TREATMENT AND REHABILITATION OF THE CHRONIC ALCOHOLIC

THE BIOLOGY OF ALCOHOLISM

Volume 1: Biochemistry

Volume 2: Physiology and Behavior

Volume 3: Clinical Pathology

Volume 4: Social Aspects of Alcoholism

Volume 5: Treatment and Rehabilitation of the Chronic Alcoholic

TREATMENT AND REHABILITATION OF THE CHRONIC ALCOHOLIC

Edited by
Benjamin Kissin and Henri Begleiter

Division of Alcoholism and Drug Dependence
Department of Psychiatry
State University of New York
Downstate Medical Center
Brooklyn, New York

PLENUM PRESS • NEW YORK-LONDON

Library of Congress Cataloging in Publication Data

Kissin, Benjamin, 1917-
　The biology of alcoholism.

　Includes bibliographies and index.
　CONTENTS: v. 1. Biochemistry.–v. 2. Physiology and behavior.–v. 3. Clinical pathology.–v. 4. Social aspects of alcoholism–v. 5. Treatment and rehabilitation of the chronic alcoholic.
　1. Alcoholism. 2. Alcoholism–Physiological effect. I. Begleiter, Henri, joint author. II. Title. [DNLM: 1. Alcoholism WM274 K61b]
RC565.K52   B5   616.8'6'1   74-131883
ISBN 0-306-37115-4 (v. 5)

First Printing – January 1977
Second Printing – April 1978

© 1977 Plenum Press, New York
A Division of Plenum Publishing Corporation
227 West 17th Street, New York, N.Y. 10011

All rights reserved

No part of this book may be reproduced, stored in a retrieval system, or transmitted, in any form or by any means, electronic, mechanical, photocopying, microfilming, recording, or otherwise, without written permission from the Publisher

Printed in the United States of America

Contributors

Frederick Baekeland, *Department of Psychiatry, Division of Alcoholism and Drug Dependence, State University of New York, Downstate Medical Center, Brooklyn, New York*
Allan Beigel, *University of Arizona College of Medicine; and Southern Arizona Mental Health Center, Tucson, Arizona*
Edward Blacker, *Division of Alcoholism, Massachusetts Department of Public Health, Boston, Massachusetts*
Howard T. Blane, *Division of Specialized Professional Development, University of Pittsburgh, Pittsburgh, Pennsylvania*
Sheila B. Blume, *Alcoholism Unit, Central Islip Psychiatric Center, Central Islip, New York; School of Medicine, State University of New York at Stony Brook, New York; Caribbean Institute on Alcoholism*
Dan W. Briddell, *Rutgers University, New Brunswick, New Jersey; now at Munson Medical Center, Traverse City, Michigan*
Morris E. Chafetz, *National Institute on Alcohol Abuse and Alcoholism, National Institutes of Mental Health, Department of Health, Education and Welfare, Rockville, Maryland; now at Johns Hopkins University, Baltimore, Maryland*
David R. Doroff, *Alcoholism Treatment Program, St. Luke's Hospital Center, New York*
Stuart Ghertner, *Southern Arizona Mental Health Center, Tucson, Arizona*
Benjamin Kissin, *Alcoholism Division, State University of New York, Downstate Medical Center, Brooklyn, New York*

Barry Leach, *Research Psychologist, New York; formerly Departments of Medicine and Psychiatry, The Roosevelt Hospital, New York*

Lawrence K. Lundwall, *Department of Psychiatry, State University of New York, Downstate Medical Center, Brooklyn, New York*

Peter E. Nathan, *Rutgers University, New Brunswick, New Jersey*

John L. Norris, *General Service Board of Alcoholics Anonymous; Associate Medical Director (retired), Eastman Kodak Company, Rochester, New York*

Earl Rubington, *Department of Sociology and Anthropology, Northeastern University, Boston, Massachusetts*

Peter Steinglass, *Center for Family Research, Department of Psychiatry and Behavioral Sciences, George Washington University School of Medicine, Washington, D.C.*

Robert Yoerg, *National Institute on Alcohol Abuse and Alcoholism, Department of Health, Education, and Welfare, National Institutes of Mental Health, Rockville, Maryland*

Preface

The present volume contains a large variety of treatment approaches to the long-term rehabilitation of the alcoholic, ranging from the biological to the physiological to the psychological to the social. The multiplicity of proposed therapies, each of which has its strong proponents, suggests that alcoholism is either a complex medical-social disease syndrome requiring a multipronged treatment approach or a very simple illness for which we have not yet discovered the remedy. The latter may, indeed, be true, but we cannot use what we do not know and must use what we do know. We do, however, have the obligation to be responsible in our treatment, to provide the best that *is* known at this time, and to be discriminating in our prescription of appropriate treatment for individual patients. If there is one conclusion we would like to offer in our preface, it is that alcoholics constitute a markedly heterogeneous population with widely disparate needs, for whom, at least at our present level of knowledge, a broad spectrum of treatment modalities is necessary. If this is true, then probably most of this book has validity.

With this volume on the treatment and rehabilitation of the chronic alcoholic, we bring to completion our five-volume series, *The Biology of Alcoholism*. As the title of the present volume indicates, we have departed from our original intention to deal solely with biological aspects of the syndrome and have attempted rather to produce a more comprehensive work. In retrospect, we find gaps and weaknesses which might have been filled or treated more adequately but, on the whole, we are satisfied that these volumes have, to a large extent, fulfilled that purpose. Our major thrust has been to present alcoholism in all of its complexities, which appear to encompass every level of human phenomenology from the biochemical to the social. This point of view, which initially generated the organization of the entire work, permeates our philosophy of treatment as well and at least has the virtue of consistency.

Reading the galley proofs and writing the preface for the last of a five-volume series is like putting the final touch of paint on a newly built house. One has a great sense of completion but also a sense of loss of a vital element of intellectual and emotional stimulation in one's life. Above all, one has a sense of obligation to those who helped plan the edifice, to those who actually built it, and to those who provided the emotional support necessary to complete a six-year project. For the first, we wish to thank Bob Ubell of Plenum Press for his constant interest in and support of this project. Among the workers, we count all of our authors, many of whom we came to know as friends as a result of this collaboration. Still among the workers, we especially thank Mildred Cohen, who nursed all of the manuscripts—many literally handwritten—from the typewriter to the publisher. Finally, among the providers of emotional support, we thank our wives, Eve and Esther, for their patience and never-ending forbearance.

Benjamin Kissin
Henri Begleiter

New York

Contents of Volume 5

Contents of Earlier Volumes
Volume 1 xix
Volume 2 xxi
Volume 3 xxii
Volume 4 xxiv

Chapter 1
Theory and Practice in the Treatment of Alcoholism
Benjamin Kissin

The Development of Alcoholism 1
   Predisposing Factors 2
   The Development of Primary Psychological Dependence 5
   The Addictive Cycle 5
   Alcoholism as "Symptom" or "Disease" 8
A Pathogenetic Classification of Alcoholics 9
   Biological Mechanisms 9
   Psychological Mechanisms 11
   Social Mechanisms 18
Implications for Therapy 22
   Breaking the Addictive Cycle 24
   Special Problems in the Rehabilitation Process 26
   The Core Problem 31

Treatment Models in Alcoholism 32
    The Medical Model of Alcoholic Rehabilitation 32
    The Behavior Modification Model 34
    The Psychological Model 35
    The Social Model 37
    The Alcoholics Anonymous Model 40
    The Multivariant Model 42
The Multimodal, Multidisciplinary Approach to the Treatment of
    Alcoholism 43
    The Multimodal Approach to the Treatment of Alcoholism 44
    The Multidisciplinary Approach to the Treatment of Alcoholism . 45
    Continuity of Care in the Treatment of Alcoholics 45
    Public Health Treatment Systems in Alcoholism 46
References 48

Chapter 2
Medical Management of the Alcoholic Patient
Benjamin Kissin

Introduction 53
The Role of the Physician in the Treatment of Alcoholism 54
    The Role of the Physician in Private Practice 55
    The Role of the Physician in General Hospitals 59
    The Role of the Physician in Special Alcoholism Programs 60
Criteria for Diagnosis 62
Criteria for Referral and/or Treatment 63
    The Therapist of Choice 68
    The Treatment Model of Choice 71
Designing a Specific Treatment Plan 75
    Special Problems and their Treatments 76
    The Core Problem and its Treatment 93
References 100

Chapter 3
Psychotherapeutic Approach
Howard T. Blane

Introduction 105
General Considerations 106
Initial Phase 113
    Relationship and Relationship Building 113
    Pretherapy Factors Affecting Treatment 115

Drinking or Not Drinking	121
Setting Goals and Confrontation	124
Therapeutic Contracts	130
The Hostile Patient	132
Middle Phase	133
General Aims and Issues	133
Homework and Other Techniques	141
Drinking Episodes	143
Late Phase and Termination	145
Relationship Issues	148
The Patient	148
The Therapist	150
Individual Psychotherapy in Relation to Other Treatment Methods	154
Summary	156
References	158

Chapter 4
Engaging the Alcoholic in Treatment and Keeping Him There
Frederick Baekeland and Lawrence K. Lundwall

Introduction	161
Detection of the Alcoholic	162
High-Risk Groups	162
Factors Hindering Identification	165
Methods of Identification	170
Engaging the Patient at the Referral Stage	174
Referral Failures: Extent and Causes	174
Remedies	175
Keeping the Alcoholic in Treatment	177
Dropping Out of Treatment: Extent and Causes	177
Remedies	184
References	187

Chapter 5
Toward a Social Model: An Assessment of Social Factors Which Influence Problem Drinking and Its Treatment
Allan Beigel and Stuart Ghertner

Introduction	197
Biological Model	198
Psychological Model	198
The Origins and Perspectives of a Social Model	199

Social Systems Theory and Its Role in the Social Model	200
Supracultural Orientation	201
Specific-Culture Orientation	203
Substructural Orientation	203
Social Etiological Factors in Drinking Behavior	203
Religious Aspects	204
Social Class Influences	207
Family Influences	209
Ethnic Aspects	212
Age, Sex, and Urbanization	213
Treatment Approaches	214
Alcoholics Anonymous	215
Therapeutic Communities	219
Halfway Houses	220
Group Therapy	221
Activity Groups	223
Family Therapy	224
The Social Learning Approach	225
Education Approaches	227
Conclusion	229
References	229

Chapter 6
Group Psychotherapy in Alcoholism
David R. Doroff

Introduction	235
Alcoholics Anonymous	237
The Psychology of A.A.	237
A.A. as a Group	238
A.A. as a Therapeutic Network	239
A Survey of Group Therapy with Alcoholics	240
Conjoint and Family Groups	254
Summary	256
References	257

Chapter 7
Family Therapy in Alcoholism
Peter Steinglass

Introduction	259
Family Therapy as a Treatment Modality	261
The Family as a System	263

The Concept of Homeostasis	264
The Concept of the "Identified Patient" or "Scapegoat"	265
Communication Patterns	265
Behavioral Context	266
Boundaries	266
Family Therapy in Alcoholism	267
Phase I: Early Interest in Family Issues and Alcoholism	268
Phase II: The Alcoholic Marriage	269
Phase III: Concurrent Therapy for Alcoholics and Spouses	270
Phase IV: The Adaptation of Family Theory to Alcoholism Therapy	275
Phase V: Conjoint Therapy with the Alcoholic Family	282
Phase VI: Multiple-Couples and Multiple-Family Group Therapy Approaches	284
Al-Anon Family Groups	289
Discussion	294
References	296

Chapter 8
Behavioral Assessment and Treatment of Alcoholism
Peter E. Nathan and Dan W. Briddell

Behavioral Theories of Alcoholism	301
Behavioral Assessment Techniques	302
Behavioral Assessment in the Laboratory	303
Behavioral Assessment in the Natural Environment	307
Behavioral Assessment: Overall Evaluation	308
Behavioral Treatment Approaches	308
Goals of Treatment	308
Modifying the Drinking Response: Aversive Conditioning	310
Modifying the Drinking Response: Operant Methods	320
Modifying the Drinking Response: Blood Alcohol Level Discrimination Training	323
Modifying Associated Behavioral Problems: Systematic Desensitization	327
Modifying the Drinking Response and Associated Behavioral Problems: Broad-Spectrum and Multifaceted Therapies	327
Modifying the Natural Environment: Community-Reinforcement Counseling and Contingency Management	335
Behavioral Treatment in Perspective	340
References	343

Chapter 9
The Role of the Halfway House in the Rehabilitation of Alcoholics
Earl Rubington

Halfway Houses	351
Origins of Halfway Houses	351
A Definition of Halfway Houses	352
One Example: The Compass Club	353
Major Characteristics of Halfway Houses	356
Some Types of Halfway Houses	357
Residents	359
Patterns of Dependence	359
Alcoholics: White-Collar, Blue-Collar, and Skid Row	360
Reaction, Impairment, and Resources	360
Halfway Houses and their Clientele	361
Rehabilitation	361
A Definition of Rehabilitation	361
Indices of Rehabilitation	362
The Social Conditions of Rehabilitation	362
Research Findings on Rehabilitation Outcomes	363
A Theory of Rehabilitation	364
Strains of Group Membership	364
Authority and the Halfway House	366
Social Types: Ways of Coping with Halfway House Authority	367
Ex-Resident Social Roles	370
Halfway House Social Types and Ex-Resident Social Roles	372
Some Hypotheses on Social Types and Ex-Resident Social Roles	373
Types of Halfway Houses and Halfway House Social Types	379
Summary and Conclusions	382
References	382

Chapter 10
Evaluation of Treatment Methods in Chronic Alcoholism
Frederick Baekeland

Introduction	385
Treatment Goals and Outcome	386
Treatment Length	388
Spontaneous Improvement, or What Happens to the Untreated Alcoholic?	389
Inpatient Treatment	390
Effectiveness of Inpatient Treatment	391

Patient and Treatment	391
Treatment Length and Outcome	393
The Effectiveness of Inpatient Psychotherapy	393
Hospital versus Outpatient Treatment	394
How Necessary Is Aftercare?	394
Outpatient Treatment	395
Dropping Out of Treatment	395
Basic Issues of Outpatient Treatment	396
Alcoholics Anonymous (A.A.)	402
General Considerations	402
A.A. Population Characteristics	404
Predictors of Success in A.A.	406
The Effectiveness of A.A.	406
A.A. Attendance as a Predictor of Success in Other Settings	408
Behavioristically Oriented Psychotherapy	409
General Considerations	409
Aversive Conditioning	410
Systmatic Desensitization	416
Operant Conditioning	416
Drug Treatment	417
Implicit Assumptions	417
Negative Results in Drug Studies with Alcoholics	418
Outcome in Drug Studies	421
Summary and Conclusions	426
References	428

Chapter 11
Factors in the Development of Alcoholics Anonymous (A.A.)
Barry Leach and John L. Norris

Introduction: Crisis for a Hungover Doctor	441
The Growth and Size of A.A.	443
In English-Speaking Lands	444
A.A. Growth Worldwide	449
The Development of A.A. and Its Structure	452
Earliest A.A. Origins	452
From "The Big Book" to the G. S. Conference (1938–1955)	456
Developments Since 1965	468
Evaluations of A.A. Effectiveness	470
Survey Method	471
Representativeness of the Sample	471
Findings of the A.A. Board Surveys	476

The Literature on A.A. 507
　　Material Published by A.A. 507
　　Bibliographies 509
　　Significant Early Publications in Mass Media 510
　　Professional, Scientific, and Technical Publications 511
　　Criticisms of A.A. 517
References 519

Chapter 12
Role of the Recovered Alcoholic in the Treatment of Alcoholism
Sheila B. Blume

Introduction 545
Definitions 546
Scope of Chapter 546
Roles of the Recovered Alcoholic, Past and Present 547
　　Independent Lay Therapists and Group Programs 547
　　Religious Programs 548
　　Independent Facilities 549
　　Medically and Psychiatrically Sponsored Programs 550
　　Industrial Programs 551
　　Antipoverty Programs 551
　　Courts and Correctional Facilities 552
　　Public Education and Information Agencies 552
The Alcoholism Counselor as Member of a Treatment Team 553
　　Unique Advantages of the Recovered Alcoholic as Counselor .. 553
　　Motivation of Counselors 554
　　Selection of Candidates to Be Counselors 555
　　Training 556
　　Job Responsibility 557
　　Problems Unique to the Recovered Alcoholic as Counselor 559
　　Remuneration and Status 561
Thoughts on the Future of Recovered Alcoholics as Counselors ... 562
References 563

Chapter 13
Training for Professionals and Nonprofessionals in Alcoholism
Edward Blacker

Introduction 567
Significant Components of Training 568
　　Objectives 568
　　Target Groups 570

Settings	575
Teaching Methods and Models	578
Subject Matter	581
Examples of Programs	583
Research Program	583
University Program	585
Community Leaders Program	586
Paraprofessionals Program	587
Evaluation	589
Guidelines for Designing a Training Program	590
Conclusion	591
References	591

Chapter 14
Public Health Treatment Programs in Alcoholism
Morris E. Chafetz and Robert Yoerg

Alcohol Problems in the United States	593
The Rise of Organizational Interest in the Problem of Alcoholism	596
Social Involvement with Problems of Alcoholism	599
Changing Legal Patterns in the Public Approach to Alcoholism	599
Involvement of the Federal Government in Alcoholism Problems	601
Patterns of Alcohol Use and Abuse within Communities	603
The Scope of Public Health Problems Involved in Alcoholism	604
Community Alcoholism Treatment Services	606
Using Available Agencies: The Experience of Industry	608
A Model Alcoholism Treatment Program	609
Prevention of Alcoholism	611
Summary	612
References	613

Index	615

Contents of Earlier Volumes

Volume 1: Biochemistry

Chapter 1
Absorption Diffusion, Distribution, and Elimination of Ethanol: Effects on Biological Membranes
by Harold Kalant

Chapter 2
The Metabolism of Alcohol in Normals and Alcoholics: Enzymes
by J. P. von Wartburg

Chapter 3
Effect of Ethanol on Intracellular Respiration and Cerebral Function
by Henrik Wallgren

Chapter 4
Effect of Ethanol on Neurohumoral Amine Metabolism
by Aaron Feldstein

Chapter 5
The Role of Acetaldehyde in the Actions of Ethanol
by Edward B. Truitt, Jr., and Michael J. Walsh

Chapter 6
The Effect of Alcohol on Carbohydrate Metabolism: Carbohydrate Metabolism in Alcoholics
by Ronald A. Arky

Chapter 7
Protein, Nucleotide, and Porphyrin Metabolism
by James M. Orten and Vishwanath M. Sardesai

Chapter 8
Effects of Ethanol on Lipid, Uric Acid, Intermediary, and Drug Metabolism, Including the Pathogenesis of the Alcoholic Fatty Liver
by Charles S. Lieber, Emanuel Rubin, and Leonore M. DeCarli

Chapter 9
Biochemistry of Gastrointestinal and Liver Diseases in Alcoholism
by Carroll M. Leevy, Abdul Kerim Tanribilir, and Francis Smith

Chapter 10
Alcohol and Vitamin Metabolism
by Joseph J. Vitale and Joanne Coffey

Chapter 11
The Effect of Alcohol on Fluid and Electrolyte Metabolism
by James D. Beard and David H. Knott

Chapter 12
Mineral Metabolism in Alcoholism
by Edmund B. Flink

Chapter 13
Alcohol–Endocrine Interrelationships
by Peter E. Stokes

Chapter 14
Acute and Chronic Toxicity of Alcohol
by Samuel W. French

Chapter 15
Biochemical Mechanisms of Alcohol Addiction
by Jack H. Mendelson

Chapter 16
Methods for the Determination of Ethanol and Acetaldehyde
by Irving Sunshine and Nicholas Hodnett

Chapter 17
The Chemistry of Alcohol Beverages
by Chauncey D. Leake and Milton Silverman

Volume 2: Physiology and Behavior

Chapter 1
Effects of Alcohol on the Neuron
by Robert G. Grenell

Chapter 2
Peripheral Nerve and Muscle Disorders Associated with Alcoholism
by Richard F. Mayer and Ricardo Garcia-Mullin

Chapter 3
The Effects of Alcohol on Evoked Potentials of Various Parts of the Central Nervous System of the Cat
by Harold E. Himwich and David A. Callison

Chapter 4
Brain Centers of Reinforcement and Effects of Alcohol
by J. St.-Laurent

Chapter 5
Factors Underlying Differences in Alcohol Preference of Inbred Strains of Mice
by David A. Rogers

Chapter 6
The Determinants of Alcohol Preference in Animals
by R. D. Myers and W. L. Veale

Chapter 7
Voluntary Alcohol Consumption in Apes
by F. L. Fitz-Gerald

Chapter 8
State-Dependent Learning Produced by Alcohol and Its Relevance to
    Alcoholism
by Donald A. Overton

Chapter 9
Behavioral Studies of Alcoholism
by Nancy K. Mello

Chapter 10
The Effects of Alcohol on the Central Nervous System in Humans
by Henri Begleiter and Arthur Platz

Chapter 11
Changes in Cardiovascular Activity as a Function of Alcohol Intake
by David H. Knott and James D. Beard

Chapter 12
The Effect of Alcohol on the Autonomic Nervous System of Humans:
    Psychophysiological Approach
by Paul Naitoh

Chapter 13
Alcohol and Sleep
by Harold L. Williams and A. Salamy

Chapter 14
Alcoholism and Learning
by M. Vogel-Sprott

Chapter 15
Some Behavioral Effects of Alcohol on Man
by J. A. Carpenter and N. P. Armenti

Volume 3: Clinical Pathology

Chapter 1
The Pharmacodynamics and Natural History of Alcoholism
by Benjamin Kissin

Contents of Earlier Volumes: Volume 3

Chapter 2
Heredity and Alcoholism
by Donald W. Goodwin and Samuel B. Guze

Chapter 3
Psychological Factors in Alcoholism
by Herbert Barry, III

Chapter 4
Interactions of Ethyl Alcohol and Other Drugs
by Benjamin Kissin

Chapter 5
Acute Alcohol Intoxication, The Disulfiram Reaction, and Methyl Alcohol
    Intoxication
by Robert Morgan and Edward J. Cagan

Chapter 6
Acute Alcohol Withdrawal Syndrome
by Milton M. Gross, Eastlyn Lewis and John Hastey

Chapter 7
Diseases of the Nervous System in Chronic Alcoholics
by Pierre M. Dreyfus

Chapter 8
Metabolic and Endocrine Aberrations in Alcoholism
by D. Robert Axelrod

Chapter 9
Liver Disease in Alcoholism
by Lawrence Feinman and Charles S. Lieber

Chapter 10
Diseases of the Gastrointestinal Tract
by Stanley H. Lorber, Vicente P. Dinoso, Jr., and William Y. Chey

Chapter 11
Acute and Chronic Pancreatitis
by R. C. Pirola and C. S. Lieber

Chapter 12
Diseases of the Respiratory Tract in Alcoholics
by Harold A. Lyons and Alan Saltzman

Chapter 13
Alcoholic Cardiomyopathy
by George E. Burch and Thomas D. Giles

Chapter 14
Hematologic Effects of Alcohol
by John Lindenbaum

Chapter 15
Alcohol and Cancer
by Benjamin Kissin and Maureen M. Kaley

Chapter 16
Alcoholism and Malnutrition
by Robert W. Hillman

Chapter 17
Rehabilitation of the Chronic Alcoholic
by E. Mansell Pattison

Volume 4: Social Aspects of Alcoholism

Chapter 1
Alcohol Use in Tribal Societies
by Margaret K. Bacon

Chapter 2
Anthropological Perspectives on the Social Biology of Alcohol: An Introduction to the Literature
by Dwight B. Heath

Chapter 3
Drinking Behavior and Drinking Problems in the United States
by Don Cahalan and Ira H. Cisin

Chapter 4
Alcoholism in Women
by Edith S. Gomberg

Chapter 5
Youthful Alcohol Use, Abuse, and Alcoholism
by Wallace Mandell and Harold M. Ginzburg

Chapter 6
Family Structure and Behavior in Alcoholism: A Review of the Literature
by Joan Ablon

Chapter 7
The Alcoholic Personality
by Allan F. Williams

Chapter 8
Alcoholism and Mortality
by Jan de Lint and Wolfgang Schmidt

Chapter 9
Alcohol and Unintentional Injury
by Julian A. Waller

Chapter 10
Alcohol and Crimes of Violence
by Kai Pernanen

Chapter 11
Alcohol Abuse and Work Organizations
by Paul M. Roman and Harrison M. Trice

Chapter 12
Education and the Prevention of Alcoholism
by Howard T. Blane

Chapter 13
The Effects of Legal Restraint on Drinking
by Robert E. Popham, Wolfgang Schmidt, and Jan de Lint

CHAPTER 1

Theory and Practice in the Treatment of Alcoholism

Benjamin Kissin

Alcoholism Division
State University of New York
Downstate Medical Center
Brooklyn, New York

THE DEVELOPMENT OF ALCOHOLISM

The American Medical Association in 1971 adopted a statement identifying alcoholism as "a complex disease with biological, psychological, and sociological components" (Todd, 1975). Like other conditions, it follows a more or less specific sequence. The susceptible individual is exposed to the causative agent and the early stages of the process begin. When that process contains self-perpetuating mechanisms which develop as a "consequence" of the condition, the syndrome is furthered. At this point the characteristics of the alcoholic have changed from their original status of "susceptibility to alcoholism" to those of alcoholism itself. This duality in which the factors perpetuating alcoholism may be not only "predisposing factors" which antedate its development but additionally mechanisms which are its "consequences" has led to much of the confusion surrounding the pathogenesis of alcoholism. A clear delineation of the interaction of these two different types of causative factors is essential to the understanding of the pathogenesis of the syndrome and to the elaboration of a rational plan of treatment.

Predisposing Factors

Our present conceptualization of the development of alcoholism is illustrated schematically in Figure 1. It follows largely on the theoretical formulations of Seevers (1968) in the psychopharmacologic area and of Jellinek (1960) in his work on the disease concept of alcoholism. Essentially, it postulates that the etiologic origins of alcoholism may be biological, psychological, or social. The biological origins may be genetic or they may be prenatal and acquired as in those infants born to alcoholic mothers. The psychological elements contributing to the development of alcoholism cover a broad spectrum of psychopathology, which may be biological or experiential, or both. The social factors contributing to the development of alcoholism add their vectorial influence to the process. The total interaction of all of these influences at the three basic levels, undoubtedly presenting widely varying patterns in different people, represents the index of predisposition toward the development of alcoholism for any given individual.

Which, if any, of these elements is most significant? At present, the most reasonable hypothesis would appear to be one which gives appropriate weight to the variables in each of these three areas—the biological, the psychological, and the social. The thrust toward the development of alcoholism is presumably a vector of these three influences so that where one or two are very great, the third may be moderate or negligible. For example, where social pressures are opposed to the development of alcoholism, either greater biological or psychological susceptibility must be hypothesized. This thesis is illustrated by the observations that women alcoholics tend to have more psychopathology than male alcoholics; Jewish alcoholics tend to have different psychopathology from Irish alcoholics; and in a different field, white middle-class heroin addicts tend to show more psychopathology than black heroin addicts from the inner city (Kaufman, 1974). This equation is particularly important in the consideration of ghetto alcoholics and heroin addicts, where alcoholism and drug addiction are almost a way of life for the unemployed, welfare-supported, social derelict. In these instances, one wonders whether the apparent sociopathic personality disorder is a cause or a consequence of the socially imposed life pattern.

On the other hand, there is some evidence that different types of alcoholics may have as their major predisposing etiological factors elements in one or another of these areas, i.e., biology, psychology, or sociology. For example,

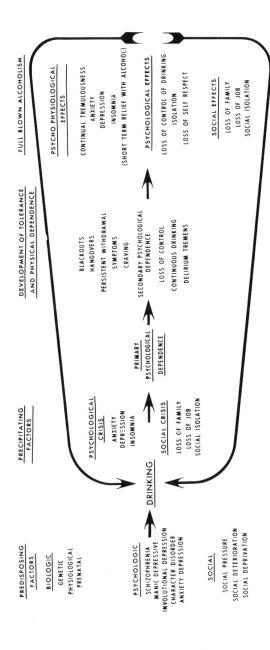

FIGURE 1. Alcoholism as symptom and disease.

some alcoholics with a strong family history of alcoholism show a special course both in their susceptibility to alcoholism and in the rapidity of its development. This phenomenon is so common among certain American Indians that the sale of intoxicating liquor to these individuals has been made illegal. The combination of a strong hereditary history, the fulminating nature of the illness, and its special prevalence in certain ethnic groups has given rise to the judgment that the dominant mechanisms in the pathogenesis of this form of alcoholism may be biological. These individuals may be called *essential alcoholics,* a term similar but not necessarily the same as that used by Knight (1937), where reference is to a specific psychodynamic mechanism rather than to a specific clinical syndrome.

An instance of alcoholism which derives mainly from psychological disorder may be the so-called situational alcoholic. Here, because of an immediate stressful situation, the individual turns to alcohol as a sedative for his extreme anxiety and depression. When the precipitating situation has been resolved, the need to take alcohol is reduced and unless the secondary effects of the alcoholic process have been superimposed, drinking may cease.

Finally, the idea has recently gained increasing credence that social forces may be among the most important in predisposing to alcoholism. The socially isolated, unemployed derelict in the ghetto area has come to be seen as almost entirely the product of social rather than biological or psychological forces. This attitude has been especially fostered by the experience in heroin addiction where the susceptibility to addiction is often more a function of the social milieu in a given place at a given time than it is of the biology or psychology of any specific individual (Chein *et al.,* 1964). There is good reason to believe that in heroin addiction—particularly during epidemic waves—social factors play the largest role, if not almost the entire role, in determining who will or will not become addicted. Probably at times of epidemics of alcoholism, as in Hogarthian England, both social conditions and economic factors—gin was cheap—are most influential. However, there is also evidence that in the more common endemic incidence of alcoholism, the biological, psychological, and social factors all play significant roles.

Particularly in the common variety alcoholic, who is neither essential, situational, or ghetto, the previously described interaction of the three major influences is more readily apparent. Even in heroin addiction epidemics, as the addiction spreads into the white middle-class and upper-class areas, it is mainly those individuals with manifest psychopathology and high accessibility to heroin who are afflicted. Similarly, where social influences leading to alcoholism are great—but not great enough to produce epidemics—the involvement of specific individuals will be a function of the interaction of the social influence with their specific biological or psychological pathology.

The Development of Primary Psychological Dependence

Given—to use Seevers's (1968) term—the *susceptible* individual, the ensuing process requires only exposure to the critical element, alcohol, for progression to occur. The first significant experience with alcohol, which many alcoholics remember vividly all their lives, is often a highly rewarding one. This reward may be in the magical relief of some underlying malignant psychological or physiological tension or it may result from a new sense of well-being and euphoria. Perhaps most often it contains elements of both. In any event, the original experience of extreme gratification which is reinforced by every subsequent experience results in the development of primary psychological dependence, the fundamental drive in all forms of drug dependence. This state involves the existence of the most powerful psychological drive for continuing alcohol-seeking behavior—the positive reinforcement conditioning paradigm. The recognition of the sufficient strength of primary psychological dependence to act as the sole drive in many severe forms of drug dependence has led some investigators to postulate that it is the only drive involved in alcohol addiction. However, as we shall attempt to illustrate in the next section, physical dependency mechanisms also probably play a major role.

The Addictive Cycle

Dynamics of the Addictive Cycle in Alcohol Addiction

The frequent ingestion of alcohol results, as it does with all depressant psychoactive drugs (alcohol, barbiturates, morphine), in the development of tolerance, both metabolic and cellular. Gradually, the individual finds that he needs ever-increasing doses to obtain the same psychological reward that he could previously obtain with smaller doses. The ingestion of continually larger doses results in the bathing of the tissues of the central nervous system in high, persistent concentrations of the drug. This, in turn, results in the gradual onset of physical dependence. With physical dependence come withdrawal symptoms and these act, both directly and indirectly, to perpetuate and to increase the rate of drug-seeking behavior (see Figure 2).

The specific role which physical dependence plays in the escalation of the addictive cycle that develops around the abuse of depressant psychoactive drugs has long been disputed. Nevertheless, it is of interest that the early definition of "addiction" (Seevers and Woods, 1953) involved the necessary presence of physical dependence. In the field of heroin addiction, Dole and Nyswander (1965) feel that physical dependence is the major driving force in heroin-seeking behavior. They describe "craving" as the psychological equivalent of

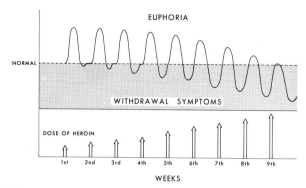

FIGURE 2. Tolerance, physical dependence, and the addictive cycle.

physical dependence manifested perhaps as persistent withdrawal symptomatology. This, according to Martin and Jasinski (1969), may continue for up to six months after withdrawal. The chemical suppression of this "tissue craving" constitutes the rationale for the generally successful Dole-Nyswander methadone maintenance treatment of heroin addiction.

The significance of physical dependence in the development of heroin addiction may be more apparent than it is in the development of alcoholism, but it is by no means more important. Tolerance and physical dependence develop more slowly and to a lesser degree with alcohol than they do with morphine or its derivatives. To that extent, their impact is less obvious in alcoholism than in heroin addiction. On the other hand, once physical dependence does develop in alcoholism, withdrawal symptomatology is far more severe and more persistent than with heroin (Kissin *et al.*, 1959; Schenker *et al.*, 1962; Tripp *et al.*, 1959). Accordingly, craving for the causative agent to relieve withdrawal symptomatology is probably as great in alcoholism as it is in heroin addiction.

The Role of Loss of Control and Craving in Perpetuating the Addictive Cycle

It is important to distinguish between craving during an ongoing drinking bout and craving when the individual is abstinent. In the first instance, withdrawal symptomatology occurs throughout intoxication with particular exacerbations in the morning after even brief periods of abstinence (Kissin, 1974). The presence of tremulousness, anxiety, depression, and insomnia constitute the conditioned stimulus (CS) for which drinking has become the conditioned response (CR). Craving during drinking is the subjective equivalent of withdrawal symptoms associated with the learned recognition that alcohol will relieve those symptoms. These reinforcing mechanisms make the craving for alcohol during a drinking binge irresistible so that the alcoholic loses his ability to control his drinking. *Loss of control* becomes then the larger descriptive term of the phenomena occurring during the compulsive drinking bout. Stein *et al.* (1968) found that uncontrolled drinkers reported "longer and more frequent blackouts, more frequent delirium tremens, 'shakes,' vague fears and phobias associated with prolonged drinking" than did controlled drinkers. Although their data also suggest more severe underlying psychopathology, these effects indicate an important role for physical dependency mechanisms in the loss of control phenomenology.

Recent work suggests that craving during abstinence may have greater psychological elements than those involved in loss of control. Ludwig *et al.* (1974) have developed an experimentally supported model which involves both physiological and psychological mechanisms. In their paper, the authors postulate that "craving for alcohol (during periods of no physical dependence) represents the cognitive—symbolic correlate of a *subclinical, conditioned withdrawal syndrome* which can be produced by appropriate introceptive and/or exteroceptive stimuli." (Periods of no physical dependence are meant to signify the absence of acute withdrawal symptoms such as would be found in an acute drinking bout.) It is the conceptualization of these authors that introceptive cues (such as emotional stress, which produces symptomatology resembling that of withdrawal) and exteroceptive cues (such as a labeled bottle associated with the previous relief of real withdrawal symptoms) interact in a meaningful paradigm which may be labeled "craving."

Hore (1974) reported on a questionnaire survey of 750 alcoholics in A.A. as to whether they had experienced moderate to severe craving for alcohol in the previous week. One-third responded positively. There was a high correlation of craving with levels of anxiety and depression. "Cravers" had a significantly higher incidence of drinking in the past month and a significantly lower incidence of a three-month period of abstinence than did "noncravers." They also reported a higher level of withdrawal symptomatology. Consequently, it may be that Ludwig *et al.* (1974) are underestimating the importance of direct physical dependency involvement even in their own experiments, where they postulate that their subjects who are "three days or more" out of an acute withdrawal episode are no longer in a state of "physical dependence." Since many studies have shown that physical dependency withdrawal phenomena to alcohol may persist for six months or longer, it is difficult to know how much of their subjects' craving was due to true withdrawal symptomatology and how much to the weaker *conditioned* withdrawal symptomatology (stress).

In either event, whatever the relative validity of these two interpretations, it is clear that some form of craving which leads to renewal of alcohol-seeking behavior does occur when the full-blown picture of physical dependency has been achieved (see Figure 1). At that point, whether the alcoholic wants to drink to relieve real symptoms of alcohol withdrawal (tremulousness, anxiety, depression, and insomnia) as in loss of control or to relieve either *real* or *conditioned* symptoms of withdrawal as in craving, it is certain that he wants to drink and does. Here, when the need to drink becomes persistent, unrelenting, and irresistible, the full-blown picture of chronic alcoholism ensues with all the stigmata of physical, psychological, and social deterioration.

Alcoholism as "Symptom" or "Disease"

Parenthetically, this conceptualization of the development of alcoholism helps illuminate the old dispute as to whether alcoholism is a symptom or a disease. It is apparent that in the early phases of alcoholism when the susceptible individual develops primary psychological dependence, alcoholism is a symptom. Here the alcoholic drinks to relieve some underlying discomfort or to satisfy some underlying need to get "high." At that stage, the alcohol is merely the instrument through which the individual achieves relief or satisfaction; the underlying pathology, whatever it may be, is the cause of the drinking. In this phase, alcoholism is a symptom of a variety of underlying pathologies, each one of which might be considered as a different illness (manic depressive psychosis, situational alcoholism, ghetto alcoholism, etc.). At the end of the process, however, when the cyclic phenomena of tolerance, physical dependence, loss of control, and craving have ensued, alcoholism becomes a disease. At this point, regardless of what may have driven the individual to drink in the first place, it is now alcohol and his physical and psychological dependence on it which are the driving motivations. With progression, the syndrome becomes more unitary both in its etiology and in its clinical manifestations. Whereas the dynamics and clinical signs of early alcoholism vary considerably, the pattern in late chronic alcoholism of physical, psychological, and social deterioration becomes only too characteristic. At this stage, alcoholism is truly a disease with a single etiologic agent, alcohol, and an essentially single diagnostic clinical picture.

This conceptualization also helps explain the apparent discrepancy between different statistics gathered on the incidence of alcoholism in the United States. Studies using the Jellinek formula usually come up with a figure of about five to six million alcoholics in the United States (Keller, 1975), while studies based on interview techniques report the incidence of "problem drinkers" as between nine and ten million (Cahalan *et al.,* 1969). Strauss (1973) and Keller (1975) have suggested that these differences are real rather than methodologic and reflect true differences between alcoholics and "problem drinkers." They differentiate between the two populations on criteria similar to those just described: The problem drinker represents the prodromal stage of alcoholism which unfolds with the development of the addictive cycle. The passage from the earlier stage to the full-blown picture is not a clear-cut episode. The entire progression lies on a continuous spectrum from social drinker to heavy drinker to problem drinker to early alcoholic to moderate alcoholic to severe alcoholic. This quantitative view is essential in defining principles of treatment which may be entirely different at different levels in the spectrum.

A PATHOGENETIC CLASSIFICATION OF ALCOHOLICS

A variety of classifications of alcoholism have been attempted, the best known of which is probably Jellinek's (1960). That classification is essentially descriptive, characterizing different types of clinical courses. This is of some value in that it illustrates the heterogeneity of the alcoholic population but it is of less value in treatment since there is no obvious relationship between the classification and pathogenetic mechanisms.

If we hypothesize that mechanisms at all three levels—biologic, psychologic, and social—contribute to the development of alcoholism in *a manner which is significant for rational therapeutic intervention,* then the description of the extent to which each of these mechanisms contributes to the whole becomes necessary to an understanding of those areas in which appropriate therapeutic intervention can and should be made. This type of classification is clearly less one of clinically descriptive categories (like Jellinek's) and is more one of pathogenetic categories. Where one or another element is dominant as in the essential, situational, or ghetto alcoholic, then the pathogenetic classification describes a discrete clinical course. On the other hand, where elements at all three levels interact with one another—as in the majority of alcoholics—only an evaluation of the specific contribution of each mechanism to the initiation and perpetuation of the disease is necessary to achieve therapeutic effectiveness.

Biological Mechanisms

Genetic and Prenatal Biological Predisposition

The existence of such a diathesis is based largely on the evidence of hereditary factors in alcoholism as demonstrated by Goodwin *et al.,* (1973, 1974) and others. It is also postulated on the clinical experience with the so-called essential alcoholic, the individual who almost with his first contact with alcohol demonstrates loss of control. One has the strong clinical impression that the alcoholic sons of alcoholic fathers often react to alcohol in a different way from the sons of nonalcoholics; the entire course of their alcoholism is more immediate, more rapid, and more violent.

The infant born to the alcoholic mother who has been drinking heavily throughout the pregnancy often shows biological effects. They may be those of acute alcohol withdrawal at birth or they may be actual organic changes such as reduced size and weight (Ulleland, 1972). More recently, a neonatal condition in the infants of alcoholic mothers known as the "fetal alcohol syndrome" has been described. This syndrome, characterized by anatomical malformations and neurophysiological dysfunction, has been reported to occur more frequently in these newborn, depending on the severity of the mother's alcoholism during the pregnancy (Jones and Smith, 1973).

The infant born with withdrawal symptoms has, by definition, developed physical dependence to alcohol, presumably involving biochemical and neurophysiological changes in the central nervous system. There is evidence in animal studies that once these changes occur, they can more readily be reactivated at a later time (Branchey et al., 1971). Whether this potential for reactivation of physical dependency persists even for years in humans is a moot question. If so, then it may be that the child of a mother who drank heavily during her pregnancy may have a residual physical dependency which can be more readily reactivated when the now adolescent child is first exposed to alcohol. If indeed this mechanism does occur, the reactivation of the addictive cycle could occur rapidly and explosively, with a clinical course similar to that of the essential alcoholic.

Physical Dependency Mechanisms as a Consequence of Alcoholism

We have described the role of physical dependence in the loss of control syndrome and its importance in perpetuating binge drinking. There seems little question either of the specific involvement of physical dependency mechanisms or of the necessity for counteracting them in order to terminate the acute binge drinking episode. More controversial is the role of physical dependence in craving. Since a method does exist for controlling physical dependency mechanisms, i.e., cross-tolerant sedative therapy, the use or nonuse of this therapeutic modality *after the acute withdrawal syndrome is over* becomes a function of whether one believes craving is psychological or physical in its origin. We believe that the dynamics of this symptom varies from patient to patient so that therapy must be adjusted accordingly.

Medical Complications of Alcoholism

These are too numerous to warrant description here and are described elsewhere (Kissin and Begleiter, 1974). It is clear that medical illnesses can interfere severely with the rehabilitation of the alcoholic, both in draining his energies and in distracting his motivation from the task of recovery. Of particular importance in developing the stigmata of alcoholism are the neurological changes in the central nervous system which occur both as a direct consequence of alcohol toxicity and as an indirect consequence of concomitant vitamin deficiency (Freund, 1973). These are significant because of clinical evidence that increasing brain damage is associated with decreased tolerance to alcohol and lower threshold for the development of acute withdrawal symptomatology.

Psychological Mechanisms

Of the three putative levels of involvement—the biological, psychological, and social—the psychological raises the greatest questions. There is little doubt that many alcoholics display severe emotional problems but these might well be the result of prolonged drinking rather than the cause. Many theoretical positions have been assumed as to the significance of predisposing psychological factors in the etiology of alcoholism. These positions cover the following range: (a) the social, which holds that sociocultural factors are preeminent in the pathogenesis of alcoholism; (b) the psychosocial which postulates an interaction of social forces with a variety of psychopathological reactivities; and (c) the psychological which postulates some basic psychopathology as the dominant element. Our own viewpoint is the psychosocial which we feel provides a bridge between the two more extreme positions, and we shall try to present evidence to support it.

Psychiatric Diagnoses Among Alcoholics

According to Sherfey (1955), the main psychiatric subdivisions into which alcoholics fall cut across the three major diagnostic categories, the neuroses, the psychoses, and the character disorders, with the character disorders accounting for the bulk of the individuals (perhaps 70–80 percent) and the other two categories some 10–15 percent each. Three other psychiatric studies, by Zwerling (1959), DeVito *et al.* (1970), and Panepinto *et al.* (1970), give similar breakdowns of the diagnostic categories, so that there appears to be strong support for Sherfey's (1955) psychiatric classification of alcoholics. Knight (1937) characterizes alcoholics along a more psychodynamically oriented parameter of essential and reactive alcoholics. The essential alcoholics resemble Sherfey's character disorders and constitute some 80 percent of the population; the reactive alcoholics appear quite similar to the neurotics in the other classifications. Knight tends to include the psychotic alcoholics in the essential category because of the similarity in their dynamics. Accordingly, the various evaluations seem not too dissimilar so that the Sherfey classification would appear to have some validity.

The Character Disorder Alcoholic. These are alternatively characterized as personality disorder, psychopathic or sociopathic personalities, and in psychoanalytic parlance, impulse disorders. The individuals in this group appear to have certain characteristics in common. Among these are:

1. A primitively structured personality. This is manifested in the fixation at the oral level as described by the psychoanalyst Fenichel (1945), by passive-dependent behavior (Machover and Puzzo, 1959), by impulsivity, by the inability to form deep, lasting human relationships, by strong denial as the dominant defence mechanism, and by marked field dependence on the Witkin et al. (1959) perception tests.

2. A tendency toward psychopathy. Alcoholics, like heroin addicts and criminals, score high on the psychopathy, depression, and mania scales of the MMPI (Hill et al., 1962). However, if the items on the Pd scale which pertain to the acute alcoholism are omitted, the level of psychopathy for alcoholics is not significantly different from normal. It is questionable whether alcoholics are as psychopathic as has generally been concluded, especially if one considers that much of the antisocial behavior occurs under the influence of alcohol, a notoriously psychotoxic drug.

3. Depression and anxiety. Alcoholics consistently score higher on depression on the MMPI (Hill et al., 1962) and on depression and anxiety on the Zung Depression and Cattell Anxiety Scales (Kissin and Platz, 1968). However, anxiety in character disorders is not internalized as it is among neurotics but is rather an external response to difficult life situations. Again, how much of this is the cause and how much the consequence of alcoholism is difficult to say.

The consistency with which these characteristics appear to hold for many alcoholics has led to the concept of the "alcoholic personality." This position, best articulated by Zwerling (1959), hypothesizes that the majority of alcoholics, at least those classified as "character disorders" all have the pattern of psychopathology just described. However, this pattern varies widely in intensity in different individuals, from mild to severe. Even more, it varies qualitatively depending on the cultural background and the specific psychopathology of the individual. In fact, however, these patterns vary so markedly that it seems arbitrary to classify all of these various patterns into a single category. Zwerling (1959) maintains that the underlying dynamics in all these groups are similar but that the interaction of cultural and psychological mechanisms are such as to produce different clinical pictures in each group.

The psychopathology of the character disorders vary widely both qualitatively and quantitatively so that a broad spectrum of clinical behaviors is visible. These may be broken down generally under the following categories which include most although by no means all types (DeVito et al. 1970).

1. The antisocial personality. These individuals show a mild to moderate level of sociopathic behavior depending on the degree of psychopathology. In

most respects they demonstrate the major characteristics or character disorders as above delineated. They are immature, impulsive, show low frustration tolerance, a high degree of hostility, and a tendency to act out. They often come from families with a high incidence of antisocial behavior, broken homes, alcoholism, and low socioeconomic status. This type of alcoholic is most commonly seen in the lowest socioeconomic group though it occurs at all levels. At its most severe, it is represented by the true psychopath.

2. The passive–aggressive personality. This type manifests itself as either (a) the passive—dependent type or (b) the passive–aggressive type. In either case, the individual appears superficially passive although in each instance the passivity is an infantile device to gain attention and affection. In the dependent type, control is attempted through helplessness, while in the aggressive type it is more often attained through "passive resistance" activities. In its most marked form, this type manifests itself as "pathological narcissism" (Kernberg, 1975).

3. Impulse disorders. These, as defined by Fenichel (1945), represent a fixation at a particularly early stage of psychosexual development and demonstrate most dramatically the characteristics of impulsivity, low frustration tolerance, grandiosity, poor reality testing, and dependency. Because of the primitiveness of the character structure, these individuals often have a particularly poor prognosis.

In its most extreme form, this pattern of pathology has become labeled "borderline" since it resembles psychotic behavior, particularly in the fantasy ideation which sometimes borders on delusion. On the other hand, these individuals seldom, if ever, develop an overt psychosis so that they probably should not be included in the latter diagnosis (Kernberg, 1975).

Borderline individuals are often highly dramatic, imaginative, creative personalities, sufficiently colorful to be portrayed in literature. Johnny Nolan of *A Tree Grows in Brooklyn,* Malcolm Lowry in his autobiographical *Under the Volcano,* and the hero of *The Iceman Cometh* typify these individuals. They are often highly verbal, extroverted in their manner although feeling inherently alienated, and given to daydreaming and fantasy. Beneath their outgoing exterior lies a deep self-destructive impulsivity. In these respects, they possess the same basic characteristic elements of the character disorders as do the other categories, but are particularly poor on reality testing.

4. Schizoid personality. These individuals are, as the name implies, schizoid in their behavior. They are withdrawn, alienated, shy, and have great difficulty in relating to other people. They are autistic, given to daydreaming, and isolated. In these individuals, the greatest problem is relating to others, and alcohol is often used as a disinhibiting "social lubricant."

These are the four major types of character disorder commonly found among alcoholics. Other so-called personality disorders are the obsessive–compulsive personality and the paranoid personality. These, however, fall less readily into the character disorder classification since the obsessive–compulsive has characteristics which are more neurotic in quality while the paranoid often borders on the psychotic. These distinctions may be valid since the prognosis of the obsessive–compulsive alcoholic is better than most while that of the paranoid is probably among the worst.

Given these four major types of character disorders which constitute perhaps 90 percent of that category, what can we say about them? Certainly there are major differences in clinical symptomatology. Equally, there are major differences in psychodynamics. Finally, there are major differences in what alcohol does for each (Russell and Mehrabian, 1975). For the antisocial personality, alcohol reduces external misery. For the passive–dependent individual, it produces infantile passivity and narcissism. For the borderline personality, it provides an escape from reality, while for the schizoid, it facilitates human interaction (DeVito et al., 1970). In what way can we postulate *an* alcoholic personality?

Perhaps not at all or perhaps only in the sense that so many of these individuals share the characteristics of the character disorder—the immaturity, impulsivity, low frustration tolerance, low level of reality testing, narcissism, grandiosity, fantasizing, repressed hostility, and so on. The high incidence of this pattern—to a greater or lesser degree—led to the concept of the alcoholic personality. Zwerling (1959) labeled it a necessary but not in itself sufficient prerequisite for the development of alcoholism. This appears to be an overstatement but to a degree the pattern does characterize many alcoholics, not only after the appearance of their alcoholism but even prior to it.

The Psychotic Alcoholic. According to Sherfey (1955), Zwerling (1959), and Panepinto et al. (1970), about 10–15 percent of all alcoholics fall under several major psychotic diagnoses: latent or overt schizophrenia, endogenous depressions, or latent or overt manic-depressive psychosis. The latter diagnoses appear to be particularly frequent among Jewish alcoholics, suggesting a psychodynamic process which is uniquely different from that of most other alcoholics. These, however, constitute a relatively small segment of the psychotic alcoholic population, most of whom fall under the latent or overt schizophrenic diagnostic category. Why these schizophrenics should resort to alcohol while so many others do not is an unanswered question. There has been speculation that alcohol may act in some as an antipsychotic tranquilizer (Irwin, 1973), while there has been equal and opposite speculation that by acting as a psychotomimetic agent, it permits the latent psychotic to act out his psychosis without concomitant anxiety. For whatever reason, some psychotics do develop alcoholism as part of their psychotic manifestation. In general, those individuals show a particularly high incidence among the skid row alcoholics with a somewhat lesser incidence among the ghetto alcoholics. Needless to say, they occur with some frequency in all other demographic categories.

The Neurotic Alcoholic. This group, according to Sherfey (1955) and Panepinto *et al.* (1970), also constitutes some 10–15 percent of the total alcoholic population and coincides with Knight's (1937) classification of the "secondary" or "reactive" alcoholic. They may also be represented by Jellinek's class of alpha alcoholics. The subtypes include the various neurotic manifestations, with anxiety neuroses and obsessive–compulsive reactions constituting the most prominent categories. This group is of particular importance in our considerations of the relationship of "theory to therapy" since the treatment strategy which is appropriate for them is quite different from that which is indicated for the others.

Also included under this category are the acute stress situations, either anxiety, depressive, or reality reactions resulting from an acute, subacute, or chronic stressful life situation (situational alcoholism). In these instances, the pattern of psychological dependence on alcohol may develop rapidly and may persist even after the acute stress situation has subsided. More often, the excessive drinking leads to a further deterioration in the life situation with exacerbation of the acute psychopathological reaction and further development of psychological dependence. Where the process progresses sufficiently for tolerance and physical dependence to develop, the addictive cycle and the phase of chronic alcoholism supervenes; then the presence or absence of the primary stressful situation becomes of secondary importance. In these instances, the disruption of the "alcoholism" vicious cycle may sometimes lead to a quite rapid recovery, particularly where the original stress is no longer operative.

The "Normal" Alcoholic. In the rehabilitation of alcoholics, one often finds that when the stigmata of alcoholism itself have subsided, there remains an essentially intact and stable individual who exhibits little or no evidence of psychopathology. This is particularly true when the individual involved is a situational alcoholic, i.e., one who uses alcohol as a temporary solution for a temporary stressful life situation or a "dyssocial" alcoholic (Keller, 1975), i.e., an overly self-indulgent member of a subculture where alcohol is an integral part of the life-style situation (advertising executives, bartenders, some inner-city alcoholics).

The idea of the dyssocial drinker (Keller, 1975) has been applied particularly to the ghetto alcoholic where, in certain inner-city areas, heavy alcohol use is almost a way of life. For this population, alcohol use and abuse become coping mechanisms for individuals who are unstable to deal with the overwhelming problems with which society has burdened them (Edwards, 1974). The concept has been developed even further in heroin addiction where the typical addict may be a boy from a broken home, a dropout from school who has learned to live in the street, physically addicted to heroin, and an outcast who must engage in illegal and antisocial activity in order to survive. Kaufman (1974) has estimated that fully 90 percent of inner-city heroin addicts fall into this category. This concept of drug addiction, like Edwards's concept of alcoholism, postulates a largely social determination where the aberrant psychological behavior is a consequence of social pathology rather than psychopathology. In these instances, where the alcoholism or drug addiction is brought under control, the individual's life-style often reverts largely to normal.

The Psychosocial Equation. The existence of these "normal" alcoholics has done much to support the sociologic position that social and cultural forces are the dominant forces in the development of alcoholism. But the clinical evidence for the prevalence of predisposing psychopathology in the majority of alcoholics is equally impressive. The presence of psychopathology in the active alcoholic is undeniable but is attributed by social advocates to the alcoholism itself; it is presumed to abate as the symptoms of alcoholism subside. Yet, in a classic study, Zwerling (1959) and Machover and Puzzo (1959) demonstrated that psychopathology, as demonstrated respectively by psychiatric interviews and psychological testing, persisted in a representative group of white middle-class ex-alcoholics who had been totally abstinent in A.A. for periods of two to ten years. Furthermore, Zwerling found strong evidence of a prealcoholic pattern of aberrant behavior of a type compatible with the postalcoholic psychopathology. These findings are consistent with the hypothesis that the premorbid personality dynamics contributed to the development of the disease.

This position is further supported by the findings of Winokur et al. (1970) and Guze et al. (1967) that there appears to be an excess of criminality, sociopathy, depression, and "abnormal personality" in the families of alcoholics, but no significant difference in the incidence of schizophrenia, mania, mental retardation, or epilepsy. Regardless of whether the association is biological or environmental in origin, the data suggest that the alcoholic's psychopathology presumably antedated the alcoholism.

Having accepted the position that individuals with a wide variety of predisposing psychopathologies become alcoholic, it becomes necessary to stress the full influence of social forces. As a general principle, which we have labeled the psychosocial equation, the general level of psychopathology in a given alcoholic population increases inversely to the prevalence of alcoholism in that subculture. For example, Jewish alcoholics show, in a statistical sense, more psychopathology than Irish or black alcoholics. This does not imply that some Irish and black alcoholics may not have severe psychopathology but a greater percentage of them might have minimal or no psychopathology than would be true among Jewish alcoholics. The case would be similar for men alcoholics as opposed to women. Consequently, many Irish and Scandinavian males, advertising executives, bartenders, and ghetto alcoholics appear to have little or no psychopathology when their alcoholism has been curtailed.

This overall conception attempts to reconcile many of the conflicting positions on the role of psychopathology in the pathogenesis of alcoholism. The question is far from resolved, yet it is not necessary to have the final answers in order to be able to institute rational treatment. The presence of psychopathology in the full-blown alcoholic is too apparent to require verification. At this point, he needs supportive psychotherapy. When the alcoholism has subsided, the astute clinician can evaluate the patient for presence or absence of psychopathology and treat or not treat accordingly.

Psychological Consequences of Alcoholism

Psychological Deterioration with Prolonged Alcoholism. As previously indicated, these effects may be organic, due to direct neurological impairment; functional, due to the disruption of personal and social relationships; or, perhaps most frequently, a combination of these. However, above and beyond these overt syndromes, clinical experience suggests that neurological deficits may be expressed in terms of increased sensitivity to the reestablishment of physical dependence. This has been described in the previous section on biological mechanisms as probably involving biochemical and neurophysiological changes in the brain as the result of the prolonged toxic effects of ethanol, with its concomitant effects of physical dependence. Here we are speaking of more macroscopic evidence of brain damage where cortical atrophy in some way increases these susceptibilities. In any event, both the behavioral and pathogenetic implications of the chronic brain syndrome in its clinical or subclinical manifestations must continually be borne in mind as an important complication of alcoholism and as one of the "special" problems requiring most intensive therapy.

The more functional elements of psychopathology which stem from prolonged alcoholism derive from the steady deterioration in personal and social relationships which accompany the alcoholic's decline into the addictive cycle. Even prior to the loss of social supports, the alcoholic finds himself continually anxious, depressed, and concerned about his loss of control and ultimately develops a complete loss of self-confidence and self-respect. When, as a result of his continued drinking, he loses first his job, then friends, then spouse, then home, the loss of self-confidence and self-respect are total. At this point, these functional deficits are translated into chronic anxiety and depression, for which the only readily available antidote is more alcohol. Accordingly, the functional psychopathologic sequence only adds to the biological addictive cycle to perpetuate drinking.

Social Mechanisms

The contributions of social disruption to the initiation of alcoholism, its perpetuation, or its reactivation, occur at every level of social, cultural and economic development (Kissin and Begleiter, 1976). Perhaps most helpful in our present analysis is a socioeconomic classification of alcoholism based on arbitrarily assigned levels of social and economic intactness, since, as we shall see, this parameter is of the highest significance in determining prognosis and treatment outcome.

Socioeconomic Classification of Alcoholism

There are several experimental findings associated with the socioeconomic classification which give that classification special significance. One is the frequently reported finding that socioeconomic stability is perhaps the single most important demographic factor in determining prognosis in the treatment of alcoholism (Kissin et al., 1970; Baekeland et al., 1975). Second is the interaction of the psychological and socioeconomic factors in determining prognosis. This may reflect the effect of level of intelligence and/or psychopathology on the level of socioeconomic achievement or conversely, the effect of socioeconomic deprivation on the development of intelligence and/or psychopathology. Third is the special importance of social rehabilitation as opposed to psychotherapy in the case of the socially disadvantaged alcoholic, particularly the dyssocial ghetto alcoholic.

The socioeconomic spectrum runs from the alcoholic executive to the skid row derelict with all levels in between. There are no adequate statistics on the exact proportions of these groups, but general estimates may be based on data extracted from Cahalan et al. (1969). The major categories, based on social, psychological, and alcohol behavior characteristics, are as follows: (a) the upper- and middle-class white-collar alcoholic, (b) the lower-middle- and upper-lower-class blue-collar alcoholic, and (c) the inner-city alcoholic and the skid row alcoholic. These categories are obviously not distinct and clear-cut since there is a fair degree of overlapping. The list is urban in nature both because socioeconomic stratifications tend to be more obvious in the city than in the country and because alcoholism tends to be more prevalent in the city. However, the corresponding suburban and rural equivalents for these classes are readily conceptualized.

The Upper-Class and Middle-Class Alcoholic (20-25 percent [?] of All Alcoholics). The demographic characteristics of this group are generally defined by their place in the socioeconomic classification. They are economically secure and socially accepted. They tend to be well educated, often of superior intelligence, and generally worldly and sophisticated (Plumeau et al., 1960). However, in other respects there is wide variability. Some are married with a personally stable environment; others are single or divorced with little personal stability despite their socioeconomic status. Some are working and productive; other are leading a parasitic existence. The similarities are the consequence of their socioeconomic status; the differences tend to reflect the wide variation in psychological dynamics and adjustment.

The psychopathology in this group runs the entire gamut described in the previous section—the character disorder, the psychotic, and the neurotic—with a not too dissimilar distribution. However, clinical experience leaves the impression that there may be a higher percentage of neurotics in this group than in any other. This impression is partly fed by the fact that this group tends to select and perhaps do best in analytically oriented psychotherapy which, as we shall discuss, appears to be most effective among neurotics. However, this may be a self-perpetuating artifact since it is only the economically secure and the intellectually sophisticated who are drawn to psychoanalysis.

The manifestation of the dyssocial reaction among the alcoholics of this class is also unique and quite obviously of a different order than the dyssocial reaction in the ghetto alcoholic. Here it tends to occur as an occupational hazard—the executive in advertising or public relations, who finds alcohol a necessary and ubiquitous social lubricant in his daily activities. In these cases, the social situation leads to an ever increasing ingestion of alcohol which may tap even minimal latent psychopathology to produce psychological dependence and, if carried far enough, ultimately tolerance, physical dependence, and chronic alcoholism. However, among these individuals where psychopathology is basically minimal, progression to the later stages is unusual.

Not uncommon are the various character disorders which we have described. These, in turn, may manifest themselves in a variety of ways, depending on which element is dominant—the hypersensitive, the schizoid, the mystical, the psychopathic, the depressed, or the self-destructive. Although all of these elements may coexist, the unique individual life history will vary considerably within the limits of the underlying personality structure.

Finally, as in all groups, a proportion of this class may have a latent or overt psychosis—either schizophrenic, depressive or manic depressive. However, because of the socioeconomic supports, these may be more successfully covered over or obscured, with the alcoholism actually helping in the process of ostensible compensation. When latency becomes overt, the compensatory mechanics break down and then the psychotic diagnosis often assumes dominance over that of the alcoholism. In these instances, the concept of alcoholism as a symptom of an underlying psychopathology is most clearly delineated.

The Lower-Middle- and Upper-Lower-Class Blue-Collar Alcoholic (40-50 percent [?] of All Alcoholics). This population, which probably constitutes the single largest segment of the total alcoholic population, is quite separate and distinct in its characteristics from the other two major groups. This middle group may be intermediate in socioeconomic terms, but psychologically, socially, and therapeutically their characteristics are in many respects not only quantitatively but also qualitatively different from both the groups above and below it on the socioeconomic scale.

Essentially, the blue-collar alcoholic differs mainly from the white-collar alcoholic in his level of psychological sophistication and from the inner-city and skid row alcoholics in his level of social stability. The first difference reflects cultural influences rather than any major difference in personality structure. It manifests itself in the blue-collar alcoholic as a special intolerance to probing insight-oriented psychotherapy. In this respect, he resembles the inner-city alcoholic (but surprisingly, not necessarily the skid row alcoholic). Accordingly, psychotherapy in both the blue-collar and inner-city alcoholics should probably be reality oriented and experiential rather than uncovering in its thrust.

In terms of his social stability, the blue-collar alcoholic resembles his white-collar counterpart more than he does the inner-city or skid row prototype. He tends to have at least a moderate level of social stability, most often having completed high school, having developed some type of vocational skill, having held a job, a home, and some family structure about him. Although his economic and social resources are not as great as those in the economic class above him, they are usually sufficient not to create a special problem.

The Inner-City and Skid Row Alcoholics (25–30 percent [?] of All Alcoholics). These two groups of alcoholics have several major characteristics which distinguish them from each other but in most respects—particularly those involving therapeutic considerations—they are sufficiently similar to warrant treatment as a single major segment. However, their differences are significant and must be commented upon.

First, the question of incidence. The fact that skid row alcoholics constitute only a small portion of alcoholics (3–5%) has been widely publicized (Cahalan *et al.*, 1969). This would leave the inner-city alcoholic to constitute some 20–25 percent of the total alcoholic population, a figure also derived from the statistics of Cahalan *et al.* (1969). Unfortunately, the fact that the skid row alcoholic constitutes such a small segment has been extensively used to support the argument that alcoholism is not a condition of the socially disrupted but rather of the socially intact. Although this is still largely true (the white-collar and blue-collar segments together constitute some 70–75 percent of all alcoholics) it is nevertheless also true that the socially disrupted alcoholic constitutes some 25–30 percent of that population. This statistic is of particular importance for public health program planning since the services and types of facilities necessary for these groups are quite different from those necessary for the more socially intact alcoholics.

Other differences between these groups are apparently mainly a function of subcultural patterns in handling sociopsychopathological behavior. The white psychotic alcoholic who deteriorates socially is no longer able to maintain his residence in his community and migrates either to a skid row community like the Bowery in New York City or to a similarly isolated, single-room occupancy, deteriorated neighborhood like the Upper West Side. The black or Puerto Rican deteriorated psychotic alcoholic, on the other hand, can continue to maintain himself in his ghetto area, sleeping either in abandoned houses or in an isolated room of his own. However, here, too, there is a significant difference in that many white skid row alcoholics have been educated and have held jobs while most black and Puerto Rican inner-city alcoholics often began drinking at a very early age and have neither education nor vocational skills, a difference which becomes significant in the rehabilitation process of these groups.

Apart from the higher incidence of psychopathology and of educational and vocational skills among the skid row alcoholics, these two groups are remarkably similar in their demographic and medical profiles. Feldman *et al.* (1974) reported in a comparative study of skid row alcoholics and inner-city alcoholics that in terms of total social, psychological, and medical pathology, the only highly significant difference was in ethnicity. Of special interest was the fact that the inner-city alcoholic showed, if anything, greater signs of malnutrition and of medical complications than did his skid row counterpart. These differences were due, probably, to the charitable activities of the Salvation Army and other social agencies on the Bowery, with benefits not readily available to the alcoholic citizen of the ghetto.

Family Disruption Among Alcoholics

Family disruption, either as a predisposing causative factor or as a consequence, is so commonly associated with alcoholism as to constitute one of the major symptoms of the syndrome. In the area of predisposition, many alcoholics come from broken homes, not infrequently because of one or two alcoholic parents. In rarer instances, an unhappy marriage of the alcoholic may itself act to initiate or aggravate the problem. Finally, once alcoholism has entered its more severe phases, some degree of family disruption almost invariably develops. This may be due to the reaction of a perfectly normal individual to the self-destructive and other destructive behavior of her spouse or it may involve a level of interaction of the special psychopathologies of the husband and wife. In any event, once family disruption has taken place, a vital element necessary for independent social rehabilitation has been lost. The plunge into chronic alcoholism becomes more rapid and complete and the necessary resources for rehabilitation less available.

Needless to say, family disruption cuts across all of the classes previously delineated in our socioeconomic classification. Here, too, possibly for social and economic reasons, the overt disruption becomes the greater as one goes lower in the socioeconomic scale. However, this is not to imply that the levels of psychological support one can hope to get, for example, from the wives of white-collar alcoholics are necessarily greater or better than those one might get from the wives of blue-collar alcoholics. The opposite is often only too commonly the case. Just as psychopathology may be greater among white-collar alcoholics, so the interaction among white-collar spouses may be "sicker" than that at lower socioeconomic levels. In general, one can state that the level of disruption is a function of the kind of pathological interaction that exists and of the common mode of handling problems in that particular subculture.

IMPLICATIONS FOR THERAPY

We have attempted to present an orderly description of the events occurring during the development of alcoholism and also of the pathogenetic mechanisms underlying those events. The susceptible individual drinks, develops primary psychological dependence which is translated clinically into craving, develops physical dependence which produces loss of control, is caught in the vicious addictive cycle, and becomes an alcoholic. This deliberately simplified account presents the basic elements of the sequence. The specific pattern for each individual alcoholic resembles the basic pattern but differs markedly according to the degree that one or another of the various pathogenetic mechanisms described is involved. Each individual falls into a different bio-psycho-social classification which describes his particular constellation of pathology and determines the outline of a rational therapeutic regimen. Yet, each individual also follows the "basic sequence" described above which is common to all.

Figure 3 is a three-dimensional model graphically illustrating the interaction of biological, psychological, and social influences. In this model, these influences are, respectively, degree of physical dependence, degree of psychological disruption, and degree of socioeconomic disruption. These are arbitrary variables and, as we shall see, others could be selected. However, these are particularly valuable for illustrating different kinds of alcoholics based on Jellinek's (1960) typology.

The heavy drinker has little or no physical dependence, only mild psychological maladjustment, and still probably little or no social disruption. The problem drinker, on the other hand, has already begun to have displacements in all three directions. The true skid row alcoholic has severe

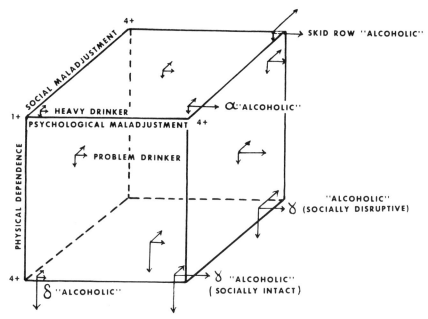

FIGURE 3. The bio-psycho-social dimensions of alcoholism.

psychological and social disruption but often very little physical dependence. The neurotic alcoholic who drinks only situationally (the alpha alcoholic in Jellinek's typology) has predominantly psychological problems and will often respond to supportive and directive psychotherapy. The delta alcoholic has developed marked physical dependence and cannot stop drinking but, as in the French peasant, maintains a fair level of psychological and social adjustment. The gamma alcoholic is usually more erratic in his drinking and psychological behavior and may or may not show social deterioration, depending on his social class and the strength of his family support. These represent some extreme types of alcoholism but obviously many other intermediate types are possible.

Within this context the rational therapeutic approach presents a series of interventions directed toward interrupting vicious cycles where they exist, extinguishing pathological conditioned reflexes and changing pathological psychosocial states. The sequence of interventions is determined by the logical relationship among these conditions, which it is necessary to relieve before one can intervene meaningfully in the next situation. There is almost a common sense logic reinforced by clinical experience which determines the sequence. To the alcoholic in delirium tremens, questions about his social stability are academic. To the dried-out derelict who has no home, no clothes, and no food, questions about his psychological immaturity are irrelevant. Some problems, particularly the medical and social, are both more acute and more amenable to treatment than the "core" problem of converting an alcoholic into an nonalcoholic. Since, as we have shown, those medical and social problems are often part and parcel of the alcoholic process, it is essential that they be attended to as rapidly as possible before a direct attack on the core problem is attempted.

Based upon this conceptualization, the natural sequence of therapeutic interventions, in order of urgency and effectiveness, is as follows: (1) breaking the addictive cycle, (2) treating special medical and social problems, and (3) treating the core problem. Breaking the addictive cycle means adequate detoxification of the acutely inebriated binge drinker or stabilization of the alcoholic during withdrawal. For such individuals, until the vicious cycle of loss of control is ruptured, no other intervention can be very meaningful. The treatment of special medical and social problems represents the next level of urgency and speed of intervention. The medical problems, both physical and psychological, are most often treated chemotherapeutically. The socioeconomic deficiencies of the patient's situation are readily approached although their complete rectification may take a very long time. More difficult to treat are the disruptions in family structure which often impinge upon the core problem. Finally, the core problem itself has two major components: (a) extinction of primary psychological dependence, the basic drive in alcoholism, and (b) modification of the individual's "susceptibility." The interaction of these elements is graphically illustrated in Figure 4.

Breaking the Addictive Cycle

The significance of the addictive cycle in perpetuating alcohol seeking behavior, either through loss of control during an actual drinking bout or through craving once active drinking has stopped, cannot be overstated. Of some importance is the question of the relative strengths of loss of control versus craving in perpetuating drinking. On both theoretical and clinical ground, one would postulate that the effects of loss of control would be greater. It is more biologically rooted, more directly related to acute withdrawal symptomatology and thus probably more potent. Craving, on the other hand, is in Ludwig's terminology a *subclinical conditioned withdrawal response* which occurs only after drinking has stopped and withdrawal symptomatology has subsided. Presumably, insofar as it is less acute, it is less powerful. Clinical experimentation appears to support this hypothesis. Platz *et al.* (1970) have shown that the single most important element in determining outcome in treatment in an outpatient setting is the drinking status of the patient at the time of admission. Whereas 69 percent of those abstinent at admission remained

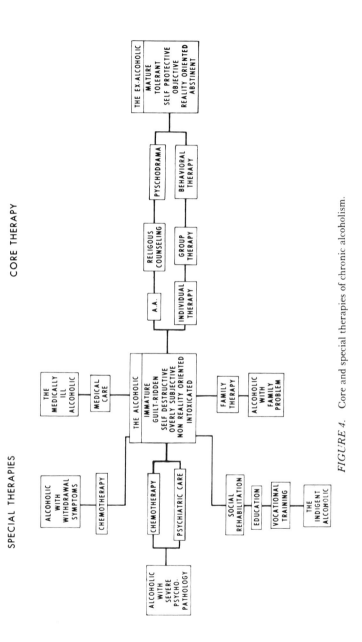

FIGURE 4. Core and special therapies of chronic alcoholism.

abstinent at the six-month follow-up period, only 19 percent of those drinking at the time of admission had been able to achieve abstinence six months later (a difference significant at $p < 0.001$). One explanation offered is that the patient who applies for admission while abstinent is better motivated than is the one who comes in intoxciated. That may be so, but it takes a great act of motivation to make the leap from intoxication to abstinence, to dry oneself out without medication. All of these reflections yield the same hard fact: it is harder to become dry than to stay dry. Stated otherwise, loss of control is an even more powerful drive to alcohol-seeking behavior than is craving.

These considerations are more than semantic. They suggest that the single most important intervention in the alcoholic on an acute binge is to help him to stop drinking *while controlling withdrawal symptomatology to the greatest possible extent*. The adequate and proper control of the acute withdrawal syndrome is not only a health- and sometimes life-saving procedure, it is the first necessary step in the rehabilitation process. Detoxification in the acute withdrawal syndrome of alcoholism follows the same principle as it does for the treatment of physical dependence to any depressant drug—the substitution of a long-acting, cross-tolerant drug for the short-acting addictive agent. In heroin addiction, the treatment of choice is methadone detoxification; in alcoholism, the treatment of choice is one of the benzodiazepines (chlordiazepoxide or diazepam). In heroin addiction, where withdrawal symptomatology is neither severe or prolonged, methadone detoxification usually takes several weeks. In the alcohol withdrawal syndrome, where, particularly in severe cases, symptoms may be extremely serious and prolonged, benzodiazepine detoxification should be continued until all symptoms are controlled. This is usually indicated for at least one week but may be necessary for two or three weeks after the cessation of drinking.

It is in the period directly following detoxification that that element of craving due to physical dependence is at its strongest (Hore, 1974). With the passage of time, it slowly subsides. Whether this reduction is due to the gradual extinction of the acute withdrawal symptoms, or rather to the gradual extinction of the *subclinical conditioned withdrawal syndrome* as postulated by Ludwig *et al.* (1974), is difficult to assess. More likely, it is due to an interaction of both of these effects. In any event, to the extent that these mechanisms do persist into the postwithdrawal stage, they constitute a powerful influence toward the continuation of drinking and must be confronted realistically. This question is considered again as the first of the special problems in the rehabilitation process.

Special Problems in the Rehabilitation Process

When the addictive cycle has been broken and the craving of the early postwithdrawal phase controlled, the primary underlying pathology which led to alcoholism in the first place surfaces. Here, now, we are faced therapeutically with alcoholism as a symptom, a symptom of many possible underlying diseases. These may be, as we have said, biological, psychological, social, or some combination thereof. It is incumbent upon the therapist at that point to attempt to delineate the specific nature of the underlying illness and, where possible, to prescribe the specific indicated therapy.

But regardless of what the underlying cause may be, operationally the process of rehabilitating the alcoholic is essentially a psychological one. This statement is based on the dictum, previously enunciated, that the main drive in all forms of drug dependence is primary psychological dependence. If, after the breaking of the addictive cycle, the role of craving due to physical dependence has been minimized, we are returned to the earlier phase where the major problem is primary psychological dependence. The core problem then is a psychological one and the core therapy must likewise be psychological.

These constructions are schematized in Figure 4. At the center of the chart is the active alcoholic who is characterized as immature, inadequate, passive, dependent, impulsive, alienated, anxious, depressed, nonreality oriented, and often intoxicated. Ideally, at the other end of the psychotherapeutic process, he emerges mature, adequate, controlled, relating, realistic, and abstinent. The core therapy is directed at achieving this metamorphosis. Obviously, it is only more or less successful. Nevertheless, until and unless some such change has been effected, treatment cannot be said to have been totally effective.

However central this psychological process may be to the ultimate rehabilitation of the alcoholic, it is often impossible even to initiate until more pressing and urgent matters are dealt with. These are the special problems enumerated in Figure 4. Just as these special problems did not become preeminent until the addictive cycle had been broken through adequate detoxification, so the core problem often cannot be addressed until some of the special problems are resolved. Again, these distinctions are not absolute. Elements of the addictive cycle carry over into the special problems; similarly, special problems may heavily influence or even determine the core problem and its treatment.

Treatment of Persistent Physical Dependency Symptomatology

As described, when the addictive cycle, with loss of control during an active drinking bout, has been broken, that element of craving due to residual physical dependence persists to a greater or lesser extent. To the extent that it does persist, it remains a potent influence toward continuing to drink. How long does that influence persist and to what extent must treatment be provided for it? The answer probably is: till that point where residual withdrawal symptomatology no longer plays a significant role in determining treatment outcome. That point is highly individual and differs from alcoholic to alcoholic, varying with the strength of physical dependence, the motivation of the individual, the available internal and external psychological and social resources, and on certain intangibles as well.

Accordingly, there appears to be a meaningful role for the use of cross-tolerant sedatives such as chlordiazepoxide, for at least a short time after detoxification in certain patients to control persistent withdrawal symptomatology and the craving associated with it. However, sedative treatment in alcoholism, although it has its place, is also susceptible to serious abuse. Accordingly, this entire question will be explored in greater depth under the section The Use of Psychoactive Drugs in the Treatment of Alcoholics in the next chapter.

Treatment of Medical Complications of Alcoholism

Where medical complications of alcoholism exist, it is essential to establish the causal relationship between the alcoholism and the medical illness in the mind of the alcoholic. Once the relationship has been definitively established, it can be used to control drinking, provided the patient is sufficiently motivated to want to recover. Most often, the discomfort of the physical malady acts as a powerful motivating influence. However, it is very important that improvement in the physical condition should accompany improvement in the drinking pattern. Continued deterioration in the medical illness, despite abstinence, only leads to a sense of discouragement which ultimately leads to drinking. Consequently, extremely active medical therapy must accompany intensive total care to ensure the overall recovery of the patient.

Medical Treatment of Psychopathology

When the biological addictive cycle has been broken, the psychological component remains. The anxiety and depression, the loss of self-confidence and self-respect persist even when the tremulousness subsides. The concomitant prolonged disturbing insomnia which characterizes the postaddictive cycle period may have its origins as much in the psychological anxiety and depression as in the biological. Accordingly, the therapist must be acutely aware of the chronic nagging sense of inadequacy which the alcoholic has in the first few weeks after the addictive cycle is broken and must use the element as an important instrument in his therapy. The possible benefits and dangers of using chlordiazepoxide and tricyclic antidepressants during this period will be discussed in the next chapter.

Equally pertinent is the medical treatment of psychopathology where the underlying condition is that of a psychosis, either schizophrenia or manic depression. The appropriate prescription of phenothiazines for schizophrenia or agitated depressions is, of course, necessary medical practice. The existence of a complicating alcoholic condition does not diminish the necessity for treating the underlying psychopathology. Recently Kline et al., (1974) have reported the effective use of lithium in the treatment of alcoholism. This would seem to constitute especially rational treatment where the alcoholic is suffering from an underlying manic depression but there may be some additional justification for its use. In a recent study, Ho and Tsai (1975) demonstrated that in mice, lithium pretreatment sharply reduces preference for alcohol. While the relevance of this to craving for alcohol in humans remains speculative, the association remains an interesting one.

Social Factors

The relevance of these factors pertain mainly to the parameter of social stability since this variable plays the major role in determining the appropriate social intervention.

Socially Stable Alcoholics—White-Collar and Blue-Collar. The individuals in these classes have several advantages going for them which are more true at the top of the socioeconomic ladder and become less so as one descends on the scale. Chief among these is the fact that socioeconomic status tends to correlate with several factors which have been shown to have positive prognostic significance: (a) intelligence, (b) job stability, (c) home stability, and (d) sources of economic, personal, and social support (Kissin et al., 1970; Gerard and Saenger, 1959). The opposite side of the coin which we have expressed is the fact that the greater the social security the alcoholic enjoys, the greater the psychopathology one must assume to be active for him to adopt the basically destructive life-style of the alcoholic. Nevertheless, because of the high correlation of social stability with positive prognosis, the individuals in this group tend to have a higher overall success rate.

Socially Unstable Alcoholics—White-Collar and Blue-Collar. The studies of Kissin et al. (1970), Mindlin (1959), and Gerard and Saenger (1959) all indicate that social instability—regardless of the socioeconomic class of the alcoholic—is a negative prognostic sign. Loss of job, home, family, and friends in a middle- or upper-class alcoholic presumably speaks for (1) greater severity of alcoholism, (2) greater psychopathology, (3) fewer available internal and external resources. Each of these factors contributes to a poor outcome. Accordingly, severe social disruption, regardless of the socioeconomic level at which it occurs, always carries an ominous significance. However, in the higher socioeconomic groups, rehabilitation is not so much a matter of providing a home, job, etc., as in the lower socioeconomic groups. It involves more (1) interruption of the disease alcoholism through thorough detoxification, (2) strong psychological intervention, and (3) reconstitution of those supports previously available (i.e., family, friends, and occupational contacts).

Of interest here is the especially successful role which Alcoholics Anonymous has played in the alcoholics of these classes (Baekeland et al., 1975). The role that Alcoholics Anonymous plays in the recovery of upper- and middle-class alcoholics is often a function of the social and psychological interaction of individuals similarly placed in the social scale. This group psychotherapeutic interaction is of a different order from that of the one-to-one level of clinical psychotherapy and may be a classical example of the greater effectiveness of "existential" and "social psychological" approaches to the treatment of character disorders in general and of alcoholism specifically.

Socially Unstable Alcoholics—Skid Row and Inner City. Therapeutically, these two groups may be treated similarly, although the previously described differences in degree of psychopathology and in educational and vocational skills obviously require appropriate differences in emphasis. However, in both of these groups, the critical element is the total absence of socioeconomic stability or even the potential for developing such a structure. In these individuals, the first and foremost thrust must be directed toward providing the primary necessities for existence—a place to sleep, food to eat, clothes to wear. Only after at least the skeleton of socioeconomic stability is provided can the core problem of psychopathology be approached. In most of these individuals social disruption has progressed to such an extent that long term inpatient care is often necessary (Kissin et al., 1970). In some cases the individuals have totally lost (or never developed) the capacity for a normal independent social existence and must be provided with "residential communities" which will act as their "family" structure for the rest of their lives. Inpatient rehabilitation wards and halfway houses are similarly needed to provide either shorter term or longer term security as needed.

On the other hand, where some semblance of social and family support remains—as is often true among the inner-city alcoholic population—these resources should be mobilized to the fullest extent possible. They serve a double function—first, to provide an additional support during rehabilitation and secondly, to provide the basis for an ultimate return to an independent social status. In this way, a fair proportion of these individuals may return to a more stable pattern of social and family relationships.

Family Problems. Family support can be one of the most helpful adjuncts in the treatment of the alcoholic where it is still available. The success of Al-Anon and Alateen—in helping the alcoholic to achieve both sobriety and social rehabilitation, and in helping the spouses and children to understand their own problems—has been substantial. Despite the anecdotal incidence of pathological wives who consciously, or more often unconsciously, seek to maintain their husbands' alcoholism, there seems to be little doubt that most alcoholics' wives wish desperately—both consciously and unconsciously—for their husbands' rehabilitation. These women (or men, where the wife is alcoholic) represent a tremendous resource where they are willing and able to be supportive and when their support is intelligently mustered. Family therapy can be one of the most effective treatment modalities in the armamentarium, particularly where the spouse is strong and intelligent and properly motivated toward aiding the alcoholic.

On the other hand, unfortunately the opposite situation also sometimes exists. Some married individuals or parents consciously, or more often unconsciously, encourage their spouses or children to continue drinking since this permits them to retain control over the psychologically and socially debilitated alcoholic. These instances require the most sensitive and expert management by the family therapist. Sometimes the individuals involved may be able to gain sufficient insight to repair their attitude and behavior. In other cases, only a complete breaking off of this destructive relationship will permit the alcoholic to achieve abstinence and self-respect. The possible pitfalls for patients, relatives, and therapists in this kind of psychological entanglement are only too apparent and only the most experienced family therapist should chance them.

The threat of a spouse to separate is one of the more powerful motivating influences for an alcoholic to stop drinking. Clifford (1960) found that the threat of separation was the single most important element bringing alcoholic men into treatment. This has variously been interpreted as a threat of total social disruption or as anxiety-related to the breaking of the last remaining barrier to latent homosexuality. In any event, the threat of separation can be used by the spouse (and the therapist) to motivate the alcoholic to make a giant step toward abstinence.

The Core Problem

When the addictive cycle has been broken and the special problems of the individual alcoholic attended to, the therapist is confronted with the core problem, the set of circumstances which underly the psychological dependence on alcoholism. Different therapists emphasize different aspects of the problem, some stressing the biological predisposition (genetic and physical dependence), some the psychological, and still others the social. As we shall see, these different emphases determine to a large extent the form of therapy offered and result in a spectrum of treatment "models"—medical, psychological, and social. However, despite their widely divergent approaches, proponents of all models tend to agree that the dynamic of most alcoholics can be described as an aberration in "coping" and it is in these terms that we shall introduce the general approach to the core problem.

It is a truism that alcoholics cannot cope. They cannot deal with the normal frustrations and irritations of the external world nor can they deal with the anxiety, depression, and sense of inadequacy which swells from within.

Accordingly, they drink, and by drinking they are able to ignore (although they do not reduce) their external and internal problems. Thus alcohol becomes their method of coping with the problems of life.

The therapy of alcoholics is largely a process of education through which alcoholics learn that drinking is an inadequate coping mechanism which can be replaced by healthier and more salutary ones. However, there are many different opinions as to how this change is to come about. The behaviorists believe that negative conditioning to drinking associated with positive conditioning to some alternative life-style is an effective technique. The psychoanalysts tend to stress the dynamics of the alcoholic's inner life which should be remolded if alcohol is to be abandoned. The existential psychiatrist believes rather in developing with the patient alternative life-styles which are more gratifying while the social psychiatrist stresses the impact of peer groups on individual behavior. The sociologist indicates that the problem may be almost entirely social requiring societal changes, while at the other end of the spectrum the biologist may advocate a more pharmacologic approach.

Accordingly, the translation of one's understanding of the patient's coping mechanism into a meaningful therapeutic approach depends on the therapist's overall view of human dynamics. This is a major element in the treatment of the core problem and will be discussed under the heading of the various treatment models.

TREATMENT MODELS IN ALCOHOLISM

The idea of developing theoretical models for a variety of psychiatric syndromes has been promulgated by Siegler and her group, who have presented different models for psychiatric illness, drug addiction, and alcoholism (Siegler et al., 1968). Although we shall use their classification for alcoholism in part, we shall base our classification more specifically on the variety of mechanisms as described thus far in this chapter and on the models which have resulted in specific treatment systems. Our list is by no means all-inclusive since the Siegler et al. (1968) classification includes models which are not based on the theoretical discussions thus far described in this chapter (e.g., their "moral" and their "family interaction" models). Nevertheless, there is a certain degree of overlap with their classification as there is with that of Pattison (1974) who has taken a similar approach.

The Medical Model of Alcoholic Rehabilitation

The medical model is based on the general thesis that alcoholism is a disease and as such is most properly treated by a physician. This conclusion, which follows logically from the disease concept of alcoholism, is so basically distasteful to the proponents of the social model of alcoholic rehabilitation that they have begun to abandon that concept (Hershon, 1974). This they have done with some reluctance and ambivalence since the disease concept has given alcoholics and alcoholism a level of acceptance and respectability which they had not previously enjoyed. Nonetheless, as we shall see, the consequences of the disease concept as represented in the medical model are so unacceptable to the social model proponents that they are often willing to throw away the baby with the bath water.

The medical model tends to stress the biological mechanisms thus far discussed in this chapter, with only little attention paid to the psychological problems and still less to the social. Since the major medical approach is chemotherapeutic (at least where surgery is not directly indicated), greater emphasis is placed upon the use of tranquilizers and deterrent medications (disulfiram) than on any other approach. Jones and Helrich (1972) found that over 65 percent of 15,000 physicians who responded to a mailed questionnaire used one or more tranquilizers as well as other medication in the treatment of chronic alcoholism.

In the medical model, the physician—most often a general practitioner or internist—is the primary therapist. Typically, he detoxifies the patient either in a hospital or on an ambulatory basis with one of the benzodiazepine tranquilizers (chlordiazepoxide or diazepam) and continues the patient on that medication for a period of at least several weeks to a few months. At first he sees the patient weekly for a 15- to 20-minute interview, during which he focuses largely on medical symptoms—nervousness, insomnia, appetite, weight, and sense of well-being. If, as is so often the case, there is a moderately enlarged liver secondary to fatty infiltration, he emphasizes the gradual decrease in size to normal which usually takes place. Psychological counseling, if any, is usually entirely supportive and encouraging, with an occasional discrete reminder of the damage which the recommencement of drinking can cause, physically, psychologically, and socially.

This regimen, which is being followed in the office of many physicians around the country, is often fairly effective. How effective, it is difficult to say, because no one appears to have made an adequate survey of the private treatment of chronic alcoholics by physicians. On the other hand, in their important study of the ambulatory treatment of chronic alcoholics in outpatient clinics, Gerard and Saenger (1959) found that medical physicians (internists and general practitioners) were the most successful of all primary therapists, followed by social workers and, significantly further behind, psychiatrists. These results may reflect some correlation with the relative effectiveness of the medical, social, and psychological models, respectively, although no real conclusions can be drawn since patients were not randomly assigned to different treatment modules.

If one considers that there are close to 75,000 general practitioners and internists in the United States, one can readily calculate that if each one were to treat ten alcoholics in his practice, this would constitute a major treatment resource for alcoholism. Clearly, this is an area of treatment potential that has remained neglected. There has been neither sufficient evaluation of the effectiveness of the medical model nor sufficient effort to motivate physicians to undertake alcoholics' treatment. More shall be said about this in the next chapter on The Role of the Physician in the Rehabilitation of Chronic Alcoholics.

The Behavior Modification Model

The behavioral model has been somewhat less comprehensive in the theoretical system which it has developed relative to the pathogenesis of alcoholism and to its treatment than have been the psychological and the social models. Essentially, it has focused more on the question of the dynamics of the conditioning paradigm involved in the development of primary psychological and less on the question of individual "susceptibility," the factor most emphasized in the other two models. Here, craving is seen entirely as a manifestation of primary psychological dependence, which is seen as a positive reward operant conditioning paradigm. In these terms, the techniques for treating alcoholism are essentially the techniques of behavior modification therapy. These include the older techniques of aversive conditioning but also the newer and theoretically more effective techniques of positive reinforcement behavior modification.

The aversive techniques include apomorphine and emetine therapy which produce nausea and vomiting when given concomitantly with alcohol, succinylcholine therapy which produces apnea, and electric shock therapy. Disulfiram (Antabuse) therapy also falls into this category, especially as it was used in its earliest clinical applications. At that time, after a sufficient build-up on disulfiram, challenge doses of alcohol were given to demonstrate the aversive effect. More recently, as the disulfiram–alcohol reaction has been more widely publicized, the challenge doses of alcohol are generally omitted. In these instances, disulfiram is thought to act as a deterrent drug rather than as an aversive agent.

A variety of modifications have been used in the aversive techniques. *Simple classical aversive conditioning* involves the administration of noxious stimuli associated with ingeston of alcohol. In *instrumental avoidance conditioning* one can avoid shock if he quickly rejects the alcohol-related symbol. In *instrumental escape conditioning,* the alcoholic escapes shock by spitting out the alcohol. Still other forms of aversive treatment are *aversive imagery techniques* and *relaxation aversion.*

The positive reinforcement version of behavioral modification is based on the concept that a conditioned reflex may be more rapidly extinguished by rewarding a nonresponse (where the response is drinking) than by punishing that response. This technique is known as operant conditioning. Its implications spread beyond simple laboratory operant conditioning situations and are thought to be applicable in every phase of therapy. For example, if maintaining abstinence and demonstrating mature behavior in a ward situation is rewarded by the compliments and respect of the staff and other patients on the ward, this is indeed a form of positive reinforcement operant conditioning. In this respect, behavior modification therapy spreads far beyond laboratory treatment; it becomes inherent in both the psychological and social models to be discussed. In this form, modern behavior modification theory has extended its techniques to a much larger arena than was true with the older aversive model.

Systematic desensitization is a technique intermediate between the other two which utilizes relaxation techniques or hypnosis to reduce the anxiety that is presumed to be causative to alcohol ingestion. A variety of new techniques which may be related to this are transcendental meditation, yoga, acupuncture, biofeedback, and muscle relaxation techniques. The rationale and the possible mode of actions of these therapeutic modalities remains to be clarified.

The Psychological Model

This model gives at least partial credence to the psychoanalytic concept of the alcoholic personality—but only partial. Although most alcoholics appear to share many of the characteristics ascribed in Figure 4 to the active alcoholic, they do this only to a greater or lesser degree. The alcoholic personality is a vectorial predisposing force, just as perhaps are genetic predisposition, social forces, and so on. Indeed, the alcoholic personality may be the indirect outcome of some of these other influences, including prolonged alcoholism itself.

Whether one subscribes to the psychoanalytic frame of reference or not, the characterization of the alcoholic as emotionally immature, passive-dependent, with strong difficulties in relating meaningfully to others, is generally accepted by clinicians of all schools. What is more important in the classical psychoanalytic model of alcoholism is the conceptualization of the basic conflictual structure and the ego defense system involved. In the neuroses, there is the central ego-alien infantile conflict which the more mature ego represses into the unconscious. "Cure" cannot occur unless and until the unconscious conflict is brought into awareness and fully experienced. In the character disorders, the major problem is not a deep-seated irritant which must be surfaced; it is more serious than that. It is a lack of development to a sufficient state of emotional maturity with which the individual can deal with the real world. Consequently, here the major ego defense system is not repression;

it is rather denial, the most primitive of defenses. Here there is no deep-seated conflict to be brought to the surface; here there is rather the need to learn to relate to others in a mature and meaningful way.

These individuals have to learn that they can derive great satisfaction and pleasure from relating in a wholesome manner to other human beings. But these other human beings must be mature, tolerant, objective, realistic, active, and relating. Since not many alcoholics (or other humans for that matter) are blessed with the presence of such model companions, the patient must turn to his therapist to learn this new experience. It is in this role that the existential psychotherapist sees himself. As therapist he is warm, tolerant, and empathetic, attempting to establish as deep and personal a relationship with the client as is possible under the professional circumstances. He accomplishes this through different techniques than does the psychoanalytic therapist. But increasingly, clinical experience seems to bear out the greater effectiveness of this type of psychotherapy in character disorders than of the analytic variety (Glasscote *et al.*, 1967).

The general thesis developed here is supported by the studies of Whitehorn and Betz (1954), who reported that in treating psychiatric patients, the interaction between therapist personality and patient psychiatric diagnosis could be a most important element in determining outcome. These workers described two types of therapist—Type A and Type B—who could be delineated by their general attitudes in treating patients. Type B were more analytic and objective in their therapeutic approach, Type A more personal, more humanistic, and more involved. Type B therapists were most effective in treating neurotic patients, Type A most successful with schizophrenics. These therapists, working in an eclectic psychiatric milieu, exercised different approaches according to their own personalities and had varying success with different kinds of patients.

Within this construction, the psychotherapeutic approach to alcoholics to a degree becomes a function of therapist personality almost as much as of therapist professional competence. This viewpoint is supported by the work of Strupp and Bergin (1969) and Luborsky *et al.* (1971), who reported that such therapist qualities as "empathy," "warmth," "acceptance," and communality of interests with the patient have great impact on the outcome of therapy. It is within this context that one can appreciate the increasingly important role that the ex-alcoholic and ex-addict counselors have come to play in the treatment of these conditions. These individuals have presumably made the transition from alcoholic to ex-alcoholic (or from addict to ex-addict) through a relearning experience with the help of some other therapist (professional or paraprofessional) or through A. A., or through personal strength of character. They present both theoretical and concrete evidence that the transition can be made. Through their very existence, they represent a social pressure upon the individual alcoholic to change his ways.

Recognition of the importance of social pressures is the contribution of another major school of psychology, i.e., social psychiatry. The social psychiatrists and psychologists stress the importance of reeducating the patient with a character disorder in the ways of society, and stress equally the power of society to teach socially acceptable behavior. The character disorder alcoholic, fixated as he is at an early stage of psychosexual development, has not only not learned how to relate to others, he has not learned the social amenities involved in human interaction. It is this inadequacy in the area of social relationships, in addition to the difficulty in personal relating, that causes the alcoholic both to feel and to be alienated. It is also this pattern of inadequate social relatedness, together with the emotional immaturity, which contributes to the sense of psychopathy—probably more implicit than real—among alcoholics.

These elements of immaturity, inadequacy of personal and social relatedness, and an unwillingness to face reality constitute a constellation of inadequacies which are particularly susceptible to the techniques of social psychiatry. Chief among these is the power of group dynamics to influence and mold the individual. One can understand why group therapy, with its various confrontation techniques, has been so effective in the treatment of alcoholics and addicts. Even more, one can appreciate why A.A. has been probably the single most effective program for the treatment of alcoholics yet developed.

Finally, there are the many other adjunctive forms of therapy, all of which are directed toward the same general end: to allow the alcoholic to learn *through actual experience* that human and social activities and relatedness are highly rewarding experiences which will substitute for the chemical nirvana which alcohol offers, or—stated otherwise—that there are other and better coping mechanisms than alcohol. Hence various activities such as recreational activity, occupational therapy, and psychodrama help the individual, through actual experience, to taste the pleasures in life that one can derive from action in a social setting. Each of these activities constitutes, in a different way, a new set of human experiences which the individual can store in his repertory. Consequently, in the psychological model the core treatment is to rehabilitate the alcoholic or—since he never really did learn these basic social qualities—to "habilitate" him.

The social psychiatric contributions, particularly group therapy and social rehabilitation through reeducation, constitute the basic psychological approaches in the social model, the next to be described.

The Social Model

In the social model approach toward the pathogenesis and treatment of alcoholism, social forces resulting in psychological dependence are seen as the major factors in the development of alcoholism. Recently, there has developed a strong tendency among workers in the field to stress these mechanisms above all others as significant elements in the pathogenesis of alcoholism. For example, Edwards (1974) in a recent presentation on the dynamics of drug addiction stated: "What we term 'characterological disorder' may perhaps more often be more accurately interpreted as an overlearned coping mechanism—something which is not really a disorder. It may instead be a learned adaptive skill which is aiding the survival of that individual in adverse circumstances. We need constantly to be alive, not only to the psychodynamics of the individual but to the social realities of that individual's past life. And to survive in the ghetto may require a range of behaviors quite other than those which the middle-class norm would suppose."

This excellent statement, which accurately relates social mechanisms to the development of primary psychological dependence in that it suggests that susceptibility to that development may be of social origin rather than of psychological, is probably somewhat more pertinent in drug addiction than in alcoholism. But only somewhat more pertinent. The social factors—particularly if we include among them the standard influences of socioeconomic status, ethnicity, and subcultural mores, as well as the more insidious elements of family interaction—are indeed extremely important in determining susceptibility.

This view tends to "depsychologize" (at least in Freudian terms) the treatment of alcoholism in a number of ways. First, it emphasizes the influence of social factors in determining susceptibility to the development of psychological dependence and stresses the need to correct the social factors if the susceptibility is to be reduced. Secondly, it sees psychological dependence as a socially dependent coping mechanism which is best extinguished through a variety of basically social techniques. Even when these techniques are psychological, they tend to be more social than either biological or psychodynamic in their orientation. For example, while the appropriate social steps are being taken to decrease basic susceptibility—getting the alcoholic a place to live, clothes to wear, educational and vocational training, and ultimately a job—the psychological approach toward extinguishing the conditioned reflex which constitutes psychological dependence is basically that of behavior modification. But behavior modification as here conceptualized is more social than biological in its application. Aversion therapies, such as those involving apomorphine, succinylcholine, or electric shock, are seen as punitive rather than as constructive. Within the framework of operant psychology as used in the social model, the dominant technique toward extinguishing the conditioned reflex is reward for nondrinking rather than punishment for drinking. However, both appropriate reward and appropriate punishment are acceptable as techniques in the psychotherapeutic situation—either group or individual—where the social approval of one's peers is utilized as a powerful reward and the social disapproval as a *powerful yet constructive* form of punishment.

This viewpoint undoubtedly has much to commend it and has been translated operationally into an effective treatment system. Its strengths lie in a variety of areas. It has stimulated a much greater awareness of the importance of socioeconomic deprivation in contributing to the development of alcoholism and drug dependence. It has emphasized the more subtle influences which ethnic and subcultural patterns play in the development of alcoholism. It has also increased our awareness of the importance of family interactions at both the pathogenetic and therapeutic levels. It has provided a rational theoretical framework in which an effective treatment system could be implemented. That treatment system includes the establishment of a variety of facilities and services to meet the various socioeconomic needs of different alcoholics. It has incorporated the use of ex-alcoholic counselors as the most effective means of demonstrating the value of social rehabilitation. It has given a rational explanation for the success of Alcoholics Anonymous and a rational basis for its incorporation into the overall treatment system. It has utilized individual therapy but even more effectively, group therapy, as an instrument of behavior modification within a social psychological framework. All in all, it probably represents the most comprehensive and coherent model which provides for the most comprehensive and coherent approach to total treatment which we now have.

Despite these real and undeniable virtues, the social model, like most models which try to be all encompassing, does suffer from several deficiencies. Among these are:

1. It does not deal sufficiently with the psychological origins of increased susceptibility. Although adherents of the social model accept the role of psychopathology in increasing susceptibility to the development of psychological dependence, they tend to feel that the psychopathology is of lesser importance than the social pathology and indeed may be a consequence of the latter. Although there is unquestionably much truth in this, more so in drug addiction than in alcoholism, there is probably much truth in the position that underlying psychopathology is also an important contributing factor.

2. It does not incorporate the facts of physical dependence and the addictive cycle into the theoretical framework. In itself this would be of little consequence if there were not the inevitable tendency to put "theory into practice." Some consequences of this shortcoming have been (a) the description of the acute withdrawal syndrome, including delirium tremens, as a psychological rather than as a biological syndrome; (b) the consequent advocacy of the nonmedical treatment of the acute withdrawal syndrome (including DTs) with its inevitable casualties (Peterson, 1973); (c) the relative unwillingness to use sedative drugs during rehabilitation regardless of the needs of the alcoholics involved or the degree of their anxiety or depression; (d) the downgrading of

the role of the medical and mental health professions in the treatment and rehabilitation of alcoholism, a position which plays into the hands of those physicians who are only too ready to leave the treatment of alcoholism to A.A. and paraprofessionals. If this attitude had been generally accepted in the field of heroin addiction, methadone maintenance, the most effective treatment modality in its field, could never have been developed.

3. Because of its emphasis on the social aspects of the pathogenesis and treatment of alcoholism, and its current de-emphasis on the medical and psychological aspects, it tends to downgrade professionalism in general. If what best reinforces socially acceptable behavior in the alcoholic is the living example of a successful ex-alcoholic, then ex-alcoholic counselors and A.A. members are better equipped to deal with the problems of an active alcoholic than are the psychologists, social workers, and professional counselors, let alone the physicians and nurses whose orientation is even further askew. Perhaps this criticism is being overstated here but it does reflect the overt position of many ex-alcoholic counselors and the covert position of many of the more militant proponents of the social model. However, these criticisms of the social model apply only to the expression of that model in its most extreme form. Certainly, most proponents of this model—particularly the more sophisticated—accept the importance of the psychological, biological, and professional aspects.

The Alcoholics Anonymous Model

The A.A. model is not generally considered as a public health model but in truth constitutes one of the major treatment systems in the country. With an estimated population of about 200,000 in attendance at any given time, the A.A. census roughly equals that of all public health programs combined—federal, state, and county—and is probably equal to the size of the population of alcoholics treated by private physicians. Since the Alcoholic Anonymous approach is unique in itself and significantly different from any other model, it requires description within the context of the mechanisms presented in this chapter.

There seems to be little doubt concerning the overall effectiveness of A.A. even though that effectiveness may not be as great as claimed by its proponents nor as meager as defined by its detractors. Here we are less concerned by how effective A.A. may be and more interested in the reasons for which it works and the population for which it is best suited.

By and large, A.A. groups seem to be most successful when they derive from a single homogeneous social and cultural group. The interaction of individuals with a common problem in a shared subculture allows a free and open exchange of confidential material not otherwise available. This shared experience of life situations within the context of positive peer pressure constitutes an imposing example of the power of social psychological mechanisms. Trice (1957) has reported on the personality structure of the individuals who tend to do best in A.A. They are those who feel most comfortable as members of a group, receiving sustenance and moral support from it—the typical "joiners." Characterologically, they fall into Reisman's (1961) "outer directed" category, a description which in other studies correlates with high field dependence (Witkin et al., 1959).

In another sense, the A.A. meeting is also an experience in existential psychology. The techniques of Alcoholics Anonymous involve confrontation, acceptance of reality, emphasis on openness and honesty, and other expressions of the existential psychiatric approach. This combination of the best elements of social and existential psychiatry—the two techniques which have proved most effective in the treatment of chronic alcoholism—probably accounts for the overall success of A.A. In addition, the existence of living models of successful change from alcoholism to sobriety fits the precepts of the social model of treatment. Finally, the absence of professionalism is in some ways an advantage although, as is also self-evident, it may be a shortcoming.

The deficiencies of Alcoholics Anonymous do relate in part to the absence of professionalism. Because it is a program which stemmed from the actual experience of successful ex-alcoholics, it is limited to that experience. A course of behavior is prescribed "because it has worked for so many." Unfortunately, it has also not worked for many. Because of the nonprofessional approach, the prescribed pattern of behavior is rigidly defined with little allowance for individual variability. (This type of approach is often true also of ex-alcoholic counselors in the social model of treatment.) More recently, this rigidity has shown signs of lessening with the recognition that many alcoholics cannot live by the precepts of A.A. alone and may need additional treatments which only professionals can supply.

However, despite its shortcomings, there can be little doubt that Alcoholics Anonymous is one of the major treatment modalities. Whatever other model may be employed, A.A. remains a valuable and important adjunct. Not the least of its contributions is the sense it gives to its members of belonging to a family. Even more, it provides human companionship to essentially lonely people on evenings and weekends when most "professional" programs and therapists are not available. These are valuable and significant services and A.A. should be offered to all alcoholics who will accept it, regardless of what the basic treatment program may be.

The additional development in Alcoholics Anonymous of the parallel organizations, Al-Anon and Alateen, has provided a similar psychosocial structure for the families of alcoholics. Consequently, A.A. has in its own way anticipated many of the professional developments in the treatment of alcoholism and has incorporated them into its overall system.

The Multivariant Model

Each of the descriptions of the five previous models has deliberately been presented in a rather extreme form. Such a presentation highlights both the strengths and weaknesses of a model. Unfortunately, it also tends to present the proponents of any model as rigid, opinionated, demagogic, and narrow. Probably these traits are less common than one often fears and most adherents of any of these five models recognize the validity of some of the claims of the other models. On the other hand, the proper treatment of chronic alcoholism is an emotion-laden subject and very often the proponents of one model or another become quite vituperative in their condemnation of all other models.

Our "prejudice" in this matter is apparent from the descriptions in this chapter. We believe that to a large degree all of the models listed are valid. It is just that not every model is valid for every alcoholic. In the middle section of this chapter, we attempted to classify alcoholics on the basis of variability of biological, psychological, and social factors. Given the wide spectrum of possible patterns, each of which could be present to a greater or lesser degree, it is apparent that all alcoholics are similar in some respects but are vastly different in others. As previously stated, we believe that alcoholics during their phase of severe alcoholism, physical dependence, and psychological and social deterioration are a largely homogeneous group, both in their pathogenetic mechanisms and in their clinical appearance. Given a unitary syndrome, the immediate treatment is unitary pharmacologic detoxification with long-acting, cross-tolerant medication. However, once the addictive cycle is broken, the previously underlying pathology—biologic, psychologic, or social—becomes dominant and alcoholics become a markedly heterogeneous population.

At that point, we believe that the appropriate treatment for any alcoholic is that which best fits his condition. This approach has two major characteristics: (a) it is eclectic and leans equally on all of the previous models and (b) it is predicated on the necessity for adequately diagnosing the condition of each alcoholic. Thus far, most treatments have been applied indiscriminately to all alcoholics as though they were a homogeneous population. More recently attempts have been made to provide "the appropriate treatment for the appropriate patient."

Some of these studies have been reviewed recently by Pattison (1974). Among those reviewed are studies by Kissin *et al.* (1970), Ludwig *et al.* (1970), Pattison (1968), Pattison *et al.* (1969), Schmidt *et al.* (1968), and Trice *et al.* (1969). Pattison (1974) concludes:

> Without reviewing these studies in detail, the following population characteristics seem, at this time, to have some empirical support:
> 1. Alcoholics who have high social competence and high psychological competence with field independence would be candidates for psychotherapy.

2. Alcoholics who have high social competence and high psychological competence with field dependency would be candidates for Alcoholics Anonymous.

3. Alcoholics who have high social competence and low psychological competence will respond to supportive medical regimes, including psychotropic drugs, and supportive types of psychotherapy.

4. Alcoholics with low social competence and high psychological competence will be excellent candidates for Alcoholics Anonymous, with participation in social and vocational rehabilitation programs of prime importance.

5. Alcoholics of both low social and psychological competence will require continuing supportive services, will not affiliate with Alcoholics Anonymous, and will not respond to solely medical or psychological regimens.

Although these conclusions are preliminary and probably inconclusive, they can serve as general guidelines to a multivariant approach to treatment prescription.

Unfortunately, because of inadequate evaluation at this time of any of the major treatment models and because of inadequate criteria as to what kind of treatment is best for what kind of alcoholic, there has been an understandable movement toward the use of that model which is easiest to develop, most acceptable to most people, and presumably most effective for most alcoholics. At the present moment the social model appears preeminent. The network of comprehensive treatment programs being developed throughout the country under the auspices of NIAAA and with the collaboration of state and city agencies fits most comfortably into that model, although major elements of the medical, psychological, and even behavior modification models are also incorporated. One might call it a modified multivariant model and it is probably the best available to us at our present level of knowledge. Nevertheless, it is important that we not lose sight of developing the criteria on the basis of which we may further individualize treatment. Only under those circumstances will we be able to provide the appropriate treatment for every individual with an alcoholism problem.

THE MULTIMODAL, MULTIDISCIPLINARY APPROACH TO THE TREATMENT OF ALCOHOLISM

As a consequence of the reconsideration of the dynamics of alcoholism, new approaches to its treatment have evolved. The general rationale for these has been described in this chapter. But from these theoretical considerations sprang specific treatment programs which, to be implemented, required a restructuring of the service delivery system. Several principles, stemming from both theoretical and practical considerations, rapidly manifested themselves as necessary to the successful implementation of a national public health program. Among these principles, which have become almost catchwords for program planners, are the terms *multimodal, multidisciplinary,* and *continuity of care.* Each of these will be described and discussed in the following sections.

The Multimodal Approach to the Treatment of Alcoholism

The term *multimodal* has indeed two different meanings which, although related, are also somewhat separate and distinct. Multimodal, in its most popular sense, means a variety of facility programs which will meet the needs of a given alcoholic in the various stages of his illness. Hence, a typical multimodal program for socially stable alcoholics might include a detoxification ward, a short-term rehabilitation ward (six to eight weeks), and an outpatient clinic. On the other hand, a typical multimodality program for socially indigent alcoholics might require all of the above facilities plus a day center, a long-term rehabilitation ward (two to three months), a halfway house (three to six months), and a residential community (six months or more).

Multimodal can also have another connotation—namely, a variety of treatment modalities for different kinds of alcoholics who require different kinds of treatment methods. For example, Pattison (1974) has presented evidence that certain types of alcoholics do best with aversion therapy programs, others with Antabuse, still others with therapeutic communities, and so on. Within this context, certain types of treatment are best restricted to a single facility. Thus, for example, it is difficult to mix aversion therapy with a typical therapeutic community. The differences are more strongly pointed up in narcotic addiction, where methadone maintenance, narcotic antagonist, and drug-free programs must almost necessarily be separated one from the other. Consequently, in this second type of multimodal program, there is less movement of patients from one facility to the next since each facility represents a different treatment modality. If the first type of multimodality could be considered vertical in structure, the second type would seem to be more horizontal.

Nevertheless, each of these types of multimodality would seem to be essential for the optimal treatment of alcoholics. As of this time, we have no proof that any one form of treatment is *the* treatment; different kinds of alcoholics appear to respond differentially to different kinds of treatment (Kissin *et al.,* 1970). Greater efforts should be made to attempt to define the kinds of alcoholic who do best in different treatment programs and then to provide the optimal treatment for each individual.

The Multidisciplinary Approach to the Treatment of Alcoholism

In our discussion of the special and core problems in the treatment of alcoholism, it became clear that not only did we need a broad spectrum of facilities to provide the necessary services, but that we also needed a broad spectrum of professionals and paraprofessionals to man them. Among the individuals needed were physicians (psychiatrists and medical physicians), nurses, nurse's aides, psychologists, social workers, professional and paraprofessional counselors, recreation therapists, occupational therapists, vocational rehabilitation counselors, and many others. All of these skills are equally necessary for a comprehensive program and no one individual skill is more important than any other.

This approach, in a domain which has traditionally been considered that of the physician, has radically de-emphasized his role—too often in the opinion of the present author, a physician—to an excessive degree. This pattern has been repeated even more strongly in the field of narcotic addiction. However this may be, it is certainly true that in few other medical fields is there the same degree of democratic sharing of authority and responsibility that one finds in the field of alcoholism and drug addiction.

This situation has been in part an historical development growing out of self-help movements by both alcoholics (Alcoholics Anonymous) and addicts (therapeutic communities) to which the mental health professions (psychology and social work) gave credence and assistance long before the medical profession. Even more, however, it is the consequence of the basic nature of the therapeutic process. As we have described the core problem in alcoholism, the central program in the therapeutic process is the process of "resocialization through living experience." Given this construction, each and every experience in the alcoholic's rehabilitation process may have almost equal weight. A home run in a baseball game with new friends or a handsomely wrought metal object may have more therapeutic value to the patient than an hour spent listening to or talking to a psychiatrist. Each therapist in the program, regardless of what his secondary services may be—medical, nursing, social, psychological, recreational, occupational, etc.—has as his first responsibility the necessity to relate to the individual and *to have the individual relate to him.* Consequently, the ideal therapeutic community is one in which all therapists and all patients are equal.

Continuity of Care in the Treatment of Alcoholics

The danger in the multimodal, multidisciplinary program is that the patient, taken care of by everybody, may end up being taken care of by no one.

In order to avoid this obvious peril, the concept of "continuity of care" has been developed. Under this design, each patient becomes the responsibility of a specific counselor who stays with and guides the patient through his multimodal, multidisciplinary experience. In this way, the patient is guaranteed continuity of care between facilities so that he does not drop into the administrative cracks between them. In addition, he experiences continuity of care in his relationship to his own counselor. This system presents certain administrative difficulties in assigning counselors to specific facilities when their responsibilities to their clients may take them to many facilities. This is, however, a small price to pay in terms of the benefits to be reaped from a continuity of care system.

The concept of the "generic counselor" has been a direct outgrowth of this system. If every therapist has as his first responsibility to his client a psychotherapeutic attitude, then he has, as secondary responsibilities, to help his client with social services, vocational and occupational services, recreational services, etc. His major responsibility is to help his client seek and obtain these services through referral to the appropriate specialist source. However, many counselors learn to provide the simpler of these services themselves, referring their clients only for those services which they themselves cannot arrange. As a result, these counselors become general practitioners or "generic" counselors, able to provide the majority of those services which an alcoholic or addict may need.

Public Health Treatment Systems in Alcoholism

It can be seen that the delivery of appropriate services for a large alcoholic population under the scheme outlined in this chapter requires the development of an entirely new, separate, and distinct service delivery system. This is, indeed, the case and this is what has actually happened. Under the direction of the NIAAA and the various state alcoholism agencies, a comprehensive system of alcoholism treatment programs has been established. These include a broad spectrum of facilities from detoxification wards to residential facilities. The activities of these programs are coordinated by federal, state, and local authorities to attempt to provide the most comprehensive system of services possible. Finally, the activities of other social agencies, both public and private, are likewise integrated in an all-embracing pattern.

The one major resource which has not yet been fully integrated into the fight against alcoholism is the medical profession. Of the 275,000 physicians in the United States, the majority do not treat alcoholism. That is not to say that they do not treat alcoholics. This they do in large numbers—but not for their alcoholism. They treat them for cirrhosis of the liver, tuberculosis, peptic ulcers, pancreatitis, and for many other alcohol-related diseases—but not for alcoholism. In a study made at the Kings County Hospital Center in Brooklyn by the author and his group (unpublished data), fully 33 percent of all males admitted to the medical and surgical wards of the hospital were alcoholic by the most rigid criteria. Yet, in less than a quarter of these was the diagnosis of "alcoholism" made and only about half of this reduced number were referred for treatment. As will be described in a later chapter, alcoholism is a treatable disease with an overall "cure rate" at least as good as that for cancer. If alcoholism is as lethal a disease as cancer and if our available treatments are at least effective as those for cancer, can there be any justification for a physician's neglect to treat an alcoholic or at least to refer such a patient for treatment?

Despite the relative indifference of the private physician to the problem of alcoholism, the medical profession remains one of the major available treatment resources in this area. Other important resources are Alcoholics Anonymous, existing health and social agencies, and the special alcoholism treatment programs recently developed. Indeed, these four resource elements constitute the major aspects of the overall public health program against alcoholism. Each forms the basis of a treatment system separate and distinct from the others, yet closely related to them. Each is based most heavily on one or more of the models previously described but each uses the concepts and techniques of the models differently.

The four major public health treatment systems are then as follows:

The Private Medical Sector. This system consists of those doctors in private practice—chiefly psychiatrists, internists, general practitioners, and osteopathic physicians—who do treat alcoholics. Mainly, they follow the medical model, though psychotherapeutic techniques are of course used by psychiatrists.

General Health, Mental Health, and Social Agencies. These include general hospitals, state mental hospitals, and a variety of social agencies. In the majority of cases, alcoholics are treated together with other types of problems. Unfortunately, many alcoholics are badly neglected in these situations. The general approaches in these programs are medical, psychotherapeutic, and social.

Specialized Alcoholism Treatment Programs. These have been set up under special legislation and with special funding from federal, state, and city agencies. Since these programs are designed for alcoholics, these individuals at least receive treatment in them. The models followed are usually interdisciplinary and may include medical, behavioral, psychotherapeutic, and social, although the latter two modalities seem to be most popular.

Alcoholics Anonymous. This, the oldest and perhaps the largest of the treatment systems is extremely effective for a significant segment of the alcoholic population. The model is informally both psychotherapeutic and social and, in nonmedical terms, highly moral.

These, then, are the four major alcoholism treatment systems which together make up the overall public health program. The interaction of the various theoretical treatment models with the actual treatment systems appears to be a function of (a) the interest of various sectors of society at a given time, (b) the availability of funds and treatment resources, (c) pragmatic evaluations of what methods appear most effective, and (d) an inexplicable rise and fall in the fashionability of certain approaches. However, in their own way, all of the above-listed treatment systems have made important contributions to the treatment of alcoholism. There is no reason why, in a more systematic and coordinated effort, they should not together constitute an overall, effective, integrated, and comprehensive attack on the problems of alcoholism.

REFERENCES

Baekeland, F., Lundwall, L., and Kissin, B., 1975, Methods for the treatment of chronic alcoholism: A critical appraisal, *in* "Research Advances in Alcohol and Drug Problems" (Gibbins *et al.,* eds.) Vol. 2, pp. 247–327, Wiley & Sons, New York.

Branchey, M., Rauscher, G., and Kissin, B., 1971, Modification in the response to alcohol following the establishment of physical dependence, *Psychopharmacologia* (Berl.) 22:314–322.

Cahalan, D., Cisin, I. H., and Crossley, H. M., 1969, American Drinking Practices: A national study of drinking behavior and attitudes, Rutgers Center of Alcohol Studies Monograph No. 6, New Brunswick, New Jersey.

Chein, I., Gerard, D. L., and Rosenfeld, E., 1964, "The Road to H; Narcotics, Delinquency and Social Policy," Basic Books, New York.

Clifford, B. J., 1960, A study of the wives of rehabilitated and unrehabilitated alcoholics, *Social Casework* 61:457–460.

DeVito, R. A., Flaherty, L. A., and Mrozdzierz, G. J., 1970, Toward a psychodynamic theory of alcoholism, *Q. J. Stud. Alcohol* 31:1009.

Dole, V. P., and Nyswander, M., 1965, A medical treatment for diacetylmorphine (heroin) addiction, *J.A.M.A.* 193:646.

Edwards, G., 1974, The Jack Donovan Memorial Lecture, Seventh Annual Eagleville Conference, Eagleville, Pennsylvania, June 7, 1974.

Feldman, J., Su, W. H., Kaley, M. M., and Kissin, B., 1974, Skid row and inner-city alcoholics: A comparison of drinking patterns and medical problems, *Q. J. Stud. Alcohol* 35:565–576.

Fenichel, O., 1945, "The Psychoanalytic Theory of Neurosis," Norton, New York.

Freund, G., 1973, Chronic central nervous system toxicity of alcohol, *Ann. Rev. Pharmacol.* 13:217–227.

Gerard, D. L., and Saenger, G., 1959, "Outpatient Treatment of Alcoholism," University of Toronto Press, Toronto.

Glasscote, R. M., Plaut, T. F. A., Hammersley, D. W., O'Neill, F. J., Chafetz, M. E., and Cummings, E., 1967, The Treatment of Alcoholism, A study of programs and problems, Joint Information Services, Washington, D.C.

Goodwin, D. W., and Guze, S. B., 1974, Heredity and alcoholism, *in* "The Biology of

Alcoholism" (B. Kissin and H. Begleiter, eds.) Vol. 3: Clinical Pathology, pp. 37–52, Plenum Press, New York.
Goodwin, D. W., Schulsinger, F., Hermansen, L., Guze, S. B., and Winokur, G., 1973, Alcohol problems in adoptees raised apart from alcoholic biological parents, *Arch. Gen. Psychiatry* 28:238–243.
Guze, S. B., Wolfgram, E., and McKinney, J., 1967, Psychiatric illness in the families of convicted criminals, A study of 519 first degree relatives, *Dis. Nerv. Syst.* 28:651.
Hershon, H., 1974, Alcoholism and the concept of disease, *Br. J. Addict.* 69:123–132.
Hill, H. E., Haertzen, C. A., and Davis, H., 1962, An MMPI factor analytic study of alcoholics, narcotic addicts and criminals, *Q. J. Stud. Alcohol* 23:411–431.
Ho, A. K. S., and Tsai, C. S., 1975, Lithium and ethanol preference, *J. Pharm. Pharmacol.* 27:58–60.
Hore, B. D., 1974, Craving for alcohol, *Br. J. Addict.* 69:137–140.
Irwin, S., 1973, A rational approach to drug abuse prevention, *Contemporary Drug Problems* 2:3–46.
Jellinek, E. M., 1960, "The Disease Concept of Alcoholism," Hillhouse Press, New Haven.
Jones, K. L., and Smith, D. W., 1973, Recognition of the fetal alcohol syndrome in early infancy, *Lancet* 2:999–1001.
Jones, R. W., and Helrich, A. R., 1972, Treatment of alcoholism by physicians in private practice, *Q. J. Stud. Alcohol* 33:117–131.
Kaufman, E., 1974, The psychodynamics of opiate dependence: A new look, *American Journal of Drug and Alcohol Abuse* 1:349–370.
Keller, M., 1975, Problems of epidemiology in alcohol problems, *J. Stud. Alcohol* 36:1442–1451.
Kernberg, O., 1975, "Borderline Conditions and Pathological Narcissism," Jason Aronson, New York.
Kissin, B., 1974, The pharmacodynamics and natural history of alcoholism, in "The Biology of Alcoholism" (B. Kissin and H. Begleiter, eds.) Vol. 3: Clinical Pathology, pp. 1–36, Plenum Press, New York.
Kissin, B., and Begleiter, H., 1974 (eds.), "The Biology of Alcoholism," Vol. 3: Clinical Pathology, Plenum Press, New York.
Kissin, B., and Begleiter, H., 1976 (eds.), "The Biology of Alcoholism," Vol. 4: Social Aspects, Plenum Press, New York.
Kissin, B., and Platz, A., 1968, The use of drugs in the long term rehabilitation of chronic alcoholics, in "Psychopharmacology: A Review of Progress, 1957–1967" (D. H. Efron, ed.), pp. 835–851, Public Health Service Publication No. 1836, Washington.
Kissin, B., Schenker, V., and Schenker, A., 1959, The acute effects of ethyl alcohol and chlorpromazine on certain physiological functions in alcoholics, *Q. J. Stud. Alcohol* 20:480.
Kissin, B., Platz, A., and Su, W. H., 1970, Social and psychological factors in the treatment of chronic alcoholism, *J. Psychiat. Res.* 8:13–27.
Kline, N. S., Wren, J. C., Cooper, T. B., Vargas, E., and Canal, O., 1974, Evaluation of lithium therapy in chronic alcoholism, *Clin. Med.* 81:33–36.
Knight, R. P., 1937, The psychodynamics of chronic alcoholism, *J. Nerv. Ment. Dis.* 86:538–548.
Luborsky, L., Chandler, M., Auerbach, A. H., Cohen, J., and Bachrach, H. M., 1971, Factors influencing the outcome of psychotherapy: A review of quantitative research, *Psychol. Bull.* 75:145–185.
Ludwig, A. M., Levine, J., and Stark, L. H., 1970, "LSD and Alcoholism: A Clinical Study of Treatment Efficacy," Charles C Thomas, Springfield, Illinois.
Ludwig, A. M., Wikler, A., and Stark, L. H., 1974, The first drink: Psychological aspects of craving, *Arch. Gen. Psychiatry* 30:539–547.

Machover, S., and Puzzo, F. S., 1959, Clinical and objective studies of personality variables in alcoholism. I. Clinical investigation of the alcoholic personality, *Q. J. Stud. Alcohol* 2:505–519.

Martin, W. R., and Jasinski, D. R., 1969, Physiological parameters of morphine dependence in man—tolerance, early abstinence, protracted abstinence, *J. Psychiatr. Res.* 7:9–14.

Mindlin, D. F., 1959, The characteristics of alcoholics as related to prediction outcome, *Q. J. Stud. Alcohol* 21:90–112.

Panepinto, W. C., Higgins, M. J., Keane-Dawes, W. Y., and Smith, D., 1970, Underlying psychiatric diagnosis as an indicator of participation in alcoholism therapy, *Q. J. Stud. Alcohol* 31:950–956.

Pattison, E. M., 1968, Abstinence criteria: A critique of abstinence criteria in the treatment of alcoholism, *Internat. J. Soc. Psych.* 14:268–276.

Pattison, E. M., 1974, Rehabilitation of the chronic alcoholic, *in* "The Biology of Alcoholism" (B. Kissin and H. Begleiter, eds.) Vol. 3: Clinical Pathology, pp. 587–658, Plenum Press, New York.

Pattison, E. M., Coe, R., and Rhodes, R. A., 1969, Evaluation of alcoholism treatment: Comparison of three facilities, *Arch. Gen. Psychiatry* 20:478–488.

Peterson, N. W., 1973, quoted in "Recovery from Alcoholism. A Social Treatment Model" (R. G. O'Briant, H. Leonard, S. D. Allen, and O. C. Ransom, eds.), Charles C Thomas, Springfield, Illinois.

Platz, A., Panepinto, W. C., Kissin, B., and Charnoff, S. M., 1970, Metronidazole and alcoholism: An evaluation of specific and non-specific factors in drug treatment, *Dis. Nerv. Syst.* 31:631–636.

Plumeau, F. E., Machover, S., and Puzzo, F. S., 1960, Wechsler Bellevue performances of remitted and unremitted alcoholics and their norma' controls, *J. Consult. Clin. Psychol.* 24:240–242.

Reisman, D., 1961, "The Lonely Crowd," Yale University Press, New Haven.

Russell, J. A., and Mehrabian, M., 1975, The mediating role of emotions in alcohol use, *J. Stud. Alcohol* 36:1508–1536.

Schenker, A. C., Schenker, V. D., and Kissin, B., 1962, Aberrations in the pulmonary respiratory patterns in alcoholics and the acute effects of ethyl alcohol and chlorpromazine upon such patterns, *Proceedings of the Third World Congress of Psychiatry*, Montreal, pp. 389–396.

Schmidt, W. G., Smart, R. G., and Moss, M. K., 1968, "Social Class and the Treatment of Alcoholism," University of Toronto Press, Toronto.

Seevers, M. H., 1968, Psychopharmacological elements of drug dependence, *J.A.M.A.* 206:1263–1266.

Seevers, M. H., and Woods, L. A., 1953, Phenomena of tolerance, *Am. J. Med.* 14:546–557.

Sherfey, M. J., 1955, Psychopathology and character structure in chronic alcoholism, *in* "Etiology of Chronic Alcoholism" (O. Diethelm, ed.), pp. 16–42, Charles C Thomas, Springfield, Illinois.

Siegler, M., Osmond, H., and Newell, S., 1968, Models of alcoholism, *Q. J. Stud. Alcohol* 29:571–591.

Stein, L. I., Niles, D., and Ludwig, A. M., 1968, The loss of control phenomenon in alcoholics, *Q. J. Stud. Alcohol* 29:598–602.

Strauss, R., 1973, Alcohol and psychiatry, *Psychiatric Annals* 3:8–107.

Strupp, H. H., and Bergin, A. W., 1969, Some empirical and conceptual bases for coordinated research in psychotherapy: A critical review of issues, trends and evidence, *International Journal of Psychiatry* 7:18–90.

Todd, M. C., 1975, How future physicians must see the alcoholic, *Rhode Island Med. J.* 75:390–401.

Trice, H. M., 1957, A study of the process of affiliation with Alcoholics Anonymous, *Q. J. Stud. Alcohol* 18:39–54.
Trice, H. M., Roman, P. M., and Belasco, J. T., 1969, Selection for treatment: A predictive evaluation of an alcoholism treatment regimen, *Int. J. Addict.* 4:303–318.
Tripp, C. A., Fluckiger, F. A., and Weinberg, G. H., 1959, Effects of alcohol on graphomotor performances of normals and chronic alcoholics, *Percept. Mot. Skills* 9:227–236.
Ulleland, C. N., 1972, The offspring of alcoholic mothers, *Proceedings of the New York Academy of Science* 197:167–169.
Whitehorn, J. C., and Betz, B., 1954, A study of the psychotherapeutic relationship between physicians and schizophrenic patients, *Am. J. Psychiat.* 111:321–331.
Winokur, G., Reich, T., Rimmer, J., and Pitts, F., 1970, Alcoholism III: Diagnosis and familial psychiatric illness in 259 alcoholic probands, *Arch. Gen Psychiat.* 23:104–111.
Witkin, H. A., Karp, S. A., and Goodenough, D. R., 1959, Dependence in Alcoholics, *Q. J. Stud. Alcohol* 20:493–504.
Zwerling, I., 1959, Psychiatric findings in an interdisciplinary study of forty-six alcoholic patients, *Q. J. Stud. Alcohol* 20:543–554.

CHAPTER 2

Medical Management of the Alcoholic Patient

Benjamin Kissin

Alcoholism Division
State University of New York
Downstate Medical Center
Brooklyn, New York

INTRODUCTION

With the advent of the multidisciplinary approach to the treatment of chronic alcoholism, a broad spectrum of professional and paraprofessional personnel have become involved in the treatment process. Among the categories involved are physicians (psychiatrists, internists, and family physicians), nurses, clinical psychologists, counseling psychologists, social workers, vocational counselors, educational counselors, recreational therapists, ex-alcoholic counselors, and doubtless many others. With the exception of the physicians, the role of each discipline is fairly well defined. On the other hand, the specific role of the physician in the treatment of alcoholics remains more ambiguous. This ambiguity stems from a variety of sources.

    1. The general ambivalence of alcoholics toward physicians, based in part on the ambivalence of physicians toward alcoholics. Almost all alcoholics have

had some experience with their private physicians. Some encounters have been positive, more negative. The unwillingness of the private physician to assume a major responsibility for the treatment of alcoholics has undoubtedly been the main source of the antipathy which many alcoholics feel toward physicians (Rathod, 1967). This underlying antipathy has been heightened by the active antagonism of Alcoholics Anonymous to the use of tranquilizers, which has become the major therapeutic modality used by physicians, i.e., internists and general practitioners (Jones and Helrich, 1972).

2. The gradual development of the "social model" as the dominant theoretical approach to alcoholism. This model opposes the "medical model" and deemphasizes the role of the physician even in the treatment of the acute alcohol withdrawal syndrome (Peterson, 1973). It is equally opposed to the use of chemotherapeutic agents, particularly tranquilizers, but in some cases, even disulfiram (Antabuse). The rise of the social model philosophy has been associated with the marked proliferation of "specialized alcoholism programs" where the physicians's role may be extremely varied.

3. The ill-defined status of the four major public health "alcoholism treatment systems." In a previous chapter we defined those systems as: (a) The private physician sector of medical care; (b) The medical–social system, including general hospitals, state mental hospitals, and social agencies; (c) The specialized alcoholism treatment programs; and (d) Alcoholics Anonymous.

The physician is an important participant in each of the first three of these systems, but his role in each one varies widely. In some cases, a physician may participate in two or three of those systems and find his functional role defined differently in each situation. This lack of definition of the specific role of the physician in each system leads to confusion in the minds of both doctor and alcoholic. Each of the two sources of ambiguity previously described contributes indirectly to this confusion and leaves unanswered the questions as to what the physician is or is not responsible for and what he should or should not be doing.

THE ROLE OF THE PHYSICIAN IN THE TREATMENT OF ALCOHOLISM

A description of the role of the physician in the treatment of alcoholism must include an exploration of several related issues: (a) What has been the role of the physician in the past, (b) how significant has been the impact of the medical profession on the larger problem of alcoholism, (c) what are the reasons that that impact has been relatively small—as it is said to be, (d) how effective is the physician in treating the individual alcoholic, and finally (e) what *should be* the role of the physician in treating alcoholics. We shall begin by examining the role of the physician in private practice, in general hospitals, and in special alcoholism programs.

The Role of the Physician in Private Practice

In speaking of physicians and their relationship to alcoholism it is necessary first to break down the broad spectrum of medical specialities into those disciplines which do or do not concern themselves with the problem, and secondly to describe the special types of involvement of those who do. On the first question is the observation—so blatant that it hardly needs validation—that with rare exceptions only psychiatrists, internists, and general practitioners show any interest in alcoholics. (A rare but significant exception is Lowenfels, 1971, who has written extensively on the implications of alcoholism in surgery.) It is not that practitioners of other specialties do not *treat* alcoholics. They see them and treat them in vast numbers but almost always for the complications of alcoholism rather than for the disease itself. That physicians in general have a blind spot for the diagnosis of alcoholism is a conclusion for which we shall shortly present evidence; nonmedical and nonpsychiatric specialists only seem to have a larger and more opaque blind spot than do the others.

On the other hand, the historical involvement of internists and general practitioners (who are really internists who do other things as well) in alcoholism has been quite different from that of psychiatrists. The more medically oriented physicians became involved through the recognition that many of the major medical illnesses (cirrhosis of the liver, polyneuropathy, acute and chronic gastritis, acute and chronic pancreatitis, etc.) were directly attributable to the social illness "alcoholism." In a related yet somewhat different pattern, psychiatrists found that several major psychiatric illnesses (delirium tremens, Korsakoff's syndrome, von Wernicke's syndrome) were similarly associated with alcoholism.

But here an interesting divergence occurred. Although both physicians and psychiatrists continued to see innumerable alcoholics as patients, the medical people saw them only for their complications while the psychiatrists began to see them for the alcoholism itself. The reasons for this are self-evident. On the negative side, because of the slowly addicting qualities of alcohol as opposed to morphine, it was not until the classic demonstration of Isbell *et al.* (1955) that delirium tremens was a withdrawal phenomenon that the role of physical dependence in alcohol addiction was clearly delineated. Consequently, physicians, seeing no evidence of specific biological involvement, were ready to view alcoholism as a moral, psychological, or social deficiency. The more readily apparent psychological origins of alcoholism were early structured by the pioneer papers of Knight (1937) and Fenichel (1945) which remained the best theoretical frame of reference available for alcoholism as a medical problem until quite recently. Since practice often follows theory, psychiatrists at first willingly undertook the treatment of alcoholics. However, as "depth analysis" and "analytically oriented psychotherapy" repeatedly proved unsuccessful, their enthusiasm waned. Because psychiatrists were the only medical specialists ever to show any interest in the condition at all, alcoholism continued to be considered predominantly as a psychiatric ailment—in the private medical sector, in health and social agencies, and largely in the new special alcoholism treatment programs as well.

Nevertheless, because the internist or family physician was almost always the first member of the medical profession to be consulted in the management of the alcoholic—either directly for the alcoholism itself or indirectly for one of its medical complications—the private physician was thrust willy-nilly into the field. Those physicians whose practices contained a large proportion of alcoholics rapidly became aware of the medical and social implications of the condition and developed some interest as a consequence. Accordingly, the major medical experience with alcoholism in the past has been by psychiatrists, internists, general practitioners, and osteopathic physicians.

Two major surveys have been conducted to study the treatment of alcoholism by doctors in private practice—the first by Bailey (1968), limited to doctors in the New York City area; the second by Jones and Helrich (1972), extended to doctors throughout the country. The two studies reinforce each other, demonstrating that the attitudes of doctors in New York City are quite similar to those throughout the United States, although there are some significant differences. The Bailey questionnaire was directed to 1815 doctors, of whom 56.6% or 1028 responded (276 internists, 187 psychiatrists, and 565 general practitioners). The Jones and Helrich questionnaire was sent to 88,302 physicians, with 15,912 responding. However, when exclusions were made for those not in private practice, the number of reliable questionnaires was 13,058 (internists 3376, psychiatrists 1674, general practitioners 6805, and osteopaths 1203).

The overall findings of the Bailey study are best stated in the summary of the report, which is here quoted verbatim.

> A 10% sample survey of New York internists, psychiatrists, and general practitioners was conducted through a mail questionnaire, in 1968, by the Alcoholism Program of the Community Council of Greater New York. The study was a partial repetition of a similar survey reported in 1962, and it is hoped that such periodic studies, by documenting the extent of medical practice and the attitudes of physicians, can provide a factual basis for community alcoholism programs planned and conducted by the medical profession for its own members.

The response rate for the 1968 survey was 56.6% an improvement over the 41.1% in 1962. One-fourth of the physicians surveyed had seen no patients with drinking problems within the preceding two months, and a few of these doctors commented that they are not willing to see alcoholics in their practices. Two-thirds of the physicians reported one to nine problem-drinking patients during the two-month period, but only a handful had seen ten or more.

Most of the physicians did not equate problem drinking with alcoholism, and some said that they had not diagnosed any of their problem drinkers as alcoholics. The majority also reported that less than half of their problem drinkers were women, and one-fourth had no such women patients. There were fewer instances of counseling with the families of problem drinkers than contacts with the drinkers themselves.

Seven-tenths of the physicians regard alcoholism as a symptom of underlying emotional problems, whereas only about two-fifths agree that alcoholism is a disease. The difficulty appears to lie in the absence of a single etiology and a specific therapy. Actually, more than one-fourth of the sample subscribe to both the disease and the symptom formulations. A minority, slightly more than one-fifth, accept the moralistic formulation that alcoholism is a self-inflicted condition resulting from personal weakness.

At least one-half of the physicians treat alcoholics for their alcoholism, although this statement is more true of psychiatrists than of internists and general practitioners, who are more likely to refer their patients for alcoholism therapy. Only about one-fourth of the physicians are affiliated with hospitals which accept alcoholics with a primary diagnosis of alcoholism. Drugs most frequently prescribed in office practice include Librium, thiamin, other vitamins, and the phenothiazines.

Three-fourths of the physicians believe that life-long abstinence is a necessary goal of treatment, and nearly as many feel that this is possible only after the alcoholic gains some insight into the origins of his drinking. Over half are pessimistic about the chances for recovery, and two-thirds find alcoholics to be difficult and uncooperative patients. Nearly two-thirds believe that alcoholics should be referred to specialized clinics for professional treatment, although only about one-third feel that helping responsibility should be turned over to Alcoholics Anonymous. Most physicians appear to regard A. A. as a valuable ally, but emphasize that the doctor must retain responsibility for treatment.

This statement in general summarizes qualitatively the experience and attitude of most physicians who treat alcoholics. Although there are quantitative differences between the Bailey and the Jones and Helrich studies, the results in the two studies are generally quite similar. On the whole, however, the Jones and Helrich data are more encouraging since they indicate a higher level of involvement on the doctor's part than in the earlier study. This might possibly be accounted for by an improvement in doctors' attitudes over the four years between the two studies. However, it is equally likely that the difference may be due to the smaller percentage of respondents in the Jones and Helrich study (18 percent vs. 56.6 percent) with presumably the most involved physicians responding. In any event, allowing for the difference between the two studies and leaning most heavily on the Jones and Helrich (1972) study since it is larger, more characteristic of the entire country, and more recent, we may conclude the following about the present treatment of alcoholics by private medical practitioners:

1. Between 70 and 75 percent of the doctors interviewed in both studies had seen three or more problem drinkers in their practice in the preceding two months.
2. More than half of these were considered alcoholic.
3. About 70–75 percent of these physicians attempted to treat these patients either alone or, more often, in conjunction with some other agency.
4. Most physicians use minor tranquilizers (chlordiazepoxide and diazepam) as the dominant form of chemotherapy, followed in frequency by phenothiazines, vitamins, antidepressants, and disulfiram.
5. Physicians appear to be evenly divided in their opinions as to whether alcoholism is a disease or a symptom, but the vast majority (about 80 percent) are agreed that it is primarily due to a personality or emotional disturbance.

Given the present state of our understanding of the pathogenesis and treatment of alcoholism, the situation among privately practicing physicians is not unencouraging. Extrapolating from the data in these articles, the average physician (internist, psychiatrist, or general practitioner) in New York City had about two alcoholics in his practice in 1968, and the average physician in the United States (Jones and Helrich, 1972) had about three alcoholics in his practice. Glasscote *et al.* (1967) estimate the average number of alcoholics in a psychiatrist's practice was about three. Using the average of these estimates, if each of the 85,000 internists, psychiatrists, and general practitioners in private practice in the United States is seeing three alcoholics a year and is treating 75 percent of them, the total of about 200,000 alcoholics in treatment at any given time represents one of the largest alcoholic patient populations under treatment within a given modality. The comparable figures for A.A. are 200,000 (Leach and Norris, 1976) while the figure for patients enrolled in specialized alcoholism treatment programs is also estimated at about 200,000. Further support for these statistics is provided by a report that private physicians provided about 3,000,000 direct services for alcoholism in their own offices each year (Todd, 1975). Given the estimate of 200,000 alcoholics in treatment, this gives an average of 15 treatments per patient per year, which average is similar to that for many outpatient alcoholism treatment programs.

These statistics suggest that the private physician has not been as negligent in treating alcoholics as has been the general impression. What appears to be truer is that most physicians do try to cope with the problem as it presents itself but often find themselves unable to diagnose, treat, or refer alcoholics appropriately because the basic standards for diagnosis, treatment, and referral have never been adequately elaborated. The recent development of criteria for diagnosing alcoholism, as developed by the National Council on Alcoholism, has been a first important step in this direction. The criteria are deficient, in the view of this author, in that they rely most heavily on medical criteria and do not differentiate sufficiently along psychosocial lines. Because of this, they do not permit a classification of alcoholics into categories related to therapeutic indications on the basis of which the private physician could decide to treat or to refer. Such a classification will be the concern of the latter part of this chapter.

The Role of the Physician in General Hospitals

Alcoholics play a major role in the dynamics of every general hospital in the United States to a degree seldom appreciated. Several surveys have reported that alcoholics constitute anywhere from 15 to 40 percent of the total admissions to the adult services of all general hospitals (the higher figures usually apply in municipal hospitals serving inner-city populations and in veterans' hospitals with predominantly male populations). These figures do not appear exorbitant when one considers that alcoholics constitute 5 percent of the total adult population in the United States and that alcoholics have a far higher incidence of medical and surgical complications than do the rest of the population.

The involvement of physicians in general hospitals occurs at four different levels: (1) diagnosis of acute and chronic alcoholism in the emergency room, (2) diagnosis of acute and chronic alcoholism in the medical and surgical wards, (3) detoxification from acute alcoholic withdrawal, and (4) referral for treatment for chronic alcoholism.

Three unpublished studies carried out by the author and his group at the Kings Country Hospital Center in Brooklyn (one of the three largest municipal hospitals in the country) illustrate the deficiencies which exist in these areas. The first involved a survey of admissions to the emergency room in the Kings County Hospital, an area which receives about 500 visitors daily. About 15 percent of these clients were diagnosed as alcoholics by the nursing staff (who are almost always more adept at diagnosing alcoholism than are the physicians). Of the 75 patients so diagnosed, only 5–10 were admitted for medical care. The remainder were given some tranquilizers and sent home. Since the results of this survey were publicized about four years ago, a counselor has been assigned to the emergency room and the rate of referral of these patients for both inpatient detoxification and outpatient treatment has increased significantly.

A similar study of admissions to the adult medical and surgical wards at the King County Hospital, using the Zimberg Interview Scale (Zimberg et al., 1971) for diagnosing alcoholism, showed 33 percent of all men and 20 percent of all women to be alcoholic as defined by those criteria, for an overall rate of about 27 percent. Of greater interest was the fact that the diagnosis of chronic alcoholism appeared in the intake or discharge diagnoses in only 20 percent of these patients. The blind spot of physicians for diagnosing alcoholism is vividly demonstrated here; 80 percent of the cases of alcoholism were not so designated. Needless to say, very few of these patients, diagnosed or otherwise, were referred for specific treatment for their alcoholism.

The third study was a five-year follow-up on patients diagnosed as delirium tremens and treated at the Kings County Hospital. Fifty consecutive patients who had been treated five years before were selected for follow-up (Perlstein, personal communication). The average age was 42 years. Of the 30 patients on whom successful follow-up was accomplished, one-third were dead. This figure agreed almost exactly with the findings of Lundquist (1961) in Stockholm and Victor and Hope (1953) in Boston in their five-year follow-ups of patients treated for DTs. In the Kings County Hospital Center study, only about 20 percent of the patients had sought treatment or been adequately referred. The others were sent home after detoxification.

It is not the purpose of this chapter to indicate what is necessary to improve the level of care for alcoholics in general hospitals. Some general actions are obviously necessary and should and are being implemented. Among these are (1) mandating the admission and treatment of acute alcoholics if the hospital is to continue to receive federal funds, (2) establishing drying-out and observation rooms for alcoholics in emergency rooms, (3) involving the social services of the hospital in referral to alcoholism programs, and (4) introducing alcoholism education in the curriculum for house staff. This last item is probably most important. Of all levels of the medical profession, house staff is probably most resistant to diagnosing and treating alcoholics (Fisher et al., 1975). They see alcoholism as "soft" medicine, if indeed as a medical problem at all. Since they are most interested in the discipline of medicine, they tend to ignore the medical–psychosocial problem of alcoholism. This attitude rubs off on medical students in their clinical years and accounts for the increasing lack of interest among students as they approach graduation (Fisher et al., 1975). But, again, house staff, like private physicians, will probably become less resistant when adequate standards for diagnosis, treatment, and referral are properly established.

The Role of the Physician in Special Alcoholism Programs

Physicians play different roles in different alcoholism programs depending upon whether the program is medically or socially oriented and if medically, whether the director is a psychiatrist, internist, or family physician. In those programs with a strong social orientation, the physician is sometimes relegated to treating solely the medical complications of the illness and is not considered to have major input into the treatment of the disease itself. Where the physician plays a more dominant role, this may be in determining overall treatment policy, in educating other staff, and in therapy itself.

However, only rarely is the physician seen as the primary therapist. The latter are more often the social worker, the counseling psychologist, or a paraprofessional ex-alcoholic counselor. And yet, in a study in which the relative effectiveness of internists, psychiatrists, and social workers in treating alcoholics in special outpatient alcoholism treatment centers were compared (Gerard and Saenger, 1966), the patients treated by internists had the highest rate of improvement, those treated by social workers next highest, and those treated by psychiatrists lowest. These results are not necessarily significant since the study was not established to test that thesis (patients were not randomized to therapists, etc.), but the data certainly do not support the popular conception that physicians do not make good primary therapists.

From the veiwpoint of cost accounting, however, ex-alcoholic counselors certainly come less expensively than do physicians and if they are equally effective, should receive preference. This position has recently gained acceptance, and more and more nonmedical professionals and paraprofessionals are being utilized predominantly as the primary therapists in alcoholism treatment programs. This development may be cost effective but there is a real question as to whether it is medically effective. The emphasis on group therapy and related methods of social psychology as the dominant approaches to treatment results in a narrow, limited-treatment program which may be effective in some patients but countertherapeutic in others. The limited success of the therapeutic community—the prototype of the social treatment model—in the field of drug addiction is well recognized. Despite this, there seems to be a determined and widespread effort to go down the same dead-end road in alcoholism.

This development is even more serious in the related field of "sobering up" or "nonmedical detoxification" centers which have developed as the result of the recent trend toward decriminalizing alcoholism in the United States. The need for medical–social detoxification centers to replace jails has become urgent. Legislators have been only too ready to accept the less expensive nonmedical detoxification model as a readily available cheap substitute for adequate medical care. Pittman (1974), the initiator of the first detoxification center in the United States, has protested strongly against this new trend of nonmedical sobering-up stations, emphasizing the dangers of inadequate medical diagnosis, care, and treatment. This is not to indicate that nonmedical sobering-up stations are not feasible but only that they require *adequate medical backup*. The alcohol counselors must be trained to recognize the early danger signs and quickly refer the patients for medical care. Adequate medical backup and adequate clinical follow-up (ambulatory outpatient clinics and halfway houses) form the essential ingredients in the public health programs which must be established to treat the alcoholics under their new decriminalized status.

The role of the physician in special alcoholism treatment centers should then be the same as his role in any other setting—to be the primary therapist where it is appropriate or the auxiliary medical resource where it appears that others may more effectively and more economically be primary therapists.

CRITERIA FOR DIAGNOSIS

In defining the role of the physician in treating alcoholics, it is necessary to distinguish between the responsibilities of the physician in private practice and those of the physician in a specialized alcoholism treatment program. In the latter situation many of the most critical issues concerning his responsibilities have already been resolved. Among these are: (1) identification of the alcoholic—this is established by the individual's presenting himself at an alcoholic clinic; (2) evaluation, development of a treatment plan, and referral—this is usually done by nonmedical professionals; and (3) definition of the physician's responsibility—this is spelled out by the program director. In the physician's private office, all of these critical decisions have to be made by the physician himself before any meaningful treatment plan can be developed and implemented.

The importance of this sequence is self-evident when one considers that it is likely that almost every alcoholic presents himself in the early stages of his problem drinking to his physician with some difficulty—physical illness, emotional disturbances, or family problems. The family physician is, with the family religious advisor, probably the first individual outside of the family to recognize the beginning of a potentially serious illness. If early recognition can lead to early treatment, there is at least the potential for early intervention and help. Consequently, the sequence of diagnosis, referral, and/or treatment by the private physician is a critical factor in the treatment of alcoholism. In this sequence, diagnosis is the first and most important step.

The most comprehensive criteria for diagnosing alcoholism have been developed by the National Council on Alcoholism Criteria Committee and have been widely published (Criteria Committee NCA, 1972). This schema describes in detail signs and symptoms which are suggestive, indicative, or actually diagnostic of the disease. The system has great value as a scientific instrument for defining a nebulous syndrome. It is, however, somewhat cumbersome and does not lend itself readily to clinical application. Nevertheless, the general principles articulated are pertinent and may be translated into a simpler formulation.

The NCA Criteria Committee report divides signs and symptoms onto two tracks: Track I—physiological and clinical, and Track II—behavioral, psychological, and attitudinal. Various findings and responses in these areas are then further categorized on three levels where level 1 is *diagnostic,* level 2 *indicative,* and level 3 *suggestive* of the presence of alcoholism. A Track I finding diagnostic of alcoholism could be a typical alcohol withdrawal syndrome; one indicative might be the presence of cirrhosis or pancreatitis, while a wide variety of pathology might *suggest* the diagnosis. In the Track II—i.e., behavioral, psychological, and attitudinal—area, a history of continued drunkenness would be diagnostic, more insidious signs such as morning shakes and morning drinking highly indicative, and the continued use of alcohol to resolve personal tensions highly suggestive.

The NCA criteria define these responses in detail and refine the physician's appreciation of the sequences involved. To that extent they remain a valuable source to which the physician can refer for specific interpretations. What is even more pertinent is the "index of suspicion" which the physician adopts. If he remembers that probably 10 percent of all the adults entering his office may have an alcohol problem, a careful history and physical examination utilizing the information which most physicians already have will provide a positive diagnosis in the majority of alcoholics. However, in this process a review of the NCA criteria is very helpful and for that purpose they are repeated in Table 1.

CRITERIA FOR REFERRAL AND/OR TREATMENT

If the major element in the early diagnosis of alcoholism is the index of suspicion, the major determinant of the course of referral and treatment is the physician's willingness to treat. If the physician is disinterested in alcoholics or even dislikes them, he is unlikely to make the diagnosis at all unless the condition is so blatant that it cannot be overlooked. In that case, having confronted the problem, the unwilling physician looks about for the easiest route for disassociating himself from the situation and refers the patient to A.A., a friendly psychiatrist, or a specialized alcoholism treatment facility in the vicinity. Having, to his way of thinking, discharged his medical responsibility, he cleanses his mind of the case and hopes that the patient never returns.

Fortunately, this type of physician represents a minority. Whereas in the Bailey study of 1968 about 52 percent of physicians were willing to treat alcoholics, in the Jones and Helrich study of 1972 the figure had risen to 70–75

TABLE 1a. Major Criteria for the Diagnosis of Alcoholism

Criterion	Diagnostic level
Track I. Physiological and Clinical	
A. Physiological dependency	
1. Physiological dependence as manifested by evidence of a *withdrawal syndrome* when the intake of alcohol is interrupted or decreased without substitution of other sedation. It must be remembered that overuse of other sedative drugs can produce a similar withdrawal state, which should be differentiated from withdrawal from alcohol.	
a. Gross tremor (differentiated from other causes of tremor).	1
b. Hallucinosis (differentiated from schizophrenic hallucinations or other psychoses).	1
c. Withdrawal seizures (differentiated from epilepsy and other seizure disorders).	1
d. Delirium tremens. Usually starts between the first and third day after withdrawal and minimally includes tremors, disorientation, and hallucinations.	1
2. Evidence of *tolerance* to the effects of alcohol. (There may be a decrease in previously high levels of tolerance late in the	
Pancreatitis in the absence of cholelithiasis	2
Chronic gastritis	3
Hematological disorders:	
Anemia: hypochromic, normocytic, macrocytic, hemolytic with stomatocytosis, low folic acid	3
Clotting disorders: prothrombin elevation, thrombocytopenia	3
Wernicke-Korsakoff syndrome	2
Alcoholic cerebellar degeneration	1
Cerebral degeneration in absence of Alzheimer's disease or arteriosclerosis	2
Central pontine myelinolysis	2
Marchiafava—Bignami's disease {diagnosis only possible postmortem}	2
Peripheral neuropathy (see also beriberi)	2
Toxic amblyopia	3
Alcohol myopathy	2
Alcoholic cardiomyopathy	2

course.) Although the degree of tolerance to alcohol in no way matches the degree of tolerance to other drugs, the behavioral effects of a given amount of alcohol vary greatly between alcoholic and nonalcoholic subjects.

    a. A blood alcohol level of more than 150 mg without gross evidence of intoxication. — 1

    b. The consumption of one-fifth of a gallon of whiskey or an equivalent amount of wine or beer daily, for more than one day, by a 180-lb individual. — 1

3. Alcoholic "blackout" periods. (Differential diagnosis from purely psychological fugue states and psychomotor seizures.) — 2

B. Clinical: Major alcohol-associated illnesses. Alcoholism can be assumed to exist if major alcohol-associated illnesses develop in a person who drinks regularly. In such individuals, evidence of physiological and psychological dependence should be searched for:

Fatty degeneration in absence of other known cause — 2
Alcoholic hepatitis — 1
Laennec's cirrhosis — 2
Beriberi — 3
Pellagra — 3

Track II. Behavioral, Psychological and Attitudinal

All chronic conditions of psychological dependence occur in dynamic equilibrium with intrapsychic and interpersonal consequences. In alcoholism, similarly, there are varied effects on character and family. Like other chronic relapsing diseases, alcoholism produces vocational, social, and physical impairments. Therefore, the implications of these disruptions must be evaluated and related to the individual and his pattern of alcoholism. The following behavior patterns show psychological dependence on alcohol in alcoholism:

1. Drinking despite strong medical contraindication known to patient — 1
2. Drinking despite strong, identified social contraindication (job loss for intoxication, marriage disruption because of drinking, arrest for intoxication, driving while intoxicated) — 1
3. Patient's subjective complaint of loss of control of alcohol consumption — 2

TABLE 1b. Minor Criteria for the Diagnosis of Alcoholism

Criterion	Diagnostic level	Criterion	Diagnostic level
Track I. Physiological and Clinical		3. Minor—indirect	
A. Direct effects (ascertained by examination)		Results of alcoholic ingestion:	
1. Early:		Hypoglycemia	3
Odor of alcohol on breath at time of medical appointment	2	Hypochloremic alkalosis	3
2. Middle:		Low magnesium level	2
Alcoholic facies	2	Lactic acid elevation	3
Vascular engorgement of face	2	Transient uric acid elevation	3
Toxic amblyopia	3	Potassium depletion	3
Increased incidence of infections	3	Indications of liver abnormality:	
Cardiac arrhythmias	3	SGPT elevation	2
Peripheral neuropathy (see also Major criteria, Track I, B)	2	SGOT elevation	3
		BSP elevation	2
3. Late (see Major criteria, Track I, B)		Bilirubin elevation	2
B. Indirect Effects		Urinary urobilinogen elevation	2
1. Early:		Serum A/G ration reversal	2
Tachycardia	3	Blood and blood clotting:	
Flushed face	3	Anemia: hypochromic, normocytic, macrocytic, hemolytic with stomatocytosis, low folic acid	
Nocturnal diaphoresis	3	Clotting disorders: prothrombin elevation, thrombocytopenia	3
2. Middle:		ECG abnormalities:	
Ecchymoses on lower extremities, arms, or chest	3	Cardiac arrhythmias; tachycardia; T waves dimpled, cloven, or spinous; atrial fibrillation; ventricular premature contractions; abnormal P waves	2
Cigarette or other burns on hands or chest	3		
Hyperreflexia, or if drinking heavily, hyporeflexia (permanent hyporeflexia may be a residuum of alcoholic polyneuritis)	3	EEG abnormalities:	
3. Late:		Decreased or increased REM sleep, depending on phase	3
Decreased tolerance	3	Loss of delta sleep	3
C. Laboratory tests		Other reported findings	3
1. Major—direct			

- Blood alcohol level at any time of more than 300 mg/100 ml — 1
- Level of more than 100 mg/100 ml in routine examination — 1
- Major—indirect — 3

Track II. Behavioral, Psychological, and Attitudinal

A. Behavioral
1. Direct effects

   Early:
   - Gulping drinks — 3
   - Surreptitious drinks — 2
   - Morning drinking (assess nature of peer group behavior) — 2

   Middle:
   - Repeated conscious attempts at abstinence — 2

   Late:
   - Blatant indiscriminate use of alcohol — 1
   - Skid Row or equivalent social level — 2

2. Indirect effects

   Early:
   - Medical excuses from work for variety of reasons — 2
   - Shifting from one alcoholic beverage to another — 2
   - Preference for drinking companions, bars, and taverns — 2
   - Loss of interest in activities not directly associated with drinking — 2

   Late:
   - Chooses employment that facilitates drinking — 3
   - Frequent automobile accidents — 3
   - History of family members undergoing psychiatric treatment; school and behavioral problems in children — 3
   - Frequent change of residence for poorly defined reasons — 3
   - Anxiety-relieving mechanisms, such as telephone calls inappropriate in time, distance, person, or motive (telephonitis) — 3
   - Outbursts of rage and suicidal gestures while drinking — 2
   - Serum osmolality (reflects blood alcohol levels): every 22.4 increase over 200 mOsm/liter reflects 50 mg/100 ml alcohol — 2

- Decreased immune response — 3
- Decreased response to Synacthen test — 3
- Chromosomal damage from alcoholism — 3

B. Psychological and attitudinal
1. Direct effects

   Early:
   - When talking freely, makes frequent reference to drinking alcohol, people being "bombed," "stoned," etc., or admits drinking more than peer group — 2

   Middle:
   - Drinking to relieve anger, insomnia, fatigue, depression, social discomfort — 2

   Late:
   - Psychological symptoms consistent with permanent organic brain syndrome (see also Major criteria, Track I, B) — 2

2. Indirect Effects

   Early:
   - Unexplained changes in family, social, and business relationships; complaints about wife or husband, job, and friends — 3
   - Spouse makes complaints about drinking behavior, reported by patient or spouse — 2
   - Major family disruptions, separation, divorce, threats of divorce — 3
   - Job loss (due to increasing interpersonal difficulties), frequent job changes, financial difficulties — 3

   Late:
   - Overt expression of more regressive defense mechanisms: denial, projection, etc. — 3
   - Resentment, jealousy, paranoid attitudes — 3
   - Symptoms of depression: isolation, crying, suicidal preoccupation — 3
   - Feelings that he or she is "losing his or her mind" — 2

percent. Consequently, most of the respondent physicians do diagnose alcoholism and do make a sincere effort to provide adequate treatment for their patients. However, the criteria for patient selection for appropriate treatment remains vague and we shall here attempt to review some of the factors to be considered in making these determinations.

The Therapist of Choice

The first decision which the physician must make is to determine who shall be the primary therapist. This decision obviously depends on the particular treatment program which is deemed most appropriate but there are also individual considerations which often influence that decision. These considerations revolve about the particular therapists who are available, the interest of the involved physicians in treating alcoholics, the particular needs of the patient, the input of the patient's family, and a variety of other factors which may determine what should be the optimal course.

The choice of primary therapist is large, ranging from the physician himself to psychiatrists, social workers, alcohol counselors, probation officers, ministers or priests, and Alcoholics Anonymous. Apart from the relative effectiveness of these various kinds of therapists, there are often individual considerations which weigh for one type of referral rather than another.

Physicians versus Psychiatrists

The terminology is not meant to imply that psychiatrists are not physicians but is merely meant to distinguish medically oriented physicians from psychiatrically oriented ones. Medical physicians (internists, family physicians, and osteopathic physicians) have different training from psychiatrists, different skills, and a different mode of practice. These differences, as among other medical specialists, enhance rather than diminish the scope of practice if there is proper appreciation of the specific indications for referral and an appropriate level of interaction among the physicians involved.

Because of the special nature of medical practice, medical physicians by and large are the first to make the diagnosis of alcoholism and consequently find it necessary to outline a course of treatment. A natural sequence of questions the physician would ask himself might be: (a) Is the patient acutely ill, i.e., intoxicated or in withdrawal or suffering from some acute medical or psychiatric condition? (b) Does the patient require hospitalization or can his symptomatology be treated on an ambulatory basis? (c) If he requires hospitalization, can it be a general hospital or need it be in some specialized treatment hospital?

Given the fact that the patient is not acutely ill and does not require hospitalization, the series of questions might continue as follows: (d) Can I treat this patient successfully myself? (e) Can I treat this patient with outside help (A.A. or a psychiatrist colleague)? (f) Is the treatment of the patient really beyond my scope and should I refer him? (g) If I should, to whom should I refer (a psychiatrist, a specialized alcoholism treatment program, A.A., etc.)? (h) If I refer him, how should I maintain a positive involvement in the patient's management?

This sequence of questions would follow just as naturally for psychiatrists if they were the first to diagnose the presence of alcoholism. Here the question of referral to an internist or family physician as a cotherapist would center largely on the presence of medical complications and on the evaluation of whether the patient could do as well under medical as under psychiatric treatment.

The question of who should be the primary therapist is a critical one and depends in a general sense on the kinds of therapeutic services that different kinds of therapists have to offer. In a more specific sense, the personal characteristics of the therapist may play a more important role than his professional qualifications; a dedicated ex-alcoholic counselor may be immeasurably more effective than an indifferent or antagonistic psychiatrist. Nevertheless, assuming that all individuals involved are of equal goodwill, it is likely that professionals with different qualifications might serve certain types of patients better than others.

By and large, the criteria which would make the medical physician the therapist of choice would include the following: (a) a close personal relationship between the physician and the patient, (b) an acute medical crisis either in the form of needed detoxification or a severe medical complication, (c) the absence of moderate to severe psychopathology, (d) a patient with that makeup which is most susceptible to medical management (this is often the socially stable blue collar worker (Kissin et al., 1970), but other characteristics of this kind of patient will be described shortly), (e) a physician who finds himself comfortable giving supportive counseling and doing family therapy.

The criteria for selecting the psychiatrist as the primary therapist might include the following considerations: (a) the absence of a suitably sympathetic family doctor, (b) the presence of moderate to severe psychopathology, (c) the need for psychotherapy—either individual or group—in the appropriate patient, and (d) the need for specialized types of treatment, e.g., lithium.

Mental Health Professionals

These are nonmedical mental health professionals—usually clinical and counseling psychologists and social workers who have taken specialized training in the treatment of alcoholism and who have set themselves up in private practice as alcohol therapists. These individuals are often extremely knowledgeable and experienced therapists who are working or have worked in specialized alcoholism treatment programs. Within the setting of the specialized treatment programs, they often provide the major therapeutic thrust either by direct therapy with patients or through their supervision of alcohol counselors.

In private practice, the effectiveness of these individuals is often limited by their inaccessibility to medical support, where the patients have medical problems or where pharmacologic support of some sort is indicated—Antabuse, tranquilizers, antidepressants, etc. Mental health professionals are without medical resource unless they are able to establish an effective liaison with some physician. When this is possible and a family physician can work closely with a mental health professional, there develops an effective therapeutic team and optimal therapy may be possible.

The question may arise as to whether what these therapists do cannot be done better by a psychiatrist. This may or may not, in fact, be the case. Many psychiatrists may have had little experience with alcoholics and some, because of their particular orientation or personality, very little success. In those circumstances, a trained psychologist or social worker with extensive experience in treating alcoholics may—working closely with a family physician—provide a more effective therapeutic milieu than a psychiatrist working alone. On the other hand, it is true that a well-trained, positively oriented psychiatrist has the advantage of being able to practice in both the medical and the psychological model, while the professional counselor tends to be restricted to the psychological model alone. Our purpose here is not to tout one primary therapist over another but merely to enumerate and evaluate the various resources available.

Paraprofessional Counselors

Paraprofessional counselors fall at this time into two major groups—ex-alcoholics and nonalcoholics (usually with a B.A. degree) who have had specialized training in the treatment of alcoholism. At present, probably all of these individuals are working in specialized alcoholism treatment units, either on their own or under the supervision of mental health professionals. However, there is a strong movement for licensing paraprofessional counselors, a movement which has great merit and one which will probably shortly be realized.

The danger in this movement, as the present author sees it, is that paraprofessional counselors will be permitted to assume professional responsibilities either in or out of specialized alcoholism treatment facilities. Although the danger of mistreatment of unsupervised paraprofessionals in a specialized treatment unit is substantial, it is at least somewhat reduced by the presence and input of other counselors. If the time should come when licensing as a paraprofessional alcohol counselor will carry with it the right to participate in private practice, the dangers of patient mismanagement could become great. Obviously, we do not feel that private referral to an alcohol counselor is a viable option.

Clergy

The clergy is in some respects similar to the mental health professional and in others to the paraprofessional counselor. Where the clergyman has had specialized training in counseling and alcoholism, he may serve in the same capacity as the professional psychologist or social worker. Where, however, he has not had such training, his approach may be moralistic or sometimes uninformed. Here the referring physician must base his referral on individual evaluation, keeping in mind at all times that zeal is not necessarily the same as knowledge.

Alcoholics Anonymous

A.A. has been found almost universally to be an effective therapy or therapeutic adjunct in the treatment of alcoholism. Almost every patient should be referred to A.A. unless there are some specific contraindications. These contraindications are few but may include the following: (a) The patient has tried it and dislikes it immensely; (b) the referring physician feels that the patient is not appropriate for the particular A.A. groups which are available; (c) the specific A.A. group involved is too narrow in its approach—as some are—and objects to the use of other treatment modalities, e.g., psychotherapy or pharmacotherapy. Under these circumstances, some patients drop out while others continue to attend, concealing their other treatment involvements; (d) the A.A. experience may interfere in the patient's family constellation (the wife may be unalterably opposed); (e) no A.A. group is readily available.

Where these contraindications do not exist, the patient should be referred to a specific group which it is thought will best fit his needs. In general, patients appear to do best in A.A. groups which are most similar in cultural and socioeconomic status to their own. Where this is not possible, one should at least attempt to approximate it. Cultural and socioeconomic compatibility also make it more possible for the wife and children to join the Al-Anon and Alateen groups, respectively—a development which may be extremely helpful in the overall treatment.

The Treatment Model of Choice

As previously stated, the criteria for selection of treatment model are often more compelling than those for selecting primary therapist and the former process will often determine the latter. As part of the question which the referring physician asks: Who should treat this patient? is the corollary: In what setting should this patient be treated? For example, if the patient is socioeconomically derelict and has no means of personal or medical support, it is obviously impossible to treat him in the private sector and he must be referred to the public support system. Since public health programs in alcoholism almost always consist of specialized alcoholism treatment units, the social status of the individual automatically determines the treatment model selected. However, the necessary course is not always so apparent and the decision to apply one or another treatment model often depends on special considerations.

In the previous chapter we have discussed the dominant available approaches under the heading of Treatment Models. These include the medical, psychological, social, A.A., and multivariant. All of these models are usually available to the referring physician and some of the criteria have been discussed under the heading of The Therapist of Choice. Here it would be of value first to describe the structure of each of these models as they exist in the private sector and secondly to indicate the criteria for selecting one model over another in a given case.

Medical Model

Characteristic Structure. This usually consists of treatment by the medical physician himself and is characterized by some of the following: (1) strong personal relationship between the physician and the patient and the patient's family; (2) stress on medical aspects of illness particularly where there is some medical complication of alcoholism; (3) strong reliance on physician often developed by initial detoxication—either in private hospital or on ambulatory basis; (4) emphasis on medical aspects of case without threat; (5) pharmacologic support—Antabuse, tranquilizers, antidepressants; (6) supportive psychotherapy or counseling; (7) support of family with some family counseling.

It should be stressed that in addition to the private office of the practicing physician, there are other elements in the medical model treatment system. These include the general hospital, and private and public specialized alcoholism programs established in this model, with both inpatient and outpatient components. However, the general approach in each of these settings is essentially similar.

Characteristics of Most Suitable Patients. (1) Socially stable—living with family and with permanent employment; (2) absence of moderate to severe psychopathology; (3) average intellectual orientation—not overly intellectual or sophisticated; (4) medically oriented—has "faith in physicians"; (5) absence of depression and self-destructive tendencies; (6) good external resources—family, friends, etc.—not alienated; (7) socially stable blue-collar and lower-echelon white-collar workers do particularly well in this setting (Kissin *et al.*, 1970; Pattison, 1974).

The Psychological Model

Characteristic Structure. This usually consists of treatment by a psychiatrist or a mental health professional (psychologist, social worker, clergyman). The setting has some of the following characteristics: (1) intensive psychotherapy, either individual or group; (2) occasional use of adjunctive medication by psychiatrists, particularly for more severe psychopathology; (3) cooperation of mental health professional with family physician where, for example, Antabuse is desirable or where medical complications exist; (4) more extensive use of family counseling.

Here, too, this model is not confined to private offices but may be found in many private treatment facilities, especially those sanitaria which deal with the long-term rehabilitation of the alcoholic.

Characteristics of Most Suitable Patients. (1) Patient inappropriate for medical model treatment; (2) patient highly intellectual and sophisticated—places less credence in "medical" explanations; (3) moderate to severe psychopathology—best treated by psychiatrists, particularly where pharmacotherapy is desirable; (4) Presence of depressive and self-destructive tendencies; (5) social stability—living at home and working full time; (6) adequate social relationships—family, friends—not alienated; (7) socially stable, sophisticated middle- to upper-class white-collar worker often does well in this setting (Kissin *et al.*, 1970; Pattison, 1974).

The Social Model

Characteristic Structure. The social model, as previously described, becomes the prototypical model used in most, but by no means all, public health programs. These programs, which often serve the socioeconomically deprived and derelict alcoholic populations, are structured to provide services which are less appropriate to those populations ordinarily served in the medical and psychological model systems, Some characteristics of the social model programs are as follows: (1) public health programs publicly supported; (2) comprehensive programs containing detoxification services, day centers, outpatient clinics, halfway houses (rehabilitation wards), and educational and vocational training services; (3) large emphasis on use of paraprofessional alcohol counselors, either supervised by mental health professionals or sometimes self-supervised; (4) medical and psychiatric input mainly for acute medical and psychiatric complications; (5) detoxification may be either medical or nonmedical with medical backup; (6) intensive use of group therapy and educational sessions; (7) emphasis on educational and vocational rehabilitation and job development; (8) provision of social services, welfare, housing, etc.; (9) heavy emphasis on patient population interaction (group therapy, etc.); (10) emphasis on community involvement; (11) use of family therapy where family relationships exist; (12) heavy involvement of A.A.

Again, it should be noted that the social model of alcohol rehabilitation is not restricted to public health programs. Some very reputable (and expensive) private programs are similarly structured even though catering to wealthier clients.

Characteristics of Most Suitable Patients. (1) The socioeconomically deprived and disrupted individual without home, job, or means of support; (2) the alienated individual without family or friends; (3) the alcoholic who needs identification with other alcoholics in order to admit to his alcoholism (similar to the "identification phenomenon" in A.A.); (4) the alcoholic who needs a broad spectrum of interactions; (5) the alcoholic without sufficient internal or external resources to make it on his own on the outside; (6) the alcoholic for whom neither the medical model nor the psychological model works; (7) the alcoholic with the dyssocial personality where social psychiatric influences (group therapy, the successful ex-alcoholic model) are most effective; (8) the alcoholic who needs prolonged, intensive inpatient or day center involvement and rehabilitation.

The Alcoholics Anonymous Model

Characteristic Structure. (1) Neighborhood groups often fairly homogeneous for ethnic, socioeconomic, and cultural parameters; (2) strong emphasis on self-acknowledgment of alcoholism and "confessional" purging; (3) emphasis on moralistic and religious elements; (4) emphasis on self-help and need to develop strength from within; (5) powerful use of successful ex-alcoholic model—"You can do it, we did"; (6) provision of social milieu for the alienated, isolated alcoholic with social interaction evenings and weekends; (7) existence of strong "identification" elements for the alienated alcoholic; (8) some antagonism to other models of treatment, e.g., pharmacotherapy and psychotherapy, though of recent times this antagonism has diminished.

Characteristics of Most Suitable Patients. (1) The alienated, isolated alcoholic who needs identification with a compatible social group; (2) the alcoholic with strong denial who needs the social pressure of peers to accept his alcoholism; (3) the outer-directed alcoholic who responds more readily to social pressure; (4) the more "mystical" alcoholic to whom abstinence in A.A. comes about as almost a religious conversion; (5) the lonely alcoholic without friends or family who finds a home in A.A.; (6) the alcoholic whose wife and children are willing to join Al-Anon and Alateen.

As previously stated, Alcoholics Anonymous offers a sufficient spectrum of opportunities for a broad enough segment of the alcoholic population to be offered to almost all of them. In some cases, A.A. alone may be sufficient therapy. Where it is not, it is often a valuable adjunct to any of the previous specific treatment models proposed. On the other hand, some groups, by taking a very strong position against psychotherapy and pharmacotherapy, make a collaborative effort between themselves and the referring physician most difficult. Usually, however, if the physician is sincere and dedicated and does not indiscriminately prescribe sedative medication, a positive working relationship becomes possible.

Selection of a Treatment Model

On the basis of the previous considerations, the physician determines which of the available treatment plans to follow. The decision as to whether to follow the medical or psychological model is frequently critical since the future prognosis of the case may be dependent upon it. On the other hand, the decision as to whether an alcoholic patient should be treated by his physician or in a special alcoholism treatment facility is often made without the conscious intervention of the physician. For one thing, relatively few patients who enter alcoholism treatment programs are referred by their physicians. Many are socially deprived and do not have private family physicians. Many others of the lower middle class and middle class who do have private physicians have found them to be unreceptive to their illness, either through ignoring the diagnosis or, having made it, ignoring the treatment. In most alcoholism treatment programs, no more than 5–10 percent of all applicants are physician-referred and then, usually, only because the physician is not interested in treating the patient or the patient cannot afford psychiatric therapy.

DESIGNING A SPECIFIC TREATMENT PLAN

Having diagnosed and evaluated the alcoholic and having determined by whom and in what general manner the alcoholic is to be treated, the physician and his collaborators must now design a specific treatment plan and implement it. Although the choice of therapist and treatment model to a degree define the treatment plan, it does so only generally and a more specific plan based on more individual considerations must be designed.

Toward that end, it is essential to consider the alcoholic patient within the context of the total array of problems. Because of the complex interaction of biological, psychological, and social mechanisms involved in the etiology, pathogenesis, and clinical manifestations of alcoholism, a practical approach toward structuring a treatment plan must be adopted. The recently developed problem-oriented approach toward medical and psychiatric treatment lends itself particularly well to the development of a treatment plan for any given alcoholic.

Table 2 illustrates what such a problem-oriented treatment plan might look like. The first column—problem areas—would, on the actual clinical form, be the only column with printed entries. The other two columns—diagnosis and treatment—would be presented vacant and would be filled out by the therapists after the first few interviews. In Table 2, as presented here, the last two columns have been filled in to illustrate some typical diagnoses which one might encounter and in column 3 some specific treatments which one might prescribe. The treatments in column 3 are not meant to correlate with the diagnoses in column 2, except in a general way.

The first six problem areas (1–6) are all fairly self-evident, i.e., they may be defined rather specifically and treatment for a given diagnosis is usually fairly specific. In a general sense they relate to the special problems described in Chapter 1 (Figure 3) and are responsive to the specific therapies described there. The last three problem areas (7–9) in Table 2, i.e., coping mechanisms, family problems, and motivation, correspond generally to the core problem described in Chapter 1 (Figure 3) and are more complex both in their constellation and in their prescribed treatment. The evaluation of the core problem involves a comprehension of the dynamics of the underlying drinking practice. It is based mainly on the understanding of the general coping mechanisms which the individual uses in his effort to adjust to the exigencies of life with particular reference to how drinking becomes the coping mechanism of choice. The treatment plan centers around evolving alternative coping mechanisms and life-styles which the alcoholic may substitute for drinking. The exploration of family relationships for both positive and negative influence is an important element in understanding the dynamics of the core problem. The specific motivation which brings the patient into treatment at a given time may be the single most significant clue as to what is bothering the patient and as to what may be used as an alternative coping mechanism to his drinking.

An analysis of each problem area in greater depth will help elucidate the process.

Special Problems and Their Treatments

As previously mentioned, the special problems are not special in the sense that they are more important than the core problem; quite the contrary is true. They are special only in that they are more apparent—more on the surface—and are usually susceptible to more immediate and direct management. They are also special in that frequently they are characterized by a greater sense of urgency than the core problem—the drinking—which usually has been going

TABLE 2. Medical Management of the Alcoholic Patient

Special problems and treatment

Problem area	Evaluation	Treatment
1. Pattern of alcohol abuse	a. Intoxication b. Periodic vs. steady c. Problem drinker vs. alcoholic d. Impulsive drinker e. Occasional breaks in abstinence	a. Detoxification b. Test for depression c. Total abstinence d. Disulfiram (Antabuse) e. Tolerant acceptance by therapist
2. Physical dependence	a. Acute withdrawal syndrome b. Residual withdrawal symptomatology c. Moderate to severe physical dependence	a. Detoxification b. Chlordiazepoxide c. Total abstinence
3. Medical complications	a. Careful evaluation and diagnosis	a. Intensive medical care
4. Psychiatric diagnoses	a. Psychoses b. Neuroses c. Character disorder d. Affective disorder e. Anxiety and agitation	a. Chemo and psychotherapy b. Dynamically oriented therapy? c. Existential psychotherapy? d. Antidepressants or lithium e. Chlordiazepoxide or phenothiazines
5. Employment status	a. Job stability b. Educational and vocational inadequacy	a. Support with employer b. Educational and vocational training
6. Socioeconomic problems	a. Homeless b. Without income c. Legal difficulties	a. Provide shelter b. Provide welfare c. Legal support

Core problems and core treatment

Problem area	Evaluation	Treatment
1. Coping mechanisms	Alcohol use: a. To suppress anxiety b. To relieve depression c. To suppress latent psychoses d. Impulsive personality e. Passive–dependent personality f. Antisocial personality g. Neurotic personality h. Dyssocial drinker "ghetto"	a. Chlordiazepoxide b. Antidepressant drugs c. Antipsychotic drugs d. Antabuse and Psychotherapy e. Psychotherapy and AA f. Psychotherapy and AA g. Psychotherapy and AA h. Rehabilitation and AA
2. Family problems	a. Supportive wife and children b. Difficult family relationship c. Destructive family relationship	a. Al-Anon and Alateen and family therapy b. Family therapy c. Separation?
3. Motivation	a. General level of motivation b. Precipitating motivation. *Threat of* 1. Loss of spouse, 2. loss of job, 3. ill health.	a. Psychotherapy and AA 1. Family therapy 2. Intervention with employer 3. Medical care

on for many years. Consequently, they are presented first in the sequence of management as aspects to be dealt with quickly and effectively so that the main business of treatment, i.e., treatment of the core problem, may be addressed.

In the private practitioner's office, some special problems are more apparent than others. For example, social destitution, which is so common a condition in public programs, is only seldom a situation with which the private physician is confronted. Nevertheless, even in private practice socioeconomic factors of this type are not to be ignored, particularly where the husband's alcoholism is causing severe economic stress in the family.

The Alcohol Abuse Pattern

Acute Intoxication. Acute intoxication occurs in three forms: (1) mild intoxication, (2) intoxication with violence, and (3) the stuporous or comatose alcoholic. For mild intoxication no therapy is necessary except perhaps some aspirin for the morning hangover. The violent alcoholic will usually subside with psychological support. Occasionally a small dose of intramuscular chlordiazepoxide may be necessary to quiet him. Alcoholic coma is a serious medical complication requiring hospitalization and acute intervention. Treatment involves general supportive measures, intravenous glucose or fructose, and occasionally dialysis.

Another form of intoxication is that known as *pathological intoxication*, which is characterized by violent behavior after relatively small quantities of alcohol. This is thought to be an atypical form of epilepsy or possibly an aberrant type of psychosis. Treatment is intramuscular diazepam (Valium) followed by phenothiazines if the former treatment is ineffective.

The Periodic Drinker and Depression. The specific pattern of alcohol abuse is usually not in itself significant except insofar as it has implications for diagnosis and treatment. For example, the periodic drinker may be suffering from a bipolar depression in which the drinking binge is often a concomitant of the depressed period. Where there is other evidence of a cyclothymic personality, either with true manic and depressed reactions, or even where the clinical symptomatology manifests itself only as periodic depressions, lithium therapy may be helpful. Although Kline *et al.* (1974) have suggested lithium therapy as a possible treatment in all forms of alcoholism with depression, clinical experience tends to indicate that alcoholics with bipolar depression do better on this treatment than do the unipolar.

The Problem Drinker and Social Drinking. The pattern of drinking which would differentiate the problem drinker from the alcoholic would be one of quality as well as quantity. The alcoholic drinks compulsively, gulps his drinks, seems to have no control over his drinking, and, because of the presence of physical dependence and residual withdrawal symptoms, "needs" alcohol.

The problem drinker as defined by Straus (1973), not having significant physical dependence, shows a lower level of necessity in his drinking pattern. The significance of the differentiation between "problem drinker" and "alcoholic" pertains mainly to the goal of total abstinence. Sobel and Sobel (1973) and others have advocated "controlled drinking" as a legitimate goal of therapy. It is highly doubtful whether the true alcoholic with moderate to severe physical dependence and withdrawal symptomatology can ever achieve controlled drinking since each drink reactivates the addictive cycle and accelerates the process. Although Sobel and Sobel (1973) did report a certain degree of improvement in gamma (i.e., true) alcoholics, what they achieved seems better described as controlled bingeing rather than controlled drinking.

On the other hand, the problem drinker with physical dependence may be a candidate for a therapeutic regimen directed toward controlled drinking. However, even here there are pitfalls. It is doubtful that the great majority of alcoholics who have been reported in the literature to have gained the capacity for social drinking were true alcoholics as here defined. On the other hand, it is possible that the early problem drinker, particularly the situational alcoholic in whom drinking is a neurotic pattern rather than the reflection of a deep-seated personality disorder, may be a candidate for such a regimen. However, before one even suggests such a goal, one must be aware of the pitfalls and one must be extremely confident of his evaluations of the patient's psychodynamics. This is sometimes possible in the setting of the private physician's or psychiatrist's office. More often, discretion is the better part of valor and one should opt for total abstinence.

The Use of Disulfiram (Antabuse). Once the decision has been made to make total abstinence the therapeutic goal, Antabuse becomes an extremely valuable clinical adjunct. The pharmacology of Antabuse is somewhat more complex than was originally considered, but its essential action is straightforward. Antabuse is a potent aldehyde dehydrogenase inhibitor which interferes with the normal metabolism of alcohol as illustrated in the following diagram:

$$\text{alcohol} \xrightarrow{\text{alcohol dehydrogenase}} \text{acetaldehyde} \xrightarrow[\text{dehydrogenase}]{\overset{\text{Antabuse}}{\downarrow\downarrow\downarrow} \text{ aldehyde}} \text{acetate} \begin{matrix} \nearrow CO_2 \\ \searrow H_2O \end{matrix}$$

Inhibition of aldehyde dehydrogenase causes a piling up of acetaldehyde resulting in the typical clinical picture of the alcohol–Antabuse reaction— nausea, vomiting, cramps, flushing, vasomotor collapse, and so on. In the past, the entire alcohol–Antabuse reaction was attributed to the acetaldehyde effect; more recently, disulfiram has been demonstrated to be an active dopamine-β-hydroxylase inhibitor as well. This latter action results in a disturbance in catecholamine metabolism which may contribute in part to the overall alcohol-Antabuse reaction (Truitt and Walsh, 1973). More significantly, it probably accounts for the occasional psychotomimetic symptomatology one sees in patients on Antabuse alone—i.e., without alcohol—and, more particularly, when the drug is given in association with an MAO inhibitor. For that reason, Antabuse should probably not be used in patients with any history suggesting a psychosis or in patients maintained on MAO inhibitors either for depression or, as with isoniazid, in the treatment of tuberculosis.

Apart from these restrictions, other contraindications to the use of disulfiram include the presence of severe medical illness (coronary heart disease, congestive heart failure, emphysema, peptic ulcer) which could lead to serious complications in the presence of an acute alcohol reaction. Serious depression in itself may be a specific contraindication since the depressed patient on Antabuse, who drinks for relief, has had his only source of respite removed. Such a patient may either stop his Antabuse to drink, or, if his depression is deep enough, may even become suicidal.

Despite this specific contraindication, the great majority of alcholics are eminently appropriate for Antabuse therapy and are benefited by it. A study by Baekeland and Kissin (1972) demonstrated that older (past 40 years of age), more socially stable, and highly motivated patients do best on Antabuse. However, these are positive factors for all forms of treatment in alcoholism so that this modality need not be restricted to this population. Another advantage of Antabuse is its prolonged action so that each daily dose ensures the patient against drinking for the next four or five days.

Abstinence and Treatment Success. Although the therapist's and patient's goal should most often be total abstinence, by no means should this be considered the sole measure of success. For one thing, other considerations such as physical health, family and social adjustment, and vocational activities are equally important to the absence of drinking. But even abstinence as a goal is not absolute. Although it is certainly true that total abstinence is usually associated with greater improvement in all areas than only partial abstinence, it is equally true that even partial abstinence can be associated with marked improvement in the patient's personal, social, and vocational relationships. Alcoholism must be considered as a chronic disease like diabetes, where, although a blood sugar of 100 mg% may be the goal, slight to moderate deviation from it would not be considered failure.

Physical Dependence

We have placed great emphasis on the presence of physical dependence in our description of the pathogenesis of alcoholism since we believe that it defines many aspects of treatment. The most evident manifestation of physical dependence occurs in the alcoholic who presents himself in acute alcohol withdrawal—a not uncommon event—since it is at that time that he is most acutely ill and best motivated to do something about his illness. The treatment of acute alcohol withdrawal is often the first step in the long road of rehabilitation for the alcoholic patient.

The Acute Alcohol Withdrawal Syndrome. The symptomatology of the acute alcohol withdrawal syndrome is variable, both quantitatively and qualitatively. In the past, various syndromes were differentiated on the basis of the predominant symptoms—delirium tremens, acute alcohol hallucinosis, and impending DTs. These different patterns are now recognized to be variants of a single syndrome with the symptomatology depending on the severity of the withdrawal state and the specific reactivities of different individuals.

The major symptoms and signs of the acute alcohol withdrawal syndrome are illustrated in Figure 1 at three hypothetical levels—mild, moderate, and severe. Some symptoms such as tremor, restlessness, anxiety, insomnia, and sweats are ubiquitous and appear in even mild withdrawal. As the syndrome deepens, the early symptoms become exaggerated and new ones are introduced:

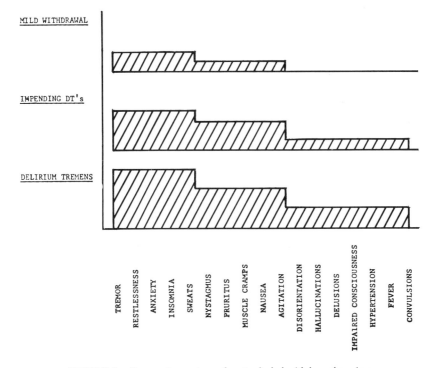

FIGURE 1. Signs and symptoms of acute alcohol withdrawal syndrome.

nystagmus, pruritus, muscle cramps, nausea, and agitation. Finally, in its most severe form, all of these symptoms worsen, and those features characteristic of delirium tremens present themselves: disorientation, hallucinations (tactile, visual, and auditory), delusions, impaired consciousness, hypertension and fever, and sometimes convulsions.

Recognition of impending delirium tremens depends on the index of suspicion of the attending physician. If the patient exhibits even mild withdrawal symptoms, he should be observed to see whether the syndrome is only beginning to develop or already abating. It must be emphasized that the withdrawal syndrome is a progressive one, so that a single observation is not sufficient to determine direction.

In any event, treatment should be prompt and sufficient. The treatment of choice are the benzodiazepines—chlordiazepoxide 75–100 mg IM t.i.d. for the first two days, followed by gradual reduction and switching to oral dosage. Even higher doses may be used where necessary since chlordiazepoxide is a relatively benign drug. Most important is that dosage be sufficient and continuous. Treatment should never be "PRN" since that indicates that medication should be administered only as the symptoms appear. Adequate treatment—i.e., regular chlordiazepoxide administration—produces suppression of the symptomatology and only rarely does the full-blown picture develop.

The treatment of alcohol withdrawal, like that of other withdrawal syndromes, is accomplished through the administration of long-acting cross-tolerant drugs. Other medications such as paraldehyde and barbiturates can also be used effectively, although detoxification appears somewhat less smooth than with the benzodiazepines. Phenothiazines are seldom indicated since they tend to increase the incidence of convulsions.

Acute fulminating delirium tremens is a serious medical condition which, when inadequately managed—particularly in the presence of medical or surgical complications—may have a mortality rate as high as 15 percent. It should be treated in a modern hospital setting with intensive medical care. On the other hand, mild to moderate cases of alcohol withdrawal may be treated in detoxification centers, but even there, ideally, with medical services available. Mild cases of withdrawal can be adequately treated by the family physician on an ambulatory basis with oral chlordiazepoxide—25–50 mg t.i.d. usually being sufficient to control symptomatology. However, the patient should be under constant surveillance at home to guard against exacerbation of the syndrome.

Necessity for Adequate Detoxification. The alcoholic in acute alcohol withdrawal is a very sick and disturbed individual. Considerate treatment at this time will create a strong sense of attachment to the therapist which will be of great value during the subsequent period of treatment. The adequacy of treatment depends on providing sufficient sedation, good nursing care, and strong psychological support. In the social model of detoxification, strong social support is substituted for sedation. This may be sufficient when the symptomatology is not too severe but can be disastrous in the presence of serious illness.

Adequate detoxification also means continuing sedation (a long-acting, cross-tolerant, nonaddictive [one hopes] drug) for as long as the patient has severe residual withdrawal symptomatology. This varies widely from patient to patient depending on the severity of physical dependence, the length and depth of the preceding binge, and individual variability. The latter, a not well-understood phenomenon, results in prolonged severe withdrawal symptoms in some individuals with a relatively moderate drinking history and relatively mild withdrawal symptomatology in patients with heavy, prolonged histories. Some of the factors which have been hypothesized to cause exaggerated and more prolonged residual withdrawal symptomatology have been (a) an asthenic as opposed to a hypersthenic body build, (b) malnutrition or intercurrent illness, (c) psychological instability. Whatever the reasons, some individuals do show moderately severe withdrawal symptomatology for six months or longer after an acute withdrawal episode (Kissin *et al.*, 1959; Schenker *et al.*, 1962; Tripp *et al.*, 1959). When this withdrawal symptomatology is associated with craving—as it so often is (Hore, 1974)—the "drive to drink" may exceed the "drive to abstinence" and treatment will end in failure.

The Use of Sedatives in the Long-Term Treatment of Alcoholics. Given craving because of persistent residual withdrawal symptomatology and the resultant drive to drink, the alcoholic patient either continues in treatment while drinking, or more often, because of a sense of failure, simply drops out. Rapid dropping out is a universal characteristic of all alcoholism treatment programs and Ditman (1966) has reported that in some programs fully 90 percent of all clients drop out during the first month. This statistic must be kept in mind when evaluating reported treatment results from various programs. Some programs do not consider clients "in treatment" until they have "shown their sincerity" by remaining in the program for at least one month. This of course markedly improves "cure" rates by excluding manifest failures.

Many techniques have been utilized to keep patients in treatment and these are discussed elsewhere (Baekeland and Lundwall, 1976). They include rapid involvement (alcoholics cannot tolerate the frustration of waiting), intensive follow-up (they have to know that the therapist "cares"), and meeting the expectations of the patient. All of these techniques are highly important and should be fully utilized. On the other hand, to the patient who is still highly tremulous, anxious, depressed, insomniac, and craving for alcohol, it is unlikely that any of these tactics will be effective unless some therapy for this residual withdrawal is provided. The simplest therapy for this symptomatology is a long-acting, cross-tolerant sedative and probably the best one available is chlordiazepoxide (Librium).

In a previous paper, we have reviewed the rationale for the use of tranquilizers (specifically chlordiazepoxide) in the rehabilitation of chronic alcoholism (Kissin, 1975). In our conclusion we wrote, "Accordingly, in the alcoholic in whom significant physical dependence has developed—and this probably represents the great majority of severe chronic alcoholics—persistent withdrawal symptomatology characterized by tremulousness, anxiety, depression, and insomnia, may persist for periods of up to six months or more after the cessation of drinking. This symptom complex, which the alcoholic knows can be at least temporarily relieved by alcohol, acts as a consistent reinforcement—both physical and psychological—to the underlying predisposing pathology which drove him to drink in the first place. Accordingly, it would seem rational—and experimental evidence tends to support this view—that to the extent that one can help to control this withdrawal symptomatology, one can help the alcoholic achieve and maintain sobriety."

The specific "experimental evidence" referred to is of two types: (1) evidence that chlordiazepoxide helps maintain people in therapy and (2) evidence that chlordizepoxide helps certain alcoholics achieve improvement. Since many studies have shown that maintenance in treatment is essential for improvement (Gerard and Saenger, 1966), these two types of evidence are not unrelated. In the area of keeping people in treatment, two studies by Ditman (1966) and Rosenberg (1974) demonstrate that chlordiazepoxide is significantly more effective than no medication in maintaining alcoholics in treatment. Another study by Kissin (1975) showed that chlordiazepoxide is more effective than phenothiazines ($p < .001$) in maintaining highly anxious alcoholics in treatment. Chlordiazepoxide also retained in treatment the same percentage of agitated patients (who tend to drop out of treatment), as did vitamins with nonagitated patients (who tend to remain in treatment).

In a series of three large studies in three different ghetto populations, Kissin (1975) reported higher improvement rates with chlordiazepoxide alone or in combination with imipramine than with placebo or other drugs. These studies, although not sufficiently controlled to allow statistical evaluation (no placebos were given concomitantly with chlordiazepoxide), do strongly suggest the effectiveness of chlordiazepoxide both in maintaining alcoholic patients in outpatient therapy and in improving prognosis in that treatment setting. A similar study by Hoff (1961) also concludes that chlordiazepoxide helps to increase long-term improvement rates.

On the other hand, the risks of prescribing sedative medication to alcoholics should not be ignored. Alcoholics are notorious in their tendency to become "addicted" to any substance with abuse potential and the sedatives—

particularly the short-acting barbiturates and some of the recently developed somnifacients (Doriden, Quaalude, and Valium)—have proven to be particularly addictive to the alcoholic. All of these drugs should assiduously be avoided in prescribing for alcoholics. These drugs, in addition to having high addiction potential, all potentiate the action of alcohol so that the synergistic interaction between them and alcohol may cause coma and death at levels at which either drug alone would be less toxic.

The prescribing physician must be keenly aware of the possible dangers in prescribing sedative medication to alcoholics and must be prepared to monitor their use closely. Contraindicated drugs, such as those described above, should not be prescribed at all. Indicated medications are prescribed in limited dosage and for short intervals, usually for no more than one week at a time. The recent federal regulation which has moved the benzodiazepines (Librium, Valium, Serax, Tranxene) to Schedule IV should be helpful.

In our own experience, chlordiazepoxide (Librium) is the drug of choice. Kissin (1975) reviews the question of risk in long-term chlordiazepoxide therapy and concludes that it is probably not too risky since (1) chlordiazepoxide is a relatively nontoxic drug with an immense gap between the clinically effective dose and the lethal dose; (2) chlordiazepoxide appears to act as an antagonist to alcohol (Goldberg, 1970; Dundee *et al.,* 1971) rather than to potentiate its effects as do almost all other tranquilizers including phenothiazines; and (3) chlordiazepoxide has a relatively low abuse potential (Irwin, 1973). These qualities combined with its demonstrated greater effectiveness, both in maintaining patients in treatment and in improving outcome, would appear to make chlordiazepoxide the drug of choice in treating chronic alcoholism.

This position, however, requires qualification. It is a well-known fact in psychotherapy in general and in the treatment of alcoholics, in particular, that a patient who is too comfortable lacks motivation to improve his status. This is in keeping with the A.A. adage that an alcoholic has to "hit bottom" before he will rebound onto the road to recovery. Unfortunately, too many alcoholics who have hit bottom lack the resiliency and the internal and external resources to rebound. They remain caught in the addictive cycle and when they are brought out of it after two or three weeks of medication, they readily slip back into it when chlordiazepoxide therapy is discontinued. However, the alternative of making them too comfortable with the chlordiazepoxide may be equally dangerous both because they become unmotivated to improve and because they may become dependent upon the drug.

The potential danger of the overeffective reduction of anxiety through the use of chlordiazepoxide is best illustrated in a paper by Shaffer *et al.* (1963). These authors reported that in a double-blind study on recently withdrawn alcoholics in an inpatient setting, chlordiazepoxide was significantly more effective than placebo in reducing anxiety and craving for alcohol. Paradoxically, however, more patients on chlordiazepoxide than on placebo drank while on weekend passes or failed to return to the ward for further treatment. When questioned as to why they had drunk, the patients on placebo implicated craving for alcohol. The patients on chlordiazepoxide on the other hand responded that their drinking was not related to craving but rather to the fact that they felt so good, they no longer thought they had an alcoholism problem! This form of self-delusion is of course only too characteristic of alcoholics but the responses do throw interesting light on the mechanisms involved.

Liebson and Faillace (1971) have explored this question from the operant psychology viewpoint. They conclude that chlordiazepoxide is an effective substitute for alcohol as a reinforcer but unfortunately, because of the mechanisms described above by Shaffer *et al.* (1963), not in itself an effective mode of treatment. They recommend rather that chlordiazepoxide and disulfiram (an effective deterrant) be combined (in a single capsule) to act both as reinforcer and blocking agent in the operant paradigm. This combination, they postulate, would have a similar combined action to that of methadone on heroin addiction, which acts as a substitute reinforcer and through its increased tolerance effects (Goldstein, 1974) blocks the euphoriant action of heroin. A preliminary experiment with the combination of chlordiazepoxide and dilsulfiram gave very promising results. This type of therapy is similar to that advocated by Kissin and Gross (1970), although there the medications were administered separately rather than in a single capsule.

The question of the use of tranquilizers in the long-term treatment of alcoholism is obviously both an important and a difficult one. The fact remains that over 65 percent of the physicians in the United States who responded to a questionnaire (there were over 15,000 respondents) did use some form of tranquilizer in the treatment of alcoholics (Jones and Helrich, 1972). We have attempted to review the evidence for and against that use elsewhere (Kissin, 1975). Nevertheless, the question still remains an open one.

Having said all this, it is important to make the following points:

1. Ideally, most alcoholics should be treated without the use of psychoactive drugs and probably many can successfully so be treated. The use of psychoactive drugs can introduce another complicating feature into treatment (drug abuse) and may also demotivate the patient. On the other hand, in those patients who demonstrate persistent anxiety-producing withdrawal symptomatology, the judicious use of chlordiazepoxide (with imipramine for depression where indicated) can help maintain them in treatment and improve prognosis.

2. Where psychoactive drugs are used, they should be used in the lowest doses necessary to control symptoms, for the shortest time possible and under the closest supervision.

3. A small number of patients may need long-term medication either because of underlying psychopathology or because of inability to abstain without it. However, every effort should be made to wean them off medication where feasible.

Other Therapeutic Implications of Physical Dependence. Given the concept of the addictive cycle and the strong implications which physical dependency mechanisms have in the pathogenesis of craving and continued drinking, there are certain conclusions which follow: (1) For the patients with moderate to severe physical dependence, controlled drinking is not a legitimate goal. (2) Early or immediate abstinence is a necessary element in the therapeutic process where there is moderate to severe physical dependence. The concept proposed by psychoanalytically oriented therapists that early abstinence is not an important element in treatment is generally unwarranted. Although it may be true for the situational alcoholic, who, at any rate, is the best candidate for such therapy, it is not true for the great majority of alcoholics.

Medical Complications

It is a truism that almost without exception chronic alcoholics demonstrate some physical evidence of prolonged alcohol abuse. One finds, in the majority of patients, some evidence of malnutrition and liver disease but, in addition, neuropathies, cardiomyopathy, brain damage, myopathies, and gastrointestinal disease (pancreatitis, peptic ulcer, colitis) are extremely common. Finally, the stigmata of the disease itself (acute and chronic manifestations of alcohol withdrawal) complete the picture of severe physical frailty.

One or another of these manifestations of physical illness or some of the concomitant psychological symptoms (anxiety, depression, tremulousness, insomnia) usually first bring the alcoholic patient to his private physician. If the physician's index of suspicion is high, he makes the diagnosis of alcoholism with complications and undertakes the total treatment of the patient. For many patients—particularly the socially stable blue-collar worker—the medical approaches in which major emphasis is directed toward the medical problem *of which alcoholism is merely one aspect* is extremely effective (Gerard and Saenger, 1966; Kissin *et al.*, 1970; Pattison, 1974). In these instances treating the medical complications becomes a major element in treating the alcoholism. For example, the present author has utilized in the treatment of such patients the technique of following the gradual shrinking of the markedly enlarged fatty liver during abstinence as a concrete sign of improvement. This technique engages the alcoholic patient, both through his concern for his health and through his need for concrete evidence of improvement.

The use of medical complications as "scare tactics" can be a double-edged sword. In some alcoholics who are relatively psychologically intact, the explanation that continued drinking truly endangers their health, if not their life, may be extremely helpful in their treatment. On the other hand, to the self-destructive alcoholic—of whom there are many—such an announcement merely reinforces his unconscious motivation and encourages him to drink further. Similarly, to the depressed alcoholic, the knowledge that he has already done himself serious and perhaps irreparable physical damage deepens his depression so that he drinks more heavily to combat it. Consequently, self destructive tendencies, depression, excessive anxiety, and other forms of severe psychopathology (psychosis) are specific contraindications to the use of scare tactics in treating alcoholic patients.

The pathogenesis, clinical manifestations, and treatment of the medical complications of alcoholism have been described comprehensively in the literature (Kissin and Begleiter 1974; Seixas *et al.*, 1975). Here it is important to stress the significance of treating the malnutrition and medical complications as part of the overall treatment of alcoholism. There is ample evidence that malnutrition and medical complication increase the severity of the acute withdrawal syndrome (Victor, 1966; Gross *et al.*, 1974). There is also suggestive evidence that these complications result in a prolongation of residual withdrawal symptomatology with all of its implications for continued drinking. Therefore, it becomes the responsibility of the physician to treat these medical complications not only for themselves but to improve the prognosis of the underlying alcoholism. Conversely, the treatment of the underlying alcoholism immeasurably improves the therapeutic course and prognosis of the medical complication.

Psychiatric Diagnoses

As previously described, it has been estimated that fully 15 percent of all alcoholics are suffering from psychosis in one form or another—schizophrenia, manic depression, involutional melancholia, or organic brain syndrome (Sherfey, 1955; Zwerling, 1959; Panepinto *et al.*, 1970). Of these, not all exhibit the full-blown manifestations of the disease, many falling rather into the borderline category. On the other hand, of the additional 10–15 percent described as "neurotic" by these authors, a fair proportion demonstrate a sufficiently high level of anxiety, depression, agitation, and hostility to warrant the categorization of "severe psychopathology." The psychopathology exhibited by the majority of patients with so-called personality disorder is more often behavioral, such as sociopathic acting-out, and although obviously of great seriousness, for the purposes of this section will not be included under the heading of severe psychiatric disorders.

Schizophrenia and Alcoholism. Patients with severe psychopathology often use alcohol—as they do other drugs of abuse—as a form of self-medication. Irwin (1973) has reported that morphine and its derivatives are potent anxiolytic, antidepressant, and antipsychotic pharmacologic agents. Alcohol has been demonstrated to have all of these properties with the anxiety-allaying effect probably the most active (Kissin and Platz, 1963). Other effects such as disinhibition of latent homosexuality and increase in grandiose tendencies in alcoholics have also been reported (Machover et al., 1961).

The stabilizing effects of drugs of abuse when used by psychotic addicts as self-medication is best illustrated in the methadone maintenance of heroin addicts. Not a few of these patients remain fairly well compensated while on standard doses of methadone—80–100 mg daily—but when it is decided to detoxify them, either at their own request or because of an ill-considered staff decision, many suffer acute psychotic breaks. This phenomenon has become sufficiently well documented to constitute one of the specific contraindications to detoxification from methadone maintenance (Kaufman, 1974).

Although the situation is not as clear-cut in alcoholism, D. Smith (personal communication) has reported a similar reaction with many abstinent alcoholics when their dosage of chlordiazepoxide and phenothiazines is reduced. She has found that many schizophrenics show severe agitation when they attempt to abstain from alcohol on first entering treatment. These symptoms may be allayed by the administration of sufficient doses of either phenothiazines or chlordiazepoxide, or, when the symptomatology is severe, both drugs in combination. These patients then make a greater or lesser adjustment to alcohol abstinence as long as they remain on medication. Attempts to reduce the level of drugs, or indeed discontinue them, are associated with acute psychotic breaks, which respond favorably to reinstitution of the medication.

One of the major symptoms in diagnosing the schizophrenic alcoholic is agitation which responds to phenothiazines. Agitation in itself is a common symptom of withdrawal in alcoholics and, as such, hardly indicative of psychosis. However, the great majority of agitated alcoholics respond poorly to phenothiazines, developing a sense of dysphoria rather than one of relief, and either discontinue the medication or drop out of treatment. As previously described, Kissin (1975), comparing the effectiveness of chlordiazepoxide and phenothiazines with agitated alcoholics, found the dropout rate to be almost three times as high with phenothiazines as with chlordiazepoxide. On the other hand, D. Smith (personal communication) found that diagnosed agitated schizophrenics preferred phenothiazines as their maintenance drug although many found chlordiazepoxide to be a valuable anxiolytic adjunct as well.

Depression and Alcoholism. Depression in one form or another is almost a universal concomitant of alcoholism. Kissin and Platz (1968) found a significant level of depression—and anxiety—in 80 percent of their patients as measured by the Zung and Cattell Scales, respectively. As with so many other findings in alcoholics, one is tempted to ascribe the depression to the prolonged alcoholism but the strong familial strain of depression in nonalcoholic relatives

(Schuckit *et al.*, 1972; Winokur *et al.*, 1970) suggests that such affective disorders may also precede rather than be caused only by prolonged heavy drinking.

The particular basis of the manifest depression varies widely and may be dichotomized along the exogenous/endogenous or reactive/psychotic parameters. Perhaps more pertinent has been the recent classification of depressions on the basis of their response to lithium versus the tricyclic antidepressants (Kupfer *et al.*, 1975). Where it now appears that both the manic and depressive phases of manic–depressive illness (bipolar depressions) respond to lithium therapy, it has recently been recognized that certain forms of unipolar depressions, previously considered unrelated to manic depressive disease, also respond better to lithium than to tricyclic antidepressants. Whether these recurrent episodes of depression actually represent a unipolar manifestation of a bipolar illness is now under investigation and must be considered in the treatment of the depressed alcoholic. Kline *et al.* (1974) have advocated the trial of lithium for most depressed alcoholics but the evidence for its efficacy is still equivocal. On the other hand, in the overt manic–depressive, its use is specifically indicated.

The tricyclic antidepressants have been widely used in the treatment of alcoholics, again with equivocal results. Kissin and Platz (1968) found that a combination of imipramine (Tofranil) and chlordiazepoxide appeared to give the best results in alcoholics generally as compared to a wide spectrum of other drugs. A defect of their study resulted from the fact that the medication was not given to depressed patients alone but rather to alcoholics at random. Oddly enough, the simple experiment of testing the effects of tricyclic antidepressants in moderately depressed alcoholics has never been reported upon in the literature—if indeed it has been done—so that the evidence for the effectiveness of this type of treatment remains circumstantial. Nevertheless, the existing experimental data supported by extensive clinical experience more than justifies the use of tricyclic antidepressants in alcoholics where depression is a significant symptom.

Agitation and Alcoholism. Where agitation is a symptom of underlying psychosis—schizophrenia or psychotic depression—phenothiazines represent the drug of choice. Where agitation is a short-lived manifestation of residual withdrawal symptomatology, chlordiazepoxide in gradually decreasing doses is rapidly effective. There is a group of alcoholics with an underlying severe anxiety neurosis, where agitation responding to the anxiolytic effect of alcohol constitutes the basic pathogenetic mechanism. This reaction, aggravated by withdrawal symptomatology, constitutes a particularly difficult vicious cycle. These patients, oddly enough, respond poorly to the phenothiazines which produce dysphoria rather then euphoria. On the other hand, both the withdrawal symptomatology and the underlying anxiety respond well to chlordiazepoxide. This group of patients who constitute a relatively small population of the total alcoholic population—perhaps 5-10 percent—are candidates for prolonged treatment with chlordiazepoxide—in a sense "Librium maintenance." Generally, these individuals do not tend to increase steadily their dose of chlordiazepoxide; however, this possibility must constantly be kept in mind and the patients closely followed.

In this section on psychiatric diagnoses, we have thus far stressed mainly the pharmacologic therapy of the various affective states. Equally pertinent is the type of psychotherapy which is appropriate in treating these different varieties of psychopathology. By and large, these questions are beyond the scope of this chapter, although some mention will be made in our discussion on coping mechanisms. As previously stated, the patient with severe psychiatric disorders should be referred to the care of a psychiatrist where the question of appropriate therapy is a routine consideration. On the other hand, where the described psychopathologies are mild or borderline, the family physician may often be able to maintain the patient in suitable equilibrium, utilizing the appropriate medication and the appropriate reality-oriented, supportive type of psychotherapy.

A final word of caution about the use of medication in alcoholics. In this section, as in the section on physical dependence, explicit indications for using pharmacotherapy have been described. What must be stressed here is the fact that many, if not most, alcoholics can be treated without any medication whatsoever. The ease of prescribing and the power of the prescription are such as to make it the path of least resistance. The physician must always keep in mind that the alcoholic is a "drug addict" who will readily substitute some other drug for alcohol. Consequently, the tendency to medicate—let alone to overmedicate—must always be guarded against. Nevertheless, the appropriate, carefully observed, and monitored prescription of needed medication can be of substantial value in the overall treatment of many alcoholics who might not be able to achieve sobriety without that help.

Employment Status

The employment status of an alcoholic has in most treatment outcome studies proven to be the single most important prognostic element (Kissin *et al.*, 1970; Mindlin, 1959). Although the level of occupational achievement is somewhat correlated with successful outcome, of much greater significance is the level of job stability. The individual who has maintained steady employment in a single job over many years has more than twice the prospect of success as the chronically unemployed alcoholic.

Whether this correlation is causative or consequential is, as with so many other correlations in alcoholism, difficult to say. The individual who is able to maintain a steady job as opposed to the one who is not, presumably (a) has a lesser level of alcoholism and (b) has greater internal stability. On the other hand, having a steady job gives the alcoholic greater motivation for sobriety since (a) the job represents economic stability, (b) it represents personal and social respectability, and (c) it presents a meaningful area of involvement. These interactions of internal and external force offer still another example of the "self-reinforcing cycles" which are positive when the individual is employed but which can become equally negative if he loses his job.

In terms of treatment—particularly in the case of men but now also with women—the importance of maintaining employment is paramount. Not infrequently the threat of losing one's job may be the critical element in motivating the alcoholic to seek treatment. Under those circumstances, it is incumbent on the therapist to work closely with the employer in the common interest of the patient. In recent years, industry has adopted a more tolerant attitude toward alcoholic employees and will make allowances for the patient if they know he is in treatment. Some major employers have developed their own industrial alcoholism programs which have high success rates, ranging from 60 to 80 percent. When such programs are available, the employee with an alcoholism problem should be encouraged to utilize them.

Special problems revolve about the chronically unemployed, socially indigent alcoholic who requires a massive rehabilitation program. These individuals are beyond the treatment scope of the family physician. However, when alcoholics at this level of social disruption are brought to the physician—as they sometimes are—he should be sufficiently familiar with public specialized alcoholism treatment programs to make a rapid referral.

Socioeconomic Factors

The impact of socioeconomic factors on the course of a given individual's alcoholism varies widely according to his socioeconomic level. In the skid row and ghetto alcoholic, these can be the most important elements in his illness. For the wealthy, they may be of no concern whatsoever. For the middle-class patients who constitute the bulk of the office practice of most physicians, these factors usually play a peripheral but often significant role.

Most often, as the middle-class alcoholic sinks further into his illness, there are associated economic problems. These result both from the deterioration of work performance and the increasing expenditures on liquor. These difficulties are first reflected in increased tension between husband and wife but ultimately result in actual financial deprivation. The shortage of funds not only creates fiscal hardship but makes it more difficult to seek the medical assistance which the patient needs. In this way an economic vicious cycle is established similar to the physiological one in which the alcoholic is caught.

At this point—or preferably prior to it—the role of the compassionate family physician becomes critical. Without adequate intercession by him, the family may shift from an independent self-supporting status to that of an indigent publicly supported one. The physician must make the decision as to whether to undertake the treatment of this individual—with the concomitant medical, personal, and financial problems—or whether to refer him to a publicly run specialized alcoholism treatment program. As previously described, where economic disruption has already progressed to a serious level, specialized treatment programs are often better equipped to deal with those types of problems. On the other hand, where the financial situation is still viable, the crisis may often provide the strongest motivation to sobriety and the skilled therapist may use it to bring about improvement.

The Core Problem and Its Treatment

The special problems of alcoholism described in the previous section are relatively discrete, readily diagnosed, and often specifically treatable. The special treatments for these problems usually constitute specific interventions, often the prescription of a medication or the referral to a specific agency. Because these problems are concrete and often urgent and because the remedies, although not always effective, are at least readily available, special treatments usually take precedence over the core treatment.

Once the special problems have been attended to, the therapist must direct his attention to the core problem. We define the core problem as the psychosocial constellation which—whatever the original cause of the alcoholism—now leads to its perpetuation and the core treatment as that psychosocial intervention which is directed toward terminating the alcoholic state. For the sake of differentiation from the previously described special treatments, we shall label the psychological interventions which compose the core treatment as "therapy." Therapy includes individual counseling, group therapy, family therapy, Alcoholics Anonymous, transactional and gestalt therapy, psychodrama, and so forth. In order to understand how each and/or all of these relate to the core problem, it is necessary for us to attempt to reconstruct the dynamics of alcoholism.

To do this, we must review the larger question of what the core problem and the core treatment are. They incorporate the process—illustrated in Figure 4 of the first chapter—by which the intoxicated, immature, impulsive, unrealistic, indecisive, anxiety- and depression-ridden alcoholic is converted (it is hoped) to the sober, mature, controlled, reality-oriented, decisive, and composed ex-alcoholic. This idyllic equation is, as the saying has it, "too good to be true" and is not always accomplished, but it remains a meaningful goal to be approached if not necessarily achieved.

A critical issue is the question of how this transition is to be brought about. Two major positions have developed as therapeutic approaches—the "existential" and the "dynamic." The existentialists, best typified by Rogers (1959) and his group, insist that whether the psychopathology of the alcoholic is antecedent or merely a consequence of alcoholism, it is a manifestation of maladaptive coping mechanisms which are best corrected by a "here and now" approach to therapy. The dynamic or Freudian approach sees the psychopathology as specifically antecedent to the alcoholism, thus essential to the alcoholism and susceptible only to dynamically oriented psychotherapy. In actual experience, the existential approach has proven more effective than the dynamic one so that psychoanalysts themselves have adopted to a degree some of the existential principles in treating alcoholics (Blane, 1976). Actually, both of these frames of reference are of value in reviewing the core problem and its treatment, and we shall refer to both throughout our discussion.

In Table 2 we have listed the three problem areas which relate to the core problem as (1) coping mechanisms, (2) family problems, and (3) motivation. These three areas do not necessarily include all the psychosocial mechanisms involved in the perpetuation of alcoholism but, together with the special problem areas, they provide a fairly comprehensive spectrum. As opposed to the special problem areas, which are generally separate and discrete, the three core areas overlap and intervene so that to a great extent there is little understanding of one except in terms of the others. It is the interaction at all three levels that provides the inner dynamics of the process and it is in those terms that we shall examine each area.

Coping Mechanisms

In either frame of reference—the existential or the dynamic—there is agreement that the alcoholic individual has developed excessive drinking as the coping mechanism with which he deals with the exigencies of life. In the existential system, this coping mechanism is seen as a learned response which the individual has acquired—in a sense a "habit"—which can be unlearned as it has been learned. This conceptualization is close to behavior therapy in its mechanisms of positive and negative reinforcement. In the dynamic frame of reference, the coping mechanism is seen as a reflection of personality problems which, in turn, are the consequence of developmental and experiential difficulties.

It is interesting to compare the spectrum of coping mechanisms hypothesized in each of the two systems since in the comparison the flavor of both systems is realized. In the existential approach, the alcoholic's coping mechanisms are usually aberrant patterns of behavior in problem solving. When the alcoholic patient has an argument with his wife, instead of working through the problem with her, he goes out and gets drunk. Similarly, when he has difficulty with the foreman on his job, he drinks. The thrust of the existential position is that behavior can be adaptive or maladaptive and much of the alcoholic's behavior is maladaptive both in social terms and in terms of the expressed goals of the patient himself. The role of the existential therapist is to deal with the immediate problem situations and to help the patient learn new ways of coping with them.

The dynamic approach is to attempt to help the patient develop insight into his own actions. The patient learns about himself—directly or indirectly: about his personality characteristics (passive–aggressive, dependent, etc.); about his defence mechanisms; about his reality testing; and perhaps even about his unconscious. Then through an understanding of his own personality and his own dynamics, he is able to construct new ways of dealing with old problems.

It is obviously impossible for us to review here the entire battery of coping mechanisms in alcoholism as described in either system, for that would involve a review of the entire field of psychotherapy. What is important is to compare the two systems in terms of their validity and then to extract those elements from both which are valuable in treating alcoholics. It is an odd paradox that while the dynamic system, possibly because of its greater theoretical richness, is more comprehensive in explaining the dynamics of alcoholism, the existential method of treatment has proven more effective in practice. This position has now been accepted by most psychoanalysts themselves (Silber, 1959, 1974) as a result of their own failures in treating alcoholics.

The major evidence for the superiority of existential techniques in treating alcoholics, drug addicts, and other character disorders has come now from general clinical experience (Glasscote *et al.,* 1967). The theoretical basis for this is clear from psychodynamic theory itself. The character disorder patient in analytic theory is a primitively structured personality with high impulsivity and low frustration tolerance. In the presence of high anxiety or depression, he seeks immediate relief—i.e., alcohol or drugs. Since dynamically oriented treatment requires a certain tolerance for the anxiety and depression created by probing psychotherapy, it is usually inappropriate for the impulsive, low-tolerance alcoholic.

Recently, a more sophisticated amalgam of existential and dynamic approaches has developed which uses dynamic mechanisms in a here-and-now frame of reference. In this system, the therapist introduces more complex concepts such as "dependence," "passive aggressive behavior," or "repressed hostility," but in such a way as to allay anxiety or depression. However, this type of dynamically oriented existential psychotherapy is appropriate only for the more sophisticated, more stable "neurotic" alcoholic and is best managed by the trained psychiatrist.

Counseling. For the family physician, a less complex, less threatening form of therapy is easier and less dangerous to handle. This less involved process may be called *counseling* as opposed to the more technical *psychotherapy* which the psychiatrist practices. Counseling as we have indicated, is directed largely toward the here and now. The counselor reviews problem areas in the client's existence and attempts to demonstrate some underlying pattern which the client will recognize as destructive. Destructive, or contrariwise, positive behaviors are defined in terms of the client's own goals and aspirations. Since the client's behavior is usually antithetical to his expressed aspirations, confrontation becomes a major therapeutic tool. However, confrontation is utilized only after a strong therapeutic alliance has developed between client and therapist, manifested mainly in a sense of trust of the client for the therapist combined with a sense of respect and a strong belief in his integrity.

Absent from the counseling situation is any probing into the deeper meanings of patient's behavior which might derive from interpretations of only indirectly related material (e.g., the psychoanalytic interpretation of dreams). Equally absent is any attempt on the part of the therapist, either overtly or covertly, to classify the client in either characterologic or psychodynamic terminology. Emphasis is always upon realistic and immediate solutions. Stress is constant on the absolute necessity for honesty, to oneself, to one's therapist, and to one's family and friends. Within this framework, the counselor attempts to work with his client, reshaping his behavior by repeatedly pointing out how certain actions are in keeping with the client's goals while others are immediately or potentially destructive.

In a theoretical sense, counseling can be viewed more as a behavior modification technique than as a dynamically oriented process. "Good" behaviors—in terms of society's and the client's own definitions—are rewarded; "bad" behaviors are condemned. Where a positive therapeutic alliance is established between client and therapist, approval becomes the positive reinforcement and disapproval the negative.

Counseling as a form of therapy is obviously a good deal less complicated and less sophisticated than psychotherapy. Yet, in some studies counselors, as exemplified by internists and social workers, have proven at least as effective as psychiatrists in the treatment of alcoholics (Gerard and Saenger, 1966). This is not necessarily to indicate that counseling is superior to psychotherapy in the treatment of alcoholics; it suggests rather than analytically oriented psychotherapy, which in the past was the psychotherapy of choice, was probably inappropriate for alcoholics. This experience also suggests that the general goals and techniques of counseling may be effective for the majority of alcoholics.

Who can provide counseling for alcoholics? Almost anyone with a stable personality of his own, adequate life experience, knowledge about the disease alcoholism, and a true desire to help. The list of individuals includes psychiatrists, physicians, psychologists, social workers, alcohol counselors, ministers, nurses, and others. As in all forms of therapy, success as a therapist depends upon one's knowledge and one's personal qualities. The "A" form of therapist, i.e., the warm, outgoing, personally responsive type (Whitehorn and Betz, 1954) generally does better with the alcoholic than the "B" type, i.e., the distant, reserved, analytical individual. Beyond that, the therapist's level of sincerity is quickly estimated by the alcoholic patient and the therapeutic response is often dependent upon that evaluation.

However, despite the apparent straightforwardness of counseling, it is not without its dangers, particularly in the hands of an inexperienced therapist. So long as counseling is supportive and directional, there is little danger of precipitating adverse reactions in an appropriate patient. But the counselor must always be on guard against trying to be overly analytic in his probings since even correct interpretations may provoke extreme anxiety or depression and cause the patient to break off treatment. On the other hand, confrontational techniques can sometimes be destructive when a patient with a fragile psychological compensation is attacked. Actue psychotic breaks were not unknown in the early days of therapeutic communities when the frontal attack by the community was the accepted treatment approach. Confrontation techniques should always be used gingerly in the hands of the less experienced therapist and consultation rapidly obtained when unexpected adverse reactions occur.

Counseling is the psychological intervention of choice for the family physician or internist. The personal rapport and level of confidence have already been established and respect is present, for both the individual and the profession. As opposed to other counselors, physicians need little or no special training in counseling because of their general knowledge of psychiatry and their general approach to the psychological problems of their patients. The counseling involved in treating alcoholics is no different from that of treating any other psychological illness with which family physicians usually deal. It entails a careful history of the drinking pattern and an attempt to correlate drinking episodes with specific types of life crises. Then, working carefully with the patient, the physician slowly brings to him the realization that life can be different and better than it was.

Inpatient versus Outpatient Treatment. Counseling is directed toward the qualitative aspects of coping mechanisms. At another level, the attending physician must make a quantitative estimate of the overall capacity of the patient to cope with the multiple problems which beset him. This estimate must be made on a series of evaluations: (1) of the seriousness of the problems with which the patient has to deal, (2) of the patient's internal resources (mainly psychological but physical as well), and (3) of the patient's external resources (particularly family support). If the problems are great and the resources small, then the patient is best referred to an inpatient facility so that he may be separated for a while from his existing environment, which at best is not helping him and at worst may be harming him.

Although, on the surface, a three to four week "rest" away from one's problem would seem always to be salutary, one study (Edwards and Guthrie, 1966) has suggested that for the working, married, living at home alcoholic, a mandated hiatus in his regular life-style may be more harmful than helpful. Consequently, the decision should be made carefully. Where, however, there is serious physical illness or where medical detoxification is necessary, referral to a general hospital for acute treatment often provides the necessary probation period during which the decision, vis-a-vis a longer term of hospitalization for rehabilitation, can be made. If that decision is made, the four- to six-week period of rehabilitation will often interrupt maladaptive coping patterns, thus permitting the patient to develop a new perspective on his illness. It is important for the physician to maintain close contact with the patient during this period, either visiting him if possible or, if not, at least writing to him to maintain the close rapport necessary for effective therapy.

Family Problems and Family Counseling

Although a larger than average percent of alcoholics are isolated and alienated human beings, the majority still manage to maintain some sort of family relationship. The strength of this relationship varies from weak to strong and the quality from extremely positive to extremely negative. The percentage of divorces among alcoholics is unusually high, running in some samples to 40–50 percent but marriages are not infrequently replaced by common-law relationships. Probably about 50–60 percent of all alcoholics, even those in publicly supported specialized alcoholism treatment programs, prove to have interested family kin, so that the percentage in private medical practice is almost certainly higher.

For most people in our society, the family is the basic unit of human interaction and the alcoholic is no exception. Since human problems are at least as often with others as with oneself, the maladaptive behavior of the alcoholic manifests itself most evidently in his/her relationship with his/her spouse. Consequently, in the search for maladaptive or destructive coping mechanisms a place to begin is at home.

In the analysis of family interactions, all relationships are significant but most important is that with the key individual—usually the spouse. The relationship between the patient and his wife often provides the clue to the dynamic of the patient's personality and of his drinking. The wife's behavior may be supportive and this can be good if it discourages his alcoholism, or bad if by supporting his dependency needs she perpetuates his drinking. On the other hand, the wife's behavior may be overtly destructive, which, so to speak, "drives him to drink." A careful exploration of the relationship between the alcoholic and his/her spouse, particularly in a joint interview, may sometimes be more revealing than prolonged individual counseling.

Family therapy is not an easy technique, is time consuming and possibly beyond the scope of most family physicians. Often if the wife is honestly involved, she will be willing to join Al-Anon, where, working through her problems with other women is similar circumstances, she may develop the coping mechanisms necessary to help her husband. Where the family problems appear more serious, referral to a more experienced family therapist is indicated.

Motivation

The question of motivation is quite possibly the most important in the entire treatment of alcoholism and is almost certainly the least understood. The internal dynamics of motivation—what it is that in one person makes him stop drinking and in another fails to do so—remains obscure.

However, what is less obscure is the immediate motivation which brings the patient to treatment at a given time. Very often it is merely the fact that physical illness has become so severe as to permit no alternative. In itself this may not reflect any real immediate motivation other than the wish to remain alive; however, even such a wish may reflect positive motivation. Often, however, the patient presents himself at the office in no apparent distress, not evidently more sick than on many previous occasions, but suddenly intent on "stopping his drinking." Superficial questioning will seldom evoke any response other than "I've just had enough," but deeper probing will often reveal the real reasons. Not infrequently this will involve the threatened loss of his home (his wife threatens to leave or actually leaves), the loss of his job, or some deeply disturbing problem involving a child (drugs, abortions, etc.). The elucidation of the real motivation which brings the patient to treatment will often provide the most meaningful insight into the patient's dynamics and will permit the physician to use that insight to hold the patient in treatment. It is interesting in this connection that Clifford (1960) reported that in an outpatient alcoholism treatment clinic, the threat of a wife to leave home was more often associated with successful treatment outcome than any other social factor.

As previously stated, these three areas—personal coping mechanisms, family problems, and immediate motivation—all interact to provide in each alcoholic patient a distinctive dynamic pattern in which is hidden the basis of his alcoholism. The skillful therapist utilizes his therapeutic sessions with his patient both in helping him to solve immediate problems and in providing diagnostic material to delineate that dynamic pattern. Where the picture is fairly clear, through constant and firm counseling and with the aid of auxiliary adjunctive treatments where necessary, the physician gently guides his patient into a more successful life-style. Where the situation is more complex and prolonged and intensive psychotherapy is indicated, the patient is directed to a skilled therapist. The details of that phase of the treatment are discussed in the following chapter on the psychotherapeutic approach.

REFERENCES

Baekeland, F., and Kissin, B., 1972, The clinical use of disulfiram in the treatment of chronic alcoholism, in "Proceedings of the First Annual Conference of NIAAA," pp. 64–78, Washington, D.C.

Baekeland, F., and Lundwall, L., 1976, Engaging the alcoholic in treatment and keeping him there, in "The Biology of Alcoholism" (B. Kissin and H. Begleiter, eds.) Vol. 5: Treatment and Rehabilitation, Plenum Press, New York.

Bailey, M. B., 1968, A Survey of Medical and Psychiatric Practice with Alcoholics, Committee on Alcoholism Community Council of Greater New York.

Blane, H., 1976, Psychotherapeutic approach, in "The Biology of Alcoholism" (B. Kissin and H. Begleiter, eds.) Vol. 5: Treatment and Rehabilitation, Plenum Press, New York.

Clifford, B. J., 1960, A study of the wives of rehabilitated and unrehabilitated alcoholics, *Social Casework* 61:457–460.

Criteria Committee National Council on Alcoholism, 1972, Criteria for the diagnosis of alcoholism, *Am. J. Psychiatry* 129:127–135.

Ditman, K. S., 1966, Review and evaluation of current drug therapies in alcoholism, *Psychosom. Med.* 28:667–677.

Dundee, J. W., Howard, A. J., Isaac, M., Taggart, J., and Howard, P. J., 1971, Alcohol and the benzodiazepines: The interaction between intravenous ethanol and chlordiazepoxide and diazepam, *Q. J. Stud. Alcohol* 32:960–968.

Edwards, E. F., and Guthrie, S., 1966, A comparison of inpatient and outpatient treatment of alcohol dependence, *Lancet* 1:467–468.

Fenichel, O., 1945, "The Psychoanalytic Theory of Neurosis," Norton, New York.

Fisher, S. C., Mason, R. L., Keeley, K. A., and Fisher, J. V., 1975, Physicians and alcoholics, *Journal of the Study of Alcohol* 36:949–955.

Gerard, D. L., and Saenger, G., 1966, "Outpatient Treatment of Alcoholism," University of Toronto Press, Toronto.

Glasscote, R. M., Plaut, T. F. A., Hammersley, D. W., O'Neill, F. J., Chafetz, M. E., and Cumming, E., 1967, The Treatment of Alcoholism: A Study of Programs and Problems, Joint Information Service, Washington, D.C.

Goldberg, L., 1970, Effects of ethanol in the central nervous system, in "Alcohol and Alcoholism" (R. E. Popham, ed.), pp. 42–56, University of Toronto Press, Toronto.

Goldstein, A., 1974, Understanding narcotic addiction, *The Journal,* Addiction Research Foundation, Toronto 3:7.
Gross, M. M., Lewis, E., and Hastey, J., 1974, Acute alcohol withdrawal syndrome, in "The Biology of Alcoholism" (B. Kissin and H. Begleiter, eds.) Vol. 3: Clinical Pathology, pp. 191–263, Plenum Press, New York.
Hoff, E. C., 1961, The use of pharmacologic adjuncts in the psychotherapy of alcoholics, *Q. J. Stud. Alcohol (Suppl.)* 1:138–150.
Hore, B. D., 1974, Craving for alcohol, *Br. J. Addict,* 69:137–140.
Irwin, S., 1973, A rational approach to drug abuse prevention, *Contemporary Drug Problems* 2:3–46.
Isbell, H., Fraser, H., Wikler, A., Belleville, R., and Eisenman, A., 1955, An experimental study of the etiology of "rum fits" and delirium tremens, *Q. J. Stud. Alcohol* 16:1–33.
Jones, R. W., and Helrich, A. R., 1972, Treatment of alcoholism by physicians in private practice, *Q. J. Stud. Alcohol* 33:117–131.
Kaufman, E., 1974, The psychodynamics of opiate dependence: A new look, *American Journal of Drug and Alcohol Abuse* 1:349–370.
Kissin, B., 1975, The use of psychoactive drugs in the long term treatment of chronic alcoholics, *Ann. N.Y. Acad. Sci.* 252-385–395.
Kissin, B., and Begleiter, H., 1974 (eds.), "The Biology of Alcoholism," Vol. 3: Clinical Pathology, Plenum Press, New York.
Kissin, B., and Gross, M. M., 1970, Drug therapy in alcoholism, *Curr. Psychiatr. Ther.* 10:135–155.
Kissin, B., and Platz, A., 1968, The use of drugs in the long term rehabilitation of chronic alcoholics, in "Psychopharmacology: A Review of Progress, 1957–1967" (D. H. Efron, ed.), pp. 835–851, Public Health Service Publication No. 1836, Washington.
Kissin, B., Schenker, V., and Schenker, A., 1959, The acute effects of ethyl alcohol and chlorpromazine on certain physiological functions in alcoholics, *Q. J. Stud. Alcohol* 20:480–492.
Kissin, B., Platz, A., and Su, W. H., 1970, Social and psychological factors in the treatment of chronic alcoholism, *J. Psychiatr. Res.* 8:13–27.
Kline, N. S., Wren, J. C., Cooper, T. B., Vargas, E., and Canal, O., 1974, Evaluation of lithium therapy in chronic alcoholism, *Clin. Med.* 81:33–36.
Knight, R. P., 1937, The psychodynamics of chronic alcoholism, *J. Nerv. Ment. Dis.* 86:538–548.
Kupfer, D. J., Pickar, D., Himmelboch, J. M., and Detre, T. P., 1975, Are there two types of unipolar depression? *Arch. Gen. Psychiatry* 32:866–874.
Leach, B., and Norris, J., 1976, Alcoholics Anonymous, in "The Biology of Alcoholism" (B. Kissin and H. Begleiter, eds.) Vol. 5: Treatment and Rehabilitation, Plenum Press, New York.
Liebson, I., and Faillace, L. A., 1971, The pharmacological reinforcement of disulfiram maintenance in chronic alcoholism, *Committee Report of the 33rd Annual Scientific Meeting Committee in Drug Dependence,* pp. 1262–1272, Toronto.
Lowenfels, A. B., 1971, "The Alcoholic Patient in Surgery," Williams and Wilkins, Baltimore, Maryland.
Lundquist, G., 1961, Delirium tremens: A comparative study of pathogenesis, course and purposes in the delirium tremens, *Acta Psychiatr. Scand.* 36:443–466.
Machover, S., Puzzo, F. S., and Plumeau, F. E., 1961, Effect of acute alcoholization on the release of homosexual trends in chronic alcoholism, *Anais do VI Congresso Inter-Americano do Psicologia,* Rio de Janiero.
Mindlin, D. F., 1959, The characteristics of alcoholics as related to prediction outcome, *Q. J. Stud. Alcohol* 21:90–112.

Panepinto, W. C., Higgins, M. J., Keane-Dawes, W. Y., and Smith, D., 1970, Underlying psychiatric diagnosis as an indicator of participation in alcoholism therapy, *Q. J. Stud. Alcohol* 31:950–956.
Pattison, E. M., 1974, Rehabilitation of the chronic alcoholic, in "The Biology of Alcoholism" (B. Kissin and H. Begleiter, eds.) Vol. 3: Clinical Pathology, pp. 587–658, Plenum Press, New York.
Peterson, N. W., 1973, quoted in "Recovery from Alcoholism. A Social Treatment Model" (R. G. O'Briant, H. Leonard, S. D. Allen, and O. C. Ransom, eds.), Charles C Thomas, Springfield, Illinois.
Pittman, D. J., 1974, The role of detoxification centers in alcoholism treatment, *Presented at North American Congress on Alcohol and Drug Programs,* San Francisco, California, December 12–18.
Rathod, N., 1967, An inquiry into general practitioners' opinions about alcoholism, *Br. J. Addict.* 62:103–111.
Rogers, C. R., 1959, A theory of therapy, personality and interpersonal relations as developed in the client-centered framework, in "Psychology: A Study of a Science" (S. Koch, ed.) Vol. 3, pp. 221–231, McGraw-Hill, New York.
Rosenberg, C. M., 1974, Drug maintenance in the outpatient treatment of chronic alcoholism, *Arch. Gen. Psychiatry* 30:373–377.
Schenker, A. C., Schenker, V. D., and Kissin, B., 1962, Aberrations in the pulmonary respiratory patterns in alcoholics and the acute effects of ethyl alcohol and chlorpromazine upon such patterns, *Proceedings of the Third World Congress of Psychiatry,* Montreal, pp. 389–396.
Schuckit, M. A., Goodwin, D. A., and Winokur, G., 1972, A study of alcoholism in half siblings, *Am. J. Psychiatry* 128:1132–1136.
Seixas, F., Williams, K., and Eggleston, S., 1975 (eds.), Medical complications of alcoholism, *Ann. N.Y. Acad. Sci.,* Vol. 252.
Shaffer, J. W., Freinek, W. R., Wolf, S., Foxwell, N. H., and Kurland, S. A., 1963, A controlled evaluation of chlordiazepoxide (Librium) in the treatment of convalescing alcoholics, *J. Nerv. Ment. Dis.* 137:494–507.
Sherfey, M. J., 1955, Psychopathology and character structure in chronic alcoholism, in "Etiology of Chronic Alcoholism" (O. Diethelm, ed.), pp. 16–42, Charles C Thomas, Springfield, Illinois.
Silber, A., 1959, Psychotherapy with alcoholics, *J. Nerv. Ment. Dis.* 12:477.
Silber, A., 1974, Rationale for the technique of psychotherapy with alcoholics, *International Journal of Psychoanalysis and Psychotherapy* 3(1):28–47.
Sobel, M. B., and Sobel, L. C., 1973, Individualized behavior therapy for alcoholics, *Behav. Ther.* 4:49–72.
Straus, R., 1973, Alcohol and psychiatry, *Psychiatric Annals* 3:8–107.
Todd, M. C., 1975, How future physicians must see the alcoholic, *Rhode Island Med. J.* 75:390–401.
Tripp, C. A., Fluckiger, F. A., and Weinberg, G. H., 1959, Effects of alcohol on graphomotor performances of normals and chronic alcoholics, *Percept. Mot. Skills* 9:227–236.
Truitt, F. B., and Walsh, M. J., 1973, The role of biogenic amines in the mechanism of action of anti-alcohol drugs, in "Proceedings of the First Annual Alcoholism Conference of the NIAAA," pp. 100–111, DHEW Publication NIH-74-675, Washington, D.C.
Victor, M., 1966, Treatment of alcoholic intoxication and the withdrawal syndrome, *Psychosom. Med.* 28:636–650.
Victor, M., and Hope, J. M., 1953, Auditory hallucinations in alcoholics, *Arch. Neurol. Psychiatry* 70:659–661.
Whitehorn, J. C., and Betz, B., 1954, A study of the psychotherapeutic relationship between physicians and schizophrenic patients, *Am. J. Psychiatry* 111:321–331.

Winokur, G., Reich, T., Rimmer, J., and Pitts, F., 1970, Alcoholism III: Diagnoses and familial psychiatric illness in 259 alcoholic probands, *Arch. Gen. Psychiatry* 23:104.
Zimberg, S., Lipscomb, H., and Davis, E. B., 1971, Sociopsychiatric treatment of alcoholism in an urban ghetto, *Am. J. Psychiatry* 127:106–110.
Zwerling, I., 1959, Psychiatric findings in an interdisciplinary study of forty-six alcoholic patients, *Q. J. Stud. Alcohol* 20:543–554.

CHAPTER 3

Psychotherapeutic Approach

Howard T. Blane
Division of Specialized Professional Development
University of Pittsburgh
Pittsburgh, Pennsylvania

INTRODUCTION

Individual psychotherapy (counseling, casework) is commonly employed in most clinical settings as part or all of the treatment of alcoholic persons. It is generally agreed that psychoanalysis is not the psychotherapeutic method of choice for alcoholics and that psychoanalytically oriented psychotherapy must be modified for effective application with them (Blane, 1968; Chafetz, 1959; Silber, 1959, 1967, 1970, 1974). Nevertheless, most individual approaches to problem drinkers have been inspired by psychodynamic principles that owe a large debt to ideas derived from psychoanalytic theory and practice (Blane, 1968; Blum and Blum, 1967). While their originators may protest, it may be argued that this is true for the currently popular schools of reality therapy, rational–emotive therapy, gestalt therapy, and transactional analysis. Exceptions are behavior therapy (which is covered in Chapter 8 of this volume) and, to a lesser estent, nondirective counseling. The literature on individual psychotherapy with alcoholics and personal observations of practices in numerous clinical settings across the country suggest that broadly eclectic approaches borrowing freely from a variety of therapeutic schools are the rule rather than the exception, although many eclectic approaches have become programmatically rigidified, thereby losing the flexibility that is the hallmark of eclecticism.

In order to outline practices and techniques of psychotherapy useful to the broadest audience of practitioners, a definition of psychotherapy which respects eclecticism and facilitates flexibility of approach is needed. Generally, psychotherapy may be thought of as an interaction between a person who seeks help and one who is trained to provide help, designed to reach mutually determined goals by means which restructure relevant aspects of the help-seeker's life. More specifically, psychotherapy is a structured emotional experience occurring in a close relationship between two persons, in which a trained individual helps another to achieve greater self-understanding, objectivity, and personal growth through a series of contacts in which relevant inner experiences and life situations of the latter are discussed. This modification of the American Medical Association (1968) definition has several advantages: It is not bound by any particular theoretical or programmatic position; it emphasizes the therapeutic relationship and the essentially emotional nature of the transactions presumed to underly therapeutic change; it makes no assumptions about alcoholism as a disease or as a symptom of an emotional disorder, while recognizing the importance of psychological and emotional factors in all drinking problems; and it is not categorical about eligibility requirements, goals, length and spacing of sessions, or other details of therapy. The definition thus provides a framework within which psychotherapy may be tailored to the unique needs and life situation of each alcoholic patient, given the constraints inherent in different clinical settings. It can encompass the heterogeneity of the alcoholic population (Canter, 1969; Curlee, 1971; Pattison, 1974), including the variety of conditions in which alcoholics present themselves at treatment facilities (Blum and Blum, 1967).

Following a section which expands on some of the themes touched upon here, techniques and practices appropriate to initial, middle, and late phases of psychotherapy are presented and discussed, allowing for variation according to individual differences among patients and institutional differences among settings. A separate section is devoted to the therapeutic relationship from the viewpoint of the patient and of the therapist. A final section examines individual psychotherapy and concurrent forms of treatment and psychotherapy in series with other treatments.

GENERAL CONSIDERATIONS

Despite a rather large body of literature on various aspects of individual psychotherapy with alcoholics, relatively little has been written about specific techniques and practices. This may be due in part to the fact that many practicing psychotherapists have been trained in psychotherapy in their generic discipline (psychiatry, psychology, counseling, social work) and have then applied their general psychotherapeutic knowledge and skills to alcoholics. The process of modifying technique until it becomes appropriate with alcoholics is usually slow and difficult, growing gradually out of frustrating trial-and-error experimentation. There is little in the way of prepared materials to make the process more rapid and less painful. An apprenticeship model for learning is sometimes not available because of a lack of trained personnel in many clinical centers. Generally speaking, it appears that social workers and counselors, with their emphasis on current reality, active development of a working treatment relationship, environmental manipulation, and strengthening of ego assets, are better equipped by their generic training to work with alcoholics than psychologists and psychiatrists, who tend by virtue of training to be more oriented toward investigation and interpretation of inner experience and inclined to be more passive in regard to the development of a therapeutic alliance. Modification of technique may therefore be more far-reaching and thus more difficult for the latter.

Regarding techniques designed for alcoholic patients, Scott (1961) presents an articulated account, with detailed clinical examples, of the handling of typical issues that occur in initial, middle, and late stages of individual psychotherapy, and Strecker (1951) makes a number of useful suggestions for psychotherapists. Blane (1968) offers several techniques for managing alcoholics according to the manner in which they manifest dependency needs. An account of the modifications of psychoanalytically oriented psychotherapy necessary for effective application to alcoholic patients may be found in Chafetz (1959). An apprenticeship model which places considerable emphasis on transference and countertransference phenomena but is rather general about technique has been described and refined over many years by Silber (1959, 1967, 1970, 1974). Bailey (1968) presents a practical account of psychotherapy from a social work perspective. Scott (1963) and Steiner (1969, 1971) discuss in detail techniques of working with patients individually prior to and during the course of group therapy. All of the above are based on experience with patients in ambulatory outpatient settings who are drawn for the most part from a special subset of the alcohol population characterized by intact marriages, employment, and middle to upper-middle socioeconomic status. Besides these sources there are numerous articles of dubious utility in the actual conduct of therapy, although they may contain useful observations about the psychodynamics of alcoholics, general statements about various aspects of psychotherapy, and isolated examples of technique.

Individual psychotherapy is only one of many forms of treatment for alcoholics, and it is usually but not always conducted in concert with other forms. The combinations of treatment are numerous, and the more typical combinations will be discussed later in a section on individual psychotherapy. Concurrent combinations include individual psychotherapy combined with group therapy, with drug therapy, with residential treatment, or with A.A.; the chain combinations include detoxification followed by individual psychotherapy or residential treatment followed by therapy. These combinations are presented here in an arbitrary manner; in practice such distinct groupings are not possible, since psychotherapeutic input certainly begins with the patient's first contacts with helping agents or institutions, and indeed the patient's prior experiences in treatment and his expectations of treatment shape the nature of those early contacts. The manner in which initial contacts are handled is often decisive in determining whether a prospective patient enters and stays in treatment (Chafetz et al., 1970).

It has been increasingly recognized that alcoholism is not a unitary phenomenon either at a clinical level (Blum and Blum, 1967; Canter, 1969; Curlee, 1971) or on the basis of research findings which have used factor analytic and other statistical grouping methods (e.g., Horn et al., 1974). This has had the effect of reinforcing the notion that individualization of care according to the unique internal and external factors that characterize each person is a must in treatment generally and psychotherapy specifically. It is also leading to the development of more complex typologies of alcoholics than those commonly used. Nevertheless, many psychotherapists and clinicians still find it useful to group alcoholics into essential (addicted) or reactive categories, a dichotomous simplification of a classification scheme originated at the Menninger Clinic in the 1930s (Knight, 1937a,b, 1938; K. Menninger, 1938; W. Menninger, 1938). According to Chafetz (1959), reactive alcoholics have relatively normal prealcoholic personality structures and are socially well adjusted with a reasonably stable pattern of family, educational, and occupational attainment and responsibility. They use alcohol to excess when temporarily overwhelmed by external stress—often a tangible loss, always a severe shock to self-esteem. Their drinking tends to be episodic, but can result in regression to a state like that of the addicted (essential) alcoholic. Reactive alcoholics tend to respond well to individual psychotherapy. The alcohol addict, on the other hand, shows "gross disturbances in his prealcoholic personality with difficulties in adjustment during the early years manifested in relationships within family, in school, at work, and in attempts at marital adjustment." Excessive drinking usually begins in late adolescence with little readily discernible stress associated with the onset of a drinking episode. Drinking is characterized by inability to abstain and loss of control. Significant modifications in psychotherapeutic technique are required for successful treatment of the alcohol addict. Other clinically useful typologies include one proposed by Blane (1968) which groups alcoholics according to the manner in which they manifest dependent behavior: dependent alcoholics, counterdependent alcoholics, and dependent–independent alcoholics. The two former categories are composed of individuals who have made fixed life adjustments in regard to dependency and who are less responsive to psychotherapy than individuals in the latter category. Steiner (1969, 1971), viewing alcoholics from the perspective of transactional analysis, has described three types of alcoholics: aggressive alcoholics ("drunk and proud of it"), psychosocially self-damaging alcoholics ("lush"), and tissue self-destructive alcoholics ("wino"). The latter two categories represent degrees of severity in self-destruction, with the wino being prepared literally to kill himself, and the lush to destroy himself psychologically and socially. Categorization of an individual into one of these classes ("games") has distinctive implications for the direction transactional therapy will take.

The use of clinical typologies may be criticized because (1) they are abstractions which neglect idiosyncracies which may be of great potential treatment value for a given individual, (2) they create a professional reality that doesn't necessarily reflect patient reality, so that all alcoholics must fit into the former regardless of self-perceived problems and needs, and (3) placement in a category of a classification system may exclude patients from therapy on the presumption that therapy for individuals in that category is ineffective. Blum and Blum (1967) assert that psychodynamic principles should be adapted "to the treatment of many kinds of alcoholics in a variety of different settings and conditions. The trend of fitting treatment to the patient, instead of rejecting patients who do not fit past notions of suitability, is a most promising one."

One way to take into account patient diversity and condition and differences in clinical setting is to classify the conditions toward which psychotherapy may be directed. These are (1) conditions which brought about the alcohol problem, (2) conditions secondary to the alcohol problem, and (3) conditions brought about by the cessation of drinking. The choice of which set of "causative" conditions to emphasize in individual psychotherapy depends on many factors (professional preparation and experience of the clinician; nature of the agency and the kinds of clientele it attracts; treatment philosophy of the agency; goals of treatment), but ideally, the choice is best made on the basis of as complete an understanding of each individual patient as possible. A brief examination of each set of conditions will make clear the potential utility of a typology which goes beyond usual diagnostic classification schemes.

Conditions which brought about the alcoholic problem may be thought of as either proximal or distal. Proximal conditions are those current events which precipitate a drinking episode, ordinarily an interaction of internal processes and an external, usually interpersonal, occurrence. An example might be a man who is unfairly accused of making a mistake by his boss but accepts the accusation and feels inadequate and guilty afterwards. Distal conditions are those developmental factors, ordinarily involving perceptions of parents and including critical incidents of childhood, that set the stage for the operation of proximal conditions. To expand on the above example, the man, as a child, was frequently belittled by his father, and the child's protestations were brusquely suppressed as talking back; the child nevertheless took great pains to please his father. He repressed his anger, turning it inward as he outwardly accepted paternal judgment. His guilt led him to seek repetitively to make up to his father by pleasing him, but he never permitted himself to feel adequate. Psychotherapy obviously addresses both proximal and distal conditions to some extent, but usually concentrates attention on one or the other. Counseling, casework, supportive, and relationship therapy tend to focus on proximal conditions in the therapeutic interaction, although a deeper historical understanding of proximal conditions may be obtained through a knowledge of distal conditions. Emphasis on distal conditions characterizes psychoanalytically oriented psychotherapy; conflicts and attendant reality distortions based on early parent–child interactions are, for example, particularly stressed in Silber's (1959, 1967, 1970, 1974) approach to alcoholics, as well as by Chafetz (1959), Strecker (1951), and most psychoanalysts (reviews of psychoanalytic contributions to an understanding of alcoholism may be found in Blum, 1966; Blum and Blum, 1967).

Conditions secondary to the alcohol problem include current physical status (chronic or acute medical complications), current psychological status, and, most importantly, reality issues, often of an interpersonal nature, that relate to family (parents, spouse, children), sexual relationships, work, and financial situation, but which also include law enforcement involvements. Generally, the approach is one of crisis intervention which can have three aims: to achieve a lasting effect in and of itself; to be a prelude to a longer term general rehabilitative effort; or to be a prelude to entrance into psychotherapy that directs itself to conditions which brought about the problem. The important effects that may be achieved by crisis intervention with alcoholics were demonstrated in a series of clinical research and demonstration studies conducted during the 1960s at the Massachusetts General Hospital in Boston (Chafetz et al., 1970).

Psychotherapeutic intervention coupled with medical intervention at times of physical crisis can, if properly handled, be extremely useful with seemingly most unlikely and intransigent patients. Individual approaches which direct themselves to immediate crises emphasize the judicious meeting of concrete needs expressed by the patient and the provision of direct services which reduce or resolve the crisis. Many clinicians express concern that rescuing patients from predicaments they have created falls so neatly within the narcissistic dynamics of the alcoholic that it will reinforce rather than ameliorate the patient's proclivity to hurt himself and others and so should be avoided. For example, Steiner (1969, 1971) discusses therapists as "rescuers" and "patsies," and Myerson (1953) examines the therapeutic dangers of getting patients out of jams. Rescue, the provision of concrete aid, should never be indiscriminate, but it can be provided in a context which has an ultimately rehabilitative thrust. Blane and Meyers (1964) describe the "constructive use of dependency needs" to foster a therapeutic relationship, and Tiebout (1949) presents a clinical example of repeated rescue which eventuated in an excellent outcome.

If the aim of crisis intervention in a particular case is to foster a relationship that will serve as a bridge to longer-term psychotherapy, the amelioration of reality problems has another advantage. It is well known that progress in psychotherapy is severely hampered if the current life situation of the patient is grossly disturbed or subject to periodic crises, first of all because much time must be spent in handling these reality issues, thus diluting the major aims of self-understanding, objectivity, and personal growth, and secondly because it is extremely difficult for the therapist and patient to disentangle the extent to which crises are self-generated or come about from external causes. Diminution, therefore, of the intensity of disturbances in the patient's life sets the stage for individual therapy that will be relatively undisturbed by reality crises.

It appears that a great deal of attention is paid in alcoholism treatment facilities to the handling of conditions that are secondary to the alcohol problem and that much of this attention is an end in itself. While crisis intervention with members of some populations (e.g., college students) can have long-lasting beneficial effects, this is rarely the case with alcoholics, so that interventive aims of continuing treatment or entering psychotherapy are most appropriate with this population. However, the possible rewards that may accrue to crisis intervention during short hospital stays (e.g., for detoxification or physical complications) seem not to be capitalized upon in many settings, but result instead in the familiar revolving-door pattern, albeit with more humane physical care.

Conditions brought about by the cessation of drinking have received a fair amount of attention in the literature, but the value placed upon abstinence itself has been so great that there is a tendency in many quarters to assume that if the patient has stopped drinking a cure has been achieved. Certainly most research and evaluation investigations have used change in drinking, particularly its cessation, as the primary if not only criterion of outcome. While many alcoholic individuals show dramatic changes in other important areas of their functioning upon cessation of drinking, research evidence indicates that this is an infrequent occurrence (Gerard *et al.*, 1962) or is short-lived without continuing intervention. Many clinicians have remarked on the feelings of well-being, optimistic certainty about the future, and sense of buoyancy and freedom that frequently follow upon a decision to remain abstinent. This flight into health, however, typically gives way in a matter of weeks or months to anxiety, depression, and irritability that may be followed by resumption of drinking or may seem little different to the individual's family and associates from his behavior when drinking excessively. Awareness by clinicians of this phenomenon is particularly important if gains already achieved are to be consolidated. While cessation of drinking minimizes reality problems, thus making therapy easier, it may also be viewed as the relinquishing of a way of behaving that binds anxiety. The temporary relief that first accompanies abstinence soon gives way to the expression of anxiety which carries with it the possibility of reaching a deeper understanding of causative factors, at the same time that it usually results in more psychological symptoms; it is thus an opportune time to begin psychotherapy.

Many clinicians (e.g., Clinebell, 1968a,b; Fleming, 1937; Pfeffer, 1958; Root, 1968; Sheldon *et al.*, 1974; Solari, 1970) see alcohol and drinking as a centralizing feature of the alcoholic's life which structures most of his daily activities. When he no longer drinks, he must be helped to acquire substitute activities to replace the emptiness of time. Bowman and Jellinek (1941) referred to this as substitutive treatment, but distinguished between the fostering of hobbies and new interests, recreational therapy, and occupational therapy, on the one hand, and "true substitutive treatment" which is largely directed to new emotional experiences, on the other. There is of course a therapeutic danger in assuming that all alcoholics have no interests or compelling concerns other than alcohol, since such an assumption goes clearly against the biographical facts of many alcoholics. Substitutive treatment is one of many resources that the clinician has to turn to, but its actual use is determined by the requirements of the individual case.

The classification of target conditions of psychotherapy into causative (proximal and distal), secondary to alcoholism, and secondary to abstinence permits us to view differences in techniques and practices of individual psychotherapy in a more systematic and general manner than has usually been the case. Many differences that have been debated in the literature disappear when viewed from this perspective; those that remain may indeed reflect basic divisions in value orientation or conceptualization, or they too may be reconciled when viewed from other perspectives, such as the model of interaction between patient type, treatment setting, and goals presented by Pattison (1974).

It must be emphasized that the classification presented here assumes that the goals of treatment are determined individually for each patient, taking into account his unique life situation and history. Goals, and thus technical parameters and the process of therapy, also vary with the nature of the clinical setting; for example, goals in a detoxification facility are quite different from those in a residential setting, and both in turn differ from those of a psychotherapist in private practice. Further, goals may change in the course of treatment as a function of shifts in the status of the patient. It is desirable that goals become more ambitious as therapy progresses, but it often turns out that goals must be more modest than was anticipated at the point therapy began. In general, psychotherapy aims to provide the individual with sufficient and flexible means of coping with conflict and feelings of shame, guilt, anxiety, depression, and a diminished sense of self so that alcohol does not have to be used as a solitary mechanism for coping.

In the following sections, technical aspects of phases and relationship issues in psychotherapy with alcoholics will be discussed primarily as they relate to the treatment of proximal and distal causative conditions of drinking problems. Where appropriate, techniques of individual therapy for conditions secondary to the drinking problem and secondary to cessation of drinking will be presented. While the three sets of conditions are discussed separately, in practice such a division does not always occur, and elements from each of the conditions are often present in a given therapeutic session. What is to be stressed, however, is that therapy is ordinarily directed primarily to one set of conditions as opposed to any of the others. When it is not, then one has reason to wonder whether the clinician has a clear sense of therapeutic direction or is simply foundering, being blown hither and yon by the moment—a situation that often occurs with beginners.

INITIAL PHASE

The initial phase of therapy consists of dealing with several interrelated issues and accomplishing several aims vital to the fate of subsequent psychotherapeutic transactions, generally within the first three to five sessions. Included in this section are (a) the relationship and relationship building, (b) pretherapy factors that may affect treatment, (c) the issue of drinking or nondrinking, (d) setting goals, (e) therapeutic contracts, and (f) handling of the resistive or hostile patient. While touched upon in the following sections, no particular stress is placed on the extremely important issue of the patient's desire for help and methods for enhancing his involvement in rehabilitation, as these are covered in Chapter 4.

Relationship and Relationship Building

It is generally accepted that the therapist must assume an active role in building a relationship with the alcoholic. He must therefore be warm, kind, nonjudgmental, and interested, but he must also be able to set limits firmly. He must be able to convey to the patient a sense of competence, respect for the patient's integrity, and tolerant acceptance. Warmth, kindness, and interest are not to be confused with overpermissiveness and niceness, nor is firmness to be identified with inflexible control (the countertransference aspects of these reactions are considered in the section on relationship issues). The therapist must be completely honest with the patient, "acknowledging . . . mistakes, errors, and feelings arising within him as soon as he becomes aware of them," as Chafetz (1959) puts it. Because alcoholics are usually extremely sensitive to rejection and because they respond positively to constancy in action, one should avoid canceling appointments without adequate notice, tardiness, and offering but not following through on suggestions.

In building a firm bond and alliance between patient and therapist, many clinicians recommend an action-oriented approach of "doing" for the patient (Becker and Israel, 1961; Blane, 1968; Catanzaro, 1968; Chafetz, 1959; Margolis et al., 1964; Sapir, 1953; Solomon, 1966). Prescriptions of drugs or the offering of coffee are tangible evidences of support. More important, however, is the capacity to provide services for real needs expressed by the patient or identified by the therapist. As Sapir (1953) states from a social work perspective: "The key feature of our approach has been to respond quickly and decisively to the needs of our patients in the initial application for help. It is out of this situation of 'asking and receiving,' where something tangible or intangible is given, that communion on a feeling level is established. If this communion can be established, it provides a favorable beginning to the development of the treatment relationship; without it nothing of any significance can happen."

This implies that actions in the patient's behalf should be natural and relevant to the patient's actual situation, not carried out simply to enhance the relationship. Otherwise, they are apt to represent little more than indiscriminate giving and therefore to be ultimately antitherapeutic. This may be one reason why some clinicians warn that concrete actions for the patient should generally be avoided (Myerson, 1953). The extent of the patient's need for concrete services will depend on his condition and on the setting. A patient admitted to a hospital facility for physical problems secondary to prolonged alcohol intake is a more likely candidate for an action approach than a sober patient who seeks help from an outpatient clinic. For the "admitted alcoholic," who is often intoxicated and unreasonably demanding, Catanzaro (1968) recommends that he be seen immediately but briefly to give necessary help and that verbal exchanges be kept to the minimum necessary for understanding the acute complication. Immediate, delimited action addressed specifically to an acute need sets the stage for a continuing relationship that can be enlarged upon when the patient is sober and not hurting. Solomon (1966) suggests a thorough physical examination, followed by advice that will improve the patient's physical health.

Some therapists, while advocating being friendly and nonjudgmental, recommend a forthright, confrontive approach. Scott (1961) says that the therapist should "be friendly and active, sprinkling the interviews with timely questions" and that he should "clarify that he is not omnipotent, and cannot be exploited, either, but is quite willing to help." Examples of timely questions are, "Do you really believe that?" "Perhaps things aren't just as you say they are," or "Are you willing to pay the price (of remaining abstinent)?" Such an approach indicates to the patient that the therapist knows his craft and understands the kinds of rationalizations alcoholics typically engage in. In a similar vein, Strecker (1951) stresses that from the first interview the therapist must always deal with the patient on a mature basis. When a patient says, "I suppose you won't let me have any liquor in the house," or similar remarks, Strecker responds, "Do whatever you think best," thus undercutting attempts to set up authority struggles while reinforcing the patient's autonomy.

Other clinicians stress essentially psychological means of advancing a positive relationship. Silber (1959) advocates making the patient aware of his fear of hostility or helplessness, which has two effects: It immediately strengthens a positive transference relationship and it tends to enhance the omnipotence with which the patient endows the therapist (note that Silber sees the therapist's omnipotence as a therapeutic asset, whereas Scott sees it as a potential danger which the therapist must disavow). Similarly, Munt (1960) indicates that the patient's fear of dependency must be dealt with immediately in order to ensure a successful treatment alliance and outcome; otherwise the patient will flee therapy early or develop an intense but therapeutically unproductive attachment.

Pretherapy Factors Affecting Treatment

Little attention has been devoted in the literature on treatment to the expectations the patient brings to treatment, how these are shaped by prior experiences in treatment, and how they affect the course of therapy in a new setting. This relative inattention is all the more surprising because alcoholics, as contrasted to other diagnostic groups, often have lengthy and varied treatment histories. There are several prior treatment experiences that are sufficiently common to merit examination of their impact on subsequent treatment and ways in which they may be handled therapeutically. They are (1) repeated episodes of detoxification without follow-through on a long-term rehabilitative regime; (2) treatment in the past, accompanied by several years of sobriety, with current fears of returning to drinking or an actual episode of drinking; (3) a long history of rehabilitative efforts with no discernible benefit and a general aura of hostility toward or discouragement about treatment; (4) transfer from one therapist to another in a context of continuing treatment; and (5) referral to individual outpatient treatment following a period of residential treatment, with the patient being very optimistic about the future. Each of these is briefly discussed below.

Repeated Detoxification

These patients are usually met in a detoxification setting. The clinician must first determine whether attempts have been made in previous hospitalizations to create a long-term rehabilitative alliance. If not, he should proceed to attempt to build a positive relationship with the patient, indicating that he is aware of missed opportunities for the patient in the past. If attempts have been made but the patient has not followed through, then the therapist should engage in a frank discussion of "What gives?" with regard to the patient's inability to follow through. If the patient does not spontaneously see the relationship between his difficulties and his use of alcohol, the technique or a variation of it, described by Catanzaro (1968) may be attempted (see Setting Goals and Confrontation for details). It often happens that detoxification occurs in a busy, impersonal medical setting where little attention is paid to the alcoholic as a person in need of more than physical help; little credence is given to the notion that patient failure to follow recommendations for continuing care is a reaction to rejection rather than lack of desire for help. The therapist who is sensitive to institutional as well as individual dynamics is in a position to help form a trusting relationship that can eventuate in a long-term treatment alliance.

Successful Past Treatment

Patients who have been doing well for some time and who slip or who are afraid of slipping often present themselves at outpatient settings. If the individual has not been drinking, it is sometimes a temptation to provide token reassurance and not explore with the patient what provokes him to seek help at this time. Overtures for help when the individual is not in overt difficulty are a distress signal that must be taken seriously. They are also a sign of strength in the patient, showing that he is able to anticipate problems and delay acting on impulse. The therapist's task is threefold: to find out what the patient perceives as being the key to his success in past treatment, to examine his relationship to his former therapist, and to determine the personal dilemma which impels him to seek aid now. Quite often, crisis intervention techniques can resolve a hazardous situation and abort regression to old defensive maneuvers. Careful use of key features in earlier treatment and the clinician's knowledge of the patient's relationship to his previous therapist may be used to make direct confrontations about the patient's current problems. It is generally unwise to attempt to improve on an earlier successful outcome by opening up and exploring new areas; such an approach usually represents an uneconomic use of the therapist's time and diminishing returns for the patient.

Unsuccessful Past Treatment

It is quite common to see patients who have previously undergone a series of relatively long-term treatments in a variety of settings by a variety of methods without receiving any long-term benefit. Often they are hostile and challenging, with challenges usually taking the implicit form of "O.K., smartypants, let's see if you can do anything for me." The patient obtains considerable secondary gains from being a champion of failures, self-destructive as this dynamic may be. Although they are provocative and challenging, these patients usually enter treatment readily, viewing it more as another contest than as help. The initial approach to a patient of this kind must be quick and firm: the therapist points out to the patient that he can help the patient only if he desires help; that he accepts the patient's strength, respects it but is not afraid of it, and is not interested in the patient's proving it to himself and the therapist again; and that he will consider treatment of the patient only if the patient shows by his behavior that he takes treatment seriously. These patients are typically afraid that if they reveal themselves, they will destroy others. They seek a figure stronger than themselves, who is unafraid of their destructive and passive wishes. Patients who accept the early confrontations and conditions of the therapist often do well in subsequent treatment. Those who refuse are better left to their own devices, because to take them into treatment represents no more than another repetition of failure. Obviously, the therapist leaves the door open, by indicating that when the patient is ready to enter treatment the therapist will be pleased to work with him.

It is necessary to distinguish the challenger from the patient who has been in many different treatment facilities but who has never received care that is addressed to his needs. The latter is usually a discouraged individual who believes or is beginning to believe that he cannot change and that there is no treatment to suit his condition; he is sometimes embittered, but more often resigned to what he sees as his fate. He is inured to rejection, which he has long since learned to expect in its grosser and to detect in its subtler forms. One must proceed slowly and carefully with these patients inasmuch as they will be suspicious of positive, optimistic approaches, searching for signs of covert rejection. A careful step-by-step review of the patient's past treatment history is indicated, starting with the first and proceeding to the most recent, with particular attention being given to shifts in the patient's perceptions of treatment settings as he progressed through them and a careful assessment by the therapist of the extent of quality of the treatment received. This review often reveals that the patient's attitude as he approached his first treatment attempt was fearful but hopeful, gradually giving way to a sense of deadened resignation as he moved from one treatment setting to another. It will also reveal such things as custodial care without a treatment program, poor quality of care, therapy that addressed itself to part but not all of the patient's relevant needs, and care given reluctantly, moralistically, angrily, or pessimistically. It is the therapist's job to help the patient restructure his perceptions of treatment so that he can regain his lost hopefulness. Rather than protecting the professional fraternity, the therapist is wise to help the patient recognize what was valuable and what was detrimental about previous treatment efforts and to support and encourage him in exploring his feelings about treatment. Also, through his own transactions with the patient, the therapist will provide a new model of treatment personnel to the patient. Naturally, the therapist must be careful not to overidentify with the patient or take sides, and he must make sure that the patient is not a hidden challenger who is attributing internal failures to past treatment agents. However, while there is something of the challenger in most alcoholics, the alcoholic who emphasizes contests is quite different from the alcoholic who has been beaten down by the treatment system.

Transfer

In settings where individual therapy is a major mode of treatment, transfer of a patient from one therapist to another is a frequent occurrence. In teaching settings, therapists in training come and go according to their training cycles, and in any setting there is always some turnover of staff personnel. For present purposes, it is assumed that an agency will have a policy about the most effective means of initiating and treating transfers of a patient from one therapist to another, and so the process of possible techniques of transfer is not discussed here. Rather, emphasis is placed on the role of the receiving therapist in effecting a smooth transition and ensuring continuity of care. Ideally, the patient will have had one or two sessions with the receiving therapist before he terminates with his old therapist, so that they can have a chance to discuss the patient's feelings about the new therapist and so that the two therapists can discuss the key dynamic points of the transfer.

In the first sessions with the patient, the receiving therapist must be alert for cues that signal feelings of loss and rejection by the former therapist or disappointment in the new therapist. The transitional period is critical, and many patients leave treatment prematurely at this point, but it also represents an opportunity to advance therapeutic aims since it recapitulates feelings engendered and reinforced by previous losses, an area that is particularly important in the psychology of most alcoholic individuals. If the patient does not bring up his feelings about the transition in the first seesion, which is often the case, then the therapist takes the initiative in a nonleading, tentative fashion, e.g., "You've covered a lot of ground with Mr. Jones; I guess you've thought about it some." If the patient demurs, it is wise not to push on, but to leave the door open: "I understand. We can come back to it when you're ready." If, toward the end of the second session, the patient has not indicated his readiness, the therapist takes a more active position, going in one of two directions, depending on his sense of the focus of the patient's feelings. When loss appears to be uppermost, the therapist may say, "You must miss Mr. Jones quite a bit, tell me about it"; if the focus is on adjusting to the new therapist, he may say, "Getting to know a new person is tough; I guess you're sizing me up against Mr. Jones. How about it?" If the patient demurs again, the therapist may indicate his awareness of how difficult it sometimes is to talk about feelings or even to identify them. If the patient is unresponsive to this kind of prodding, the therapist may then briefly outline how change and loss typically affect people and show his understanding and appreciation of feelings the patient has. Whether the patient picks this up or not (and many alcoholic patients will not), he will nevertheless receive the message that the new therapist is on his side, understands him, and accepts his feelings.

When the transfer is between therapists from different professional backgrounds, the patient may often express his feelings by making invidious comparisons between the professions or by asking the therapist how his profession differs from that of the previous therapist. Clinicians are often put off by such comparisons and questions. Similarly, in transfers between therapists in training or between a staff member and a trainee, the student therapist will often take comments about himself in relation to the previous therapist personally. In either case, the patient's observations should not be taken as slights or commendations of one's profession or personal abilities, but should be understood, responded to, and pursued as reflecting the patient's feelings about loss. When the therapist fails to do so, he runs the risk of losing the patient in this crucial phase because of the press of unfinished business on the patient's part and his consequent feelings that the therapist does not understand him because the all-important feelings about loss were not addressed.

Referral upon Discharge from a Residential Program

In recent years there has been a proliferation of residential treatment facilities with four-week to three-month programs which stress group and milieu approaches to cutting through the rationalizations most alcoholics hold about drinking and to helping patients restructure their lives without alcohol (residential programs are not to be confused with community-based halfway houses which tend to be of variable length, have less structured programs, and serve homeless persons). The longer the program, the greater the chance of working through problems and consolidating gains, but many programs serve a clientele with jobs and family responsibilities who rely primarily on third-party payment for their treatment, so that four-week programs predominate. When patients leave the residential facility to return to the community, they are typically on the crest of a wave of self-understanding, full of optimism and confident about their control over alcohol. Without continuing outpatient care, however, the danger of regression among these patients within a few months after discharge is great. Usually patients are referred to an appropriate outpatient setting, but engaging them in continuing treatment is often difficult because they perceive themselves as cured. Once he is engaged in therapy, however, the ultimate outcome can be excellent because the patient has already passed through the initial, and most difficult, phase of treatment.

It is usually not possible to distinguish genuine progress toward self-understanding during residential treatment from the sense of well-being, almost rebirth, that commonly accompanies a period of abstinence. Many observers have commented on the latter. In the 1930s, for example, Fleming (1937) recommended an initial phase of medically oriented hospital treatment that might last from two weeks to several months and serve as a prelude to a second phase of substitutive treatment lasting from six months to a year. He states: "It is interesting to note that almost without exception near the close of this period [the first phase] the patient's mental state is characterized by great optimism, the inevitable good resolutions, and expressions of a firm conviction that the cure is successful." Fleming warns that this is a crucial time for the patient, who often leaves the hospital "with the almost certain result of his starting to drink again." Diethelm (1936) attributed the insight shown by patients at the beginning of treatment to "alcoholic euphoria"; this period is followed by a revolt of a few weeks and only after this can cooperation be expected. A contemporary view is provided by Steiner (1971), who describes a biphasic sequence of withdrawal sickness lasting about a week, followed by a lull of about two weeks during which the patient feels strong, confident, and "on the top of the world." This phase is followed by "withdrawal panic" in which the alcoholic is frightened and obsessed about the "disconcerting changes of consciousness" which, according to Steiner, result from the continued lack of alcohol. Withdrawal panic may result in resumption of drinking, but if it does not, the next phase is the "honeymoon," during which the alcoholic enjoys his adult state and feels genuine relief and well-being. This phase can be expected to last as long as three months, after which it subsides and previous conflictual behavior patterns, with or without alcohol, begin to reemerge. It should be noted that Steiner is quite explicit that this sequence of events shows wide individual differences. For individual patients, some phases may be shorter or longer or may not even occur.

Whether the patient returning to the community from a residential program is in a transitory state or has begun to make real headway in discovering himself is somewhat academic, because the same technical approach is adopted for both. Because the patient has been primed, so to speak, for therapy by the residential experience, the therapist may be direct and straightforward in explaining to the patient that while he has made extensive gains, experience has shown that these gains are lost without continuing treatment. The patient's response usually is some variant expression of his newfound strength, such as,

"I know, but I would like to try things on my own for a while and I know I can do it." It is useful at this point if the therapist, either on the basis of actual meetings with the patient before he completed the residential program or clinical material obtained from the facility, can pick up on a crucial bit of unfinished business to demonstrate to the patient, with due respect for his autonomy, that there are areas of his life which still require understanding. Another approach is to bring up reality problems attendant upon transition from treatment facility to home, where different modes of adjustment will be required both by the patient and by significant others in his social environment. In the event the patient still wishes to go it alone, which often happens, the therapist must make every attempt to leave the door open for the patient to return to treatment if trouble occurs. Experience shows that transfer from residential to outpatient individual therapy is the most hazardous of pretherapy situations. At this time, resistance is at its peak and patients sincerely perceive themselves to be cured, which may be symptomatically true. Careful prior preparation while the patient is still resident at the facility, combined with appeals to reason and experience, is necessary to establish a continuing relationship with these patients.

Drinking or Not Drinking

Three points of view run through the history of individual psychotherapy with regard to whether the patient should drink during treatment. The most commonly expressed position states that the patient must achieve abstinence as soon as possible, that abstinence should continue throughout therapy, and that a major goal of treatment is that the patient remain abstinent after therapy is completed. A variant of this position involves no stand on lifelong abstinence but demands abstinence during treatment. A second position does not make drinking an issue, but implicitly assumes that patients are not drinking or are drinking in ways which do not affect the course of therapy deleteriously. In this view, drinking is seen as a symptom of an underlying emotional disturbance, and therapy follows fairly traditional notions that self-understanding of conflicts which gave rise to the symptom will result in symptom removal, while direct attacks will prove to be of no avail. The third position is closely allied to the second but is explicit about prohibiting drinking during therapeutic hours on the grounds that meaningful emotional experiences cannot occur when the patient's state of consciousness has been altered by a drug.

Present-day thinking generally favors cessation of drinking during treatment, with an explicit or implicit emphasis on abstinence as an ultimate treatment goal. It appears to be generally felt that injunctions about drinking should be flexible and that each situation that violates the general rule should be dealt with on its own merits. Thus, it may be better to have an active alcoholic in treatment, aiding him with conditions secondary to the alcohol problem, than for the person to have no treatment at all. It also appears that some segments of the alcoholic population do not require injunctions about drinking. Since the decision to enter psychotherapy is preceded by giving up drinking, drinking in treatment is not a relevant treatment topic for this group.

Silber's approach (1959, 1967, 1970, 1974) illustrates the position which makes no rules about drinking during the course of therapy. As he points out, the patients he deals with are highly motivated, having entered therapy either after being on a waiting list for one to ten months or having indicated a desire to know more about their inner reasons for drinking, in combination with being curious about themselves and possessing the "intellectual potential" for psychotherapy. In therapy conducted with hundreds of patients by therapists in training under Silber's supervision, "hardly any" patients had to be terminated because drinking interfered with keeping their appointments (Silber, 1970). That drinking does not become an issue in this approach unless it interferes with therapy is illustrated by this example (Silber, 1959): "The patient gave up his drinking when therapy began, although more recently there has been a tendency to drink sporadically. This drinking has not been a problem thus far in treatment." Therapists with other approaches to alcoholic patients would see the resumption of drinking as a danger signal and would actively explore it with the patient.

Flexibility about abstinence during treatment is exemplified by Bailey's (1968) suggestions. Although she advances the general rule that abstinence is the first goal of therapy, she states that "many, perhaps most, alcoholics cannot achieve it at the beginning of treatment. To demand it as an initial condition is to deprive the individual of the help he needs." Steiner (1971) holds a similar but somewhat more uncompromising point of view. In the first sessions he indicates that he expects the patient to stop drinking completely as soon as possible and to agree to remain abstinent for at least one year. However, he realizes that "while a majority of alcoholics are willing to stop drinking, it is not always true that they will do so immediately upon being advised of this necessity." He identifies two types of patients who continue. The first continues to drink in his usual pattern and comes to therapy sessions under the influence, while the second reduces or reorganizes his drinking, confining it to weekends or periods between sessions. A third category may also be identified: the patient who takes a drink or two immediately before coming to the session; that is, who has been drinking but is not intoxicated. These patients should not be grouped with those who have a drinking episode during the course of treatment; "relapses" are discussed in the Middle Phase section under Drinking Episodes.

Handling of the patient who continues to drink in a treatment context which focuses on drinking and seeks abstinence as a condition of treatment has been discussed by Bailey (1968) and Steiner (1971), among others. Bailey, in common with most clinicians, views intoxication in early sessions as a form of testing the relationship, the patient's way of seeing whether the therapist can accept him and yet be firm. While each case must be understood and dealt with on its own merits, she recommends that the intoxicated patient should be seen on the first occasion to reassure him of the therapist's interest and acceptance. On a second occasion, the therapist may explore with the patient why he finds it necessary to drink before the session and this may in turn strengthen the relationship. It may be added that it is usually helpful to aid the patient to see that drinking during treatment is counterproductive, but that the therapist's nonacceptance of drinking during the sessions is distinct and separate from his acceptance of the alcoholic as a person. Steiner recommends that the patient who comes to therapy sessions while drinking be told that therapy "is useless under these conditions, and that because it is a waste of time, the therapist will not indulge in it at all."

Steiner (1971) observes that the patient who drinks between sessions may keep this from the therapist or lie about drinking bouts. He recommends that the therapist take the responsibility to make himself aware of the quantity and extent of the patient's extracurricular drinking by questioning the patient and by gaining information from relatives, friends and employers, with the patient's permission. The therapist should continue to focus on the drinking issue until the patient gives it up or decides to leave therapy. According to Steiner, most of these patients cease drinking within three months or terminate treatment. He finds that about 5 percent of patients who enter treatment do not want to stop drinking and eventually discontinue therapy. It is interesting to note the similarity between this figure and Silber's (1970), the one obtained in a drinking-oriented approach, the other in a context in which drinking is hardly mentioned. One implication is that emphasis or nonemphasis on drinking in treatment is a less salient issue than the quality of the relationship between patient and therapist. Certainly in other clinical contexts it has been observed that the content of therapeutic interactions is less crucial to outcomes than what is qualitatively conveyed by the relationship (see, e.g., Milmoe et al., 1967; Harford et al., 1970).

Another way of dealing with the patient who continues drinking between sessions but conceals it from the therapist does not involve questioning the patient or gaining information from informants. This view holds that the patient may at first gain secret gratification from "putting one over" on the therapist. The advantages, however, will in a matter of a few weeks begin to fade because the therapist is not aware that he is, in the patient's mind, being bilked. The patient will at this point either give up drinking between sessions or tell the therapist about it. In the latter instances, its meaning and significance for therapy is examined, and the usual result is that the patient gives up drinking, although he may, like some of Steiner's patients, decide that he is not ready for therapy and discontinue it. A general rule of thumb is that if drinking outside therapeutic contacts is affecting treatment, the patient will bring it up sooner or later if he is committed to therapy. If he is not, then he will in any event leave treatment. The problem with a continued focus on drinking behavior by the therapist is that it can be extremely difficult to prevent it from degenerating in the patient's view into surveillance which will drive him from treatment, despite its being part of the original agreement between patient and therapist.

Handling of drinking in the therapy situation will also vary with the goals of treatment, a topic discussed in the following section. But, as Bailey (1968), Blane (1968), Blum and Blum (1967), Chafetz (1959), and others have pointed out, there are many instances in which the goal of cessation of drinking is unrealistic, where the emphasis instead is on limited restructuring of the patient's life situation so it is more tolerable for him and those in his immediate environment. Generally, in these instances, drinking during treatment is not an issue.

Setting Goals and Confrontation

There are two broad sets of goals in psychotherapy with alcoholics which reflect basic differences in beliefs about the best way to treat alcoholics. The first set of goals has been succinctly stated by Silber (1970): "The object of the psychotherapy (is to help) the patient to live and function more comfortably and with greater awareness of himself and his surroundings. . . . The treatment is aimed at making the patient more familiar with his fears, differentiating his wishes from reality, and stressing the differences between impulse, thought, and action." Note the absence of any necessary emphasis on cessation of drinking as a goal. The second set of goals emphasizes the attainment of abstinence first, followed by a period of substitutive treatment or a period of treatment that focuses on proximal or distal causative conditions. It is likely that those who subscribe to the first set of goals deal primarily with patients who are not drinking when they are first seen, whereas those subscribing to the second set of goals first see the patient when he is still actively alcoholic. This in part accounts for the differences in goals. More critical differences between goals involve whether treatment following abstinence is substitutive, focuses on proximal causative conditions, focuses on distal causative conditions, or is some combination of these. These differences will depend on the philosophy of the facility and the background and interests of the clinician conducting the treatment. Ideally, all types of treatment should be available and patients should be assigned to them according to their expectations and individual assessment by the staff. In practice, this ideal is rarely attained.

In settings where abstinence is an explicit goal, it is usually recommended that the patient be made concretely aware that he has an alcoholic problem or suffers from alcoholism in order to recognize the importance of abstaining from the use of alcohol. The techniques for creating this awareness are lumped together under the term *confrontative,* and may range from rather brusque, unrelenting scare tactics in the hands of novices or poorly trained personnel to carefully worked-out psychoeducational procedures that provide a smooth entry into therapy. Bailey (1968) clearly indicates that effective confrontation involves careful attention to method and training.

> Confrontation does not mean lecturing the client, or even a primarily didactic approach, but rather a frank and open discussion of drinking and the problems it creates for him.... The main purpose of confrontation is to bring the realities and consequences of drinking into the open, whether or not the word alcoholism is used.... A good drinking history is helpful as a basis for discussion between client and worker.... Confrontation fundamentally means a realistic evaluation of his drinking problem with the client. It aims to cut away denial and omnipotent phantasy and thus usually mobilizes some anxiety.... Confrontation may have to proceed at a slower pace when ... the alcoholism is masked by some other difficulty (pp. 81–83).

Catanzaro (1968) offers a detailed outline of a method for confronting alcoholics by focusing on the symptoms of alcoholism and by avoiding derogatory or accusatory statements. He divides questions into those relating to early-stage symptoms (e.g., "Do you frequently drink more than you intend?") and later-stage symptoms (e.g., "Do you lose time from work due to drinking?"), with the former asked prior to the latter. The point at which Catanzaro recommends that one confront the patient with his alcoholism depends on the information gathered and an assessment of the strength of the relationship that has been established. This may not be until the second or third interview, when any of the following kinds of statements may be made: "On the basis of the evidence we've gathered, it appears that you have a drinking problem," or "It looks like drinking too much is definitely part of your problem," or "It looks like you may be hooked on alcohol." Catanzaro finds, contrary to some professional folklore, that the patient who is approached about his drinking behavior in a sincere, understanding manner—with avoidance of terms like *alcoholic* and *alcoholism* until used by the patient—and with whom rapport has been established will usually be cooperative and responsive.

Programmatic approaches such as this are useful in many settings and are particularly comforting to beginning clinicians. They are, however, subject to several disadvantages, all related to the need for a flexible, open-minded stance on the part of the therapist so that he can appreciate individual differences in the kinds of problems patients bring with them. The first is a feeling pervasive in many clinical settings that, if the patient doesn't admit that he has an alcoholic problem, treatment has nowhere to go. This often results in a sense of urgency on the part of clinicians, especially novices, which leads to poor timing of confrontation and a heavy-handed and overly persistent reaction when the patient takes exceptions to the worker's remarks. Under a steady barrage of confrontations, most patients, alcoholic or not, will either become angry or, more often, agree with the clinician in order to preserve peace. During a forceful first interview, which was tape-recorded, the patient finally said, "You know, if it means that much to you, sure, I'm an alcoholic." But it was clear he was keeping his own counsel. Anger or pro forma agreement do not advance treatment, and most patients approached in this manner soon discontinue treatment.

A second disadvantage is a common tendency to feel that the patient who admits that he can't control his drinking behavior is in a state of resistance and that he doesn't "really" accept his alcoholic problem. Here the clinician may attempt to lead the alcoholic to a deeper level of awareness, or "surrender," in Tiebout's terms (1949, 1953), that goes beyond mere "compliance." While these patients may indeed be in a transitory phase or in a state of resistance, the clinician may be wiser to accept them as they are now, rather than implicitly reject their self-perception by focusing on the need for greater recognition of the problem. Anyone who has worked therapeutically with alcoholics or other kinds of patients knows that seeing oneself clearly is not a steady, constant state, but fluctuates with one's level of anxiety and guilt.

Another disadvantage of a programmatic approach to confrontation is that it focuses on the drinking problem in instances where the drinking problem, as severe as it may be, is not of central importance in the patient's life situation. Distinguishing the patient who is drinking excessively and steadily in response to a life crisis or series of crises from the one who has had a lifelong love affair with alcohol is simply a matter of taking a good history from the patient. Reactive alcoholics represent one class of alcoholics for which an emphasis on drinking behavior slows rather than advances the course of therapy. Since it is not central, discussion of the drinking problem may be used by the client as a means of resisting examination of the core problem(s), just as many essential alcoholics focus on emotional conflicts as a means of resisting a closer look at the all-encompassing role alcohol plays in their lives. This begins to explain why some clinicians recommend staying away from the drinking issue while others indicate that emphasizing it is absolutely necessary: Both may have been dealing primarily with different groups of clients and have begun to believe that no other kinds of alcoholics exist. Also, it may be that in many instances in which a drinking problem erupts suddenly in response to overwhelming stress, end goals do not necessarily include lifelong abstinence.

This observation leads to a fourth disadvantage, i.e., the extent to which confrontation is necessary when goals include regularization of certain aspects of a patient's grossly disturbed life situation without holding out for abstinence.

Many alcoholics in their 50s and 60s who have always lived in a social environment that values the heavy use of alcohol and who have been drinking excessively since adolescence often show a tendency to modify but not stop drinking as they get older. Here, for instance, unambitious but useful aims may be sought, such as maintaining a continuing relationship with a single agency rather than "shopping around" among treatment facilities without any coordination of care, reducing the number of episodes of alcohol abuse, controlling drinking so that it occurs at times that do not affect occupational functioning, or switching the patient from hard liquor to beer (this is not to suggest that drinking beer makes one less of an alcoholic, but merely to indicate that, perhaps because of socially defined expectations, aggressive and bizarre behavior occur more commonly among whiskey drinkers). The desirability of confrontation in cases such as these will vary from individual to individual and while it may be necessary for a patient to know and understand the "chronic disease" he is suffering from in order to achieve therapeutic goals, in most instances confrontation will serve little therapeutic purpose, when change in the patient's life situation is the most critical goal.

Given a nondrinking patient, what goals will the therapist strive for in continuing therapy? As indicated above, these may be generally classed as substitutive treatment, treatment which focuses on proximal causative conditions, and treatment which focuses on distal causal conditions, and whether one or some combination of these will guide therapy depends on the philosophy of the setting and the preparation of the therapist. All have the general aim of helping the patient to arrive at a point where he can function without pain and without the destructive use of alcohol. All hope that this general goal will be accompanied by heightened self-esteem, a sense of well-being and growth, appreciation and use of one's assets and acceptance of one's limitations, rewarding and self-respecting relationships with others, and capacity to handle and accept difficulties as they arise. The extent to which these are achieved will of course depend largely on what the patient brings with him to therapy in the way of basic psychological, social, emotional, educational, and occupational achievement and promise.

An ever-present danger is that the clinician will set goals which are beyond the patient's capacity to achieve. In cases of doubt it is a useful rule of thumb to set goals lower rather than higher; then as treatment progresses, new, more ambitious goals may be introduced as appropriate. To find it necessary to lower goals as therapy progresses is difficult for both patient and therapist and can result in discontinuance when therapy is relatively far advanced. The process of setting goals must be a mutual one, but from his point of view the therapist must strive to make goals as concrete as possible within a time framework that takes into account the alcoholic's characteristic difficulty in delaying gratification. As this latter capacity improves in treatment, longer-term planning with the patient will become more relevant to his personal growth. Major life goals initiated by the patient must be viewed tentatively, and decisions about them should be delayed until a full exploration of their meaning to the patient and his rehabilitation has been conducted. Often, for example, the patient will view his spouse or his job as major causative factors in his alcoholism and may entertain divorce or a change in career as a solution. To encourage or permit such steps in the absence of full understanding may be a major disservice to the patient because they often reflect resistance and thus have no therapeutic advantage and may even damage the patient's life further. Strecker (1951), for example, indicates that the widely held belief that a change in occupation will solve the problem is a vain hope: "There are no occupational dry docks." Sometimes alcoholics, able and well trained in major professions and their occupational functioning little marred by drinking, give up their professions during treatment with no personal advantage to themselves or to society.

The establishing and communication of concrete goals is more readily accomplished in substitutive treatment than in therapy which deals with proximal causative conditions, and it is more difficult in therapy which emphasizes distal causative conditions. This is because substitutive treatment deals largely with the identification, trying-out, and consolidation of specific behaviors and activities in the patient's immediate life. There are, however, techniques that can be employed in both forms of causative treatment that may be viewed as goals but are primarily adjuncts to facilitate the progress of therapy. These techniques may be subsumed under the term *homework*, discussed in detail in a later section.

The basic initial step in setting goals consists of some form of the questions "What sort of things do you expect to get out of therapy?" and "How do you expect to get them?" First responses are often global and unrealistic, and it is the therapist's job by asking questions, reflecting, and making gentle evaluative comments to help the patient differentiate goals and make them concrete and short term by identifying subgoals. It is useful to try to make as explicit as possible with the patient the steps whereby each minigoal will be accomplished. No matter how fantastic the patient's goals may be, they should not be dismissed or belittled, although their relation to reality may be highlighted by the therapist. If the patient says that he wants everyone to leave him alone and not to pester him at work and at home, yet be Johnny-on-the-spot when he wants them, the therapist may reply, "Who doesn't?" or "I'd like that, too. How would you work it out?"

When the patient arrives at a set of attainable, concrete goals that the therapist perceives as being within the patient's sphere of competence, he may summarize them and indicate that he can help the patient to achieve what he wants to achieve. It should be noted that while the initial period of goal setting may take anywhere from a single session to several, goals are constantly under review by both therapist and patient. It frequently happens that early-established goals disappear from view to be replaced by other, usually more meaningful aims. The therapist must be sensitive to the spontaneous expression of goals that have never been mentioned. It should be understood that while goals are not inflexibly pursued, they form a framework within which therapy proceeds, and it is also useful for therapist and patient to have some understanding that major life decisions are not to be made during the course of therapy without thorough review and exploration by the patient in the treatment setting.

Goals also vary according to the therapist's purpose, i.e., substitutive, proximal causative, or distal causative. Patient goals relevant to the therapist's objectives are reinforced and pursued, while irrelevant goals are not and may even be actively discouraged. The substitutively oriented therapist encourages the discussion of goals related to activities and behavior patterns that serve to occupy time formerly devoted to alcohol, such as group activities (including A.A.), work, hobbies, athletic activities, religious participation, and joint family projects. He considers proximal causative conditions only to the extent that they are related to the major goal of substitution and is usually indifferent to goals related to distal causative factors. The therapist who concentrates on goals related to an understanding of proximal causative conditions encourages specification of goals that have to do with current interpersonal relations and the manner in which the patient copes with emotional states that are followed by drinking. He usually accepts personal historical material but does not pursue it when the patient frames it in terms of goals of treatment. In a sense, he is interested in substitutive goals, but only to the extent that they represent coping with or resolving psychological factors that brought about drinking episodes. The distal causative therapist is receptive to personal historical material, responsive to perceptions of parents and childhood memories, and interested in continuities between present and past. He ordinarily devotes a considerable amount of time in early interviews to obtaining a family history, including the patient's perceptions of each family member, his earliest memories, and recurrent dreams. This tends to set a tone for the subsequent conduct of therapy just as a drinking history does for the clinicial who plans to focus on drinking behavior and its cessation.

To the extent that a therapist holds a developmental view of alcoholic problems, he will seek to form a psychodynamic understanding of each patient and this will inform his technical approach whether he has substitutive, proximal causative, or distal causative objectives. His understanding of the psychogenesis of a patient's alcohol problem does not, however, mean that a therapist with substitutive or proximal causative goals will explicitly focus on the relationship between childhood conflicts and substitutive activities. Even the therapist whose stated aims center on distal causative conditions will usually not make the kinds of interpretations to the patient about early childhood conflicts that characterize some forms of intensive psychotherapy. These factors are more fully addressed in the sections on the middle phase of therapy.

Therapeutic Contracts

Implicit in much of what was presented in the preceding section on goals and confrontation is the nature of the terms under which the patient and therapist proceed in treatment. Most commonly, there is only the most general understanding between the two about their respective roles and expectations, and while each may feel that a mutual agreement exists, it frequently turns out that the specific terms of the agreement are not known and may be contradictory. The usual procedure when this is discovered (if it is) is to negotiate around the specific situation that impelled the discovery. This process may be repeated many times in the course of therapy. Each episode is a danger point that can result in discontinuance of treatment by the patient or even termination by the therapist. On the other hand, such negotiations can be a potent tool for emotional reeducation of the patient, helping him learn how to resolve interpersonal conflicts openly and to take each situation on its own merits. Many clinicians prefer a more explicit approach, feeling that it may be unfair for the patient not to know specifically the terms of a therapeutic contract and that nonspecific agreements may result in (unconscious) abuse of the therapeutic relationship by the therapist. Some feel that general agreements are inefficient because the working frame of reference they establish is so loosely defined that much time may be lost in following unproductive paths. A corollary to this is that the patient may find a general agreement conducive to resistance and acting-out, both counterproductive for attaining psychotherapeutic goals. More specific therapeutic contracts are thought by many to be particularly necessary in treatment of alcoholic patients, who are prone to manipulation and acting-out.

The nature of contracts varies widely; they may focus on a few specific items (e.g., abstinence, length and frequency of sessions, length of therapy) or may be more general and comprehensive. Time-limited therapy is sometimes advisable in dealing with patients where there is substantial reason to believe that the patient's dependence on the relationship will eventually block treatment; in such an instance the patient should agree to the time limit before entering treatment. Steiner (1971) is an advocate of a general but explicit contract that has the characteristics of legal contracts. These include mutual consent, valid consideration, competence, and lawful object. Mutual consent is central to therapeutic contracts and consists of an offer by the therapist, explicitly communicated with definite terms, that is accepted by the patient. Valid consideration refers to benefits conferred by the therapist (making good on his offer, achieving treatment goals) and to benefits conferred by the patient (fees, efforts toward achieving treatment goals). Competence in the case of alcoholics usually means that an intoxicated individual is not competent to enter a contract, while lawful object means that nothing in the contract should violate laws or public morals. Steiner gives an abbreviated example of a request for treatment, an offer of treatment by the therapist, and an acceptance by the patient. The offer of treatment, which outlines the terms of the contract, is given below with minor modifications.

> O.K., Mr. Jones, I will accept you in treatment. While you are in therapy with me . . . I will expect you to stop drinking entirely as soon as possible and will expect you to continue to abstain for at least one year, since it is my experience that individuals who do not abstain for a year tend not to recover from alcoholism. If you remain abstinent for a year while in therapy, you will probably be cured, that is to say, you will gain control over your drinking to the point that drinking will no longer be of concern to you. This treatment implies that you will be actively pursuing not only sobriety but any number of other states of affairs which might be conducive to a cure. I, as a therapist, will be guiding you but the responsibility for your situation has always been and will always be yours. I will be able to see you on Monday at 10:00 A.M.

This contractual statement is specific, optimistic, supportive, and forthright. It gives the patient a clear but not overdetailed idea of his responsibilities and the length of stay in therapy for a successful outcome. It indicates the therapist's role and his limitations and indicates the shape which therapy will assume.

Blane (1968) suggests that general therapeutic contracts be phrased in ways that address themselves to the characteristics of the individual patient. Thus the terms of entrance into therapy will be different for a patient who is anxious and fearful and needs support than for one who has a history of subtle interpersonal manipulation or who is subject to outbursts of anger when threatened. Perry *et al.*, (1970) suggest the following as major ingredients in any treatment agreement: If the patient wants to cease drinking, the counselor has the knowledge, skill, legitimated authority, and commitment to help him to do so; the therapist will give total commitment to the patient; he will always accept the client himself, although he may not always accept his actions; and he will not withdraw his support.

Whether the terms of entrance into treatment are spelled out or left general is a question each therapist will decide for himself. Either approach has its advantages and disadvantages, and the choice has more to do with how comfortable the therapist feels with one or the other than with any inherent benefit of either. Therapists comfortable with both approaches, however, may have the luxury of being able to apply either, according to what appears indicated for a given patient.

The Hostile Patient

Angry, hostile, and resentful patients are often seen in settings they have been forced to attend under pressure from the courts, employers, or relatives. Therapy programs for offenders on probation have been described by Margolis et al. (1964) and Mills and Hetrick (1963). More recently, these have largely been replaced by public inebriate programs. Most of these programs use the kinds of techniques described in preceding sections.

Scott (1963), describing his experiences in an outpatient setting, outlines a method for initial contacts with alcoholics who are subject to outbursts of anger and use projection as a characterological mode of defense. These patients typically open the first interview by angrily blaming their spouse for any problems they have (e.g., "She's always cutting me down," "She's trying to destroy me"). Rather than attempting to redirect the patient toward himself, Scott recommends that the patient be encouraged to elaborate on his feelings of projected blame. Just at the end of the hour, when the patient has no time to protest, it is suggested that the patient be told that he has a problem with alcohol and his temper but that he has not talked about himself. He is encouraged to try harder to do this in the next session.

The patient will, according to Scott, characteristically start the second session in the same vein as the first, emphasizing the injustices he has collected in the past week. "After a few minutes of listening, the therapist hands the patient some notes on hostility in which the basic dynamics are explained. . . . The notes are read over slowly, and the patient is asked to question any and all remarks. This stimulates some very defensive questioning on the patient's part . . . [and he] will want to know again and again if the therapist thinks he is hostile. An affirmative answer is given." The patient typically handles the anxiety this technique generates by angrily stating that he is not an angry person. He is than confronted with this reaction. At the end of this session, "the patient is asked to take the notes home, reread them again, and return with as many questions as he likes." He is also requested to bring his wife with him.

The third session, with both patient and wife attending, typically goes in one of two directions: a verbal battle between the two or a tearful and depressed reaction by the wife to the patient's accusations. For the latter, Scott recommends urging the wife to express herself with the therapist's support and backing. For verbal battles, the therapist serves as a clearinghouse for all remarks, relaying accusations made by the patient to the wife, and relaying her countercharges to him in turn. Usually, the couple simmers down in a short while and begins to converse in a more adult fashion, and eventually both want an evaluation from the therapist as to whether they can hope to change their situation. Scott finds that this technical plan works well and leads to successful entrance into group therapy in a surprisingly large number of cases.

Scott does not discuss the dynamics of change that underly this apparently little-used but promising technique. Some of the elements that may be identified include the initial acceptance of projected hostility, the therapist's lack of fear or anger as indicated by his confrontations at the end of the first session and during the second, his provision of intellectual means of understanding anger as another method of handling it, and his recognition of the ambivalent importance and value that the marital relationship has for the patient. Scott's success indicates that this technique merits further attention.

MIDDLE PHASE

Three aspects of the middle phase of therapy are considered: (a) general aims and issues common to this stage of treatment; (b) homework and other techniques; and (c) the therapeutic handling of drinking episodes that occur in treatment. As noted earlier, the division of therapy into initial, middle, and late phases is an arbitrary device useful for purposes of presentation and training. Each case has its unique characteristics, and the criteria presented below for ascertaining when the initial phase ends and the middle phase begins are not meant to apply to all cases. As we have seen, therapy begins before the patient sees a therapist and if it is successful it never ends, even though the patient has long since stopped seeing the therapist. One must never lose sight of therapy as a process when examining its "stages."

General Aims and Issues

The middle phase of treatment begins when the patient is abstinent or his drinking does not grossly interfere with the conduct of therapy, the groundwork for a positive relationship has been established, a therapeutic agreement has been made, and the goals of treatment have been provisionally arrived at. This state of affairs may be reached as early as the end of the first session or may not be attained for a period of two or three months. The major and most time-consuming work of therapy is conducted during this middle phase, and in its most general form it consists of exploration, understanding, trying-out, and working through. As Chafetz (1959) puts it, the therapist "must be prepared to encourage his patient to carry out tasks, to make decisions and, consistent with the patient's abilities, to meet and deal with reality in a mature manner." For the psychoanalytically oriented therapist, this involves several modifications in standard technique, even when the therapy focuses on distal causative conditions underlying the alcoholic problem. While some discussion of process, technique, and dynamics of the middle phase of treatment with alcoholics may be found in the literature, there are few available comprehensive accounts of this stage of therapy and, in general, less is available than for the initial phases of treatment. Silber (1959, 1967, 1970, 1974), from a psychoanalytic position, Scott (1961, 1963), from an eclectic psychodynamic point of view, and Steiner (1971), from a transactional analytic orientation, have had the most to say about what occurs in treatment after the initial phase. Interestingly, much of the material presented by Scott and Steiner is based on group therapy, but nevertheless can be directly applied to individual psychotherapy with some exceptions.

The middle phase of therapy may be profitably examined in terms of whether it is essentially substitutive, focuses on proximal causative conditions, or on distal causative conditions. While there is a substitutive element in all three orientations (discussed in the next section), substitutive treatment itself tends to center on getting the patient involved in a supportive social and interpersonal network that emphasizes the values of abstinence. The therapist may encourage the patient to become involved in this social support system or may assume a more active role in introducing him to it by making referrals and introducing him to representatives of the system. These latter typically include clergymen, former alcoholics, members of Alcoholics Anonymous, and other interested individuals. In therapy itself, a primary function of the therapist is to have the patient report his experiences and feelings about involvement in the substitutive system, with the therapist being sensitive to difficulties the patient may be having in shifting from an alcoholic to a nonalcoholic life. Emphasis on drinking continues, with the therapist exploring situations and circumstances in which the patient drinks, with an eye to environmental manipulation if it appears warranted. The therapist also indicates that the patient may, or should, call him any time he has an impulse to drink; membership in A.A. also fills this support function. There is relatively little concern with more than superficial understanding or focus on psychological realities. However, the therapist takes care to observe the emergence of problems, complaints, or rationalizations that precede drinking and brings these up directly with the patient. The positive aspects of family life, work, and interpersonal functioning are reinforced and conflictual elements are handled by helping the patient to repress them. For example, a patient may say that his wife has been bugging him lately. The therapist, aware that this feeling often precedes a drinking bout, may ask the patient about the difficulty with his wife, but he simultaneously stresses the strengths of the marriage and encourages the patient to see things from the wife's point of view.

The therapist's stance with regard to abstinence is to make it as important in the patient's life as alcohol once was. As Shea (1954) puts it, the most important therapeutic ally is "the substitution of one obsession for another. . . . The patient must come to the state where he is now as obsessed with the idea of being sober as he was formerly obsessed by the notion of being drunk. . . . He must become consumed with the notion that he is now a nondrinker. . . . The easiest way to tackle alcoholism directly is to make nonalcoholism an obsessive issue with the patient." The social support system that the therapist with a substitutive orientation attempts to create for the patient is strongly directed toward abstinence, extolling its values and benefits.

The middle stage of therapy focusing on proximal causative conditions will take one of two broad directions depending on whether the patient is a reactive or essential alcoholic. In the first instance, the therapist centers on recent losses which precipitated excessive drinking and takes up the issue of drinking only when it interferes with treatment. This is particularly the case when the onset of the alcohol problem can be precisely dated in association with a loss. The therapy may consist of grief work, as outlined by Bellwood (1975), when the loss is clearly identifiable and specific. More often, however, there are a number of severe threats to self-esteem that are not losses in the usual sense, i.e., death of a relative or close friend. These may be a series of business reverses, family illnesses and problems, or an exaggeration of the loss of youth as reflected in a mid-life depression. Here much work has to be devoted to helping the patient see the emotional impact these events have had upon him. Affective reactions to loss are usually repressed and isolated and the drinking symptom, while relatively sudden in its onset, appears as an inexplicable phenomenon to the patient and his relatives and friends. They often rationalize it as being caused by fatigue, overwork, drinking necessary for the job, and so on. It is rarely perceived as being a means of handling anxiety, depression, and anger.

The major therapeutic task is to help the patient understand affectively the relationship between his anxiety and depression and the guilt and anger he experiences towards the lost objects. The patient is encouraged to ventilate his depressed feelings and his fears. This will usually be accompanied or followed by self-recriminations ("If I had listened to him when he first complained about the pain, he would be here today"; "He quit college because I wasn't a good father"; "I didn't get the foreman's job because I don't know how to handle men."). The therapist helps the patient to disentangle reality from feelings, to separate out feelings of love, anger, and guilt, and to see that wishes are not deeds. When the patient is able to discharge the deep-seated anger he feels toward the lost object with little anxiety, a marked shift in his feelings about himself typically occurs. He begins to reevaluate himself, to appreciate his assets, to feel less depressed, and to decrease his alcohol intake markedly. Links between current losses and the patient's early psychological history are typically marked and obvious to the therapist. It is a temptation to shift therapy to an understanding of these early causative conditions, but with this type of patient, this is generally avoided because it may prolong treatment to little therapeutic advantage. Reactive alcoholics have usually been reasonably successful and problem-free until the onset of the alcohol problem. In delimited conditions like these it is generally enough to resolve the conflict that resulted in the drinking problem so that the patient can return to his previous state of equilibrium. Generally speaking, the more specific the loss, the more recent its occurrence, and the shorter the period of destructive drinking, the shorter the therapy. Sometimes an excellent result may be obtained in a matter of months. When losses are diffuse and the drinking problem has begun to have some of the typical sequelae of chronic excessive drinking (marital disruption, impaired job functioning, trouble with the police), therapy is complicated and may take much longer.

The middle phase of treatment for essential alcoholics in which proximal causative conditions are the focus is quite different from that outlined above for reactive alcoholics. The major differences consist of a greater focus on drinking behavior itself, emphasis on increasing the range and strength of coping mechanisms other than alcohol, and a constant analysis of salient psychological themes as they occur in concrete daily situations. As Blane (1968) has put it: "The therapeutic focus . . . is on strengthening personality assets and shoring up areas of previous strength now in danger of collapse. . . . Current conflicts and the patient's solutions to them are understood by the worker to be variations of earlier-life conflicts and previously learned solutions, but the worker ordinarily does not seek out memories and fantasies of childhood. . . . The goal is to help the patient respond to present conflict flexibly, so that he can select the most relevant and appropriate solution."

During this phase, the therapist is alert to evidences of rationalization, denial, and projection, which are among the more common defenses used by alcoholics. Painstaking work is necessary to help the patient see how he avoids facing himself by use of these defenses. In the literature there are many exhortations to "expose defensive maneuvers" or to "block the alcoholic's pathological personality defense mechanisms," but there is little suggested in the way of techniques to use. Some techniques for dealing with projection have been presented in an earlier section. Margolis *et al.* (1964) suggest that many alcoholics live in a state termed *suspension of thought,* which is related to discharge of affect through acting-out. Suspension of thought may also be thought to accompany denial, and to a lesser extent rationalization and projection, since the major characteristic of suspension of thought lies in an inability to reflect on one's own behavior, feelings, and fantasies. Margolis *et al.* (1964) recommend that the therapist encourage patients to pay greater attention to and express their feelings and thoughts. This may be accomplished by reflection, nondirective questioning, or direct questioning. Encouragement of expression with alcoholic patients, many of whom are verbally passive and unspontaneous in therapy, serves not only as training for therapy but also as an indirect and nonthreatening means of getting beyond defenses.

Scott (1961) suggests a more active technique with passive and anxious patients, including confrontative questions such as "You're afraid of me aren't you?" or "Are you always so shy?" or "What do you hope to accomplish by your anxiety?" Such questions and comments are designed to investigate the hostility that presumably underlies the passive–anxious exterior. The expression of hostility in a neutral setting in itself advances therapeutic progress. Once expressed, the task then becomes to help the patient to distinguish between projected hostility and realistic assertiveness and between wish and reality. Because rationalization is more accessible to conscious understanding than denial and projection, the therapist should center his direct confrontive efforts on it rather than the latter. Often the patient has at least some consciousness that he is presenting a masking reason for drinking. Usually a gentle but direct challenge to rationalization is sufficient and has the added benefit of strengthening the therapeutic alliance because the patient feels that the therapist understands him in a way that few people do. Once the fact that the patient tends to rationalize his behavior rather than communicate his affective experience has been established, the therapist may develop a kind of shorthand with the patient when he engages in it by, for example, assuming a skeptical expression with a raised eyebrow or by making a brief comment ("There you go again," "Are you kidding yourself again?"). These techniques are useful for keeping the patient on an emotional topic and diluting resistance without affecting the relationship.

Denial in treatment is generally accepted and not challenged because it is a danger signal that the anxiety aroused is more than the patient can handle at the time. For example, on the basis of what a patient has said in previous sessions about authority figures and his emphasis in the current session about how he likes his boss, how well they get on together, and how helpful the boss is, the therapist might comment that it seems that the patient is afraid of the strength of persons in power and handles the fear by adopting a submissive position with them. The patient may respond by saying that the connection made by the therapist is simply untrue; the therapist may then give the evidence for making the assertion. If the patient still denies the association, the therapist may say, "Well, perhaps I was wrong; if it's important it will come up again," or "We can look at it when you're ready." Such statements have the effect of reducing immediate tension but leaving the item on the agenda, so to speak. Sometimes the statement stimulates the patient's curiosity enough that he will pursue the topic on the spot.

Projection, too, should generally be handled gradually, although some of Scott's direct techniques may be in order when there is a strong therapeutic alliance and when the therapist is sure of his ground. A potential difficulty that therapists should be aware of and attempt to avoid (although this is frequently not possible) is becoming a central object of a patient's projective system as either a persecutor or a benefactor. The best way to avoid being placed in either situation is to maintain a consistently honest, open and nonjudgmental attitude toward the patient. The therapist must take pains to clarify with the patient that acceptance of him as a person doesn't mean that he necessarily approves of all his actions—and this distinction must be returned to again and again in the course of treatment. For patients with strong and fixed projective positions, the therapist's neutrality will be of little import, for the patient will cast him according to an inner mold that is little regulated by interpersonal reality. Fortunately, most alcoholics do not have such fixed positions. From a transactional analytic standpoint, Steiner (1969, 1971) discusses the alcoholic's proclivity to see therapists as "persecutors" or "rescuers"; he gives several technical recommendations for avoiding being forced into either of these roles (Steiner, 1971), most of which may be subsumed under the general category of adoption of a neutral, firm, consistent stance that avoids unthinking responses to patient provocation (the treatment issues involved in the interaction between the patient's perceptions of the therapist, and vice versa, are dealt with in detail in the section on Relationship Issues).

When therapy has as its centralizing theme the understanding of distal causative conditions, the middle phase of treatment will be distinguished by an interest in childhood memories, fantasies, and dreams and the manner in which these interrelate psychodynamically. Among its aims is the reconstruction of the patient's past with resolution of distortions of reality learned by the patient in order to reduce the experience of emotional pain brought about by conflict. This involves a thorough understanding with the patient of his perceived past, the types of crippling reality distortions he uses, and the kinds of painful feelings that activate particular distortions and the expression of those feelings. In addition, the therapist helps the patient to achieve a sufficiently detached awareness of conflicts engendering painful affect so that the level of affect at least diminishes to a point where it no longer overwhelms the patient and can be experienced without resort to defensive maneuvers that seriously disrupt the individual's perception of reality. With regard to alcoholic patients, this means that there may be relatively little emphasis on the patient's drinking behavior itself, except insofar as some of the patient's attitudes about drinking represent distortions of reality (e.g., "Drinking is no problem for me—I can take it or leave it alone").

It is generally accepted, even by its advocates, that this form of treatment—psychoanalytically oriented psychotherapy—is suitable for only a small percentage of alcoholic individuals who come to treatment settings. Silber (1974) states that only about 5 percent of the alcoholic patient population are sufficiently motivated to be amenable to psychoanalytic psychotherapy. As indicated earlier, control over drinking and drinking during treatment are usually not problems in the treatment of these patients; as Silber (1967) puts it, "the drinking itself frequently slipped into the background, superseded by attempts to understand anxiety, aggression, or any number of other symptoms or problems of adjustment."

Even though psychoanalytic psychotherapy is considered the treatment of choice for a "select group" of the "best-motivated" alcoholics, virtually all of its practitioners recommend some variations from its typical application. The major variation resides in the therapist's active efforts to establish a positive relationship in the opening phase of treatment, as we saw in an earlier section, by being open and warm and immediately fulfilling appropriate patient requests. Modifications of technique in the middle phase of treatment are not as specific to the therapy of alcoholics as the literature suggests but rather represent variations in technique that have been generally adopted for nonintensive (i.e., once a week), short-term, and/or goal-limited forms of individual therapy that proceed from a psychoanalytic model. Thus, positive transference is encouraged, transference is interpreted only if it threatens to disrupt treatment, the therapist is more active in keeping the patient on topics germane to the goals of treatment, reconstructions of the past are made in accordance with the goal-limited nature of treatment, and ego-assets are reinforced rather than analyzed, while areas of ego-weakness are avoided. If it becomes necessary to examine transference, negative reactions are interpreted before positive reactions. These variations from psychoanalysis and intensive psychoanalytically oriented psychotherapy characterize much of the individual psychotherapy focusing on distal causative conditions provided for a wide range of problems, including drinking problems, in mental health and other settings.

While many have made contributions to therapy dealing with distal causative conditions (e.g., Blum and Blum, 1967; Chafetz, 1959; Knight, 1937a. 1938; Margolis *et al.,* 1964; Strecker, 1951), Silber (1959, 1967, 1970, 1974) has presented the most coherent and contemporary picture of technical aspects of the middle phase of individual psychotherapy with alcoholics. Following the initial stage of treatment, which consists mainly of fostering a positive transference by the therapist, who presents himself as "an all-knowing figure in a psychological sense" with special knowledge about what is going on in the patient's mind, the patient is encouraged to recognize, experience, and accept his feelings. The expressing of feelings, wishes, or thoughts is discouraged. During this period the therapist shifts from the role of an omnipotent figure to that of a benevolent, consistent, and firm figure who clarifies the distinction between wishes and actions. Much of the middle phase of therapy is devoted to facilitating and strengthening the patient's identification with the therapist and to strengthening existing defenses. This is accomplished by several means: (1) The therapist conveys to the patient the idea that the more he is able to understand himself and his problems, the more he will be able to control concerns that cause him pain. The therapist helps by listening and remembering, and thereby the patient learns to deepen self-awareness. "The patient is offered the expectation that through increased knowledge and understanding comes increased control over his person and his functioning. Such increased knowledge becomes the touchstone of the therapy" (Silber, 1974). (2) The therapist "lends" the patient his ego-functions in an attempt to strengthen the patient's unstable sense of identity by providing guidance and information in learning how to function interpersonally and intrapersonally. Suggestion is freely used by the therapist and readily seized upon by the patient, who, according to Silber, is typically quite open to it. Precepts, psychologically sound information and knowledge, which aid in evaluating what the patient observes in himself and in others, are freely given on relevant occasions. The therapist is viewed and in many ways acts as the parents the patient wishes he had. (3) A related and critical aspect of Silber's approach is to point out and work through with the patient the reality of his parents' psychological and behavioral abnormalities. The approach is based on Silber's clinical observation that "every alcoholic patient seen had at least one parent who manifestly displayed psychotic behavior," usually during the patient's early childhood. Because parental bizarreness has been developmentally normative for the patient, the therapist's placing disturbed parental behavior against the background of an average environment enables the patient to see himself in a new and different context. It further extends the process of differentiating oneself from the parents, thereby facilitating identification with the therapist. Confrontation with parental abnormality is preceded by a period in which the patient's perceptions of his parents are obtained and during which the therapist disentangles contradicting, confused, bizarre, or psychotic parental behavior and attitudes from those which are more normal.

Silber, in common with others who have applied psychoanalytic principles and techniques to nonintensive therapy, recommends that the therapist generally avoid transference interpretations because they tend to arouse early unconscious material, whereas treatment emphasizes current functioning. Transference phenomena—usually negative—that threaten the viability of therapy may be interpreted. This appears to contradict this therapy's focus on distal causal conditions by inhibiting rather than eliciting childhood material. Silber's approach involves seeking sufficient early material to help the therapist make a reasonable assessment of the patient's perceptions of parental figures and to permit an essentially intellectual reconstruction of these perceptions with the aim of facilitating new and sounder identifications. The therapy does not seek to open the psychogenetic field more broadly because it is not designed to provide the continuity and length required for such an endeavor and because there is some question whether alcoholics benefit from it.

If the therapist becomes a parental substitute in the patient's mind, "new identification with a healthy, reality-oriented figure is possible." The process, which occurs gradually and without outward sign, accounts for marked and sudden shifts in patient behavior. Because many alcoholics have an unstable sense of self, they are often able to form temporary identifications easily and have a distinct "as if" tendency, i.e., a capacity to behave as if they were someone else. Silber suggests that these tendencies are useful in helping to foster more permanent identification with a new parental figure, mediated by the relationship with the therapist. These new identifications gradually come to replace the diffuse and uncertain identity the patient brought to therapy.

Homework and Other Techniques

Homework—that is, the assignment of psychological, behavioral, or interpersonal tasks to be accomplished between sessions—is a commonly recommended technique, especially in substitutive treatment and therapy emphasizing proximal causative conditions. Homework, which may assume many forms, has been usually thought of an an adjunct to therapy to be used according to the requirements of each individual case, but Steiner (1971) considers it an integral, formal component of his approach. He defines it as "assigned work that the patient does between therapy meetings toward the fulfillment of his contract." Sometimes homework is self-assigned, a good prognostic sign. Kinds of homework that Steiner recommends assigning include (1) overcoming social anxiety by systematic desensitization, (2) asserting oneself with significant others, (3) writing an essay countering one's negative self-image, and (4) structuring time by following a tight schedule of activities, having fun, or looking up old friends. A particular assignment is determined by material the patient brings up. Steiner makes the point that the therapist should check to see if it was done; if not, it should be reassigned and the patient challenged if he again fails to accomplish it. Homework that is too difficult should be avoided because failure may be a negative effect on treatment. While assignments should not be markedly below what a patient is capable of, it is better from a learning standpoint to have them too "easy" than too "hard."

Overcoming social anxiety and other phobias by systematic desensitization simply means that the individual is given a series of tasks, graded from "easier" to "more difficult" in terms of achieving a desired goal. As each easier task is accomplished, and consolidated by repetition, the patient is assigned the next more difficult task. Steiner gives the example of a shy patient given the following series: asking the time of day on a busy street, then asking for directions, smiling at people in the street, complimenting someone on his appearance, and making small talk.

Miscellaneous types of homework have been recommended by several therapists. Strecker (1951) finds it useful to suggest that the patient make a schedule each night of the next day's activities, with special attention being paid to times when the temptation to drink is aroused. Departures from the schedule are discussed by patient and therapist. Scott (1961) favors an approach that blends uncovering with a focus on immediate reality situations. One technique of maintaining a focus on daily situations is to assign homework, the success or failure of which is discussed in therapy along with the patient's feelings about it. Like Steiner, Scott makes interpersonally relevant assignments, including acceptance of blame when it is realistic, being more charitable to one's spouse, or controlling anger at work. Behavioral successes may be far more important in enhancing a patient's positive image of himself than self-understanding without behavioral feedback. Sheldon et al. (1974) also stress the importance of specifically assigned homework. Less specific "assignments" include suggestions that during the week the patient think about a theme that has emerged in the course of therapy. Therapists often employ such a device with the patient who finds it difficult to stay on a topic. Sometimes it is useful to suggest that the patient record his dreams and the thoughts and feelings that he has about them.

Among other techniques that may be used during the middle phase of therapy are role-playing, "armchair" problem solving (Sheldon et al., 1974), reversal of roles, and "protection" (Steiner, 1971). Most of these techniques are best suited to substitutive therapy and causative therapy focusing on current reality, and many of them may be used in connection with the assignment of homework, serving as prelude and practice for trials outside therapy.

While role-playing is ideally used in group settings, it also has its uses in individual therapy. Real or relevant artificial situations may be employed in which the patient plays both himself and the role of the other (e.g., the spouse, parent, child, boss) or in which the therapist assumes the role of the patient or the other figure. Stress is placed on what is going on cognitively and affectively in the mind of each player. The situation may be role-played several times with alternate outcomes and with discussion afterwards about the merits of each alternative. Armchair problem solving bears some similarity to role-playing in that it focuses on alternative ways of dealing with interpersonal situations that are problematic for the patient; it differs in that problems are discussed by the patient rather than played out. A verbal problem is presented to the patient, who suggests several ways of handling it. As in role-playing the alternatives are discussed, and a clear series of alternatives is left with the patient. Some therapists continue this technique with a variety of different but related problems, while others attempt to help the patient to see a guiding principle in the selection of the "best" alternative that he may apply to diverse situations as they occur.

Reversal of patient–therapist roles is helpful when the patient feels the therapist has let him down in some way or when the therapist is aware that the patient is having difficulty looking at a topic of importance. In the first instance, the therapist asks the patient how he would deal with the situation in question and suggests that the patient act as the therapist and he, the therapist, act as the patient. Sometimes patient and therapist switch seats. The therapist may play out the patient's feelings of disappointment in him, along with the consequent anger, while the patient may alternatively play the "good" therapist and the "bad" therapist. Role reversal tends to give the patient perspective on himself in relation to the therapist, to bring up new material, and to permit the therapist to introduce indirectly feelings that the patient is resisting. When the patient is having difficulty in achieving a deeper awareness of himself, role reversal is often helpful because it permits the therapist-as-patient to bring up threatening material in a way that the patient is able to accept.

Steiner (1971) gives the name "protection" to the commonly needed and commonly used technique of supporting the patient in times of psychic crisis brought about by taking new emotional or interpersonal steps. For example, a patient may have always been able to express love only when certain conditions were met, these conditions magically meaning that he wouldn't be rejected by the loved one. He reaches a point in therapy where he is prepared to love unconditionally, but upon trying it out, he becomes anxious and feels abandoned, overwhelmed, and lost. The therapist must be available to the patient when such crises occur. Ordinarily, the support, reassurance, and understanding of the therapist is all that is required to help the patient through the crisis; most often the crisis is brought to the therapist's attention by a telephone call and may be handled then. Occasionally a special session is required, and this is nearly always sufficient. When the therapist is away he should leave the patient the telephone number of another therapist to contact if a crisis should occur (he rarely does, for the "protection" implied in the therapist's making alternate arrangements is sufficient to permit the patient to postpone the crisis until the therapist returns or to handle it on his own).

Drinking Episodes

The most critical juncture during the middle phase of treatment typically occurs when the patient has a drinking episode; the manner in which the episode is handled often determines whether treatment will be successful or not. Generally speaking, the drinking episode(s) is an extreme but hardly unusual way of testing the relationship with the therapist: "Will he still love me if I drink?" Secondarily, it may represent the patient's anger and hostility toward the therapist, usually occurring in a context where the patient feels the therapist has let him down or unjustly attacked him. The episode is critical, not because of the return to drinking, but because it may result in a premature discontinuance of treatment. For the patient who has idealized the therapist, the guilt and shame of letting him down by drinking is so great that the patient cannot face him and does not return. For the therapist who is proud of his "success," the anger and disappointment about the patient's drinking may result in the therapist pushing the patient out of treatment in more or less subtle fashion. Both of these are essentially relationship issues (discussed in the section so titled) that are readily avoidable by the therapist's anticipation that the relationship will be tested in many ways, including drinking, and structuring therapy to take this into account from the beginning. As indicated earlier, therapists who at the beginning specifically make nondrinking a condition of therapy are inevitably asked testing questions about it, a common one being, "What happens if I do drink?" If the question is asked as a point of information, the therapist, after exploring what the patient thinks might happen, may explain that it would become a topic for discussion and they would look at it together. If the question is a challenge, the therapist must first explore and help to resolve what lies behind it and then give a similar explanation. The basic idea is to convey to the patient that a drinking episode doesn't automatically mean the end of therapy, but is an issue to be understood and resolved by patient and therapist working together.

Some therapists recommend that the patient should be made explicitly aware from the beginning of treatment that he may have a relapse but that if he does, treatment will not be discontinued. The difficulty with this policy is that it doesn't take into account the patient who will not have a relapse; furthermore, it can easily be interpreted by the patient to indicate that the therapist condones drinking. The alcoholic may see this as a lack of the firmness he requires from a person before he can trust him. In such an instance, therapy may founder.

If the stage has been properly set at the beginning of treatment, drinking episodes can be readily dealt with in treatment as another part of the relationship between the patient and therapist. This will usually result in greater awareness and understanding on the patient's part of his interpersonal and intrapersonal reasons for drinking and engaging in other self-destructive behaviors. Steiner (1971) finds that such episodes have a beneficial effect and also are useful prognostic signs. If the episode is as long and severe as earlier ones, then the patient is typically not progressing in therapy, despite compliance with it. For most patients, however, drinking episodes during treatment are shorter and less severe then previous episodes and give them the opportunity of contrasting drinking with nondrinking feelings, behavior, and sense of self in a general context of increased control and self-consciousness. This phenomenon is commonly found in general psychotherapeutic practice, where during the course of treatment patients will repeatedly but less intensely "touch base," as it were, with regressive behaviors they have all but discarded.

LATE PHASE AND TERMINATION

Characteristics of the late phase of therapy and criteria for deciding when treatment is successful and should be terminated have been barely touched upon in the literature on individual psychotherapy with alcoholics. Later stages of treatment involve consolidation of gains made in the middle phase, the expansion and generalization of new behavior and coping pattern to diverse situations, and preparation of the patient for termination. Some of the criteria for deciding when the treatment has reached this indeterminate late stage are that the patient (1) reports spontaneous engagement in new activities, interests, and behavior; (2) handles unique, potentially conflictual situations in an adult and self-satisfying manner; (3) accepts setbacks without becoming anxious or depressed, or without acting-out; (4) knows and experiences feelings as they occur; (5) when conflicted, examines and works through the conflict himself or in a nondefensive way with the therapist; and (6) shows a growing desire to try things on his own and be independent of the therapist. To expect these criteria to manifest themselves as baldly as this listing indicates would be unrealistic; the therapist must be alert to signs and cues that indicate their presence and he may explore these signals with the patient as they occur. Nor need a patient meet all these criteria before therapy can be closed as a success. Because the criteria are interrelated, a patient showing solid evidence of any one or two is likely to possess the others nascently. Finally, it is understood that these criteria can only genuinely be met in a therapeutic context where major behavioral and psychological gains have already been made during the middle phase of treatment. In other words, meeting the criteria for deciding upon terminating therapy assumes that treatment has been going on for some time.

The length of treatment has not received much specific attention in the literature, but it appears generally accepted that, to be successful, therapy must continue for at least a year. Steiner (1971), for example, recommends that the contract made at the beginning of treatment be made for a year of therapy. As early as 1937, Fleming stated that the first six months was only a prelude to a lengthy period of substitutive rehabilitation. Optimal length is probably somewhere between one and two years. The one-year minimum holds for substitutive and both forms of causative treatment, with longer periods being more appropriate for treatment emphasizing distal causative conditions. A year minimum is indicated because it is during the first year that the initial glow of sobriety wears off and the patient is apt to start drinking again. Further, it takes time to work through some of the defenses that prompt the patient to believe that he can resume drinking again without harm. Finally, the patient needs considerable time to try out and consolidate substitutive coping mechanisms and alternative life-styles.

Length of treatment, then, may be used as a very general guide to the late phase and a temporal framework within which to begin to consider termination. When any of the criteria listed above are in evidence some time before the completion of a year of treatment, the therapist must view them with suspicion for they may reflect resistance rather than arrival at a new level of mental health. The dangers of early termination are well illustrated in case material presented by Peltenburg (1956). Sometimes, flights into health are permanent, but they are more often short-lived. When a patient discontinues treatment because he believes himself "cured," the therapist should make every effort to maintain continuing contact, even if minimal, with the patient so that it will be easier for him to resume treatment should he subsequently desire it.

The late phase of successful therapy is generally an easy time for the therapist. The patient does most of the work himself, in a sense, acting the role of both therapist and patient, with the therapist acting as an observer who may be occasionally helpful. The therapist's job becomes more difficult once the subject of termination has been broached and a decision has been made. It is ideal, of course, for the patient himself to initiate termination, assuming that termination seems appropriate at the time. Many patients bring up the possibility of ending treatment indirectly by speaking, for instance, of how good it feels to be doing things on their own that they've never done before. The therapist may respond by agreeing with the patient and making a general comment about how rewarding it is to feel the master of one's own fate. The patient may or may not pick up on this. If he does, then the topic of termination can be tentatively introduced. If not, the patient will ordinarily bring up similar perceptions in succeeding sessions; it may then devolve on the therapist to suggest to the patient that he seems to be thinking about when therapy should end. Other patients, quite aware of the strides they have made and confident of themselves, sense that their changed status implies the ending of therapy and panic at the thought of being on their own. These patients, rather than hinting at ending treatment, suddenly regress to previous self-destructive behaviors in a striking fashion. The therapist in this situation may be quite direct with the patient, interpreting his fantasies of abandonment while reassuring him. Regressive behavior will ordinarily subside, but the subject of termination has been brought into the open, albeit in dramatic fashion. The therapist must convey to the patient the notion that an episode of regressive behavior, while signaling distress, does not invalidate the genuineness and stability of gains the patient has made.

Once termination becomes a public topic between patient and therapist, they can make a plan for phasing-out, with the mutual understanding that it is flexible and can be changed as needed. Typically, frequency of sessions will be reduced to every other week for a while, then to once a month, and then to an on-call basis. The door is never closed. The therapist makes it clear that he welcomes seeing the patient from time to time and that he is always there should the patient need him. When a plan has been decided upon, it is not unusual for patients, even those who have themselves introduced the subject of termination, to engage in old behaviors. These are interpreted to the patient, and if it seems indicated the plan is revised but not discarded. Some patients, usually those with a history of traumatic abandonment by a parental figure, will terminate therapy abruptly, although not necessarily prematurely. The process of gradual termination revives too many intensely painful feelings for the patient to go through with it. Such abrupt terminations, in a context of solid therapeutic gains, are probably best for these patients and not necessarily antitherapeutic.

Earlier, the dangers of early termination were mentioned. The continuance of therapy when it has outlived its usefulness is equally hazardous and represents an uneconomic use of the therapist's and patient's time. Considering the strength of many alcoholics' need for dependency attachments, it is not surprising that this is a frequent occurrence. Often, the patient's needs find a ready response in the therapist's fantasy of himself as nonexpendable or as turning out perfect therapeutic products. It is a temptation to many therapists to prolong therapy after a patient has had an episode of regressive behavior as part of a panic response to termination. Unless there are definite indications of serious unfinished business, this is generally an unwise course to follow and may even encourage the patient to "cry wolf!" each time the ending of therapy is in sight. Therapists who emphasize distal causative conditions often permit a patient to become obsessively involved in early childhood memories, fantasies, and dreams. This tends to lengthen therapy with no corresponding increment in attaining treatment goals and should be avoided. Silber's injunctions (1974) in this regard are useful. Another element that contributes to the prolongation of therapy is an inattention to setting and resetting goals in the early stages of treatment, thereby providing few points of reference for deciding when treatment should end. While it is true that some patients may have to be carried supportively for many years, or even for life, this can be done on an episodic basis (e.g., once a month) with visits of five to 15 minutes. In general, when treatment has been going on for two years or more without much in the way of observable gains in the past few months, it is wise for the therapist to review the entire course of therapy and to take initiative with the patient about termination.

RELATIONSHIP ISSUES

Considerable attention has been devoted in the literature to relationship issues in therapy with alcoholics, presumably because the quality and nature of the relationship is thought to be a major factor mediating high rates of early dropout and antitherapeutic discontinuance of therapy, both of which have been repeatedly demonstrated to characterize alcoholism treatment facilities. Early explanations which held that the alcoholic is typically "unmotivated" or that the helping professional is hostile and rejecting have given way to a greater emphasis on typical psychodynamics that patient and therapist respectively bring to the therapeutic situation and how these may interact to produce a treatment failure. Much of this work has been conducted from a psychoanalytic perspective in which patient dynamics which affect the relationship are termed *transference* and therapist dynamics which affect the relationship are termed *countertransference*. In a general sense both processes refer to the transfer of early patterns of relating to parental figures and significant others to the contemporary therapeutic relationship. Both processes distort the current relationship because they attribute to the other characteristics he does not possess. In the following paragraphs, typical patient and therapist dynamics important for the fate of the therapeutic relationship are examined, along with techniques for reducing incidence of failed relationships. However, it should be noted that factors other than the psychological aspects of interpersonal relations may explain high dropout rates. These include organizational imbalances and poor coordinative mechanisms among agencies such that patients "get lost" in the system; interlocking social systems which may be viewed as depending on failures in treatment to keep the systems going and so encouraging them (see, e.g., Wiseman, 1970, for an astute interactional analysis of hospital, jail, and skid row as interdependent systems); and pejorative social labeling of alcoholics which heightens patients' resistance to identifying themselves as having problems with alcohol. With regard to the latter, Garitano and Ronall (1974) distinguish individual from cultural life-style, arguing that cultural life-style, i.e., living out a social definition of alcoholism, must be addressed in treatment along with the patient's individual life-style, i.e., the psychodynamics that motivate excessive drinking.

The Patient

Alcoholics, in common with other persons having behavioral and impulse disorders, display a variety of relational styles which many therapists view as countertherapeutic, manipulative, and hard to work with in a therapeutic setting. It may be difficult for the therapist intuitively to penetrate this facade to sense the fear, helplessness, and inadequacy which he has been instructed are certainly there. It is, however, important to remember that the patient learned these modes of reacting to new, potentially significant interactions in early childhood in order to ward off overwhelming feelings of abandonment and psychic pain experienced in relation to parental figures. He is doubly threatened when he comes into treatment: first, by the possibility of giving up alcohol and the associated removal of one barrier to getting close to painful feelings, and second, by the possibility of developing an emotionally significant relationship with all the attendant dangers of reactivating early childhood relational wishes and fears. The therapist who appreciates these threats is less apt to be put off when the patient initially presents himself strongly on the defensive.

Any clinician who has worked with alcoholics is aware of the many types of alcoholics who present themselves. These include the anxious, passive, nonverbal patient who speaks only when spoken to and volunteers virtually nothing; the glib, often rambling, sometimes seductive patient, full of anecdotes, who never stops to examine himself; the patient who borrows self-esteem by association, impressing others with his friendly, working relations with prestigious figures, often alcoholism professionals; the hostile patient who challenges the therapist's competence and his ability to resolve a problem on which no one has made any headway, a person gifted in provocation and brinksmanship; the injustice collector who finds all blameworthy but himself; and the dependent patient who puts himself completely at the disposal of the therapist whose job it is to "cure" him, a patient talented in prolonging the hour. Scott (1961, 1963), as noted earlier, makes suggestions for handling some of these styles—notably the passive and projective patients—designed to get beyond defensiveness and to advance self-awareness. Steiner (1969, 1971) emphasizes the challenging quality some alcoholics bring to treatment in his discussion of the "drunk and proud of it" game, and several of these interpersonal styles and their general therapeutic management have been discussed by Blane (1968).

Certainly a steady diet of patients with only these relational styles would try the soul of a martyr, but this need not be the case. First, many alcoholics present themselves initially in quite a different fashion, seeking help and guidance in ways that are more compatible with many therapists' notions of the "good" patient. This is little talked about, probably because it presents no problem or conflict to the therapist, but also because it is incongruent with stereotypes about alcoholics that many clinicians possess. Second, when the therapist initially focuses on the defensive nature of relational styles and addresses himself to the fears that underlie them, as Silber (1974) recommends, these thorny modes of relating often subside or even disappear. Some clinicians (e.g., Chafetz, 1959) suggest that one of the values of a team approach to alcoholics lies in the diffusion of the target for difficult patient behaviors, so that no one staff member is the sole recipient. In diversified health settings, it is often the policy that only a portion of a therapist's caseload be devoted to alcoholic patients. This has obvious advantages in broadening the clinician's experience and will ordinarily have a salutary effect on his work with alcoholics. With the recent increase in the number of independent alcoholism treatment facilities, this policy becomes less workable, but nonetheless it should be seriously considered. Even where policy does not permit it, it appears that team approaches, which are commonly used in alcoholism treatment facilities, diffuse patient–staff relationships and that strong emphasis on treating spouses, children, and other relatives of alcoholics serves the same function as a diversified caseload in a multipurpose setting.

Another aspect of transference that has been noted in the literature to characterize the initial phase and much of the middle phase of therapy is the alcoholic's tendency to perceive the therapist as an omnipotent figure (Becker and Israel, 1961; Scott, 1961; Silber, 1959, 1967, 1974). Scott (1961) recommends confronting the patient when he shows evidence of expecting the therapist to effect a magical cure by pointing out that he, the therapist, can help the patient, but that the responsibility for change is the patient's; a similar stance is generally adopted by Steiner (1971). Becker and Israel (1961) reflect Scott's belief that the therapist's omnipotence in the patient's eye is an obstacle to therapeutic progress, but at the same time point out its usefulness in consolidating a therapeutic alliance in the early stages of treatment. Silber (1959, 1967, 1974), on the other hand, recommends actively utilizing the patient's fantasies of the therapist's power and strength not only to consolidate the relationship but to further the aims of therapy. Thus, he recommends the use of early confrontations about preconscious fears in order to enhance the therapist's all-knowing qualities in the patient's mind. Once a firm relationship is established, Silber recommends a shift to a more psychoeducative role on the therapist's part, but he does not suggest that the patient's fantasies of the therapist's omnipotence be interpreted during the course of treatment. Rather the therapist remains a strong, consistent, knowledgeable figure, serving as an identification model for the patient.

The Therapist

Countertransference phenomena have been extensively examined in the literature, particularly by Moore (1961, 1965), Selzer (1957), and Silber (1959, 1967). Within a transactional analysis context, Steiner (1971) makes valuable suggestions for handling countertransference responses. Blane (1968) has contributed a social and psychological analysis of professional reactions to alcoholics in treatment. The importance of the therapist's attitude toward the patient has nowhere been more stressed than in the literature on alcoholic clients; as we noted earlier, the attitude assumed by the therapist in the initial phase of treatment is critical for the continuance of treatment. An open, warm, interested, firm, but nonjudgmental stance is a prerequisite to the establishment of a genuine working alliance between patient and therapist. In this section, the interest resides in reactions aroused in the therapist as the interaction progresses over time. A number of these reactions may be identified, including hostility, overpermissiveness, inability to see the importance of the parental role in the patient's life, and subtle encouragement of the patient to act upon the therapist's own forbidden wishes. All these have their ultimate genesis in hostility aroused by the therapist's personalizations of the patient's situation.

Selzer (1957) presents examples of both overt–conscious and covert–unconscious hostility toward alcoholic patients by professionals. In particular, he views unrealistic treatment goals—for example, immediate demands for abstinence—and a preoccupation with whether the patient is drinking as evidences of unconscious hostility. He, like Moore (1961, 1965), traces the source of the hostility to the therapist's unwitting envy of the pleasurable aspects of excessive drinking. Moore further observes that therapists frequently sense the anger they hold toward the patient and find it incongruent with the socially valued ideal of the accepting therapist. The arousal of anxiety is defended against "by establishing a reaction formation in the form of an overly indulgent and permissive attitude toward the alcoholic" (Moore, 1961). He believes this attitude "is destructive of the patient's chance of recovery as it impairs his reality testing and encourages denial of the severity of the drinking problem." Overpermissiveness is characterized by being overly kind, granting the patient any request he makes, and aligning oneself with the patient, including the patient's projective, injustice-collecting system.

How can these kinds of countertransference phenomena be handled? Knowledge that they are common clinical experiences should of course be part of the basic training of any therapist, whether he be a psychiatrist, psychologist, social worker, rehabilitation counselor, or alcoholism counselor. This is a necessary component of training and a major way of sensitizing beginners to what they may expect. A most effective way of dealing with countertransference phenomena is on a case-by-case basis according to an apprenticeship model of supervision, and another is by means of group processes, whereby therapists meet for the sole purpose of exploring their attitudes toward alcohol, problem drinking, and alcoholism and determining how these attitudes may be linked to their therapeutic activity.

Much of Silber's residency training program (1959, 1967) is devoted to modifying attitudes residents bring with them and to dealing specifically with countertransference reactions as they occur in individual cases. He finds that the alcoholic's masochistic provocation designed to get the therapist angry at him and thus reduce his guilt constitutes a critical juncture in therapy. He handles this in supervision by helping the therapist to an intellectual understanding that alcoholics often have strong wishes to be punished and are skilled in stimulating aggressive and sadistic wishes in others. In provocative situations, the therapist points out that the patient seems to want to be punished and is perhaps confusing thoughts and actions. Another frequent and therapeutically important manifestation of countertransference resides in the manner in which the therapist handles the patient's early parental situation. Silber (1967) believes that the therapist's need to deny the pathology of the parent can affect the whole course of treatment unless it is worked out in supervision. As indicated previously, alcoholic patients often report incidents of bizarre parental behavior without any realization of its strangeness, and in supervision therapists often recite the incidents with an equal lack of awareness. Upon questioning, therapists typically will find "reasons" to make the parental behavior appear plausible because they identify with the patient's need to preserve the integrity of the pathological parent. According to Silber (1967):

> The therapist identifies with his patient in this instance and struggles not to see the irrational parent. At the same time, some of the doctors start to attack the patient rather vigorously at this point. The therapist "prefers" to see the patient as an external threat rather than face the internal threat stirred up by anxiety in relation to identification with the patient. It would seem then that the pathological parent stirs up early childhood anxieties in the therapist (fear of an overpowering, irrational parent or fantasies of this nature). The therapist reacts by attacking the patient, and is unable to interpret to the patient the manifestations of his pathological parental environment.

Therapists whose own wishes for infantile indulgence are strong may obtain indirect but painless (for them) gratification by subtly encouraging the alcoholic to continue his acting-out behavior. This may be suspected when a therapist seems overly interested in the patient's recital of impulsive, thoughtless escapades or behavior during drunken debauches, when he actively elicits such memories, or when he rewards or reinforces the pleasurable aspects of these memories by enjoying or condoning them. The therapist who identifies with a patient's impulses can be of little help in distinguishing thoughts from actions and in strengthening the patient's control functions. Again, such tendencies are best resolved in supervision or through small-group discussion.

Steiner (1971) examines countertransference within a transactional analysis framework, using its vocabulary. Typical roles the therapist should avoid are those of persecutor, rescuer, and patsy. By virtue of their training, therapists are rarely openly persecutive, but Steiner gives an example of the therapist who prolongs treatment when the patient has indeed recovered. A rescuer is a therapist who becomes overinvolved emotionally with his patients and who finds it necessary to maintain an image of himself as an omnipotent rescuer. Such overinvolvement, if unchecked, can result in damage not only to the patient but to the therapist as well. The role of patsy cannot be understood as countertransference per se, but rather refers to technical weaknesses on the part of the therapist, such as failure to blend reality with psychodynamic understanding, not checking to see if confrontations are accepted, or assuming automatically that what a person says about his behavior actually reflects his behavior. Steiner also speaks of the gallows transaction in which the therapist approves of or condones self-destructive behavior; this is one of the criteria for identifying the therapist who attempts to discharge his own impulses through the patient.

One area with implications for transference and countertransference and, indeed, the general conduct of psychotherapy with alcoholic patients is the sex of therapist and patient. Virtually nothing has been written about differences in treating men and women alcoholics and absolutely no attention has been paid to same-sex or cross-sex combinations of patient and therapist. Parenthetically, a fairly large number of published case vignettes on therapy describe women alcoholics, even though the ratio of male to female alcoholics in treatment is typically 5:1 or higher. It may be speculatively presumed that women therapists find it easier than men to accept the patient's infantile self-indulgence and his provocations to aggression, and this would probably be the case with patients of both sexes. Men therapists find it more difficult to accept provocation by male patients, not only because of the dynamics discussed above but also because such provocations are often phrased in terms of challenges to engage in machismo competition, to which males in our society are particularly prone to respond. With women patients, men are apt to accord provocation a higher psychogenetic quality than it usually has in reality and interpret it in sexual terms. Alternatively, they may not recognize its dynamic importance and dismiss it as "playing games"; the dismissal is good technical practice, but the failure to recognize and address the masochistic dynamic may adversely affect treatment. Women therapists, on the other hand, find it more difficult than men to be firm and set limits, and this may be of special import in therapy with impulse-ridden patients like alcoholics.

Male patients who tend to blame their wives for much of their difficulties will attempt to recruit the male therapist as an ally in reinforcing this perception. With a woman therapist, the patient may attempt a similar kind of alignment, introducing a seductive, sexual element, or he may polarize therapist and wife into a good–bad dichotomy. Female patients who feel oppressed by their husbands are likely to respond to a male therapist in a manner similar to the male patient with a female therapist, and will seek alignment with a female therapist.

It should be noted that the above ideas are partly speculative and partly based on supervision of beginning therapists who worked with alcoholics and other types of patients. The area has received no systematic attention in the clinical or research literature on psychotherapy with alcoholics and is a fertile field for inquiry that will illuminate and, one hopes, improve psychotherapeutic practice.

INDIVIDUAL PSYCHOTHERAPY IN RELATION TO OTHER TREATMENT METHODS

In the foregoing discussion of aspects of individual psychotherapy, the focus has necessarily, but arbitrarily, been limited to transactions between patient and therapist. It would be not simply misleading but incorrect to leave an impression that therapy of alcoholics is limited to these two participants or that psychotherapy and other forms of treatment are mutually exclusive. The first point refers to the extremely important role that family members may play in facilitating individual therapy and the second to the multiple interrelations between individual psychotherapy and other forms of treatment.

Therapists are unanimous about the importance of the participation of family members in rehabilitation of alcoholic clients. In intact nuclear families, the nonalcoholic marital partner is the most central figure and therapeutic involvement may include early contributions to the alcoholic's and the family's history with occasional contacts thereafter to see how things are going with the patient and the family from the spouse's point of view. Or—depending on the philosophy of the clinical setting and the role the spouse plays in maintaining the patient's problems—the marriage partner may become actively engaged in treatment. Parents and children may also actively participate in treatment. With the recent emphasis on the importance of family dynamics in symptom-formation and maintenance and the family as an interdependent system, many clinical settings now offer couples' therapy, couples groups, and family therapy. With regard to individual therapy, family members are least encouraged to become actively involved in therapy that focuses on distal causative conditions, while there is apt to be considerably greater active participation in therapy which emphasizes proximal causative conditions and in substitutive treatment. This is the case because the latter two modes of therapy are devoted to a careful scrutiny and reworking of current reality, while distal causative therapy tends more to view current reality manifestations as distortions that reflect early relationships with parental figures and that will be modified as the patient comes to differentiate himself from parental bonds. This is not to say that distal therapy is unconcerned with current reality but merely that it views the impetus for change as coming from within the patient, whereas the other therapies, particularly substitutive, see it as emanating more from the patient's interaction with his environment.

Individual psychotherapy may be conducted as the sole form of treatment, in series with other forms of treatment, or concurrently with other modalities. Some of the more common combinations will be briefly described, along with a discussion of the considerations they raise for the conduct of individual therapy. When individual therapy is conducted as the sole form of treatment, as in private practice or in an outpatient setting devoted partly or completely to the treatment of alcohol problems, the therapist is wise to have formed relationships with emergency facilities for immediate inpatient care when it may be required. Furthermore, he should have established cooperative working relationships with the alcoholism treatment network in his area so that he may refer patients unsuited to individual psychotherapy to the most appropriate resources in the community. Personal relationships with one or more interested members of A.A. may be particularly useful adjunctively.

A frequent treatment pattern is referral for individual psychotherapy following a short-term hospitalization for detoxification, ordinarily within the same institution. It is important that personnel in the detoxification and outpatient units develop coordinative mechanisms that permit the initiation of the idea of continuation of treatment with the patient as early in his hospital stay as possible. Ideally, this allows the therapist to have two or three sessions with the patient in the outpatient setting prior to discharge. Since many patients who are being detoxified view it as confinement rather than treatment, an early task for the therapist is to help the patient clarify his feelings about treatment in and out of the hospital. A variation of this series is when the patient comes to outpatient individual psychotherapy after detoxification followed by a period of residential care (usually one to three months, with the shorter time being more common). The ideal situation here is for the patient to have been seen concurrently with residential treatment by the same therapist who will continue to see him upon discharge. This unfortunately occurs only rarely. The danger, of course, in continuing or instituting therapy after a period of residential treatment is that the patient at this time is ordinarily at the height of a euphoric state of "goodness" brought on by abstinence and self-discovery, and is often confident that he can make it on his own. Another way of putting it is to say that denial is then at its peak. This makes continuity of care critical. When the residential treatment facility is located in an institution geographically separated from the locale for postdischarge therapy, the problem is especially acute and the danger of losing the patient is great. The most careful coordinative and cooperative relationships, including a policy and associated mechanism for actively reaching out, are indicated. These may include sharing of staff, provision for the patient to make one or two visits to the outpatient facility prior to discharge, tapering down the intensity of residential care prior to discharge, and making it explicit to patients upon admission that residential treatment is only a first step toward successful attainment of treatment goals.

Individual psychotherapy provided on an inpatient or outpatient basis is often concurrent with other forms of treatment. Perhaps the most common concurrent treatments are group therapy, couples therapy, drug therapy, and A.A. The most important consideration here is that the treatments be integrated so that they are complementary rather than competitive. Such integration may be assured by structuring and open communication. In settings where staff members hold fixed positions about the relative value of particular treatment forms, it is probably best to avoid the use of concurrent treatment, although it may be a good idea to hold meetings of the staff for self-examination and resolution of differences that may exist. Agencies which have a policy of tailoring the treatment to the requirements of the individual will use substitutive, proximal causative, and distal causative therapy accordingly, and will recommend concurrent forms of treatment as appropriate to each case. This situation does not always obtain in practice, however, and many settings will favor one type of approach over the other. Common concurrent combinations are substitutive therapy with A.A. and proximal causative therapy with group therapy or couples therapy. Distal causative therapy, as noted above, is frequently used in isolation or after detoxification. There is, however, no necessary incompatibility between A.A. and proximal or distal causative therapy and in many settings they will be used concurrently. Usually different therapists are provided for each form of treatment, thus requiring a mechanism to ensure regular exchange of information about what is going on in the respective treatment modalities that might affect the others. At the beginning of treatment, the patient must be informed and give his consent to this exchange of information; while such an arrangement is usually suitable to the patient, it may be a source of difficulty later in therapy when highly sensitive material is being brought up. If it seems therapeutically warranted, the therapist may discuss the situation with the other treatment agents to obtain a temporary waiver, especially when the material may be presumed to be unrelated to what is occurring in the other modalities. While neither Steiner (1971) nor Scott (1963) discuss concurrent group and individual therapy, both address the question of introducing the patient to group forms of treatment and offer helpful suggestions for introducing the patient to a concurrent treatment modality.

SUMMARY

Individual psychotherapy, to be of maximum benefit, must meet the unique characteristics of each client and must be based on a thorough assessment and understanding of his total life situation, including his immediate needs, interpersonal relationships, social and drinking history, and personality strengths and weaknesses. Therapy may direct itself to conditions—either proximal or distal—which brought about the alcoholic problem, those secondary to the problem, or those following cessation of drinking. When therapy is viewed within these categories along with patient characteristics, the therapist's clinical orientation and the type of treatment setting, parochial differences about such factors as goals, confrontation, and drinking in relation to treatment largely disappear.

The initial phase of therapy is especially important in the treatment of alcoholic patients because of the high risk of early discontinuance. The active development of a positive relationship and an appreciation of the client's previous treatment experiences are critical to the future therapeutic course. Confrontation, setting goals, and making a therapeutic contract proceed according to the therapist's orientation and his understanding of the patient and his situation. Too quick an approach heightens defensiveness and drives the patient from treatment; too slow an approach undermines the patient's confidence in the therapist and results in a degenerating and unproductive treatment situation.

Once the initial phase of therapy is completed, which may take a short time or as much as several months, the middle phase of exploration, self-understanding, working through, and trying-out commences. While the process will vary according to whether treatment is essentially substitutive, focuses on proximal causative conditions, or on distal causative conditions, common techniques include listening, reflecting, confronting, and interpreting. Other techniques include the assignment of homework (interpersonal tasks, time structuring, systematic desensitization), role-playing, armchair problem solving, reversal of roles, and "protection." Drinking episodes during the middle phase are often positive occurrences when handled properly by the therapist, but can be critical for the continuance of therapy when they are not. The late phase of therapy, characterized by rapid personal growth and minimal therapist activity, may include transitory regressive behavior related to fears about termination. Criteria for assessing readiness for termination can aid the therapist to avoid premature termination or unnecessary prolongation of therapy.

Transference and countertransference issues are nowhere more apparent than in the treatment of alcoholic clients, and clinicians must become aware through training, supervision, and experience of the manner in which their emotional responses, if unchecked, can interact detrimentally with the alcoholic's dynamics. Psychotherapy does not occur in isolation but includes the participation of family members and other significant persons and is often conducted, in various combinations, with other forms of treatment.

ACKNOWLEDGMENTS

The preparation of this chapter was supported in part by grants AA00057 and AA00491 from the National Institute on Alcohol Abuse and Alcoholism.

REFERENCES

American Medical Association, 1968, "Manual on Alcoholism," American Medical Association, Washington, D.C.
Bailey, M. B., 1968, "Alcoholism and Family Casework; Theory and Practice," Community Council of Greater New York, New York.
Becker, G. S., and Israel, P., 1961, Integrated drug and psychotherapy in the treatment of alcoholism, *Q. J. Stud. Alcohol* 22:610.
Bellwood, L. R., 1975, Grief work in alcoholism treatment, *Alcohol Health and Research World*, Spring, pp. 8–11.
Blane, H. T., 1968, "The Personality of the Alcoholic," Harper & Row, New York.
Blane, H. T., and Meyers, W. R., 1964, Social class and establishment of treatment relations by alcoholics, *J. Clin. Psychol.* 20:287.
Blum, E. M., 1966, Psychoanalytic views of alcoholism; a review, *Q. J. Stud. Alcohol* 27:259.
Blum, E. M., and Blum, R. H., 1967, "Alcoholism; Modern Psychological Approaches to Treatment," Jossey-Bass, San Francisco.
Bowman, K. M., and Jellinek, E. M., 1941, Alcohol addiction and its treatment, *Q. J. Stud. Alcohol* 2:98.
Canter, F. M., 1969, The future of psychotherapy with alcoholics, in "The Future of Psychotherapy" (C. J. Frederick, ed.), pp. 253–296, Little, Brown, Boston.
Catanzaro, R. J., 1968, Basic principles of treatment, in "Alcoholism" (R. J. Catanzaro, ed.), pp. 83–89, Charles C Thomas, Springfield, Illinois.
Chafetz, M. E., 1959, Practical and theoretical considerations in the psychotherapy of alcoholism, *Q. J. Stud. Alcohol* 20:281.
Chafetz, M. E., Blane, H. T., and Hill, M. J., 1970, "Frontiers of Alcoholism," Science House, New York.
Clinebell, H. J., Jr., 1968a, Pastoral counseling of the alcoholic and his family, in "Alcoholism" (R. J. Catanzaro, ed.), pp. 189–207, Charles C Thomas, Springfield, Illinois.
Clinebell, H. J., Jr., 1968b, "Understanding and Counseling the Alcoholic," Abingdon Press, New York.
Curlee, J., 1971, Attitudes that facilitate or hinder the treatment of alcoholism, *Psychotherapy* 8:68.
Diethelm, O., 1936, "Treatment in Psychiatry," Macmillan, New York.
Fleming, R., 1937, The treatment of chronic alcoholism, *N. Engl. J. Med.* 217:779.
Garitano, W. W., and Ronall, R. E., 1974, Concept of life style in the treatment of alcoholism, *Int. J. Addict.* 9:585.
Gerard, D. L., Saenger, G., and Wile, R., 1962, The abstinent alcoholic, *AMA Arch. Gen. Psychiatry* 6:83.
Harford, T., Blane, H. T., and Chafetz, M. E., 1970, Language predictability and psychiatric interviews, *Psychol. Rep.* 31:725.
Horn, J. L., Wanberg, K. W., and Adams, G., 1974, Diagnosis of alcoholism; factors of drinking, background and current conditions in alcoholics, *Q. J. Stud. Alcohol* 35:147.

Knight, R. P., 1937a, The dynamics of treatment of chronic alcohol addiction, *Bull. Menninger Clin.* 1:233.
Knight, R. P., 1937b, The psychodynamics of chronic alcoholism, *J. Nerv. Ment. Dis.* 86:538.
Knight, R. P., 1938, The psychoanalytic treatment in a sanitarium of chronic addiction to alcohol, *J. Am. Med. Assoc.* 111:1443.
Margolis, M., Krystal, H., and Siegel, S., 1964, Psychotherapy with alcoholic offenders, *Q. J. Stud. Alcohol* 25:85.
Menninger, K. A., 1938, "Man Against Himself," Harcourt, Brace, New York.
Menninger, W. C., 1938, The treatment of chronic alcohol addiction, *Bull. Menninger Clin.* 2:101.
Mills, R. B., and Hetrick, E. S., 1963, Treating the unmotivated alcoholic, *Crime and Delinquency* 9:46.
Milmoe, S., Rosenthal, R., Blane, H. T., Chafetz, M. E., and Wolf, I., 1967, The doctor's voice: Postdictor of successful referral of alcoholic patients, *J. Abnorm. Psychol.* 72:78.
Moore, R. A., 1961, Reaction formation as a countertransference phenomenon in the treatment of alcoholism, *Q. J. Stud. Alcohol* 22:481.
Moore, R. A., 1965, Some countertransference reactions in the treatment of alcoholism, *Psychiatric Digest* 26:35.
Munt, J. S., 1960, Fear of dependency; a factor in casework with alcoholics, *Social Work* 5:27.
Myerson, D., 1953, An active therapeutic method of interrupting the dependency relationship of certain male alcoholics, *Q. J. Stud. Alcohol* 14:419.
Pattison, E. M., 1974, Rehabilitation of the chronic alcoholic, *in* "The Biology of Alcoholism" (B. Kissin and H. Begleiter, eds.) Vol. 3, pp. 587-658, Plenum, New York.
Peltenburg, C. M., 1956, Casework with the alcoholic patient, *Social Casework* 37:81.
Perry, S. L., Goldin, G. J., Stotsky, B. A., and Margolin, R. J., 1970, "The Rehabilitation of the Alcohol Dependent," D. C. Heath, Lexington.
Pfeffer, A. Z., 1958, "Alcoholism," Grune & Stratton, New York.
Root, L. E., 1968, Social casework with the alcoholic and his family, *in* "Alcoholism" (R. J. Catanzaro, ed.), pp. 208-222, Charles C. Thomas, Springfield, Illinois.
Sapir, J. V., 1953, Relationship factors in the treatment of the alcoholic, *Social Casework* 34:297.
Scott, E. M., 1961, The technique of psychotherapy with alcoholics, *Q. J. Stud. Alcohol* 22:69.
Scott, E. M., 1963, A suggested treatment plan for the hostile alcoholic, *Int. J. Group Psychother.* 13:93.
Selzer, M. L., 1957, Hostility as a barrier to therapy in alcoholism, *Psychiatr. Q.* 31:301.
Shea, J. E., 1954, Psychoanalytic therapy and alcoholism, *Q. J. Stud. Alcohol* 15:595.
Sheldon, R. B., Davis, H. G., and Kohorn, R. L., 1974, Individual counseling and therapy with the alcoholic abuser, *in* "Alcohol Abuse and Rehabilitation Approaches" (J. G. Cull and R. E. Hardy, eds.), pp. 137-154, Charles C Thomas, Springfield, Illinois.
Silber, A., 1959, Psychotherapy with alcoholics, *J. Nerv. Ment. Dis.* 129:477.
Silber, A., 1967, Psychodynamic therapy in alcoholism, *in* "Alcoholism" (R. Fox, ed.), pp. 145-151, Springer, New York.
Silber, A., 1970, An addendum to the technique of psychotherapy with alcoholics, *J. Nerv. Ment. Dis.* 150:423.
Silber, A., 1974, Rationale for the technique of psychotherapy with alcoholics, *International Journal of Psychoanalytic Psychotherapy* 3:28.
Solari, G. C., 1970, Psychotherapeutic methods in alcoholism, *in* "Alcohol and Alcoholism" (R. E. Popham, ed.), pp. 165-169, University of Toronto Press, Toronto.
Solomon, P., 1966, Psychiatric treatment of the alcoholic patient, *Int. Psychiatry Clin.* 3:159.
Steiner, C. M., 1969, The alcoholic game, *Q. J. Stud. Alcohol* 30:920.
Steiner, C., 1971, "Games Alcoholics Play: The Analysis of Life Scripts," Grove Press, New York.

Strecker, E. A., 1951, Psychotherapy in pathological drinking, *J. Am. Med. Assoc.* 147:813.
Tiebout, H. M., 1949, The act of surrender in the therapeutic process with special reference to alcoholism, *Q. J. Stud. Alcohol* 10:48.
Tiebout, H. M., 1953, Surrender versus compliance in therapy; with special reference to alcoholism, *Q. J. Stud. Alcohol* 14:58.
Wiseman, J. R., 1970, "Stations of the Lost," Prentice-Hall, Englewood Cliffs, New Jersey.

CHAPTER 4

Engaging the Alcoholic in Treatment and Keeping Him There

Frederick Baekeland and Lawrence K. Lundwall

Department of Psychiatry
Division of Alcoholism and Drug Dependence
State University of New York
Downstate Medical Center
Brooklyn, New York

INTRODUCTION

Although heroin addiction is more dramatic and newsworthy because of its stronger association with crimes against property, it is generally agreed among professionals in the field of alcoholism and drug dependence that alcoholism is by far the more extensive public health problem of the two. Four major stumbling blocks have always faced those who try to treat alcoholics: (1) The vast majority of alcoholics remain undetected and undiagnosed and do not receive treatment for their alcoholism, if they ever get it, until their condition is far advanced. (2) Once referred for treatment, a high percentage of alcoholics fail to negotiate successfully the jump from referring source to treatment facility. (3) Once in treatment, the alcoholic patient is likely to quickly drop out of it. (4) Among the variety of treatment approaches available, it is by no means clear which is most appropriate for a given patient.

DETECTION OF THE ALCOHOLIC

High-Risk Groups

It has recently been estimated that there are 9 million alcoholics in the United States (National Council on Alcoholism Fact Sheet, 1972a), and alcoholism accounts for almost a quarter of all first state mental hospital admissions (Cahn, 1970). In 1971, the last year about which complete information is available, 278,000 patients with a primary diagnosis of alcoholism were being treated in psychiatric facilities (Redick, 1973). (The number with a secondary diagnosis of alcoholism is not known.) Jones and Helrich (1972) estimated that about 70,000 alcoholics were being seen by private practitioners. Hence, only about 350,000 alcoholics receive medical and/or psychiatric treatment for alcoholism per se. On the other hand, A.A.'s current membership is about 250,000 (Alcoholics Anonymous, 1972), so that in terms of primary treatment it reaches almost twice as many alcoholics as the medical profession. Taken together, A.A. and medical and psychiatric treatment sources thus reach about 6.7 percent of the alcoholic population, for the most part briefly. Surely, this must be one of the most dismal failures to detect and treat a serious and widespread medical–psychiatric condition known to the field of public health. That this is not simply an American problem but rather an international one probably conditioned by prevalent attitudes about alcoholism is suggested by a recent study conducted in London by Edwards *et al.* (1973a). They reported that the likely ratio of cases in need of attention for their alcoholism to those in contact with an appropriate agency probably lay between 4:1 and 9:1.

Clearly, then, there has been a general failure to detect alcoholics. Their discovery can be facilitated if they are looked for in certain facilities, occupations, and medical and psychiatric categories where they tend to be concentrated, but are most often not labeled as such.

The internal medical complications of alcoholism are well known. Accordingly, their mortality rates are higher than those of nonalcoholics for a wide variety of internal medical conditions, such as cancer of the upper digestive and respiratory tracts, arteriosclerotic heart disease, pneumonia, cirrhosis of the liver, and gastric or duodenal ulcer (Schmidt and de Lint, 1972). The internist and general practitioner hence should be on the lookout for alcoholism among patients with cirrhosis or ulcers. Because of their poor nutrition, alcoholics have a higher than average risk of contracting tuberculosis. In fact, it has been estimated that as many as half the patients in tuberculosis hospitals are alcoholics (Cahn, 1970). Hence, the medical man should be on the lookout for alcoholism among his tuberculosis patients. Both he and the psychiatrist should also be aware that alcoholism is a common but often overlooked cause of impotence (Lemere, 1973). It should also be considered in the differential diagnosis of insomnia as a recent study of alcoholics starting outpatient treatment showed that about 35.7 percent of them suffered from moderate or severe sleep disturbance (Baekeland *et al.*, 1974a).

Alcoholics are also concentrated in general medical services of urban hospitals. Thus, in an urban ghetto hospital it was reported that 60 percent of male and 34 percent of female medical inpatients abused alcohol, while 18 percent of admissions to the medical service were for alcoholism-related conditions (Zimberg *et al.*, 1971). Similarly, six other studies of general medical wards in urban hospitals have shown that from 12 to 60 percent of the male patients surveyed were alcoholic (Barcha *et al.*, 1968; Green, 1965; McCusker *et al.*, 1971; Moore, 1971; Nolan, 1965; Pearson, 1962). The wide variation in the prevalence of alcoholism reported in these studies doubtless reflects population differences, but it may also be due to differences in detection techniques, which ranged from simple history taking to specially designed questionnaires or rating scales. It is interesting that one of these studies (Barcha *et al.*, 1968) found that one-third of the alcoholic patients seemed to be in remission. Despite the fact that 16.3 percent of the sample were thought to be alcoholic on the basis of a 17-item questionnaire, very few were given that diagnosis by the medical staff, and in the absence of DTs or obvious liver disease, very few of the charts contained an adequate drinking history. Similarly, in the McCusker *et al.* (1971) study at Harlem Hospital, in New York City, which reported a 60 percent alcoholism rate in male medical ward patients, less than half the alcoholics were diagnosed as such by the resident staff, who failed to detect only 28 percent of the severe cases but missed 92 percent of the moderate ones.

Alcoholism is also significantly associated with a variety of psychiatric disorders. Thus, in a general psychiatric facility which was diagnosis- and research-oriented, 37.5 percent of male admissions received a diagnosis of alcoholism (de Groot and Adamson, 1973). In particular, there is a clear association between alcoholism and both schizophrenia and drug abuse. In public psychiatric clinic or hospital populations, higher than expected rates of schizophrenia have been found among alcoholics and have been reported to range from 15 to 39 percent (Button, 1956a; Kalant and Hawkins, 1969; Panepinto *et al.*, 1970; Zimberg *et al.*, 1971; Zwerling, 1959). There is even a stronger association between drug abuse and alcoholism. Accordingly, in one general psychiatric hospital alcohol use and abuse was found to contriubte to one-fourth of the admissions, and there was a significant correlation between recent alcohol use and cocaine, sedative hypnotic, and antianxiety drug use (Crowley *et al.*, 1974). Heroin addicts have a high incidence of alcoholism. In four studies of addicts, this ranged from 20.5 to 30 percent (Anchersen, 1947; Kieholz and Battegay, 1967; Kolb, 1925; Retterstöl and Sund, 1965), with the wide variation in prevalence probably accounted for by the different times and countries in which they were conducted. Furthermore, as many as 20 percent of unselected heroin addicts entering a methadone program may become problem drinkers (Baden, 1971; Goldstein, 1972), while even in programs that screen out alcoholics 10 percent may subsequently be identified as alcoholics (Perkins and Bloch, 1970).

In recent years there has been growing evidence of an association between alcoholism and depression, both in men (Mayfield and Coleman, 1968) and women (Schuckit *et al.*, 1969; Winokur *et al.*, 1970, 1971). Indeed, on the basis of epidemiological evidence, it has recently been suggested that there may be at least three types of alcoholism, one of which is associated with a unipolar affective disorder (Schuckit *et al.*, 1969; Winokur *et al.*, 1971). Accordingly, those who commit or attempt to commit suicide have a higher than average risk of being alcoholics, According to a recent review, 6-21 percent of alcoholics commit suicide versus 1 percent of the general population, and in 11 studies of suicides, 21.7 percent were alcoholic (Goodwin, 1973). A recent study by Mayfield and Montgomery (1972) reported that 29 out of 34 patients who tried to shoot or stab themselves were alcoholic and that of these 29, 26 were intoxicated at the time of the suicide attempt. Hence, depression in a patient should raise the physician's index of suspicion that he may be dealing with an alcoholic. Indeed, Woodruff *et al.* (1973) found that the major difference between alcoholics who seek clinic psychiatric treatment and those who do not was that the former were depressed while the latter were not.

Alcoholics are also concentrated in courts, jails, and prisons. They contribute disproportionately to highway accidents and deaths (Brenner and Selzer, 1969; Selzer, 1969; Selzer and Ehrlich, 1967; Smart and Schmidt, 1967; Waller, 1965, 1968), and drunkenness is the most common basis for arrest in the United States. A large proportion of other arrests, such as those for disorderly conduct and vagrancy, involve drunkenness (Glaser and O'Leary, 1966). However, about 40 percent of arrests for public drunkenness involve either infrequent offenders or those not well known to the police or courts (Cahn, 1970). In one study of men with sentences of 30 days or more, the average arrestee had 16.5 prior arrests, of which 12.8 were for public intoxication (Pittman and Gordon, 1958). In another study, intoxication rates among those arrested amounted to 64 percent of those charged with carrying concealed weapons or committing knifings or shootings, 60 percent of those arrested for robbery, burglary, larceny, forgery, and auto theft, and 45 percent of those charged with rape (Shupe, 1954). Among prisoners at Sing Sing, 22 percent were drinking and 15 percent intoxicated at the time they committed their crimes (Banay, 1942). Along similar lines, Guze and co-workers (1963, 1972) and Goodwin *et al.* (1971) reported alcoholism in 43 percent of the felons they studied. However, according to a recent review, most crimes of violence, including murder, are committed by nonalcoholics (Goodwin, 1973).

Lest we overlook the obvious, alcoholics tend to be concentrated in bars, which satisfy the needs of lonely, homeless men whose needs are not being met by health and welfare agencies. Such alcoholics, alienated from their families, not part of the regular work force, on welfare and chronically depressed, find the bar their only chance for socialization (Dumont, 1967).

Alcoholics are likely to be found in certain occupations rather than others. Thus, bartenders, cooks and restaurant workers, longshoremen and stevedores, musicians, newspaper workers, housepainters, and policemen run a higher than average risk of alcoholism (Hitz, 1973). Furthermore, certain kinds of occupational risk factors have been found to characterize high-risk jobs in industry. These include absence of clear goals, freedom to set work hours, low structural visibility, work addiction, occupational obsolescence, novel work status, required on-the-job drinking, mutual benefits, reduction of work controls, and, most important of all, absence of supervision (Roman and Trice, 1970). It is not clear what percentage of industrial employees are alcoholic, but it is considered an important problem there, where it is a major cause of absenteeism. From 1 to 10 percent, with a mean of 3 percent, have been estimated to be alcoholic (Trice, 1959a). In industrial settings, early-stage deviant drinkers may be especially hard to detect. They can "dry out" for considerable periods of time, during which they usually show excellent job performances.

Summing up, patients with a higher than average risk of being alcoholic, and who should hence be routinely screened for alcoholism, are those with gastric or duodenal ulcers, cirrhosis, or tuberculosis, those on general medical wards of urban hospitals, admissions to general psychiatric hospitals, depressed patients and those who attempt suicide, schizophrenics, and heroin addicts. Nonpatients who should be screened for alcoholism include persons arrested for drunken driving or involved in automobile accidents, bartenders, cooks and restaurant workers, longshoremen and stevedores, musicians, newspaper workers, housepainters, policemen, and the habituees of bars.

Factors Hindering Identification

If alcoholism is so prevalent and, moreover, especially concentrated in institutional settings where medical and psychiatric expertise is so high, why, then, is failure to identify the alcoholic so widespread even in such settings? First of all, as those who deal with alcoholics know all too well, these patients are very apt either to deny their affliction or else to minimize the extent of their drinking (Chafetz et al., 1970; Hayman, 1966). Cahalan et al. (1969), in a survey study which depended on subjects' self-reports and defined heavy drinkers as those who drink five or more times a week or three or more drinks several times a week, found that while 12 percent of alcohol users were classified as heavy drinkers and 6 percent of them as problem drinkers, only 16 percent of them described themselves as heavy drinkers. Another survey study (Bailey *et al.,* 1966) reported that a quarter of those who admitted drinking problems in a first survey failed to do so in a second one, while of those first labeled as nonalcoholic, 8.5 percent now acknowledged a drinking problem. Similarly, in another study, 10 out of 35 presumed alcoholics failed to meet both quantity and frequency criteria for heavy drinking (Mulford and Miller, 1960). Along similar lines, Summers (1970), who interviewed 15 skid row alcoholics on admission and again two weeks later with a 14-item questionnaire, found that only 1 patient out of 15 gave the same answers as on admission, while 14 of the 15 changed 50 percent of their responses. More important, although only 4 out of 15 changed their estimate of the number of years of heavy drinking, in the second interview half gave a more severe history, half a better one than they had in the first place—a finding that suggests denial may not have been the only source of unreliability in this population. Confusion and memory impairment in the wake of acute alcohol withdrawal may have played a hand in the results. Certainly, a number of studies suggest variable impairment of brain function in hospitalized alcoholics. Thus, Weingartner *et al.* (1971) found a reversible memory impairment involving hold formation (storage) which disappeared with three weeks of abstinence. Reversible impairment of intellectual and cognitive function have also been reported by others (Jenssen *et al.,* 1962; Kish and Cheney, 1969; Tomsovic, 1968). It may be dangerous to generalize from Summers's (1970) small sample study for a second reason: She studied a special and atypical patient (the skid row alcoholic), who could be expected to be especially unreliable. A final indication of the potential unreliability of the alcoholic's drinking history is given by the observation of Schaefer *et al.* (1972) that in a group of patients rated on a multifactorial outcome measure, 50 percent were improved at one year in the opinion of the investigators as opposed to only 19.2 percent in the eyes of the patients' relatives. In contrast to all the preceding is the careful study of Guze *et al.* (1963), who were able to interview both subjects and their relatives in the case of 90 of 221 felons and detected among them 39 alcoholics, 26 percent of whose responses disagreed with those of their relatives with respect to a 17-item history. However, in only 6 percent of questions did the subject give a negative response and the relative a positive one, but since the sample was biased in favor of more socially stable felons (those with contactable relatives), the most severe alcoholics, those who would be totally rejected by their families and might be most unreliable in their answers, were likely missed. On the other hand, Guze and Goodwin (1972), who reinterviewed these felons eight to nine years later, found that the more extensive the original drinking problem, the more likely the subject was to be consistent in his answers at follow-up. They concluded that the original felon–relative inconsistency was largely a manifestation of mild or borderline alcoholism. The results of Guze and his collaborators notwithstanding, it appears that quantity–frequency items are quite unreliable in alcoholics, but that they may not be due entirely to denial.

Failure even to consider alcoholism in the differential diagnosis or to take the kind of history which would reveal it must rank as a major culprit in the widespread failure of physicians to detect alcoholism among their patients. Thus, Cahn (1970), in a survey of 28 university hospitals and clinics, found that the training of personnel in the treatment and handling of alcoholic patients was limited and highly restrictive. The average physician's ignorance about alcoholism is illustrated by a study done in a large urban hospital's emergency room, where it was found that the physicians there tended to think of alcoholism as a disorder of derelicts and were apt preferentially to diagnose it in lower-class patients while missing it in those who were married, employed, and not socially isolated (Chafetz et al., 1962). Alcoholics who were missed were also apt not to have medical or surgical complications of their alcoholism (Blane et al., 1963; Wolf et al., 1965). Although not specifically studied in other medical settings, such as private practitioners' offices, it would be surprising if the erroneous stereotype of the alcoholic as a socially deteriorated person did not impede efficient diagnosis there as well.

We will recall that two studies conducted on the medical wards of big city hospitals found that the diagnosis was missed in a high percentage of cases (Barcha et al., 1968; McCusker et al., 1971). Since they dealt with largely indigent populations, it seems that another factor besides ignorance was at work, namely, unwillingness to treat the alcoholic. The average physician's aversion to working with alcoholics has been amply documented. For example, general hospitals usually will not admit patients if they are drunk or have a hangover, while patients in impending DTs may be given medication and sent away or sent away with nothing at all, and among general practitioners, only the physical damage caused by alcoholism is treated (Glasscote et al., 1967). Similarly, Pittman and Sterne (1962) found that 40 percent of the hospitals they studied would not admit alcoholics with a primary diagnosis of alcoholism while a third of them would do so only under certain circumstances (DTs or, paradoxically, no signs of agitation). Many of the reasons they gave, which included statements that alcoholics are apt to sign out against medical advice, often disturb other patients, don't return for outpatient care, are repeatedly admitted and are combative, seem patently specious since they apply to so many other patients as well. Psychiatrists and psychologists, though perhaps a little more enlightened about alcoholism, are just as unwilling as internists and general practitioners to treat the alcoholic patient. Thus, 90 percent of psychiatrists surveyed in one study found alcoholics harder to treat than other patients, and 16 percent saw no patients with serious drinking problems. Among psychiatrists who treated them, alcoholics amounted to only about 5 percent of their caseload (Glasscote et al., 1967). Similarly, Hayman (1956), in a survey of members of the Southern California Psychiatric Association, also reported that only a low percentage of respondents were willing to treat alcoholics. Oddly enough, it is possible that nonpsychiatric physicians may be more willing to treat alcoholics (State of Tennessee Legislative Council, 1962).

Why are physicians so unwilling to treat alcoholics? A number of studies indicate that the answer to this question is not a simple one. Thus, Knox (1971), who studied VA psychiatrists and psychologists, found that they rejected the disease concept, preferred to characterize alcoholism as a behavior problem, symptom complex, or escape mechanism, only inconsistently advocated neuropsychiatric hospitalization, and considered treatment benefits very limited. Freed (1964) also found that psychiatric hospital staff were nonaccepting of alcoholism as an illness, and in another study it was found that the psychiatrists studied had reported negative feelings when working with alcoholics (Robinson and Podnos, 1966). Other investigators have found that 60–75 percent of psychiatrists dealing with alcoholics thought them unmotivated (Moore and Buchanan, 1966; Sterne and Pittman, 1965). Even the professional who specializes in the treatment of alcoholism tends to want only to treat motivated problem drinkers who will abstain, and he is unlikely to accept patients in legal trouble (Einstein et al., 1970). Knox's (1971) study suggests that most professionals tend to think of alcoholism as an incurable condition. On the other hand, while they underestimate the utility of medical treatment, they also may overestimate the efficacy of A.A., which has generated much more publicity for itself, this notwithstanding the fact that alcohol clinics seem to do better than A.A. with patients of equivalent socioeconomic status (Baekeland et al., 1974b). Above and beyond their acceptance of A.A.'s inflated claims of success, it is not surprising that physicians and other professionals tend to downgrade the efficacy of medical treatment in comparison with it. They have been found to be more pessimistic about alcoholism than those in other occupations, and almost half of them have a moralistic orientation toward alcoholism which varies inversely with professionalization (Mogar et al., 1969; Pittman and Sterne, 1962). Alcoholics Anonymous, on the other hand, has an evangelistic emphasis and stresses the willpower of the alcoholic. Several studies have related physicians' authoritarianism to their unwillingness to treat alcoholics. Thus, Mendelson et al. (1964) found that the more authoritarian doctor shuns an interpersonal treatment program for alcoholics and recommends custodial treatment, while Gray *et al.* (1969) reported that physicians high on authoritarianism both prefer not to, and do not, treat alcoholic patients as often as less authoritarian physicians. It is not only physicians in hospitals or office settings who tend to have negative biases against alcoholics and fail properly to diagnose alcoholism in their clients. Although alcoholics probably constitute a sizable percentage of the caseload of welfare workers, probation officers, and public health nurses, studies of family service, welfare, and mental health clinics show that such agencies tend not to accept diagnosed alcoholics as part of their case load (Cahn, 1970; Cumming, 1968; Pittman and Sterne, 1962). Pittman and Sterne (1962), in a study done in St. Louis, found that half of the public and voluntary agencies they looked at had less than 3 percent of alcoholics in their case load. Known alcoholics were ineligible in 19 percent and eligible in another 36 percent only on the condition that they were well motivated and/or got treatment elsewhere. Similarly, Cumming's (1968) data show that alcoholics are very rarely included among referrals to agencies. That this is not just due to laxness on the part of alcoholism treatment facilities is suggested by the finding of Cohen and Krause (1971), who studied a counseling agency and found that the staff was unwilling to work with identified alcoholic husbands although they would deal with their wives.

In summary, then, physicians, other professionals and agencies tend to be unwilling to deal with alcoholics. Some of the reasons for this apparently include misinformation about the nature of alcoholism, undue pessimism about its prognosis under medical treatment, and a moralistic orientation toward alcoholism.

What can be done to change professionals' negative attitudes toward alcoholics? The impact of large-scale community programs to change public opinion about any kind of psychiatric illness has rarely been scientifically studied. However, Cumming and Cumming's (1957) study of a large-scale attempt to change public attitudes toward patients with major psychiatric illness in a Canadian community suggests that they may be of limited value. Similarly, one attempt to change industrial supervisors' attitudes about alcoholism showed little success (Trice and Belasco, 1968). On the other hand, Hill (1970), who administered a questionnaire about attitudes toward alcohol and alcoholism to public health nurses given varying amounts of a training program about alcoholism consisting of both a series of case presentations with group discussions and group seminars, found that the program was effective. She identified five factors, all of which changed: I. Authoritarian Caretaker, II. Unsympathetic, III. Social Drinking (accepts it), IV. Skid Row Stereotype, and V. Moral Relativism (not moralistic about alcoholism). Factor IV, Skid Row Stereotype, was easiest to change, while Factor III, Social Drinking, changed most as a function of length of training. In any case, it is clear that there is an inverse relationship between amount of experience in working with alcoholics and pessimism about alcoholism (Pittman and Sterne, 1962). It is also clear that the introduction of alcoholism treatment programs into a community of itself will not change community or professional attitudes toward alcoholism (Dersch and Talley, 1973). It is likely that the answer to the problem of negative professional attitudes toward alcoholics lies in intensive exposure to and education about alcoholism early in professional education. It seems plausible that professionals who are high on authoritarianism are also those who reject the disease concept of alcoholism and take a moralistic point of view toward it. Are they necessarily those in whom there is the least hope of attitude change? That the answer to this question may be no is suggested by Adorno's (1950) findings that authoritarianism and education are negatively correlated.

In summary, then, in any attempt to increase the alcoholic's acceptability as a patient, the following seem indicated: (1) Physicians, social workers, and psychologists should receive didactic education about alcoholism and practical training in handling alcoholics *early* in their careers and training. (2) In propaganda and publicity programs, the disease concept of alcoholism should be used, the condition's treatability stressed, and the efficacy of medical–psychiatric treatment underscored.

Methods of Identification

There are two approaches to the problem of detecting the alcoholic. One, much more intensively studied, involves directly observing and interviewing the subject or else giving him a questionnaire. The other, less investigated but nonetheless promising, entails evaluation of third-party information.

Most workers in the field of alcoholism treatment and research would not cavil with Trice's (1959b) definition of alcoholism as a condition where the individual's use of alcohol deviates from the norms of the patient's key social groups, role performance (home, work, social) is impaired, the individual suffers emotional and/or physical damage, and he is unable to stop drinking excessively for extended periods of time. Inevitably, however, in extenso definitions of alcoholism may vary from facility to facility and from researcher to researcher. Such definitional differences in turn affect the construction and application of whatever instrument is used to detect alcoholism. Recently, the National Council on Alcoholism (1972b,c) published an extensive set of physiological and behavioral criteria for the diagnosis of alcoholism. Jellinek (1946) and Jackson (1957) have also reported extensive sets of criteria. At the other end of the scale, Edwards *et al.* (1972) asked only, on the morning after a night's drinking, how often do the individual's hands shake and how often he takes a drink, while Selzer (1967) proposed a set of only five questions. Complex and extensive sets of criteria have the disadvantage of being time-consuming to administer and requiring the ready availability of specialized medical skills and tests. On the other hand, they may make possible detection of mild or moderate alcoholism which might otherwise be overlooked. By contrast, short tests are insensitive and are likely only to pick up more severe cases while missing mild or moderate ones. On the other hand, they take little time to administer and score. We shall see that there appears to be an acceptable middle ground between these two kinds of instruments. In any case, as is true of other diagnostic tests, a useful test for alcoholism should detect a high percentage of alcoholics but misidentify as alcoholic only a small fraction of those who are nonalcoholic. It goes without saying that it should also be reliable.

Most instruments designed to detect alcoholics, like Cahalan's (1970), include a number of measures of loss of control over drinking such as frequent intoxication, binge drinking, symptomatic drinking, and psychological dependence on alcohol, as well as various kinds of problems due to drinking such as those with spouses or relatives, friends and neighbors, superiors or colleagues at work, the law, or with health or money. Such tests can be quite efficient even if they have a relatively small number of questions. Thus, Guze *et al.* (1962) with 17 questions were able correctly to detect 38 of 39 independently diagnosed alcoholics in a group of 90 felons and later validated their instrument in a follow-up study (Guze and Goodwin, 1972). Selzer (1971) also reported excellent results with the brief 25-item Michigan Alcoholism Screening Test (MAST), while failing to identify only 15 out of 264, or 5.7 percent of patients whose alcoholism was verified in a validation study which reviewed all local medical and agency facility records. However, it diagnosed as alcoholic only 50 percent of those arrested for drunken and disorderly conduct. This inventory may be too heavily weighted on items based on long-range consequences of heavy drinking so that it underdiagnoses alcoholics with a relatively short history of alcoholism. Its ability to detect alcoholics was also validated by Moore (1972) among psychiatric inpatients, where it failed to pick up only 2 of 128 (1.6 percent) problem drinkers. Other alcoholism inventories which appear to have face validity but have not been subjected to reliability and validity studies have been reported by Auerback (1966), Steinhilber *et al.* (1967), Shelton *et al.* (1969), and McCusker *et al.* (1971). By contrast, Mortimer *et al.* (1973) studied a self-administered yes/no questionnaire which took 35 minutes to administer. They did reliability and validation studies and item scores were weighted according to the findings of a multiple regression analysis. Only 8 percent of known alcoholics were missed while only 1 percent of controls were identified as alcoholics.

Another approach to the detection of the alcoholic has been to try to identify him via items drawn from the Minnesota Multiphasic Personality Inventory (MMPI), such as the alcoholism scales developed by Hampton (1953), Holmes (Button, 1956a), and Hoyt and Sedlacek (1958) and reviewed by MacAndrew and Geertsma (1964), who criticized three of these scales because they had been developed by comparing alcoholics with normals, so that they tended to differentiate on the basis of general psychiatric disorder rather than alcoholism per se. Furthermore, they criticized all of these scales as measures of personality factors because they included two items directly relating to alcohol consumption, items whose inclusion would, on the other hand, strengthen them if used for purposes of detection. MacAndrew (1965) subsequently developed a 49-item scale by comparing alcoholics with general psychiatric outpatients. Its detection accuracy was 81 percent and it had good test–retest reliability (Primo *et al.*, 1972; Rohan et al., 1969). Others have found the MacAndrew scale less effective than its maker did (Uecker, 1970; Vega *et al.*, 1971; Whisler and Cantor, 1966) and have reported problems with false positives. Recently, however, de Groot and Adamson (1973) have developed a new 39-item MMPI derived clinical scale that was 81 percent accurate with 11.5 percent false negatives and 7.5 percent positives. They point out that inclusion of the two alcohol-related items would improve detection efficiency.

It would appear, then, that instruments derived directly from clinical experience with alcoholics are more efficient than those based on the MMPI. Of those so far developed, the scales of Guze *et al.* (1963), Selzer (1971), and Mortimer *et al.* (1973) seem best to combine simplicity, efficiency, reliability, and validity.

As already discussed, in any detection instrument too much weight should not be put on alcohol quantity–frequency items, which are apt to be among the most unreliable. Although it may be easy for alcoholics to prevaricate about such items, it may be hard for them to do so consistently about all the items in an alcoholism detection inventory. Thus, Selzer and Ehrlich (1967), who instructed 99 hospitalized alcoholics to lie about their drinking problems to an interviewer who used the MAST, reported that 92 percent of them got scores greater than or equal to the cutting score of 5. This result, however, cannot be taken strictly at face value since these patients had already been labeled as alcoholics and were in treatment, and hence had nothing much to gain by successfully deceiving the interviewer.

The detection of the alcoholic via a third party's information—although inevitably somewhat hampered by the fact that relatives, even spouses, who presumably know the patient best, may be poorly informed about details such as alcohol intake, blackouts, and the like (Albertson, 1971), as alcoholics themselves are about psychopathology in their first-degree relatives (Rimmer and Chambers, 1969)—has the advantage of facilitating his discovery in the earlier stages of his affliction. That third-party interviewing can be a practical and workable approach is suggested by the results of Cohen and Krause's (1971) study, which was conducted in a Family Service agency. They found that they could train caseworkers to identify alcoholics among the husbands of women applying for agency help in the course of a first contact telephone history which stressed, inter alia, the use of cues such as indebtedness, frequent job changes or absences from work, gambling, infidelity, isolation, and avoidance of responsibility.

Third-party information also plays an important role in the discovery of the alcoholic in industry, where it may also be combined with direct observation of the possible alcoholic employee. Alcoholism is a serious problem in industry, where alcoholics have been found to have two to three times the normal absenteeism rate. Industry is the only place where systematic detection efforts have been made. Company alcoholism programs have reported alcoholism rates ranging from 1 to 10 percent, and 50–70 percent success rates have been reported in company treatment programs (Trice, 1959a), where patients have the double advantage of high social stability and the motivating leverage of threat of dismissal. Those who deal with the problem of alcoholism in industry stress the importance of being able to detect the alcoholic employee before the late middle stage when he begins to report to work intoxicated. Besides absenteeism (usually reported as minor illnesses or as family emergencies), other signs of alcoholism among employees are spasmodic work, increased off-the-job accidents, and financial difficulties which may lead to garnished pay. The higher the SES of the employee, the less likely his absenteeism record is to be helpful, so that in upper management alcoholism is an obvious problem only at a late stage. As already noted, in a controlled study of an extensive training program for supervisors consisting of lectures and case discussions involving causes of deviant drinking, identification, rehabilitation, and courses of action open to supervisors, the changes associated with training were small. Completion of a two-hour questionnaire, however, was a far more potent change agent, particularly with respect to supervisors' attitudes and actions (Trice and Belasco, 1968). The authors of this study emphasize that many industrial training and educational programs are unsuccessful because concentrating on the illness aspect of alcoholism may result in increased acceptance and tolerance of the alcoholic by his supervisor. Presnall (1967) lists the major drawbacks of training supervisors to discover alcoholism among employees as the following: (1) Some supervisors never learn to talk tactfully and effectively with employees. (2) The supervisor is made into an amateur diagnostician. (3) He may have to argue with the employee about whether or not he is an alcoholic. (4) Some supervisors are reluctant to invade the employee's privacy. The National Council on Alcoholism has recommended that (a) supervisors be familiar with the company's policy and procedures for referring suspected cases of alcoholism; (b) that they be trained to observe work behavior that indicates performance problems and changing work patterns; (c) that they be told that the company wants corrective discipline consistently applied and with increasing pressure.

In summary, taking an alcoholism history should be routine among first admissions in medical settings. If possible, however, the application of an alcoholism detection instrument like those of Guze *et al.* (1962), Selzer (1971), and Mortimer *et al.* (1973) is preferable. Furthermore, alcoholics can and should be detected via third-party information in agency, clinic, hospital, and industrial settings.

ENGAGING THE PATIENT AT THE REFERRAL STAGE

Referral Failures: Extent and Causes

As already noted, a major problem in the treatment of alcoholism is that so many alcoholics remain unidentified. Moreover, even if detected, more often than not they are not given appropriate referrals. Beyond this, if referred, they often fail to show up at the facility to which they were directed. Careful studies in this area are rare. It is known, however, that the main sources of referral to alcohol clinics are the patient's family, physicians, hospitals and service agencies, and the legal/police/penal establishment (Cahn, 1970). All of these sources fall down in the job of identifying and referring the alcoholic to an appropriate treatment facility. Thus, Cumming (1968) found that a family agency sent only 1 of 40 clients referred to A.A. while a Catholic welfare agency sent only 1 out of 102 there, a police complaint desk referred only 1 in 55, and the social service department of a dispensary made no referrals to an alcoholism treatment facility. Similarly, Pittman and Sterne (1962) found that only 16 percent of the agencies they studied consistently took the initiative in providing help for their alcoholic clients. A.A., followed by outpatient clinics, were the most frequent referral targets. The importance of the police in the identification and referral of the alcoholic is underscored by the paucity of lower-class patients in specialized alcoholism treatment wards, so that jails have in fact become their treatment agencies (Cahn, 1970). The record of hospitals in referring alcoholics is not much better than that of agencies. Pittman and Sterne (1962) found that 55 percent of those studied had no standard system of referral for the alcoholic, while 10 percent referred patients to A.A. and 35 percent either to A.A. or to public psychiatric clinics. Only 15 percent made adequate routine preparations for the alcoholic patient's return to the community.

The bulk of alcoholics referred to an alcoholism treatment facility never show up there, or if they do, stay for only a visit or two. Thus, in one study, only 5 percent of alcoholics seen in a big-city hospital emergency room that served a lower-class population were successfully referred to an alcoholic clinic (Chafetz et al., 1962). Data are not available on success rates in referrals to A.A., but it would be surprising if they were not also low.

Alcoholism information and referral centers have been established in many cities, but there has been little systematic study of them. Nonetheless, it appears that if lay staffed, as they so often are, they function as little more than referral centers for A.A. Besides A.A., they may refer patients to halfway houses, most of which, because they are usually staffed and run by nonprofessional A.A. members, have only the most nebulous relationship with other kinds of alcoholism treatment facilities (Cahn, 1970).

A number of factors have been found to contribute to referral failures. In an emergency room setting these include anger in the voice of the referring physician (a reflection of his negative attitudes toward the patient) (Milmoe et al., 1967) and dilatory scheduling of first appointments (Mayer, 1972). (The latter is certainly much easier to control than the former, which points up the need for intensive educational efforts to rid the medical profession of its false stereotype of the alcoholic as a skid row individual.) That such variables should be crucial is quite understandable in view of some of the typical features of the alcoholic's personality. As so well summarized by Blane (1968), these include feelings of inferiority and worthlessness, fear of new situations which present challenges, in particular of relationships which make demands on him, ambivalence, denial, and low frustration tolerance. No wonder, then, that the alcoholic is especially sensitive to others' negative attitudes and that he will not wait a long time for an appointment.

Remedies

There has been relatively little scientific study of remedies for referral failures among alcoholics. However, in a piece of research intended both to test the effect of casework with wives of alcoholics on their husbands' behavior and to engage the latter in treatment, a group of experimental counselors focused on confirming the existence of both the husband's alcoholism and the stage reached in it and tried to clarify its effects on each family member while they tried to convince the client that the worker was competent to help deal with the problem. Factors noted that affected the husband's engagement in treatment included the following: his fear that he would be controlled or lose something he valued and his wife's fear of criticism and of his retaliation (Bateman, 1971). Significantly more husbands were seen by the experimental counselors, and their drinking was reduced (Cohen, 1971).

Other, more direct approaches to engaging the alcoholic in a first clinic interview have been tried. Thus, in the study previously noted where only 5 percent of alcoholics seen in a big-city hospital emergency room were successfully referred to an alcohol clinic (Chafetz et al., 1962), the simple expedient of having the patient seen by a psychiatrist and social worker increased the rate of successful referrals to 65 percent. Similarly, one preliminary visit of a social worker to female alcoholics in a correctional institution increased the yield of successful referrals from 1 percent to 50 percent (Demone, 1963). Of the emergency room patients, 27.3 percent ended up in continued treatment (i.e., at least five sessions), of those in the correctional institution, 18 percent. Finally, writing or telephoning the patient to remind him about his first appointment will increase the rate of successful referrals (Koumans et al., 1965, 1967). The success of such expedients may be due in part to their giving the alcoholic the feeling that someone is genuinely interested in him (something he often deeply doubts) and that he is not being lost in the shuffle. That the referral phase can be confusing and bewildering for him is highlighted by the emergency room study cited above where it was found that he had to see 8 to 16 people before getting treatment at an alcohol clinic (Chafetz, 1967).

Although no specific studies have been done of other remedies for unsuccessful referrals, it seems plausible that describing A.A. methods to a patient and getting an A.A. member to visit him before he leaves a hospital or emergency room, for example, would increase the rate of successful A.A. referrals. It also seems likely that staffing information and referral centers with professional personnel would result in a wider variety of referrals which might better suit the varied needs of patients.

The picture in the referral stage may not be as bleak as it seems on first glance since it cannot be assumed that the alcoholic patient who is referred to an alcoholism treatment facility and fails to show up there never gets treatment for his alcoholism. The evidence for this is indirect and comes from studies done in general psychotherapy clinics, where it has been found that at least 21–46 percent of patients who do not keep their first appointments end up in treatment elsewhere within the next year (Brandt, 1964; Chameides and Yamamoto, 1973).

Summing up, medical and agency facilities generally fail both to identify alcoholics and to refer them to appropriate treatment facilities. If referred, most patients don't show up for their first appointments. Negative staff attitudes toward the alcoholic and dilatory scheduling of appointments are important contributors to referral failures; they can be partly alleviated by first engaging the spouse in treatment, by having a professional orient the patient about the nature of the treatment program he is about to enter, and by telephoning or writing him to remind him of his first appointment.

KEEPING THE ALCOHOLIC IN TREATMENT

Dropping Out of Treatment: Extent and Causes

Inpatient, outpatient, and halfway house alcohol treatment programs are all plagued by so many of their patients dropping out of treatment. It is not surprising that reported dropout rates have tended to be lower (13.7 percent to 39.2 percent, with a mean of 28 percent) in inpatient settings (Bowen and Androes, 1968; Gross and Nerviano, 1973; Hoy, 1969; Kamin and Caughlin, 1963; Knox, 17972; Moore and Ramseur, 1960; Pokorny et al., 1968; Rathod et al., 1966; Rhodes and Hudson, 1969; Rohan, 1970; Simpson and Webber, 1971; Tomsovic, 1968; Wilkinson et al., 1971) than in outpatient treatment programs, where 52 percent to 75 percent of patients drop out before the fourth session (Baekeland et al., 1973; Blane and Meyers, 1964; Ditman and Cohen, 1959; Gerard and Saenger, 1966; Storm and Cutler, 1968; Wilby and Jones, 1962). The lower attrition rates found in inpatient programs may stem both from the higher motivation of patients who are willing to engage themselves in such lengthy programs (typically 60 to 90 days) and from the fact that they offer more intensive treatment and support than do outpatient programs. Halfway house programs, which usually offer little beyond room, board, and participation in A.A. and many of which cater largely to class 4 and 5 persons, also have very high dropout rates. Thus, in one halfway house where it was felt that maximum benefit was to be derived from a 90-day stay, the mean stay was 40 days, the median stay 28 days, and 86.6 percent of patients stayed less than 90 days (Rubington, 1970). Similarly, in an urban ghetto day care program, 26 percent dropped out after the first session (Zimberg et al., 1971), and in a local residence-work program run under the auspices of a national religious organization, 96.8 percent stayed less than one year (considered the optimal length of stay) and over 50 percent stayed less than three weeks, the average length of stay found by Cahn (1970) in his survey of halfway house programs.

Inpatient studies have been rarer than those pursued in outpatient settings, but they have generally been far better designed and have used psychological tests and measures rather than retrospectively obtained demographic data (Goldfried, 1969; Pisani and Motansky, 1970; Zax et al., 1961) or one psychological measure (Blane and Meyers, 1963; Karp et al., 1969; Mozdzierz et al., 1973). Miller et al. (1968) and Wilkinson et al. (1971) studied 90-day inpatient programs which relied heavily on group psychotherapy. They reported that younger patients with an earlier onset of alcoholism and more unstable marriages were most likely to leave treatment. A number of psychological differences were found to distinguish dropouts from program completers in the two studies. Although neither examined the same parameters and both used somewhat different tests, their results were quite consistent. Dropouts had higher MMPI scores, were more hostile and aggressive, less mature, responsible, and emotionally controlled, had less self-esteem and more self-doubt, and were more socially dependent although more socially isolated and unaffiliated. Miller et al. (1968) examined the precipitants of drinking bouts. Dropouts turned out to be more likely to react to a tough problem with hostility and a tendency to lose control of it and either leave the situation or get drunk. They were also more likely to report that alcohol helped them relax and gave them confidence, that they drank when guilty or remorseful or for spite, that their drinking bouts were longer, and that they always got drunk when drinking. Mozdzierz et al. (1973) in an MMPI study confirmed the common clinical impression that patients who do worse exercise more denial than those with better outcomes. They found that dropouts were more defensive and denied their feelings of hostility and suspiciousness, as well as their problems with interpersonal relationships. Finally, Voth (1965) found that alcoholic inpatients who scored high on autokinesis (the apparent motion of a spot of light projected on a dark ground) were more likely to elope. Persons with high scores have greater ego autonomy, while those with low scores are more suggestible, responsive to external stimuli, socially active, and emotionally labile and impulsive, a dichotomy which resembles that of field independence and field dependence, variables found to be strongly related to dropping out in a variety of outpatient situations, as we shall see later.

In summary, then, the alcoholic who drops out of an inpatient treatment program seems to be one who is in a more advanced stage of alcoholism, has more passive–aggressive and psychopathic features, is more apt to deny his hostility, suspicion, and interpersonal problems, and depends on alcohol for relief of feelings of resentment, anxiety, or depresssion.

Despite many methodological differences, the ten studies of outpatient alcoholism treatment found suitable for review agreed on a number of points. Thus, three studies reported that lower socioeconomic status or factors related to it (low education, income, and occupational status) were associated with defecting from treatment (Baekeland et al., 1973; Kissin et al., 1968; Pisani and Motansky, 1970). However, Baekeland et al. (1973), who divided dropouts into immediate dropouts (1 visit only), rapid dropouts (>1 visit but ≤ 1 month of treatment), and slow dropouts (>1 but <6 months of treatment), found that education predicted dropping out only in the case of slow dropouts. On the other hand, Kissin et al. (1968) reported that lower SES favored dropping out among patients assigned to psychotherapy (as did Pisani and Motansky, 1970) but that patients with higher occupational status tended to drop out of a long-term rehabilitation program while, on the other hand, dropping out and SES were unrelated among patients assigned to drug treatment. The last finding may have been due to too small a range in SES among drug patients.

Social isolation was found to be predictive of dropping out in several studies. Thus Zax *et al.* (1961) reported that dropouts were more likely to be single or, if married, not living with their wives, while Goldfried (1969) found that they were apt to be single or separated. On the other hand, Baekeland *et al.* (1973) reported that a patient's living alone was predictive of dropping out only if he was an immediate dropout.

Three studies implied that negative or ambivalent attitudes toward treatment favored dropping out. Thus Blane and Meyers (1963) discovered that counterdependent individuals (those who are resistant to proffered help, insist on their ability to do things for themselves, feel that they do not need anything and do not believe in the utility of trusting others) are more likely to drop out than overtly dependent patients, who request psychological or material help. Similarly, Baekeland *et al.* (1973) reported that immediate dropouts are most likely previously to have made an appointment and then never to have shown up for it, presumably an indication of their overt ambivalence about treatment. They also found that rapid dropouts endorsed questionnaire items that both favored and rejected supervision by authority figures, presumably an indication of covert ambivalence about treatment. Along somewhat the same lines, Zax (1962) found that dropouts were more likely than remainers to have a previous history of dropping out, and Allen and Dootjes (1968), who used the Gough Adjective Checklist, found that the number of visits patients made was negatively correlated with autonomy but positively correlated with deference. It seems, then, that the more independent, less compliant patient is more likely to drop out of treatment.

Four studies implicated poor motivation as a cause of dropping out, either on the basis of global clinical impressions (Robson *et al.*, 1965) or on that of related variables. Thus, both Zax *et al.* (1961) and Goldfried (1969) found that institutionally referred patients (who by and large tend to be coerced into treatment) are more likely to drop out than patients who are self-referred. On the other hand, patients on disulfiram are much less likely to drop out of treatment if they take it under supervision, a procedure that directly shores up weak motivation (Gerrein *et al.*, 1973). Along related lines, Gertler *et al.* (1973) found that dropouts were less likely than remainers to have been abstinent for at least one year at some time before admission.

The predictive value of personality traits in outpatient alcoholics has received relatively little attention but nonetheless research studies have suggested important roles for sociopathy and behavioral and/or perceptual dependence. Thus, Goldfried (1969) reported that a history of arrests was related to dropping out of treatment, a finding consonant with the inpatient studies already discussed and one which suggests an association between poor impulse control and dropping out. That two other outpatient studies (Baekeland *et al.*, 1973; Kissin *et al.*, 1968) failed to replicate this finding may be due to their use of less stringent cutoff points. Perceptual and behavioral dependence were implicated by Karp *et al.* (1970), who found in a relatively small patient sample that the more field dependent (Witkin *et al.*, 1962) the alcoholic patient in psychotherapy, the more likely he was to drop out of it. On the face of it, the findings of Blane and Meyers (1963) and Karp *et al.* (1970) seem to contradict each other, since overt behavioral dependence is a personality feature that has been found to characterize highly field dependent subjects (Witkin *et al.*, 1954). On the other hand, counterdependence seems to be a reaction formation against dependence (Blane, 1968). In any case, before either finding can be accepted with much conviction, both measures should be studied in the same group of patients. Karp *et al.* (1970) also reported that field dependence was unrelated to dropping out in patients in drug therapy. However, recent work in our clinic (Baekeland and Lundwall, unpublished results) shows that in a one-month double blind drug study of the effects of an oxazepam–protriptyline combination on outpatient alcoholics (primarily lower-class patients), field dependent subjects were more likely to miss appointments and to drop out of treatment. The two studies' findings may not be as discrepant as they seem since the field dependent patient's nurturant and succorant needs are less well served in a situation where he is receiving an admittedly experimental medication which can only be seen by the prescribing physician as one of questionable value since he does not know whether the patient is receiving drug or placebo.

There is a good deal of evidence, some of it direct, some of it indirect, that affiliation with A.A. is relatively incompatible with continued attendance at an alcoholism clinic. Thus, Gertler *et al.* (1973) found that dropouts were more likely to be regular A.A. attenders. Furthermore, as already noted, Allen and Dootjes (1968) discovered that clinic attendance correlated negatively with autonomy and positively with deference as measured by the Gough Adjective Checklist. Subjects who agreed with A.A. were higher on autonomy and lower on deference. The implication is hence that there was an inverse relationship between A.A. affiliation and clinic attendance. Moreover, persons who join A.A. seem to be more field dependent than those who don't (Karp *et al.*, 1965) and are more apt to be physically stable (Trice and Roman, 1970). It hence seems likely that since A.A. does not offer medical treatment, patients with high levels of psychiatric and somatic symptoms may drop out of it for the sake of more specific treatment. Finally, alcohol clinics serve largely lower-class populations, while few class 4 and 5 persons attend A.A. (Edwards *et al.*, 1967; Jones, 1970; Pittman and Gordon, 1958). It hence seems probable that A.A. and clinics serve very different populations. It should be kept in mind that A.A. tends to have an exclusivistic quasireligious sectarian character (Jones, 1970), so that it is apt to downgrade all other forms of long-term treatment for alcoholism. Typically, it discourages its members' visits to alcohol clinics because it is adverse to the use of medication and looks askance at psychotherapy (Cahn, 1970).

In sum, the composite picture of the alcoholic outpatient who is most likely to drop out of treatment is that of a field dependent, counterdependent, highly symptomatic, socially isolated lower-class person of poor social stability who is highly ambivalent about treatment and has psychopathic features. The skid row alcoholic is the most extreme example of this kind of patient. If not a class 4 or 5 individual, the dropout is more likely to be an A.A. affiliate than a clinic attender.

Although there has been considerable study of the features of the alcoholic patient that help determine whether or not he stays in treatment, there has been almost no formal investigation of institutional factors responsible for his abandoning it. Nonetheless, it is clear that a built-in feature of many outpatient alcoholism treatment programs is a series of hurdles designed to eliminate "undesirable" or "poorly motivated" patients. Often there is some kind of an initial screening interview to determine acceptability, followed by from one to seven orientation lectures attended with or without the spouse, all this finally leading to one or two intake interviews (Cahn, 1970). No wonder, then, that the alcoholic, who is so apt to interpret ordinary frustrations as personal rejections, is so likely to drop out before he is really engaged in treatment.

How the first interview in psychotherapy is conducted, especially if the patient is of lower SES, may be crucial in whether or not he understands his and the therapist's prescribed roles and whether or not he subsequently drops out of treatment (Gill et al., 1954; Orne and Wender, 1968). Experimental support for this idea comes from two studies. In one, the patients of two social workers of similar background and training but with different personalities (and, presumably, different ways of orienting new admissions to an alcoholism clinic) and who did not subsequently see them again (they were turned over to an internist for further treatment) differed in the rate at which they dropped out after the first month of treatment (Baekeland et al., 1973). Stronger support comes from another study in which an initial "anticipatory socialization" interview significantly reduced dropping out of psychotherapy (Hoehn-Saric et al., 1964). The aims of this interview were: (1) to provide a rational basis for patients to accept psychotherapy as a way of helping them, (2) to clarify the roles of patient and therapist, (3) to provide a general outline of the course of treatment and its vicissitudes with particular emphasis on clarification of the development of angry or otherwise negative feelings toward the therapist.

Many clinics offer only psychotherapy and/or counseling. What, however, of the patient who does not understand or want psychotherapy? First of all, it should be recalled that the typical alcohol clinic patient is a lower SES individual and that therapists usually are middle-class people. Hence, they may only very imperfectly understand many facets of the life of lower-class (class 4, 5) persons, who make up the bulk of hospital and clinic populations in public facilities. Therefore, their values and hence their implicit (if not explicit) explanations about the patient's life goals and the conduct of his treatment may differ greatly from the ideas the patient himself has about such matters. For example, the lower-class individual puts much more emphasis on the present than on the future (Gursslin et al., Winter 1959–60; Hollingshead, 1949; Hyman, 1953; Seward and Marmor, 1956), is more concrete and task-oriented (Gursslin et al., Winter 1959–60), and is less other-directed and less likely to conform to social and expert opinion (Hyman, 1953). He is also more apt to have physical, as opposed to psychological, symptoms and is less psychologically minded (Hollingshead and Redlich, 1958). Finally, he is more poorly motivated, less patient, and less discontented and dissatisfied with himself (Schmidt et al., 1968) than the middle-class patient. Hence, it is not hard to understand why forms of treatment which emphasize long-range goals and self-understanding via psychological constructs may seem to him bewildering, if not downright irrelevant or nonsensical, or that a discrepancy between patient and therapist treatment expectations has been shown to aid and abet dropping out in at least five studies conducted in psychotherapy clinics (Freedman et al., 1958; Garfield et al., 1963; Goldstein, 1960; Heine and Trosman, 1960; Rickels, 1968). Orne and Wender (1968) have pointed out that the usual ground rules for psychotherapy include the following: (1) The patient must participate verbally and actively. (2) The therapist's task is to help him to understand himself. He is not supposed to give him advice or tell him what to do. (3) The course of therapy will be stormy. (4) Causality is complex and unconscious. They contrast these assumptions with those of medical and surgical treatment, where (a) the patient is relatively passive, (b) the doctor's task is to make him well and, although treatment is sometimes quickly effective and sometimes prolonged, (c) the patient's personal feelings may have little to do with its results, and (d) causality is often simple and generally physical. No wonder, then, that the lower-class patient is more likely to drop out of any kind of treatment that uses the psychotherapeutic model sketched above rather than the medical one he is used to. Hence, insistence on fitting the alcoholic patient, willy-nilly, to the Procrustean bed of psychotherapy, can only result in high dropout rates unless patients have been extensively screened and oriented.

The importance of promptly attending both to the alcoholic's general medical problems and to his psychological symptoms is suggested by the findings of Baekeland et al. (1973) that outpatients who drop out in the first month of treatment are apt to have especially high levels of anxiety, depression, and somatic symptoms and that they experience less amelioration of their symptoms than those who stay longer. The clinic in which their study was conducted, like so many others, did not offer internal medical services but had to refer patients for them to medical clinics, a time-consuming and frustrating procedure calculated to encourage discouragement and resentment in the easily frustrated and angered alcoholic patient. The conclusion is unavoidable that the services of a general practitioner or internist should be available on the premises of outpatient alcoholism treatment facilities. That patients who dropped out in the first month had higher levels of anxiety and depression and less reduction in these symptoms also points up the importance of rapid, aggressive treatment of psychological symptoms. The desirability of early aggressive treatment of both somatic and psychological symptoms is suggested by Gibson and Becker's (1973) finding that changes in alcoholics' depression scores in the first 14 days of treatment are mainly a function of changes not only in mood but also in somatic symptoms. An important although little investigated but relevant problem in this area is that of the prevalence of unrecognized withdrawal symptoms in alcoholic outpatients.

In view of the fact that so many patients are offered only psychotherapy but find it uncongenial, it should not be surprising that the number of treatment options offered the alcoholic patient seems to be an important factor whether or not he will stay in treatment. Thus, by using a design in which alcoholic outpatients were randomly assigned to drug treatment or to psychotherapy, but either had or did not have the option of switching to an alternative form of treatment than that originally assigned, Kissin et al. (1970, 1971) found that treatment acceptors did better than treatment rejectors, and also that the greater the number of treatment choices, the better the outcome. The results obtained by Kissin et al. (1970, 1971) make all the more sense in view of the report from Pattison et al. (1973) that patients who attended an aversion-conditioning hospital, an outpatient clinic, a halfway house, and a police work center constituted rather different clinical populations and that each facility tended to have quite different treatment approaches adapted to the special populations they dealt with.

If offering the patient treatment options is one side of the coin of keeping him in treatment, offering him the most appropriate kind of treatment within the general kind of treatment he has been offered and will accept is surely the other side of that same coin. Fundamentally, this is but another aspect of the general problem of spelling out the determinants of effectiveness of treatment outcome, since dropping out of treatment is closely related to other measures of outcome (Baekeland et al., 1973). The effects of within modality differences in treatment approach on patient retention have been best studied in drug treatment. Thus, in a quadruple blind study, Ditman (1961) found in a largely skid row population of outpatient alcoholics that patients given chlordiazepoxide had a lower short-term dropout rate than those patients on placebo, imipramine, thioridazine, or diethylpropion. Recently, Rosenberg (1974) reported similar findings in a somewhat better prognosis population of outpatient alcoholics. On a double blind basis he randomly assigned new patients to group psychotherapy plus one of four treatment conditions: disulfiram, chlordiazepoxide, multivitamins, or no medication. Since only 44 percent of those assigned to disulfiram would accept it, the other 56 percent were randomly assiged to the other three conditions. Dropout rates were lowest in patients on chlordiazepoxide or disulfiram, followed in order by those on multivitamins and no medication. Although there was not a significant difference between dropout rates in chlordiazepoxide and disulfiram patients, it must be kept in mind that the latter were self-selected and probably better motivated, so that it seems likely that had these patients been matched with those on chlordiazepoxide for motivation, dropout rates would have been lower in the latter. After five months, between-group differences disappeared, a finding that could be ascribed to the final disappearance by this time of withdrawal symptoms, which chlordiazepoxide is effective in countering (Kaim et al., 1969). A recent study in our program tends further to support the idea that chlordiazepoxide is more effective in holding outpatient alcoholics in treatment than currently used medications in other pharmacological groups (Baekeland et al., unpublished). On admission and at each visit the clinic physician evaluated each patient on a 4-point scale for anxiety, depression, sleep disturbance, and current alcohol intake. He then prescribed medication as he saw fit, usually prescribing phenothiazines for severe agitation, chlordiazepoxide or diazepam for moderate anxiety, and vitamins where there was little evidence of agitation or anxiety. Phenothiazine group and chlordiazepoxide group patients were both equally symptomatic and significantly more so than those given only vitamins. However, dropout rates were significantly higher in the phenothiazine group than in the other two groups. Since the chlordiazepoxide group was more symptomatic than the no-medication group, patients in it would have been expected to be more likely to drop out (Baekeland et al., 1973). Nonetheless, in the two groups, dropout rates were the same, a further indication that chlordiazepoxide and oxazepam are more effective than other kinds of medication in keeping outpatients from dropping out of treatment.

Remedies

What of the patient, who, once having started treatment, then drops out of it? Research on remedies for this problem has been scanty, but it seems that simple measures can be very effective. Thus, Panepinto and Higgins (1969) found that they were able to reduce first-month dropout rates in an alcohol clinic from 51 percent to 28 percent by the simple expedient of sending patients appointment letters whenever they missed a scheduled visit. In the early phase of treatment, it has been found that group as opposed to individual intake sessions (Gallant et al., 1969) and giving additional services to the patient, such as having him seen early by the social worker and seeing relatives (Wedel, 1965), were effective in reducing immediate dropout rates.

Another plausible approach, though one that has not been studied, is to try reengaging the dropout in treatment by dint of a home visit. Equally plausible is trying to work in cooperation with the referring source, be it physician, hospital, or agency. All too often, it appears that they get no feedback about the patient's progress in clinic treatment, and vice versa.

Alcoholics, most of whom come from disturbed families, tend to marry someone from a disturbed background (Rimmer and Winokur, 1972). In view of such assortative mating, it should occasion no surprise that engaging the alcoholic's spouse in treatment also seems a promising approach to keeping him in treatment. Granted that staying in treatment is a strong, if crude, measure of outcome, there is a good deal of evidence to support such an approach in selected cases. That it is not applicable to even the bulk of married alcoholics is indicated by Edwards et al. (1973b) in a thorough and critical review of the literature which showed that the classical description of the alcoholic's wife, advanced between 1937 and 1959, as an aggressive, domineering woman who married her husband to mother or control him is without foundation. Nevertheless, there seems to be a good rationale for couple treatment. Rae and Drewery (1972) have shown that if wives of alcoholics are divided into those above and below the median of the MMPI psychopathic deviate (Pd) scale and are compared with wives of normals, the former see themselves as more masculine, as do their husbands, who see themselves as feminine. However, high Pd wives deny that their husbands are feminine, an attitude that may work counter to therapists' efforts to get patients to recognize problems in this area. Pd scale scores also seem connected with alcoholic patients' wives' perception of their husbands' independence or dependence. Control husbands scored much more independent than dependent, control wives much more dependent than independent, as one might expect of those who conform to conventional sex-role stereotypes. However, alcoholic husbands saw themselves as equally independent and dependent. High Pd wives recognized this conflict, but low Pd wives denied it, something that may have negated these patients' efforts to achieve independence if they had high Pd wives or enhanced them if they had low Pd wives. Hence, it is not surprising that in another study (Rae, 1972) successfully treated inpatients had wives with lower Pd scores. However, having a low Pd wife was a prognostic indicator only in marriages marked by employment instability and sexual disturbance. In line with Rae's (1972) findings, it is interesting that Smith (1967), who asked the wives of married inpatients to attend a wives' group (65.2 percent of them did so), found that at 4–9-month follow-up, the husbands of wives who cooperated were doing significantly better. However, since wives were not randomly assigned to treatments, it is not clear whether the results are due to wife group therapy or to the fact that cooperative wives would have proved, for example, to have had lower Pd scores on the MMPI than uncooperative ones. Conversely, Pemberton (1967) found that assistance and support of alcoholic women by their husbands promoted better results. Gallant *et al.* (1970) reported on treatment of 118 couples, of whom at least one in each was alcoholic, with group psychotherapy at the rate of two hours every two weeks. Their high reported success rate of 44.9 percent at 2–20 months might in part be attributed to the fact that these were highly selected patients discharged from their inpatient service, where patients are not discharged until the spouse visits the unit for a family session. Similarly, Burton and Kaplan (1968), who studied group counseling at the rate of 1½ hours a week with alcoholics and their spouses, focusing on husband–wife interactions, at follow-up found fewer marital problems in 61.7 percent, drinking improvement in 46.8 percent, and reduced social pathology in 38.2 percent. It should be kept in mind that this was a highly selected group of patients. One controlled study (Corder *et al.*, 1972) bears out these reports of the effectiveness of alcoholic couple group psychotherapy, though in a hospital setting. Controls received the usual four weeks of daily group therapy, didactic lectures, RT, and OT. Experimental subjects got the same program for three weeks, then an intensive four-day workshop with wives and their husbands which consisted of two-couple therapy sessions, session videotape analyses, group lectures and discussions about alcoholism, discussions of game-playing in alcoholism, joint recreational counseling, A.A. and Al-Anon meetings, and meetings with representatives from follow-up treatment agencies. At 7-month follow-up, experimental subjects were significantly more abstinent (55 percent), and a higher proportion of them were attending follow-up treatment and had been employed one month. Obviously, in this experiment, couple treatment was effective, but it is difficult to know which aspect of it was most responsible for the observed treatment effects or to what extent subject selection had a hand.

Clearly, more systematic research needs to be done to pinpoint clinic and therapist factors that drive alcoholic patients out of treatment and methods effective in keeping him in treatment, or, once he has left it, in reengaging him in treatment. However, even now it seems that the following measures are effective in keeping the patient in treatment or in getting him back: offering him a variety of treatment options, group intake sessions, offering him ancillary services, such as social work and internal medical services (on the premises), and engaging his spouse in treatment. Hence, the clinician dealing with the alcoholic would do well to heed the following seven commandments: (1) Eliminate waiting lists and offer immediate admission. (2) Satisfy the patient's dependency needs by offering as wide a range of ancillary services as possible. (This is especially important in the case of the class 4 or 5 patient, who is often severely in need of everything that the social worker can offer him.) (3) Offer a variety of treatment modalities to the patient and apply that which is best suited to him rather than that which is easiest or simply happens to be available. (4) Explain clearly to the patient the aims, scope, probable results, side effects, and duration of the kind of treatment to which he is assigned, and make sure he understands his role in it. (5) Find out whether a patient has previously dropped out of treatment. If he has, explore the reasons for it with him thoroughly and right away. Do not allow him to store up unverbalized resentment, and let him know that he can express negative feelings toward the clinic or therapist without fear of retribution. (6) Maintain contact with the significant other and engage his or her help. Engage him or her in treatment if necessary. (7) In more symptomatic lower-class patients, put major emphasis on rapid symptom relief. Do not withhold medication where it might help.

REFERENCES

Adorno, T. W., 1950, "The Authoritarian Personality," Harper, New York.
Albertson, R., 1971, Identifying alcoholism as a problem, in "Casework with Wives of Alcoholics" (P. C. Cohen and M. K. Krause, eds.), pp. 33-46, Family Service Association of America, New York.
Alcoholics Anonymous, 1972, "The Fellowship of Alcoholics," New York.
Allen, L. R., and Dootjes, I., 1968. Some personality considerations of an alcoholic population, *Percept. Mot. Skills* 27:707-712.
Anchersen, P., 1947, On the prognosis of narcomania (euphomania), *Acta Psychiatr. Scand.* 22:153-193.
Auerbach, A., 1966, Office management of the alcoholic, *Amer. Acad. Gen. Practice* 33:102-104.
Baden, M. M., 1971, Methadone-related deaths in New York City, in "Methadone Maintenance" (S. Einstein, ed.), pp. 143-152, Marcel Dekker, New York.
Baekeland, F., Lundwall, L., Shanahan, T. J., 1973, Correlates of patient attrition in the outpatient treatment of alcoholism, *J. Nerv. Ment. Dis.* 157:99-107.
Baekeland, F., Lundwall, L., Shanahan, T. J., and Kissin, B., 1974a, Clinical correlates of reported sleep disturbance in outpatient alcoholics, *Q. J. Stud. Alcohol* 35:1230-1241.
Baekeland, F., Lundwall, L., and Kissin, B., 1974b, Methods for the treatment of chronic alcoholism: A critical appraisal, in "Research Advances in Alcohol and Drug Problems" (Y. Israel, ed.) Vol. 2, John Wiley and Sons, New York.
Bailey, M. B., Haberman, P. W., and Sheinberg, J., 1966, Identifying alcoholics in population surveys; a report on reliability, *Q. J. Stud. Alcohol* 27:300-315.
Banay, R. S., 1942, Alcoholism and crime, *Q. J. Stud. Alcohol* 2:686-716.
Barcha, R., Stewart, M. A., and Guze, S. B., 1968, The prevalence of alcoholism among general hospital ward patients, *Am. J. Psychiatry* 125:681-684.

Bateman, M. E., 1971, Engaging the alcoholic man and his wife in casework treatment, *in* "Casework With Wives of Alcoholics" (P. C. Cohen and M. K. Krause, eds.), pp. 59–78, Family Service Assocation of America, New York.

Blane, H. T., 1968, "The Personality of the Alcoholic: Guises of Dependency," Harper & Row, New York.

Blane, H. T., and Meyers, W. R., 1963, Behavioral dependence and length of stay in psychotherapy among alcoholics, *Q. J. Stud. Alcohol* 24:503–510.

Blane, H. T., and Meyers, W. R., 1964, Social class and the establishment of treatment relations by alcoholics, *J. Clin. Psychol.* 20:287–290.

Blane, H. T., Overton, W. F., and Chafetz, M. E., 1963, Social factors in the diagnosis of alcoholism. I. Characteristics of the patient, *Q. J. Stud. Alcohol* 24:640–663.

Bowen, W. T., and Androes, L. R., 1968, A follow-up study of 79 alcoholic patients: 1963–1965, *Bull. Menninger Clin.* 32:26–34.

Brandt, L. W., 1964, Rejection of psychotherapy, *Arch. Gen. Psychiatry* 10:310–313.

Brenner, B., and Selzer, M. L., 1969, Risk of causing a fatal accident associated with alcoholism, psychopathology, and stress: Further analysis of previous data, *Behav. Sci.* 14:490–495.

Burton, G., and Kaplan, H. M., 1968, Marriage counseling with alcoholics and their spouses—II. The correlation of excessive drinking behavior with family pathology and social deterioration, *Br. J. Addict.* 63:161–170.

Button, A. D., 1956a, A study of alcoholics with the Minnesota Multiphasic Personality Inventory, *Q. J. Stud. Alcohol* 17:263–281.

Button, A. D., 1956b, The psychodynamics of alcoholism: A survey of 87 cases, *Q. J. Stud. Alcohol* 17:443–460.

Cahalan, D., 1970, "Problem Drinkers," Jossey-Bass, San Francisco.

Cahalan, D., Cisin, I. H., and Crossley, H. M., 1969, "American Drinking Practices: A National Survey of Behavior and Attitudes," Monograph No. 6, Rutgers Center of Alcohol Studies, New Brunswick, New Jersey.

Cahn, S., 1970, "The Treatment of Alcoholics: An Evaluation Study," Oxford University Press, New York.

Chafetz, M. E., 1967, Motivation for recovery in alcoholism, *in* "Alcoholism: Behavioral Research Therapeutic Approaches" (R. Fox, ed.), pp. 110–117, Springer, New York.

Chafetz, M. E., Blane, H. T., Abram, H. S., Golner, J., Lacy, E., McCourt, W. F., Clark, E., and Meyers, W., 1962, Establishing treatment relations with alcoholics, *J. Nerv. Ment. Dis.* 134:395–409.

Chafetz, M. E., Blane, H. T., and Hill, M. J., 1970, "Frontiers of Alcoholism," Science House, New York.

Chameides, W. A., and Yamamoto, J., 1973, Referral failures: A nine-year follow-up, *Amer. J. Psychiatry* 130:1157–1158.

Cohen, P. C., 1971, Outcome of the project and next steps, *in* "Casework With Wives of Alcoholics," (P. C. Cohen and M. K. Krause, eds.), pp. 113–117, Family Service Association of America, New York.

Cohen, P. C., and Krause, M. K. (eds.) 1971, "Casework With Wives of Alcoholics," Family Service Association of America, New York.

Corder, B. F., Corder, R. F., and Laidlaw, N. D., 1972, An intensive treatment program for alcoholics and their wives, *Q. J. Stud. Alcohol* 33:1144–1146.

Crowley, T. J., Chesluck, D., Dilts, S., and Hart, R., 1974, Drug and alcohol abuse among psychiatric admissions, *Arch. Gen. Psychiatry* 30:13–20.

Cumming, E., 1968, "Systems of Social Regulation," Atherton, New York.

Cumming, J., and Cumming, E., 1957, "Closed Ranks: An Experiment in Mental Health Education," Harvard University Press, Cambridge.

de Grott, G. W., and Adamson, J. D., 1973, Responses of psychiatric inpatients to the MacAndrew Alcoholism Scale, *Q. J. Stud. Alcohol* 34:1133-1139.
Demone, H. W., Jr., 1963, Experiments in referral to alcoholism clinics, *Q. J. Stud. Alcohol* 24:495-502.
Dersch, G., and Talley, R., 1973, Responses to alcoholics by the helping professions in Denver, *Q. J. Stud. Alcohol* 34:165-192.
Ditman, K. S., 1961, Evaluation of drugs in the treatment of alcoholics, *Q. J. Stud. Alcohol Suppl. No. 1:* 107-116.
Ditman, K. S., and Cohen, S., 1959, Evaluation of drugs in the treatment of alcoholism, *Q. J. Stud. Alcohol* 20:573-576.
Dumont, M. P., 1967, Tavern culture: The sustenance of homeless men. *Am. J. Orthopsychiatry* 37:938-945.
Edwards, G., Hensman, C., Hawker, A., and Williamson, V., 1967, Alcoholics Anonymous: The anatomy of a self help group, *Soc. Psychiatry* 1:195-204.
Edwards, G., Gattoni, F., and Hensman, C., 1972, Correlates of alcohol-dependence scores in a prison population, *Q. J. Stud. Alcohol* 33:417-429.
Edwards, G., Hawker, A., Hensman, C., Peto, J., and Williamson, V., 1973a, Alcoholics known or unknown to agencies: Epidemiological studies in a London suburb, *Br. J. Psychiatry* 123:169-183.
Edwards, P. Harvey, C., and Whitehead, P. C., 1973b, Wives of alcoholics: A critical review and analysis, *Q. J. Stud. Alcohol* 34:112-132.
Einstein, S., Wolfson, E., and Gecht, D., 1970, What matters in treatment: relevant variables in alcoholism, *Int. J. Addict.* 5:43-67.
Freed, E. X., 1964, Opinions of psychiatric hospital personnel and college students toward alcoholism, mental illness and physical disability: An exploratory study, *Psychol. Rep.* 15:615-616.
Freedman, N., Engelhardt, D. M., Hankoff, L. D., Glick, B. S., Kaye, H., Buchwald, J., and Stark, P., 1958, Dropout from outpatient psychiatric treatment, *AMA Arch. Neurol. Psychiatry* 80:657-666.
Gallant, D. M., Bishop, M. P., Stoy, B., Faulkner, M. A., and Paternostro, L., 1969, The value of a "first contact" group intake session in an alcoholism outpatient clinic: Statistical confirmation, *Psychosomatics* 7:349-352.
Gallant, D. M., Rich, A., Bey, E., and Terranova, I., 1970, Group psychotherapy with married couples: A successful technique in New Orleans alcoholism clinic patients, *J. La. State Med. Soc.* 122:41-44.
Garfield, S. L., Affleck. D. C., and Muffly, R., 1963, A study of psychotherapy interaction and continuation of psychotherapy, *J. Clin. Psychol.* 19:473-478.
Gerard, D. L., and Saenger, G., 1966, "Outpatient Treatment of Alcoholism," University of Toronto Press, Toronto.
Gerrein, J. R., Rosenberg, C. M., and Manohar, V., 1973, Disulfiram maintenance in outpatient treatment of alcoholism, *Arch. Gen. Psychiatry* 28:798-802.
Gertler, R., Raynes, A. E., and Harris, N., 1973, Assessment of attendance and outcome at an outpatient alcoholism clinic, *Q. J. Stud. Alcohol* 34:955-959.
Gibson, S., and Becker, J., 1973, Alcoholism and depression; the factor structure of alcoholics' reponses to depression inventories, *Q. J. Stud. Alcohol* 34:400-408.
Gill, M. M., Newman, R., and Redlich, F. C., 1954, "The Initial Interview in Psychiatric Practice," International Universities Press, New York.
Glaser, D., and O'Leary, V., 1966, "The Alcoholic Offender," Department of Health, Education, and Welfare, Washington, D.C.
Glasscote, R. M., Plaut, T. F. A., Hammersley, D. W., O'Neill, F. J., Chafetz, M. E., and Cum-

ming, F., 1967, "The Treatment of Alcoholism," Joint Information Service of the American Psychiatric Association and the National Association for Mental Health, Washington, D.C.

Goldfried, M. R., 1969, Prediction of improvement in an alcoholism outpatient clinic, *Q. J. Stud. Alcohol* 30:129–139.

Goldstein, A., 1972, Blind controlled dosage comparisons with methadone in 200 patients, in "Proceedings of the Third National Conference on Methadone Treatment," Public Health Service Publication No. 2172, Washington, D.C., U.S. Government Printing Office, pp. 31–37.

Goldstein, A. P., 1960, Therapist and client expectation of personality change in psychotherapy, *J. Counsel. Psychol.* 7:180–184.

Goodwin, D. W., 1973, Alcohol in suicide and homicide, *Q. J. Stud. Alcohol* 34:144–156.

Goodwin, D. W., Crane, J. B., and Guze, S. B., 1971, Felons who drink: An 8-year follow-up, *Q. J. Stud. Alcohol* 32:136–147.

Gray, R. M., Moody, P. M., Sellars, M., and Ward, J. R., 1969, Physician authoritarianism and the treatment of alcoholics, *Q. J. Stud. Alcohol* 30:981–983.

Green, J. R., 1965, The incidence of alcoholism in patients admitted to medical wards of a public hospital, *Med. J. Aust.* 1:465–466.

Gross, W. F., and Nerviano, V. J., 1973, The prediction of dropouts from an inpatient alcoholism program by objective personality inventories, *Q. J. Stud. Alcohol* 34:514–515.

Gursslin, O. R., Hunt, R. G., and Roach, J. L., Winter 1959–60, Social class and the mental health movement, *Social Problems* 7:200–218.

Guze, S. B., and Goodwin, D. W., 1972, Consistency of drinking history and diagnosis of alcoholism, *Q. J. Stud. Alcohol* 33:111–116.

Guze, S. B., Tuason, V. B., Gatfield, P. D., Stewart, M. A., and Picken, B., 1962, Psychiatric illness and crime with particular reference to alcoholism: A study of 223 criminals, *J. Nerv. Ment. Dis.* 134:512–521.

Guze, S. B., Tuason, V. B., Stewart, M. A., and Picken, B., 1963, The drinking history: A comparison of reports by subjects and their relatives, *Q. J. Stud. Alcohol* 24:249–260.

Hampton, P. J., 1953, The development of a personality questionnaire for drinkers, *Genet. Psychol. Monogr.* 48:55–115.

Hayman, M., 1956, Current attitudes to alcoholism of psychiatrists in Southern California, *Am. J. Psychiatry* 112:484–493.

Hayman, M., 1966, "Alcoholism: Mechanism and Management," Charles C Thomas, Springfield, Illinois.

Heine, R. W., and Trosman, H., 1960, Initial expectations of the doctor–patient interaction as factor in continuance in psychotherapy, *Psychiatry* 23:275–278.

Hill, M. J., 1970, Public health nurses and the care of alcoholics: a study of attitude change, in "Frontiers of Alcoholism" (M. E. Chafetz, H. T. Blane, and M. J. Hill, eds.), pp. 278–291, Science House, New York.

Hitz, D., 1973, Drunken sailors and others; drinking problems in specific occupations, *Q. J. Stud. Alcohol* 34:496–505.

Hoehn-Saric, R., Frank, J. D., Imber, S. D., Nash, E. H., Jr., Stone, A. R., and Battle, C. C., 1964, Systematic preparation of patients for psychotherapy: I. Effects on therapy behavior and outcome, *J. Psychiat. Res.* 2:267–281.

Hollingshead, A. B., 1949, "Elmtown's Youth," John Wiley and Sons, New York.

Hollingshead, A. B., and Redlich, F. C., 1958, "Social Class and Mental Illness," John Wiley and Sons, New York.

Hoy, R. M., 1969, The personality of inpatient alcoholics in relation to group psychotherapy, as measured by the 16-PF, *Q. J. Stud. Alcohol* 30:401–407.

Hoyt, D. P., and Sedlacek, G. M., 1958, Differentiating alcoholics from normals and abnormals with the MMPI, *J. Clin. Psychol.* 14:69-74.

Hyman, H. H., 1953, The value systems of different classes: A social psychological contribution to the analysis of stratification, in "Class, Status and Power" (R. Bendix and S. M. Lipset, eds.), pp. 426-441, Free Press, Glencoe, Illinois.

Jackson, J. K., 1957, The definition and measurement of alcoholism, *Q. J. Stud. Alcohol* 18:240-262.

Jellinek, E. M., 1946, Phases in the drinking history of alcoholics. Analysis of a survey conducted by the official organ of Alcoholics Anonymous, *Q. J. Stud. Alcohol* 7:1-88.

Jenssen, C. O., Cronholm, B., Izikowitz, S., Gordon, K., and Rosen, A., 1962, Intellectual changes in alcoholics; psychometric studies in mental sequels of prolonged intensive abuse of alcohol, *Q. J. Stud. Alcohol* 23:221-242.

Jones, R. K., 1970, Sectarian characteristics of Alcoholics Anonymous, *Sociology* 4:181-195.

Jones, R. W., and Helrich, A. R., 1972, Treatment of alcoholism by physicians in private practice, *Q. J. Stud. Alcohol* 33:117-131.

Kaim, S. C., Klett, C. J., and Rothfeld, B., 1969, Treatment of the acute alcohol withdrawal state: A comparison of four drugs, *Am. J. Psychiatry* 125:54-60.

Kalant, H., and Hawkins, R. D. (eds.), 1969, "Experimental Approaches to the Study of Drug Dependence," University of Toronto Press, Canada.

Kamin, I., and Caughlan, J., 1963, Subjective experience of outpatient psychotherapy, *Am. J. Psychother.* 17:660-668.

Karp, S. A., Witkin, H. A., and Goodenough, D. R., 1965, Alcoholism and psychological differentiation: Effect of achievement of sobriety on field dependence, *Q. J. Stud. Alcohol* 26:580-585.

Karp, S. A., Winters, S., and Pollack, I. W., 1969, Field dependence among diabetics, *Arch. Gen. Psychiatry* 21:72-76.

Karp, S. A., Kissin, B., and Hustmyer, F. E., 1970, Field dependence as a predictor of alcoholic therapy dropouts, *J. Nerv. Ment. Dis.* 150:77-83.

Kieholz, P., and Battegay, R., 1967, Vergleichende untersuchungen über die genese un den verlauf der drogabhängigkeit und des alkoholismus, *Schweiz. Med. Wochenschr.* 97.893-898.

Kish, G. B., and Cheney, T. M., 1969, Impaired abilities in alcoholism: Measured by the General Aptitude Test Battery, *Q. J. Stud. Alcohol* 30:384-388.

Kissin, B., Rosenblatt, S. M., and Machover, S., 1968, Prognostic factors in alcoholism, *Amer. Psychiat. Ass. Psychiat. Res. Rep.* No. 24, pp. 22-43.

Kissin, B., Platz, A., and Su, W. H., 1970, Social and psychological factors in the treatment of chronic alcoholism, *J. Psychiat. Res.* 8:13-27.

Kissin, B., Platz, A., and Su, W. H., 1971, Selective factors in treatment choice and outcome in alcoholics, in "Recent Advances in Studies of Alcoholism" (N. K. Mello and J. H. Mendelson, eds.), Publication # (HSM) 71-9045, U.S. Government Printing Office, Washington, D.C.

Knox, W. J., 1971, Attitudes of psychiatrists and psychologists toward alcoholism, *Am. J. Psychiatry* 127:1675-1679.

Knox, W. J., 1972, Four-year follow-up of veterans treated on a small alcoholism ward, *Q. J. Stud. Alcohol* 33:105-110.

Kolb, L., 1925, Types and characteristics of drug addicts, *Ment. Hyg.* 9:300-313.

Koumans, A. J. R., and Muller, J. J., 1965, Use of letters to increase motivation for treatment in alcoholics, *Psychol. Rep.* 16:1152.

Koumans, A. J., Muller, J. J., and Miller, C. F., 1967, Use of telephone calls to increase motivation for treatment in alcoholics, *Psychol. Rep.* 21:327-328.

Lemere, F., 1973, Alcohol-induced sexual impotence, *Am. J. Psychiatry* 130:212–213.
MacAndrew, C., 1965, The differentiation of male alcoholic outpatients from nonalcoholic psychiatric patients by means of the MMPI, *Q. J. Stud. Alcohol* 26:238–246.
MacAndrew, C., and Geertsma, R. H., 1964, A critique of alcoholism scales derived from the MMPI, *Q. J. Stud. Alcohol* 25:68–76.
Mayer, J., 1972, Initial alcoholism clinic attendance of patients with legal referrals, *Q. J. Stud. Alcohol* 33:814–816.
Mayfield, D. G., and Coleman, L. L., 1968, Alcohol use and affective disorder, *Dis. Nerv. System.* 29:467–474.
Mayfield, D. G., and Montgomery, D., 1972, Alcoholism, alcohol intoxication, and suicide attempts, *Arch. Gen. Psychiatry* 27:349–353.
McCusker, J., Cherubin, C. E., and Zimberg, S., 1971, Prevalence of alcoholism in general municipal hospital population, *N.Y. State J. Med.* 71:751–754.
Mendelson, J. H., Wexler, D., Kubzansky, P. E., Harrison, R., Leiderman, G., and Solomon, P., 1964, Physicians' attitudes toward alcoholic patients, *Arch. Gen. Psychiatry* 11:392–399.
Miller, B. A., Pokorny, A. D., and Hanson, P. G., 1968, A study of dropouts in an inpatient alcoholism treatment program, *Dis. Nerv. Syst.* 29:91–99.
Milmoe, S., Rosenthal, R., Blane, H. T., Chafetz, M. E., and Wolf, I., 1967, The doctor's voice: A postdictor of successful referral of alcoholic patients, *J. Abnorm. Psychol.* 72:78–84.
Mogar, R. E., Helm, S. T., Snedeker, M. R., Snedeker, M. H., and Wilson, W. M., 1969, Staff attitudes toward the alcoholic patient, *Arch. Gen. Psychiatry* 21:449–454.
Moore, R. A., 1971, The prevalence of alcoholism in a community general hospital, *Am. J. Psychiatry* 128:638–639.
Moore, R. A., 1972, The diagnosis of alcoholism in a psychiatric hospital: A trial of the Michigan Alcoholism Screening Test (MAST), *Am. J. Psychiatry* 128:1565–1569.
Moore, R. A., and Buchanan, T. K., 1966, State hospitals and alcoholism: Nationwide survey of treatment techniques and results, *Q. J. Stud. Alcohol* 27:459–468.
Moore, R. A., and Ramseur, F., 1960, Effects of psychotherapy in an open-ward hospital on patients with alcoholism, *Q. J. Stud. Alcohol* 21:233–252.
Mortimer, R. G., Fillins, L. D., Kerlan, M. W., and Lower, J. S., 1973, Psychometric identification of problem drinkers, *Q. J. Stud. Alcohol* 34:1332–1335.
Mozdzierz, G. J., Macchitelli, F. J., Conway, J. A., and Krauss, H. H., 1973, Personality characteristic differences between alcoholics who leave treatment against medical advice and those who don't, *J. Clin. Psychol.* 29:78–82.
Mulford, H. A., and Miller, D. E., 1960, Drinking in Iowa. V. Drinking and alcoholic drinking, *Q. J. Stud. Alcohol* 21:279–291.
National Council on Alcoholism Fact Sheet, 1972a, "Facts on Alcoholism," August 1972.
National Council on Alcoholism, 1972b, Criteria for the diagnosis of alcoholism, *Am. J. Psychiatry* 129:127–135.
National Council on Alcoholism, 1972c, Criteria for the diagnosis of alcoholism, *Ann. Intern. Med.* 77:249–258.
Nolan, J. P., 1965, Alcohol as a factor in the illness of university service patients, *Am. J. Med. Sci.* 249:135–142.
Orne, M. T., and Wender, P. H., 1968, Anticipatory socialization for psychotherapy: Method and rationale, *Am. J. Psychiatry* 124:1202–1212.
Panepinto, W. C., and Higgins, M. J., 1969, Keeping alcoholics in treatment: Effective follow-through procedures, *Q. J. Stud. Alcohol* 30:414–419.
Panepinto, W. C., Higgins, M. J., Keane-Dawes, W. Y., and Smith, D., 1970, Underlying psychiatric diagnosis as an indicator of participation in alcoholism therapy, *Q. J. Stud. Alcohol* 31:950–956.

Pattison, E. M., Coe, B., and Doerr, H. O., 1973, Population variation among alcoholism treatment facilities, *Int. J. Addict.* 8:199–229.
Pearson, W. S., 1962, The "hidden" alcoholic in the general hospital, *N.C. Med. J.* 23:6–10.
Pemberton, D. A., 1967, A comparison of the outcome of treatment in female and male alcoholics, *Br. J. Psychiatry* 113:367–373.
Perkins, M. E., and Bloch, H. I., 1970, Survey of a methadone maintenance treatment program, *Am. J. Psychiatry* 126:1389–1396.
Pisani, V. D., and Motansky, G. U., 1970, Predictors of premature termination of outpatient follow-up group psychotherapy among male alcoholics, *Int. J. Addict.* 5:731–737.
Pittman, D. J., and Gordon, C. W., 1958, Criminal careers of the chronic police case inebriate, *Q. J. Stud. Alcohol* 19:255–268.
Pittman, D. J., and Sterne, M. W., 1962, "The Carousel: Hospitals, Social Agencies and the Alcoholic," Missouri Division of Health.
Pokorny, A. D., Miller, B. A., and Cleveland, S. E., 1968, Response to treatment of alcoholism: A follow-up study, *Q. J. Stud. Alcohol* 29:364–381.
Presnall, L. F., 1967, Alcoholism in industry, in "Behavioral Approaches" (R. Fox, ed.), Chapter 10, pp. 118–133, Springer, New York.
Primo, R. V., Terrell, F., and Wener, A., 1972, An aversion-desensitization treatment for alcoholism, *J. Consult. Psychol.* 38:394–398.
Rae, J. B., 1972, The influence of wives on the treatment outcome of alcoholics: A follow-up study at two years, *Br. J. Psychiatry* 120:601–613.
Rae, J. B., and Drewery, J., 1972, Interpersonal patterns in alcoholic marriages, *Br. J. Psychiatry* 120:615–621.
Rathod, N. H., Gregory, E., Blows, D., and Thomas, G. H., 1966, A two-year follow-up study of alcoholic patients, *Br. J. Psychiatry* 112:683–692.
Redick, R. W., 1973, Utilization of psychiatric facilities by persons diagnosed with alcohol disorders, *National Institute of Mental Health, Mental Health Statistics, Series B., No. 4.*, U.S. Dept. of Health, Education, and Welfare, Rockville, Maryland.
Retterstöl, N., and Sund, A., 1965, "Drug Addiction and Habituation," Hogfeldt; Kristiansand, Norway.
Rhodes, R. J., and Hudson, R. M., 1969, A follow-up study of tuberculous skid row alcoholics, I. Social Adjustment and drinking behavior, *Q. J. Stud. Alcohol* 30:119–128.
Rickels, K., 1968, Non-specific factors in drug therapy of neurotic patients, in "Non-specific Factors in Drug Therapy" (K. Rickels, ed.), pp. 3–26, Charles C Thomas, Springfield, Illinois.
Rimmer, J., and Chambers, D. S., 1969, Alcoholism: Methodological considerations in the study of family illness, *Am. J. Orthopsychiatry* 39:760–768.
Rimmer, J., and Winokur, G., 1972, The spouses of alcoholics: An example of assortative mating, *Dis. Nerv. Syst.* 33:509–511.
Robinson, L., and Podnos, B., 1966, Resistance of psychiatrists in treatment of alcoholism, *J. Nerv. Ment. Dis.* 143:220–225.
Robson, R. A. H., Paulus, I., and Clarke, G. G., 1965, An evaluation of the effect of a clinic treatment program on the rehabilitation of alcoholic patients, *Q. J. Stud. Alcohol* 26:264–278.
Rohan, W. P., 1970, A follow-up study of hospitalized problem drinkers, *Dis. Nerv. Syst.* 31:259–267.
Rohan, W. P., Tatro, R. L., and Rotman, S. R., 1969, MMPI changes in alcoholics during alcoholization, *Q. J. Stud. Alcohol* 30:389–400.
Roman, P. M., and Trice, H. M., 1970, The development of deviant drinking behavior: Occupational risk factors, *Arch. Environ. Health* 20:424–435.
Rosenberg, C. M., 1974, Drug maintenance in the outpatient treatment of chronic alcoholism, *Arch. Gen. Psychiatry* 30:373–377.

Rubington, E., 1970, Referral, past treatment contacts, and length of stay in a halfway house: Notes on consistency of societal reactions to chronic drunkenness offenders. *Q. J. Stud. Alcohol* 31:659–668.

Schaefer, H. H., Sobell, M. B., and Sobell, L. C., 1972, Twelve month follow-up of hospitalized alcoholics given self-confrontation experiences by videotape, *Behav. Ther.* 3:283–285.

Schmidt, W.., and de Lint, J., 1972, Causes of death in alcoholics, *Q. J. Stud. Alcohol* 33:171–185.

Schmidt, W., Smart, R. G., and Moss, M. K., 1968, "Social Class and the Treatment of Alcoholism," University of Toronto Press, Toronto, Canada.

Schuckit, M., 1972, The alcoholic woman: A literature review, *Psychiatry Med.* 3:37–43.

Schuckit, M., Pitts, F. N., Jr., Reich, T., King, L. J., and Winokur, G., 1969, Alcoholism: I. Two types of alcoholism in women, *Arch. Gen. Psychiatry* 20:301–306.

Selzer, M. L., 1967, Problems encountered in the treatment of alcoholism, *Univ. Mich. Med. Cent. J.* 33:58–63.

Selzer, M. L., 1969, Alcoholism, mental illness, and stress in 96 drivers causing accidents, *Behav. Sci.* 14:1–10.

Selzer, M. L., 1971, The Michigan Alcoholism Screening Test (MAST): The quest for a new diagnostic instrument, *Am. J. Psychiatry* 127:1653–1658.

Selzer, M. L., and Ehrlich, N. J., 1967, A screening program to detect alcoholism in traffic offenders, *in* "The Prevention of Highway Injury," (M. L. Selzer, P. W. Gikas, and D. F. Huelke, eds.), pp. 44–50, University of Michigan Highway Safety Research Institute, Ann Arbor.

Seward, G., and Marmor, J., 1956, "Psychotherapy and Culture Conflict," Ronald Press, New York.

Shelton, J., Hollister, L. E., and Gocka, E. F., 1969, The drinking behavior interview, *Dis. Nerv. Syst.* 30:464–467.

Shupe, L. M., 1954, Alcohol and crime: A study of the urine alcohol concentrations found in 882 persons arrested during or immediately after the commission of a felony, *J. Criminal Law, Criminology and Political Science*, 44:661–664.

Simpson, W. S., and Webber, P. W., 1971, A field program in the treatment of alcoholism, *Hospital and Community Psychiatry* 22:170–173.

Smart, R. G., and Schmidt, W., 1967, Responsibility, blood alcohol levels and alcoholism, *in* "The Prevention of Highway Injury," (M. L. Selzer, P. W. Gikas, and D. F. Huelke, eds.), pp. 38–43, University of Michigan Highway Safety Research Institute, Ann Arbor.

Smith, C. G., 1967, Marital influences on treatment outcome in alcoholism, *J. Irish Med. Ass.* 60:433–434.

State of Tennessee Legislative Council, 1962, "Final Report—Alcoholism Study," Nashville, Tennessee.

Steinhilber, R. M., Kuluvar, V. D., Anderson, D. J., Heilman, R. O., and Hansen, P. L., 1967, Symposium on the problem of the chronic alcoholic, *Mayo Clin. Proc.* 42:705–723.

Sterne, M. W., and Pittman, D. J., 1965, The concept of motivation: A source of institutional and professional blockage in the treatment of alcoholics, *Q. J. Stud. Alcohol* 26:41–57.

Storm, T., and Cutler, R. E., 1968, "Systematic Desensitization in the Treatment of Alcoholics: An Experimental Trial," Alcoholism Foundation of British Columbia, Vancouver.

Summers, T., 1970, Validity of alcoholics' self-reported drinking history, *Q. J. Stud. Alcohol* 31:972–974.

Tomsovic, M., 1968, Hospitalized alcoholic patients. I. A two-year study of medical, social and psychological characteristics, *Hospital and Community Psychiat.* 19:197–203.

Trice, H. M., 1959a, "The Problem Drinker on the Job," Ithaca, Cornell University, Bulletin 40, New York State School of Industrial and Labor Relations, April, 1959.

Trice, H. M., 1959b, The affiliation motive and readiness to join Alcoholics Anonymous, *Q. J. Stud. Alcohol* 20:313–320.
Trice, H. M., and Belasco, J. A., 1968, Supervisory training about alcoholics and other problem employees: A controlled evaluation, *Q. J. Stud. Alcohol* 29:382–398.
Trice, H. M., and Roman, P. M., 1970, Sociopsychological predictors of affiliation with Alcoholics Anonymous: A longitudinal study of "treatment success," *Soc. Psychiatry* 5:51–59.
Uecker, A. E., 1970, Differentiating male alcoholics from other psychiatric inpatients; validity of the MacAndrew Scale, *Q. J. Stud. Alcohol* 31:379–383.
Vega, J. C., Smith, D. F., Skeeters, D. E., and Auvenshine, C. D., 1971, MMPI alcoholism scales; factor structure and content analysis, *Q. J. Stud. Alcohol* 32:1055–1060.
Voth, A. C., 1965, Autokinesis and alcoholism, *Q. J. Stud. Alcohol* 26:412–422.
Waller, J. A., 1965, Chronic medical conditions and traffic safety, *N. Engl. J. Med.* 273:1413–1420.
Waller, J. A., 1968, Patterns of traffic accidents and violations related to drinking and to some medical conditions, *Q. J. Stud. Alcohol Suppl.* 4:118–137.
Wedel, H. L., 1965, Involving alcoholics in treatment, *Q. J. Stud. Alcohol* 26:468–479.
Weingartner, W., Faillace, L. A., and Markley, H. G., 1971, Verbal information retention in alcoholics, *Q. J. Stud. Alcohol* 32:293–303.
Whisler, R. H., and Cantor, J. M., 1966, The MacAndrew Alcoholism Scale: A cross-validation in a domiciliary setting, *J. Clin. Psychol.* 22:311–312.
Wilby, W. E., and Jones, R. W., 1962, Assessing patient response following treatment, *Q. J. Stud. Alcohol* 23:325–334.
Wilkinson, A. E., Prado, W. M., Williams, W. O., and Schnadt, F. W., 1971, Psychological test characteristics and length of stay in alcoholism treatment, *Q. J. Stud. Alcohol* 32:60–65.
Winokur, G., Reich, T., Rimmer, J., and Pitts, F. N., Jr., 1970, Alcoholism: III. Diagnosis and familial psychiatric illness in 250 alcoholic probands, *Arch. Gen. Psychiatry* 23:104–111.
Winokur, G., Rimmer, J., and Reich, T., 1971, Alcoholism: IV. Is there more than one type of alcoholism? *Br. J. Psychiatry* 118:525–531.
Witkin, H. A., Dyk, R. B., Faterson, H. F., Goodenough, D. R., and Karp, S. A., 1962, "Psychological Differentiation," John Wiley and Sons, New York.
Witkin, H. A., Lewis, H. B., Hertzman, M., Machover, K., Meissner, P. B., and Wapner, S., 1954, "Personality Through Perception," Harper, New York.
Wolf, I., Chafetz, M. E., Blane, H. T., and Hill, M. J., 1965, Social factors in the diagnosis of alcoholism in social and nonsocial situations. II. Attitudes of physicians, *Q. J. Stud. Alcohol* 26:72–79.
Woodruff, R. A., Guze, S. B., and Clayton, P. J., 1973, Alcoholics who see a psychiatrist compared with those who do not, *Q. J. Stud. Alcohol* 34:1162–1171.
Zax, M., 1962, The incidence and fate of the reopened case in an alcoholism treatment center, *Q. J. Stud. Alcohol* 23:634–639.
Zax, M., Marsey, R., and Biggs, C. F., 1961, Demographic characteristics of alcoholic outpatients and tendency to remain in treatment, *Q. J. Stud. Alcohol* 22:98–105.
Zimberg, S., Lipscomb, H., and Davis, E. B., 1971, Sociopsychiatric treatment of alcoholism in an urban ghetto, *Am. J. Psychiatry* 127:1670–1674.
Zwerling, I., 1959, Psychiatric findings in an interdisciplinary study of forty-six alcoholic patients, *Q. J. Stud. Alcohol* 20:540–554.

CHAPTER 5

Toward a Social Model: An Assessment of Social Factors Which Influence Problem Drinking and Its Treatment

Allan Beigel

University of Arizona College of Medicine
and
Southern Arizona Mental Health Center
Tucson, Arizona

and

Stuart Ghertner

Southern Arizona Mental Health Center
Tucson, Arizona

INTRODUCTION

Most contemporary models used in assessing and understanding the phenomena of problem drinking are based on the belief that alcoholism is an individual problem. These approaches to care are based upon an understanding of physiological and/or psychological characteristics and their sequelae. Treatment usually is directed toward eliminating individual physical or psychological dysfunction and principally involves chemotherapeutic and/or psychotherapeutic techniques.

Biological Model

Early theorists who adopted this approach defined problem drinking and its results primarily by its physiological characteristics and followed a disease model (Martin, 1962). According to Pattison (1966), this "heuristic–cultural" need to define alcoholism as a disease has sprung from a medical–pathophysiological model which views disease as resulting from an alteration in body physiology. This requires the demonstration of a pathophysiological process such as an allergenic or endocrine disorder. To qualify alcoholism as a disease, it is also necessary to demonstrate a physiological addiction, an habituation, or a compulsion to drink.

This biological model, however, fails to take into account the importance of individual differences or social roles. It presupposes that the problem drinker is only "physically sick" and sets forth to "cure him without paying attention to his interaction with either his external (sociological) or internal (psychological) environment" (Enelow, 1974).

Psychological Model

Like the biological model, the psychological approach also focuses on the individual by attempting to understand and treat the developmental personality disorder with which the drinking problem is associated. Usually, through a one-to-one encounter between therapist and patient, treatment is designed to alleviate the personality disorder of the individual problem drinker.

This approach encourages the development of insight by promoting an increased awareness of those heretofore unconscious motivations which have led to excessive drinking.

The origins of this model may be traced back to the late 1800s, when Freud and Breuer proposed their psychodynamic formulation of hysteria. They hypothesized that psychologically traumatic events were associated with the onset of observed hysterical symptoms. If these forgotten events could be brought into consciousness, along with their accompanying feelings, Freud and Breuer observed that patients lost their symptoms and improved.

Although this concept has been expanded and modified by many theorists since Freud, Enelow (1974) points out that all intrapsychic models share in common the assumption that symptoms and disordered behavior are produced by unconscious conflicts between opposing tendencies or interests of the individual or as a result of opposing forces generated when personal wishes come in conflict with demands imposed by society and internalized at an early age.

The Origins and Perspectives of a Social Model

Despite long-standing adherence to a biological or psychological model by many theorists, there have been no specific metabolic or psychic markers identified which allow for a conclusive prediction that serious problem drinking will appear in a specific individual. This "failure" of the biological and psychological approaches led researchers in the mid-1950s to begin to focus their attention on a model which utilized social systems theory as a theoretical base.

Rather than emphasizing biological or psychological factors, proponents of the social model viewed problem drinking as resulting from phenomena derived from all aspects of an individual's life pattern. Rather than de-emphasizing the social dimensions of behavior and considering them as unessential to the origin and course of problem drinking, these theorists proposed that the social context in which drinking behavior begins and continues is a primary source of the problem and must be addressed if rehabilitation is to occur.

The early concept of alienation expressed by Durkheim et al. (1956) was a forerunner of later approaches to the formulation of a social model. By suggesting that an individual's social and economic environment defines and shapes his goals and the roles which he assumes in attempting to attain them, Durkheim placed individual behavior in an appropriate social context. He believed that if a person is successful in pursuing goal-directed behavior or if he perceives that success is possible, he will behave in a socially acceptable manner. However, when an individual no longer views these goals or aspirations as potentially attainable, the rules and values of society lose their regulatory value. This leads to a feeling of alienation and subsequent deviant behavior.

Roebuck and Kessler (1972) later summarized the historical development of the sociological approach by emphasizing three empirical events: (1) the failure to find a unique personality type or a unique psychiatric nosological group associated with problem drinking; (2) the importance of both sociological and social–psychological variables in problem drinking as reported by the large number of studies on drinking behavior patterns and on sociocultural factors which influence its occurrence; and (3) the development of operant conditioning theory and its implications for the relationship between the environment and changes in behavior patterns.

Other examples of recent studies which reflect these events include the survey by Cahalan et al. (1969) of national drinking behavior. He and his co-workers concluded that drinking behavior is primarily a result of sociological and anthropological rather than psychological variables. Support for this conclusion was drawn not only from an examination of the life histories of problem drinkers but also from the abundant research regarding drinking patterns and the manner in which these patterns are influenced by age, family history, ethnicity, sex, social status, religion, marital relationships, and extent of urbanized living.

In addition, the social consequences of problem drinking behavior often perpetrate its continuance. Higher rates of mortality, poverty, promiscuity, illegitimate pregnancies, and physical and mental abuse have been noted. Furthermore, the severe disruption of both family and social life, as revealed through data regarding divorce, desertion, and child abuse rates, often leads to further social consequences.

It is recognized that the problem drinker faces conflicts and problems which are not necessarily a result of social forces. However, the social model maintains the theoretical position that problem drinking is a phenomenon affecting *all* aspects of an individual's life interdependently, and that psychological or biological aspects cannot be viewed independent of the sociological.

In accord with this growing recognition of the role of social dysfunction in the development of alcoholism, the World Health Organization's definition (1952) notes: "Alcoholics are those excessive drinkers whose dependence upon alcohol has attained such a degree that it shows a noticeable mental disturbance or an interference with their bodily and their mental health, their inter-personal relations, *and their smooth social and economical functioning*; or who show the prodromal of such developments. They therefore require treatment."

SOCIAL SYSTEMS THEORY AND ITS ROLE IN THE SOCIAL MODEL

The major thesis to be developed in this chapter is that social systems theory provides a different and more realistic orientation to understanding problem drinking and developing therapeutic approaches. In a systems analysis, the problem drinker is considered neither a victim nor a victimizer, but a product of system disruption. The total system is viewed as the troubled unit, whether it is a marriage, a family, a society, or a culture, and effective treatment can only be provided at a system level.

Smoyak (1973) suggests that a social systems analysis entails the viewing of persons within a single social unit as well as the examination of their roles in the multiple subsystems of the larger system. Society is composed of internal dyads, triads, and more complex groups which are constantly shifting and changing. When a social system is in equilibrium, all subsystems are working and the negotiation and decision-making processes proceed smoothly. When a disruption of the operating order occurs, outside help may be needed to restore some parts to a workable condition.

Therefore, as Bowen (1974) suggests, systems theory attempts to focus on the functional facts of these relationships: what happened and how it happened, as well as when and where. An essential difference between traditional biological or psychological models and social systems theory is the lack of emphasis by the latter on the "why" of human behavior. According to Bowen:

> Why thinking has also been a part of cause and effect thinking because ever since man became a thinking being, he began to look around for causes to explain events that affected him. In reviewing the thinking of primitive man, we are amused at the various evil forces he blamed for his misfortunes or the benevolent forces he credited for his good fortunes. We can chuckle at the causality that man in later centuries assigned to illness before he knew about germs and micro-organisms. We can smugly assure ourselves that scientific knowledge and logical reasoning have now enabled man to go beyond the erroneous assumptions and false deductions of past centuries and that we now assign accurate causes for most of man's problems. However, an assumption behind systems theory is that man's cause and effect thinking is still a major problem in explaining his dysfunctions and behavior. A major effort in systems theory is to get beyond cause and effect thinking and to concentrate on facts which are the basis for systems thinking (p. 116).

Bowen also points out that cause-and-effect thinking can lead to a tendency to blame one's fellow man for problems. The greater the degree of anxiety within a system, the greater the tendency for even the most reasonable person to resort to blaming others.

In addition, a discrepancy often exists between what man does and what he says he does or will do. Systems research, by attempting to isolate observable facts about man and his relationships, avoids falling into the trap of looking for "why" explanations which contribute to circular reasoning and little progress.

In his book, *The Female Alcoholic,* Kinsey (1966) points out that most social model theorists have generally tended to view social systems as operating at three structural levels: (1) the "supra-culture orientation" which applies to all cultures or societies; (2) the "specific cultures orientation" which applies to a single culture, related cultures, or to a comparison between specific cultures; and (3) the "sub-structural orientation" which involves social and cultural systems and institutions such as the family and religion as well as demographic variables such as age, sex, and ethnicity.

Supracultural Orientation

Horton (1943) postulated that the common primary role of alcoholic beverages in the cultures which he studied was anxiety reduction. As a result, he hypothesized that: (1) drinking tends to be accompanied by a release of sexual and aggressive impulses; (2) the strength of the drinking response in any society tends to vary directly with the level of anxiety in that society; and (3) the strength of the drinking response tends to vary inversely with the strength of the counteranxiety elicited by painful experiences during and after drinking.

Horton speculated that the sources of painful experiences could be classified as either (a) actualization of real dangers which are a result of impairment of physiological function, (b) social punishment which result from the impairment of function, (c) social punishment invoked by the inappropriate release of sexual impulses, or (d) social punishment invoked by the inappropriate release of aggressive impulses.

By examining the files of a cross-cultural survey completed by the Yale Institute of Human Relations, Horton was able to review 77 societies. He rank ordered the degree of drinking behavior in these societies based on information regarding the degree of insobriety commonly reached by adult male drinkers. He was able to substantiate his first and second hypotheses but not his third, although he tentatively verified statement (c).

Bales (1946) also utilized systems theory to show how social and cultural organizations influence the rate of problem drinking in three ways: (1) the degree to which the culture influences the adjustment of individual inner tensions in its members (e.g., culturally induced anxiety, guilt, conflict, suppressed aggression, and various sexual tensions); (2) the attitude concerning drinking which the culture produces in its members (the pivotal factor is whether these attitudes suggest drinking as a means of reducing inner tensions or instead create a strong counteranxiety to drinking); and (3) the degree to which the culture offers substitute means of satisfaction (if inner tensions are sufficiently acute, individuals will be compulsively habituated to alcohol in spite of opposed social attitudes unless substitute means of satisfaction are provided).

Ullman (1958) has emphasized another aspect of the supracultural framework. He suggests that when societies have well-established drinking customs, values, sanctions, and attitudes which are consistent among all segments of the society and accepted by all, a low rate of alcoholism will prevail. He maintained that the development of the addictive process depends on a psychological state which is precipitated, in part, through sociological variables and attitudes. Specifically, Ullman suggests that problem drinking will occur when (1) there is emotional arousal with regard to drinking; (2) drinking is accompanied by a stress situation; and (3) these aforementioned circumstances occur frequently in a setting where alcohol is used to produce a tension-reducing effect. Furthermore, if ambivalence characterizes this emotional arousal, conflict will ensue and increase guilt.

Jellinek (1959), while attempting to describe individual typologies (alpha, beta, gamma, delta, and epsilon) of drinking behavior, recognized that their occurrence varies according to the specific sociocultural context. For example, in those societies which prescribe large daily intakes of alcohol, only those individuals with high psychological vulnerability run the risk of alcohol addiction. Among those societies and cultures which accept only small daily amounts of alcohol, a small psychological vulnerability suffices for exposure to a high risk of becoming a serious problem drinker. Therefore, the societal rate of problem drinking is more properly ascertained by studying the interaction of two basic factors: (1) the degree of psychological vulnerability among a society's members and (2) the amount of alcohol intake which is socially acceptable in that society.

Specific-Culture Orientation

Many studies have investigated problem drinking within specific societies and cultures. Several investigators have concentrated their efforts on the negative sociocultural conditions to which problem drinking may be a response. For example, Clinebell (1963) suggests that a major factor in the prevalence of problem drinking is the attempt to satisfy religious needs through a nonreligious means (i.e., alcohol). Merton (1957) has focused on those societies which maintain culturally prescribed aspirations (ends) and socially structured methods for realizing these aspirations (means). Finally, numerous investigators have focused their attention on the American culture and these studies will be described in more detail in the next section.

Substructural Orientation

Those social subsystem influences which are thought to play an important role in the development of problem drinking in specific societies and cultures are religion, family relationships, social class, ethnicity, sex, age, and the degree of urbanization. Because these are diverse topics which have been extensively researched, the intent of the following section is to present a brief profile of how each influences the drinking behavior of individuals.

SOCIAL ETIOLOGICAL FACTORS IN DRINKING BEHAVIOR

Various aspects of the social system are important variables among all societies in defining the expected behaviors and the role performances of their members. These variables also relate to drinking practices and those behaviors which are basic to understanding the development of problem drinking. Studies of racial, familial, and other differences which influence drinking patterns are important not only in understanding how problem drinking occurs, but also in planning effective treatment modalities.

Religious Aspects

Most members of a society receive a religious identification at birth even though their parents may have no formal affiliation or involvement with a church. This religious identification usually provides an individual with a frame of reference for viewing the universe and social reality as well as a system of beliefs which often offer sociological mechanisms for the relief of guilt and anxiety. While nominal affiliation may reduce the individual's identification with a given belief system, information about a religious affiliation may still be useful in predicting a given individual's drinking behavior. Variables such as the intensity of belief and the influence of religious ideas are also crucial. However, these factors often make it difficult to judge the impact of religious orientation and affiliation on drinking behavior.

Three American studies correlating drinking behavior and religious affiliation have concluded that the highest percentage of drinkers (87-92%) of any major religious group is found among Jewish-Americans (Riley and Marden, 1947; Mulford, 1964; Cahalan and Cisin, 1968). However, these authors and others (Cahalan and Cisin, 1968; Cahalan, 1970) point out that heavy drinking among Jews is below the national average and that problem drinking among Jews suggests that the rate may be as low as 2 per 1000 (Bailey et al., 1965) to zero (Roberts and Myers, 1967).

Snyder (1962) concluded that, whereas Jews may have a greater usage of alcoholic intoxicants than any other major religious group in America, "both in this country and abroad, rates of alcoholism and other drinking for Jews are very low." He proposed three hypotheses for these findings. First, that a majority of Jews view moderate drinking and sobriety as a "Jewish virtue" while intoxication and hedonistic consumption of alcohol are seen as a "drunken vice." Secondly, he speculated that the Jew is motivated toward moderation not by fear from any out-group, but by a concern about the reactions of his peers who strongly adhere to Jewish norms and customs. The relative effectiveness of this group's control may be attributed in part to the group's solidarity arising from its reaction to tensions between Jews and non-Jews. When in-group relationships are weakened and out-group contacts increase, the presence of heavy drinking increases among Jews. Finally, as originally suggested by Bales (1946) and expanded upon by Keller (1970), the "Jewish ritual" is important, since members are taught how to drink in a "controlled" manner through the religious-cultural integration of alcohol into daily life.

Catholics have almost as high an incidence of drinking behavior, which has been varyingly estimated at 79 percent (Riley and Marden, 1947), 83 percent (Cahalan et al., 1969), and 89 percent (Mulford, 1964). Mulford (1964) found that approximately 12 percent of Catholics were "heavy drinkers" whereas Cahalan et al. (1969) placed this figure at 19 percent. Bailey et al. (1965) estimated that the rate of severe problem drinking among Catholics in New York City was approximately 24 per 1000, which exceeded the rate for Protestants (20 per 1000) and Jews (2 per 1000). Only Negro Baptists had a higher rate (40 per 1000). Unfortunately, little theoretical discussion of Catholic drinking patterns is available. In contrast to the other major religious denominations, Catholic dogma appears to be associated with a more tolerant attitude toward problem drinking behavior.

Investigations regarding the relationship between drinking behavior and Protestantism usually have focused on Lutherans, Episcopalians, Methodists, and Baptists. The percentage of Protestants estimated to be consumers of alcoholic beverages has been varyingly placed at between 59 percent (Riley and Marsden, 1947) and 63 percent (Mulford, 1964). These lower percentages are related to the increased role of an abstinence philosophy among many Protestant denominations.

In his study on religious affiliation and drinking behavior, Skolnick (1958) noted that the Methodist Church is the largest single religious denomination to espouse a total abstinence position and that it has historically been a strongly organized and influential group in the "temperance movement." Mulford's (1964) national survey reported that whereas 61 percent of all Methodists were involved in drinking, only 5 percent could be considered heavy drinkers. Similar figures were noted by Cahalan et al. (1969), who estimated that approximately 66 percent of this religious denomination were drinkers, with only 10 percent being heavy drinkers.

Another Protestant denomination which is strongly opposed to alcohol use and maintains an organized effort against drinking is the Baptist Church. Mulford (1964) found that 48 percent of all Baptists were drinkers but only 9 percent of those surveyed were heavy drinkers. Cahalan et al. (1969) reported that 4 percent of Baptists surveyed drank, but only 7 percent were heavy drinkers. These data refer only to white Baptists. Negro Baptists have a much higher rate of serious problem drinking (Bailey et al., 1965).

Compared to these two Protestant groups, Lutherans and Episcopalians have a more permissive view of drinking behavior. The Episcopal Church sanctions moderate forms of drinking, but is opposed to drunkenness. Cahalan and Cisin (1968) found that Episcopalians had a drinking prevalance (91 percent) exceeded only by the Jews, with 14 percent being heavy drinkers; and 81 percent of the Lutherans sampled in this study were drinkers, but only 6 percent were heavy drinkers.

Cahalan (1970) continued to study differences between Protestant denominations by dividing them into two groups, liberal and conservative, based on their philosophy about abstinence. He concluded that conservative Protestant groups (Methodists and Baptists) had more than twice as many abstainers and only about half as many heavy drinkers.

Strauss and Bacon (1953) have also reported a relatively high prevalence of heavy drinking among Mormons which they interpreted as resulting from rebellious behavior against the prohibitive policies of the church. They concluded that when drinking is not controlled through an approved intake norm, as is the case in religions which preach abstinence, serious problem drinking is more likely to occur and a greater range of drinking patterns may be present.

In contrast, Skolnik (1958) and Preston (1969) have suggested that abstinence itself is an "intake" norm, but that it is the frequent and emotionally charged teachings associated with the abstinence norm which foster "extreme drinking" behavior. Furthermore, since nominal use is prohibited, the role model of the heavy consumer is more prominent. Preston (1969) also hypothesized that the prohibitionistic religious philosophy fosters ambivalent attitudes toward alcohol and that this ambivalence also plays a role in encouraging heavy drinking behavior.

In summary, hypotheses which have been proposed regarding the relationship between religious affiliation and drinking behavior are as follows:

1. Controls over drinking behavior which are derived from a religious group affiliation have a noticeable impact on the membership of that religious group. Those religious organizations which have a more permissive attitude toward alcohol consumption (Jewish, liberal Protestant, and Catholic) report the highest percentages of nonabstainers. In contrast, Methodists, Baptists, and Mormons, with a restrictive attitude regarding alcohol consumption, report the lowest percentages of nonabstainers.
2. Those religious denominations with a higher percentage of nonabstainers tend to report a lower incidence of problem drinkers.
3. These investigations have only presented *theoretical* relationships between religious affiliation and drinking practices. Indeed, Roebuck and Kessler (1972) have suggested that, because of a growing secularism and materialism within the memberships of many religious organizations, as well as a pattern of increasing assimilation, it is likely that religious differences will become less important as determinants of drinking practices.

Social Class Influences

The socialization process as it reflects social class assignment has a tremendous impact upon human behavior. A newborn child begins a process of learning covert and overt patterns of b havior which are associated with the social class of his or her parents. Social class influences values, norms, and standards which govern daily activities. Each class level possesses unique characteristics which may conflict with the cultural pattern of the total society or with other subcultural groups within the society.

The determination of workable criteria for assignment to social class is an age-old problem which continues to confront social scientists who are investigating the relationship between social class and drinking behavior. Researchers have utilized different techniques and methodologies when investigating this problem; but, for the purposes of this brief review, only their theoretical conclusions will be presented.

One of the most comprehensive studies regarding social class and drinking behavior was completed by John Dollard (1945). He found that, within the "true" upper class of society, the level of alcohol consumption was high and more readily condoned. Only antisocial behavior which occurred while inebriated received critical attention. Meanwhile, it was the "newly rich" upper class which maintained a "party" image, drinking with greater abandon than the traditional families of the upper class. According to Dollard, those individuals with relatively "new money" were more insecure and constantly preoccupied with meeting the standards of the more "established" families. He also contended that parental controls were more tenuous in the newly rich upper class and that the marks of social competition had a damaging impact on their offspring, who attempted to escape from their discomforts through alcohol. True upper-class males have a minimal need to strive for wealth and position, resulting in little class pressure. They drink only on formal social occasions or in casual gatherings.

These observations led to a general hypothesis that the attitude toward drinking among class members is strongly influenced by the particular stresses of membership in that class. For example, Dollard also described middle-class males as desperately striving for recognition and status to maintain the disparity between themselves and those whom they consider to be from a lower class. In addition to this pressure, there are also strong taboos against drinking, especially for women, since excessive drinking is a behavior which is associated with the lower class. Among the lower class, frequent drinking is observed both in homes and in taverns, with the latter taking on an atmosphere of a "social club." Drunkenness is not viewed as a disgrace and drinking is socially unrestrained with weekend binges and chronic drunkenness frequent.

In contrast to Dollard, McCord *et al.* (1959) concluded that the upper class, in addition to the lower class, was most likely to produce confirmed serious problem drinkers. The authors concluded that social control of drinking was the weakest among these two classes, with excessive drinking being viewed as an accessible outlet for anxiety and aggression.

Those variables most frequently cited as indicators of social class (both singularly and in combination) such as education, occupation, and income have also been examined with regard to their specific relationship to drinking behavior.

Educational achievement is a major means by which upward social mobility, whether through financial rewards and/or social recognition, is realized. Education is more than the formal learning process by which facts and techniques related to subject matter are learned. It is also a preparation for living and the opportunity to develop human potential. Nonachievement of educational skills may handicap an individual in many ways. Although the research does not suggest a definitive relationship between education and prevalence of problem drinking, authors such as Riley and Marsden (1947), Mulford and Miller (1960), Mulford (1964), and Cahalan and Cisin (1968) have concluded that, in general, as the years of education increase, so does the percentage of drinkers within that educational group.

Findings regarding problem drinking are less clear. Cahalan and Cisin (1968) reported that individuals with a grammar school education or less had the lowest percentage of drinkers (53 percent), but the highest percentage of heavy drinkers (20 percent). However, a high percentage of problem drinkers is also found among those who have completed high school or some college. Therefore, the results of these studies and others suggest that there is no consistent relationship between problem drinking and educational experience.

Occupational activity also serves as a major index of economic and social status in a community. Occupations can be arranged in a hierarchy from those with low self-esteem and few monetary rewards to those with high self-esteem and great financial rewards. Mulford (1964) and Cahalan and Cisin (1968) have indicated that there is a slight, although inconsistent, tendency for the percentage of drinkers to increase with occupational status. However, as with educational experience, this trend is not consistent among problem drinkers. The lowest-ranking occupational group in Mulford's study (laborers and janitors) had the lowest percentage of drinkers (69 percent) and the second lowest percentage of problem drinkers (10 percent). While the highest percentage of drinkers was found in the two highest occupational categories (lawyers and physicians), the percentage of problem drinkers in these groups was the lowest. In contrast, the second highest percentage of problem drinkers (scientists, college professors, and engineers) had the highest percentage of problem drinkers (24 percent).

As one might expect from the above discussion, a positive correlation exists between education, occupation, status, and income. The results of several studies investigating the relationship between drinking behavior and income have shown that there is a general pattern of increased numbers of both drinkers and heavy drinkers as income rises (Riley and Marsden, 1947; Strauss and Bacon, 1953; Cahalan et al., 1969). Similarly, one might also suspect that there is a general pattern of increased numbers of both drinkers and heavy drinkers as education, occupation, and status rise.

In summary, while drinking behavior appears to be more common the higher one proceeds in the social stratification system, it is groups at both the upper and lower ends that have a greater proportion of presumed problem drinkers. We emphasize "presumed" as it is most difficult to obtain an accurate count of those at the upper end of the scale with serious problem drinking because of the lack of reporting from private physicians and medical facilities.

Family Influences

Most social and behavioral scientists stress the importance of the nuclear family as a primary socializing influence. To understand the social etiological factors which play a role in the onset of problem drinking, attention must be paid to the family because of its role as a provider of an early sense of social unity and relatedness to others. Furthermore, the inclusion of the family in this review is justified by the important role which family members serve in modeling behavior for each other. Finally, investigations of familial influences on drinking behavior have been an integral aspect of many prevention and treatment approaches. Only those investigations focusing on the parents of the severe problem drinker will be reviewed.

Early research regarding the families of problem drinkers produced similar results and is best exemplified by the findings of Whitman (1939), who compared the family histories of 100 male problem drinkers with a similar number of nonproblem drinkers matched for age, education, and nationality. He reported that alcoholic offspring had received a disproportionate amount of love from their mothers in comparison to their fathers. In these family environments, the presence of a domineering, oversolicitous mother and a stern, distant, and autocratic father created fear in the child, resulting in feelings of insecurity and dependency.

Subsequent studies have investigated several situational factors in the families of male problem drinkers and these results have both confirmed and repudiated Whitman's early findings. For example, Pittman and Gordon (1958) and Moore and Ramseur (1960) found that while approximately one-third of the male alcoholics felt rejected by their fathers, none had felt overprotected or overindulged.

In contrast, Moore and Ramseur (1960) found that 26 percent of the mothers in their studies of male problem drinkers were overindulging, while 13 percent were viewed as rejecting. Pittman and Gordon (1958) characterized the typical mother–son relationship among alcoholics as follows: "The general thread that runs through the cases is an emotionally impoverished relationship between mother and son, with consequent deprivation of social and psychological ramifications which are usually found in the primary group of the family" (p. 87).

McCord *et al.* (1960) emphasized that both maternal and paternal influences play a role in the development of problem drinking. In over a dozen categories, the authors found statistically significant differences between the familial environment of the problem drinker and the nonproblem drinker. The most critical variables differentiating these two groups were maternal alternation of affection and rejection, maternal deviance, paternal antagonism, parental escapism, and the presence of an "outsider" in conflict with parental values. The authors found that boys who were strongly encouraged by their mothers to be dependent were no more likely to become problem drinkers than those who were encouraged to be independent.

In contrast, the adolescent male who is deprived of a clear conception of the male role due to paternal antagonism and escapism is more likely to become a problem drinker. Deprived of a positive male role model, the adolescent assumes the role of an aggressive and independent male based on social stereotypes. At the same time, strong dependency needs (fostered by the mother) conflict with this self-image. Therefore, the development of problem drinking is often a result of an unsuccessful attempt to compromise between these two conflicting personality elements (independence versus dependence).

The role of parental absence is also commented upon frequently in the research literature. For example, Kinsey (1966) noted that 50 percent of the subjects in his study of hospitalized female alcoholics came from homes where the parents had not lived together regularly during the patient's childhood; furthermore, in 35 percent of these cases, either the mother, the father, or both parents had died during the problem drinker's childhood. Similar data have been reported by Pittman and Gordon (1958) as well as Moore and Ramseur (1960). The specificity of this finding for problem drinkers is called into question by Holtman and Friedman (1953) who compared 500 hospitalized problem drinkers with 600 schizophrenics and four other diagnostic categories. The authors concluded that parental absence was found no more frequently in problem drinkers than in the five other diagnostic groups.

The behavior modeling of parents also plays an important role in the development of problem drinking. Cahalan *et al.* (1969) observed a high correlation between the drinking patterns of parents and those of their children. They also pointed out that, as the attitudes of mothers or fathers toward drinking became more favorable, the percentage of drinkers and heavy drinkers in the entire family increased.

Similarly, Jackson and Connors (1953) theorized that the parents of problem drinkers would have different attitudes toward drinking than would the parents of nonalcoholics. They found that nonproblem drinkers more frequently came from homes in which neither parent drank and where both parents disapproved of drinking. On the other hand, they also discovered that moderate drinkers came equally from nondrinking homes and homes where both parents drank. The authors speculated that those who were raised in nondrinking homes would either develop a negative attitude toward drinking or would be more likely to become problem drinkers. Those children who were raised in homes in which the parents were moderate drinkers were more likely to learn specific and detailed definitions of appropriate drinking behavior. Despite this hypothesis, they found that problem drinkers more frequently came from homes in which one parent (usually the father) engaged in some form of drinking. As a result, Jackson and Connors concluded that it is the presence of an "ambivalent" environment which is most likely to contribute to the development of a problem drinker because of the absence of a well-organized family structure with firmly held attitudes that could lead to restraint from excessive drinking.

However, Robins, Bales, and O'Neal (1962) point out that fathers who drink excessively are no more likely to produce offspring who become problem drinkers than those fathers who have been arrested, unemployed, deserters of their families, or sexually or physically abusive to wives or children. The authors concluded that the critical factor in the father's behavior which influences the development of problem drinking is the antisocial component rather than the father's drinking habits.

In summary, research about the families of problem drinkers has frequently centered on the lack of appropriate role models due to parental deprivation, parental absence, antisocial behavior including problem drinking, and the various constellations of paternal and maternal attitudes regarding alcohol which influence their children. The research literature suggests that the parents of problem drinkers deviate from the norm and fail to relate to their prealcoholic children in a manner which is conducive to successful adult adjustment. However, it is important to note that, as Roebuck and Kessler (1972) emphasize, sufficient statistical data are not available to define more specifically the nature of those relationships so that accurate prediction of later problem drinking can be made.

Ethnic Aspects

The ethnic group affiliation is a major aspect of the cultural milieu in which an individual is socialized. Recent research has examined the possible critical role which ethnic background plays in the development of problem drinking and its impact on treatment approaches.

Despite some contradictory evidence, a majority of the research suggests that the incidence of problem drinking is generally higher among Negroes than whites (Cahalan and Cisin, 1968; Bailey et al., 1965; King et al., 1969; Maddox and William, 1968). This higher incidence of problem drinking among Negroes is most often attributed to the frustration which they experience living in a white American society (Frazier, 1947; Strayer, 1961; Locke et al., 1960).

McCord (1969) and Cahalan (1970) suggest that the Negro experiences a higher degree of stress because of the unfavorable outlook toward future relationships with family, friends, and life goals. McCord also hypothesized that, in many Negro families, the father does not reinforce or exemplify a positive male role. Therefore, male offspring are faced with a confused role definition and react by creating a pseudoindependent facade. In addition, the white society (with its attitudes) tends to undermine the tenuous conception of manhood to which the young Negro male is exposed during childhood. Finally, alcohol becomes a source for acting out the resulting frustration because of its easy availability and cheapness.

This increased incidence is not limited to Negro males since Negro women also have a higher rate of problem drinking than white women. However, they also have a higher rate of total abstinence than white women (Cahalan et al., 1969; Bailey et al., 1965; Locke et al., 1960). Strayer (1961) explains that historical and economic factors have made the Negro family strongly matriarchal and this places additional burdens on the female. These responsibilities may give Negro women more motivation for sobriety, leading to the higher abstinence rate. On the other hand, Knupfer (1963) suggests that, when women are less economically dependent on men, their behavior more closely resembles that of men—possibly explaining the higher rate of problem drinking in Negro females. Their explanation is supported by data which show that, while fewer than 50 percent of Negro females drink, the rate of problem drinking among those who drink is relatively high.

Robins et al. (1968) have also pointed out that the adult drinking pattern of the Negro is associated with an increased incidence of other deviant behaviors such as family instability, irregular work habits, and trouble with the law. These are also factors which play a role in the higher incidence of problem drinking which is seen in their children after they become adults. Consequently, excessive drinking, poor parental performance, and childhood behavior problems are connected in a vicious cycle.

The Indian Health Service believes that no other condition adversely affects so many aspects of Indian life in the United States as does problem drinking. The death rate from cirrhosis of the liver among Indians has been reported as three times that of the remaining U.S. adult population (Harris and Harris, 1968); all alcohol-related deaths are 12 times the national average (Harris and Harris, 1968); the arrest rate for alcohol-related offenses is 5 times greater than in the rest of the population (Ferguson, 1968; Whittaker, 1963; Stewart, 1964); and the rate of alcohol-related traffic fatalities is similarly higher (Stage and Keast, 1966).

Observation of Indian drinking patterns indicates that lone drinkers are less common than in other ethnic groups. In most Indian communities, the solitary drinker is considered a deviant by both drinkers and nondrinkers alike (Ferguson, 1968; Kunitz *et al.*, 1971; Littman, 1970; Dutoit, 1964).

Littman (1970) notes that American Indians more readily admit their unwillingness to give up drinking than non-Indians because they do not feel as stigmatized by their problem drinking. This resigned attitude toward their drinking practices appears to tie in with certain religious beliefs.

However, differences in intrafamilial relationships, tribal backgrounds, and the degree of acculturation noted in various tribes make it difficult to find a single set of dynamics which applies to all Indian problem drinkers. Kunitz and Levy (1974), summarizing the research literature, have suggested the following reasons for the higher rate of problem drinking among Indians: (a) the learning of drinking behavior by Indians was taught by whites; (b) the use of intoxication to achieve aboriginal goals with the survival of those goals often coming into conflict with the stress for social change and cultural contact; and (c) the adoption of deviant rules of behavior by those Indians whose access to positions in white society has been blocked through the traditional system.

Age, Sex, and Urbanization

Age, sex, and urbanization represent important variables which all societies use in defining the expected behaviors and role performances of their members. These characteristics have also been studied in relationship to drinking practices. However, these variables overlap significantly with those previously discussed.

The concept of age grading is a phenomenon observed in all societies and implies that expected behaviors change as an individual moves from childhood to adolescence, to adulthood, and into old age. For example, as applied to problem drinking, Maddox (1962) concluded that the drinking practices of youth are highly predictable from the drinking patterns of their parents and that first drinking experiences tend to occur in the home under parental supervision.

In a longitudinal study surveying men between the ages of 18 and 59, Cahalan and Room (1972) found that drinking problems were most prevalent in the youngest group of males (18–24) and that most drinking problems diminish rapidly after age 50 among those who survive. This had been previously suggested by Drew (1968), who also concluded that problem drinking disappears with increasing age. However, these authors (Cahalan and Room, 1972; Drew, 1968) also pointed out that environmental and sociological factors are more predominant correlates of problem drinking than age.

Women usually develop drinking problems at a later age, in their 30s and 40s. Relatively few women reported drinking-related problems prior to this age group. Several explanations have been advanced for this later onset including ambivalent conflicts about marital dependence and increased feelings of alienation as children grow up (Kinsey, 1966).

Cahalan (1970) reported that drinking-related problems are higher in larger cities than in small towns or rural areas. Cahalan et al., (1969) also reported that the highest proportion of abstainers and infrequent drinkers were among those age 45 and over who live in low urbanization areas. The authors hypothesized that these variations reflect different social pressures regarding drinking behavior with greater social control being available in smaller cities, towns, and rural areas. However, Cahalan (1970) pointed out that, although there is a higher prevalence of drinking problems among men in urbanized areas, this may be partially a result of people at risk migrating to the cities where support resources are more available.

While these variables are probably the least significant in influencing the onset of problem drinking behavior, they should not be overlooked because of their value as part of larger demographic studies of other social factors which contribute to the onset and continuance of problem drinking.

TREATMENT APPROACHES

During the past 25 years, the abundant research into those social factors which influence the development of problem drinking has also had an important impact on treatment approaches to this problem, leading to a dramatic spiral of interest in social treatment approaches.

The social model of alcohol rehabilitation may be defined as a complex of theoretical approaches specifically designed to assist those whose problem drinking behavior cannot be modified without a change in the social context in which function (Enelow, 1974). Its guiding principle is that dysfunctional alcohol use cannot be altered by attention only to the individual since a large portion of behavior is determined by external forces. Accordingly, any attempt to change problem drinking behavior requires paying significant attention to social influences upon that behavior. The social rehabilitation model abandons as a primary goal trying to change an individual problem drinker so that he will fit into the society in which he lives. Instead, it focuses on changing the social context in which he lives.

Prior to examining specialized social treatment approaches, it is necessary to comment upon two generic aspects of the model. First, a more precise definition of "rehabilitation" is needed. Traditional definitions do not adequately express the ultimate goal of social treatment since, by definition, rehabilitation implies a restoration to a former state of well-being. In contrast, the social rehabilitation model for problem drinkers stresses the achievement of a state of mental and physical health not previously experienced. Furthermore, therapy is considered a social learning process following which the successful "graduate" no longer needs alcohol and has made a personality adjustment that allows for a more mature and objective management of his or her daily life encounters. This leads to improved socialization with a more meaningful relationship between the individual, family, friends, and the community.

Second, although alcohol abuse has a significant effect on the individual's ability to adequately adapt to his environment, treatment outcome in this model is assessed along dimensions of social adjustment. Utilization of social outcome measures is consistent with the theoretical position that problem drinking is a phenomenon which affects all aspects of an individual's life pattern. While Foster *et al.* (1972) suggest that these could be equivalent to the "traditional" criterion of abstinence, a composite measure of sociological functioning is preferable to a single measure which uses only abstinence.

Within the context of this theoretical framework, many social treatment approaches have been developed and are currently in use. In the remainder of this section, we will examine the prominent ones.

Alcoholics Anonymous

Alcoholics Anonymous (A.A.), a nationwide network of voluntary support groups, composed of recovering and recovered problem drinkers, is one of the foremost social treatment approaches. Although it is directly available only to problem drinkers, it also provides support to the families of problem drinkers through associated organizations such as Al-Anon and Alateen. Founded in 1935, A.A. now serves more than 300,000 problem-drinking men and women who are dedicated to maintaining sobriety through total abstinence. The organization consists of more than 9000 active groups, of which about two-thirds are in North America and the remainder in 80 foreign countries (Blum and Blum, 1967).

The major goal of A.A. is to assist members in attaining and maintaining sobriety, with the only qualification for membership being the desire to stop drinking. Each member pledges to help any problem drinker, whenever called upon, either by being a constant companion, by providing understanding and sympathetic help at any hour of the day or night, by assisting with domestic–financial–legal problems, and/or by sharing shelter. There are no membership dues or fees.

The A.A. approach to the treatment of alcohol dependency requires the provision of equivalent gratification without the use of alcohol. This is commonly attained through the substitution of gratifying interpersonal relationships in place of alcohol.

Through these "new" interpersonal relationships, role models are provided which help an individual problem drinker to see how others with similar problems in the past have attained sobriety. If a problem drinker meets a member of A.A. with whom he has something in common, particularly during the early stages of his struggle to attain abstinence, this can greatly facilitate acceptance of the difficulty and bring about what Tiebout (1944) has called the "surrender phenomenon." This phenomenon involves an increased acceptance of the essential role which drinking has played in causing living problems. A key feature is the role modeling arising out of the identification that takes place between the problem drinker and the "recovered" A.A. member.

This "initial surrender" begins the recovery process, but it is probably in the continuing, long-term "treatment" of the problem drinker that A.A. has proven most successful. Since one of the problem drinker's biggest frustrations has often been the severe social constrictions which the use of alcohol has placed on his or her life, by welcoming members and encouraging them to participate in social interactions, A.A. can result in a problem drinker becoming part of a group for the first time in many years. Although on the periphery at first, but at least in contact with others who have had similar experiences, the problem drinker gradually finds it easier to build and maintain trust than he would if contact was only with the general population. The problem drinker learns that there are people who understand him and are interested in him. If these initial contacts are successful, improved self-esteem and a new willingness to take chances with others may quickly lead to further improvement in interpersonal functioning. If the new member has difficulty in accepting his new behavior and new relationships, the "safe" environment of A.A. can help him to tolerate those normal aspects of human interactions which he had found so troubling in the past.

Two essential programs assist in the implementation of this philosophy. Although an extensive explanation of their features may be found elsewhere (Alcoholics Anonymous Comes of Age, 1957; Alcoholics Anonymous, 1960;

The Twelve Steps and Twelve Traditions, 1953). A brief summary follows:

The Twelve Steps are a series of statements set forth to describe what successful members must overcome to remain sober. The steps are guideposts for the problem drinker attempting to reorganize his life. These 12 steps may be summarized briefly: (1) accepting oneself as an alcoholic, i.e., as unable to drink in a controlled situation because of physiological and emotional pathology; (2) a willingness to change one's personality, relationships, and way of life in a manner conducive to healthier adjustment; (3) accepting daily spiritual help from a self-defined God; (4) taking an honest and continuing "personal inventory"; (5) taking action to change dysfunctional personality attributes and behaviors as they are recognized; and (6) helping other problem drinkers to recover.

There are also the Twelve Traditions, a series of rules governing the behavior of A.A. groups and individual A.A. members in relationship to outsiders. The traditions make it clear that A.A. exists for the purpose of achieving only one goal, helping an individual to maintain sobriety. They specify that A.A. will become involved in no other activity and that members will use A.A. facilities only for achieving and maintaining sobriety.

In most large cities, a problem drinker may attend a formal A.A. meeting at almost any hour of the day or night or become involved in an impromptu meeting at one of the many club headquarters. A majority of A.A. members actually make initial contact as a result of a decision made during an emergency. When an emergency occurs, A.A. immediately sends out two members before the problem drinker has a chance to rebuild his defenses. This "emergency team" devotes its energies to establishing rapport, providing emotional support, assisting the problem drinker in defining the nature and extent of the problem, and helping him to decide on a course of action once the alternatives are clear. The atmosphere is permissive, accepting, and unemotional. The crisis team members often use personal stories to illustrate their own ambivalences at the onset of their membership and how they were resolved. This enables a potential member to recognize that his callers are people with similar problems. Because he is informed about their problem-drinking background, he is able to identify with them and feel that his own actions are not particularly unique, reprehensible, or hopeless. The helpers agree that a problem exists and give the problem drinker an opportunity to discuss these problems in a matter-of-fact atmosphere. The team members avoid being drawn into the individual's feelings about himself, his rationalizations, or his searching for causes. Their attitude is that a problem exists whether or not the specifically claimed rationalizations and punitive causes are valid. The A.A. program is outlined to the problem drinker along with other alternatives. He is left with information on how to reestablish contact with A.A.

Should an individual choose to join A.A., he is assigned a sponsor who undertakes voluntarily to work with the new member. The sponsor believes that the effort (whether or not it succeeds) to help another problem drinker furthers his own sobriety. The sponsor assumes the role of an "older" experienced individual who can provide emotional support, understanding, and guidance, assisting the new member to clarify his own thinking as well as to work toward sobriety. It is emphasized to the new member that he has no obligations to his sponsor and can terminate the relationship whenever he chooses.

A.A. offers four types of meetings: (1) general meetings open to the public, (2) meetings open to select nonproblem drinkers, (3) closed meetings for A.A. members, and (4) study group meetings. Group meetings are the primary therapeutic instrument. The group, acting like a family, allows new opportunities for its members to develop healthy interpersonal relationships with others and to arrive at a more realistic and satisfying self-conception.

Most members begin by visiting different groups until they find one in which they feel comfortable. Groups cannot refuse to accept a member, nor can they ask a member to leave. However, it is understood that any member who feels uncomfortable will leave and look for another group which is more congenial. The groups have no formal authority structure with the only authority being derived from the prestige of maintaining sobriety and the ability to help others resolve their problems through diminishing anxiety and enhancing self-esteem. Group meetings are informal and there are no tabooed or insignificant subjects.

From the research literature (Maxwell, 1962; Trice, 1957, 1959; Mayer and Black, 1974) it is possible to formulate some conclusions regarding the individual who is most likely to experience some success in Alcoholics Anonymous. He or she is most likely from the middle class. Mayer and Black (1974) suggest that the skid row population is unable to respond to the demands for regular attendance or to the verbal interchange regarding problems necessary at formal meetings. Furthermore, the necessary social and economic support which these individuals require is not within the scope of A.A. A.A. also has difficulty attracting problem drinkers from the upper class, possibly because of the upper class's traditional rejection of abstinence from alcohol as a treatment goal as well as a desire to avoid public exposure and social interaction with other levels of society.

For the middle-class problem drinker, the social contacts provided, the structured guidelines concerning what should be accomplished in treatment, and the opportunity to identify and associate with other problem drinkers are attractions of this therapeutic modality. The intellectualization in the A.A. system provides a conceptual scheme based upon a belief that problem drinking is a physical, emotional, and spiritual disease. This is a comfortable and familiar mode of interaction for the middle-class person and also allows for alleviation of social guilt through helping other problem drinkers.

Therapeutic Communities

The original model for the "therapeutic community" was established by Maxwell Jones (1962) in London at the Belmont Hospital. Because of his experience in using large groups in the rehabilitation process of various psychiatric disorders, Jones was commissioned to establish a similar program for people with "social" disorders including alcoholism and drug addiction.

The 24-hour therapeutic community for problem drinkers initially was located in a hospital setting. This presented some problems because the environment needed was substantially different from that required for the physically ill. When a patient enters a hospital for a physical illness, a specific treatment such as surgery is involved. To maintain that individual within the hospital, he is given a bed and three meals a day, provided with a television, and encouraged to move about the hospital should he be ambulatory. However, these adjuncts are not considered part of treatment, but rather amenities necessary to maintain good morale.

However, with the alcoholic patient, except in an acutely intoxicated or withdrawal state, there are no signs or symptoms which require confinement to a bed or to a room for treatment. Rather, the entire hospital is the treatment environment and becomes a laboratory for observing, influencing, learning, or testing. For example, mealtime is no longer considered an adjunct to treatment, but rather a critical aspect of treatment since the patient's attitude toward food and his behavior in the dining room may provide opportunities for the staff to understand and treat him more effectively.

In the therapeutic community, the total experience is essential to therapy. Jones and his followers proposed that friendliness and stimulation are essential components of a therapeutic atmosphere with the total staff being involved with the total patient. In the therapeutic community, the patient examines his feelings and reactions to all situations and discusses them with staff and other patients. He is encouraged to try new methods of behavior and to test new roles with other patients and staff.

A major effort is made within the treatment setting to treat everybody, including patients, nurses, physicians, and other staff, in an equal and democratic fashion. Whenever possible, staff participate freely in social situations and at social functions. This informality communicates to the patient that the professional staff (especially the physician) is a benign authority and counteracts unrealistic notions about its magical powers.

The treatment program of a therapeutic community involves a variety of sessions. During these sessions, many opportunities are provided for interaction and discussion of problems. Among these meetings are: (1) a daily community meeting of patients and staff, often called a patient government meeting; (2) social and recreational meetings with staff and/or volunteers; (3) discussion groups which often focus on social problems and treatment for individual patients (aside from the intellectual and emotional stimulation these discussion groups provide, their aim is to settle the various problems associated with group living, promote patient participation in unit activities, and allow individuals to discuss with the group their personal problems); (4) psychodrama; and (5) follow-up groups in which planning takes place for vocational roles to be assumed later.

A therapeutic community has three goals: (1) establishing an egalitarian atmosphere; (2) transmitting a therapeutic orientation staff to patients to new arrivals; and (3) promoting daily discussions involving the entire treatment unit. It encourages a patient to demonstrate his attitudes, abilities, and behavior and provides opportunities to observe behavior which may indicate therapeutic directions that will substantially aid in recovery. The ultimate goal of a 24-hour therapeutic community is to establish ways in which problem drinkers can learn, through their social experiences in the program, new ways of approaching those social problems which have been handled, in the past, through the abuse of alcohol.

Halfway Houses

Halfway houses provide the problem drinker with another experience in group living. Generally, they are used as an intermediate stage from hospital or jail to the community. Although houses differ in their mode of financial support and other minor characteristics, the basic structural features, as described by Rubington (1967), are common to most of these environments.

Most halfway houses are managed by recovered problem drinkers who serve as successful role models to the new residents. The primary goal for all members is a return to a higher level of functioning without reliance on alcohol for support. Following an initial drying-out period, usually done outside the halfway house (in a hospital or detoxification center), most houses require that residents find some kind of gainful employment and/or share in household duties. Members are encouraged to form their own discussion groups within the house to help each other resolve personal and group problems. Many halfway houses also require attendance at A.A. meetings. As Weppner (1973) suggests, most halfway houses are oriented toward a living mode more vigorous than any which exists in the external middle-class society. For example, a daily routine in many houses is to rise early, have a morning meeting, work (either in or out of the house), attend an educational seminar in the afternoon, and then return to work. Evenings are devoted to reading or listening to approved books or tapes as well as to more discussion and group therapy sessions. Conversations which do not relate to problem drinking, group activities, or the house philosophy are considered "negative" and are not encouraged.

In the halfway house environment, all ties to the outside are initially severed in an effort to strengthen the "primary group nature" of the self-help organization (Cooley, 1909). This encourages the new resident to submerge himself totally into the activities and philosophy of the group and to subscribe completely to its rules and values. The newcomer quickly learns that peers will inform "old-timers" in the house about any deviation from the norm. Group therapy sessions may take several different approaches, but will generally begin with a confrontation emphasis. As the individual adapts to the halfway house situation, more support-oriented groups take place to assist him in leaving the environment and returning to the "outside" world.

Group Therapy

Many of the same methods utilized in individual therapy have been applied to the treatment of problem drinkers in groups. Group psychotherapy has the practical advantage of providing treatment to many persons exhibiting similar problems. Many variations are possible and have been described at length by Fox (1962) and Weiner (1965). Within these groups the methods often vary considerably, depending upon the theoretical orientation of the group leader. In addition to the varieties of group therapy, there are also different techniques for conducting the group sessions.

The origins of group therapy can be traced back to ancient Greece and the Dionysian courses. Aristotle described these day-long performances as emotionally cathartic. More recently, formal group psychothcrapy was described in the early 1700s by Anton Mesmer, who conducted group hypnotic sessions in France.

In the early 1900s, a modern adaptation of Aristotle's principles surfaced through Moreno's adaptation of the classical Greek drama which he called psychodrama. Moreno utilized a stage on which the protagonist reenacted psychological situations, problems, and conflicts under the watchful eye of a "director." Among problem drinkers, this technique has proved popular and now constitutes one of the most extensively used treatment modalities.

Group psychotherapy theory is based on the belief that social behavioral disorder arises from difficulties with interpersonal interactions. Consequently, attempts to understand and ameliorate these disorders can be carried out best in a group setting. A member can discover how his self-image is distorted by observing others and their reactions to him while operating in a safe environment. Consequently, a primary benefit of the group method in the treatment of the problem drinker is broadening and deepening the limited base of identification previously held (Mullan and Sangiuliano, 1966).

The rationale for placing a problem drinker in group psychotherapy is derived from the isolationism and sense of nonbelonging which is often behind the facade of "sociability." The problem drinker functions precariously in society and carefully selects friends and acquaintances who will accept his behavior. As a corollary, he withdraws from those groups which will not tolerate his behavior. This leads to a giving up of important relationships with family, friends, and associates. In the end, the problem drinker is only able to relate to other problem drinkers since they have few expectations or demands for him.

Group psychotherapy confronts and attacks this social behavior pattern. Murphy (1963) suggests three basic advantages of the group model: (1) The group can supply a warmth and cohesion which resembles family solidarity; (2) the group can prepare members for the future by giving them opportunities in the group to experiment with various forms of social adaptation such as love, cooperation, and disagreement, which can be later utilized outside the group; (3) individuals in the group are allowed to experience giving as well as receiving.

A group usually consists of approximately 7 to 12 members who, together with a trained leader, meet to explore their interpersonal relationships within the group. The group focuses on the immediate feelings of members towards themselves and/or others and places a great value on their honest expression. A heightened level of effective expression, properly directed, leads to the development of a group atmosphere of trust and support. The group leader's task is to create a group structure which encourages the development of this atmosphere.

Yalom (1974) suggests that the major forces which promote change within the group are interpersonal learning, group cohesiveness, catharsis of emotions, the development of socializing techniques, increased altruism, the instillation of hope, the imparting of information, the corrective recapitulation of the primary family group, and imitative behavior. All of these factors do not operate in every group and some may be more relevant and helpful to a particular group member than others. As Yalom conceptualizes the value of groups for problem drinkers, the goal is not to help an individual attain sobriety but to assist him in working through those underlying socialization conflicts which have produced the need for alcohol. The group's task is to assist members in overcoming feelings of self-contempt, loneliness, alienation, and disengagement; to understand and alter abrasive, maladaptive, and self-defeating styles of self-presentation; and to help members involve themselves meaningfully in a peopled world.

Activity Groups

There are many activities which are used to promote increased mental and physical health in problem drinkers. Through activity group therapy, new interests and outlets for self-expression are offered to engage the problem drinker in activities such as fishing, organized sports, dancing, arts and crafts, and hiking, or structured social activities such as dancing, bowling, and bridge playing.

The activity group provides a vehicle for bringing people together; for providing positive reinforcement for acceptable social intercourse; for offering nondestructive and nonthreatening outlets for aggressive and competitive drives; for stimulating the growth of affection; and for reducing fears of intimacy and tenderness (Blum and Blum, 1967). The therapist in these groups can structure at his discretion the organization of the activity to increase or suppress spontaneity and emotional expression.

Slavson (1943) points out that the group should be a substitute family which incorporates positive elements and counteracts many of the destructive or undesirable relations that problem drinkers have had with their own families. Furthermore, he suggests that an important phase of treatment is to redirect aggression, through sublimation, into socially or spiritually significant activities. In activity groups, individuals are encouraged to vent interpersonal hostility and emotions on materials rather than on each other.

Another virtue of activity groups is that nonverbal individuals who are inhibited in communicating their problems (as many problem drinkers are) need not resort to words in order to express themselves. This relieves the pressure to relate to people until the problem drinker is ready. This particular social treatment has significant implications for those problem drinkers who come from a social class which does not encourage intellectualizing or verbal communication or who are from different cultural backgrounds and have language difficulties.

Slavson (1943) points out two important elements of these groups which are essential to the development of an improved self-image in its members. First, the group must be permissive, allowing the individual to discover that the world is not necessarily frustrating, denying, and punitive. Second, because power is diffused within the group, the patient is subject to group control rather than individual therapist control. Since the group is involved in real activities which are of interest to its members, the manner in which authority is exercised is both vital and concrete. The problem drinker learns to control his impulses and to adjust to the exercise of power.

Family Therapy

The social treatment model also encourages giving help to the families of problem drinkers since they play an important role in the recovery of the problem drinker. Having often been raised within a disturbed family background, the problem drinker in turn creates disturbance in his own family. Sometimes the impact of the drinking problem on the spouse and children may be more destructive than on the problem drinker. Spouse and children may also need help to learn ways in which they may be contributing to the problem and how they can assist in the recovery. Although a spouse may not be responsible for the problem drinking of his or her partner, the reaction to the behavior may make the situation worse. Without the family's involvement in the therapeutic process, the prognosis is poorer. Although the identified problem drinker may wish to deny his negative impact on the family, it is important to engage the entire family in meaningful self-evaluation and dialogue.

Family therapy for problem drinking usually focuses on marital communication, family equilibrium, and pathological family interactions (Meeks and Kelly, 1970; Esser, 1971). The goal of therapy is not to single out the problem drinker for separate treatment as an identified patient, but rather to study and treat the whole family and to uncover the individual problems of all family members. Grandparents or brothers and sisters may play an important role in maintaining family homeostasis and, in these cases, an extended family approach is appropriate.

The importance of the family relationship—not only in bringing about or sustaining patterns of alcohol abuse, but also in being involved in improvement—has been recognized for many years (Jackson, 1958). There are many complex mechanisms by which improvement or lack of it is mediated by family relationships. Work with the family as a therapeutic unit received its initial impetus in 1949 when Malzberg published his study of alcoholic admissions without psychosis to licensed hospitals in New York State (Malzberg, 1949). These nonpsychotic alcoholics exhibited a much higher degree of social and marital stability than the alcoholics with psychosis which he had previously described, but lower than the general non-problem-drinking population. Similar results were noted by Strauss and Bacon (1953), Lisansky (1957), and Rosenbaum (1958).

A review of the literature reveals a progression from consideration of the spouse as a "part" of the problem drinker's environment to a concern about the spouse as a person to his or her own right and, finally, to the current focus on the interaction between marital partners. The hypothesis is that improvement in the area of marital conlict is correlated highly with improvement in drinking behavior. Counseling which focuses on the marital pathology of problem drinkers may, therefore, be therapeutic not only for marital problems but for the drinking problem as well.

Other investigators have studied the manner in which the family reacts to the problem drinker in its midst. For example, Lemert (1961) described the progressive social isolation of the alcoholic family and how the impoverishment of the male role leads to decreased economic status, an impaired decision-making process, poor child-rearing behavior, and inadequate sexual performance.

While the family treatment method chosen often depends on the philosophy of the therapist, a significant factor in bringing about effective treatment participation is the establishment of a pattern of communication among family members and a commonly accepted goal between all participants. Only in this way can the role adaptations which have been thrust on family members as a reaction to the condition be examined effectively.

The Social Learning Approach

The social learning approach utilizes specific techniques based on the assumption that problem drinking is a product of social experience rather than organic or inherited factors. Application of this approach arises out of an understanding of the influence of learning on behavior (Ullman, 1958; Shoben, 1956; Conger, 1956; Kingham, 1958).

Within the social learning framework, alcohol abuse is viewed as a socially acquired, habitual behavior pattern maintained by specific reinforcement contingencies. Excessive drinking enables the alcoholic to avoid or escape from unpleasant, anxiety-producing situations; to exhibit more varied, spontaneous social behavior; to gain increased social reinforcement (either positive or negative attention) from a spouse, family, friends, or business associates; or to avoid the withdrawal symptoms associated with termination of drinking behavior. The apparent lack of influence of the unpleasant consequences of excessive alcohol consumption (i.e., loss of family or job, physical disorders, hangovers, etc.) may be related to the length of delay between the onset of drinking behavior and the occurrence of these aforementioned events (Vogel-sprott and Banks, 1965).

In learning theory, problem drinking is also described as a pattern of learned behavior in which the learned response, or symptom, is inappropriate to the situation. Because of individual differences involved in the ability to learn, generalize, and distinguish required responses, different people are likely to have different patterns of symptom behavior. Also, in a specific individual, some responses and sensory modalities may be more or less resistant to the processes of acquisition, generalization, and extinction than others.

Dollard and Miller (1950) have offered a brief explanation of learning theory. For an individual, four basic requirements must be met: (1) He must want something; (2) he must notice something; (3) he must do something; and (4) he must get something. There must also be (a) either a primary drive (i.e., hunger, sex, thirst) or learned drive (fear); (b) a cue which serves as a distinctive stimulus from the environment; (c) a response, either behavioral or psychological; and (d) a reward (reinforcement) which acts as a positive reinforcer causing the act to be reproduced or a punishment which acts as a negative reinforcer leading to avoidance behavior.

Kepner (1964) suggests that the use of alcohol is a source of two important rewards: (1) The physiological changes produced by alcohol-induced states may be experienced as intensely pleasurable, thereby acting as positive reinforcers and (2) alcohol may also provide temporary relief from unpleasant or punitive stimuli such as anxiety and guilt. Each time alcohol is used to achieve one of these aims, the drinking response is reinforced and the tendency to repeat the behavior is strengthened. With time, an individual may use alcohol to avoid all anxiety, thereby becoming a problem drinker.

The treatment of problem drinking based on learning principles is not a simple reconditioning process designed to remove the symptom. A comprehensive social learning model involves a twofold approach (Kepner, 1964). First, somatic complaints must be eliminated before any emotional and personal problems can be successfully treated. Continued drinking while in treatment reinforces the addiction, complicates the already existing problems, and interferes with the new learning which must take place if treatment is to be effective. Kepner suggests setting a series of graded tasks leading to sobriety. If closer approximations to the final goal (sobriety) are consistently rewarded, new patterns of behavior may eventually be established.

Kepner (1964) proposes the following rewards for encouraging sobriety among patients. The therapist should adopt an accepting attitude which values the problem drinker as a human being rather than devaluates him as a "drunk." Drinking behavior should be regarded as a serious but not a moral problem and the individual should not be made to feel sinful, woeful, ashamed, or guilty. Although medication may be used to relieve anxiety experienced by the problem drinker when problems are overwhelming, the therapist should also encourage other necessary steps toward sobriety by offering support, understanding, and encouragement. By continuing to reinforce the achievement of sobriety, the problem drinker begins to experience a sense of accomplishment. The miseries of early sobriety may be alleviated in part by encouraging an individual to find substitute satisfactions to replace drinking such as hobbies, physical recreation, community activities, intellectual pursuits, or employment, or even relearning the art of being husband and father or wife and mother.

In the social learning approach, the therapist offers a nonpunitive audience and permits the previously learned maladaptive attitude and responses of the problem drinker to be expressed, examined, and changed. Treatment of the problem drinker begins with eliminating the urge to drink and is followed with a treatment of subsequent symptoms and exploration of the problem. A careful investigation of the frequency, timing, and locale of drinking behavior uncovers the "punishment" which he has been trying to avoid through the use of alcohol. In some cases, abstinence alone can restore the patient psychologically and adjusted behavior can be maintained without further treatment. However, abstinence alone is not enough for most problem drinkers and many individuals have a previously learned repertoire of responses that are still inhibited or blocked by continuing responses of anxiety and guilt to social situations.

In attempting to rearrange social reinforcements in a problem drinker's environment, significant persons in the environment are also taught ways of reinforcing new patterns of sobriety behavior and of punishing or extinguishing excessive drinking behavior. Sulzer (1965) made peer companionship and spouse attention contingent upon non-alcohol-drinking behavior. The patient's friends were instructed to meet him periodically for a "drink." At this meeting, if he ordered an alcoholic beverage the friends were to leave immediately. If he ordered a nonalcoholic beverage, they were to remain and provide social reinforcement through conversation. Furthermore, continued sober behavior was socially reinforced by verbal remarks made by the wife and therapist. Cheek *et al.* (1971) trained wives to apply social learning techniques to those behaviors of their problem-drinking husband which were most disruptive to the family. Miller (1972) applied behavioral contracting to a problem drinker and his wife in an effort to alter reinforcement contingencies which maintained drinking behavior.

It appears that elimination of drinking through chemical or psychological means is an insufficient solution to this complex problem. In the future, social learning techniques aimed at changing reinforcement contingencies in the environment (home, work, social, recreation) offer greater promise for more effective treatment.

Educational Approaches

An educational approach assists the problem drinker in acquiring information which will help to improve both his physical and mental health. In this approach, usually conducted in a group setting, the problem drinker first learns the inappropriateness of his present way of life, then acquires a realization that alcohol is an important cause of that inappropriateness and that there are certain tolls which can be used to change this inappropriate behavior. The group functions to educate the problem drinker as to how alcohol leads to family, social, and vocation disruptions and then orients him to community facilities and agencies which can assist in treatment.

As in group therapy, the educational group also traces its origins to ancient Greece, but its composition and goals differ considerably from group psychotherapy. The educational group enrolls students rather than patients. Those joining the group may be individuals with alcohol problems, relatives, or those merely interested in learning more about problem drinking. The group is generally organized as a course with lectures and discussion.

A program developed by Marsh (1935) best exemplifies the traditional educational approach. Classes may be held in a variety of settings. A classroom environment helps maintain a pleasant atmosphere and encourages optimism and light humor. Students have an opportunity to become involved in classroom management (arranging chairs, taking attendance, and greeting newcomers) to increase their sense of personal involvement. Efforts are made to impress upon the students from the beginning that they are not different and this subject matter is not unique to them. In an effort to develop powers of attention and concentration, Marsh encourages note taking, which also assists in creating an educational atmosphere.

He also suggests using didactic techniques (or classroom exercises) to pursue appropriate goals. These include the development of peer support through assigning one student the task of aiding or befriending another classmate. This assistance can be helpful not only to the individual receiving help but also to the person giving help. Once the group has been well established, the subject matter can move to more personal discussions involving family history, social assets and liabilities, and feelings of inferiority. As in group psychotherapy, the sharing of information enables an individual to realize that his feelings and aspirations are not unique.

Other techniques involved in the didactic educational model include the presentation of special films which deal with specific problems associated with problem drinking. These films often depict typical family interaction problems and can be a starting point for discussions of more personal concerns by the class members. The direction of any discussion depends on the skills and desires of the group leader but, whenever possible, emphasis is placed on problems in daily living that are linked to excessive drinking.

The educational approach serves as an excellent vehicle for helping problem drinkers because it may be the first step toward treatment and lead to a deeper therapeutic involvement by providing sufficient support to allow for a regrouping of forces.

CONCLUSION

In this chapter, we have attempted to concisely describe the theoretical, etiologic, and therapeutic aspects of the social model of problem-drinking behavior. In contrast to the more traditional biological and psychological models, the social model attempts to integrate various external influences which impact on the problem drinker and to recognize that these influences play a major role in determining the onset, nature, and course of problem drinking.

The control of problem drinking is a long-term public health goal. In striving to achieve this end, the perspective of the social model also offers a useful approach to the introduction of preventive strategies. Within the social model, primary prevention may be geared toward changing cultural attitudes which effect individual adoption of problem drinking as a solution to the stresses of everyday life. A secondary prevention impact may be obtained through continued analysis of and intervention into family dynamics which precipitate and support problem drinking. Early contact with couples and families surrounding their conflict can lead to their resolution and a decrease in the use of alcohol as a mechanism for escaping from them. Finally, tertiary prevention can be obtained by the expansion of current social treatment approaches, as described in this chapter, which are designed to rehabilitate the problem drinker and return him to a functioning role in society.

The increased influence of social issues as important factors in understanding problem drinking will continue to have a major impact on the development of strategies to combat this public health problem. The material presented in this chapter has tried to provide the reader with a fuller understanding of this methodological approach to problem drinking.

REFERENCES

"Alcoholics Anonymous Comes of Age: A Brief History of A.A.," 1957, Alcoholics Anonymous Publishing, New York, 1957.
"Alcoholics Anonymous," 1960, Alcoholics Anonymous World Service, New York (First edition, 1939).
Bailey, M. P., Haberman, P. W., and Alksne, H., 1965, The epidemiology of alcoholism in an urban residential area, *Q. J. Stud. Alcohol* 26:19–40.
Bales, R. F., Cultural differences in rates of alcoholism, 1946, *Q. J. Stud. Alcohol* March: 480–499.
Blum, E. M., and Blum, R. H., 1967, "Alcoholism: Modern Psychological Approaches to Treatment," Jossey-Bass, San Francisco.
Bowen, M., 1974, Alcoholism as used through family systems theory and family therapy, *Ann. N.Y. Acad. Sci.* 233:115–122.
Cahalan, D., "Problem Drinkers," Jossey-Bass, San Francisco.

Cahalan, D., and Cisin, I. H., 1968, American drinking practices: Summary of findings from a national probability sample, I. Extent of drinking by population sub-groups. *Q. J. Stud. Alcohol* 29:130–151.
Cahalan, D., Cisin, I. H., and Crossley, H. M., 1969, "American Drinking Practices, Rutgers Center for Alcohol Studies, New Brunswick, New Jersey.
Cahalan, D., and Room, R., 1972, Problem drinking among American men aged 21–59, *Am. J. Public Health* 62:1473–1482.
Cheek, F. E., Franks, C. M., Laucius, J., and Burtle, B., 1971 Behavior modification training for wives of alcoholics, *Q. J. Stud. Alcohol* 32:456–461.
Clinebell, H. J., 1963, Philosophical–Religious factors in the etiology of the treatment of alcoholism, *Q. J. Stud. Alcohol* 24:473.
Conger, J. J., 1956 Reinforcement theory and the dynamics of alcoholism, *Q. J. Stud. Alcohol* 17:296–305.
Cooley, C. H., 1909, "Social Organization: A Study of the Larger Mind," Scribner's, New York.
Dollard, J., 1945, Drinking mores of the social classes, *in* "Alcohol, Science and Society," Greenwood Press, Westport, Conn.
Dollard, J., and Miller, N., 1950, "Personality and Psychotherapy," McGraw-Hill, New York.
Drew, L., 1968, Alcoholism as a self limiting disease, *Q. J. Stud. Alcohol,* 29:957–967.
Durkheim, E., Cabot, H., and Kahl, J., 1956, "Human Relations," Harvard University Press, Cambridge.
Dutoit, B. M., 1964, Substitution: A process of cultural change. *Hum. Organ.* 23:61–66.
Enelow, Allen J., 1974, Social Models in the Therapy of Alcoholism, Paper presented at the Hawaiian Medical Association Annual Convention (March, 1974).
Esser, P. H., 1971, Evaluation of family therapy with alcoholics, *Br. J. Addict.* 66:251–255.
Ferguson, F. N., 1968, Navajo drinking: Some tentative hypotheses. *Hum. Organ.* 27:159–167.
Fox, R., 1962 Children in the alcoholic family, *in* "Problems in Addiction: Alcoholism and Narcotics" (C. Bier, ed.) Fordham University Press, New York.
Foster, F. M., Horne, J. L., and Wanberg, K. W., 1972, Dimensions of treatment outcome. *Q. J. Stud. Alcohol,* 33:1079–1098.
Frazier, E. F., 1947, "Black Bourgeoise," Colliers Brooks, New York.
Harris, F. R., and Harris, L., 1968, "Indian Health Sources: A Blue Cross report in the Health Problems of the Poor," Blue Cross, Chicago.
Holtman, J. E., and Friedman, S., 1953, A consideration of parental of probation and other factors in alcohol addiction, *Q. J. Stud. Alcohol* 14:49–57.
Horton, D., 1943, The function of alcohol in primitive societies: A cross cultural study, *Q. J. Stud. Alcohol* 4:199, 320.
Jackson, J. P., 1958, Alcoholism and the family, *Annals of American Academy of Political and Social Sciences* 315:90–98.
Jackson, J. K. and Connors, R., 1953, Attitudes of parents of alcoholics, moderate drinkers and non-drinkers toward drinking, *Q. J. Stud. Alcohol* 14:596–613.
Jellinek, B., 1959 "The Disease Concept of Alcoholism," Yale University Press, New Haven.
Jones, M., 1962, "Social Psychiatry," Charles C Thomas, Springfield, Illinois.
Keller, M., 1970, The great Jewish drinking mystery, *Br. J. Addict.* 64:287–296.
Kepner, E., 1964, Application of learning theory to the etiology and treatment of alcoholism, *Q. J. Studies Alcohol* 25:279–291.
King, L., Murphy, G., Robins, L., and Darvish, H., 1969, Alcohol abuse: A crucial factor in the social problems of Negro men, *Am. J. Psychiatry* 125:96–104.
Kingham, R., 1958, Alcoholism and the reinforcement theory of learning, *Q. J. Stud. Alcohol* 19:320–330.

Kinsey, B. A., 1966, "The Female Alcoholic," Charles C Thomas, Springfield, Illinois.
Knupfer, G., 1963, Factors related to amount of drinking in an urban community. *California Drinking Practices Study Report No. 6.,* Berkeley, Division of Alcohol Rehabilitation, California Division of Public Health.
Kunitz, S. J., and Levy, J., 1974, Changing ideas of alcohol use among Navaho Indians, *Q. J. Stud. Alcohol* 35:176-195.
Kunitz, S. J., Levy, J. E., Odoroff, C. S., and Bollinger, J., 1971, The epiodemiology of alcoholic cirrhosis in two Southwestern Indian tribes, *Q. J. Stud. Alcohol* 32:706-720.
Larkins, J. R., 1965, "Alcohol and the Negro," Record Publishing, Zebulon, Ohio.
Lemert, E. M., 1961, Alcoholism and the Family: The Research Problem. Unpublished paper read at a meeting of the North American Association of Alcoholism Programs, October, 1957, *in* "Alcoholism and Marriage: A Review of Research and Professional Literature," (M. B. Bailey) *Q. J. Stud. Alcohol* 30:81-94.
Lisansky, E. S., 1957, Alcoholism in women: Social and psychological concomitants. I. Social History Data, *Q. J. Stud. Alcohol* 18:588-623.
Littman, G., 1970, Alcoholism, illness and social pathology among American Indians in transition, *Am. J. Public Health* 60:1769-1787.
Locke, B. Z., Kramer, M., and Pasamanick, B., 1960, Alcoholic psychoses among first admissions to public mental hospitals in Ohio, *Q. J. Stud. Alcohol* 21:473.
Maddox, G. L., 1962, Teenage drinking in the United States, *in* "Society, Culture and Drinking Patterns" (D. J. Pittman and C. R. Snyder, eds.), Wiley, New York.
Maddox, G. L., and William, J. R., 1968, Drinking behavior Negro collegians. *Q. J. Stud. Alcohol* 29:117-129.
Malzberg, B., 1949, First admissions with alcoholic psychoses in New York State, year ended March 31, 1948. With a note on first admissions for alcoholism without psychoses, *Q. J. Stud. Alcohol* 10:361-470.
Marsh, L. C., 1935, Group therapy and the psychiatric clinic, *J. Nerv. Ment. Dis.* 82:381-392.
Martin, C. G., 1962, Use of "disease" for chronic alcoholism, *J. Am. Med. Assoc.* 179:742.
Maxwell, N., 1962, Alcoholics Anonymous: An interpretation, *in* "Society, Culture, and Drinking Patterns" (D. J. Pittman and C. R. Snyder eds.), Wiley, New York.
Mayer, J., and Black, R. M., 1974, A description of some selected treatment approaches in alcohol abuse, *in* "Alcohol Abuse and Rehabilitation Approaches" (J. G. Cull and R. E. Hardy, eds.), Charles C Thomas, Springfield, Illinois.
McCord, W., 1969, "Life Style in the Black Ghetto," W. W. Norton, New York.
McCord, W., McCord, J., and Gudeman, J., 1959, Some current theories on alcoholism: A longitudinal evaluation, *Q. J. Stud. Alcohol* 20:727-744.
McCord, W., McCord, J., and Gudeman, J., 1960, "Origins of Alcoholism," Stanford University Press, Stanford.
Meeks, D. E., and Kelly, C., 1970, Family therapy with families of recovered alcoholics, *Q. J. Stud. Alcohol* 31:399-413.
Merton, R. K., 1957, "Social Theory and Social Structure," The Free Press, New York.
Miller, P. N., 1972, The use of behavioral contracting in the treatment of alcoholism: A case study, *Behav. Ther.* 3:593-596.
Moore, R. A., and Ramseur, F. A., 1960, A study of a background of 100 hospitalized veterans with alcoholism, *Q. J. Stud. Alcohol* 21:51-67.
Mulford, H. A., 1964, Drinking and deviant drinking, U.S.A., 1963, *Q. J. Stud. Alcohol* 25:634-648.
Mulford, H. A., and Miller, D. E., 1960, Drinking in Iowa, V. Drinking and alcoholic drinking, *Q. J. Stud. Alcohol* 21:483-499.

Mullan, H. and Sangiuliano, I., 1966, "Alcoholism: Group Psychotherapy and Rehabilitation," Charles C Thomas, Springfield, Illinois.

Murphy, G., 1963, Group psychotherapy in our society, in "Group Psychotherapy and Group Function," (M. Rosenbaum and M. Berger, eds.) Basic Books, New York.

Pattison, E. N., 1966, A critique of alcoholism treatment: With special reference to abstinence, Q. J. Stud. Alcohol 27:49–71.

Pittman, D. J., and Gordon, C. W., 1958, "Revolving Door," Free Press, Glencoe, Illinois.

Preston, J. D., 1969, Religiosity and adolescent drinking behavior, Sociological Quarterly, Summer: 372–383.

Riley, J. W., and Marsden, G. F., 1947, The medical profession and the problem of alcoholism, Q. J. Stud. Alcohol 7:265–273.

Roberts, B. H., and Myers, J. K., 1967, Religion, national origin, immigration and mental illness, in "Sociology and Mental Disorders" (F. K. Weinberg, ed.), Aldine, Chicago.

Robins, L. N., Bales, W. M., and O'Neal, P., 1962, Adult drinking patterns of former problem children, in "Society, Culture and Drinking Patterns" D. J. Pittman and C. K. Snyder (Eds.): Wiley and Sons, New York.

Robins, L., Murphy, G. E., and Breckenridge, M. B., 1968, Drinking behavior of young urban Negro men, Q. J. Stud. Alcohol 29:657–684.

Roebuck, J., and Kessler, R., 1972, "The Etiology of Alcoholism: Constitutional, Psychological, and Sociological Approaches," Charles C Thomas, Springfield, Illinois.

Rosenbaum, B., 1958, Married women alcoholics at the Washingtonian Hospital, Q. J. Stud. Alcohol 19:79–89.

Rubington, E., 1967, The halfway house for the alcoholic, Ment. Hyg. 51:552–560.

Shoben, E. J., Jr., 1956, Views on the etiology of alcoholism. III. The behavioristic view, in "Alcoholism as a Medical Problem," (H. D. Kruse, Ed.), Harper, New York.

Skolnick, J. H., 1958, Religious affiliation in drinking behavior, Q. J. Stud. Alcohol 19:452–470.

Slavson, S. R., 1943, Group therapy special section meeting, Am. J. Orthopsychiatry 13:648–690.

Smoyak, A., 1973, Therapeutic approaches to alcoholism based on systems theory, Occupational Health Nursing 21:27–30.

Snyder, C., 1962, Culture and Jewish sobriety: The ingroup–outgroup factor, in "Society, Culture and Drinking Patterns" (D. J. Pittman and C. K. Snyder, eds.), Wiley and Sons, New York.

Stage, T. B., and Keast, T. J., 1966, A psychiatric service for Plains Indians, Hospital and Community Psychiatry 17:74–76.

Stewart, O., 1964, Questions regarding American Indian criminality, Hum. Organ. 23:61–66.

Strauss, R., and Bacon, B., 1953, "Drinking in College," Yale University Press, New Haven.

Strayer, R., 1961, A study of the Negro alcoholic, Q. J. Stud. Alcohol 30:111–123.

Suicide, Homicide, and Alcoholism among American Indians: Guidelines for Help, 1973, *HEW Publication No. (ADM) 74-42* U.S. Government Printing Office, Washington, D.C.

Sulzer, E. S., 1965, Behavior modification in adult psychiatric patients, in "Case Studies in Behavior Modification" (L. P. Ullman, and L. Krasner, eds.), Rinehart and Winston, New York.

"The Twelve Steps and Twelve Traditions," 1953, Alcoholics Anonymous Publishing, New York.

Tiebout, H. M., 1944, Therapeutic mechanisms of Alcoholics Anonymous, Am. J. Psychiatry 100:468–473.

Trice, H. M., 1957, A study of the process of affiliation with Alcoholics Anonymous, Q. J. Stud. Alcohol 18:39–54.

Trice, H. M., 1959, The affiliation motive and readiness to join Alcoholics Anonymous, Q. J. Stud. Alcohol 20:313–320.

Ullman, A. D., 1958, Sociocultural background of alcoholism, Annals of the American Academy of Political and Social Science 351:22–30.

Vogel-sprott, M., and Banks, R. K., 1965, The effect of delayed punishment on an immediately rewarded response in alcoholics and nonalcoholics, *Behav. Res. Ther.* 3:69–73.

Weiner, H. B., 1965, Treating the alcoholic with psychodrama, *Group Psychotherapy* 18:27–49.

Weppner, R. S., 1973, Some characteristics of an ex-addict self-help therapeutic community and its members, *Br. J. Addict.* 68:243–250.

Whittaker, J. O., 1963, Alcohol and the Standing Rock Sioux Tribe: Psychodynamic and cultural factors in drinking, *Q. J. Stud. Alcohol* 24:80–90.

Whitman, M. P., 1939, Developmental characteristics and personalities in chronic alcoholics, *J. Abnorm. Soc. Psychol.* 34:361.

World Health Organization, 1952, Expert Committee on Mental Health, Alcoholism Sub-Committee, Second Report, World Health Organization Technical Representatives Service, No. 48 (August, 1952).

Yalom, D., 1974, Group therapy and alcoholism, *Ann. N.Y. Acad. Sci.* 233:85–103.

CHAPTER 6

Group Psychotherapy in Alcoholism

David R. Doroff

Consultant, Alcoholism Treatment Program
St. Luke's Hospital Center
New York, New York

INTRODUCTION

Whether one views alcoholism as a disease entity or whether, as some have suggested, it is seen as an adaptational effort, there is little question that it has been of concern as long as history has been recorded. Both its longevity and the number of cures that have been in vogue at one time or other leave no doubt as to the difficulty in controlling it. Perhaps no one has phrased it as eloquently as the late comedian W. C. Fields, who, during the delivery of his famous "Temperance Lecture," remarked, "Don't say you can't swear off drinking—it's easy. I've done it a thousand times!"

The proffered cures have ranged from the exercise of willpower to rendering alcohol illegal, with a variety of measures falling between these extremes. As often as not, the cure has produced difficulties as severe as those it sought to remedy. The persistence of the problem and the extent to which it has intruded into the social and economic fabric of our lives have at last brought it out of the realm of fear, superstition, and moralizing and into the focus of science. As science turned its attention to alcoholism it was only natural that psychology should also become involved. With psychology came the attempts to apply psychological theory to the treatment of alcoholism. These efforts have included the use of hypnosis, psychoanalysis, and conditioning techniques, as well as variations on more traditional modes such as exhortation and guilt inducement. Yet there has been a notable lack of success in the treatment of alcoholism based upon a dyadic relationship. Clinical data have resulted and some measure of theory construction has ensued, but precious few cures.

In recent years there appears to have emerged a consensus among the scientific and professional community to the effect that among the various psychotherapies a group approach seems to offer the brightest prospect.

Group psychotherapy has, in the past three decades, been viewed by increasing numbers of authorities as the psychotherapeutic modality of choice in the treatment of alcoholism. The early promise of psychoanalysis and other individual therapies did not prove out, and despite its contributions to both the clinical and theoretical literature, psychoanalysis has been disappointing as a method of treatment for alcoholism. At the same time as these failures were being experienced, Alcoholics Anonymous was beginning to experience its successes. Thus, there was, and has continued to be, a powerful thrust in the direction of increasing reliance on the group.

Group is a relatively new aspect of psychotherapy and has still a great deal of controversy centered about it. There was, until the Second World War created a need, no group psychotherapy to speak of. It was only when military necessity created a demand for an economic and speedy way of returning to duty those military casualties who suffered from combat neurosis that group psychotherapy was given its impetus. In the last decade it appears to have gained a measure of legitimacy, but even now, as Durkin (1964) has pointed out, there is no *theory* of group. Thus, there are extant today a variety of techniques and technical approaches to group psychotherapy, some of which assume a theoretical stance drawn from a theory of the psychopathology of the individual, and others which appear to proceed ad hoc and post hoc. This is certainly the case in the attempt to treat the alcoholic through the group medium. In fact, if one considers the goals of the various approaches, the discrepancies are so wide that one is led to the inescapable conclusion that there is little or no agreement as to what constitutes the problem. At one end of the spectrum is A.A. claiming that alcoholism is a disease and sobriety alone constitutes the cure. At the other end are the psychoanalytic theories which often imply that there is no disease, and that alcoholism is merely the stage on which a more fundamental personality conflict is being played out.

The focus of this chapter will be to present as wide an overview of group therapy in alcoholism as possible, with an emphasis upon the historical development and to provide an examination of the underlying premises of the various approaches. Additionally, consideration will be given to certain clinical and theoretical issues.

ALCOHOLICS ANONYMOUS

To date, as far as can be determined, Alcoholics Anonymous has established itself as the foremost extant treatment agency for alcoholism. Yet, it does not claim to be an organization of professionals and its methods seem almost naively simple in this era of psychological sophistication.

A.A.'s success, after millennia of failures, is impressive and, regardless of whatever limitations may be decried and whatever objections may be raised, commands attention to its methods and means.

The story of its founding and subsequent growth are too well known to warrant repetition here. However, its *psychology,* its functioning as a *group,* and its functioning as a *network* appear to have implications for a theory of alcoholism as well as for the clinical treatment of alcoholism.

The Psychology of A.A.

A.A. has been variously described. Some critics contend that it harbors an essentially antipsychotherapy attitude and, depending upon the definition of psychotherapy that is employed, this criticism may well have validity. Others have characterized A.A. as using a repressive–inspirational type of approach that involves only a limited utility of confrontation, which is, after all, a cornerstone of psychotherapy.

Still other critics have commented that an approach which involves a definition of alcoholism as a *disease* and restricts itself to the arrest of alcoholic behavior misperceives and misconstrues the problem, substituting symptom treatment for an understanding of the alcoholic as a person who was deeply troubled long before he became an alcoholic.

Nonetheless, despite both its critics and its own stated intensities, there is a psychology that A.A. employs, and it merits careful evaluation. Confrontation, as described by Buie and Alder (1972) takes a number of forms. Corwin (1972) had similarly noted a variety of confrontation techniques which are described as ranging from "routine" to "heroic." Routine confrontation is the everyday business of psychotherapy. It may take the form of a comment such as "How does that strike you?" or "You must have been angry to react that way." It may be an inquiry into a dream, or calling to a patient's attention that he has forgotten something. Heroic confrontation consists of stance taken by the therapist that challenges the very nature and conduct of the patient in regard to the therapeutic process, and may come in the form of an ultimatum. Freud (1937/1959) himself seems to have been the first to employ it when he told a patient that unless he (the patient) was more productive, the analysis would have to be terminated at the end of a year.

In the psychology of A.A., confrontation of striking dimension plays a central role. Consider, for example, the demand that the alcoholic acknowledge the fundamental fact of his alcoholism, or the relentless confronting of the various rationalizing and denying mechanisms employed by alcoholics to maintain the secret hope that drinking can be resumed. A.A. appears to have managed to combine its exhortations and inspirations with a confrontational approach which, however limited its scope, is clearly on the same plane of intensity as its basic commitment to struggle against alcoholism. Combined with the aspect of confrontation is that particular attitude of hostility toward the introspective. This seems to have grown out of recognition that the brooding, depressive, self-hating, and remorseful attitudes in alcoholics are part of a cycle that almost inevitably leads to a resumption of drinking. Thus, rather than dismiss the attitude of A.A. as simply another antipsychotherapy stance, it seems more appropriate to regard it as an effort to mitigate against the guilt-laden depression cycle that moves the alcoholic inexorably toward his next drink.

In addition to the continuing confrontation and the efforts to lessen the burden of guilt, loss of self-esteem, and loss of hope, A.A. makes extensive use of ego-enhancing techniques such as almost total acceptance of the individual. The atmosphere into which the alcoholic walks for his first meeting is eloquently described by a recovered alcoholic who remembers her first meeting: "I was very frightened when I walked in, but I was received with such warmth that I felt totally accepted. No one put any pressure on me. They told me that I didn't have to say anything. It was all right to just sit there. But when I saw how people were toward each other, I wanted to be a part of it. So I was able to talk openly about my drinking. At the end of the meeting people came up to me and gave me their phone numbers—nobody asked me for mine. They just gave me theirs and told me to call any time I needed to."

The effect of A.A.'s inspirational stance seems to be a powerful one. The same woman further describes her experience: "I had felt totally hopeless. But I seemed to get hope from the fact that these people had been drunks just like me and now they seemed so close. I really wanted to be part of it. It was like a loving family."

A.A. as a Group

The literature on alcoholism contains many references to the emotional hunger of the alcoholic. It also describes the alcoholic as a person who has severe difficulties in interpersonal relationships. To the extent that A.A. allows a newcomer to cathect the *group*, so it allows him to experience a slightly diffused source of nourishment and as a further consequence *not* to feel as though he stands in danger of being swallowed by a single person to whom he might be too closely drawn.

The description offered above of an A.A. group being experienced as a "loving family" characterizes the kind of transference that A.A. groups seek to promote and which appears to be of crucial importance in the achievement of sobriety. The issue of acceptance appears to be central. The individual is offered acceptance *no matter what he is* and *no matter what he has done*. The only demand made in exchange is that he has a desire to stop drinking. This acceptance almost certainly touches the alcoholic deeply and must be of extreme importance to him. Being able to experience himself as accepted for what he is stands in marked contrast to the demands that are made upon a person throughout his life, and most particularly with the "false self" that emerges out of the matrix of the early mothering experience in which the individual learns to disguise his feelings and attitudes and, in order to survive, to adopt a make-believe attitude that will please mother (Winnicott, 1965). Now he finds a mother who says in effect, "I love you no matter what."

Thus, there is added to the alcoholic's motivation to stop drinking an enormous reward. Not drinking is no longer simply a massive and painful deprivation (as it is in analytic therapy) but brings with it the kind of rewards for which the individual has yearned all his life. It seems both accurate and fair to characterize A.A. as a group that legitimizes certain regressive features. The alcoholic is both allowed and encouraged to experience himself as the infant who is the apple of his mother's eye. He can call upon anyone at any time. He need not long experience the frustration and stress of having to curb his desires. If he calls instead of drinking, he will almost certainly be rewarded for having done so. In short, the alcoholic seeking sobriety is offered therapeutic regression as an alternative to intoxication. Further, he is allowed to control the degree of involvement with any single individual by spreading his attachment to the group itself, or to others in it. Thus the potentially explosive issue of the loss of one's individuality through absorption by another as the price of attachment is quickly neutralized. Similarly diffused and, through the dispersion, lessened, is the reaction of intense anger to the frustration of infantile–narcissistic demands. Here again the opportunity to maintain the dependent tie is enhanced far beyond what might be capable in a dyadic relationship.

A.A. as a Therapeutic Network

If, as some authorities believe, most alcoholics are individuals whose character structure has remained fixated at an infantile level, then there is a corollary that implies that for A.A. to be successful in its work, it must somehow be appropriate to the struggles of infancy. At the risk of oversimplification, a parallel might be useful here. The needs of an infant (or infantile personality) are relatively few, though intense, and, during the period of infancy only very gradually relenting. Of the mother is demanded 24-hour availability. She must sacrifice her own wishes, disregard the promptings of her own body, and, in short, devote herself entirely to the care of the infant. Winnicott (1965) has described this and refers to it as "primary maternal pre-occupation." With the alcoholic there is a strikingly similar intensity of demands, which appears to also have an unrelenting quality. The capacity of A.A. to diffuse responsibility among its members enables it to form a *network* which can see the alcoholic through his crisis period far more effectively than a single person. Thus the excessive quality of the alcoholics' demands need not become a problem of such magnitude as to exhaust the resources and limits of a single person. The alcoholic who drinks because he cannot cope with life's refusal to honor his narcissistic demands has the opportunity to feel cared for and to replace the emotional gratification of alcohol during his period of withdrawal.

Another aspect of A.A.'s group psychology that needs commentary is its demand of the alcoholic that he commit himself to *life* membership. If this is viewed as an offer of life support rather than a demand, it is easier to understand the effect it has in aiding the alcoholic in maintaining sobriety. There is some evidence that often enough a spouse or parent may need the alcoholic to be "sick" and to remain alcoholic. As A.A. is an organization of individuals rather than a single individual, it allows for a diffusion of the hostility that inevitably emerges out of the context of an eternally dependent relationship. Further, A.A. provides a balance to this; the alcoholic, once he has achieved sobriety now has the opportunity to become the sponsor of a fledgling member. Thus, as in a family, the bond is perpetuated by means of identification and role expectation and fulfillment. A new avenue of gratification is open to the alcoholic who has achieved sobriety—an avenue which can aid him in the maintenance of his sobriety through supporting him in his new role as a strong and helping person. Even though a change in the character structure has not been effected, the self-esteem is strengthened and renewed through narcissistic reward and is experienced as an inner resource upon which the recovered alcoholic may draw in times of stress.

A SURVEY OF GROUP THERAPY WITH ALCOHOLICS

Even the most casual perusal of the literature on alcoholism makes it abundantly clear that no single theory of alcoholism exists. There is a range that encompasses views of alcoholism as a medical problem, defining it as a disease entity, as well as viewing it as a mere symptomatic expression of underlying severe psychopathology. Although there are a number of technical aspects relative to the treatment of alcoholism in which rather substantial agreement exists across the theoretical range, there are significant differences in overall approaches to treatment which are dictated by the theoretical stance one takes.

At one end of the spectrum A.A. stands, defining alcoholism as the problem in itself, and as a consequence, having sobriety as the single goal. Without saying as much, A.A. appears to have taken the view that alcoholism is an addiction and at the same time essentially ruled out questions relating to the premorbid personality of the individual. At the other end, one finds some of the essentially classical analytic statements that seem to imply that once the underlying causes (e.g., oral deprivation) are removed, the problem ceases to exist. Between these extremes are a number of alternative approaches, some of which appear to have a theoretical base, but many of which seem to be rather ad hoc efforts at "doing something" and hoping that it will have a curative effect, though what is meant by "cure" is, as often as not, left unarticulated.

The relevant literature on group psychotherapy in alcoholism is not a large body. But despite its relatively small size it is not easy to classify it. Consequently, this section will present an essentially chronological view with commentary wherever it is deemed relevant.

Group psychotherapy is a relatively young phenomenon, the still gawky child of a forced marriage between psychoanalysis and necessity as occasioned by the Second World War. Thus it is that one of the very first papers to appear was that of Heath (1945). Heath worked with a population of merchant seamen and attempted to adapt some of the principles of A.A. to his patients. He felt that many of them refused to become involved with A.A. because they felt social class differences created a much too stressful situation. Heath's work included efforts to establish a group support and also attempted to make use of hypnosis, "the purpose being to attempt to destroy the pleasant associations which often come to the fore with the taking of the first drink." Heath did not provide any statistical data to evaluate his work, a condition of omission obtaining in virtually all of the published work to be described here.

Evseeff (1948) attempted to develop a group format within the setting of a state hospital which would be highly structured and well organized. He selected those patients who he believed would be most responsive. These were individuals with "psychoneurotic personalities or with cyclothymic personality reactions," and "with an IQ of 90 or higher." He designated three stages of therapy. The first is "designed to stress that there is a serious problem behind his (the alcoholic's) drinking." The second phase consisted of group sessions that were structured by the therapist as discussions focused around such topics as "What is personality?" and included such others as the concept of psychosexual development. The purpose of the second phase appears to have been an educative one, designed to help the individual to think psychologically. The final phase was the uncovering phase. Included in it were such auxiliary techniques as modified psychodrama and the use of the Thematic Apperception Test.

It is interesting to note in passing that both of these early workers in the field introduced techniques that have survived to this date and have a great deal of credibility. Heath's work with a homogeneous group as opposed to heterogeneous is generally the accepted procedure. Although his patients were even more homogeneous than being simply alcoholics, virtually all authorities agree that alcoholics respond best to therapy when in an all alcoholic group. Evseef's structured sessions have been duplicated by a large number of workers and appear to serve as a way of avoiding the resistance to therapy that is often manifest in the alcoholic's long rambling monologue that has both a frustrating and angering effect on the other group members. Haber *et al.* (1949) present a detailed description of the inpatient treatment of alcoholics. They argue persuasively for a closed and segregated ward, careful patient selection (eliminating the frankly psychotic, psychopathic, deteriorated, and those with no *real* desire for treatment [italics added]), patient preparation, A.A. involvement, individual therapy as well as group, and a therapeutic milieu. In addition to the foregoing, these writers include a preparatory phase of three educative sessions. The issue of preparation appears to have been rather neglected by most writers in the field or at best been given a rather casual notation. Mullan and Sangiuliano (1966) are the notable exception, and their work deserves special mention. It will be taken up at a later point. Another item of interest described by Haber *et al.* (1949) is the issue of the frequency of sessions. Relatively few writers have seen fit either to comment as to the frequency of sessions or to go beyond one or, at most, two sessions weekly. These writers describe their groups as meeting five times weekly. It is the opinion of this writer that the frequency of sessions is extremely important, the group session substituting for the emotional gratification achieved through alcohol. More likely than not, the extent of the alcoholic's involvement with A.A., particularly the meeting every night, serves to enable him to maintain sobriety on a day-to-day basis, rather than endure long periods of time during which he would experience only stress. The final note on Haber *et al.* is their use of role-playing. This seems to be a variation of psychodrama and as such to be the first attempt at applying it to working with an alcoholic population. It is an ancillary technique that appears to have been used with favorable results by others.

Brunner-Orne and Orne (1954) presented a paper dealing extensively with the earlier-mentioned approach of Evseeff (1948). They provide an elaborate rationale for the use of the technique (which they call "directive group therapy") as well as enough clinical illustration to make clear exactly *how* the technique is applied. For the rationale they draw upon two sources. The first is the now classic study of Lippert and White (1947) of the effects of leadership styles (laissez-faire, authoritarian, democratic) upon group functioning. The authors characterize their approach as "directive, . . . yet very permissive." They attempt to keep the group goal directed, though they emphasize that in their judgment the atmosphere of permissiveness is "the factor which determines therapeutic efficacy."

This is then tied to a psychoanalytically based view of the character structure of the alcoholic. "In a sense, alcoholism is analogous to character neurosis. The character neurotic also controls current anxiety, but by means of acting out. He also is immature and has low frustration tolerance. Further, he, as well as the alcoholic, is in conflict with society." They note that withdrawal from alcohol will produce a higher level of anxiety. This, they feel, provides the point at which therapeutic intervention is possible. "Provided the patient is given adequate support at this point and is made to feel sufficiently secure, he will gradually acquire other more adequate defense mechanisms." Unfortunately, there is no effort made to provide a theoretical underpinning as to exactly how this change takes place. The authors are not clear as to what is meant by "acquire." However, due to the profound nature of defense mechanisms, one would have to assume structural changes. Defenses are, after all, a manifestation of character structure. Yet, it is clear from the description, and from the author's assessment, that "our group essentially is not a therapeutic group, but a 'work group.'" They go on to indicate that the therapeutic effects of the group are based upon its providing "substitutive emotional satisfactions," lessening of guilt, neutralizing some of the rationalizing defenses, and an apparent cathecting of sobriety. Thus it would appear that what has taken place has been the achieving of sobriety by adopting behavioral alternatives (rather than new defenses) through means that are essentially a replication of A.A.'s approach.

Greenbaum (1954) attempted to combine Antabuse with analytic group therapy. He began with four basic assumptions as to the psychological functions of alcoholism. These were: (1) for self-punishment and for the satisfaction of instinctual drives, (2) an aggressive and hostile act against significant persons, past and present, (3) to preserve and enhance self-esteem, and (4) to meet and satisfy narcissistic needs.

The patients in Greenbaum's group were those culled from a pool of alcoholic patients by virtue of being deemed the best motivated. The acceptance of Antabuse was part of the treatment program.

Prior to the beginning of group therapy, each patient was given a psychological test battery and the results were discussed with him. Greenbaum structured the group by having it include only alcoholics (though it was otherwise heterogeneous) and by having it meet twice weekly. He articulated his approach by characterizing it as analytically oriented, noting little difference from groups of nonalcoholics and having as its aim the analysis of the total personality and not concentrating on the alcoholic symptom. In addition to a brief introduction into certain concepts relative to both psychoanalysis and group psychotherapy, Greenbaum took an active role in encouraging the patients to openly reveal shameful and secret material, as well as encouraging them to act as adjunct therapists in addition to being patients. Finally, he stressed that increased understanding was by itself no cure and that the patient would have to make efforts to apply his insights.

Although his data are unsophisticated, Greenbaum is one of the very few writers who provide any data. He noted that of those patients who were in a group that met twice weekly, 80 percent showed improvement as measured by the maintainance of sobriety during the period of treatment. Of those in group on a once-weekly basis, 50 percent showed improvement. He also noted that his best motivated patients achieved the best results.

The work of Brunner-Orne and Orne and of Greenbaum warrants special consideration because of the theoretical implications. Though both papers view the alcoholic from a psychoanalytic stance, the conclusions they reach as to treatment are remarkably different. Brunner-Orne and Orne offer nonanalytic treatment. It is treatment that focuses upon alcoholism per se, defining it as the problem and leaving untouched the character problems. Greenbaum, on the other hand, considers the analytic work of the group to be paramount. What seems surprising is that two papers written from a psychoanalytic standpoint should have such differing views as to the treatment process. It is the opinion of this writer that the issue is centered around the narcissistic problem that is so frequently encountered in the alcoholic patient. An earlier paper by Pfeffer *et al.* (1949) alludes to this when the authors note that they believe that group is the modality of choice for the alcoholic because it involves "less dangerous transferences to the therapist." Freud wrote many years earlier (1920) of the problem, indicating that the lack of a stable transference made analysis impossible. It was not until very recently that a detailed discussion of the narcissistic character was presented, together with an exhaustive description of the treatment approach. Kohut (1971) makes clear that what one is confronted with is the particular quality of the narcissistic transference. The difficulties are of an order that makes intense countertransference reactions highly probable. Giovacchinni (1972) has written of this, suggesting it is the attitude of the narcissistic person which essentially fails to recognize the existence of others that provokes a reaction of anger in the therapist. It seems to this writer that the decision of Brunner-Orne and Orne to confine their goals to symptom relief reflects a decision to avoid the "dangerous transferences." Unfortunately, it has also an implication that an alcoholic is essentially uncurable. This is doubly problematic since these are widely known and respected authorities whose opinions carry great weight.

The issue of transference was discussed again by Martensen-Larsen (1956). He ascribed the lack of success in individual therapy with alcoholics to the inability of the patient to handle the intense ambivalence in the transference without breaking off treatment. Group is suggested as the treatment of choice because it vitiates the transference. He attributes other benefits to group therapy as helping to strip away the denial and rationalizing, as well as a "mutualistic community feeling." His concern with the hostile aspects of the transference is responsible for a technique he introduces. Midway through a two-hour session, tea and pastry are served.

Despite an evidently well-considered concern for the effect of the alcoholic's hostility on his ability to remain in treatment, Martensen-Larsen's paper gives the impression that his efforts to mitigate acting out through forestalling hostility have been only partially successful. He describes irregular attendance, membership in more than one therapy group, and absence to the point of having a two-person group. It is clear enough that whatever measures a therapist may take to forestall hostility, by themselves they cannot be sufficient, though perhaps they are necessary. Psychological problems are obviously both wide enough and deep enough to make any course of therapy hazardous and unpredictable. Martensen-Larsen further responds to the problems encountered in the transferences by advocating the use of groups having cotherapists. He believes that this will further enable the patient to split the transference, keeping one therapist as the benign figure by displacing onto the other the anger that was originally felt toward the significant parent.

Generally, Martensen-Larsen appears to have responded to the impulsive and infantile characteristics of the alcoholic by a willingness to act protectively in various situations. He describes, for example, the sort of situation that often arises in groups of alcoholics when tempers flare. "As alcoholics sometimes may be very hostile to each other or may indulge in disclosing each others' problems too fast, the therapist should not hesitate to change the subject." Similarly he describes the use of mixed-sex groups as "intermediate steps in treatment." He feels that premature exposure of the male alcoholic to the hostility that women often feel toward them would almost surely result in a regressive response. It may appear that Martensen-Larsen acts overprotectively, but this writer concurs with his assessment of the pitfalls that line the route of outpatient therapy with alcoholics.

The problem of the alcoholic who does not want treatment is taken up by Brunner-Orne (1956) in a paper describing the treatment of involuntary subjects. These are court-referred alcoholics who must attend clinic or be faced with incarceration. The methods and means of treatment are the same as those described in the earlier work of Brunner-Orne and Orne (1954). A claim of "encouraging" results is made, but no data are available for purposes of assessment. It is indeed unfortunate, for even modest success with resistive or poorly motivated alcoholics would be a most significant event. One need only consider the nearly unanimous opinion of workers in the field that only the best motivated alcoholics are treatable to realize its importance. Scott (1956) presented what appears to be an adaptation of the late Fritz Perls's "hot seat" technique. This involves an essential departure from the group format for what appears to be a kind of individual therapy with the other group members acting as ancillary therapists for the one patient who is designated as "being on the pan," to use Scott's term. Scott indicates that the therapist has the responsibility of coming "to the aid of the patient" if the questioning becomes too antagonistic, as well as refocusing if the questions are "aimless." The therapist also determines when to end the turn.

Preston (1960) presents what appears to be a shotgun approach as she suggests that individual, joint (husband and wife), and group therapy combined offers the treatment plan of choice for the alcoholic. Although she advocates a combination of approaches, her bias seems to be in favor of the treatment of what might best be designated as the alcoholic couple. Regarding the rationale for an approach so varied, Preston can only comment, "Perhaps it is because of the severity of the personality difficulties and interpersonal relationship problems in these patients that a varied intensification of therapy appears as a possible solution." Such a lack of specificity is impressive only in a negative manner. However, she does have thoughts as to the basis for conjoint therapy. Essentially she feels that the marriage represents a gestalt in which there is a powerful reciprocal effect from one spouse to the other such that any disequilibrium will be met by a counterforce from the untreated spouse attempting to restore the previous balance. In a variation of this basic theme she notes that often the alcoholic will begin to make gains subsequent to abstinence and the previously unrecognized psychopathology of the spouse will emerge and attempt to vitiate the gains. Like several others, she offers selection criteria which, again like others, suggest that the healthiest, best motivated, and most insightful patients have the best chances for success. It is noteworthy that there has generally been an avoidance of the issue raised by the poorly motivated patient, namely, how to treat him in spite of himself.

Feibel (1960) draws upon psychoanalytic theory to establish a rationale for the eminence of certain therapeutic processes. She feels that the group situation provides opportunities for the patient to learn to substitute speaking for acting-out; to strengthen his ego via group membership which is in itself an identification process; to reduce the tendency to rely upon denial as anxiety is diminished; to dilute the transference to the therapist, thus enabling its positive side to be strengthened; and to enable him to accept interpretation without feeling attacked. These aspects given, certain tasks fall to the therapist including: (1) to help the patient verbalize his life history; (2) to help him to begin working through the mechanism of denial, allowing for a gradual emergence of reality; (3) to use interpretation so as to enable the patient to connect his past life with his present actions and to recognize the inappropriate aspects of his behavior; (4) to accept the rage of the patient that emerges out of his frustration in the transference; (5) to turn the patient toward further exploration, understanding, and reorganization; (6) to allow for a "beginning incorporation of the 'good' therapist which develops toward a later identification with the therapist."

Although her thinking is analytic, particularly as she draws upon it to assess some of the character problems, her technique is modified, and she considers that it differs from classical analysis in that (1) the content of the material obtained is more directed than produced by free association; (2) there is not a detailed working through of unconscious conflicts; and (3) the transference is diluted.

Although, once again, no data are provided, Feibel's work is one of the very few that is specific in making explicit the connections between theory and practice, and in more fully articulating both the clinical and theoretical aspects of the alcoholic's character structure. Unfortunately, it is a paper that is bound by the limits of its time and consequently unable to more deeply and specifically describe some of the issues that *various* alcoholics struggle with. Such material has only very recently emerged (Horner, 1975) and will be referred to further on in this paper.

Esser (1961) has received a great deal of attention, though his work does not appear to contain anything that adds to or offers alternatives to existing knowledge of theory or of techniques. Working in the Netherlands, he has concentrated upon those alcoholics who are apparently sufficiently stable as to have remained married. His approach has been to focus upon sobriety as the single goal and to draw upon the A.A. philosophy. Once again, he appears to conclude that from a characterological point of view, the alcoholic is incurable. "Deeply repressed unconscious conflicts, particularly in alcoholics with middle-aged fixed character structures, are but rarely reached through the group approach."

Fox (1962) offers a rather detailed, yet concise description of the role of group therapy in a total treatment program. She feels there is considerable need indicated for such services as hospitalization, medication, Antabuse, and A.A. Additionally she feels that abstinence during treatment is crucial.

The kind of group therapy she espouses is very much a now classical psychoanalysis-in-group approach. She also makes a point that appears to be crucial to the treatment program and to an understanding of the resistance that is encountered. The alcoholic patient's resistance to therapy is based upon a resistance to *permanent sobriety* and to the giving up of all that alcohol does for him. For inexplicable reasons, this very simple and very basic concept appears to have been overlooked in the literature. It is the opinion of this writer that a failure on the part of the therapist to properly identify and articulate this resistance will almost invariably lead to a failure in treatment, regardless of what theoretical or clinical approach is taken.

Psychodrama seems to be an ancillary technique favored by a number of writers. Weiner (1966) presents an overview of some techniques, but unfortunately does not offer a clearly articulated rationale for the use of psychodrama. Like other treatment phenomena, psychodrama appears to be a technique whose adherents seek to make of it a major *movement* in psychotherapy. Unfortunately this loss of perspective results in a lack of substance and rather overblown rhetoric. Though she makes sweeping claims as to what psychodrama can accomplish for the alcoholic, there is no indication of *how* it does so, and there is a lack of articulation of cause–effect relationships, as well as the usual absence of data. However, the essential purpose of the psychodrama appears to be an intensification of the affect experienced by the patient in the context of the reenactment of a significant event in his life. It is believed that the experience of intense affect has an ego-strengthening effect, as well as providing insight into the nature of the conflict. It is the experience of this writer that psychodrama does have the capacity to evoke powerful emotions. But this should not be taken as an indication that a careful exercise of judgment as to when and with whom the technique may be employed is no longer necessary.

The work of Mullan and Sangiuliano (1966) represents the most comprehensive, fully articulated work encountered by this writer. Strangely, the work is rarely mentioned by other writers and seems not to have made much of an impact on the field. It is worth hazarding a guess that the approach itself raises issues that many authorities would prefer to avoid, such as its emphasis that patient and therapist are engaging in "*mutual* change" through the therapeutic process (italics added). Similarly, detailed account is given of how the therapist shares his own dreams with the group, allegedly for the purpose of "facilitating." What seems to be the tragedy here is that the work has been rather neglected in its entirety—an apparent case of both baby and bath water being disposed of.

The work itself is a book which describes a total treatment program. However, only the last six chapters, which are devoted to group psychotherapy, will be discussed here.

The first issue taken up is that of patient selection. Generally there is agreement with previous writers as the crucial factors are designated as (1) age (the patient should not be too aged); (2) mental capacity such as to rule out those who are retarded; (3) the extent of the affective defect—that is, elimination of the collapsed or burned-out patient; and (4) including only those who are reasonably well motivated. In addition to these criteria, there is offered one other, namely, the therapist's "subjective response." This doesn't appear to be anything new, but has not been articulated in the literature on group therapy with alcoholics. In general psychotherapeutic practice it is an accepted practice for a therapist to disqualify himself on the basis of subjective or anticipated countertransference difficulties but this appears, at least by implication, to be unacceptable in working with alcoholics.

In addition to these inclusion criteria, several others are offered. These are that (1) the patient acknowledge that he is an alcoholic, (2) that he seek psychotherapy, (3) that he accept other forms of rehabilitation such as Antabuse and/or A.A. membership, (4) that he show some degree of insight, dynamic activity, and commitment to keeping appointments, and (5) that he combine anxiety with efforts at sobriety.

Under the rubric of selection criteria are exclusion criteria. These are listed as (1) psychotic behavior, (2) inability to form relationship with any staff member, (3) psychopathic behavior, and (4) severe physical impairment.

Mullan and Sangiuliano consider how and when a patient enters group to be as critical as the selection process. They state that failure to properly prepare a patient is the primary reason for dropping out of group. They feel that the single most important aspect of preparation is the development of a trusting relationship between patient and therapist. The state of the group at the time of a new patient being introduced is also considered to be of extreme importance. Certain conditions are felt to be optimal including the group being not full, free of emergencies, and not too deeply involved in unconscious material. Placement of the alcoholic patient in nonalcoholic group is considered possible only when the patient has been sober for a long time, has had considerable therapy, and has formed a strong bond with the therapist.

The issue of alternate sessions (meetings of the group without the therapist being present) is considered and is strongly felt to be contraindicated.

A separate chapter is devoted to what is called "quasi-group cohesion." This is identified as an early form of resistance and manifested by a superficial camaraderie among the patients for the purposes of excluding the therapist and developing an atmosphere of mutual support so as to preclude confrontation.

Other early resistances are noted as involving generalizations so as to avoid the impact of the specific, drinking, and misinterpretation of the real purpose of the group.

The next chapter is given over to a description of the therapist's efforts to develop what is called "authentic cohesion." It is within this context that the therapist introduces the tactic that is designed to counter the group's effort to exclude him. This involves his becoming a quasi-patient. Although much is given over to articulation of the need for judicious use of this approach, the meaning is clear. "A therapeutic gestalt eventually forms between patients and therapist and at this point the therapist is able to permit himself a fuller range of response which include his feelings and fantasies" (p. 267). Thus, slightly ahead of their time, the writers have brought the so-called humanistic movement into the group treatment of the alcoholic. It is worth speculating that it is upon this rock that their ship has foundered—thus the relative obscurity into which the work has fallen. Unfortunately, politics and therapeutic issues appear to have become interwoven. It is the opinion of this writer that the approach that is espoused raises serious questions. But at the same time, one can only look with regret upon the disdain with which the work in total has been treated by the professional community.

There is a question that is paramount in evaluating such an approach, and the authors themselves allude to it in proscribing alternate sessions (patient's meeting without the therapist), as the dangers of acting-out are warned against. It seems clear enough that in dealing with essentially infantile individuals (whose proclivity for acting-out conflict is so well established), the introduction of highly personal material on the part of the therapist may well stimulate fantasies that the patient may be unable to handle without acting-out and the course of therapy be disrupted beyond repair. On a clinical basis it is easy to see how a "personal" approach would feed into and further stimulate the fantasies of a narcissistic patient, so as to lead him to expect from the therapist gratification of all his infantile yearnings. When such is not forthcoming, the eruption of rage as a response to the narcissistic wound may well result in a premature termination of therapy. Often the unreal expectations of the narcissistic patient lead him to make demands on the therapist of such magnitude and in such a manner as to provoke severe countertransference reactions (Giovacchinni, 1972). This almost inevitably leads to a destruction of the therapeutic relationship.

Withal, a distinct service has been done in pointing out the problems of resistance through pseudocohesion and exclusion of the therapist.

Regarding the central problem in the treatment of alcoholic persons, namely, acting-out as the basic resistance, a view is taken that characterizes what is called the treatment paradox. In order to forestall acting-out, the therapist may attempt to keep anxiety to a minimum. Having done so, he finds that little or nothing therapeutic happens. When he attempts to inject more anxiety into the treatment scene, acting-out occurs. The solution that is offered suggests that (1) the anxiety must reach a minimal level in order that there be any genuine motivation; (2) the tendency to act-out may be limited by the involvement of the clinic staff; (3) the involvement itself must be on the therapist's terms.

On a more theoretical level what is implied is that the therapist avoid attacking the patient's narcissism and make use of the positive aspects of the symbiotic transference to "join forces" with him. This is done by working within the transference and with a consistent expression of concern. It is in this manner that the concept of involvement on the therapist's terms seems most clear. This writer strongly supports the contention that in working with the narcissistic (alcoholic) individual, focusing upon the transference is crucial if there is to be any real hope of forestalling acting-out.

Another chapter is devoted to the use of dreams. It is contended that dreams are particularly useful in working with the kind of alcoholic who is likely to use superrationality as a defense. Further, bringing in the dream increases the depth of contact between the dreamer and the group. "The therapy group, through the dream, becomes engaged in the non-rational aspects of the dreamer's and the therapist's lives, thus circumventing the patient's penchant for abstractions and rationalizations" (p. 268).

Like the previous chapter, the one on the utilization of dreams contains a highly controversial element, namely, the therapist's use of his own dreams. As described, the therapist will selectively share all or parts of his own dreams as well as personal reactions to the patients' dreams for the purpose of "facilitating." However, upon examination of some of the clinical data that are provided, it seems appropriate to question whether or not the therapist's manner of personalized response reflects *his* "needs" as well as therapeutic interventions. In fairness to the authors, it is noted that warning is given against naive use of this approach, listing some of the dangers inherent in it. If one takes such an approach it is crucial that the therapist (1) must be secure in his own boundaries, (2) must be almost totally aware of countertransferential material and attitudes, and (3) must use it only to report his percept of the individual patient or the group.

The last chapter is devoted to issues of termination of the patient from the group, beginning with the observation that among alcoholics there is a high frequency of premature termination initiated by the patient. Although the chapter offers a wealth of clinical detail on acting-out by leaving the group, there is no attempt to provide a theoretical basis for it. It would seem appropriate in such a chapter to devote some time to a discussion of the problems relative to yearnings for attachment, fears of engulfment, fears of abandonment, and some of the crucial problems that are the hallmark of the alcoholic's life. Despite the fact that much material of both descriptive and prognostic value is included, there is no tying of the clinical data to the concepts of the narcissistic personality with its developmental and characterological implications. This comment is not meant to bring the authors to task, but rather to make note of what is lacking, since the theoretical material referred to has begun to emerge only in the last few years. (Kohut, 1971, 1972; Horner, 1975).

The patient who drops out is felt to be reacting to a combination of factors which include (1) poor selection (should the patient have been selected for group in the first place?); (2) insufficient preparation; (3) incorrect timing of entry into the group; and (4) placement in an inappropriate group.

The patient who leaves group prematurely poses problems for the group as well as for himself. "Precipitous dropouts militate against an effective group functioning. Great care should be taken to select patients who are more likely to remain in treatment. The group therapist must become adept at selecting patients who can tolerate at least minimal anxiety" (p. 305).

With regard to the therapist's response to premature termination, a description is given of efforts to reach out and reengage the patient via phone calls or letters.

A final note to the chapter is a bit puzzling as an apparent contradiction with an earlier statement appears. Notation is made that severe underlying pathology, such as psychosis or psychopathy, is not a crucial determinant as to whether or not a patient remains in therapy. This appears to be at odds with an earlier statement in the same book which states that psychopathic behavior is a contraindication to inclusion in a group.

Overall, Mullan and Sangiuliano have produced a most worthy book. There are, to be sure, at least four serious shortcomings in the mind of this writer:

1. Insufficient character differentiation among the alcoholic population. Little attention is given to treatment considerations for various kinds of individuals such as schizoid personalities, as though all alcoholics are essentially alike.
2. An inconsistency with certain issues, such as the therapeutic stance in regard to the psychopath.
3. The issue of the manner of the therapist's involvement, which seems dangerously close to the acting-out of countertransference reactions.
4. Failure to go beyond traditional analytic thinking as to theory and technique relative to the problems of the treatment of individuals fixated early (pre-oedipal) in psychosexual development.

These objections notwithstanding, the work must be considered a major contribution to the literature on the group therapy of alcoholism, and is by far the most comprehensive, specific, and fully articulated work on the subject extant.

There has been almost nothing in the literature regarding the *compulsory* treatment of the alcoholic. Other than the previously noted paper of Brunner-Orne (1956), only the work of Gallant (1968) seems to have addressed itself specifically to that problem. "Revolving door" alcoholics known to the court are compelled to attend group as a condition of probation. Gallant reports some trends toward success as measured by sobriety, at least during the period of attendance. One inference that might be drawn is that to the degree to which the environment can be controlled, so is the alcoholic increasingly treatable. If the manifest anxiety can be turned to talking, rather than acting-out, then a therapeutic engagement is increasingly possible.

A fairly recent book which has drawn attention to itself (though somewhat less among the professional community) is Steiner's *Games Alcoholics Play* (1971). As the title suggests, it is an attempt at applying Berne's conceptualizations to the alcoholic individual. Generally it is a rather flamboyant, almost messianic opus in which claims for success appear nearly without reservation, and the work of other theorists and practitioners is dismissed in a cavalier manner. Despite this, Steiner does make some interesting and notable comments. He takes up the problem of the alcoholic who appears in a state of intoxication. His solution is to allow the patient to attend the group session, but not to overtly participate. The rationale is "based on the fact that a person under the influence of alcohol is unable to cathect the properly functioning Adult ego state which, after all, is the fundamental tool of therapeutic improvement" (p. 116). Further, "the reaction of the group and the therapist to the intoxicated patient, who is basically in his child state, will be clearly remembered by the patient. Thus, the therapist avoids playing the persecutor . . . and he avoids playing the role of Rescuer or Patsy . . ." (p. 117).

Despite the efforts at Bernian jargon, the concepts themselves are sound and essentially in the same dimension as the remarks of Mullan and Sangiuliano (1966) regarding the importance of engagement on the therapist's terms.

On the other hand, Steiner makes statements that contradict virtually the entire body of opinion, and offers little or no rationale for them. For example, "selection of patients for groups is seen as unnecessary and undesirable . . ." (p. 159).

Further, Steiner alone among authorities and writers in the field seems to imply that the alcoholic is perfectly well suited to inclusion in a group of nonalcoholic persons. The issue of the alcoholic's sense of isolation in a group of nonalcoholics is not taken up. Nor does he deal with the disruptive effects of the psychopathic alcoholic, or the excess of demands of the narcissistic alcoholic, but rather issues edicts as though therapy were a matter of cure by decree. Despite a fair amount of good clinical material, it is a work that must be read with serious reservations about many of its claims.

An even more recent work that must be included is a book called *The Drinking Man,* by McClelland *et al.* (1972). Although it is primarily a research work, it merits inclusion here because at its conclusion it suggests that group has a crucial role in solving the alcoholic problem. The book is an investigation of drinking among males and reaches some interesting conclusions. Foremost among these has to do with what drinking among males is *not.* The authors contend that drinking is not (1) to reduce anxiety, (2) to meet dependency needs, or (3) to be free of responsibility. Rather, it is their contention that men drink to feel stronger. They argue that alcoholic men are driven by a wish for aggrandizement and for a wish to feel powerful. This, they feel, is so fundamental an issue as to defy attempts at root cures. A solution is proposed via the socialization of the power need. A.A. is given as an example of meeting the power need by channeling it into doing something someone else. The alcoholic meets his own needs to feel powerful by becoming the sponsor of another alcoholic.

These contentions are remarkably simple, and, in this writer's opinion, overly simple. No issue is taken with the idea of a power need, but rather with the elimination of other considerations, and secondly, with the stopping at that point, as though a power need were itself an irreducible element. Readers of McClelland *et al.* are further recommended to the work of Kohut on narcissistic rage (1972).

CONJOINT AND FAMILY GROUPS

In addition to the more traditional groups that have been described in the preceding section, efforts have been made in the direction of both family and conjoint groups.

A family group would consist of all those persons living in the same household. Thus, in many cases, the group would comprise an extended family and might include a mother-in-law, for example, whose role in the family and its modes of functioning would likely be of significance.

A conjoint group would consist of number (3–6) of alcoholic persons and their spouses.

The rationale for such approaches is exemplified in a remark made by John Warkentin. After some 20 minutes of listening to a woman berate her alcoholic husband, Warkentin turned to her and remarked, "It's odd, but after listening to you, I have a feeling that I've just *got* to have a drink" (personal communication).

Unfortunately, the literature yields very little on either the conjoint or the family group. Bailey (1961), in an excellent review, noted that most of the work to that date had consisted of descriptions of the wife's pathology. Other workers in the field have felt that the attitudes of the spouse of the alcoholic were such that made these persons highly resistant to therapy and little long-range therapeutic engagement could be reported. Some writers found that it was possible to work with wives of alcoholics as a group (Ingersheimer, 1959), but the potential volatility was noted recently by Esser (1971), who emphasizes the need to avoid fights in the family. He is one of the very few who reports success in family therapy with alcoholics. It should be noted however, that Esser's patient families have already qualified as well motivated since the therapist is introduced to the family by an A.A. member.

Another recent paper by Ablon (1974) describes the work of the Al-Anon family group. These groups consist of the relatives of the alcoholic who meet as a self-help group to discuss the problems of living with an alcoholic. The point is made that under these circumstances, which are nonthreatening, progress can be achieved.

Both Ablon and Esser present a position that suggests that one of the aspects of any successful working with the alcoholic or his relatives is the maintainance of a placid, nonthreatening atmosphere.

Evidence in support of that conception is not lacking but is rather strongly supported by many writers who describe the breaking off of treatment at even the mildest of confrontations. However, some papers are extant that describe success. Ingersheimer (1959) reports on a group which had as its avowed purpose the exploration of whether the wives behavior and/or attitudes might contribute to the husband's drinking. Here, too, the sample was obviously biased in the direction of the best motivated. Berman (1968) reports on conjoint therapy in an inpatient setting. His groups met twice weekly and were analytic. He attributes his success to the group's capacity to challenge as well as accept and encourage. He also notes that those persons who had previously unsuccessful therapy had little success in his groups. Scott (1963) reports of his approach to the treatment of the "hostile" alcoholic through a conjoint group, with confrontation taking place from the beginning.

To this writer it seems likely that when an individual is involved in a deeply dependent relationship which tends to reinforce his pathological behavior, then inclusion of the spouse (and the family, where applicable) is indicated. The crucial issues seem to be those of technique. If there is to by any therapy, as such, then there must be some confrontation. Unfortunately, there is relatively little information available as to the techniques that offer the greatest promise of therapeutic engagement of any but the very best motivated of alcoholics or their spouses and families. What seems probable is that confrontation is both possible and necessary when the alcoholic (and/or spouse) is well motivated. When he is not well motivated then efforts with and without confrontation are likely to fail.

SUMMARY

This chapter has attempted to present an overview of the group therapy approach to the treatment of alcoholism. Various modalities have been viewed; various papers have been described; various issues have been considered. What seems clear to this writer is that the present state of affairs is a highly imperfect one with much controversy. However, issues of technique are not all the controversy. Even more basic is the still unresolved problem of a definition of alcoholism. Added to this is the lack of adequate research. There are still the various issues of technique, including the problems of character diagnosis, patient selection, patient preparation, the length of treatment, treatment frequency, spouse and/or family involvement, abstinence, and the network concept. The list might well be expanded.

Essentially this writer feels that a disservice has been done in the overall failure to pay sufficient attention to two diagnostic issues. These are the distinction between the primary and the reactive alcoholic, and attention to the various character types that constitute the spectrum of primary alcoholics. A theory or practice of treatment for alcoholics that fails to distinguish between the alcoholic who is a psychopathic personality and one who is a narcissistic personality and one who is a schizoid personality is likely to be far less than helpful.

There is also a need for considerable theory-based research. Failure to develop such research results in a proliferation of papers that take various positions but provide no assessable data and generate little useful research. One theoretical approach that might be adaptable to the alcoholic is the work of Horner (1975), which utilizes a psychoanalytic-object relations theory of character development, relating failure of specific developmental tasks to specific psychopathologies.

Finally, there is a need to assess the spectrum of group psychotherapy approaches as they range from the extreme group focus of Bion (1961) to the highly individual focused work of Wolf and Schwartz (1971). It seems to this writer that the statements one makes about the problems that alcoholics have will surely merit consideration in deciding what philosophy of group to employ in their treatment.

REFERENCES

Ablon, J., 1974, Al-Anon family groups, *Am. J. Psychother.* 28(1):30–45.
Adler, G., and Buie, D. H., 1972, The misuses of confrontation with borderline patients, *International Journal of Psychoan. Psychotherapy* I(3):109–120.
Bailey, M., 1961, Alcoholism and marriage: A review of research and professional literature, *Q. J. Stud. Alcohol* 22:81–97.
Berman, K., 1968, Multiple conjoint family groups in the treatment of alcoholism, *J. Med. Soc. N.J.* 65:6–8.
Bion, W., 1961, "Experiences in Groups," Basic Books, New York.
Brunner-Orne, M., 1956, The utilization of group psychotherapy in enforced treatment programs for alcoholics and addicts, *Int. J. Group Psychother.* 6:272–279.
Brunner-Orne, M., and Orne, M., 1954, Directive group therapy in the treatment of alcoholics: Technique and rationale, *Int. J. Group Psychother.* 4:293–302.
Brunner-Orne, M., and Orne, M., 1956, Alcoholics, *in* "Fields of Group Psychotherapy" (S. R. Slavson, ed.), International Universities, New York.
Buie, D. H., and Adler, G., 1972, The uses of confrontation with borderline patients, *International Journal of Psychoan. Psychotherapy* I (3):90–108.
Corwin, H., 1972, The scope of therapeutic confrontation from routine to heroic, *International Journal of Psychoan. Psychotherapy* I (3):68–89.
Durkin, H., 1964, "Group Psychotherapy," International Universities, New York.
Esser, P. H., 1961, Group psychotherapy with alcoholics, *Br. J. Addict.* 57:105–114.
Esser, P., 1971, Evaluation of family therapy with alcoholics, *Br. J. Addict.* 66:251–255.
Evseeff, G. S., 1948, Group psychotherapy in the state hospital, *Dis. Nerv. Syst.* 9:214–218.
Feibel, C., 1960, The archaic personality structure of alcoholics and its indications for group therapy, *Int. J. Group Psychother.*, 10:39.
Fox, R., 1962, Group psychotherapy with alcoholics, *Int. J. Group Psychother.* 12:56–63.
Freud, S., 1920, "Introductory Lectures of Psychonalysis," Boni and Liveright, New York.
Freud, S., 1959, Analysis terminable and interminable (1937), "Collected Papers" Vol. 5, pp. 316–357, Basic Books, New York.
Gallant, D., 1968, A comparative evaluation of compulsory (group therapy and/or Antabuse) and voluntary treatment of the chronic alcoholic municipal court offender, *Psychosomatics* 9:306.
Giovacchinni, P., 1972, Technical difficulties in treating some characterological disorders: Counter transference problems, *Int. J. Psychoan. Psychother.* 1(1):112–128.
Greenbaum, H., 1954, Group psychotherapy with alcoholics in conjunction with antabuse treatment, *Int. J. Group Psychother.* 4(30):30–45.
Haber, S., Paley, A., and Block, A., 1949, Treatment of problem drinkers at Winter Veterans Administration Hospital, *Bull. Menninger Clin.* 13:24–30.
Heath, R., 1945, Group psychotherapy of alcohol addiction, *Q. J. Stud. Alcohol* 5:555–562.
Horner, A., 1975, Stages and processes in the development of early object-relations and their associated pathologies, *Int. Rev. Psycho-Anal.* 1(1):95–105.
Ingersheimer, W., 1959, Group psychotherapy for nonalcoholic wives of alcoholics, *Q. J. Stud. Alcohol* 20:77–85.
Kohut, H., 1971, "The Analysis of the Self," International Universities, New York.
Kohut, H., 1972, Narcissism and narcissistic rage, *in* "Psychoanalytic Study of the Child," p. 27, Quadrangle Books, New York.
Lippert, R., and White, R. K., 1947, An experimental study of leadership and group life, *in* "Readings in Social Psychology," (T. Newcombe and E. Hartley, eds.), Henry Holt, New York.

Martensen-Larsen, O., 1956, Group psychotherapy with alcoholics in private practice, *Int. J. Group Psychother.* 6:28–37.

McClelland, D., Davis, N., Kalin, R., and Wanner, E., 1972, "The Drinking Man," Free Press, New York.

Mullan, H., and Sangiuliano, I., 1966, "Alcoholism: Group Psychotherapy and Rehabilitation," Charles C Thomas, Springfield, Illinois.

Pfeffer, A. Z., Friedland, P., and Wortis, H., 1949, Group psychotherapy with alcoholics; preliminary report, *Q. J. Stud. Alcohol.* 1:217–251.

Preston, F., 1960, A combined independent, joint, and group therapy in the treatment of alcoholism, *Mental Hygiene* 44:522–528.

Scott, E., 1956, A special type of group therapy and its application to alcoholics, *Q. J. Stud. Alcohol.* 17:288–290.

Scott, E., 1963, A suggested treatment plan for the hostile alcoholic, *Int. J. Group Psychother.* 13:93–100.

Steiner, C., 1971, "Games Alcoholics Play," Grove Press, New York.

Weiner, H., 1966, An overview of the use of psychodrama and group psychotherapy in the treatment of alcoholism in the United States and abroad, *Group Psychotherapy* 19:159–165.

Winnicott, D. W., 1965, Ego distortion in terms of true and false self, *in* "The Maturational Process and the Facilitating Environment," International Universities, New York.

Wolf, A., and Schwartz, E., 1971, Psychoanalysis in groups, *in* "Comprehensive Group Psychotherapy," (H. Kaplan and B. Sadock, eds.), Williams and Wilkins, Baltimore.

CHAPTER 7

Family Therapy in Alcoholism

Peter Steinglass

Center for Family Research
Department of Psychiatry and Behavioral Sciences
George Washington University School of Medicine

INTRODUCTION

In 1972 a husband and wife were recruited as research subjects for an experimental treatment program being carried out under the auspices of the National Institute on Alcohol Abuse and Alcoholism (Steinglass *et al.*, 1975). The couple, an alcoholic woman and her concerned husband were an attractive, well-educated, middle-class couple with three children.

For the past five years Mrs. D had been returning home from her job as an elementary school teacher and on a thrice-weekly schedule had been consuming up to one-fifth of scotch over a late afternoon and early evening time frame. Her husband would characteristically return home to find her extremely intoxicated, depressed, voicing suicidal intentions, and histrionic. He reported experiencing anger and frustration in response to his wife's intoxication, and an increasing sense of helplessness in the face of the chronicity of the problem. Mrs. D took the position that alcoholism was an inherited trait in her family, and that she would inevitably meet a tragic fate.

Mr. and Mrs. D had, for the previous five years, made contacts with a variety of traditional resource professionals in an increasingly desperate attempt to find a solution for Mrs. D's pernicious drinking problem. These contacts had included visits to a family doctor who felt Mrs. D was an anxious person under stress and prescribed minor tranquilizers, an internist who diagnosed Mrs. D's alcoholism and suggested a course of Antabuse which she refused, and the pastor of a local church who suggested Alcoholics Anonymous and made an initial contact for Mrs. D. However, after attending two meetings she refused to return, telling her husband that she was clearly hopeless.

As Mrs. D's drinking increased in severity, she began experiencing difficulties at work. Supported by the opinions of her husband and her family physician, she accepted a referral to a psychiatrist. The psychiatrist, a specialist in alcoholism, recommended in-hospital detoxification to break the drinking cycle, followed by a four- to six-week stay at a residential treatment center. Mrs. D accepted the recommendation for detoxification, but refused the residential treatment program. There followed a series of brief, usually abortive, contacts with community-based alcoholism programs offering group therapy with medical backup. A social worker in one of these alcoholism treatment programs was impressed, on taking a history, with the extent to which Mrs. D focused on difficulties with her husband and, after seeing them together, recommended them to the NIAAA program.

As a result of this referral, Mr. and Mrs. D were seen jointly for an evaluation. The evaluation uncovered long-standing patterns of interactional conflict. It seemed clear that Mrs. D's abuse of alcohol had become a central issue between the couple, but it was unclear exactly what role Mrs. D's alcohol consumption and states of intoxication were playing in the couple's interactional life. However, it did seem apparent that although only one member of the marriage was drinking excessively, both husband and wife were abusively using this drinking pattern and its behavioral consequences to level a variety of pernicious and destructive charges at each other. Lastly, it was noted that despite a steadily worsening pattern of alcohol consumption, behavior within the marriage seemed remarkably stable, with one week's interaction looking very much like another.

As a result of this evaluation, it was suggested to Mr. and Mrs. D that they were engaged in what the research team called an "alcoholic marriage" (Steinglass et al., 1971a). This term implied that both members of the marriage had an equal stake in the perpetuation of the alcoholic symptom, even though the symptom was limited to the wife alone. It was recommended to Mr. and Mrs. D that they volunteer for a treatment program in which they would be seen in a therapy setting along with other couples, but at no time would they be seen individually. They were told that the goal of treatment would be to gain a better understanding of the role abusive alcohol consumption was playing in their marriage, on the assumption that this improved understanding would lead to improvement in their interpersonal relationship. Once again it was suggested that a period of hospitalization would be necessary, but this time it was to be the couple that would be hospitalized together. They were also informed that during the period of hospitalization they would actually be encouraged to use alcohol so the therapists could observe their interactional behavior during states of intoxication. The problem, they were told, was a dysfunctional marriage in which alcohol had been inappropriately used to perpetuate certain patterns of behavior.

Until recently the existence of such a treatment program was unimaginable and probably unthinkable. Certain aspects of the recommendations made to Mr. and Mrs. D are clearly at this point experimental in nature. However, the conceptualization of alcoholism as a family problem and the clinical interest in conjointly interviewing married couples and family members as part of a diagnostic evaluation and treatment program has increased dramatically in the past decade. In recognition of this trend, the *Second Special Report to the U.S. Congress on Alcohol and Health* (Keller, 1974) called family therapy "the most notable current advance in the area of psychotherapy [of alcoholism]."

The gap between the treatment approaches recommended to Mr. and Mrs. D in their initial contacts with professionals and the approach recommended by the NIAAA program is a vast one. The closing of this gap has resulted from a series of clinical and research findings over a 40-year period which have provoked increasing interest in the role of the family in the development and perpetuation of chronic alcoholism. In this chapter we will review this process historically and present the detailed data that are currently available regarding family treatment approaches to alcoholism.

FAMILY THERAPY AS A TREATMENT MODALITY

The recognition that disturbed family life may play a significant etiological role in the genesis of psychopathology dates back to the earliest psychoanalytic notions of human behavior. All psychodynamic theories of human functioning have viewed family relationships as the emotional substrate out of which the adult personality is formed, and disturbed family relationships therefore are thought to increase the likelihood of adult neurosis, psychosis, and personality disorder. The notion of simultaneously interviewing two or more family members as part of a psychotherapeutic intervention, however, did not occur until the mid-1950s, and has only become firmly established in the decade of the 1960s. Nevertheless, although relatively slow in starting, family therapy has rapidly established itself as an exciting and innovative therapeutic approach, and has become part of the standard repertoire in most community mental health centers and residential treatment programs.

Family therapy developed in response to two fundamental clinical observations. The first was the repetitive experience of mental health professionals working with psychotic patients, that despite the intensity or format of their therapeutic work with the individual patient, abnormal functioning invariably reappeared when the patient returned home and resumed his preexisting family life. The second clinical observation, long known and understood by playwrights and novelists but not as well appreciated by clinicians because of their concentration on the individual patient, was that each family member has a unique and often disparate version of family events and family relationships. The first clinical observation suggested that progress for the individual patient might only be secured if simultaneous and complementary progress also was achieved in the individual's social environment, namely his or her family. The second observation suggested that individual differences in perceptions of family life could only be resolved for the clinician if he or she observed the family directly. That meant bringing the entire family in for a simultaneous interview.

These clinical observations received considerable support from concurrent research findings regarding the relationship between schizophrenia and family life. This work, especially the notions of Bateson et al. (1956) on the "double bind," the notion of "pseudomutuality" by Wynne et al. (1958), Wynne and Singer's (1963; Singer and Wynne, 1965a,b) work on thought disorder, and the family "skew" notion of Lidz et al. (1965) all suggested that schizophrenia seems to develop in conjunction with a particular pattern of disturbed family functioning. The critical implication of these theoretical notions was that in certain instances it may be more profitable to view the family rather than the individual as the basic unit of pathology. A short additional step brings us to the conclusion that therapeutic interventions should be directed toward the whole family rather than its symptomatic member alone. This evolution from individual to family is summarized by Ackerman (1962) as he contrasts psychoanalytic and family treatment methods: "Psychoanalytic treatment focuses on the internal manifestations of disorder of the individual personality. Family treatment focuses on the behavior disorders of a system of interacting personalities, the family group."

At the present stage of its development, family therapy is a term used to describe a wide variety of therapeutic techniques. Bowen (1976), for example, in his review of the first two decades of family therapy, lists 12 distinct therapeutic approaches or modalities all qualifying as bona fide forms of family therapy. Foley (1974) narrows these approaches down to four basically distinct forms of therapy: conjoint family therapy, multiple impact therapy, network therapy, and multiple family therapy.

This diversity in form and structure of the therapeutic setting led the GAP Committee (1970) on the family to conclude in their report, *The Field of Family Therapy*, that family therapy is "not a treatment method in the usual sense." Pointing out that there is "no generally agreed upon set of procedures followed by practitioners who consider themselves family therapists," they conclude that the shared base is the common conviction about the relationship between individual and family psychopathology, and the belief in the therapeutic benefits of seeing the family together.

In other words, despite the continuing debate over the structural components of the therapeutic situation (for example, who should the therapy be addressed to: the parental couple, the whole family, the extended family, the social network; or what schedule should be adopted for the sessions: the traditional weekly sessions of conjoint therapy, the intensive two-week schedule of multiple-impact therapy, the marathon quality of network therapy), family therapists share a body of core concepts that apply to all family therapy approaches.

As we shall see in our review of family approaches to alcoholism, most of the popular family therapy techniques have at one time or another been applied to families containing alcoholic members. These techniques, however, have often been implemented by therapists who are relative novices to the family field. The result has been a somewhat haphazard application of poorly understood techniques resulting in confusion about outcome and significance of the therapeutic venture.

Let us therefore spend a few moments reviewing six key concepts that differentiate family therapy from other therapeutic modalities. Although a unified or comprehensive theory of family functioning, family pathology, and therapeutic change has yet to be developed, the following core concepts have gained widespread acceptance, and therefore deserve our attention.

The Family as a System

Drawing on concepts from general systems theory, family therapists have found it profitable to think of the family as an operational system. This view treats the whole family as the primary organizational unit. Individuals within the family represent component subsystems of this primary organizational unit. The emphasis is on patterns of interrelationship between these component parts, hence the focus on interactional behavior, structural patterning within the family, and the balance or stability of the system as a whole. Any single piece of behavior for the family systems therapist has to be understood first in terms of how all the component parts (individuals) are contributing to or making the behavior possible, and secondly how the behavior is affecting all the individuals in the family. Pathology becomes redefined as a structural or functional imbalance in the family rather than difficulties being experienced by any single individual within the family. Therapy is focused on the improved understanding of the structure and patterns of functioning within the family, and on correcting those imbalances that have led to stress or strain within the "system."

The Concept of Homeostasis

First introduced by Don Jackson (1957), this concept utilizes a term introduced by Cannon to describe regulatory mechanisms in physiological systems, and applies it to family systems. The notion is that families tend to establish a sense of balance or stability, and have built-in mechanisms to resist any change from that predetermined level of stability. This stability does not necessarily imply a healthy state of affairs. The family, for example, might include as part of this stabilization pattern a piece of chronic psychopathology, such as chronic alcohol abuse. But Jackson's notions imply that regardless of the quality of stabilization, families have built-in mechanisms to restore their specific sense of balance whenever events occur that tend to disrupt this balance.

The *feedback loop* is the primary mechanism for the maintenance of homeostasis. In physiology, the *negative feedback loop* was a term coined by Cannon to describe the pituitary–endocrine gland axis controlling hormonal levels in the body. As the hormone product of one component part of the axis increased, the hormonal product at the other end of the axis decreased, and vice versa. In family terms, the negative feedback loop is most frequently used to describe two interrelated aspects of behavior, usually carried out by two or more individuals in the family. The behaviors are interrelated but conflict with each other, in that their presence tends to inhibit each other. A pattern is established in which behavior A is invariably followed by behavior B, which in turn inhibits or prevents the further expression of behavior A. The end result is a return to the prior level of functioning (the homeostatic level) of the family. A clinical example of such a negative feedback loop is the spouse who encourages socializing or brings alcohol into the home or demands to be taken to a restaurant where drinks are ordered, every time the alcoholic member of the family makes a try at abstinence.

The feedback loop is one of a series of mechanisms postulated by family therapists to explain the intricate series of checks and balances they observe occurring in families, the end result of which is the remarkable stability of interactional behavior exhibited clinically by most families. *Cybernetics*, a term introduced by Norbert Wiener (1961), has become an attractive theoretical model utilized to explain this process of self-regulation in family systems. Although originally intended to explain processes of control in physical systems, and utilized most profitably to develop concepts of logic in computer science, Wiener's concepts have proven highly adaptable to the field of human interactional behavior. Cybernetic regulation is the key concept that attempts to explain the perpetuation or maintenance of chronic patterns of behavior in family systems. In this sense, it becomes the key concept when family theorists attempt to explain interactional factors maintaining chronic alcohol abuse in families.

The Concept of the "Identified Patient" or "Scapegoat"

Perhaps the most revolutionary impact of family therapy on psychiatry has been the redefinition of psychopathology in family terms. The schizophrenic individual becomes the schizophrenic family. The alcoholic individual becomes the alcoholic family. The antisocial individual becomes the antisocial family. This transformation occurs via the concept of the identified patient. According to this concept, the symptomatic member of the family—be he or she schizophrenic, alcoholic, or psychopathic—is not merely a disturbed individual who would be clearly symptomatic in his own right regardless of the behavioral setting. Instead, he or she is the labeled or identified patient selected by the family system of which he or she is a member to express for the entire family the particular piece of disturbance represented by the symptom selected.

For example, the antisocial adolescent is viewed as acting out antisocial fantasies shared by all family members, and the schizophrenic individual is seen as manifesting the stresses and strains created by the psychotic pattern of interaction within the family as a whole. An extension of this concept implies that the selected individual might, through his or her symptom expression, be protecting or stabilizing the level of functioning of other family members. In this view, the alcoholic member of the family might, through his or her drinking, be protecting the family from overwhelming depression or intolerable levels of aggression. Such a model would be used to explain, for example, the clinical appearance of significant depression in a nonalcoholic spouse when the alcoholic stops drinking.

Communication Patterns

Family therapists have focused on communication patterns within the family with the same enthusiasm that analysts have reserved for dream material. Communications, both verbal and nonverbal, are viewed as reflecting the basic structural and interactional patterns governing the family's behavior, and therefore frequently become the primary focus of attention during therapy sessions. Some family therapists even contend that improvement in patterns of communication should be the only therapeutic goal. According to this view, once the family employs healthy channels for communication, they will be able to tackle and resolve any conflicts that come their way.

In alcoholism, interest in communication has focused on the nature of communicational patterns during intoxication. An appreciation of the contrast between sober and intoxicated interaction of the entire family has been viewed by some family therapists as critical to their understanding of the dynamics of the alcoholic family.

Behavioral Context

A cornerstone of individually oriented dynamic psychiatry has been the role of conflict as a determinant of behavior. Behavior is seen as internally motivated, resulting from the conflict between individual wishes and external reality. Although motivating factors may be outside of conscious awareness, the process leading to the final expression of behavior is a process that resides in the individual and is therefore under individual control. The behaviorist model, although postulating a different mechanism for behavior, also views the process as internal to a single individual.

Family therapists, because they work with more than one individual, have been interested not only in internal processes but also in the relationship between individual behavior and the interactional field within which the behavior is expressed. This relationship has been called the "context" for behavior. The context, which is usually thought of as a combination of setting, and cast of characters in the setting, can predetermine the behavior of any single individual in that contextual field by limiting the possible choices or range of behaviors the individual can successfully or appropriately express. The family therapist is particularly interested in combinations of behaviors and contexts for behavior which occur repeatedly in a particular family. These combinations often evolve into characteristic patterns of interaction. These patterns of interaction, if potentially destructive to the family unit as a whole, may then become primary targets for investigative work in the therapeutic setting.

Drinking behavior for the family therapist is therefore inadequately described if only the quantity and frequency of alcohol consumption is considered. The drinking must also be examined in terms of the behavioral context within which it occurs, and the characteristic pattern of relationships between family members that emerge when alcohol is present.

Boundaries

Because family therapists are interested in interactional *fields,* they must also pay attention to the quality of the boundaries separating participants in the field, and the boundary surrounding or separating the entire field from the outside world. Family therapists are therefore concerned both with the nature of the boundaries separating individual members of the family (e.g., generational boundaries) and with the relationship of the nuclear family to the outside world. Rigid and impermeable boundaries between the family and the outside world tend to isolate the family, preventing them from utilizing extrafamily resources to benefit individuals within the family. Excessively permeable boundaries, on the other hand, destroy the family's sense of group integrity and connectedness, often preventing them from behaving as an effective unit. Alcoholic families have been characterized as having extremely rigid boundaries, leading to a characteristic sense of isolation within the community. Therapeutic interventions might therefore involve intensive examination of such boundary phenomena.

FAMILY THERAPY IN ALCOHOLISM

As we shall see in our review of the literature, alcoholism therapists have come relatively late to the family field, and family therapists have only very recently begun to view alcoholism as an area of interest. This mutual disregard is frankly not at all surprising.

From the perspective of the traditional establishment in the alcoholism field, the priority issue has been the transformation of alcoholism from a moral problem into a medical problem. This conversion has been seen as a necessary prerequisite for the transfer of responsibility for alcoholism treatment from the judicial system into the medical establishment. With this goal in mind, the emphasis has been on the medical model. Alcoholism is viewed as a disease process with an etiology, a set of symptoms, a typical course, and a predictable prognosis. However, the medical model is designed primarily to describe disease processes as they affect an individual. Hence family therapy feels strange and foreign.

From the perspective of the family therapist, on the other hand, clinical interest focused on disturbed communicational patterns and structural dissonance within the family. Although these phenomena are hardly absent in the family with an alcoholic member, the abusive consumption of alcohol, and its attendent behavioral and physical consequences, appeared at first glance to be so overwhelming that it was hard to imagine successful treatment being achieved in any way other than by intensive work with the individual who was doing the drinking.

Despite these obstacles, family therapy techniques have been used with increasing enthusiasm in alcoholism treatment. We will review this development historically, paying attention both to theoretical trends, and to those innovative experimental studies that have advanced the field.

Phase I: Early Interest in Family Issues and Alcoholism

In 1937 Robert Knight published a classical paper which raised, for the first time, curiosity about the role of family factors in the etiology of alcoholism. Based on his intensive psychoanalytic case history work with chronic alcoholic men, Knight's report hypothesized that alcoholism arises in a family constellation composed of a domineering mother and a passive father, a notion that subsequently gained widespread acceptance. The following year Chassel (1938) published another case report. He also postulated a two-generational model leading to the production of an identified alcoholic. But in contrast to Knight, Chassel emphasized the presence of an abusive and domineering but somewhat capricious father as the key element.

These early papers continue to be of interest to us because they established a particular framework for exploring family factors in alcoholism. This framework incorporated perspectives which in retrospect actually retarded the growth of family therapy as an acceptable treatment approach for alcoholism at the same time that it focused on family factors as etiological agents. Since these perspectives are still very much with us, they deserve brief mention.

First, for Knight and Chassel family factors were of interest only from the etiological perspective. The behavior demanding explanation was the abusive drinking of the alcoholic individual, and little attention was paid to interactional behavior or object relations. Family factors were therefore viewed as the germinative medium nourishing the alcoholic's psychopathological development. The logical conclusion was that knowing the family helps to understand the individual, but treatment remained individually oriented. Second, the emphasis on a specific family constellation associated with the production of alcoholism presages the notion that alcoholic individuals are more alike than different, that a specific alcoholic "personality" can be demonstrated to exist, and that families with alcoholic members are also more alike than different. These families should therefore be lumped together and dealt with by "alcoholism" professionals rather than family experts in general family treatment centers. Third, the emphasis on family constellation led inevitably to an emphasis on structural components of family life rather than communicational styles or functional components of family living. This emphasis also later isolated alcohol therapists from developments in the family therapy field which tended to be oriented toward issues of interaction and communication.

So even though the family therapy field didn't really develop until the decade 1955–1965, existing notions and already established orientations in alcoholism therapy left alcohol therapists unprepared to adopt or experiment with these new techniques.

Phase II: The Alcoholic Marriage

In the 1950s interest turned to the clinical study of marriages between male alcoholics and their wives (Bailey, 1961). The primary concern centered on the role of the wife in initiating and perpetuating her husband's drinking. A debate arose between two factions: One was represented primarily by psychiatrists and psychiatric social workers who viewed the wife of the alcoholic as a person with severe, long-standing psychopathology antedating marriage, which led her to choose an alcoholic husband as a way of satisfying and stabilizing intrapsychic needs (Futterman, 1953; Bergler, 1949); the other faction was represented primarily by sociologists who explained the behavior of these wives as directly resulting from having to deal with the repetitive pressures and stresses placed upon the marriage by the husband's drinking (Kogan and Jackson, 1965; J. K. Jackson, 1962, 1958, 1954). In retrospect this debate was probably artificial. The most recent review of this literature (Edwards et al., 1973) concludes that no convincing evidence has emerged suggesting a single personality "type" characteristic to wives of alcoholics, or a theoretical explanation of their behavior.

In any event, although interactional models were being proposed to explain behavior in an alcoholic marriage, most of the clinical data stimulating these ideas came from individually oriented therapy or research. For example, a clinician would be impressed with repetitive stories of inconsistent behavior on the part of his alcoholic patient's wife in which the wife is described as keeping the liquor cabinet well stocked, pouring drinks for her husband, and making excuses for him in his work situation, at the same time that she is complaining bitterly about his excessive drinking and threatening to leave if he doesn't stop. Only rarely were these clinical data verified via a clinical interview with the wife as well (MacDonald, 1958; Whalen, 1953). Sociologists, on the other hand, obtained much of their data directly from wives, and had little opportunity to substantiate these reports via direct observation or collateral interviewing (Kogan and Jackson, 1965; J. K. Jackson, 1954).

However, these studies were important in providing a changing focus for therapy. Whereas earlier studies focused on family issues only from a historical perspective, the focus on the alcoholic marriage was a focus on the here and now. As long as the alcoholic individual was viewed in isolation, and explanations for his or her abusive drinking were related only to individual psychodynamics or pathophysiology, then the only logical treatment approaches would be individually oriented. If, however, questions were raised about the extent to which an interactional relationship between a husband and wife might either cause or perpetuate abusive drinking, then logic would dictate that a place had to be found for the spouse in the treatment plan. As Joan Jackson (1962) has noted, "Once attention had been focused on the families of alcoholics, it became obvious that the relationship between the alcoholic and his family is not a one-way relationship. The family also affects the alcoholic and his illness. The family can either help or interfere with the treatment process." Jackson therefore concludes that significant family members must be taken account of, if not actively involved, in treatment in order to achieve success.

Phase III: Concurrent Therapy for Alcoholics and Spouses

In 1954 a project was instituted in the outpatient department of the Henry Phipps Psychiatric Clinic at the Johns Hopkins Hospital involving concurrent group meetings of male alcoholics and their wives (Gliedman, 1957; Gliedman et al., 1956a,b). This project represented the first attempt to adapt the most successful psychological therapy approach to alcoholism—group therapy—to a family orientation. Nine male alcoholics and their wives were recruited and placed in two separate groups, one for the alcoholics, one for the wives. Thus once they had volunteered for the study, husband and wife went their separate ways and entered into a group which developed its own schedule, therapy format, rules, and group process issues. Although the intent was to involve the marital couple concurrently in the treatment process, Gliedman and his associates assumed that different therapeutic issues would exist for alcoholic husbands versus wives. Each spouse would therefore need a different group in order for appropriate therapy to occur. For example, the wives' group was envisioned as "the usual analytically oriented therapy group," while the alcoholics' group was viewed as more structured, with specific techniques adapted to control anxiety levels in the group.

Despite the very small patient sample, this study is a pivotal one in the development of family techniques for the treatment of alcoholism. Although the specific results of the study were equivocal (marginal but not convincing improvement in most of the patients treated), ground was broken in a number of important areas that have subsequently become characteristic of family approaches. Perhaps the most important of these areas is the issue of outcome variables.

Most treatment programs have focused almost entirely on a diminution of drinking as the sole outcome variable of merit. Although the wisdom of this approach has been questioned on occasion, the majority of treatment programs continue to be judged against a standard of percentage of patient population abstinent within a specified time frame. Gliedman, by including wives as potential clientele for the treatment program, significantly expanded the scope of appropriate outcome variables against which successful treatment was to be judged. Symptom reduction, for example, applied to the wife as much as it applied to her alcoholic husband. If symptoms such as depression are applicable for the spouse, then they must also be applicable for the identified alcoholic. Thus reduction in depression is added to reduction in drinking as an acceptable criterion for successful treatment. Secondly, the concurrent treatment of both members of a marriage naturally leads to an examination of marital satisfaction and marital interactional behavior as target criteria for therapeutic change. Although Gliedman is primarily a group therapist who explains his technique as an extension of group therapy to a natural collateral group, a subtle change has begun in which the marriage itself has become an appropriate focus for the therapist's attention.

Patients were therefore evaluated before and after treatment by means of four measures: (a) a drinking checklist to measure the severity of drinking; (b) a symptom checklist to indicate the amount of distress from psychological symptoms; (c) the mutual satisfaction or dissatisfaction experienced by the alcoholic husbands and their wives with each *other* during sobriety as contrasted with intoxication; and (d) a social ineffectiveness scale.

Within this widened perspective, Gliedman and his associates found that although there was some reduction in drinking behavior, the greatest changes in behavior resulting from the concurrent group therapy technique were in the areas of "marital milieu" (defined as satisfaction of alcoholic husband and his wife with each other) and "personal morale" (the alcoholic individual's satisfaction with himself). A significant change also was felt to occur in reduction in psychological symptomatology, especially irritability and depression, on the part of both alcoholic husband and nonalcoholic wife. The least change seemed to occur in the area of social effectiveness, which was judged to be poor at the start of therapy and showed little improvement as a result of the group experience. Gliedman's conclusion was that his concurrent group therapy technique was most effective in its ability to improve or elevate self-esteem in a patient group that tended to be demoralized prior to therapy.

Following the Johns Hopkins study, several clinical papers appeared in the literature describing group techniques for working with spouses of alcoholics (Westfield, 1972; Pattison *et al.*, 1965; Pixley and Stiefel, 1963; Burton, 1962; MacDonald, 1958). These papers indicated a growing interest in the development of techniques for changing the treatment focus from the alcoholic individual alone to the alcoholic individual in a marital context. It also reflected the conviction that the inclusion of the nonalcoholic spouse in the treatment of those alcoholic individuals who retained a stable marriage was a necessary prerequisite for successful therapy. Pixley and Stiefel (1963), for example, state, "There is no question at this point that if psychotherapy is to be effective for a larger proportion of the alcoholic population the wife must also be treated."

The most ambitious study of concurrent group psychotherapy was carried out by Ewing and his colleagues (1961) at the University of North Carolina School of Medicine. For a period of four years, starting in 1955, a program was established offering an optional concurrent group therapy program for spouses of alcoholic individuals already in treatment. Although the program was offered to male and female alcoholics alike, only wives of male alcoholics volunteered for the program.

During the first 18 months of the program's inception, 32 still-married alcoholic men were accepted into the group therapy program offered by the Department of Psychiatry for alcoholism treatment. Of these 32 men, 16 wives volunteered to participate in concurrent group psychotherapy sessions. In contrast to the Johns Hopkins group, similar schedules were adopted for the husbands' group and the wives' group. Both groups met weekly in different rooms in the same building.

Although the basic technique was described as "dynamically oriented group psychotherapy," Ewing concurs with Gliedman in noting the particular needs of alcoholic patients for more structured experiences stemming from a lower capacity to withstand tension and anxiety. However, in contrast to Gliedman, Ewing noted considerable resistance in the wives' group to traditional psychotherapeutic techniques, and an unwillingness to develop any real therapeutic orientation. Their clinical description of group process in the wives' group indicates a general resistance to any discussion of intragroup feelings, but a much greater facility in talking about attitudes toward alcohol and drinking and their feelings about their husbands' abusive use of alcohol (extragroup feelings).

Long-term follow-up data provided by Ewing's group is impressive on two scores. First is the finding of a significantly greater persistence in therapy for those male alcoholics whose wives were attending a concurrent group psychotherapy session (see Table 1). Second, long-term follow-up (a minimum of three years after the inception of group therapy) indicated significantly improved control of drinking and considerable improvement in marital harmony for those men engaged in concurrent group therapy with their wives, as opposed to men coming alone to the therapy program. The question of whether this improvement was due to the specific working-through of marital issues in the therapy sessions, or merely due to the increased longevity of treatment (because the engagement of the wives assisted in keeping their alcoholic husbands in treatment for a longer period of time), is raised but left unanswered by this study. However, since engagement of alcoholics in long-

TABLE 1. University of North Carolina Concurrent Therapy Study[a]

Comparison of Length of Attendance of Patients on the Basis of Wives' Participation in Group Meetings[b]

Patients	Attended less than 6 months	Attended more than 6 months	Total
Married men			
Coming alone	13	3	16
Coming with wife	7	9	16

Comparison of Improvement in Patients on the Basis of Wives' Attendance in Group Meetings[c]

Patients	Totally sober or very much improved	Only slight or no improvement	Uncontacted at follow-up	Total
Married men				
Coming alone	3	7	6	16
Coming with wife	8	3	6	16

[a] From Ewing et al. (1961).
[b] Chi square = 4.80; $p < 0.05$.
[c] Chi square = 3.46; $p\ 10 \to 0.05$.

term therapy has in and of itself been a major obstacle to successful treatment, the results of the Ewing study have to be viewed as impressive.

Ewing's findings were strongly supported in a study carried out by Smith (1969) at the University of Edinburgh. Despite the fact that the treatment program was radically different (alcoholics were hospitalized for up to a six-week stay as opposed to being treated on an outpatient basis in the Ewing study), the institution of a separate therapeutic group for wives of alcoholic men led to a significantly greater rate of improvement as contrasted with men whose wives did not attend (see Table 2). Of the 14 patients who were treated in concurrent groups with their wives, 6 remained abstinent at 16-month follow-up, 3 showed considerable improvement in drinking, whereas only 5 demonstrated no change in drinking behavior. Of the 8 patients whose wives did not attend concurrent group meetings, only 1 was abstinent at 16-month follow-up, 3 showed improvement, 2 showed no change, and 2 had died from alcoholism-related difficulties (1 suicide and 1 death from pneumonia). Smith also demonstrated that the difference between the "wives attending" and "wives not attending" groups was not related to a differential level of social stability in the marriages prior to treatment.

TABLE 2. University of Edinburgh Concurrent Therapy Study[a]

Treatment Outcome and Group Attendance[b]				
	Abstinent	Improved	No change	Dead
Wives attending	6	3	5	0
Wives not attending	1	3	2	2

Treatment Outcome and Social Stability[c]				
	Abstinent	Improved	No change	Dead
Socially stable	5	3	1	0
Socially unstable	2	3	6	2

Group Attendance and Social Stability[d]		
	Socially stable	Socially unstable
Husbands of attending wives	5	9
Husbands of nonattending wives	4	4

[a] From Smith (1969).
[b] Tau = 0.381, Z = 2.49, p = 0.006. The two dead included one suicide and one death from pneumonia, presumably related to the patient's alcoholism. Both men were in their forties and both were socially unstable. Treatment outcome is better in those patients whose wives attend the group meeting.
[c] Tau = 0.414, Z = 2.70, p = 0.0035. The socially stable do better following treatment than those patients who are unstable socially.
[d] Tau = 0.132, Z = 0.86, p = n.s. A patient's social stability is not related to his wife's attendance or nonattendance at the group.

These clinical papers have therefore been by and large enthusiastic about the concurrent group psychotherapy technique. Although the emphasis remains on the effectiveness of the technique as an adjunct to the treatment of the alcoholic husband, wives are reported to be engaged in treatment for their own needs, having demonstrated independent issues of concern that could benefit from therapeutic examination.

A pessimistic note, however, is sounded by Pattison et al. (1965) in their work with lower-class families. Attempts to involve lower-class families troubled by alcoholism problems in family-oriented treatment programs have been infrequent. Alcoholism in these families tends to present as just one of the wide range of social and psychological problems. Since alcohol abuse is often a socially acceptable norm in these socioeconomic groups, the notion of alcoholism as a distinct disease process is alien, and such families are resistant to treatment approaches specifically directed at alcoholism as a symptom.

Pattison attempted to involve wives of male alcoholics referred to the Cincinnati Alcoholism Clinic by the courts and social agencies in an "orientation class" which was viewed as an adjunct to the husbands' treatment program. In contrast to Ewing and Smith, the Cincinnati program was viewed as only marginally effective in involving the wives, and of undemonstrated value in facilitating the husbands' treatment. Pattison's conclusion is not that family approaches should be abandoned. In fact, he argues strongly that in lower socioeconomic groups, just as in middle-class groups, alcohol problems often exist in a family context and may be aggravated by the family context as well as impacting on the family's ability to cope with its multiple problems.

But therapeutic approaches must be adapted to social realities. He therefore specifically suggests the use of the public health nurse (Pattison, 1965) and the home visit as an appropriate technique of intervention with alcoholic families. Such a program contrasts dramatically with more traditional public health approaches to alcoholism, in which the lower-class alcoholic is usually isolated from his or her family, treated in a residential setting with a heavy emphasis on detoxification and individual rehabilitation, and then returned to a socioeconomic situation usually totally out of the awareness of the therapists who have been working with him or her.

Phase IV: The Adaptation of Family Theory to Alcoholism Therapy

The studies discussed up to this point, although taking cognizance of family factors in alcoholism, were by and large adaptations of existing individual and group therapy techniques. During this same time period, however, a body of clinical theory dealing with family pathology, family concepts of symptom formation, and family-oriented therapeutic interventions was being developed. The strong clinical orientation of these notions separated these emerging theoretical ideas from earlier social scientific approaches to the family. Whereas earlier sociological and anthropological approaches had concentrated on family structure, cultural influence, and role theory, clinically oriented theorists such as D. D. Jackson (1965a,b), Bowen (1966), Ackerman (1966), Minuchin *et al.* (1967), and Boszormenyi-Nagy and Framo (1965) concentrated on general systems theory, communications theory, cybernetics, and game theory, as well as the familiar foundations of psychodynamic–psychoanalytic theory.

Since most of these clinical thinkers were psychiatrists, their attention was naturally drawn toward new explanations for traditional psychiatric conditions such as schizophrenia, psychosomatics, and adolescent dysfunction. Although somewhat puzzling in retrospect, alcoholism and drug abuse was almost totally ignored both theoretically and clinically.

In the late 1960s and early 1970s some cracks began to appear in this wall

TABLE 3. Major Studies Employi

Investigator	Treatment location	Principal mode of therapy	Goals of treatment	Number of patients treated
J. Ewing, V. Long, and G. Wenzel	University clinic	Concurrent therapy for alcoholics and spouses	Reduce drinking behavior via involvement of spouse in therapy.	32 alcoholics 16 spouses
L. H. Gliedman, D. Rosenthal, J. D. Frank, and H. T. Nash	University hospital outpatient clinic	Concurrent therapy for alcoholics and spouses	Improvement of the couples' marital relationships through the use of group treatment.	9 alcoholic men and their wives
C. Smith	Community hospital	Concurrent therapy for alcoholics and spouses	Education on the process of alcoholism. Better understanding of marital relationship. Overcome rivalrous feelings about therapy of spouse.	22 alcoholics and their wives
P. H. Esser	Community clinic	Family therapy	Reduction of drinking behavior and improved family functioning via involvement of family as a whole in the treatment process.	14 families
D. Meeks and C. Kelly	Posthospital clinic	Conjoint family therapy	Improve family communication and problem-solving ability.	5 families
G. Burton and H. Kaplan	University clinic	Multiple-couples group therapy	Improvement in the area of marital conflict with focus on excessive drinking only to the extent that it had a role in the conflict.	39 couples
D. Cadogan	Posthospital clinic	Multiple-couples group therapy	Reduction in drinking behavior via emphasis in therapy on marital communication, family equilibrium, dormant conflict, and pathological family interaction.	20 couples—control 20 couples—therap*
B. F. Corder, R. Corder, and N. Laidlaw	State alcoholism clinic	Multiple-couples group therapy	Improve interaction between spouses. Encourage participation in follow-up services offered in patient's community. Improve existing program without spending more money.	20 alcoholics—contr 19 alcoholics and spouses—therapy
D. M. Gallant, A. Rich, E. Bey, and L. Terranova	State alcoholism clinic	Multiple-couples group therapy	Improvement of the marital relationship. Abstinence for the alcoholic.	118 couples
P. Steinglass, D. Davis, and D. Berenson	State hospital	Multiple-couples group therapy	Improvement in family functioning through emphasis on examination of the relationship between alcohol use, intoxicated behavior and interaction.	10 couples

Family Therapy Techniques for Alcoholism

low-up time	Outcome measurements	Critical Results
onths and years	Effect of wife's attendance at group therapy meetings on the length of time in treatment and amount of drinking of her alcoholic spouse.	

	Coming Alone	Coming with Wife
Six-month follow-up		
Attending less than 6 months	13	7
Attending more than 6 months	3	9
Three-year follow-up		
Abstinent/Decreased drinking	3	8
Slight/No change in drinking	7	3
Uncontacted	6	5

low-up time	Outcome measurements	Critical Results

nediately ollowing reatment	1. Drinking checklist.
	2. Symptom checklist: Assesses patient's psychological difficulties.
	3. Adjective checklist: Assesses satisfaction and dissatisfaction alcoholic and spouse receive from sobriety and intoxication.
	4. Social ineffectiveness scale: Rates alcoholic spouse on interpersonal ineffectiveness.
		Greatest changes:
		  Satisfaction of patient and spouse with each other
		  Satisfaction with self
		Least changes:
		  Social ineffectiveness

months	Statistical measurement of the relationship between patient's social stability, treatment outcome, and group attendance of wife. Abstinence was rated optimal and death was worst outcome.	Patient's social stability and wife's attendance at a spouse's meeting related to favorable treatment outcome. Patient's social stability not found to be related to attendance of wife.
onths ɔ 2 years	Evaluation of general family integration and functioning of each family member and amount of drinking by alcoholic.	Abstinent—satisfactory home life 9 No change 2
luated uring various oints in reatment	Evaluated progress by reviewing selected tape recordings on an ongoing basis and focusing on changes in: (1) drinking behavior; (2) family interaction and equilibrium; (3) family involvement in treatment; (4) "identified patient" effects; (5) therapist involvement.	All families involved showed improved relating, healthier communication, and increased mutual support. Two of the alcoholics maintained their sobriety and the other three reduced their drinking substantially.
77 onths	Questionnaire given before and after treatment measured changes in (1) drinking amount; (2) family pathology (areas of considerable disagreement between the couple); (3) social deterioration (consequences of alcoholic's drinking, i.e., hospitalization, etc.).	

	Decreased	No change	Increased
Drinking changes	20/36	10/36	6/36
Family pathology	18/24	3/24	3/24
Social pathology	18/38	20/38	—

low-up time	Outcome measurements	Critical Results
onths	Before and after treatment—administration of Primary Communication Inventory (PCI) and the Conjugal Life Questionnaire (CLQ) to measure communication in marriage and how much trust and acceptance there is in the marriage; assessment of the amount of drinking.	

Drinking Behavior	Decreased	No change	Increased
Therapy group	9	4	7
Control group	2	5	13

In the therapy group, spouses of patients who were still drinking had significantly lower scores on the CLQ than spouses of abstinent patients, implying greater difficulty with acceptance and trust.

low-up time	Outcome measurements	Critical Results
onths	Determined current drinking pattern, treatment, marital and employment status, and compared the two groups by means of Fisher's Exact Test applied to the resulting contingency table for each category.	

	Therapy	Control
Resumed drinking	8	17
Improved treatment status	15	4
Unemployed more than 1 month	1	10

low-up time	Outcome measurements	Critical Results
20 onths	Success: improvement in the marital relationship and abstinent or no more than 2 drinking episodes. Failure: unhappy home life, problems with drinking Unknown: discontinued treatment	53 definite successes 41 failures 24 unknown
nths	1. Contextual drinking history 2. Individual assessment (SCL-90, Connors, SSIAM). 3. Dyadic Interaction Assessment (Olson & Ryder, IMC, Drewery & Rae Interpersonal Perception Technique, Ravich Interpersonal Game/Test). 4. Marital Satisfaction Inventory. 5. Structured interview evaluating individual, marital, family, and social functioning.	*Drinking behavior* Decreased 5 No change 2 Increased 2 Marked improvement clinically in marital interaction. Analysis of formal testing of interaction not yet completed.

of indifference. The first marriage of family theory and alcoholism therapy appeared in an article by Ewing and Fox (1968), in which they adapted theoretical concepts associated with both Bateson's and Jackson's work with families, especially Don Jackson's notion of homeostasis in family systems (1957). The alcoholic marriage is viewed as a "homeostatic mechanism" which is "established . . . to resist change over long periods of time. The behavior of each spouse is rigidly controlled by the other. As a result, an effort by one person to alter his typical role behavior threatens the family equilibrium and provokes renewed efforts by the spouse to maintain the status quo."

Specifically addressing marriages between male alcoholics and their wives, they suggest a process in which these two people strike and elicit an "implicit . . . interpersonal bargain," a marital "quid pro quo," to use Jackson's (1965a) terminology, in which the male alcoholic's passive dependency needs implicitly encourage his wife's protective nurturing needs. A sexual bargain is also struck engaging an undemanding alcoholic husband in a behavioral pattern which complements the behavior of his sexually unresponsive wife. Both of these interactional pacts are played out within the context of a cyclical system, in which the alcoholic marriage alternates between periods of sobriety and periods of intoxication. "By alternating between suppression of impulses and direct expression of them, he can maintain the conflict surrounding impulse gratification for a lifetime."

Ewing and Fox recommended family therapy for such families for two reasons: It increases the likelihood that a drinking problem will be acknowledged by a patient population (middle-class gamma-type male alcoholics) who are usually resistive to such self-labeling procedures, and it stimulates motivation toward change within the alcoholic himself.

The specific therapeutic modality suggested is, as has already been mentioned, concurrent group therapy. A husband group and a separate wife group meet at the same time but in different rooms and with different therapists for dynamically oriented group psychotherapy meetings adapted to the shared life experiences and problematic interests of the group members. Both groups are seen as going through an initial resistance phase in which they attempt to pin the rap on their spouse as the troublemaker in the marriage, followed by a period of insight into their own role in the maintenance of marital homeostasis. Of particular note are clinical examples from the wives' groups underscoring their ambivalence about their husbands' achieving sobriety. Although sobriety brings with it the desired goal of behavioral stabilization, it often carries with it seemingly undesirable aspects for both husband and wife, such as unfamiliarity with the new pattern of the marital relationship, new demands for intimacy, increased depression, etc.

Based on their extensive clinical experience, Ewing and Fox conclude that "alcoholism can no longer be seen purely in terms of intrapsychic dynamics. ... It is the family emotional homeostasis which seems to perpetuate the drinking and it is this behavior which must be changed if the drinking is to be controlled." Their therapeutic approach emphasizes the need for reciprocal work with husband and wife in order to coordinate change in both halves of the homeostatic dyad. The corollary prediction is that working within an individual framework might increase the drive to change in the individual but would also increase the pressure toward resistance on the part of the spouse. Therapeutic efforts in one direction are therefore countermanded by resistances in the other direction, minimizing the opportunity for a positive therapeutic outcome.

Steinglass and his co-workers (Steinglass et al., 1975, 1971a,b; Davis et al., 1974; Wolin et al., 1975) have incorporated many of the same concepts (homeostasis, marital bargain, complementary role functioning) in a more comprehensive interactional model of alcoholism, developed in response to clinical observations of family interaction made during states of experimentally induced intoxication. These observations suggested that interactional behavior during intoxication is highly patterned and often dramatically different from the behavior predicted by the family during sobriety. As one example, a family that claimed drinking by their "identified alcoholic" caused depression, fighting, and estrangement was observed to show increased warmth toward each other, increased caretaking, and greater animation when the "alcoholic" was permitted to drink.

The interactional model proposed by Steinglass is based on general systems concepts of family functioning. These concepts posit that families are operational systems, obeying laws general to all systems, including the importance of organization, drive toward homeostasis, circularity of causal events, and feedback mechanisms as factors determining the quality of interaction between the component parts of the systems (in this case members of the family plus alcohol).

Alcohol ingestion and intoxicated behavior is then viewed from the perspective of the extent to which and the manner in which it affects the interactional life of the members of the family. Steinglass also suggested that alcohol, by dint of its profound behavioral, cultural, societal, and physical consequences, might assume such a central position in the life of some families as to become an organizing principle for interactional life within these families. He labeled such a family an "alcoholic system." In such a system the presence or absence of alcohol becomes the single most important variable determining the interactional behavior not only between the identified drinker and other members of the family, but between nondrinking members of the family as well.

This model implied that an intricate and delicate balance exists between drinking and the day-to-day functioning of the family. In fact, it was suggested that in certain instances alcohol might be unconsciously viewed by the family as a *stabilizing* rather than a disruptive influence on their interactional life. Although superficially disruptive, from a different vantage point, the abusive use of alcohol seemed to produce extremely patterned, predictable, and rigid sets of interactions which dramatically reduced uncertainties about the family's internal life and its relationship to the external society.

The opportunity to directly observe intoxicated interactional behavior led not only to unique theoretical proposals but also to quite different conclusions about therapeutic intervention. If, in fact, alcohol might be aiding "system maintenance," which in clinical terms means serving some important dynamic function in the interactional life of the family, then the first role of the therapist dealing with the drinking symptom in a family context is an appreciation of the relationship between alcohol and family life. In certain situations it seems clear that the identified patient's drinking behavior emerges de novo in a family situation at a time of stress or strain. In these situations the drinking behavior might well be viewed as a signal or symptom reflecting this stress or strain, and crisis intervention is called for. On the other hand, if alcohol consumption is part of an ongoing interactional pattern within the family system, then the traditional therapeutic intervention aimed toward abstinence is totally inadequate to the task.

A logical extension of this theoretical model is to view family therapy not so much from the point of view of involving family members as a mechanism for improving treatment with the identified alcoholic, but rather to view the entire family or the marriage itself as the patient. Therapeutic intervention becomes interactionally oriented rather than intrapsychically oriented, and goals for treatment center around an improvement in the functioning, flexibility, and growth potential of the family system as a whole rather than the more limited focus on reduction in drinking on the part of the identified alcoholic.

A paper by Davis *et al.* (1974) expands on this theoretical model in two significant directions. First, it incorporates behavior theory, and second, it underscores the importance of focusing on maintenance factors rather than etiological factors at this very primitive stage of our understanding of chronic alcoholism. Pointing out that historically there have been two major premises underlying therapeutic approaches to alcoholism—the notion that excessive drinking is maladaptive and the belief in the existence of ultimate causes as explanations of why alcoholism develops—Davis notes that these premises have given rise to a wide variety of therapeutic approaches. These range from moralistic exhortations and aversive behavioristic approaches deriving from the maladaptive premise to the uniform psychodynamic or psychobiological approaches based on ultimate cause theories. Clinical experience, however, suggests that alcoholic behavior is more profitably thought of as a final common pathway. Incorporating behavioral concepts into the systems model allows for clinical diversity while at the same time suggesting new therapeutic strategies.

Davis *et al.* postulate the following: (a) that the abuse of alcohol has certain *adaptive consequences,* (b) that these adaptive consequences are sufficiently reinforcing to serve as the primary factors maintaining the habit of drinking, regardless of what underlying causation there may be, and (c) that the particular adaptive consequences or "primary factors" for each individual may differ and might be operating at a number of different levels including intrapsychic, intracouple, or the level of maintenance of homeostasis in a family or wider social system, but that the final common pathway is the reinforced chronic abuse of alcohol.

Two major implications for therapy are suggested. First, it is necessary for the therapist to determine the specific manner in which drinking behavior is serving an adaptive function for an individual or family. The maladaptive consequences are obviously readily apparent. Search for the adaptive consequences requires more clinical skill. Second, it is suggested that once the adaptive consequences of drinking have been ascertained, therapy may be structured around helping a patient to manifest the adaptive behavior while sober instead of only during drinking, and to learn effective alternate behaviors.

Bowen (1974), using similar concepts, also views alcoholism as potentially explainable in the language of family systems theory. Pointing out that alcoholism is one of the common human dysfunctions, Bowen contends that as a dysfunction, alcoholism must "exist in the context of an imbalance in functioning in the total family system." In this context, every family member is seen as contributing to the dysfunctional behavior of the alcoholic member. In fact, Bowen would contend that the dysfunction of the alcoholic can only continue with the support of his or her family. Treatment that alters the behavior patterns of these other family members will therefore, by definition, eliminate the necessary substratum for the existence of alcoholism. Bowen therefore states that "when it is possible to modify the family relationship system, the alcoholic dysfunction is alleviated, even though the dysfunctional one may not have been part of the therapy."

Bowen's statement represents the most undiluted justification for family therapy of alcoholism currently in the literature. It is probably, at this stage, too strong a pill for most alcoholism therapists to swallow. A more integrative approach, particularly one that takes account of the profound behavioral consequences of alcohol consumption, is probably more useful at this stage of our knowledge. However, insofar as it represents the interest and concern of an influential family therapist in the problems of alcoholism, Bowen's views deserve our close attention.

Phase V: Conjoint Therapy with the Alcoholic Family

The previous section has reviewed the growing *theoretical* literature on family therapy for alcoholism. The literature reporting results of the use of conjoint family therapy for alcoholism has to date been limited to infrequent clinical papers describing case histories offered in support of the use of family therapy techniques. By conjoint family therapy, we are now talking about techniques involving conjoint interviewing of both members of a marital pair, or conjoint interviewing of two or more members of a nuclear or extended family. Concurrent interviews occuring in separate locations, or therapy techniques involving multiple families in a group format are discussed in separate sections. Of the limited number of reports on family therapy with alcoholics currently available, none has appeared in a journal or publication primarily addressed to family issues (e.g., *Family Process* or *Journal of Marriage and the Family*).

The state of the literature, however, is in all likelihood unrepresentative of the extent to which family therapy techniques are actually being utilized for the treatment of alcoholism. In many alcoholism treatment centers it is routine for therapists to insist on the inclusion of other family members in the initial evaluation, and conjoint interviewing techniques are often included as one option available to the treatment team. However, it is not yet the case that family treatment centers routinely view the treatment of alcoholism as within the scope of their expertise. Traditional family agencies will often refuse to work with families containing an identified alcoholic member, even when the families present themselves because of problems other than alcoholism. This is particularly true with agencies working with lower-middle-class and lower-class families where a rapid referral to the "alcohol" center is the preferred disposition regardless of the nature of the presenting complaint.

An interesting study by Meeks and Kelly (1970) evaluating the efficacy of family therapy techniques introduced during the recovery phase of the treatment of the alcoholic member of the family is the most representative and influential of the clinical studies of conjoint family therapy. Although only five families were treated and studied, as a pilot study this report is of considerable interest to us.

Meeks and Kelly adhere firmly to the theoretical orientation of the family therapist and proceed to apply treatment techniques developed by one of the founders of the family therapy movement, Virginia Satir. They accept a "basic premise" common to all family therapists that the family itself, rather than individual members within the family, is the unit of treatment, and that the family as a psychological unit has internal processes which help to establish an emotional balance characteristic of the family as a whole. They also accept the premise that dysfunctional behavior on the part of one family member may actually be viewed as functional by the family as a whole insofar as this behavior helps to restore the critical psychological balance or equilibrium that the family desires.

In their study, conjoint family therapy was begun following an intensive 7-week program of individual and group psychotherapy in a day treatment program. During this 7-week program, family members were seen separately from the "alcoholic patient." At the beginning of the aftercare phase, however, family members were seen in conjoint interviews only; the alcoholic member was never seen separately from his or her family during the aftercare phase. Families were seen for periods ranging from 10 to 12 months.

The treatment techniques employed were modeled on guidelines established by Satir (1967). The focus was on the traditional family therapy interest in interaction, communication, role performance, and redefinition of problems in family rather than in individual terms. Therapeutic goals were derived from Ackerman (1966), another family therapy pioneer, and included achievement of a clearer definition of interactional conflicts, improved and more open communications about these conflicts, a greater understanding of intrapsychic determinants of interpersonal conflicts, and an improved level of complementarity in family role relations. Treatment evaluations included an interest in the drinking behavior of the identified alcoholic, but focused more intensely on issues of improved family interaction and family equilibrium. Such issues as problem definition, communication, patterns of relating, and methods of problem solving, are included as possible variables indicating improved family functioning. The study also attempted to assess the extent of the family involvement in the treatment process, the extent to which the therapists were able to remove the alcoholic member from the "identified patient" status, and issues of therapist involvement.

Although Meeks and Kelly underscored the exploratory nature of their report, they were enthusiastic about their experience. They concluded that techniques geared toward redefining alcoholism issues in family terms are quite profitable. The more drinking behavior can be seen as merely one aspect of family interaction, the greater the likelihood that the "alcoholic" member of the family will be able to shed his or her label and establish new patterns of interaction within the family.

Esser (1968, 1970, 1971) reached similar conclusions in reports stemming from his experience with conjoint family therapy in the Dutch city of Haarlem. Once again the emphasis is on the recovery or aftercare phase of treatment. Family therapy is seen as potentially expanding the scope of treatment from hospitalization and clinical care for the identified alcoholic, to a more sociotherapeutically oriented approach to the entire family. The family of an alcoholic is viewed as a "group under stress," but this stress is related as much to disturbed interactions as it is to the behavioral effects of alcohol. Restoration of communication, concentration on role conflicts, and the removal of the alcoholic from the role of the "identified patient" are again seen as the central issues that the therapist must approach.

These clinical reports can be best characterized as promising but unsubstantiated; enthusiastic but primarily impressionistic. They seem to reflect the level of optimism attached to family therapy for alcoholism (e.g., the *Alcohol and Health* statement, Keller, 1974; Chafetz et al., 1974), but leave unanswered questions about the verifiable efficacy of these techniques.

Phase VI: Multiple-Couples and Multiple-Family Group Therapy Approaches

Multiple-couples group therapy is a particularly popular form of family therapy currently being utilized in alcoholism treatment. This technique uses the group setting and group process to assist couples in examining marital interaction patterns and the relationship between these patterns and drinking behavior. Its increasing popularity as a treatment modality in alcoholism treatment programs is perhaps related to the assumption that it represents the "best of all possible worlds." It retains, in format at least, a group therapy structure and is therefore attractive to many alcoholism therapists who have viewed group therapy as the treatment of choice. However, it also acknowledges and takes account of the importance of family factors in the exacerbation of alcoholism and is responsive to the growing feeling that alcoholism treatment is less effective if significant family members are not involved in therapy.

In the family therapy field, multiple-couples group therapy has evolved as a specialized form of multiple-family therapy (Laquer, 1972), in which anywhere from 3 to 20 families are convened for therapy sessions which involve not only the adult generation, but also children, and at times three-generational extended families. Multiple-couples groups, from the point of view of the family therapist, represents an attempt to work within a group setting with the most significant dyad in the family, on the assumption that improvement in the working relationship between husband and wife will have profound effects on all relationships within the family.

In the alcoholism field, in contrast to the historical trend in family therapy, multiple-couples groups stem primarily from the desire to include spouses in group therapy, as opposed to working with couples in a group context. This distinction, although perhaps subtle at first glance, has significant implications for the type of therapy that evolves. If the therapist's primary training and orientation is in group process and group therapy, then multiple-couples groups will retain this emphasis on the group as a whole. If the therapist's training is in family therapy, however, the major focus would be on the couple as an interactional unit, and the relationship between the couple and other couples in the group will be seen as a model of the couple's interface and interactions with the outside world. Since alcoholic couples and families are often socially isolated, especially as abusive drinking increases, the multiple-couples group offers a unique setting to work on problems of social isolation and interaction with the outside world.

A growing body of experimental and clinical literature now exists concerning multiple-couples group therapy approaches to alcoholism. This literature includes traditional treatment outcome studies (Cadogan, 1973; Corder et al., 1972), reports of experimental treatment techniques (Steinglass et al., 1975), and summaries of clinical experiences (Sands and Hanson, 1971; Gallant et al., 1970; Berman, 1968). We will examine three reports more extensively: an outcome study of multiple-couples therapy based on group techniques (Cadogan, 1973); an experimental study based on family therapy principles (Steinglass et al., 1975); and a clinical report of the extensive use of multiple-couples groups in an operational alcoholism treatment program (Gallant et al., 1970).

Cadogan presented the first controlled study in the literature of multiple-couples group therapy in alcoholism treatment. Forty alcoholics (both men and women) and their spouses were recruited while the alcoholics were still inpatients at a traditional alcoholism unit, and asked to volunteer for a "new and effective method of treatment" in which "an attempt would be made to improve family problem-solving patterns, to encourage the expression of feeling in marital communications and to develop a new awareness of the effect of their behavior on others." The study group represented the first 40 couples volunteering for this new outpatient multiple-couples group therapy program. Subjects were then randomly assigned to one of two groups: an immediate treatment group, or a waiting list in which they continued with the traditional treatment program but did not engage in the outpatient multiple-couples group. Ultimately, 20 couples were assigned to each group. Groups proved to be comparable in age, socioeconomic status, severity of alcoholism, and involvement with other treatment programs (especially A.A.).

The treatment group engaged in open-ended multiple-couples group therapy sessions (90-minute sessions on a once-weekly schedule). The average group was composed of five couples and membership was fluid, with dropouts being replaced by newly interviewed recruits. Follow-up evaluation occurred six months after the couple was recruited for the study. Follow-up results were striking. At six months, 9 alcoholic members in the therapy group remained abstinent, 4 were doing some drinking, and 7 had relapsed completely. Amongst the control group, however, only 2 were abstinent, 5 were drinking moderately, and 13 had demonstrated complete relapse.

Gallant *et al.* have provided a report of the most extensive application to date of multiple-couples group therapy as an integral phase or component of an ongoing alcoholism treatment program. Their program, the New Orleans Alcoholism Clinic, is comprised of two integrated units, a 36-bed inpatient unit, and an outpatient alcoholism clinic. Gallant has been routinely assigning every discharged married patient who is returning home to live with his or her spouse to a multiple-couples therapy group in the outpatient clinic as the major form of ongoing treatment. These therapy groups, composed of four to seven couples meeting every two weeks for a two-hour session, have a traditional alcoholism treatment goal of total abstinence for the alcoholic combined with an interpersonal goal of improvement in marital interaction. Treatment techniques combine both family therapy orientations toward analysis of interactional behavior, and group therapy techniques of encouraging direct exchange and feedback between all members of the group.

Gallant has reported the results of 118 couples assigned to the clinic's multiple-couples groups. Follow-up data was not systematically gathered, but most couples were contacted following treatment and drinking history and quality of family life was explored. (The follow-up period varied from 2 months to 20 months.) Fifty-three of the 118 couples were considered to be definite successes at the time of follow-up (either complete abstinence or no more than two brief drinking episodes, and "reasonable" marital relationship) and 41 were considered definite failures (unhappy family life, frequent drinking episodes, or sobriety felt by the treatment team to be temporary and without satisfaction or contentment). Twenty-four couples were lost to follow-up. Based on these findings Gallant *et al.* "conclude that marital couples group therapy is the treatment of choice at this time for married alcoholic patients. The denial and projection mechanisms, exaggerated in the alcohol-marital problem, are more easily approached and treated in a group."

Steinglass and his colleagues have carried out work with multiple-couples therapy groups as part of their research studies examining interactional behavior in alcoholic families. An experimental treatment program was established at NIAAA's Laboratory for Alcohol Research in which couples with one or two alcoholic members were placed in an intensive, six-week multiple-couples group therapy program. Although the treatment program was conceptualized primarily as an experimental model which permitted the establishment of a rich clinical field allowing for the examination of interactional behavior, the treatment process itself was highly unusual and proved to be quite fascinating in its own right. In contrast to Cadogan's work, where the emphasis was on the desirability of involving the spouse in a group process, the NIAAA group was firmly based in family therapy.

The experimental treatment program was divided into three phases: an initial two-week outpatient phase in which groups met for three sessions per week; a ten-day inpatient phase during which time three couples were simultaneously admitted to an inpatient facility; and finally a posthospitalization three-week outpatient phase of two group meetings per week. Following the six-week intensive treatment program, groups reconvened at six-week intervals for follow-up sessions over a six-month follow-up period.

The core of the program, and clearly its most innovative feature, was the hospitalization period. The hospital setting itself was a redesigned inpatient unit in a traditional state hospital. This unit was described by the research team as a "simulated apartment setting" which was supposed to provide a homelike atmosphere for the couples, allowing them to reproduce as accurately as possible their usual interactional behavior. Of greatest importance, however, was the fact that alcohol was made freely available during the first seven days of the hospitalization period and couples were asked to engage in their usual drinking patterns while they were on the ward. This last feature of the treatment program was an extension of the use of experimentally induced intoxication as a potential adjunct to therapy, and was based on theoretical notions about the role alcohol can play in maintaining fixed interactional patterns within families (these notions are discussed in Phase IV). The specific rationale provided to the couples for this free availability of alcohol was that the therapist, by being able to directly observe intoxicated behavior, could gain a better understanding of the role that alcohol consumption was playing in the couples' lives.

The treatment program utilized a variety of techniques to examine patterns of interaction exhibited by each couple, and to contrast the difference between interaction during sobriety versus interaction during intoxication. These techniques included videotape recording and feedback, role-playing techniques, use of one-way mirror observation and feedback from observers, analysis of speech and communication patterns, emphasis on nonverbal behavior and postural analysis, and use of three-generational family genograms. All of these techniques have been used extensively by family therapists in more traditional settings.

The multiple-couples groups were conceptualized by the researchers as a societal system composed of three distinct elements or levels: individuals, couples, and whole group. This type of group was therefore viewed as an excellent vehicle for observing the relationship between individual dynamics and intracouple dynamics, while at the same time having an opportunity to observe the couple's behavior in negotiating its position in a group of strangers (perhaps analagous to the relationship between family and the outside society). However, the therapeutic target was always the couple, and individual dynamics or whole group behavior was viewed from the vantage point of its relationship to each of the three couples.

Although this NIAAA program obviously represented a radical departure from traditional alcoholism treatment, it also is the purest example in the literature of an approach to alcoholism treatment based on family principles. Let us therefore summarize the main features that made this program unique: First, the program recruited middle-class, intact couples who displayed a substantial degree of economic and interactional stability despite the chronic abuse of alcohol by one of its members. Second, the program not only did not insist on the usual abstinence model of treatment; it actually suggested that intoxicated behavior can be utilized by the therapist as an adjunct to treatment. Third, instead of viewing the individual alcohol abuser as the "problem," therapy was directed at the couple. Fourth, both drinker and spouse enjoyed similar status as inpatients in a psychiatric hospital, and treatment goals and techniques were based on examining the relationship between alcohol use, intoxicated behavior, and interaction. And last, the therapists insisted on improved family functioning rather than a reduction of drinking behavior as the primary target. Because each of these features represented a natural extension of family theory into alcoholism treatment, they have profound implications regarding potential consequences of the more generalized application of family therapy techniques in alcoholism treatment. These implications will be more fully discussed at the close of this chapter.

Although Steinglass and his colleagues have advised caution regarding outcome results from this experimental study, emphasizing the highly experimental, pilot nature of the program, it was found that couples responded quite positively to the treatment approach. Although only ten couples were treated, they all completed the study despite its strenuous demands, and reported a profound emotional impact deriving particularly from the in-hospital experience. The enthusiasm of patients for the therapeutic work was particularly impressive to the therapists in light of the fact that all couples had failed repeatedly in previous therapeutic efforts. The therapists were also most enthusiastic about the in-hospital experience as a mechanism facilitating the rapid clinical understanding of the relationship between drinking behavior and interactional life for each of the couples worked with.

Reports of six-month follow-up data have not yet been completed and it is therefore at this time unclear to what extent the intensity of this therapeutic experience was meaningfully integrated, and had a lasting effect in changing the interactional patterns of the couples involved. In all likelihood this experimental study will be more valuable in suggesting directions for therapy rather than establishing a definitive therapeutic approach.

Two additional studies, although not focusing specifically on multiple-couples therapy, will be mentioned here because they also involved simultaneous work with spouses in a traditional hospital setting. Corder et al. (1972) carried out a pilot project at a residential alcoholism treatment center. Wives of male alcoholics were included in a four-day intensive workshop which followed a traditional three-week inpatient program. The workshop program included group therapy and videotape analysis of the sessions, didactic lectures, group discussions of "game-playing" and role-playing, recreational activities, and A.A. and Al-Anon meetings. A six-month follow-up performed on the pilot group of 20 alcoholics indicated a significant reduction in drinking for the experimental group as compared to a control group that had gone through the traditional treatment program alone.

Paolino and McCrady (1976) have been experimenting with the "joint admission" of a nonalcoholic spouse as a "guest" of the hospital. The couple lives on a psychiatric ward that includes patients of all diagnoses and ages over 12. The nonalcoholic spouse participated in ward activities as much as possible while retaining his or her job. The patient and spouse also participated in three types of weekly therapy groups: a group for *problem drinkers only,* a group for *spouses only* of problem drinkers, a group for *couples* in which one member is a problem drinker and the other member does not have a problem with alcohol. These groups all continued after the couples left the hospital and were considered an essential part of the treatment program.

Paolino and McCrady conceptualized the goals of such a joint admission as follows: (a) to give the staff the opportunity to observe the couple's interactions; (b) to provide comprehensive feedback for the couple about their patterns of interacting; (c) to integrate the spouse into the milieu so that the spouse has the same opportunity as the problem drinker to experience the closeness and caring of the unit; (d) to integrate the spouse into the milieu so that the spouse may incorporate the approach to handling problems which is taught in the milieu. They plan extensive experimental studies based on this treatment approach.

AL-ANON FAMILY GROUPS

Al-Anon is an indigenous self-help movement which arose spontaneously as a parallel but separate movement to Alcoholics Anonymous in the late 1940s. Over 5000 Al-Anon Family Groups now exist with a world-wide distribution.

Al-Anon Family Groups are modeled after Alcoholics Anonymous, that is, a group fellowship of peers sharing a common problem. In the case of Al-Anon the peers are spouses, children, and close relatives of alcoholics who are usually but not necessarily part of an A.A. group. The typical member is the wife of an alcoholic man in an upper-middle-class family with a strong religious orientation. Although infrequently studied systematically by social scientists, Al-Anon Family Groups have been characterized as "a remarkable, self-help, nonprofessional modality of group therapy and group education" (Ablon, 1974).

Al-Anon closely parallels A.A. in structure and function. It exists for the most part outside the traditional medical or social service framework within which alcoholics are usually "treated" by the community, with membership recruited through word of mouth, newspaper advertisements, and church group support. The group meetings correspond to A.A. meetings in their religious flavor, their emphasis on anonymity, and their structured format for achieving self-help. The operational philosophy of the program includes the Twelve Steps, adapted from A.A., supplemented by the Twelve Traditions, a series of policy statements that outline the nature of Al-Anon and the goals and limitations of a particular group (see Table 4.) The group meeting itself follows a semistructured format, chaired by a group leader, and usually centers around ways in which different group members can adapt or incorporate the Twelve Steps into their lives.

Ablon (1974) has attempted the most comprehensive analysis of the group process that occurs during Al-Anon meetings. She feels that all successful Al-

TABLE 4. The Twelve Steps of Al-Anon

1. We admitted we were powerless over alcohol—that our lives had become unmanageable.
2. Came to believe that a Power greater than ourselves could restore us to sanity.
3. Made a decision to turn our will and our lives over to the care of God as we understand Him.
4. Made a searching and fearless moral inventory of ourselves.
5. Admitted to God, to ourselves, and to another human being the exact nature of our wrongs.
6. Were entirely ready to have God remove all these defects of character.
7. Humbly asked Him to remove our shortcomings.
8. Made a list of all persons we had harmed and became willing to make amends to them all.
9. Made direct amends to such people whenever possible except when to do so would injure them or others.
10. Continued to take personal inventory and when we were wrong, promptly admitted it.
11. Sought through prayer and meditation to improve our conscious contact with God as we understand Him, praying only for knowledge of His will for us and the power to carry that out.
12. Having had a spiritual awakening as a result of these Steps, we tried to carry this message to others, and to practice these principles in all our affairs.

Anon members must accept one basic didactic lesson and three principles for operating in the groups themselves. The fundamental didactic lesson is an acceptance of A.A.'s concept of alcoholism, that is, "an obsession of the mind and an allergy of the body." This concept of alcoholism implies that the alcoholic member is behaving neither out of irresponsibility nor out of a perversity toward other family members. Instead, he or she has a disease which is totally outside of the alcoholic's control, and in this sense other family members are advised that they shouldn't "take it personally." A disease analogy is usually made, the Al-Anon family member being told that the alcoholic can no more control his or her drinking than the cancer patient or diabetic can control the organic changes associated with their illnesses.

The three operational principles stressed by Ablon are first, a "loving detachment from the alcoholic," in which the family member accepts the fact that they have control only over their own behavior and must accept the fact that only the alcoholic can work out his or her own solutions; second, the reestablishment of self-esteem and independence; and third, a reliance on a "higher power," a religious emphasis that of course also closely parallels A.A.

Although Al-Anon Family Groups are difficult to compare to traditional therapy groups, they would probably be most appropriately characterized as supportive or introspective groups analogous to other religiously oriented groups such as Quaker meetings. The group member is encouraged to find him/herself in the many similar stories shared by other group members, and adjust or adapt his/her life to the operational principles followed by the group. Interpersonal interaction is for the most part discouraged, and direct confrontation between group members is usually strictly censured. In this last regard, the groups tend to be more softly supportive than their A.A. counterparts.

Although Al-Anon Family Groups have often been dismissed by professional therapists because of their quasi-religious framework and their anti-intellectual and antiscientific tradition, they are the single largest "treatment" program involving families with alcoholic members currently available. Although for the most part they fly in the face of the theoretical notions that underlie the family therapy movement (the view of the alcoholic as the helpless victim of a disease process being the most flagrant example of this), several of the operational principles bear a remarkable resemblance to principles advocated by many family therapists. For example, the principle of "loving detachment" is quite similar to a principle advocated by many family therapists who insist that families will show movement or growth only when individual members concentrate on changing their own behavior rather than attempting to manipulate or change others in the family system.

Most studies of Al-Anon Family Groups to date have included enthusiastic anecdotal reports of favorable clinical outcomes for the families involved. The one study that carried out more systematic assessments (Bailey, 1965) concluded that most Al-Anon members were more satisfied with this experience than any other therapy contact they had had, and these members related this feeling particularly to their new and changed understanding of alcoholism. Parenthetically, male alcoholics involved in A.A. were found to have had an easier time achieving and maintaining sobriety if their wives were simultaneously attending Al-Anon fellowship meetings.

DISCUSSION

Our review of the existing literature leaves us with a sense of guarded optimism about the application of family therapy techniques in the treatment of alcoholism. Although every study we have mentioned concludes with an enthusiastic statement encouraging greater use of family therapy, it is also apparent that very little hard evidence exists at this point demonstrating either the efficacy of family therapy by itself or the comparative value of family therapy versus more traditional forms of therapy in the treatment of alcoholism.

Considerable controversy exists concerning the adequacy of methodologies currently available for psychotherapy research. However, even using existing standards for research, the studies reviewed in this chapter are problematic. In fact, virtually every study should be viewed as a pilot or exploratory venture, rather than a definitive attempt to validate a treatment method. Nevertheless, despite the absence of solid experimental data to support their clinical convictions, therapists in the alcoholism field seem increasingly convinced that family therapy techniques represent a powerful new addition to the therapeutic armamentarium. We therefore should anticipate that the use of these techniques will increase dramatically over the next decade.

It is less clear, however, exactly what form these family therapy techniques will take. Our survey of the existing data indicates that no single family therapy technique has gained a dominant position or demonstrated superior credentials regarding treatment of alcoholic families. Instead, for alcoholism therapy just as for family therapy in general, a wide variety of approaches is currently being advocated and should be deemed supportable. We can think of these family approaches to the treatment of alcoholism as falling into the following categories: (a) "pure" family therapy, based on family systems theoretical formulation of alcoholism; (b) group or individual approaches designed specifically to fulfill criteria suggested by a family theory of alcoholism; (c) techniques involving concurrent therapeutic work with other family members in addition to the identified alcoholic, but using a more traditional indiviudally oriented psychodynamic theoretical base; (d) specialized techniques developed for working with spouses of alcoholics, usually eclectic in nature; and (e) supportive approaches geared to assist spouses and children of alcoholics to deal with common difficulties they face as a result of having an alcoholic member in their family.

It also seems clear that family approaches have not been confined to one particular location, such as an outpatient clinic. The fact that several investigators have been experimenting with the simultaneous admission of alcoholic and spouse to an inpatient setting indicates that virtually every therapeutic setting can be adapted to incorporate family issues. In fact, the only generally agreed upon prerequisite for family therapy in alcoholism is the presence of an intact family.

If the current enthusiasm for family therapy continues, it is not unlikely that future therapy programs for alcoholism will be split into two distinct approaches: a family-oriented psychosocial approach applied primarily to middle- and upper-class alcoholics with intact families, and a biomedically-oriented approach combining pharmacotherapy, behavioristic techniques, and group therapy applied primarily to single alcoholics. The former approach would structure itself on a family systems theoretical framework, while the latter approach would remain within the more traditional framework of the medical model of alcoholism.

Although from one vantage point the incorporation of family therapy techniques into alcoholism treatment programs represents a logical commonsense extension of clinical experience, it is important to recognize that certain aspects of the family therapy movement have stemmed from radical underpinnings. Therapists not aware of these underpinnings often move somewhat naively into work with families rather than individuals, only to discover much later the long-term implications of their change in therapeutic focus. Therefore, although it certainly seems warranted to encourage the continuing expansion of family treatment techniques in the alcohol field, we also need to discuss potential implications of this trend.

Three major implications must be underscored. The most basic, although not necessarily the most obvious implication is a shift in theoretical thinking and conceptualization, what Bateson (1972) has called an "epistemological" change. This change, embodied in the theoretical notions connected with family therapy, redefines notions of disease processes, dysfunctioning, causal relationships, motivation, etc. The most obvious consequence of this change in thinking is the redefinition of alcoholism in family terms discussed in the previous section labeled Phase IV. These notions, logically extended, have led several psychiatrists to postulate the existence of a subpopulation of families in which the use of alcohol actually *stabilizes* interaction for the family and is therefore viewed as *adaptive* rather than dysfunctional (Davis et al., 1974; Steinglass et al., 1971a). At the very least, however, the theoretical underpinnings of family therapy suggest that the alcoholic is merely the labeled victim, set apart by his family as a scapegoat and protecting his or her family through the repetitive cycling of intoxicated behavior from the family's difficulties in coping with each other and with the outside world.

The second major implication is the changing definition of the patient population. This aspect not only involves the obvious change that nonalcoholic family members are also to be actively involved in the treatment process; it also means that, since the family therapist sees the whole family as his patient, therapeutic efforts are equally directed toward alcoholic and nonalcoholic patients. The shift toward family therapy has also meant a significant change in the socioeconomic status of the patient population. Whereas in previous years traditional alcoholism treatment centers were working with lower-middle-class and lower-class alcoholics, many of whom were already separated from family and work, while A.A. and Al-Anon provided services for middle-class and upper-class individuals and families, family therapy techniques have usually been applied to the intact, middle-class family.

In addition to the change in the socioeconomic level of the patient population, there has been a change in the timing for intervention in what must be thought of as a chronic disease process. The availability of family therapy techniques has increased the likelihood that therapists will intervene at a much earlier stage in the alcoholic process. In fact, many families seek therapy, not because of a devastating alcohol problem, but because of a devastating problem in family communications, parent–child conflict, sexual difficulties, etc. The existence of a pernicious drinking pattern in one or more members of the family might only emerge as history-taking proceeds.

The third implication arises from changing outcome goals. The traditional goal of abstinence as an isolated behavioral change makes very little sense in the context of family therapy (Pattison, 1968). Substantial clinical evidence now suggests that abstinence as an isolated behavioral change often carries with it attendant increases in pressure on the family system, frequently with dissolution of the marriage. It was the recognition that family members other than the identified alcoholic had to be actively involved in the treatment process and undergo change, in order to successfully reduce drinking while maintaining the family structure, that has led to the increasing use of family techniques. Although most of the studies reported in this chapter retained a reduction in drinking as a primary indication of successful treatment outcome, increasingly such family variables as communication, social functioning, relationships with extended family, evidence of symptomatology in children, etc. were used as alternative outcome variables.

If one looks at families with alcoholic members being treated in family therapy clinics, then the shift is perhaps even more dramatic. Traditional family therapists might even view alcoholism as a symptom of secondary importance, with treatment goals being primarily directed toward issues of family functioning and a reduction in drinking viewed as a desirable secondary gain following the primary change in the quality of family life.

These implications add up to a quite striking shift in one's concept of alcoholism. In family therapy terms, alcoholism is merely one of a wide range of behaviors used by families in their ongoing functioning. Although it obviously is a dramatic and highly visible piece of behavior, it is the tip of the iceberg. The important issues, the family therapist would maintain, exist in the structure, communicational style, and relationship system of the family. Any typological or diagnostic system should be based on these latter qualities, rather than on the existence or absence of a drinking symptom. Alcoholism, by this analogy, is the equivalent of fever, or adolescent delinquency. Whereas it makes some sense to view such a symptom or piece of behavior as primary in the diagnostic and treatment process (the fever work-up or the therapy group for adolescent runaways), increasing knowledge directs us toward the antecedents of this behavior.

The logical extension of family theory therefore argues against the artificial grouping of families by behavioral symptoms such as abusive drinking. Extended further it means that families with alcoholic members who view themselves as dysfunctional and seek help should be seen in family therapy centers rather than alcoholism treatment centers. Furthermore, if family therapy proves to be the treatment of choice for middle-class and upper-class alcoholics with intact families (as suggested in a still very preliminary way by the studies reported on in this chapter), then our task will be to integrate such families into family therapy programs and eliminate entirely the separate alcoholism treatment program that has up to this point claimed such families as part of its natural clientele.

The above issues underscore the importance of our gaining a clearer understanding of family techniques before we make sweeping claims about its applicability. The initial enthusiasm for family therapy approaches to alcoholism, although clearly worth supporting, must continue to be evaluated objectively rather than being embraced in an uncritical fashion. Although this is obviously sound advice regarding any new treatment technique, it is particularly relevant here because family therapy carries with it the potential for radical changes and shifts in our approaches to the treatment of alcoholism.

ACKNOWLEDGMENT

The preparation of this chapter was supported by Grant No. RO1 AA 01441 from the National Institute on Alcohol Abuse and Alcoholism. The author wishes to thank Ms. Lydia Tislenko for her invaluable assistance in the preparation of this chapter.

REFERENCES

Ablon, J., 1974, Al-Anon family groups, Impetus for change through the presentation of alternatives, *Am. J. Psychother.* 28(1):30.
Ablon, J., 1976, Family structure and behavior in alcoholism: A review of the literature, in "The Biology of Alcoholism" (B. Kissin and H. Begleiter, eds.) Vol. IV, pp. 205–242.
Ackerman, N. W., 1962, Family psychotherapy and psychoanalysis: Implications of difference, *Family Process* 1:30.
Ackerman, N. W., 1966, "Treating the Troubled Family," Basic Books, New York.
Bailey, M., 1961, Alcoholism in marriage: A review of research and professional literature, *Q. J. Stud. Alcohol* 22:81.
Bailey, M., 1965, Al-Anon family groups as an aid to wives of alcoholics, *Social Work* 10:68.
Bateson, G., 1972, "Steps to an Ecology of Mind," Ballantine Books, New York.
Bateson, G., Jackson, D. D., Haley, J., and Weakland, J. H., 1956, Toward a theory of schizophrenia, *Behav. Sci.* 1:251.
Bergler, E., 1949, "Conflict in Marriage," Harper & Row, New York.
Berman, K. K., 1968, Multiple conjoint family groups in the treatment of alcoholism, *J. Med. Soc. N.J.* 65:6.
Boszormenyi-Nagy, I., and Framo, J. L. (eds.), 1965, "Intensive Family Therapy," Hoeper, New York.
Bowen, M., 1966, The use of family therapy in clinical practice, *Compr. Psychiatry* 7:345.
Bowen, M., 1974, Alcoholism as viewed through family systems theory and family psychotherapy, *Ann. N.Y. Acad. Sci.* 233:115.
Bowen, M., 1976, Family therapy after twenty years, in "American Handbook of Psychiatry," (S. Arieti, ed.).
Burton, G., 1962, Group counseling with alcoholic husbands and their wives, *Marriage and Family Living* 24:56.
Burton, G., Kaplan, H. M., and Mudd, E. H., 1968, Marriage counseling with alcoholics and their spouses, *Br. J. Addict.* 63:151.
Cadogan, D. A., 1973, Marital group therapy in the treatment of alcoholism, *Q. J. Stud. Alcohol* 34:1187.
Chafetz, M., Hertzman, M., and Berenson, D., 1974, Alcoholism: A positive view, in "American Handbook of Psychiatry" (S. Arieti and E. B. Brody, eds.) Vol. 1, pp. 367–392, Basic Books, New York.
Chassell, J., 1938, Family constellation in the etiology of essential alcoholism, *Psychiatry* 1:473.
Corder, B. F., Corder, R. F., and Laidlaw, N. L., 1972, An intensive treatment program for alcoholics and their wives, *Q. J. Stud. Alcohol* 33:1144.
Davis, D. I., Berenson, D., Steinglass, P., and Davis, S., 1974, The adaptive consequences of drinking, *Psychiatry* 37:209.

Edwards, P., Harvey, C., and Whitehead, P. C., 1973, Wives of alcoholics: A critical review and analysis, *Q. J. Stud. Alcohol* 34:112.
Esser, P. H., 1968, Conjoint family therapy for alcoholics, *Br. J. Addict.* 63:177.
Esser, P. H., 1970, Conjoint family therapy with alcoholics—a new approach, *Br. J. Addict.* 64:275.
Esser, P. H., 1971, Evaluation of family therapy with alcoholics, *Br. J. Addict.* 66:251.
Ewing, J. A., and Fox, R. E., 1968, Family therapy of alcoholism, *in* "Current Psychiatric Therapies" (J. H. Masserman ed.) Vol. 8, pp. 86–91, Grune & Stratton, New York.
Ewing, J. A., Long, V., and Wenzel, G. G., 1961, Concurrent group psychotherapy of alcoholic patients and their wives, *Int. J. Group Psychother.* 11:329.
Ferber, A., Mendelsohn, M., and Napier, A., 1972, "The Book of Family Therapy," Science House, New York.
Finley, D. G., 1966, Effect of role network pressures on an alcoholic's approach to treatment, *Social Work* 11:71.
Foley, V. D., 1974, "An Introduction to Family Therapy," Grune & Stratton, New York.
Futterman, S., 1953, Personality trends in wives of alcoholics, *Journal of Psychiatric Social Work* 23:37.
Gallant, D. M., Rich, A., Bey, E., and Terranova, L., 1970, Group psychotherapy with married couples: A successful technique in New Orleans alcoholism clinic patients, *J. La. State Med. Soc.* 122:41.
Gliedman, L. H., 1957, Concurrent and combined group treatment of chronic alcoholics and their wives, *Int. J. Group Psychother.* 7:414.
Gliedman, L. H., Nash, H. T., and Webb, W. L., 1956a, Group psychotherapy of male alcoholics and their wives, *Dis. Nerv. Syst.* 17:90.
Gliedman, L. H., Rosenthal, D., Frank, J. D., and Nash, H. T., 1956b, Group therapy of alcoholics with concurrent group meetings with their wives, *Q. J. Stud. Alcohol* 17:655.
Gray, W., Duhl, F. J., and Rizzo, N. D. (eds.), 1965, "General Systems Theory and Psychiatry," Little Brown, Boston.
Group for the Advancement of Psychiatry Report, 1970, "The Field of Family Therapy," Vol. 7, Report No. 78.
Haley, J., 1963, "Strategies of Psychotherapy," Grune & Stratton, New York.
Haley, J., and Hoffman, L., 1967, "Techniques of Family Therapy," Basic Books, New York.
Howells, J. G. (ed.), 1971, "Theory and Practice of Family Psychiatry," Brunner/Mazel, New York.
Jackson, D. D., 1957, The question of family homeostasis, *Psychiat. Q. Supplement* 31:79.
Jackson, D. D., 1965a, Family rules: Marital quid pro quo, *Arch. Gen. Psychiatry* 12:589.
Jackson, D. D., 1965b, The study of the family, *Family Process* 4:1.
Jackson, J. K., 1954, The adjustment of the family to the crisis of alcoholism, *Q. J. Stud. Alcohol* 15:562.
Jackson, J. K., 1958, Alcoholism and the family, *Annals of the American Academy of Political and Social Science* 315:90.
Jackson, J. K., 1962, Alcoholism and the family, *in* "Society, Culture and Drinking Patterns" (D. J. Pittman and C. R. Snyder, eds.), pp. 472–492, Wiley, New York.
Keller, M. (ed.), 1974, Trends in treatment of alcoholism, *in* "Second Special Report to the U.S. Congress on Alcohol and Health," pp. 145–167, Department of Health, Education and Welfare, Washington, D.C.
Kelly, D., 1973, Alcoholism and the family, *Md. State Med. J.* 22(1):25.
Knight, R., 1937, The dynamics and treatment of chronic alcohol addiction *Bull. Menninger Clin.* 1:233.

Kogan, K., and Jackson, J., 1965, Stress, personality and emotional disturbance in wives of alcoholics, *Q. J. Stud. Alcohol* 26:486.

Laquer, P., 1972, Mechanisms of change in multiple family therapy, in "Progress in Group and Family Therapy" (C. J. Sager and H. S. Kaplan, eds.) pp. 400–415, Brunner/Mazel, New York.

Lidz, T., Fleck, S., and Cornelison, A. R., 1965, "Schizophrenia and the Family," International Universities Press, New York.

MacDonald, D. E., 1958, Group psychotherapy with wives of alcoholics, *Q. J. Stud. Alcohol* 19:125.

Meeks, D. E. and Kelly, C., 1970, Family therapy with the families of recovering alcoholics, *Q. J. Stud. Alcohol* 31:399.

Minuchin, S., 1974, "Families and Family Therapy," Harvard University Press, Cambridge, Massachusetts.

Minuchin, S., Montalvo, B., Guerney, B. G., Rosman, B. L., and Schumer, F., 1967, "Families of the Slums," Basic Books, New York.

Paolino, T. T. J., and McCrady, B., 1976, Joint admission as a treatment modality for problem drinkers: A case report. *Am. J. Psychiatry* 133:222.

Pattison, E. M., 1965, Treatment of alcoholic families with nurse home visits, *Family Process* 4:75.

Pattison, E. M., 1968, A critique of abstinence criteria in the treatment of alcoholism, *Int. J. Soc. Psychiatry* 14:268.

Pattison, E. M., Courless, P., Patti, R., Mann, B., and Mullen, D., 1965, Diagnostic therapeutic intake groups for wives of alcoholics, *Q. J. Stud. Alcohol* 26:605.

Pattison, E. M., DeFrancisco, D., Wood, P., and Frazier, H., 1975, A psychosocial kinship model for family therapy. Presented at the American Psychiatric Association Annual Meeting, May 1975.

Pittman, D. J., and Tate, R. L., 1969, A comparison of two treatment programs for alcoholics, *Q. J. Stud. Alcohol* 30:888.

Pixley, J. M., and Stiefel, J. R., 1963, Group therapy designed to meet the needs of the alcoholic wife, *Q. J. Stud. Alcohol* 24:304.

Sands, P. M., and Hanson, P. G., 1971, Psychotherapeutic groups for alcoholics and relatives in an outpatient setting, *Int. J. Group Psychother.* 21:23.

Satir, V., 1967, "Conjoint Family Therapy," Science and Behavior Books, Palo Alto, California.

Singer, M., and Wynne, L., 1963, Thought disorder and family relations of schizophrenics: II. Classification of forms of thinking, *Arch. Gen. Psychiatry* 9:199.

Singer, M., and Wynne, L., 1965a, Thought disorder and family relations of schizophrenics: III. Methodology using projective techniques, *Arch. Gen. Psychiatry* 12:187.

Singer, M., and Wynne, L., 1965b, Thought disorder and family relations of schizophrenics: IV. Results and implications, *Arch. Gen. Psychiatry* 12:201.

Smith, C. J., 1969, Alcoholics: Their treatment and their wives, *Br. J. Psychiatry* 115:1039.

Steinglass, P., Weiner, S., and Mendelson, J. H., 1971a, A systems approach to alcoholism: A model and its clinical application, *Arch. Gen. Psychiatry* 24:401.

Steinglass, P., Weiner, S., and Mendelson, J. H., 1971b, Interactional issues as determinants of alcoholism, *Am. J. Psychiatry* 128:275.

Steinglass, P., Davis, D. I., and Berenson, D., 1975, In-hospital treatment of alcoholic couples, Presented at the American Psychiatric Association Annual Meeting, May 1975.

Weiner, S., Tamarin, J. S., Steinglass, P., and Mendelson, J. H., 1971, Familial patterns in chronic alcoholism: A study of a father and son during experimental intoxication, *Am. J. Psychiatry* 127:1646.

Westfield, D. R., 1972, Two years' experience of group methods in the treatment of male alcoholics in a Scottish mental hospital, *Br. J.,Addict.* 67:267.

Whalen, T., 1953, Wives of alcoholics: Four types observed in a family service agency, *Q. J. Stud. Alcohol* 14:632.

Wiener, N., 1961, "Cybernetics" (2nd ed.), M.I.T. Press, Cambridge, Massachusetts.

Wolin, S., Steinglass, P., Sendroff, P., Davis, D. I., and Berenson, D., 1975, Marital interaction during experimental intoxication and the relationship to family history, in "Experimental Studies of Alcohol Intoxication and Withdrawal" (M. Gross, ed.), Plenum Press, New York.

Wynne, L., and Singer, M., 1963, Thought disorder and family relations of schizophrenics: I. Research strategy, *Arch. Gen. Psychiatry* 9:191.

Wynne, L., Ryckoff, I., Day, J., and Hirsch, S., 1958, Pseudomutuality in the family relations of schizophrenics, *Psychiatry* 21:205.

CHAPTER 8

Behavioral Assessment and Treatment of Alcoholism

Peter E. Nathan and Dan W. Briddell*
Rutgers University
New Brunswick, New Jersey

This comprehensive review of behavioral efforts to assess and treat alcoholism begins by considering very briefly the present status of behavioral theories of the etiology of alcoholism. What follows then is a detailed review of recent efforts to assess levels and kinds of alcohol consumption with behavioral techniques. Finally, the remainder and bulk of the chapter highlights the manifold behavioral methods and programs that have been developed in the effort to modify alcoholic drinking.

BEHAVIORAL THEORIES OF ALCOHOLISM

Behavioral views of the etiology of alcoholism have not been well articulated. Most current behavioral theories are variations of a basic theme—first developed by Conger (1951, 1956) in the animal laboratory—that drinking is a

---

* Present address: Munson Medical Center, Traverse City, Michigan.

learned means of reducing conditioned anxiety. Most variations on this basic theme derive from disagreement over the precise mechanisms responsible for the presence of the conditioned anxiety. Franks (1970) offers a detailed historical treatment of this body of research and the resultant conflict over its interpretation.

Implicit in the hypothesis that drinking is a reinforced behavior is the assumption that alcohol eases prevailing high levels of anxiety in most or all alcoholics (that is, technically, negative reinforcement). However, one recent study comparing the behavior of alcoholics and matched nonalcoholics during a period of experimental intoxication (Nathan and O'Brien, 1971) reported that alcohol actually *increases* levels of anxiety and depression in alcoholics following an initial 12- to 24-hour period of drinking during which levels of anxiety decreased modestly. Another recent behavioral study which drew the same conclusion (Okulitch and Marlatt, 1972) also speculated that alcohol may actually function as a discriminative stimulus with both positive (reinforcing) and negative (punishing) properties for the alcoholic. An excellent review of the animal literature on tension-reduction and alcohol was published by Cappell and Herman in 1972; research in the area with humans was carefully reviewed by Mello (1972) and Marlatt (1975b); while a review of both research areas was recently published by Cappell (1974).

All three reviews conclude that the simple tension-reduction model of alcoholism has proven to be insufficient to account for continued drinking by chronic alcoholics. In an effort to extent the behavioral view of etiology beyond this model, Miller and Eisler (1975) suggest that "social learning theory," a more complex and inclusive theory of behavior, is more useful for understanding the development and maintenance of excessive drinking by alcoholics. Their explication of this view, excerpted here, may now represent the majority view of behavioral researchers:

> Within a social-learning framework alcohol and drug abuse are viewed as socially acquired, learned behavior patterns maintained by numerous antecedant cues and consequent reinforcers that may be of a psychological, sociological, or physiological nature. Such factors as reduction in anxiety, increased social recognition and peer approval, enhanced ability to exhibit more varied, spontaneous social behavior, or the avoidance of physiological withdrawal symptoms may maintain substance abuse (Miller and Eisler, 1975, p. 5).

BEHAVIORAL ASSESSMENT TECHNIQUES

Behavioral assessment instruments and procedures for use specifically with alcoholics have recently been developed in recognition of the extensively documented unreliability of alcoholics' self-reports of their own drinking (Summers, 1970). Traditional personality assessment methods also fail to permit specification of the kind and severity of a given individual's immoderate drinking for treatment-planning purposes (Franks, 1970), as well as being inappropriate for the pre- and post-treatment assessments that are necessary components of all behavioral treatment efforts (Miller, 1972).

Behavioral Assessment in the Laboratory

Three distinct kinds of behavioral measures of drinking in laboratory settings have been developed in recent years. The first of these, from the historical point of view, enabled Mello and Mendelson (1965, 1966, 1970), Nathan et al. (1970), and Nathan and O'Brien (1971) to evolve objective operant indices in the experimental laboratory of the relative reinforcement value of alcohol, money, and socialization/interpersonal isolation to the male alcoholics they studied. Among their findings were that alcohol is profoundly reinforcing for male alcoholics during times of drinking, that many male alcoholics demonstrate a biphasic pattern of drinking (a week-long "spree" followed by a much more lengthy period of "maintenance" drinking), and that many male alcoholics choose to drink in interpersonal isolation even when socialization is freely available to them. The latter observation has been questioned by the results of more recent studies by Bigelow (1973) and Griffiths et al. (1974a), who reported that most of their male chronic alcoholic subjects were significantly *more* sociable on drinking days than on days they were not permitted to drink; by a later study by Griffiths et al. (1975), who observed that other chronic alcoholic subjects, confronted with a forced choice between socialization and money during periods of sobriety and drunkenness, more often chose socialization over money during periods of experimental intoxication; and by a study by Thornton et al. (1975), which revealed that heavy drinkers in Gottheil's Fixed-Interval Drinking Decision research model (detailed below) displayed a dramatic increase in social behavior when they began programmed drinking on the research ward setting in which they were living. Though their subjects may have consumed less alcohol, been less chronic in their alcoholism, or of different socioeconomic status than the earlier subjects in Mello and Mendelson's and Nathan's studies, the nature of the effects of alcohol on interpersonal behavior remains an open question.

A recent operant study of the behavior of women alcoholics in an experimental laboratory setting similar to those used by Mello and Mendelson and by Nathan brings into question the extent to which these earlier findings have relevance to alcoholism among women (Tracey et al., 1976). Tracey's study revealed that the women alcoholics in her investigation demonstrated a singular (rather than biphasic) pattern of alcohol consumption in that they remained "maintenance" drinkers throughout the period programmed for drinking. Further, these women chose to drink more during periods of socialization than periods of isolation, another characteristic of their drinking at considerable variance with drinking by prior male alcoholic subjects. Because the female subjects may have differed from earlier male subjects in socioeconomic status as well as in chronicity of alcoholism, the role of gender identity in the phenomenon of differential drinking patterns has yet to be fully elucidated.

Another recent operant study of alcoholism showed that alcoholic subjects were more inclined than nonalcoholics to work to obtain alcohol in the presence of interpersonal stress (Miller, Hersen, Eisler, and Hilsman, 1974), a study which directly addressed the etiologic role of stress in alcoholism.

The operant approach has also been employed to predict and/or assess the efficacy of behavior change procedures, a most promising application of behavioral assessment methods. In the only one of these studies to be reported thus far, Miller, Hersen, Eisler, and Elkin (1974) observed that the pretreatment operant rates emitted by 20 male alcoholics to earn alcohol was predictive of their ultimate response to an eight-week behavioral treatment regimen. Ten alcoholics who were judged treatment successes after treatment were found to have emitted fewer operant responses (hand-switch presses) to earn points worth beverage alcohol before treatment began than another ten matched alcoholics who were treatment failures. The potential of this method—and others currently being developed—to predict the chances an individual alcoholic will benefit from behavioral treatment *before it begins* is very encouraging.

A second behavioral assessment strategy currently under development involves the use of "taste tests." Taste tests require subjects to taste a variety of available alcoholic and nonalcoholic beverages and to rate the beverages along taste dimensions. However, the real purpose of this procedure is surreptitiously to measure actual alcohol consumption. Taste tests, like the operant methods reviewed above, have been employed both to examine environmental determinants of drinking by alcoholics and to reflect changes in drinking following treatment.

Marlatt recently employed an "alcohol taste rating task" to explore the effects of stress (Higgins and Marlatt, 1973), expectancy (Marlatt, 1973b; Marlatt *et al.*, 1973), and modeling influences (Caudill and Marlatt, 1975) on consumption of alcohol. In the first of these studies (Higgins and Marlatt), alcoholics did not drink more beverage alcohol than nonalcoholics after the imposition of experimental stress (the threat of painful electric shock). This result suggests that stress may not play the direct role in the decision to drink by chronic alcoholics it has previously been presumed to play. In the second and third of these studies, both alcoholics and nonalcoholics drank more of the beverages which they were "testing" when told they were testing alcoholic beverages; however, alcoholics drank more than nonalcoholics of both alcohol and a mixer and the mixer alone under this experimental condition. By contrast, the *actual* beverage given subjects (alcohol and tonic or tonic alone) did not affect their consumption. Accordingly the "loss of control" hypothesis— that alcoholics cannot stop drinking once they begin—was not supported by these studies. The effects of modeling influences upon social drinking behavior was assessed by means of the taste test in the most recent of the studies in this series (Caudill and Marlatt, 1975). Groups of male college students who were heavy social drinkers drank with a confederate of the experimenters who either drank heavily or lightly during the taste test; another group of subjects were not exposed to a drinking model. Subjects who were exposed to the heavy drinking model drank significantly more alcohol than subjects exposed either to the model who drank lightly or to no model, suggesting that modeling, a basic learning paradigm, may indeed play a role in the etiology of alcoholism.

Miller and Hersen have employed a similar technique to assess changes in motivation to drink following behavioral treatment efforts. In the first of these studies, Miller and Hersen (1972a) employed Miller's taste test to evaluate the effects of two kinds of aversive conditioning agents (loud noise and imagined aversive agents) on drinking during the taste test. In Miller and Hersen's second study (1972b), the same assessment technique was used to evaluate the effects of electrical aversion on taste test consumption levels. In both instances, subjects drank less beverage alcohol (from 100 cc containers containing bourbon or vodka in water) during treatment than either before or after treatment. Beyond suggesting that aversive treatment apparently has no lasting inhibitory effects on alcohol consumption, these data also suggest that this behavioral measure of alcohol consumption may provide a first-order index of treatment efficacy that is quick, easy to employ, and reliable in its interpretation.

The final behavioral assessment approach dates to the mid-1960s, when drinking at experimental bars established within laboratory settings first permitted objective, reliable, and continuous observation of drinking behavior by alcoholics and nonalcoholics. While one of these experimental bars (that outfitted by Nathan at Boston City Hospital) only dispensed beverage alcohol when subjects "paid" with operant tokens, others dispensed alcohol ad libitum, up to liberal limits, as part of a specific measurement system. In one of the most significant reports from such an experimental bar, Schaefer *et al.* (1971) reported reliable differences between alcoholic and nonalcoholic subjects in drinking topography: Alcoholics chose to gulp straight drinks without ice while nonalcoholics preferred mixed drinks with ice which they sipped. Williams and Brown (1974), employing a similar experimental bar setting, recently replicated this study in New Zealand. They reported that New Zealand and North American alcoholics and nonalcoholics drink in strikingly similar ways. For the most part, however, free-drinking assessment methods have been used less often recently to gather this kind of normative data, more often to predict or assess treatment efficacy, a use to which the taste tests have also recently been put.

The Fixed-Interval Drinking Decision (FIDD) model developed by Gottheil and his colleagues for research and treatment purposes is an ad lib measure that has been employed for predictive purposes. The six-week FIDD program, developed by Gottheil and his coworkers (1971, 1972) for use at the Coatesville (Pennsylvania) Veterans Administration Hospital, incorporates a predrinking orientation week, a subsequent four-week period during which fixed amounts of alcohol are available at hourly intervals from 9:00 a.m. through 9:00 p.m. on week days only, and a postdrinking week. When data on drinking patterns during the four-week FIDD drinking period were compared to posttreatment drinking behavior, it was reported that drinking during the program predicted subsequent posthospital drinking more reliably than patients' past drinking behavior or expressed desire to stop drinking. Specifically, patients who did no drinking during the FIDD program more often maintained abstinence on leaving it than patients who either began to drink and then stopped or those who drank through the entire program (Gottheil, Alterman, Skoloda, and Murphy, 1973; Gottheil, Crawford, and Cornelison, 1973; Skoloda et al., 1975).

The "probe" sessions interposed three times during the 17-session behavioral treatment program reported by Sobell and Sobell (1973a) also represent an application of ad lib drinking as a measure of therapeutic progress. The Sobells' probe sessions permitted subjects to drink their choice of an alcoholic beverage in the absence of the aversive contingencies operative during preceding treatment sessions. This assessment revealed that the aversive conditioning procedures imposed during prior treatment sessions had no apparent effect on drinking during the probe sessions. However, in view of the positive outcome data the Sobells reported at one- and two-year follow-ups (1973b, 1976), one must conclude either that the probe data were unable to reflect a positive therapeutic change that was taking place or that they were correctly reflecting absence of therapeutic change *during* treatment that occurred only after active treatment ceased.

Treatment evaluation research recently conducted at the Alcohol Behavior Research Laboratory at Rutgers University also employed interposed ad lib drinking periods (formally, ABA and ABACA single-subject designs) to assess behavioral treatment efficacy before, during, and immediately after periods of experimental treatment. To this end, Silverstein *et al.* (1974) employed an ABA assessment strategy in their study of variables determining ability of alcoholics to discriminate their own blood alcohol levels and then to maintain them at moderate levels. Wilson *et al.* (1975) used a similar evaluative design to assess the efficacy of aversive conditioning as a treatment for alcoholism while Steffen (1975) and Steffen *et al.* (1974) chose the same design to investigate the extent to which biofeedback of frontalis muscle action potential heightens the effects of relaxation training on subsequent drinking by chronic alcoholic subjects.

Behavioral Assessment in the Natural Environment

The techniques for behavioral assessment of drinking discussed above were developed and used in laboratories studying both the environmental determinants of excessive drinking and experimental behavioral treatments to modify it. By contrast, most long-term assessment of alcoholic drinking "in the field" is distinctly nonbehavioral in nature. Such measures have ranged from general purpose psychological tests like the MMPI and the California Psychological Inventory to tests specifically designed for use with alcoholics like the Alcadd Test and the Manson Evaluation Test. These assessment measures typically suffer from the same serious problems of reliability, validity, and utility as personality tests used in other contexts.

Several recent efforts to develop and refine new behavioral assessment procedures for use in conjunction with long-term follow-up of behavioral treatment in the natural environment are worthy of brief mention here. The first of these is the Sobells' intensive follow-up of their comprehensive comparative behavioral treatment of 70 chronic alcoholic patients (1973a,b, 1976). Two important methodological advances characterized their follow-up. The first was the novel way in which outcome data were organized around the *drinking disposition*. The second was the intensive nature of the follow-up: Almost 100 percent of original subjects participated in that follow-up and supplementary data from many previously unsolicited sources was utilized to maximize the reliability of data on drinking disposition. Gathering follow-up data within the drinking disposition format requires coding and assigning drinking behavior during the days after each follow-up contact to several distinct outcome categories: *controlled drinking* (drinking at or below defined moderate limits), *abstinence, not incarcerated–drunk,* and *incarcerated–alcohol-related*. Follow-up contacts were made by telephone or in person at least monthly. With this procedure judgments about number of days of drinking and patterns of consumption could be made with considerable assurance.

Besides his alcohol taste rating task, Marlatt has recently developed the Drinking Profile (1975a) a 19-page questionnaire completed during a standardized interview and designed to yield a thorough behavioral profile of drinking preferences, rates, patterns, and settings as well as motivational and reinforcement factors associated with the drinking. The Drinking Profile contains all the data necessary for a complete behavioral assessment of a given individual's drinking problem. Further, readministration of the drinking profile questionnaire during follow-up may be useful in the assessment of long-term treatment effects.

Behavioral Assessment: Overall Evaluation

Because alcoholism involves a single "target behavior"—uncontrolled drinking—whose dimensions can be specified unequivocally, efforts to develop behavioral assessment procedures to describe drinking by alcoholics have proliferated.

Strengths of assessment methods that actually involve alcohol consumption (the laboratory assessment procedures) are numerous. They are reliable, analogous to drinking behavior outside the laboratory, salient to the problem of alcohol abuse, and quantifiable within prescribed limits. For these reasons, the operant, taste test, and ad lib drinking assessment methods have proven of considerable value to investigators studying environmental, psychological, and interpersonal variables affecting drinking in the laboratory. An important shortcoming these methods share, however, is that they cannot be used to assess abstinent alcoholics because all require alcohol consumption. As a consequence, these methods cannot be used for abstinence-oriented therapy outcome research and evaluation.

BEHAVIORAL TREATMENT APPROACHES

Goals of Treatment

The only acceptable treatment goal for alcoholics, until very recently, has been total abstinence. This sole treatment aim stemmed from the widely held conviction that alcoholism is a progressive, irreversible disease characterized by loss of control over drinking during periods of drinking and profound craving for alcohol during periods of sobriety (Jellinek, 1960; Ludwig, 1972; Ludwig and Wikler, 1974; Ludwig et al., 1974; Williams, 1948). The belief that a single drink by the "dry alcoholic" inevitably leads to loss of control over drinking has been reinforced through the years by Alcoholics Anonymous, which considers total abstinence to be its only treatment goal.

Several investigators have recently challenged this single orientation to alcoholism treatment by demonstrating that one drink—or several—does not always unleash irresistable craving and loss of control over drinking in the laboratory (Cutter *et al.*, 1970; Gottheil, Crawford, and Cornelison, 1973; McNamee *et al.*, 1968; Merry, 1966; Williams, 1970). Further, Bigelow, Griffiths, Cohen, and their co-workers have reported that volunteer chronic alcoholics will choose to drink moderately, despite the availability of large quantities of alcohol, in order to live in an enriched rather than an impoverished ward environment (Cohen *et al.*, 1971) or to spend time with other alcoholics rather than alone (Griffiths *et al.*, 1974b). From these and similar results these researchers conclude that alcoholics can moderate their drinking voluntarily after once having begun. They suggest, as a consequence, that controlled drinking may not be an appropriate treatment goal for the alcoholic. In addition, other investigators (Pattison, 1966; Pattison *et al.*, 1968) have recently observed that some alcoholics do return voluntarily to moderate levels of drinking even after little or no specific therapeutic intervention. Finally, critics of abstinence as the sole treatment objective point to data (National Institutes of Mental Health, 1969) suggesting that less than 20 percent of treated alcoholics ever successfully attain abstinence.

With these findings in mind, several clinical researchers have initiated experimental treatment programs with controlled drinking the prime treatment objective. Several of these studies are reviewed in detail below. What is important to note here, however, is that on the basis of admittedly preliminary outcome data, controlled drinking may well be an appropriate treatment objective for some individuals—but an inappropriate objective for others. What is not now known is the nature of reliable predictors of response to abstinence-oriented and controlled drinking-oriented treatment. In other words, it is time that we ask what factors must be considered when client and clinician formulate treatment goals with respect to abstinence or controlled drinking.

We believe that development of predictors of response to treatments with such varying goals will require attention to at least the following crucial factors: (1) evaluation of the functional relationships between excessive drinking and choice of treatment goal; for example, does alcohol serve as a tension-reducer or a social lubricant in the patient's life; (2) assessment of pretreatment consumptive behavior as a predictor of treatment response; for example, how much does the patient drink when he is exposed to one of the recently-developed "taste tests"; (3) assessment of the patient's social support systems as an adjunct to posttreatment planning; that is, is someone in the patient's environment committed to abstinence (controlled drinking) as a treatment goal and will he or she help the patient maintain that goal; (4) evaluation of prior treatment experiences; for instance, does a history of failure to achieve abstinence following abstinence-oriented treatment mean that the patient should work towards controlled drinking as a future treatment goal; (5) assessment of patient's expectations about chances for successful treatment as a predictor of treatment choice; does the patient, for example, come to treatment accepting only the legitimacy of an abstinence-oriented treatment? Detailed consideration of issues related to treatment goal clarification is given elsewhere (Briddell and Nathan, 1976).

We believe that a *thorough behavioral assessment* must be conducted for each individual alcoholic and that, given the preliminary nature of controlled-drinking research, extreme caution should be exercised before encouraging a problem drinker to attempt to "control his drinking" rather than to attempt to stop it altogether. At this stage we feel it unwise to choose control drinking as a treatment goal for a patient unless that patient has previously tried and failed in abstinence-oriented programs. We also believe that no currently abstinent patient should experiment, with or without a therapist's help, with controlled drinking as an alternative to abstinence.

Modifying the Drinking Response: Aversive Conditioning

Electrical Aversion

Almost 50 years ago Russian physician N. V. Kantorovich treated 20 Russian alcoholics by pairing the sight, smell, and taste of a variety of alcoholic beverages with repeated electric shock. Follow-up of this dramatic new "Pavlovian" treatment lasted from 3 weeks to 20 months and revealed that fully 70% of patients receiving the treatment remained abstinent through the follow-up period. Since Kantorovich, unlike many present-day alcoholism researchers, also followed up a control group of 10 alcoholics given hypnotic suggestion or medication (they returned to alcohol within a few days of their release from hospital), one must be impressed by the design and results of this first reported application of aversive conditioning to alcoholism.

Despite the encouraging thrust of Kantorovich's findings electrical aversion was not employed again to treat alcoholism until the 1960s, perhaps because of widespread belief that the remediation efforts of Alcoholics Anonymous required no supplementary help.

In any event, encouraged by theoretical papers by Eysenck (1960) and Rachman (1961) claiming the efficacy of electrical aversion as treatment for a variety of behavioral disorders, McGuire and Vallance (1964) reported their attempts to modify drinking by electrical aversion. Viewing their method as straightforward aversive conditioning and reporting in the same paper on use of the technique with other behavioral problems in addition to alcoholism, McGuire and Vallance concluded that the behavior of alcoholics responded less well than other forms of deviant behavior to the treatment procedure.

Shortly after the McGuire and Vallance study, Blake (1965, 1967) initiated an extensive series of studies of electrical aversion at the Crichton

Royal Hospital in Scotland. A more sophisticated application of behavioral principles, Blake's paradigm combined training in progressive relaxation—to reduce stress and anxiety—with electrical aversion to eliminate alcohol consumption. This simultaneous attack on conditions associated with chronic alcoholism (anxiety and stress) as well as on the uncontrolled drinking itself anticipated the more recently developed broad-spectrum approaches to the disorder that are discussed below. In the initial developmental study of this procedure, Blake (1965) recruited 37 alcoholics in their middle 40s drawn from higher socioeconomic classes who suffered from what seems to have been only moderate alcoholism. All Ss received relaxation training (averaging about 12 sessions), "motivation arousal" (designed to increase Ss' treatment motivation by forcing them to focus on the consequences of their alcohol abuse), and aversion conditioning (lasting about 15 sessions). The latter procedure programmed shock to begin when a sip of alcohol was taken and to terminate when the alcohol was expectorated. Technically, this conditioning method should be termed *aversion–relief,* since it was intended to instill both aversion to alcohol and subsequent feelings of relief from the aversion when the alcohol was expectorated. Follow-up data on these Ss were elicited at 1-, 3-, 6-, 9-, and 12-month intervals. They indicate that about half of the 34 Ss located for follow-up were sober at the 6-month mark; whether these same Ss had remained abstinent through the entire 6 months was uncertain. An additional follow-up of the same patients (Blake, 1967) described a new treated group of 22 alcoholic Ss against whom the original 37 aversion–relaxation Ss were compared. Matched with the original 37 Ss for age, sex, socioeconomic status, chronicity of alcoholism, previous hospitalization, concurrent psychiatric diagnosis, and intelligence, these new Ss received aversion–relief therapy but not concurrent relaxation training. Tables 1A and 1B summarize 12-month follow-up data for

TABLE 1A. Twelve Months' Follow-Up of 37 Cases Treated by Relaxation and Electrical-Aversion Therapy

	Cases			
Outcome	Male	Female	Total	Percent
Abstinent	12	5	17[a]	46
Improved	3	2	5[a]	13
Relapsed	8	3	11	30
Others	4	—	4	11
Total	27	10	37	100

[a] Includes 3 subjects (1 male, 2 female) relapsed, readmitted for further relaxation–aversion therapy and have been abstinent or improved for 12 months since last discharge.

TABLE 1B. Twelve Months' Follow-Up of 22 Cases
Treated by Electrical-Aversion Therapy Alone

	Cases			
Outcome	Male	Female	Total	Percent
Abstinent	3	2	5	23
Improved	6	—	6[a]	27
Relapsed	4	2	6	27
Others	5	—	5	23
Total	18	4	22	100

[a] Includes one subject readmitted for booster treatment during follow-up (Blake, 1967, p. 91).

both groups. They show that alcoholics receiving both relaxation training and aversion therapy (Table 1A) were more apt to be judged either abstinent or improved at follow-up than those given electrical aversion therapy alone (Table 1B). However, neither "time in treatment," the reliability of follow-up data, nor the comparability of Blake's two groups of Ss were dealt with in either of Blake's reports. Further, Blake failed to employ an untreated control group for comparative purposes, preventing us from knowing how many Ss would have moderated their drinking over the 12 months *without* treatment. As a consequence, while Blake's results encourage the view that alcoholism treatment that aims to alter "associated problems" in addition to excessive drinking itself may be more useful than treatment attempting only to modify the drinking response, they do not provide unequivocal endorsement for electrical aversion as a treatment of choice for chronic alcoholism.

In a study designed to evaluate the efficacy of a behavioral treatment paradigm similar to that developed by Blake, MacCulloch *et al.* (1966) employed a procedure previously used with homosexual Ss called "anticipatory avoidance learning" to treat a group of four alcoholics. MacCulloch and his colleagues described the alcohol-related stimuli used for the purposes of aversive conditioning in the following words: "Ss are presented with a range of photographs of beer and "spirits," the sight of an actual bottle of alcohol (stoppered), the sight of an open bottle of alcohol, and alcohol in a glass. We also use a tape recording which consists of a repeated invitation to have a drink of whatever alcohol is the patient's preferred choice" (pp. 187–188). Slides of orange squash were counterposed as relief stimuli.

A complete battery of alcohol-related and nonalcohol-related stimuli (pictures of an orange drink) was shown randomly to each Ss. When the alcohol-related stimuli were presented, they were followed eight seconds later by a shock unless switched off by the S, at which point the nonalcohol-related (aversion–relief) stimuli appeared. This sequence was intended to promote an increase in the reinforcement value of the nonalcohol-related stimuli and a concomitant decrease in the reinforcement value of the alcohol-related stimuli.

In commenting on their essentially negative findings—all four Ss returned to drinking shortly after treatment ended—the authors observed that three of the four Ss had consistently failed to avoid shock during conditioning by consistently switching off alcohol-related stimuli (in other words, they had presumably failed to permit conditioning to occur). They also noted that unknown biochemical factors in alcoholism may have played a role in this failure of "anticipatory avoidance learning" with alcoholics, as against its apparent success with homosexuals. Finally, MacCulloch and his co-workers speculated that social factors associated with alcohol consumption (e.g., the conviviality with which it is associated, the interpersonal facilitation it permits) may also be stronger maintainers of excessive drinking than similar factors in the social milieus of homosexuals.

Though other investigators (Hallam et al., 1972; Hsu, 1965; Miller and Hersen, 1972b; Sandler, 1969) have reported on the use of electrical aversion conditioning with alcoholics—with decidedly mixed results—only one additional direct clinical trial of electrical aversion with chronic alcoholics will be reviewed here. A study by Vogler et al. (1970) will be discussed because it demonstrated sophisticated design and methodology, programmed an extensive follow-up with apparent care, and, less favorably, presented its positive results in a manner which may unintentionally have disguised a central conceptual/methodological error.

Vogler's study was designed to build on Blake's "encouraging" outcome data by adding appropriate control groups. Vogler's treatment procedure also programmed booster reconditioning sessions in line with suggestions by other behavioral theorists that the efficacy of aversive conditioning and other behavioral procedures might be extended beyond the immediate posttreatment period in that way. Seventy-three male chronic alcoholics were randomly assigned to one of four treatment groups: (1) Booster Conditioning: Ss who received 20 in-patient aversion–relief conditioning sessions and 3 outpatient reconditioning sessions. (2) Conditioning Only: Ss who received only aversion–relief conditioning when they were inpatients; they did not return for booster reconditioning sessions. (3) Pseudoconditioning: Control Ss who sipped beverage alcohol, like Booster Conditioning and Conditioning Only Ss, but then received random rather than contingent shock upon alcohol ingestion. (4) Sham Conditioning: Ss who received the same treatment as had Ss in the two conditioning groups except that they received no shocks. A Ward Control Group which received only "routine hospital treatment" was added to the study later.

An extensive series of follow-up contacts with all Ss revealed that the

Booster Conditioning Group had maintained abstinence significantly longer than any of the other groups at an eight-month follow-up ($p < 0.002$). Unfortunately, these results could deceive unless the data on which the significance level was computed are carefully inspected. Those data reveal that 13 Booster Conditioning and Conditioning Only Ss and one Pseudoconditioning S dropped out of the study during its treatment phase while one Pseudoconditioning and seven Ward Control Ss were lost during follow-up. If these Ss had been included in the statistical computations—if they are viewed as the treatment failures they likely were—the apparent efficacy of the behavioral treatment employed by Vogler and his colleagues would be significantly attenuated. A one-year follow-up of the same Ss (Vogler et al., 1971) revealed that the encouraging eight-month follow-up data reported in the 1970 article had become less encouraging at the one-year mark. Specifically, they indicated that members of the two conditioning groups did not differ from pseudoconditioning Ss on any "rehospitalization criteria," the measure employed at this follow-up to reflect treatment efficacy.

Against this backdrop of conflicting findings, Miller et al. (1973) recently attempted more fully to evaluate the efficacy of electrical aversion with alcoholics in the context of an analog drinking study. The methodology of earlier investigations was improved by including appropriate control groups and an objective pre–post behavioral measurement of treatment efficacy. Thirty hospitalized males free from organic brain disease, cardiovascular disorder, or other major medical problems were matched and assigned to one of three groups. Aversive Conditioning Ss received escape conditioning according to Blake's 1965 procedure. They were instructed to sip but not swallow the drink, with shock being delivered concurrent with the sip and terminated when S spat the alcohol into a pan. Shock was delivered to the forearm; its intensity was adjusted between 3 and 8 ma until the subject "reported pain and a distinct flexion of the arm was observed." Subjects received 500 shock trials over a ten-day treatment period. Control Conditioning Ss were exposed to exactly the same conditioning sequence except that they received a 0.05 ma shock; the majority of these Ss reported that they felt no sensation at this shock intensity level. This group served as a control for "expectancy" factors implicit in the aversive conditioning procedure. Group Therapy Ss received six one-hour group sessions designed to help them achieve a "better understanding of the social and emotional precipitants of excessive alcohol consumption." All Ss received Miller and Hersen's (1972a,b) taste test before and after treatment. It revealed that all groups reduced posttreatment alcohol consumption, the Aversion Conditioning group by 36%, the Control Conditioning group by 37%, and the Group Therapy group by 30%. In short, the study revealed no significant differences in treatment efficacy among the treatment groups. These results discourage the view that aversive conditioning per se is responsible for the positive treatment effects that some investigators have observed following aversive conditioning with alcoholics. Miller and his colleagues suggest, in fact, that such treatment effects may be a function instead of therapeutic instructions, demand characteristics, or other nonspecifics of the conditioning procedure (see Hallam et al., 1972 or Hallam and Rachman, 1972, for a further discussion of this issue).

Two studies recently completed at our own laboratory, designed to evaluate the effects of electrical aversion by a more direct behavioral measure (Wilson et al., 1975), will be noted here. Treatment effects were assessed by free operant drinking baselines in a seminaturalistic experimental setting. Attitudes toward alcohol were measured daily by a semantic differential technique. Six gamma-type male alcoholics (Jellinek, 1960) served as subjects in the first two studies (1A and 1B). These subjects had previously tried and failed to achieve abstinence in other treatment programs; all had been screened to rule out major medical and psychiatric disorders. Subjects were treatment-motivated and participated in the study on a voluntary basis.

During an initial baseline period (which lasted three days in study 1A and four days in study 1B), Ss were allowed to consume 86-proof blended whiskey or bourbon to the maximum of 18 one-ounce drinks a day (1A) or 30 drinks (1B). Drinks were served at a simulated bar and dispensed by a computer-operated console in each S's bedroom. In both cases, consumption was computer-recorded during all baseline periods of the study. After one postbaseline "drying out" day, during which blood alcohol levels were allowed to return to zero, three Ss received electrical escape conditioning modeled after the procedures used by Blake (1965) and Vogler et al. (1970) (one S in 1A, two in 1B) while the other three Ss received backward (control) conditioning such that they were shocked immediately prior to being instructed to sip but not swallow the alcohol. Both treatments were administered twice daily, with 15 conditioning trials constituting a session. Treatment lasted three days during study 1A and four days during 1B. The range of shock intensities was similar to that reported by Miller et al. (1973). Shock levels were increased whenever a S appeared to be habituating to the shock. A second baseline followed the first treatment phase of the study. The treatments were reversed for all subjects during the second treatment phase. This phase, in turn, was followed by a final baseline period.

Results of both studies are summarized in Table 2.

The table shows that neither escape nor backward conditioning was effective in reducing alcohol consumption. Only subject #3 in study 1B demonstrated a substantial reduction in alcohol consumption following escape conditioning. However, it was later learned that he returned to drinking within a week of his discharge from the laboratory.

These results are consistent with findings by Miller and his co-workers

TABLE 2. Mean Number of Ounces of Alcohol Consumed[a]

	Baseline 1	Treatment 1	Baseline 2	Treatment 2	Baseline 3
Study 1A					
Subject 1	10.7	Escape conditioning	9.7	Backward conditioning	10.3
Subject 2	11.0	Backward conditioning	11.3	Escape conditioning	11.7
Study 1B					
Subject 1	21.3	Escape conditioning	21.5	Backward conditioning	23.0
Subject 2	13.7	Escape conditioning	15.0	Backward conditioning	15.5
Subject 3	28.0	Backward conditioning	25.5	Escape conditioning	1.5
Subject 4	24.3	Backward conditioning	25.5	Escape conditioning	26.5

[a] From Wilson et al., 1975.

(1973) and reviews by Nathan (1976) and Briddell and Nathan (1976) to the effect that electrical aversion by itself does not cause a significant reduction in alcohol consumption or any permanent attitude change toward alcohol in alcoholics. Whether electrical aversion *combined with* other behavioral methods (e.g., Sobell and Sobell, 1973) is of value in reducing alcohol consumption, however, remains to be determined.

Chemical Aversion

Chemical aversion for alcoholism involves the repeated pairing of nausea and vomiting, produced by emetic chemicals like Emetine or Apomorphine, with the sight, taste, smell, or touch of alcohol. The most extensive use of the chemical aversion paradigm with alcoholics was reported by Voegtlin (1940, 1947), Lemere and Voegtlin (1950), and Lemere *et al.* (1942) in a series of investigations. They describe the most comprehensive program of aversion therapy with alcohol abusers to date. Still in use now at the Shadell Sanatorium in Seattle, the hospital at which these data were originally gathered, as well as at other private hospitals located mostly in the western United States, the method Voegtlin and his associates developed in the late 1930s involves the use of Emetine as the aversive agent. In general, the chemical aversion method requires, first, ingestion or injection of the emetic, followed shortly thereafter by ingestion of a sip of the patient's favorite alcoholic beverage. Ingestion of the beverage is timed to coincide with the prolonged and severe vomiting that invariably follows. Two or three such trials constitute a typical 45-minute treatment session. A treatment session every other day, up to the usual four to six sessions during the patient's hospital stay, is the standard chemical aversion paradigm.

Lemere and Voegtlin reported in 1950 on an extensive follow-up of over 4000 patients given chemical aversion conditioning 10 to 13 years after treatment. Overall, 60% of these patients were abstinent after 1 year and about 50% were abstinent at the 13-year follow-up. These figures, however, include patients who relapsed but were successfully retreated. As a result, their comparability to therapeutic outcomes generated by other treatment methods remains uncertain.

Other investigators have also employed nausea-producing chemicals to establish conditioned aversion to alcohol. Thimann (1949), for example, used Emetine approximately as did Voegtlin and Lemere to treat 245 alcoholic patients. At a four-year follow-up, 51% of his patients had remained abstinent, but his exclusive reliance on self-report data detracts from the overall significance of his findings. Similarly Raymond (1964), in the context of a choice situation in which patients could choose either alcoholic or nonalcoholic beverages as a behavioral measure of response to chemical aversion, reported that patients receiving chemical aversion chose alcoholic beverages significantly less often than patients not receiving it. These results also lend support to the view that chemical aversion has promise in this application.

The most favorable data on chemical aversion since publication of Voegtlin and Lemere's results are contained in a recent paper by Wiens and his colleagues (1975). Summarizing treatment outcome data from 261 alcoholic patients treated at a private hospital outside Portland, Oregon, Wiens and his co-workers report that 63% of these patients had apparently remained abstinent during the 12-month period following an inpatient program of chemical aversion in 1970. Since close contact with these patients was maintained throughout the entire follow-up period, there is good reason to have confidence in the reliability of these data.

It must be noted, at the same time, that "expectancy"—belief that a therapeutic method will work—may have played an important role in the Shadell/Voegtlin and Portland/Wiens treatment paradigms. Building on a lengthy history of documented treatment success, those who developed treatment programs at these two hospitals clearly encouraged the expectation in their patients that chemical aversion is a proven treatment method. And, in fact, as it was employed at these treatment facilities in conjunction with other modes of intervention—including inpatient group therapy conducted along Alcoholics Anonymous lines as well as irregular personal and family counseling—chemical aversion does appear to have been quite effective. The specific role of chemical aversion per se in these "treatment packages," however, remains undefined since experimental analyses of the separate components of the two total treatment programs were not undertaken. Finally, most of the patients treated by Voegtlin and Lemere and by Wiens and Montague were better motivated and of higher educational and socioeconomic level than patients receiving other forms of behavioral treatment by other researchers. One cannot interpret the apparent success of chemical aversion without considering the prognostic significance of these demographic variables. That is, since it is well known that motivation, education, and socioeconomic status play important roles in predicting outcome of treatment for alcoholism, one must consider the possibility that the favorable prognosis of patients given chemical aversion in these two hospitals played as important a role in their recovery from alcoholism as the chemical aversion they received.

It is clear, then, that research designed specifically to "unpackage" total treatment programs which include chemical aversion and comparative studies of patients from differing socioeconomic and educational levels receiving such treatment are necessary before the promise of chemical aversion can be fully evaluated.

Another chemical aversion procedure—employing succinylcholine chloride dehydrate (Scoline) to induce total apneic (respiratory) paralysis as the aversive event—was described by Sanderson *et al.* (1962, 1963) and Laverty (1966) as having yielded equivocal outcome data. As this aversive procedure also involves unconscionable ethical, psychological, and medical risks, it will not be reviewed further here. A detailed consideration of Scoline conditioning is given, however, in Nathan (1976).

Covert Aversion

Covert aversion is an aversion conditioning method which employs aversive images instead of aversive shocks or chemicals. Also termed *covert sensitization,* this procedure has been put forth by Cautela (1966, 1967, 1970), its creator, as a promising treatment approach for obesity, smoking, obsessions and compulsions, stealing, and homosexuality as well as alcoholism. For use with alcoholics, Cautela suggests combining covert sensitization with standard relaxation training and systematic desensitization in order to treat the anxiety component of the patient's drinking behavior as well as the drinking response itself.

Central to covert sensitization are the aversive scenes the therapist provides the patient. Cautela gives the following example:

> You are walking into a bar. You decide to have a glass of beer. You are now walking toward the bar. As you are approaching the bar you have a funny feeling in the pit of your stomach. Your stomach feels all queasy and nauseous. Some liquor comes up your throat and it is very sour. You try to swallow it back down, but as you do this, food particles start coming up your throat to your mouth. You are now reaching the bar and you order a beer. As the bartender is pouring the beer, puke comes into your mouth. . . . As soon as your hand touches the glass, you can't hold it down any longer. You have to open your mouth and you puke. It goes all over your hand, all over the glass and the beer. You can see it floating around in the beer . . . (Cautela, 1970, p. 87).

On theoretical and practical grounds, covert sensitization offers advantages over chemical or electrical aversion. It permits patient and therapist to adjust the topography of the aversive imagery to the specifics of the patient's drinking behavior; it allows direct (imaginal) association between the aversive image, the behavior (drinking), and the environment associated with the drinking (including the people and places with which it is associated). Covert sensitization is also sufficiently "mobile" that it can be independently employed by the patient when he or she feels tempted to drink.

Despite these advantages, covert sensitization is still not widely used to treat alcoholics. One reason for this is the equivocal nature of outcome data following its use. Cautela (1967) reported that a 29-year-old female alcoholic given covert sensitization remained abstinent at an 8-month follow-up and Anant (1967) reported that 96% of 26 patients receiving the same treatment were abstinent at follow-up periods ranging from 8 to 15 months; neither study included control groups designed to assess the differential efficacy of the separate components of the total treatment package (which also included relaxation training and systematic desensitization). The one covert sensitization study (Ashem and Donner, 1968) which did employ a control group—to compare treatment and no treatment effects—revealed that only 6 of 15 subjects receiving covert sensitization remained abstinent at a 6-month follow-up. None of 8 control subjects was similarly abstinent. These data suggest that while the covert sensitization package may exert some effect on drinking by alcoholics, that effect is probably a modest one whose overall clinical utility must be considered peripheral.

Aversive Conditioning in Perspective

We conclude, with others (e.g., Franks, 1970; Wilson *et al.*, 1975), that none of the aversive conditioning procedures reviewed above can by itself constitute the treatment of choice for chronic alcoholism.

We are struck by the results of a variety of studies which suggest that electrical aversion by itself has little impact on excessive drinking. On the other hand, recent research reviewed below indicates that electrical aversion in combination with other behavioral procedures may have promise. These studies merit replication and extension. Though much less research has been conducted with covert aversion, we would suggest that essentially the same conclusions about its potential efficacy singly and in combination with other behavioral procedures can be drawn from its small body of research literature.

By contrast, chemical aversion with nausea-inducing drugs may be of more significant value by itself than the other aversion conditioning methods, perhaps because nausea, unlike electrical stimulation or imagined aversion, relates much more directly to the natural consequences of excessive alcohol consumption. Garcia *et al.* (1974) and Wilson and Davison (1969) have written interesting reviews of this issue and drawn similar conclusions. But despite the apparent empirical and theoretical superiority of chemical aversion, its efficacy alone has not been tested. As a consequence, we do not maintain that it should be used in other than a multifaceted behavioral treatment program until its utility in isolation has been established.

Modifying the Drinking Response: Operant Methods

Aversive conditioning aims to modify or eliminate the drinking response of the alcoholic by pairing it with an aversive event such that the drinking response itself will acquire aversive properties. By contrast, operant approaches aim to modify the drinking response by manipulating its consequences in order to reduce its frequency. Despite their recent origin, pilot operant treatment programs designed to modify the target behavior of excessive drinking by alcoholics show considerable promise.

Cohen and her colleagues (1971) and Bigelow, Griffiths, and their coworkers (1973, 1974) describe such a series of pilot investigations undertaken at the Alcoholism Behavior Research Unit, Baltimore City Hospitals. In the first of these projects, Cohen and her colleagues reported that five of five chronic alcoholic subjects voluntarily moderated their drinking in a laboratory setting when moderate drinking (no more than 5 ounces of beverage alcohol a day) resulted in an enriched living environment and not doing so (drinking from 6 to 24 ounces) was associated with an impoverished environment. The enriched environment permitted subjects to work in the hospital laundry four hours a day for one dollar an hour, to have use of a private telephone, to have access to a fully-equipped recreation room, to participate in daily group therapy, to eat a regular hospital diet, to possess a bedside chair, and to receive visitors. The impoverished environment allowed none of these "extras."

In another of these pilot investigations, Bigelow *et al.* (1974) explored the effects on drinking of a brief period of isolation contingent on drinking. An "isolation" phase, during which 10 minutes of isolation in a small booth followed each ingestion of an ounce of 95-proof ethanol up to 12 ounces a day, preceded and followed baseline drinking periods of indeterminate length. The ten alcoholic subjects of this brief study consumed 95% and 92%, respectively, of the alcohol available to them during pre- and postisolation phase periods. Contingent time-out from reinforcement during the isolation phase suppressed drinking to 52% of the alcohol available, suggesting that isolation was sufficiently aversive that it was avoided even at the expense of alcohol consumption.

Alcoholics who appeared to possess adequate social reinforcers outside the hospital were selected for inclusion in a study reported by Bigelow, Liebson, and Griffiths (1973). Opportunities to earn a weekend pass to attend a family gathering or to acquire special privileges to visit a girlfriend were the reinforcers made available to the two subjects of this study. These privileges were made contingent upon their drinking 5 or fewer of the 12 ounces of alcohol that were available to them each day. Drinking beyond 5 ounces a day occurred on only 1 of 16 days for one subject for whom these contingencies were in force and not at all for the other. Both of these studies demonstrate that alcoholics will drink in moderation when the contingencies for doing so are directly tied to the availability of social reinforcers.

In a second experiment detailed in their 1973 paper, Bigelow and his colleagues tested the utility of alcohol itself as a reinforcing consequence for *controlled drinking* by alcoholics. The research revealed that not one of five alcoholics drank excessively (beyond 8 oz/day) when alcohol for a subsequent day of drinking was made contingent upon not exceeding the predetermined moderation. By contrast, when no contingencies were attached to drinking, all five subjects drank as much as they could.

Miller, Hersen, Eisler, and Watts (1974) describe an unusual single-subject operant treatment study designed to evaluate the effectiveness of contingent reinforcement for reduced blood alcohol levels of a single outpatient alcoholic. The method used to obtain reliable measures of drinking in the natural environment is of special interest. Blood alcohol level was assessed in the home or at the place of employment by a research assistant dispatched twice weekly on an unpredictable basis to collect breath samples for later analysis by gas chromatograph. Contingent reinforcement—three dollars' worth of VA hospital commissary coupons—was given for each zero BAL reading following a lengthy initial baseline period during which breath samples were also collected. After each chromatographic analysis, once the three-week contingent reinforcement period began, the research assistant returned to the patient's home or job either to deliver reinforcement or to inform him that he had failed to meet the criterion for reinforcement. After three weeks of this, a three-week noncontingent reinforcement phase was introduced during which the patient received commissary coupons following each breath test regardless of BAL. Contingent reinforcement was then reinstated for the final three weeks of the study. The two contingent reinforcement periods were much more effective than either noncontingent reinforcement or no reinforcement (during the baseline period) in reducing drinking by this patient. The practical utility of this mode of treatment, however, might be questioned on the basis of cost effectiveness—the high cost in paraprofessional time spent locating the patient, taking breath samples from him, analyzing them, and then delivering the reinforcement or announcing its absence. This procedure, however, may have practical application as an intermittent follow-up procedure—assessing BAL in the natural environment.

Also within the operant paradigm, Wilson *et al.* (1975) recently reported that drinking by alcoholics can be effectively suppressed, as expected, by strong contingent electric shock. When a painful experimenter-administered shock routinely followed consumption of a one-ounce drink of beverage alcohol, a substantial decrease in drinking from baseline (nonshock) levels was observed; when contingent shock was subsequently withdrawn, subjects returned to much higher alcohol consumption levels. Of greater interest are results from a second phase of the same study. During this phase, shock was *self-administered* rather than *experimenter-administered*. Over the course of nine treatment days, the self-administered punishment schedule was "thinned" from continuous shock to a light variable ratio schedule (FR 1 to VR 2 to VR 4 to VR 20). The results of this punishment procedure are summarized in Figure 1. It shows that drinking by *S* #2 was completely suppressed during and after the contingent self-

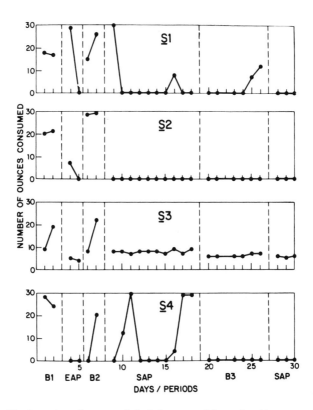

FIGURE 1. Total number of ounces of alcohol consumed by each subject across baseline (B) drinking, experimenter-administered (EAP), and self-administered punishment (SAP) conditions (Wilson *et al.*, 1975, p. 20).

shock sequence, while S #3 maintained a controlled pattern of consumption during the same period. These findings, admittedly preliminary, suggest that self-punishment, like the other methods of externally-imposed contingent control reviewed here, may effectively modify drinking by some alcoholics. We believe that the self-punishment model of aversive control warrants intensive additional study.

Modifying the Drinking Response: Blood Alcohol Level Discrimination Training

In 1970, Australians Lovibond and Caddy described the first research program designed specifically to investigate the therapeutic potential of blood alcohol level discrimination training. Their program contained two elements, BAL discrimination training, followed by BAL control training. During BAL discrimination training, their 31 alcoholic subjects drank ethanol in fruit juice to a BAL of 80 mg/% and then observed their subjective—visceral reactions to the changing intoxication levels. For 2 hours following drinking, subjects estimated their own BALs every 15–20 minutes, receiving accurate feedback on actual BAL after every estimation. Each subject received one such 2-hour training session prior to the next, control conditioning, phase of the study. During this phase, subjects were required to drink to a moderate level of intoxication (65 mg/%) within 1.5 hours of its start. Subjects also continued to estimate BAL, for which they continued to receive feedback on accuracy. On reaching, then exceeding, the target BAL of 65 mg/%, subjects began to receive painful shocks via chin electrodes. Thirteen control subjects received random, noncontingent shocks (rather than contingent ones) both before and after the target BAL had been reached.

Twenty-one of 28 experimental subjects who completed this treatment sequence were "regarded tentatively as complete successes in that they were drinking in a controlled fashion, exceeding 70 mg/% BAL only rarely" four or more months following treatment. These encouraging results must be regarded with some caution, however, since the authors had to rely, at least in part, upon subjects' self-reports of treatment effectiveness, a source of follow-up data many consider to be unreliable. Further, the absence of explicit efforts to assess BAL estimation accuracy once the two-hour discrimination training session ended makes it impossible to accept the authors' assertion that the "complete success" they observed stemmed largely from retention of this skill during the control conditioning and follow-up periods.

In an effort to retest and extend the BAL discrimination-controlled drinking treatment paradigm, Silverstein et al. (1974) then undertook an experimental analysis of what they considered to be the central components of BAL discrimination training with four alcoholic subjects. The first of two phases of the study comprised a 10-day *BAL discrimination training* period which included 2-day pre- and posttraining baseline periods. Discrimination training itself—which lasted 6 days—was designed to provide subjects with feedback on the accuracy of their BAL estimations; feedback was accompanied by social and token reinforcement for increases in estimation accuracy.

Data from this phase of the study (graphed in Figure 2) revealed that subjects could not estimate BAL accurately before discrimination training: Their pretraining estimates averaged 112 mg/% from actual BAL during the 2-day pretraining baseline period. Estimation accuracy by all four subjects improved dramatically (to an ultimate 14 mg/% mean discrepancy) at or slightly after the introduction of BAL feedback. The addition of social and token reinforcement for accuracy resulted, surprisingly, in no additional gain in accuracy. Accuracy began to deteriorate, however, when the posttraining baseline period (which no longer provided for feedback) was reintroduced. Estimation accuracy deteriorated further (to 53 mg/% mean discrepancy) when subjects moved from the programmed drinking of this first phase of the study to ad lib drinking during the first baseline period of the study's second phase, *control training*. During this phase, which lasted another 12 days, three of the original subjects were trained to maintain their BALs within a delimited range. Techniques for

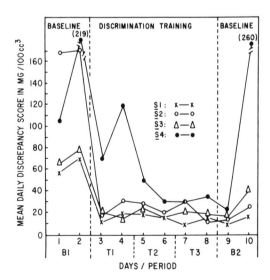

FIGURE 2. Accuracy of BAL estimation during Phase I discrimination training. All four subjects improved their estimation accuracy when feedback on BAL was introduced at the beginning of T1 (Silverstein *et al.*, 1974, p. 8).

bringing their drinking behavior under control included the following: (1) Control over time of and upper limits on drinking was gradually shifted from experimenter to subject; (2) the range of positively reinforced BALs was gradually narrowed from 30–130 mg/% to 70–90 mg/%; (3) reinforcement for controlled drinking and feedback on accuracy were gradually faded out. Figure 3 shows that subjects learned to maintain their BALs within the prescribed target range but failed to maintain this control when external BAL feedback was withdrawn during the second baseline period of this phase of the study.

This study suggests that alcoholics can learn to estimate BAL accurately and can be helped to achieve control over their drinking within narrow limits when they are provided occasional external feedback on the accuracy of their estimates. It also indicates that the four alcoholics studied could not estimate BAL from internal cues, contrary to Lovibond and Caddy's earlier findings.

In fact, there was reason to ask at the end of this study whether *any* person—alcoholic or nonalcoholic—could be taught to discriminate BAL accurately on the basis of internal–visceral cues. This open question derived from the fact that the studies just reviewed, in common with others in the same research area (Bois and Vogel-Sprott, 1974; Paredes *et al.*, 1974a,b; Vogel-Sprott, 1975), failed to control for the possible influence on discrimination accuracy of all-important exteroceptive and/or proprioceptives cues (e.g., prior exposure to sample BAL curves, awareness of strength of drink, knowledge of elapsed time since last drink, and BAL feedback after previous drink). For this reason, Huber *et al.* (1975) designed a study controlling these variables that would evaluate the differential effectiveness of BAL discrimination training focused on internal cues, on external cues, and on a combination of the two. All 36 subjects of the study (social drinkers) gave no-feedback BAL estimates during an initial baseline session. Subjects were then matched on the basis of the accuracy of their pretraining BAL discriminations and assigned to one of the separate three training groups. Each group of 12 subjects then received discrimination training during a subsequent training session according to the following plan: (1) *Internal* cue subjects were trained via checklists and self-report instruments to associate specific internal sensations and feelings with specific BALs; (2) *external* cue subjects were trained with a programmed learning booklet to calculate their BALs on the basis of a drink count, recognition of the passage of time, and knowledge of their own alcohol metabolism constants; (3) *internal plus external* cue subjects received both kinds of training. All subjects received feedback following their BAL estimates during Session 2. They were then retested for BAL estimation accuracy during the third and final session. All subjects gargled with an anesthetic mouth wash during Sessions 2 and 3 to ensure their inability to discriminate the alcoholic content of their drinks on the basis of taste and olfactory cues.

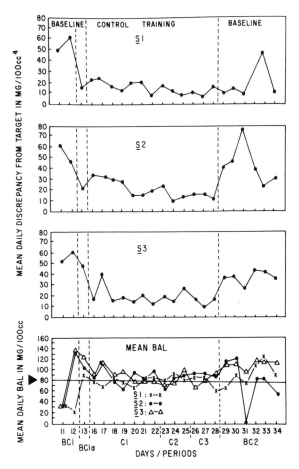

FIGURE 3. Discrepancy from target BAL during control training. All three subjects demonstrated controlled drinking by the end of C1. It was maintained until reinforcement for controlled drinking was terminated at the beginning of BC2 (Silverstein *et al.*, 1974, p. 11).

Analysis of the results of the study revealed that, *unlike* prior *alcoholic subjects*, (1) these nonalcoholic subjects did learn to estimate BAL solely on the basis of internal cues, (2) all three sets of training procedures increased discrimination accuracy by approximately 50%, and (3) the three training groups did not differ in ultimate BAL estimation accuracy. The lasting value of this study may ultimately derive from the development it permitted of effective methods for training accurate BAL discrimination. What is needed now are rational criteria for judging which BAL estimation methods are most appropriate for which kinds of subjects. The answers to this question, still open, are important because BAL estimation from internal cues almost certainly has greater ultimate therapeutic potential than discrimination from external cues. What remains to be established, however, is whether alcoholics can discriminate BAL from internal cues, like the nonalcoholics in the study of Huber *et al.* (1975). Or has the phenomenon of tolerance guaranteed that no alcoholic can learn to associate a particular blood alcohol level with a particular set of internal, subjective cues to intoxication?

Modifying Associated Behavioral Problems: Systematic Desensitization

Systematic desensitization, considered by many behavior therapists to be the treatment of choice for the neuroses, has only rarely been utilized as the sole ingredient of a behavioral prescription for alcoholism. Only Kraft and Al-Issa (1967) and Kraft (1969), in fact, have reported using systematic desensitization by itself for this purpose. Both clinical reports describe the successful use of systematic desensitization to reduce the anxiety which interfered with the individual social functioning of eight young alcoholic subjects—anxiety severe enough to have played an apparent causal role in the development of their alcoholism. Of the eight alcoholic subjects treated by systematic desensitization alone, all were thought to have improved their social functioning and reduced their alcohol consumption by the end of treatment. Extensive follow-up data on these patients, however, was not reported.

Modifying the Drinking Response and Associated Behavioral Problems: Broad-Spectrum and Multifaceted Therapies

In 1965 Lazarus, then and now a major figure in the behavior therapies, proposed a "broad-spectrum" behavioral approach to alcoholism that combined systematic desensitization with a variety of other behavioral techniques designed to modify both the alcoholic's excessive drinking and behavioral problems associated with the drinking. Among the components of Lazarus's broad-spectrum behavior therapy for alcoholism were medical care to treat alcohol-related physical disabilities, aversion conditioning to directly modify the patient's drinking behavior, tests and interviews to identify "specific stimulus antecedents of anxiety" in order to enable construction of anxiety hierarchies for systematic desensitization, assertive training, behavioral rehearsal, and hypnosis to "counter-condition anxiety-response habits," and development of a relationship with the patient's spouse, to help her see and alter her role in the patient's alcoholism. This early commitment to a multifaceted behavioral approach to the alcoholic's associated behavioral problems as well as his excessive drinking itself foreshadowed subsequent developments both in behavioral approaches to alcoholism and in Lazarus's own approach to innovation in behavior therapy. Virtually all current efforts to treat alcoholism are multifaceted—a multitechnique intervention designed to confront the full range of the individual's problems having identified or assumed functional relationships to his excessive drinking.

A recent example of this approach to individual behavioral treatment for alcoholism is contained in a report by Wilson and Rosen (1975) on the successful treatment of a 30-year-old male alcoholic. Deciding early in treatment that the most appropriate treatment goal for this patient was controlled drinking, Wilson and Rosen proceeded to employ blood alcohol level discrimination training, aversion conditioning, behavioral rehearsal, assertive training, self-monitoring of alcohol consumption, and contingency contracting to attenuate the patient's drinking and to improve his ability to deal with stress in familial and social settings. At a six-month follow-up, these researchers report that the patient had been able successfully to maintain a stable pattern of controlled drinking.

Of more relevance to the readers of this chapter, in all likelihood, would be detailed descriptions of large-scale demonstrations of the efficacy of multifaceted behavioral approaches to alcoholism and its associated personal and interpersonal consequences. Miller *et al.* (1974) describe such a program in force on a 15-bed alcoholism ward at the Veterans Administration Center, Jackson, Mississippi. Miller and his co-workers present a coherent behavioral rationale for their choice of the following three-fold behavioral approach to alcoholism treatment: (1) techniques "to decrease the immediate reinforcing properties of alcohol" (including electrical aversion, chemical aversion, and covert sensitization); (2) development of "new ways of coping with life . . . incompatible with alcohol abuse" (involving instruction in more appropriate ways to cope with stressful situations); (3) alteration of his social and vocational environment so that the patient "receives increased satisfactions from life" (these include family therapy and vocational counseling). Pomerleau and Brady (1975a) describe a similarly multifaceted treatment package, this one for use with outpatient alcoholics who have sufficient motivation to change their drinking habits to invest $300 of their own money in treatment. A small portion of that sum is returned each treatment session, contingent upon the patient's cooperation with a multifaceted treatment package that includes self-monitoring of alcohol intake, contingency management techniques, explicit efforts to shape more adaptive drinking behavior, and identification of and practice in interpersonal behaviors inimical to uncontrolled drinking. Six of seven patients treated in a pilot exploration of this treatment approach reduced their drinking following involvement in the program; four of seven maintained this gain at the six-month mark. Data from a more lengthy follow-up of substantially more patients, currently under way (Pomerleau and Brady, 1975b), are required, however, before this promising behavioral treatment package can be recommended for wider distribution.

One of the most ambitious applications of a multifaceted behavior modification program to alcoholism was that of M. B. Sobell, L. C. Sobell, Schaefer, and Mills at the Patton State Hospital in California. The separate components of this program, as outlined in reports by Mills *et al.* (1971) and Sobell and Sobell (1973a), were embedded within a 17-session treatment plan which "dealt directly with the inappropriate behavior of excessive drinking and emphasized a patient's learning alternative, more appropriate responses to stimulus conditions which previously functioned as setting events for his heavy drinking." (Sobell and Sobell, 1973a, p. 53). Individually tailored to take into account each patient's prior treatment history, the specifics of his drinking problem, and the treatment goals established for him, the treatment program at Patton State Hospital also varied as a function of whether the patient's treatment goal was abstinence or controlled drinking. Either treatment goal could be accommodated within the following general treatment plan:

Sessions 1 and 2 (Videotaping). Subjects drank to drunkenness. While in that state, they engaged in intensive discussions with the staff about the roots of their alcoholism, their behavior during intoxication, and their apprehensions about the treatment program upon which they had embarked. Videotapes of these sessions were made.

Session 3 (Treatment Plan). The treatment plan was explained in detail to all subjects. At the same time, controlled drinking subjects were trained to recognize and identify the separate components of mixed drinks, a skill in which most alcoholics are deficient.

Session 4 and 5 (Videotape Replay). The videotapes of drunken behavior previously recorded were replayed, both to demonstrate to subjects how inappropriate and maladaptive their behavior often is during such times and to help increase motivation for treatment as prior research had suggested it might.

Session 6 (Failure Experience). Subjects were required to complete a series of tasks 20 minutes before the session began that were, in fact, impossible to complete. The therapy session which followed focused then on subjects' maladaptive responses to this stress as well as prior stresses as this general category of responses was likely a part of their chain of alcoholism.

Sessions 7 through 16 (Stimulus Control). During portions of each of these 10 sessions, controlled drinking subjects were shocked whenever they behaved like uncontrolled drinkers (as defined by Schaefer *et al.*, 1971) and reinforced with alcohol (up to a predetermined limit) when they drank like social drinkers; subjects whose treatment goal was abstinence were shocked whenever they drank at all. During other portions of the same sessions, treatment subjects were helped to identify crucial stimulus variables (stressors) associated with their individual decisions to drink; they were also aided in evolving a set of potentially effective responses to cope better with those stressful situations in the future. Modeling and role-playing these alternate coping responses also took place.

Session 17 (Summary; Videotape Contrast). Selected replays of drunken behavior from Sessions 1 and 2 were contrasted with videotapes of sober behavior taken during Session 16. Progress during therapy was discussed and each subject was given a wallet-sized list of *dos* and *don'ts* specific to his own drinking history and response to treatment. Discharge from hospital usually followed completion of the program by no more than two weeks.

In an extensive formal trial of this multifaceted behavioral treatment program (Sobell and Sobell, 1973a,b, 1976), assessment of the drinking, social, personal, and vocational behavior of 70 male gamma alcoholic patients who had completed the full treatment regimen took place at six-week, six-month, one-year, and two-year follow-up intervals. The 70 patients had been assigned to one of two treatment groups before treatment began on the basis of prior treatment history; 40 patients were assigned to a group whose treatment goal was controlled drinking, 30 patients were assigned to a group whose goal was abstinence. These two groups were each further divided into experimental and control groups to yield a study with four groups in all: Controlled Drinker, Experimental (CD-E); Controlled Drinker, Control (CD-C); Nondrinker, Experimental (ND-E); Nondrinker, Control (ND-C).

Follow-up data at six weeks and six months, summarized in Table 3, show that subjects in all four groups were more likely to be abstinent or drinking in a controlled fashion at six weeks than at six months, that patients in the two experimental treatment groups (CD-E, ND-E) hewed closer to their respective treatment goals (controlled drinking, abstinence) than did the control groups, and that subjects in the treatment groups also maintained significantly better "drinking dispositions" during both follow-up intervals (that is, they were more often abstinent or drinking in controlled fashion) than men in the two control groups. Evaluation of the vocational status and general adjustment of the same 70 subjects revealed that members of the experimental treatment groups were also functioning better in these areas than control subjects.

Eighteen-month and two-year follow-up data on the same subjects are shown in Table 4. These data confirm observations made from six-week and six-month data showing that experimental treatment subjects (CD-E, ND-E) had remained abstinent or controlled in their drinking a significantly greater percentage of time than control subjects. The same table also shows that treatment subjects in the controlled drinker group (CD-E) spent fewer drunk days, days in hospital, or days in jail and more abstinent or controlled drinking days

TABLE 3. Mean Percentage of Days Spent in Different Drinking Dispositions by Subjects in Four Experimental Groups for 6-Week and 6-Month Follow-Up Intervals

Drinking disposition	Experimental condition[a]			
	CD-E	CD-C	ND-E	ND-C
Six-week follow-up				
Controlled drinking	41.80	10.70	7.20	12.93
Abstinent, not incarcerated	30.95	39.32	60.33	42.13
Drunk	17.55	42.70	23.20	41.60
Incarcerated, alcohol-related				
Hospital	9.15	2.00	6.94	3.20
Jail	0.55	5.35	2.33	0.44
Total	100.00	100.00	100.00	100.00
Six-month follow-up				
Controlled drinking	27.33	9.10	2.87	14.54
Abstinent, not incarcerated	37.89	29.40	62.63	16.55
Drunk	20.33	50.50	19.38	40.91
Incarcerated, alcohol-related				
Hospital	12.12	4.10	12.25	8.09
Jail	2.33	6.90	2.87	19.91
Total	100.00	100.00	100.00	100.00

[a] Experimental conditions were controlled drinker, experimental (CD-E), controlled drinker, control (CD-C), nondrinker, experimental (ND-E), and nondrinker, control (ND-C) (Sobell and Sobell, 1973a, p. 65).

than treatment subjects in the abstinence group (ND-E). In other words, behavioral treatment for controlled drinking appears to have been more efficacious than behavioral treatment for abstinence.

Because these outcome data are more favorable than those reported from virtually any other alcoholism therapy project, they are correspondingly more encouraging. They are encouraging because they suggest that implementation of a multifaceted, rationally derived, behavioral treatment program with a large group of alcoholics, combined with lengthy, thorough follow-up, yields experimental treatment subjects who drink less and drink less often than control subjects. Further—and as important—these results also encourage the view that controlled drinking may well be a viable treatment goal for some alcoholics.

Unfortunately, the Sobells' data also give pause because important design and data analysis problems prevent their unequivocal interpretation. Among these problems are the following:

1. The experimental treatment procedures effected changes in drinking behavior on follow-up and not during treatment. Why? Specifically, why did

TABLE 4. Mean Percentage of Days Spent in Different Drinking Dispositions by Subjects in Four Experimental Groups Displayed Separately for the Third and Fourth 6-Month (183 Day) Follow-Up Intervals

Drinking disposition	Experimental condition[a]			
	CD-E	CD-C	ND-E[b]	ND-C[c]
Follow-up months 13–18				
Controlled drinking	21.58	4.03	2.89	1.13
Abstinent, not incarcerated	61.56	36.67	59.40	39.35
Drunk	12.81	51.77	19.67	36.85
Incarcerated, alcohol-related				
Hospital	2.71	2.59	6.48[d]	11.59
Jail	1.34	4.94	11.56	11.08
Total	100.00	100.00	100.00	100.00
Follow-up months 19–24				
Controlled drinking	23.55	7.59	4.41	1.99
Abstinent, not incarcerated	63.64	36.24	61.59	43.99
Drunk	11.72	46.73	21.08	38.09
Incarcerated, alcohol-related				
Hospital	0.46	2.39	5.78	6.05
Jail	0.63	7.05	7.14	9.88
Total	100.00	100.00	100.00	100.00
Follow-up year 2 (months 13–24)				
Controlled drinking	22.57	5.81	3.65	1.56
Abstinent, not incarcerated	62.60	36.46	60.50	41.67
Drunk	12.27	49.25	20.37	37.47
Incarcerated, alcohol-related				
Hospital	1.58	2.49	6.13	8.82
Jail	0.98	5.99	9.35	10.48
Total	100.00	100.00	100.00	100.00

[a] Experimental conditions were controlled drinker experimental (CD-E), $N = 20$; controlled drinker control (CD-C), $N = 19$; nondrinker experimental (ND-E), $N = 14$; and, nondrinker control (ND-C), $N = 14$.
[b] Does not include data for one ND-E subject who died of a subdural hematoma at the start of his second year follow-up interval.
[c] Does not include data for one ND-C subject who died of drug-related causes about 8 weeks after hospital discharge.
[d] Nondrinker group incarceration data tend to reflect extreme scores of a few individual subjects (Sobell and Sobell, 1976, p. 200).

subjects drink to excess during nonshock "probe" test periods introduced during Sessions 8, 12, and 16 to test for treatment efficacy as treatment was taking place, only to modify their drinking patterns to embrace abstinence or controlled drinking on completion of the treatment program?

2. What separate component(s) of the treatment program, of the many that were used, was (were) responsible for the treatment gains reported? Since

the Sobells' "individualized behavior therapy" for alcoholics drew on a variety of treatment methods, each designed to focus on a different component of the maladaptive behavior of its alcoholic subjects, it is important that the most active of these methods (e.g., aversive conditioning, training in social skills, videotape replay) be identified. Likewise, it would seem most useful to assess the extent to which the lengthy, intensive follow-up, which required almost weekly contact between a single follow-up worker and the study's 70 subjects, contributed to the positive results that were reported. Specifically, one wants to know whether this follow-up procedure comprised "booster" therapy, intensive record-keeping, both, or neither.

3. Follow-up data were generated from self-reports of drinking behavior, a species of information many investigators consider to be unreliable. Recently, however, the Sobells have reported interesting data questioning the widespread acceptance of the invalidity of alcoholics' self-reports (Sobell and Sobell, 1975; Sobell *et al.*, 1974). As a result, the question probably ought to remain an open one.

4. All follow-up data were gathered by a single person who was, in fact, one of the two principal investigators. While intentional bias on this person's part can, of course, be ruled out, unintentional bias cannot be. Fortunately, this problem is in the process of being dealt with. Two independent teams of investigators are now following up the 70 patients on whom follow-up data has been reported by the Sobells in an effort to provide independent validation for their findings. Until then, of course, the specifics of these data must be considered promising but tentative.

5. The format within which the Sobells reported follow-up data—"drinking dispositions" expressed as percentage of controlled-drinking, abstinent, drunk, and incarcerated days within successive six-month follow-up periods—makes excellent sense. Follow-up data cast in this way probably represent a truer picture of drinking behavior over a circumscribed follow-up period than data from a patient's post hoc assessment of his behavior during a similar time period. On the other hand, because this format is a new one, previous outcome data cannot be compared with these data. As a result, it is impossible to know whether the Sobells' patients actually did better posttreatment than pretreatment, just as it is impossible to compare the Sobells' success rate with that achieved by other researchers using other outcome indices.

A recent study by Hedberg and Campbell (1974) addressed itself to the relative efficacy of different behavioral methods for the treatment of alcoholism, an important and necessary task the design of the Sobells' study did not permit. Of the four separate methods investigated by Hedberg and Campbell (behavioral family counseling, electrical aversion, covert sensitization, and systematic desensitization), two (family counseling and electrical aversion) were similar to components of the Sobell and Sobell treatment program. After randomly assigning 49 alcoholics to one of four treatment groups and treating members of each group on an outpatient basis for a year, Hedberg and Campbell reported: Behavioral family counseling was most effective in modifying alcohol consumption; systemic desensitization, while effective, was less effective than behavioral counseling; covert sensitization and electrical aversion were, respectively, relatively and completely ineffective in this regard. The design of this study did not permit assessment of the *interactive* effects of multifaceted alcoholism treatment program; it nonetheless represents a step in the direction of enabling necessary and important assessment of the separate components of such behavioral programs.

Another recent study, this one by Vogler *et al.* (1975) compared the efficacy of two broad-spectrum treatment packages that differed markedly in their composition. Forty-two chronic alcoholic patients completed the inpatient treatment portion of the study and a subsequent one-year follow-up. The 23 patients in Group 1 received comprehensive behavioral treatment while they were inpatients; this treatment included videotaped recording and replay of intoxicated behavior, alcohol education, blood alcohol level discrimination training, aversion training both for overconsumption (drinking above a BAL of 50 mg/%) and for "alcoholic drinking" (gulping instead of sipping, drinking straight liquor instead of mixed drinks, etc.), and behavior counseling. The 19 patients in Group 2 received only alcohol education and behavior counseling along with the standard hospital milieu treatment both groups received. Both groups also received booster treatment sessions over the one-year follow-up period once a week for 4 weeks, then once a month for 11 months.

Follow-up data revealed that the two treatment groups did not differ in rates of abstinence or controlled drinking: Sixty-two percent were either abstinent or drinking in a controlled fashion at the one-year mark. This rate of treatment success is clearly better than that achieved by most other nonbehavioral clinicians working with comparable groups of chronic alcoholic males. In like fashion, the two groups did not differ in degree of change in four of five specific drinking and drinking-related behaviors after treatment; both groups achieved positive changes in preferred beverage consumed, drinking companions chosen, drinking environment, and number of days per month lost from work because of drinking. The one behavior which did differentiate the two groups was pre–post alcohol intake: Group 1 decreased from 17.2 to 3.5 gallons a year of absolute alcohol while Group 2 decreased from 11.8 to 6.9 gallons (the national average is 2.6 gallons per adult per year). An a posteriori comparison made between posttreatment consumption residual scores revealed a significant difference between the two groups beyond the 0.01 level.

One explanation for these differences in treatment outcome relates simply to time in treatment. Because Group 1 received both *more* as well as *different* treatment, its favorable treatment outcome could have stemmed entirely from this fact. However, a correlational analysis of treatment outcomes and time in treatment failed to reveal a single significant correlation, effectively removing this argument from further consideration.

The results of this study merit serious discussion for the following reasons: (1) Some of the subjects in both groups were able to maintain a pattern of controlled drinking throughout the follow-up year; (2) many subjects in both groups did better than expected through the year in terms of a variety of alcohol-related behaviors; and (3) only one index of treatment outcome, amount of alcohol consumed, differentiated the two treatment groups. These results suggest, then, that continued research on behavioral treatment approaches to controlled drinking be encouraged. They also point to the undoubted importance that booster treatment sessions possess in determining successful treatment outcome; it may be that both treatment groups did well during follow-up largely because booster sessions were programmed for both. The Sobells reached similar conclusions about the therapeutic value of intensive follow-up in their treatment research reports (1973b, 1976). Finally, the results of this study reemphasize that the manner in which outcome data are cast in alcoholism treatment studies profoundly affects their conclusions. To this end, if Vogler and his colleagues had adopted either the usual criterion of change employed in alcoholism treatment studies—rate of abstinence—or even the Sobells' more sophisticated index of change—drinking disposition—they would have been unable to make the valid outcome distinction between their treatment groups they did make on the basis of alcohol consumption.

Modifying the Natural Environment: Community-Reinforcement Counseling and Contingency Management

Alcoholics, like almost everyone else, live in the midst of persons who are important to them—spouses, parents, children, employers, and friends. Reciprocal relationships are maintained between the alcoholic and other individuals within his social–interpersonal environment; the alcoholic influences and is influenced by this matrix of relationships. In recognition of this uniquely human phenomenon, behavioral researchers have begun to trace the impact of reciprocal influence processes on the development and maintenance of alcohol abuse. For example, Goldman et al. (1973) contrasted the effects on drinking by four alcoholics of group and individual decision-making (about whether or not to drink) while Hersen et al. (1973) examined functional relationships between the verbal and nonverbal interactions of an alcoholic and his wife. Miller and Hersen (1975), in a more applied project, observed that the modification of disordered relationship patterns between an alcoholic and his wife resulted in a beneficial impact on the husband's drinking. The increasing emphasis behavior researchers and therapists have begun to place on helping alcoholics develop more adaptive responses to stressful interpersonal situations reflects their growing realization of the intimate relationship between interpersonal competence and alcohol abuse. The Sobells and Miller and his colleagues pioneered this new thrust. Most recently, these efforts have led to attempts to extend behavioral interventions to "alcohol abuse in the natural environment," as against developmental-analog investigations within laboratory or hospital settings.

Hunt and Azrin's (1973) "community-reinforcement" approach to alcoholism represents one of the first efforts at intervention on this level. In preparation for the study, 16 alcoholics hospitalized in an Illinois State mental hospital were divided into two matched groups on the basis of marital status, age, education, vocational status, and number of previous hospitalizations. A control group of 8 patients received standard hospital milieu therapy along with counseling on the health hazards and interpersonal problems associated with continued drinking and detailed information about local Alcoholics Anonymous groups. An experimental group of 8 patients received "community-reinforcement counseling" along with the standard hospital treatment. Community-reinforcement counseling actually involved direct modification of the alcoholics' interpersonal and environmental support systems. To this end, an experienced behavioral clinician helped each of the patients in this group to find employment, improve family and marital relations, enhance social skills, and structure reinforcing social activities. Although precisely how the clinician undertook these rehabilitative efforts is not spelled out in Hunt and Azrin's report, his or her labors must have been arduous indeed, in view of the limited resources these patients must have brought to the treatment situation.

Since earlier research has amply demonstrated that alcoholics with social, family, and vocational ties have better prognoses for successful rehabilitation than alcoholics without such environmental supports, the component of community-reinforcement counseling in the Hunt and Azrin study clearly heightened prospects for successful treatment outcome. Beyond this, however, the experimental treatment program specifically built upon these newly developed "natural reinforcers" (e.g., job, wife's sustained attention, effective social relationships) by ensuring that their continued availability depended upon continued sobriety. This contingency arrangement was coordinated during weekly visits by a counselor for the first month following hospital discharge; subsequent visits continued less frequently during the remainder of a six-month follow-up period.

At the six-month follow-up, it was found that the community-reinforcement group as a whole had spent significantly less time drinking, unemployed, away from home, or institutionalized than the control group. These data encourage the view that direct environmental manipulation, combined with contingent access to environmental reinforcers like those employed in this study, may be of value for the treatment of chronic alcoholics. Despite the promising nature of these data, however, two cautions should be observed. The first of these stems from the apparent costs of this program. Specifically, it seems clear that few communities could afford the expense of the continuing individual supervision and counseling this program provided on a community-wide basis. Second, as noted above, a clear picture of the therapeutic utility of the unique "active ingredient"—contingent access to natural reinforcers—employed in this study is not possible because of its design. That is, before treatment began, each experimental patient received intensive help in finding a job, improving social skills, and increasing marital communication. Since control patients did not receive this help, a comparison between their posttreatment drinking behavior and the more moderate drinking shown by members of the experimental group does not give an unencumbered view of the specific therapeutic activity of contingent access to these natural reinforcers.

Complementing this ambitious study, several recent case studies suggest that management of behavioral contingencies by "significant others" (rather than by professionals, as above) can reliably alter alcoholic drinking in the natural environment. Differential social reinforcement from peers (Sulzer, 1965) and wives (Cheek *et al.*, 1971) aided in the successful management of alcoholic drinking. Likewise, contingency contracts between an alcoholic and his wife served to establish and maintain—at a six-month follow-up—a stable pattern of controlled drinking (Miller, 1972).

Contingency management extending from the laboratory to the natural environment continues to be a prime focus of research by Bigelow, Liebson, Griffiths, and their co-workers at the Baltimore City Hospitals. This research direction received initial impetus from therapeutic success with a heroin addict who was simultaneously abusing alcohol (Liebson and Bigelow, 1972). Treatment of this patient proceeded by making his maintenance on methadone (a potent reinforcer) contingent upon ingestion of disulfiram (Antabuse). Regular ingestion of Antabuse, in turn, prevented the patient from drinking since alcohol–disulfiram reactions are extremely aversive. Encouraged by their initial success in using methadone as an environmental reinforcer to moderate alcoholic drinking, these researchers then treated nine more multiply addicted men precisely the same way (Liebson *et al.*, 1973). Again, treatment was successful—with drinking taking place on 1.4% of patient days when methadone maintenance was contingent on sobriety, but on 19.2% of days it was not.

A similar treatment program (Bigelow, Liebson, and Lawrence, 1973) involved four hospital employees who had been referred for treatment because they were in danger of being fired for "work decrements related to excessive drinking." These patients were given the opportunity to work contingent upon regular Antabuse ingestion. The four employees were required to report to the hospital's alcoholism clinic for their Antabuse each day before work. Failure to report resulted in no work and no pay. At the time the 1973 report was written, the four had accumulated over 1000 treatment days; "*none* of the patients has *ever failed* to satisfy the contingency" during variable follow-up periods. A tenfold reduction in number of sick days and unexcused absences from work was also reported. The three investigators describe another outpatient contingency program in the same report. Seven alcoholics contracted with the therapist to make cash security deposits to ensure their continued attendance at an alcoholism clinic to receive Antabuse. The security deposit could be earned back bit by bit by regular clinic attendance; failure to report on an occasion resulted in the loss of a small portion of the deposit to charity. At the time of the report, the seven patients in the program had accumulated 835 days under contract. Six of the seven had achieved the contract-specified goal, sobriety. Overall, drinking had taken place on only 2.6% of treatment days, a figure which represents the single patient who failed to achieve abstinence. At the conclusion of an initial contract, four of the seven patients then agreed to contract for additional periods of sobriety, despite the "heavy response requirements involved in participation in this kind of program" and the sizable financial commitment involved in the initial security deposit.

Miller and colleagues have also begun to explore the utility of contingency contracting in both analog and actual "natural environments." The first of two papers reporting on their work in this area contrasted the effects on analog drinking by inpatient alcoholics of four different kinds of behavioral contracts (Miller, Hersen, and Eisler, 1974). Treatment effects were assessed by comparing drinking before and after the contracts were introduced; drinking was reflected via an operant analog drinking task in which lever pressing was reinforced with alcohol on a fixed-ratio (FR50) reinforcement schedule.

The study began with an initial 10-minute practice trial at the operant console, followed by four daily 10-minute operant sessions. Operant drinking behavior during the latter four sessions constituted the baseline (preintervention) measure of consumption. At this point, a "treatment goal" of approximately half the pretreatment consumption rate was set. To help attain this goal, one of the following four behavioral contracts was given each of the study's 40 subjects immediately after the end of the baseline drinking sessions: (1) Verbal instructions: Subjects were given standard instructions to limit their drinking to a predetermined number of "shots" of beverage alcohol; limits were set at about half baseline drinking levels; (2) Written agreement: Subjects received precisely the same standard instructions but in written form. The instructions were signed by both subject and psychologist, with a copy retained by both parties; (3) Verbal instructions plus reinforcement: To the same verbal instructions was added the information that successful completion of a 10-minute operant session (that is, drinking no more than the predetermined limit) would result in reinforcement of $1 while an unsuccessful session (drinking beyond the limit) would result in loss of $2; (4) Written agreement plus reinforcement: The same reinforcement information was written down, together with written, signed instructions.

Following the establishment of these four contracts, four postintervention operant drinking sessions were administered. Verbal instructions alone resulted in 2 of 10 subjects achieving their treatment goals while written agreement resulted in 4 of 10 subjects doing so. By contrast, all but 1 of 20 subjects in Groups 3 and 4 achieved the same goal. A chi-square analysis revealed a significant relationship between the experimental conditions and goal attainment.

Miller (1975), encouraged by these findings that behavioral contracts "empowered" by reinforcement and punishment contingencies can modify alcoholic drinking (albeit in an analog setting), designed a behavioral intervention program along similar lines for alcoholic outpatients. Selecting a most difficult target population, skid row alcoholics who had repeatedly and unsuccessfully been institutionalized for public drunkenness, Miller chose ten such men for admission to an outpatient behavioral intervention program and ten more for allocation to a control group receiving routine services from the city's helping agencies. The groups were matched on a variety of relevant variables.

For two months before behavioral intervention began and two months after it was in force, data on number of public drunkenness arrests and number of hours worked per week for all 20 subjects were gathered. Behavioral intervention provided the 10 subjects in the intervention group who maintained blood alcohol levels at or below 10 mg/% with an array of goods and services including meals, jobs, clothing, cigarettes, counseling services, elective medical care, and shelter. Once every five days, these subjects were tested for blood alcohol level by breath sample. They were also observed at random times through successive five-day periods for gross signs of intoxication. A blood alcohol reading above 10 mg/% or observation of the subject in a grossly intoxicated condition resulted in immediate cessation of access to the contingent reinforcers for five days. At the end of that five-day period, the subject was again given access to all reinforcers *if* he met the same drinking criteria. Behavioral intervention lasted for two months.

Mean number of arrests and mean number of hours employed per week for subjects in the behavioral intervention group changed once the intervention commenced. Number of arrests decreased significantly ($p < 0.025$) while number of hours worked per week increased significantly ($p < 0.05$) during invention. No change in the same behaviors was observed for the control subjects. Analyses of random blood alcohol readings revealed comparable changes, with BALs during contingent reinforcement significantly lower ($p < 0.001$) than during baseline for experimental treatment subjects. No comparable change was observed for the control subjects.

Though Miller points out that these promising pilot data may well have been functions of nonspecific as well as specific treatment variables (differences in "time in treatment" and BAL feedback given experimental and control subjects), he also admits to encouragement with the positive nature of his findings. We are similarly encouraged by his results, though we believe that an additional posttreatment follow-up of patients in both experimental and control groups would have enabled important assessment of the long-term effectiveness of this new intervention package. Overall, the results of the work reviewed in this section are noteworthy for the following reasons: (1) Behavior modification in and of the natural environment should permit large numbers of alcoholics to be treated simultaneously by more or less standard procedures while they remain in their home environments; (2) reliable follow-up of treatment should be enhanced since patients report to a treatment facility on a regular basis for contingency management; (3) treatment has considerable flexibility—initial behavioral contingencies can be left in force permanently or if appropriate, faded out and then reinstated if necessary; (4) the desirable participation of third parties in treatment (e.g., an employer or a spouse monitoring contingent behavior) is facilitated by the systematic nature of the treatment program.

It should be clear, however, that fundamental to the success of programs like these is the availability of natural reinforcers in the environment. That is, the alcoholic must have, for example, gainful employment, money in the bank, and an interested spouse before any or all of these reinforcers can be employed on a contingent basis for management of his or her problem drinking. For this reason, effective behavioral intervention on the community level, as reported by Hunt and Azrin, may well involve *first* helping the alcoholic secure the incentives and support that are necessary to provide the therapeutic leverage which can enable ultimate modification of his problem drinking.

Behavioral Treatment in Perspective

Does the mass of research data reviewed above enable us to reach state-of-the-art conclusions about the relative merits of the multitude of developing behavioral approaches to alcoholism? We think that it does, and that the following general statements can be made regarding treatment strategy:

1. Aversive conditioning—electrical, chemical, or covert—has not been proven effective *in isolation* as a means for modifying or eliminating excessive drinking. When tested for therapeutic efficacy by themselves, neither electrical

nor covert aversion led to significant changes in drinking behavior on either a short-term or long-term basis. Untested in this way, chemical aversion may nonetheless possess greater therapeutic potency in view of its more direct relevance to the drinking sequence.

2. Of the other behavioral methods which aim specifically to modify the drinking response itself, including the several operant procedures as well as blood alcohol level discrimination training methods, none has progressed beyond the pilot, exploratory phase of development. As a result, we know that systematic reinforcement and punishment contingencies enforced in a laboratory setting will eventuate in more moderate drinking by alcoholics; we know, too, that with appropriate feedback alcoholics can also learn to discriminate blood alcohol levels and to use that skill to control their drinking, again in the laboratory. However, we have no assurance that stimulus control acquired in the laboratory will extend to settings in which expectancy variables and reinforcement contingencies cannot so easily be managed.

3. It seems clear that multifaceted behavioral programs which attack alcoholic drinking and the alcoholic's associated behavioral problems with a variety of techniques and procedures simultaneously have proven more efficacious overall than unitary methods with a more limited purview. We conclude, as a result, that such broad-spectrum approaches are now to be recommended, at least until additional research enables specification of the active ingredient(s) within them.

4. Attempts to extend behavioral control over excessive drinking from the laboratory to the natural environment are most promising. Whether such efforts involve actual manipulation of the environment—as in Hunt and Azrin's community-reinforcement therapy—or depend upon contingency management and contingency contracting to achieve the same ends, all encourage the view that the natural environment can be (perhaps *must* be) utilized as a natural reinforcer to maintain abstinence or induce controlled drinking in the alcoholic.

Is controlled drinking a viable goal for some alcoholics? Perhaps. *Is continued research on controlled drinking a legitimate pursuit for the behavioral researcher?* Certainly. *Do we now know enough to select patients for abstinence-oriented or controlled-drinking-oriented treatment programs on a rational basis?* No. But what we do know now about controlled drinking encourages us to believe that it may be almost the only viable treatment goal for some alcoholics—those who have tried and failed to achieve abstinence on repeated occasions. What we also believe, parenthetically, is that alcoholics who have been able to maintain abstinence should certainly not attempt to become social drinkers with these techniques as their therapeutic efficacy in this context has not been established.

Efforts to develop controlled-drinking-oriented treatment programs as alternatives to abstinence-oriented treatment have met with enormous resistance from well-intentioned persons who are convinced that the "disease of alcoholism" cannot be cured by providing the alcoholic continued access to his or her principal poison. Others, just as emphatically, reject this view and predict that controlled drinking may be the most appropriate treatment goal for many or most alcoholics. Our view of this matter, a bit more temperate, reflects our memory of the repeated disappointments alcoholics have suffered each time a miracle cure for their affliction has had to be debunked. As a result, we maintain our plea for continued federal, state, and local support for research exploring treatment approaches to alcoholism from every perspective—and let the data that evolve from these investigations be the final arbiter of the matter.

Can future directions for behavioral research on alcoholism be defined? Indeed. They include the following:

1. Experimental analysis of the active components of multifaceted behavior therapy for alcoholism. Such a study might compare three multifaceted programs that differ from each other only in that the aversion component of each employs electrical aversion, chemical aversion, or covert aversion. To this end, a recent study by Caddy and Lovibond (1976) might serve as a model in this regard. The study compared treatment outcome for three groups of 20 chronic alcoholic patients each. One group received behavioral treatment like that developed by Lovibond and Caddy (1970), consisting of blood alcohol level discrimination training, aversive conditioning, alcohol education, and "psychotherapy" emphasizing self-regulation training. The second group was provided the same treatment package minus the aversive component, while the third received the initial treatment program minus training in self-regulation. At 6- and 12-month follow-up marks, patients in group one were doing better than those in the other two groups. But beyond the specifics of these outcome data, the research model employed by these investigators deserves support and encouragement.

2. Identification of psychophysiologic correlates of successful and unsuccessful behavioral treatment. This research would ask whether changes in the character of physical tolerance or physical dependence must accompany short-term changes in drinking behavior for these changes to be permanent.

3. Experimental analysis of those components of a multimodal approach to controlled drinking which contribute to long-term abstinence or to long-term successful controlled drinking. This research would include alcoholics who do *not* accept controlled drinking as a viable treatment goal as well as those who do, since studies of controlled drinking to this time have employed only alcoholics who failed to achieve abstinence and, for that reason, believed that controlled drinking might be the only appropriate treatment goal for them.

4. Identification of objective correlates of adaptive change in drinking behavior. This research could permit prediction of the long-term outlook for a

given patient's alcoholism before or after treatment begins. Such a correlate might be drinking behavior during a standard ad lib period or in a taste test immediately preceding or following a course of behavioral treatment. Such an assessment pretreatment may enable a more rational decision on the nature of treatment. This assessment posttreatment would aid in making decisions about the nature and extent of booster therapy to be arranged during the posttreatment period. In short, analog assessment procedures (e.g., taste tests, ad lib drinking) must be systematically validated against comprehensive, long-term follow-ups.

5. Development of the most appropriate follow-up procedures and periods to assess the outcome of behavioral treatment for alcoholism. At present, researchers use such a variety of follow-up procedures employed over such a range of follow-up periods that direct comparison of the treatment efficacy of different methods over time is impossible. Development of standard behavioral procedures for follow-up—rather than reliance on test scores or self-reports on drinking behavior—to be administered at systematic intervals post-treatment would enable this direct comparison. Cost-effectiveness studies of intensive versus intermittent follow-up and long-term versus short-term follow-up are also called for. They would answer important questions like whether five years of follow-up predict permanency of change in drinking behavior more validly than a two-year follow-up or whether four different behavioral outcome measures at six months versus one or two measures at two years are more predictive of final treatment status.

6. Continued development of procedures for extending stimulus control over drinking acquired in the laboratory to the natural environment. More cost-effective means for heightening the reinforcement value of natural reinforcers (as in the community-reinforcement paradigm), combined with more sophisticated contingency management systems, may be the path to this goal.

ACKNOWLEDGMENT

Preparation of this chapter was aided by U.S. Public Health Service Grant AA 00259-05, National Institute on Alcohol Abuse and Alcoholism, to P. E. Nathan. The authors are grateful to Barbara Honig and David Lawson, whose help in the preparation of this chapter was most significant.

REFERENCES

Anant, S. S., 1967, A note on the treatment of alcoholics by a verbal aversion technique, *Can. Psychol.* 1:19–22.

Ashem, B., and Donner, L., 1968, Covert sensitization with alcoholics: A controlled replication, *Behav. Res. Ther.* 6:7–12.

Bigelow, G., 1973, Experimental analysis of human drug self-administration, Paper read at Eastern Psychological Association (May, 1973).

Bigelow, G., Liebson, I., and Griffiths, R. R., 1973, Experimental analysis of alcoholic drinking, Paper read at American Psychological Association (August 1973).

Bigelow, G., Liebson, I., and Lawrence, C., 1973, Prevention of alcohol abuse by reinforcement of incompatible behavior, Paper read at Association for Advancement of Behavior Therapy (December 1973).

Bigelow, G., Liebson, I., and Griffiths, R. R., 1974, Alcoholic drinking: suppression by a behavioral time-out procedure, *Behav. Res. Ther.* 12:107–115.

Blake, B. C., 1965, The application of behavior therapy to the treatment of alcoholism, *Behav. Res. Ther.* 3:78–85.

Blake, B. G., 1967, A follow-up of alcoholics treated by behaviour therapy, *Behav. Res. Ther.* 5:89–94.

Bois, C., and Vogel-Sprott, M., 1974, Discrimination of low blood alcohol levels and self-titration skills in social drinking, *Q. J. Stud. Alcohol* 35:87–97.

Briddell, D. W., and Nathan, P. E., 1976, Behavior assessment and modification with alcoholics: Current status and future trends, *in* "Progress in Behavior Modification" (M. Hersen, R. M. Eisler, and P. M. Miller, eds.) Vol. 2, Academic Press, New York.

Caddy, G. R., and Lovibond, S. H., 1976, Self-regulation and discriminated aversive conditioning in the modification of alcoholics' drinking behavior, *Behav. Ther.* 7:223–230.

Cappell, H., 1975, An evaluation of tension models of alcohol consumption, *in* "Research Advances in Alcohol and Drug Problems" (Y. Israel *et al.* eds.), Wiley, New York.

Cappell, H., and Herman, C. P., 1972, Alcohol and tension reduction—A review, *Q. J. Stud. Alcohol* 33:33–64.

Caudill, B. D., and Marlatt, G. A., 1975, Modeling influences in social drinking: An experimental analogue, *J. Consult. Clin. Psychol.* 43:405–415.

Cautela, J. R., 1966, Treatment of compulsive behavior by covert sensitization, *Psychol. Rep.* 16:33–41.

Cautela, J. R., 1967, Covert sensitization, *Psychol. Rep.* 20:459–468.

Cautela, J. R., 1970, The treatment of alcoholism by covert sensitization, *Psychotherapy: Theory, Research and Practice* 7:86–90.

Cheek, F. E., Franks, C. M., Laucius, J., and Burtle, V., 1971, Behavior modification training for wives of alcoholics, *Q. J. Stud. Alcohol* 32:456–461.

Cohen, M., Liebson, I. A., Faillace, L. A., and Allen, R. P., 1971, Moderate drinking by chronic alcoholics, *J. Nerv. Ment. Dis.* 153:434–444.

Conger, J. J., 1951, The effects of alcohol on conflict behavior in the albino rat, *Q. J. Stud. Alcohol* 12:1–29.

Conger, J. J., 1956, Alcoholism: Theory, problem and challenge. II. Reinforcement theory and the dynamics of alcoholism, *Q. J. Stud. Alcohol* 17:291–324.

Cutter, H. S. G., Schwab, E. L., and Nathan, P. E., 1970, Effects of alcohol on its utility for alcoholics, *Q. J. Stud. Alcohol* 30:369–378.

Eysenck, H. J. (ed.), 1960, "Behaviour Therapy and the Neuroses," Pergamon Press, London.

Franks, C. M., 1970, Alcoholism, In "Symptoms of Psychopathology" (C. G. Costello, ed.), Wiley, New York.

Garcia, J., Hankins, W. G., and Rusiniak, K. W., 1974, Behavioral regulation of the milieu interne in man and rat, *Science* 185:824–831.

Goldman, M., Taylor, A., Carruth, M., and Nathan, P. E., 1973, Effects of group decision-making on group drinking by alcoholics, *Q. J. Stud. Alcohol* 34:807–822.

Gottheil, E., Corbett, L. O., Grasberger, J. C., and Cornelison, F. S., Jr., 1971, Treating the alcoholic in the presence of alcohol, *Am. J. Psychiatry* 128:475–480.

Gottheil, E., Corbett, L. O., Grasberger, J. C., and Cornelison, F. S., Jr., 1972, Fixed interval drinking decisions: 1. A research and treatment model, *Q. J. Stud. Alcohol* 33:311–324.

Gottheil, E., Alterman, A. I., Skoloda, T. E., and Murphy, B. F., 1973, Alcoholics' patterns of controlled drinking, *Am. J. Psychiatry* 130:418–422.

Gottheil, E., Crawford, H., and Cornelison, F. S., Jr., 1973, The alcoholic's ability to resist available alcohol, *Dis. Nerv. Syst.* 1973, 34:80–84.

Griffiths, R. R., Bigelow, G., and Liebson, I., 1974a, Assessment of effects of ethanol self-administration on social interactions in alcoholics, *Psychopharmacologia* 38:105–110.

Griffiths, R. R., Bigelow, G., and Liebson, I., 1974b, Suppression of ethanol self-administration in alcoholics by contingent time-out from social interactions, *Behav. Res. Ther.* 12:327–334.

Griffiths, R. R., Bigelow, G., and Liebson, I., 1975, Effects of ethanol self-administration on choice behavior: Money vs. socializing, *Pharmacol., Biochem. Behav.* 3:443–446.

Hallam, R., and Rachman, S., 1972, Theoretical problems of aversion therapy, *Behav. Res. Ther.* 10:341–353.

Hallam, R., Rachman, S., and Falkowski, W., 1972, Subjective, attitudinal and physiological effects of electrical aversion therapy, *Behav. Res. Ther.* 10:1–14.

Hedberg, A. G., and Campbell, L., 1974, A comparison of four behavioral treatments of alcoholism, *J. Behav. Ther. Exp. Psychiatry* 5:251–256.

Hersen, M., Miller, P., and Eisler, R., 1973, Interactions between alcoholics and their wives: A descriptive analysis verbal and nonverbal behavior, *Q. J. Stud. Alcohol* 34:516–520.

Higgins, R. L., and Marlatt, G. A., 1973, The effects of anxiety arousal upon the consumption of alcohol by alcoholics and social drinkers, *J. Consult. Clin. Psychol.* 41:426–433.

Hsu, J., 1965, Electroconditioning therapy with alcoholics: A preliminary report, *Q. J. Stud. Alcohol* 26:449–459.

Huber, H., Karlin, R., and Nathan, P. E., 1976, Blood alcohol level discrimination by non-alcoholics: The role of internal and external cues, *Q. J. Stud. Alcohol* 37:27–39.

Hunt, G. M., and Azrin, N. H., 1973, The community-reinforcement approach to alcoholism, *Behav. Res. Ther.* 11:91–104.

Jellinek, E. M., 1960, "The Disease Concept of Alcoholism," College and University Press, New Haven.

Kraft, T., 1969, Alcoholism treated by systematic desensitization: A follow-up of eight cases, *Journal of the Royal College of General Practice* 18:336–340.

Kraft, T., and Al-Issa, I., 1967, Alcoholism treated by desensitization: A case study, *Behav. Res. Ther.* 5:69–70.

Laverty, S. G., 1966, Aversion therapies in the treatment of alcoholism, *Psychosom. Med.* 28:651–666.

Lazarus, A. A., 1965, Towards the understanding and effective treatment of alcoholism, *S. Afr. Med. J.* 39:736–741.

Lemere, F., and Voegtlin W. L., 1950, An evaluation of the aversion treatment of alcoholism, *Q. J. Stud. Alcohol* 11:199–204.

Lemere, F., Voegtlin, W. L., Broz, W. R., and O'Hallaren, P., 1942, Conditioned reflex treatment of alcohol addiction. V. Type of patient suitable for this treatment, *Northwestern Medicine, Seattle* 4:88–89.

Liebson, I., and Bigelow, G., 1972, A behavioural-pharmacological treatment of dually addicted patients, *Behav. Res. Ther.* 10:403.

Liebson, I., Bigelow, G., and Flame, R., 1973, Alcoholism among methadone patients: A specific treatment method, *Am. J. Psychiatry* 130:483.

Lovibond, S. H., and Caddy, G. R., 1970, Discriminated aversive control in the moderation of alcoholics' drinking behavior, *Behav. Ther.* 1:437–444.

Ludwig, A. M., 1972, On and off the wagon: Reasons for drinking and abstaining by alcoholics, *Q. J. Stud. Alcohol* 33:91–96.

Ludwig, A. M., and Wikler, A., 1974, "Craving" and relapse to drink, *Q. J. Stud. Alcohol* 35:108–130.

Ludwig, A. M., Wikler, A., and Stark, L. H., 1974, The first drink, *Arch. Gen. Psychiatry* 30:539–547.

MacCulloch, M. J., Feldman, M. P., Orford, J. F., and MacCulloch, M. L., 1966, Anticipatory avoidance learning in the treatment of alcoholism: A record of therapeutic failure, *Behav. Res. Ther.* 4:187.

Marlatt, G. A., 1973a, A comparison of aversive conditioning procedures in the treatment of alcoholism, Paper read at Western Psychological Association (April 1973).

Marlatt, G. A., 1973b, Determinants of alcohol consumption in a laboratory taste-rating task: Implications for controlled drinking, Paper read at American Psychological Association (August 1973).

Marlatt, G. A., 1975a, The drinking profile: A questionnaire for the behavioral assessment of alcoholism, *in* "Behavior therapy assessment: Diagnosis and evaluation" (E. J. Mash and L. G. Terdal, eds.), Springer, New York.

Marlatt, G. A., 1975b, Alcohol, stress, and cognitive control, Paper read at NATO-sponsored International Conference on Dimensions of Stress and Anxiety (June–July 1975).

Marlatt, G. A., Demming, B., and Reid, J. B., 1973, Loss of control drinking in alcoholics: An experimental analogue, *J. Abnorm. Psychol.* 81:233–241.

McGuire, R. J., and Vallance, M., 1964, Aversion therapy by electric shock, a simple technique, *Br. Med. J.* 1:151–152.

McNamee, H. B., Mello, N. K., and Mendelson, J. H., 1968, Experimental analysis of drinking patterns of alcoholics. Concurrent psychiatric observations, *Am. J. Psychiatry,* 124:1063–1069.

Mello, N. K., 1972, Behavioral studies of alcoholism, *in* "The Biology of Alcoholism" Vol. 2 (B. Kissin and H. Begleiter, eds.), Plenum Press, New York.

Mello, N. K., and Mendelson, J. H., 1965, Operant analysis of drinking patterns of chronic alcoholics, *Nature* 206:43–46.

Mello, N. K., and Mendelson, J. H., 1970, Experimentally induced intoxication in alcoholics; a comparison between programmed and spontaneous drinking, *J. Pharmacol. Exp. Ther.* 173:101–116.

Mendelson, J. H., and Mello, N. K., 1966, Experimental analysis of drinking behavior of chronic alcoholics, *Ann. N. Y. Acad. Sci.* 133:828–845.

Merry, J., 1966, The "loss of control" myth, *Lancet* 1:1267–1268.

Miller, P. M., 1972, The use of behavioral contracting in the treatment of alcoholism: A case report, *Behav. Ther.* 3:593–596.

Miller, P. M., 1975, A behavioral intervention program for chronic public drunkenness offenders, *Arch. Gen. Psychiatry* 32:915–918.

Miller, P. M., and Eisler, R. M., 1975, Alcohol and drug abuse, *in* "Behavior Modification Principles, Issues, and Applications" (W. E. Craighead, A. E. Kazdin, and M. J. Mahoney, eds.), Houghton Mifflin, Boston.

Miller, P. M., and Hersen, M., 1972a, A quantitative measurement system for alcoholism treatment and research, Paper read at Association for Advancement of Behavior Therapy.

Miller, P. M., and Hersen, M., 1972b, Quantitative changes in alcohol consumption as a function of electrical aversive conditioning, *J. Clin. Psychol.* 28:590–593.

Miller, P. M., and Hersen, M., 1975, Modification of marital interaction patterns between an alcoholic and his wife, in "Counseling Methods" (J. D. Krumboltz and C. E. Thoresen, eds.), Holt, Rinehart & Winston, New York.

Miller, P. M., Hersen, M., Eisler, R., and Hemphill, D. P., 1973, Electrical aversion therapy with alcoholics: An analogue study, *Behav. Res. Ther.* 11:491–497.

Miller, P. M., Hersen, M., and Eisler, R. M., 1974, Relative effectiveness of instructions, agreements, and reinforcement in behavioral contracts with alcoholics, *J. Abnorm. Psychol.* 83:548–553.

Miller, P. M., Hersen, M., Eisler, and Elkin, T. E., 1974, A retrospective analysis of alcohol consumption on laboratory tasks as related to therapeutic outcome, *Behav. Res. Ther.* 12:73–76.

Miller, P. M., Hersen, M., Eisler, R., and Hilsman, G., 1974, Effects of social stress on operant drinking of alcoholics and social drinkers, *Behav. Res. Ther.* 12:67–72.

Miller, P. M., Hersen, M., Eisler, R., and Watts, J. G., 1974, Contingent reinforcement of lowered blood alcohol levels in an outpatient chronic alcoholic, *Behav. Res. Ther.* 12:261–263.

Miller, P. M., Stanford, A. G., and Hemphill, D. P., 1974, A comprehensive social learning approach to alcoholism treatment, *Social Casework* 55:279–284.

Mills, K. C., Sobell, M. B., and Schaefer, H. H., 1971, Training social drinking as an alternative to abstinence for alcoholics, *Behav. Ther.* 2:18–27.

Nathan, P. E., 1976, Alcoholism, in "Handbook of Behavior Modification"(H. Leitenberg, ed.), Appleton-Century-Crofts, New York.

Nathan, P. E., Titler, N. A., Lowenstein, L. M., Solomon, P., and Rossi, A. M., 1970, Behavioral analysis of chronic alcoholism, *Arch. Gen. Psychiatry* 22:419–430.

Nathan, P. E., and O'Brien, J. S., 1971, An experimental analysis of the behavior of alcoholics and nonalcoholics during prolonged experimental drinking, *Behav. Ther.* 2:455–476.

National Institutes of Mental Health, Alcohol and Alcoholism, USPHS Publication No. 1640, Washington D.C.

Okulitch, P. V., and Marlatt, G. A., 1972, Effects of varied extinction conditions with alcoholics and social drinkers, *J. Abnorm. Psychol.* 79:205–211.

Paredes, A., Jones, B. M., and Gregory, D., 1974a, An exercise to assist alcoholics to maintain prescribed levels of intoxication, *Alcohol Technical Reports* 2:24–36.

Paredes, A., Gregory, D., and Jones, B. M., 1974b, Induced drinking and social adjustment in alcoholics, *Q. J. Stud. Alcohol* 35:1279–1293.

Pattison, E. M., 1966, A critique of alcoholism treatment concepts with special reference to abstinence, *Q. J. Stud. Alcohol* 27:49–71.

Pattison, E. M., Headley, E. B., Gleser, G. C., and Gottschalk, L. A., 1968, Abstinence and normal drinking: An assessment of changes in drinking patterns in alcoholics after treatment, *Q. J. Stud. Alcohol* 29:610–633.

Pomerleau, O. F., and Brady, J. P., 1975a, Behavior modification in medical practice, with an example from the treatment of problem drinking, *Pa. Med.* 78:49–53.

Pomerleau, O. F., and Brady, J. P., 1975b, Behavioral treatment of problem drinking, Paper read at Association for Advancement of Behavior Therapy (December 1975).

Rachman, S., 1961, Sexual disorders and behaviour therapy, *Am. J. Psychiatry* 118:235–240.

Raymond, M. J., 1964, The treatment of addiction by aversion conditioning with apomorphine, 1964, *Behav. Res. Ther.* 1:287–291.

Sanderson, R. E., Campbell, D., and Laverty, S. G., 1962, Traumatically conditioned responses acquired during respiratory paralysis, *Nature* 196:1235–1236.

Sanderson, R. E., Campbell, D., and Laverty, S. G, 1963, An investigation of a new aversion conditioning treatment for alcoholism, *Q. J. Stud. Alcohol* 24:261–275.

Sandler, J., 1969, Three aversive control procedures with alcoholics: A preliminary report, Paper read at Southeastern Psychological Association (April 1969).
Schaefer, H. H., Sobell, M. B., and Mills, K. C., 1971, Baseline drinking behavior in alcoholics and social drinkers: Kinds of drinks and sip magnitude, *Behav. Res. Ther.* 9:23–27.
Silverstein, S. J., Nathan, P. E., and Taylor, H. A., 1974, Blood alcohol level estimation and controlled drinking by chronic alcoholics, *Behav. Ther.* 5:1–15.
Skoloda, T. E., Alterman, A. I., Cornelison, F. S., and Gottheil, E., 1975, Treatment outcome in a drinking-decisions program, *J. Stud. Alcohol* 36:365–380.
Sobell, L. C., and Sobell, M. B., 1975, Self-reports by alcoholics, *J. Nerv. Ment. Dis.* 161:32–42.
Sobell, M. B., and Sobell, L. C., 1973a, Individualized behavior therapy for alcoholics, *Behav. Ther.* 4:49–72.
Sobell, M. B., and Sobell, L. C., 1973b, Alcoholics treated by individualized behavior therapy: One year treatment outcome, *Behav. Res. Ther.* 11:599–618.
Sobell, M. B., and Sobell, L. C., 1976, Second-year treatment outcome of alcoholics treated by individualized behavior therapy: Results, *Behav. Res. Ther.* 14:195–215.
Sobell, M. B., Sobell, L. C., and Samuels, F. H., 1974, Validity of self-reports of alcohol-related arrests by alcoholics, *Q. J. Stud. Alcohol* 35:276–280.
Steffen, J. J., 1975, Electromyographically induced relaxation in the treatment of chronic alcohol abuse, *J. Consult. Clin. Psychol.* 43:275.
Steffen, J. J., Nathan, P. E., and Taylor, H. A., 1974, Tension-reducing effects of alcohol: Further evidence and some methodological considerations, *J. Abnorm. Psychol.* 83:542–547.
Sulzer, E. S., 1965, Behavior modification in adult psychiatric patients, *in* "Case Studies in Behavior Modification" (L. P. Ullman and L. Krasner, eds.), Holt, Rinehart, & Winston, New York.
Summers, T., 1970, Validity of alcoholics' self-reported drinking history, *Q. J. Stud. Alcohol* 31:972–974.
Thimann, J., 1949, Conditioned reflex treatment of alcoholism. II. The risk of its application, its indications, contraindications, and pscycotherapeutic aspects, *N. Eng. J. Med.* 241:408–410.
Thornton, C. C., Alterman, A. I., Skoloda, T. E., and Gottheil, E., 1975, Drinking and Socializing in "introverted" and "extraverted" alcoholics, Paper read at National Council on Alcoholism annual meeting (April 1975).
Tracey, D., Karlin, R., and Nathan, P. E., 1976, Experimental analysis of chronic alcoholism in four women, *J. Consult. Clin. Psychol.* (in press).
Vogel-Sprott, M., 1975, Self-evaluation of performance and the ability to discriminate blood alcohol concentrations, *J. Stud. Alcohol* 36:1–10.
Voegtlin, W. L., 1940, The treatment of alcoholism by establishing a conditioned reflex, *American Journal of Medical Science* 199:802–809.
Voegtlin, W. L., 1947, Conditioned reflex therapy of chronic alcoholism: Ten years' experience with the method, *Rocky Mount. Med. J.* 44:807–812.
Vogler, R. E., Lunde, S. E., Johnson, G. R., and Martin, P. L., 1970, Electrical aversion conditioning with chronic alcoholics, *J. Consult. Clin. Psychol.* 34:302–307.
Vogler, R. E., Lunde, S. E., and Martin, P. L., 1971, Electrical aversion conditioning with chronic alcoholics: Follow-up and suggestions for research, *J. Consult. Clin. Psychol.* 36:450.
Vogler, R. E., Compton, J. V., and Weissbach, T. A., 1975, Integrated behavior change techniques for alcoholics, *J. Consult. Clin. Psychol.* 43:233–243.
Wiens, A. N., Montague, J. R., Manaugh, T. S., and English, C. J., 1975, Pharmacologic aversive counterconditioning to alcohol in a private hospital: One year follow-up, Unpublished manuscript, University of Oregon Medical School.
Williams, R. J., 1948, Alcoholics and metabolism, *Scientific American* 179:50–53.

Williams, R. J., and Brown, R. A., 1974, Differences in baseline drinking behavior between New Zealand alcoholics and normal drinkers, *Behav. Res. Ther.* 12:287–294.

Williams, T. K., 1970, The ethanol-induced loss of control concept in alcoholism, Ed.D. dissertation, Western Michigan University.

Wilson, G. T., and Davison, G. C., 1969, Aversion techniques in behavior therapy: Some theoretical and metatheoretical considerations. *J. Consult. Clin. Psychol.* 33:327–329.

Wilson, G. T., and Rosen, R. C., 1975, Training controlled drinking in an alcoholic through a multifaceted behavioral treatment program: A case study, *in* "Counseling Methods" (J. D. Krumboltz and C. E. Thoreson, eds.), Holt, Rinehart & Winston, New York.

Wilson, G. T., and Tracey, D. A., 1975, An experimental investigation of the effects of covert sensitization on excessive drinking by chronic alcoholics, Unpublished manuscript, Rutgers University.

Wilson, G. T., Leaf, R., and Nathan, P. E., 1975, The aversive control of excessive drinking by chronic alcoholics in the laboratory setting, *J. Appl. Behav. Anal.* 8:13–26.

CHAPTER 9

The Role of the Halfway House in the Rehabilitation of Alcoholics

Earl Rubington

Department of Sociology and Anthropology
Northeastern University
Boston, Massachusetts

HALFWAY HOUSES

Alcoholism and its rehabilitation result from experiences in groups. Certain kinds of experiences in groups are more likely to result in alcoholism. Similarly, certain forms of experiences in groups are more likely to result in rehabilitation. This chapter theorizes on the kinds of experiences in halfway houses most likely to produce successful treatment outcomes.

Origins of Halfway Houses

Alcoholics, late in their drinking careers, experience patterned crises. Heavier drinking means more frequent and prolonged intoxications, and, necessarily, more detoxications. Each intoxication constitutes a rupture in group membership, if the alcoholic is still a member of a family, work, or friendship group. Each intoxication constitutes a barrier to subsequent admission to group membership if the alcoholic is not currently a member of any group.

At the same time, the need for detoxication persists. And, after detoxication, relations with groups are still in abeyance. Thus, for example, existing groups are hesitant if not reluctant to accept the alcoholic back. But, regardless of whether or not his groups are willing to take him back, he is in no condition to be taken back.

Alcoholics without any significant group memberships prior to detoxication face the problem of locating a primary group, once the "drying out" period is over (Rubington, 1960). While they may or may not want any group affiliations, subsequent to detoxication, they generally lack the resources for making an independent sober adult life for themselves. Most of the time, they lack job and money, sponsors and references, friends and backers.

In addition, alcoholics, whether affiliated or unaffiliated, suffer numerous impairments. After detoxication, they may still be shaky, nervous, or with some kind of tranquilizing drug still in their bloodstream. There are immediate aftereffects of a drinking bout, as well as other physical sequelae of prolonged drinking. Besides these physical effects, there are such psychic effects as guilt, remorse, and self-rejection.

For affiliates and nonaffiliates alike the postdetoxication situation poses a dilemma. The crisis is compounded because of the tendency to relapse. Many, if not most alcoholics respond to the crisis by drinking again. Many alcoholics, after drying out, simply relapse for lack of any better alternative. This only activates the drinking cycle and increases alienation from significant social groups.

Halfway houses came into being to solve the problems of the postdetoxication situation. In time they developed into interim groups. Originally they were only intended to tide the alcoholic over until such time as he could "get back on his feet again." Later, they tried to help residents stay sober long after they had left the halfway house. In effect, then, halfway houses came into being to offer support during transition—when alcoholics are between groups (Blacker and Kantor, 1960).

A Definition of Halfway Houses

The halfway house may be defined as *a transitional place of indefinite residence of a community of persons who live together under the rule and discipline of abstinence from alcohol and other drugs*. Transition refers to the condition of being between groups, of being in passage from the "dry" to the "sober" state. The community of persons are bound together by the common affliction of alcoholism and the common interest in abstinence. Residence refers to sharing bed and board and living under one roof. And most important of all is the rule of abstinence which is imposed and enforced by authority of the resident staff.

A number of organizations are halfway houses in name only (Baker, 1972). Thus, absence of a common residence, of community, of a period of indefinite residence, and of either imposition of an abstinence rule, its enforcement, or both, disqualifies an organization from being called a halfway house.

One Example: The Compass Club

In recent years, halfway houses for alcoholics have increased in the United States. Literature on halfway houses has not, however, kept pace with this growth. This literature consists of official statements, surveys, follow-up researches, and field studies, in that order.

Official statements include brochures, state regulations, manuals of procedures, proceedings, and the like (A Body of Knowledge, 1975). There have been a handful of important surveys of halfway houses (Blacker and Kantor, 1960; Cannon, 1973). Data for these surveys come from mail questionnaires or field visits. Follow-up studies, of which there have been but a handful, focus on treatment outcomes (Blumberg et al., 1973; Katz, 1966; Myerson and Mayer, 1966; Nash, 1962). Finally, there have been a few field studies of halfway houses for alcoholics (Albrecht, 1969; Rubington, 1957, 1960, 1963, 1964, 1965, 1968, 1970). The description of the Compass Club which follows below is based on one of these field studies. It is included here to give some idea of how halfway houses work.*

The Connecticut Commission on Alcoholism established the Compass Club as a pilot facility in 1956. It was primarily intended to serve chronic drunkenness offenders, persons who had been arrested for public intoxication three or more times in a given year. The Commission wanted to extend sobriety begun in either the state's inpatient alcoholism treatment center, Blue Hills Hospital, or in custody while serving time for public drunkenness, and to cut costs of processing offenders through the criminal justice system.

The Commission, after a six-month search for a site, leased space in what was then the Yale Hope Mission, a three-story building serving homeless and transient men in downtown New Haven. The Compass Club occupied a front office adjoining the front door to the mission, and a large administrative office

---

* To protect the anonymity of staff, residents, commission, and others, I have changed the name of the Compass Club in some of my publications. Thus, in a 1963 paper it is called simply The Club. In four other publications (1964, 1965, 1968, 1970), it is called Shelter House. In two other publications (1957, 1960) and in the present piece it is referred to by its right name. For the record, all of these papers deal with the Compass Club.

in the rear of the building on the first floor. Downstairs the Club leased a lounge, a portion of the dining room for eating, and a segment of the kitchen for preparing meals. On the second floor, it rented five private single rooms. On the third floor, it rented a small ten-bed dormitory plus a large single room adjoining it for a staff member.

Save for a secretary and cook, all Compass Club staff were recovered alcoholics and members of Alcoholics Anonymous. In the beginning there was only the director and his assistant. Later, the staff expanded to include the director and five counselors. About that same time it increased in size, renting 10 more private rooms on the second floor for a total capacity of 25 residents. After 30 days, a man was eligible to move downstairs to occupy his own private room. He was also given a front door key to the mission. Both room and key were intended to mark his rising status in the Club.

When counselors felt certain residents showed signs of attitude and behavior change, they said: "It looks like Mickey Shaughnessy is 'getting the program.'" The "program" consisted of a simple body of rules summarized below:

1. One drink and you're out.
2. Eat morning and evening meals in the Club.
3. Observe the 11:00 p.m. curfew nightly.
4. After the 72-hour restriction is up, go out and find work.
5. Pay $3 a day rent or $21 a week.
6. Go to all mandatory meetings and group discussions.
7. Talk to counselors about drinking problems.
8. Attend outside A.A. meetings.
9. Perform cleanup and maintenance details as assigned.
10. Be cooperative.

When the Club first opened, staff felt 90 days would give the men enough time to get back on their feet. With time, they changed their minds about how long members should stay. In all, three changes in optimum stay took place. First, they increased the maximum stay to 120 days. Later, they extended the time to six months. Finally, they changed maximum stay to an indefinite period to be decided on mutually by staff and resident. During the period of the field study, some six years in all, there were three different directors, and these changes, for the most part, usually followed changes in directorships. Also, the Club finally settled on two mandatory group meetings a week. The "big meeting" was held on Monday nights right after dinner in the basement lounge and all staff and members present attended. Tuesday, Wednesday, and Thursday nights small "group discussions" were held around a table in the administrative office. Men were assigned to one of these meetings according to the day of the week. Whichever counselor was on duty chaired the meeting. It was a form of lay group therapy and all present were expected to talk about their drinking problems during the hour they sat around the table.

Besides learning how to run a halfway house, the staff kept records. One important item of information that concerned them was how long residents stayed and how they left. Staff recorded discharges as follows: A—according to plan; B—with notice; C—without notice; and D—disciplinary discharge. These discharges represented different stay-benefit ratios. Residents who left "according to plan" had stayed either the limit or longer and then had left sober. Residents who left "with notice" usually left a few weeks or a month before the end of the desired length of stay. They were sober at the time and moved to an apartment or took a job in another city. Those who left "without notice" were considered AWOL. Staff assumed that most AWOLs had gotten drunk and had not bothered to return or to notify the Club. And, finally, most disciplinary discharges were for on-premise violations of the no-drinking rule.

During the study period, the median length of stay held at around 28 days; the average was 40. During that time, approximately 15 percent of the residents were discharged according to plan. A study of the first 179 admissions revealed that 24 or 13 percent had stayed the limit and left according to plan. Staff believed that duration of membership was important. They reasoned that the longer a resident maintained sober associations, the longer he would be sober after discharge. A pilot follow-up study of the first 100 admissions bore them out. Their proposition, of course, held only for those who were discharged according to plan or with notice. Those discharged according to plan were found to be sober longer after discharge than those who were discharged with notice.

Residents came to the Compass Club principally from three sources of referral. Approximately one-third came directly from the county jail. Another third came from Blue Hills. The last third were miscellaneous, including such diverse sources as the Yale Hope Mission, walk-ins, A.A., probation, state hospitals, etc. Restrictions on admission affected the social characteristics of residents. In addition to a minimum of three arrests for public drunkenness annually, residents had to be able to work, be between 21 and 55, and not be overly psychotic. As a result, the average age of residents was 45. Half of the residents had never married; the other half had been married one or more times prior to their admission. Of the ever-married, less than five were still married at the time of their admission. Staff allowed only two readmissions. Over the course of the study, the readmission rate rose considerably (Rubington, 1960).

For the first year and a half, the turnover was rather high. Eight residents would have been discharged and eight new ones admitted in the course of a month. Since the bulk of discharges were mainly for drinking, the staff did become discouraged from time to time. Given the fact that all of them were new to counseling in a halfway house and that all of them desired long-term sobriety for their residents, their discouragement is quite understandable.

Major Characteristics of Halfway Houses

The Compass Club study suggests that halfway houses may be distinguished by the following major characteristics: culture, social organization, physical plant, and ecology.

Culture

Halfway houses share an antidrinking culture. Their core beliefs, values, and norms can be summed up as follows: (a) belief—you only get drunk from drinking; (b) value—sobriety is better than drunkenness; (c) norm—whatever you do, don't drink.

Social Organization

Halfway house statuses include director, manager, counselor, cook, and residents. When houses first open, it is not uncommon for one person to fill all staff statuses. In those circumstances, there is probably one staff person for about 15 residents. As houses mature, they divide staff labors, add personnel, and increase the number of residents.

Physical Plant

Most halfway houses are converted dwellings, formerly private homes, boarding houses, or apartment houses. But if most are free-standing, others occupy sections of a larger building, e.g., a wing of a hospital, hotel, or YMCA. And, in other instances, a halfway house may occupy an entire building on the grounds of a hospital.

Ecology

Generally, halfway houses for alcoholics are found in one of three areas. Close to the central business district, in lower-middle-class residential neighborhoods, or, infrequently, on the city's border or in a suburban location. Other things equal, the larger the physical plant, the closer it is to the central business district.

Prior to 1950, most transitional facilities for alcoholics were operated by rescue missions, the Salvation Army, and other church organizations, and were found in or close to the downtown section. After 1950, privately run halfway houses appeared. These houses sought to remain small to keep costs down and to create and maintain a homelike or family atmosphere. As the halfway house movement grew, houses tended to copy one another. Simple in organization and in ideology, halfway houses came to exhibit four features: small size, simple rules, elementary statuses, and informal atmospheres.

Size. Alcoholics, particularly those from skid row, have considerable experience with state hospitals and jails. Experience in these big organizations only makes them feel smaller and alienates them from rehabilitation. So halfway houses stay small to make residents feel bigger.

Rules. Big organizations, run on bureaucratic principles, have many specific rules. Halfway houses, as small organizations, have few rules. Some claim to have only one rule: "One drink and you're out."

Elementary Statuses. Big helping organizations have layers of personnel. They deal with patients according to the principle of specialization. But this impairs meaningful communication because most task specialists lack personal experience with or understanding of alcoholism. Halfway houses reverse these principles of organization. Halfway house staff of manager, director, counselors, and cook are recovered alcoholics, not task specialists. Their expertise comes from recovering from alcoholism. This gives them credibility with those alcoholics who believe that "only an alcoholic can understand the alcoholic."

Informality. Big organizations run like impersonal machines. The larger the scale, the colder the atmosphere. As bureaucracy grows in size, in statuses, and in rules, it all but drains out human warmth. So halfway houses simply reverse the cooling action of the large-scale organization, infusing all activities with informality.

Some Types of Halfway Houses

It is possible to classify halfway houses according to sponsorship and philosophy. Thus, there are privately run, state-subsidized, and church-operated halfway houses.

Culture

The antidrinking culture of the privately run halfway house centers exclusively on the program of Alcoholics Anonymous and its famous Twelve Steps. The antidrinking culture of the state-subsidized halfway house is based upon the principles of social rehabilitation. And, of course, the antidrinking culture of church-operated houses is based upon the principles of religious redemption. Total and permanent abstinence is the sole criterion of the privately run halfway house and is only possible if a person follows the Twelve Steps. State-subsidized halfway houses include other objectives along with abstinence as criteria of success. Thus, for example, working, paying taxes, paying one's rent, being off welfare rolls, and not being in a hospital or a jail constitute measures of success. Being publicly financed, the state-subsidized halfway house seeks to help residents become independent, self-supporting, nondrinking adults. As a pure type, the privately run halfway house tries to change the way its residents think about drinking. The state-subsidized halfway house works on behavior change in areas that can be measured and counted. The church-operated house focuses on residents' relationship with God. The central emphasis is upon the quality of spiritual feeling among residents.

Social Organization

Any halfway house is an amalgam of its pattern of staff–resident and resident–resident contacts. In the privately run halfway house, the manager interacts frequently with residents. As a result, he (or she) gets to know them very well. In a sense, residents are naked and find it extremely hard to hide. The manager commands more activities because he gleans information through direct contacts with residents. Thus, his span of control is great and is total. Numerous halfway houses are one-man operations. Given these conditions, residents quickly decide to follow the leader or leave. There is no third alternative. This may well explain why one survey reveals that half the residents of the halfway houses studied left within two weeks (Cahn, 1970).

The state-subsidized halfway house is intermediate in size, having somewhat more residents and also more staff. A manager, two or three counselors, and a cook are about average. Here the pattern of contacts are somewhat different. To be sure, there are contacts between one staff member and one resident. But given the tendency for state-subsidized houses to have more of a "structured program" during the day and night, there is a greater round of activities and meetings in which staff meets and makes contact with residents in small groups. The amount of influence any staff member can have is muted and, at the same time, residents are somewhat less exposed, less vulnerable. In turn, they have more allies among fellow residents. A hierarchy of statuses dilutes staff authority, particularly that of the counselors. Similarly, contacts among the residents increase the chances of clique-formation. If the pattern of control in the privately run houses is closer to *domination,* then the pattern of staff control in the state-subsidized houses is closer to *manipulation.*

Generally speaking, church-operated halfway houses are the largest. Their staff is a little larger than the state-subsidized halfway houses and shows a tendency towards specialization. As a consequence, staff–resident contact is one-to-a-batch, mostly for a specific rather than a general purpose. The larger size cuts down the information obtained in staff–resident contact. Apparent conformity to organizational norms then becomes of great importance. Specific items of behavior rather than a total pattern or configuration, as in the case of the privately run halfway house, come to symbolize conformity to or dissent from the halfway house's program. The pattern of control in the church-operated halfway house is closer to *exhortation*. There are also elements of what might be called *regimentation*.

Physical Plant

Privately run halfway houses are usually in converted dwellings. State-subsidized halfway houses more often occupy leased space in a mission or shelter, a municipal building, a hotel, or a YMCA. Occasionally, they occupy part or all of a building on the grounds of a county or a state hospital. By contrast, church-operated halfway houses more often occupy part or all of a mission, shelter, or parish structure.

Ecology

Privately run halfway houses most often are found in lower-middle-class residential neighborhoods, less frequently in working-class neighborhoods. State-subsidized halfway houses, when located in a converted dwelling, are most often found in working-class neighborhoods, less often in lower-middle-class districts. When they are housed in municipal buildings, hotels, missions, or shelters, they are usually located in the downtown business district. Those in hospital buildings are usually on the outskirts of the city. The church-operated halfway houses are sometimes found in working-class neighborhoods, frequently in or near the central business district, but most of the time right in the homeless men's section.

RESIDENTS

Alcoholics vary according to what *stages* they are in their drinking careers; the extent and degree of physical, psychological, and social *dependence* on alcohol; their social *class* background; the nature of the *societal reaction* they have experienced; the degree of their *impairment* because of drinking; and the nature and number of social *resources* available to them for their recovery.

Patterns of Dependence

To begin with, alcoholics are either early, middle, or late stage (Trice, 1960). Similarly, all are dependent upon alcohol but not necessarily to the same degree. To be sure, late-stage chronic alcoholics tend to resemble one another and appear to manifest an identical configuration of physical, psychic, and social troubles. But, at the outset of their drinking careers, they may well have gotten started under diverse conditions. And, at midpoint, one form of dependence may well stand out as stronger than another. Thus, physical dependence is the outstanding characteristic of the addictive alcoholic while psychological dependence is the major feature of the symptomatic alcoholic (Jellinek, 1960). What may not be as clear is the fact that social dependence is the essential characteristic of the typical nonaddictive alcoholic. All varieties of dependence commingle after a while in the life history of most alcoholics. Nonetheless, singling out the dominant pattern may be useful for analytic purposes.

Alcoholics: White-Collar, Blue-Collar, and Skid Row

For persons who become alcoholics only two kinds of mobility are possible—lateral or downward. Thus, taking alcoholics at their current social class face value, they are either white-collar, blue-collar, or skid row. These class distinctions affect chances of recovery and forms of treatment (Schmidt et al., 1968). They work on three areas: societal reaction, degree of impairment, and social resources.

Reaction, Impairment, and Resources

Societal reaction encompasses patterned ways in which agencies of social control have typically dealt with the alcoholic. The pattern varies in severity. By the time any alcoholic has decided to enter a halfway house of whatever kind, he has already experienced some degree of severity. One index of severity is number of arrests for public drunkenness (Pittman and Gordon, 1958). For purposes of classification, alcoholics who have become halfway house residents can be distinguished by whether they have experienced intense, heavy, moderate, or mild penalties for their excessive drinking.

While societal reaction takes into account negative sanctions applied to alcoholics, impairment measures the damage that excessive drinking has done to their minds and bodies. Thus, alcoholics can be classified by whether they have been heavily, substantially, or only moderately impaired by their drinking.

Finally, social resources means the number of people, groups, and social agencies still willing to offer support to an alcoholic seeking recovery. And so alcoholics can be classified by whether they have many, some, few, or no group memberships left.

In each of the categories of alcoholic as distinguished by class, all possible combinations of societal reaction, impairment, and social resources will appear. Nonetheless, there will be central tendencies characteristic of each. Thus white-collar alcoholics will more often be characterized by relatively mild societal reactions, minimal impairment, and many social resources. Blue-collar alcoholics will more often have experienced moderately severe societal reactions, will manifest substantial impairment, and will vary from some to few resources. Skid row alcoholics will have experienced the most severe societal reactions, will be heavily impaired, and will for the most part, have few or no social resources.

Halfway Houses and Their Clientele

Halfway houses, through sifting and sorting, tend to draw a clientele which is either predominantly white-collar, blue-collar, or skid row. All houses may very well have some proportions of each type. But over the course of their existence they try to fit candidates to their program and its objectives. Thus, white-collar alcoholics will predominate in privately run halfway houses, blue-collar alcoholics will predominate in state-subsidized houses, and skid row alcoholics will be in the majority in church-operated halfway houses.

As the detoxication–halfway house treatment system grows and expands, there may well be changes in this distribution of the population of alcoholic types among the differing kinds of halfway houses. But, for the present, it appears that there are five major sources of referral for halfway houses: walk-ins, detox centers, criminal justice system, hospitals (state hospitals and specialized inpatient alcoholic treatment centers), family members, and A.A. Once again, all types of halfway houses obtain referrals from all sources. But church-run halfway houses will have a preponderance of walk-ins, while state-supported halfway houses will have a growing preponderance of detox referrals. Finally, the privately run halfway house will obtain its major share of residents from A.A. referrals, family, and friends.

REHABILITATION

A Definition of Rehabilitation

Rehabilitation means reinstatement of former rank and privileges, restoration to previous condition and status, and vindication of moral character and reputation. When an alcoholic resumes his position in his family or at his place of employment, it is fair to say that he has been rehabilitated. However, rehabilitation is not always predicated on resumption of old group memberships. In many instances, old groups no longer exist. And, if some still do, most no longer want a former alcoholic member back on any terms. Furthermore, even in those instances where a group might welcome back their prodigal son, it often happens that he does not care to return. What is crucial to the definition of rehabilitation then, is the return of self-control. Now able to restrain himself, the person no longer requires regulation by agencies of social control whether in the system of criminal justice or in the system of health care. Thus, it is not the fact of group membership which signifies rehabilitation but rather the fact of self-control.

Indices of Rehabilitation

In the interaction between helper and alcoholic, helpers look for signs of rehabilitation. Alcoholics, in turn, seek to present these signs to their helpers for validation, support, and reward. The major sign, accepted by both parties, in some evidence of compliance with the rule of sobriety, regardless of its source, whether internal or external. Restitution for prior damages of whatever kind is often accepted as a sign of rehabilitation. Also important is an expressed willingness to change one's attitudes and one's actions. More important, of course, is actual change: The person stops drinking. Propriety in word and in deed is also helpful. Additional signs of rehabilitation are having an address, an established or fixed place of residence, and, since people are known by the company they keep, sober associations are additional signs of rehabilitation.

The Social Conditions of Rehabilitation

If self-control is the definition of rehabilitation, then group membership is one of its social conditions. The necessary condition for the rehabilitation of alcoholism is membership in a group seeking abstinence. The sufficient condition for rehabilitation is submission to the group's authority. Hence, residence in a halfway house may be instrumental in the rehabilitation of alcoholics. Compliance with the discipline and rule of abstinence imposed by the staff, in turn, may be activated and supported further by intimate association with other persons seeking the goal of sobriety, namely other residents as well as the staff. Numerous members of the halfway house movement will say that they are primarily trying to teach their residents how to live without alcohol. The first step towards learning that lesson is to acquire ways of conforming to the abstinence norm in the halfway house itself. For abstinence is the capital of rehabilitation and it can only be acquired in the course of special group membership.

Research Findings on Rehabilitation Outcomes

As more and more agencies assume public health responsibility, standards for judging rehabilitation change. Just as state-subsidized halfway houses have to have strict financial accountability, so also will more and more of them have to report the results they have achieved to justify investment of public monies in rehabilitation. With change in sponsorship, funding, and organizational support goes change in evaluative criteria. The change includes judges, systems of collecting and reporting data, and areas of behavior to be included in the concept of treatment outcome. These have necessarily led to more conflict over who can or should treat the alcoholic and what treatment success means.

Two schools of thought, lay and professional, have been ranged against one another for some time now. Lay judges of rehabilitation usually are members of Alcoholics Anonymous who get their data on rehabilitation from observed or reported attendance at A.A. meetings, and self-reports of treatment status. Generally, there is only one area of behavior worth considering, namely, is the person sober or isn't he? Professional judges look for objective accounts, including, wherever and whenever possible, evaluation studies. They seek measures of changes in behavior and attitudes in a number of areas (employment, arrests, drinking patterns, and attitudes toward the self, to mention only a few) before and after exposure to treatment (Pattison, 1974).

There is a vast literature on treatment evaluation, primarily when the treatment is medical, psychological, or a combination of both (Emrick, 1974, 1975). There are only a few studies of the results of halfway house rehabilitation. Though methods of evaluation are generally much cruder when compared with the psychological literature, one generalization seems to hold. Using long-term sobriety as a criterion of successful treatment, follow-up studies report on the average that only 20 percent of ex-residents are still sober six months or more after their discharge. For example, one study following up ex-residents a year after treatment found that only 12 percent had been continuously sober during that time (Blumberg et al., 1973). Myerson and Mayer (1966) following up 101 ex-residents ten years after residence found 22 percent sober, working, and living in families. Katz (1966) found 25 percent of 293 ex-residents still sober six months after residence in a Salvation Army facility. Finally, Nash (1962) found that 22 percent of a sample of 79 ex-residents were sober and working more than six months after they had left the halfway house.

Generally, then, these studies do not give the halfway house high marks when grading their efforts as rehabilitative agencies. Their main conclusion would seem to be that halfway houses achieve significant results and lasting changes (using the abstinence criterion) with a minority of their residents. Quite obviously there is a problem here. Since the 1950s, there has been a great increase in the number of halfway houses. Conservatively, there probably are over 1000 in all of North America right now. But what little research there is suggests that the effectiveness of halfway houses is inversely proportional to their number. The rapid growth of halfway houses may simply mean that more and more people in the alcoholism treatment complex believe in their effectiveness.

On the other hand, research reporting on relatively low rates of effectiveness in rehabilitation may have taken an overly simple view of rehabilitation. Most researchers hold a statistical theory, namely, that exposure to halfway houses ought to make a greater contribution to successful rehabilitation when compared with chance. Since most studies lack a control group, they do not test this theory at all. These studies have neither tested the halfway house theory of rehabilitation nor their own statistical theory. Moreover, they are silent on residents' backgrounds, the kind of halfway house they entered, the nature of their experiences there, and the relationship of their experience to the outcome. Hence, it is rather hard to arrive at any important theoretical conclusions based on these negative results.

A THEORY OF REHABILITATION

At present, there is no theory on the halfway house rehabilitation process. As things now stand, results are either equivocal or negative. In neither instance do they tell why halfway houses do or do not work. Nor do they tell with which kinds of residents they are most likely to work. A theory of rehabilitation would make it possible to answer both of these important questions. We turn now to such a theory.

Strains of Group Membership

As noted earlier, a necessary condition of rehabilitation is membership in a group that requires sobriety. In the course of such membership, the alcoholic person experiences rewards for sobriety, punishment for inebriety. In effect, then, he learns the rules and norms of a group that demands abstinence as one of its terms of relationship. But since people do not give up their ways without a struggle, living under these terms imposes considerable strain.

When a stringent rule of no drinking is added, the alcoholic is almost at his breaking point. Membership in a halfway house also requires submission to its rules, rules that are not of the resident's making. Thus, there is exposure to an external discipline, to a body of imposed rules, topped off by the severe abstinence requirement. Finally, there is the strain that results from close contact with a number of involuntary associates—staff and residents.

Thus, if the constant in halfway houses is subjection to the rule of group membership, then the important variable is the specific social response to these strains of group experience. Thus, any social theory of halfway house rehabilitation must explain variations in treatment outcomes as a consequence of adaptations to the strain of sober group membership.

Rehabilitation, as noted, is a social process in its own right. An index of its achievement is the degree of self-control the person has, the mastery over his own impulses—in particular, the impulse to drink and to get drunk. In this process of regaining self-control, the first state is comparable to the "surrender" process which members of Alcoholics Anonymous speak of. This surrender entails submission to the rules of a group.

To become abstinent, the person has to learn to live without alcohol. He has to learn how to become sober and how to stay sober. He begins by assuming the social status of alcoholic. Being in that position, which has its rights and duties like any other social position, he is expected to take positive action on that status, such as, for example, attending meetings where sobriety is the subject of discussion. In the course of attending these meetings in the company of people very much like himself, he comes to define himself as a person who cannot drink. He says to himself: "I cannot drink." He adopts the self-conception of an abstainer. The product of these activities is sobriety. A mutually stable set of expectations is slowly reproduced over the long course of time, the result being that the person and his actions are now predictable. Where, previously, family, friends, and acquaintances had come to expect inebriety from him, they now come to expect, just as he does, abstinence as a taken-for-granted aspect of everyday life.

Naturally, this does not come about overnight. It takes time. Translated into social action, it means that over the course of sober associations, the resident of the halfway house takes the group's norms and makes them his own. As he becomes attached to the staff, he similarly becomes committed to their norms, particularly the rule of abstinence. Thus, during the initial phases of membership, he learns how to become sober and to stay sober while in residence in the halfway house. Then, in the later stages of his residential career, he acquires the skills and the motives for staying sober after he has completed his period of residence.

But how he accomplishes sobriety during the initial phases and how he acquires the motives and the skills for postresidential abstinence depends, to a very large extent, on how he comes to terms with the authority of the halfway house. The manner of exercise of authority and the manner of response to it change during the resident's career. But authority never goes away entirely; it is always a fact of the halfway house experience. Authority produces the major strains of group experience. As such, it is important to understand how residents handle their problems of authority.

Authority and the Halfway House

Because of its small size and informality, the halfway house thrusts people into close contact. Two social segments constitute the community in contact—the staff and the residents. Staff defines and enforces the authority of the halfway house. Residents interpret and respond to that authority. Ultimately, then, when questions of the definition of any specific situation come up, a resident faces the problem of deciding whose side he will take, the staff's or the resident's, in appraising how he will think, feel, and act accordingly in that specific situation. As is true in most other groups, the interests of one social segment are not wholly identical with the interests of another social segment. Consequently, the resident has to establish for any given situation what his interests are and which side seems closest to his interests as he understands them. Given the fact of a difference in interests, the chances of conflict are quite good. It thus becomes necessary to make choices on issues that come up in the course of halfway house life. The mode in which these issues are resolved tend to fall into a set of patterns. Social types in the halfway house are defined by the patterned way they manage conflicts produced by the ever-present problem of authority.

The main interest of residents is in reducing staff authority. The main interest of the staff is in the maintenance of their authority. It is here that the issue of authority in the halfway house is joined.

Since there are only two social segments in the halfway house, the resident may or may not affiliate with either in an attempt to cope with the ever-present problem of authority. Thus, a resident can side with staff in the hope that they will minimize their control over him. A resident may elect to side mainly with other residents, hoping through this affiliation to mitigate staff authority. A resident may elect to side with neither segment in the hopes of reducing staff authority. And finally, a person may decide to affiliate with both social segments, depending upon the situation, so as to again reduce if not mitigate exposure to staff authority. Inside the halfway house there are mainly three kinds of social situations. These situations differ by the number and the mixture of the company. Thus there are situations in which the resident finds himself solely in the company of one or more residents. Then there are situations in which the resident finds himself solely in the company of one or more staff. Finally, there is the situation of mixed company when the resident is in the presence of one or more residents as well as one or more staff members.

Definition of and response to authority varies with these three social situations. For example, when residents alone interact, the principal common factor to their situation is their residence in the halfway house. And, the main feature of that residence which they share is exposure to staff authority. With staff members absent, that authority is, although potential or latent, not manifestly present. When, on the other hand, the resident is alone in the company of one or more staff members, that authority is manifestly present and he is on his own as to how he will come to terms with that staff authority. Finally, when the resident is in the company of other residents and staff members are also present, he has a set of potential allies in mitigating, if not reducing, the scope of staff authority if he so desires.

Staff members set down, define, and enforce the rules of halfway house life. Residents cannot easily avoid these constraints of group membership. Halfway house discipline requires discharge of these obligations. Residents do not police one another. That task falls to the staff. Residents, however, pay attention to evasion of norms by fellow residents (Rubington, 1965). Any evasion without notice or punishment constitutes a failure in enforcement. All residents experience costs of membership, abstinence being one of the high personal costs. Should they find one of their number paying less, literally and figuratively, it will be to their advantage to find out if they too can reduce their costs in similar fashion. Thus, depending upon how the entire staff exercises its authority, there could well be a contest over who controls the organization, the staff or the residents.

Social Types: Ways of Coping with Halfway House Authority

There are, then, three social situations in which the resident faces the problem of authority in the halfway house—in the company of other residents only, in the company of staff only, or in the company of other residents and the staff. In the course of a resident's membership, he will face these three social situations in varying degrees. Patterned styles of coming to terms with the dilemmas these situations pose have been developed. These patterns form four social types: mixer, company man, regular guy, and loner.

The Mixer

Mixers are the diplomats of social relations in a halfway house. They make choices with an eye toward reducing conflict with the particular people with whom they are interacting. For instance, when in the company of fellow residents only, they define most situations in accordance with the residents' views. If these views are somewhat antagonistic toward the staff and its program, then mixers will go along with whatever sentiments the residents express as a group. In those situations of mixed company, formal or informal, mixers are careful to avoid being put on a spot, of having to make a public choice for the view of one social segment as opposed to another. For instance, a mixer will enter the dining room before anyone else, and take a seat at an empty table. Thus, he does not have to decide with whom to sit. Similarly, during a group discussion chaired by an alcoholism counselor with a half dozen other residents around the table, a mixer will not volunteer any comments if he can help it. Should the counselor solicit his opinion or feelings on some matter under discussion, the mixer is most careful to respond with an observation which does not offend the sentiments of the counselor or of those residents present and which is widely shared by both staff and residents. Finally, when a mixer is in the company of staff members only, he will be most careful to talk about matters which are of some personal importance to him and which do require some degree of subjection to the authority of the staff member or members present. In these situations, he complies with the rule of talk on alcohol problems. And, in as much as he behaves like a well-motivated resident, he makes the staff member's task easier and more rewarding.

The Company Man

In all situations, the company man gives the impression that he looks at things the way the staff does. He defines situations the way they do. In situations where only residents are present, he behaves like a staff spokesman. Some or all staff members have become his role-models. Taking this line sometimes brings the company man into antagonistic relations, if not actual overt conflict, with other residents. If this should prove costly in reducing his chances of forming close ties with some of the residents, it can be used to cement relations with some, if not all, staff members when he is in contact with them alone. With or without subtlety, directly or indirectly, he can make it plain that he, in contrast to some of the other residents of the halfway house, is a strong supporter of the current staff regime and its ways of defining and enforcing the halfway house's rules. Finally, in situations of mixed company, whether formal or informal, he will go out of his way to make his choices public. He has considerable skill in finding out which residents seem to be publicly against certain staff members, whether it be on personal or ideological grounds. Once he learns this important social information, he sets himself against that person. His public behavior thus indicates the point of view he embraces as his own. If, for instance, the person whom he has observed to be opposed to the staff voices public sentiments that are anti-A.A. then he will go out of this way to make it clear that he holds pro-A.A. sentiments. Thus, by defining himself as opposed to those who have voiced opposition to the staff, he defines himself as a public supporter of the staff and their definition of the situation.

The Regular Guy

Regular guys are company men turned inside out. Their own choices and alignments are quite clear in all social situations. When in the company of residents only, they talk like residents. Their basic loyalties lie with their fellow residents as opposed to members of the staff. By virtue of the fact that they are alike, equal in the sense of their exposure to and subjection to the authority of the staff, they are bound by these quasi-kinship ties to all others in the same situation, their fellow-residents in the halfway house. Their orientation is quite simple, obvious, and pervasive for all three halfway house situations. Alone with residents, they side with them whenever a discussion of events taking place in the house comes up. They carry over the residents' view in those situations of mixed company, be they meals or meetings, in which residents and staff come together. In these circumstances, they publicly support the residents' view. Depending on current atmosphere in the halfway house, the view ranges all the way from harsh, overt antagonism to friendly, but kidding, relations. Here the joking relationship makes it possible for the regular guy to express negative sentiments toward the staff in a socially acceptable fashion. The joking relationship, however, is most likely to manifest itself when the regular guy finds himself alone in the presence of one or more staff members. The joking relationship permits the regular guy to express his social distance from the staff's view. By so doing, the regular guy makes it plain that it is not so much that he is hostile to staff, whether in particular or in general, so much as he is much friendlier to other residents and their sentiments, to their definition of the halfway house situation.

The Loner

The loner is an enigma to both staff and residents alike and for a very simple reason. Overtly, it is absolutely impossible to know whose side he is on, what point of view he favors, which segment he identifies with. The loner gives no sign to any of the participants in halfway house life where he stands. Most of the time in the three social situations he is noncommittal. When in the company of fellow residents only, he remains silent whenever a tentative loyalty test is given. When in the company of staff members only, he will be equally, if not more, impenetrable. And whenever he finds himself in the situation of mixed company, he adjusts by hiding self. As a result, residents may assume that he is loyal to the staff and mistake him for a company man. And, staff may assume that he is loyal to the residents and mistake him for a regular guy. In view of the fact that most loners are guarded and most incommunicative, it is easy to see how these mistaken assignments can happen.

These social types are ways of coping with authority. All are patterns of overt social behavior that can be observed in the halfway house. The inner states that accompany these patterns are inferred from the behavior. The more frequently and the more openly that residents talk, the easier it becomes to assign them to one of the four possible social types. But just as people vary in their drinking patterns, so do they vary in the ways they talk. The ways in which residents talk or do not talk helps to assign them to their social types. Some men are reticent, somewhat reluctant to say too much during their initial settling-down period in the halfway house. And, if some men, despite the close quarters they share, never become intimate with one another, still others only become friendly and close with some of their fellow residents. And, it others only become, in time, close with some staff members, there are, finally, those residents who can be equally at home and at ease in the single company of residents or of staff or in the mixed company of both residents and staff. The ease with which they can be in the company of other men and talk about a variety of subjects without recourse to alcoholic beverages and without flaring up or expressing hostility is an acute measure of their adjustment to the strains of the experience of sober group membership.

Each of these social types has come into being as a means of reducing the costs of membership in the halfway house. The major cost, of course, centers around subjection to the authority of the halfway house and its rule and discipline of abstinence. And each of these social types varies in its ability to reduce the costs of group membership. It is precisely here where a link between social type (a behavioral means of coping with the process of rehabilitation) and the outcome of the selfsame rehabilitation process may be established.

On admission to a halfway house, ways of coping with its built-in social strains are available, then. A new resident in time can become a mixer, a company man, a regular guy, or a loner. These social types encompass alternate ways of coping with the problem of halfway house authority. In effect, each is not only a way of coming to terms with the strains of sober associations but is also a precursor, a link, and a predictor of post-halfway house adjustment.

Ex-Resident Social Roles

Residents have careers in halfway houses. These are defined and classified by such items as duration of membership, conformity to halfway house rules, social type, etc. These careers fit ex-residents to one of four post-halfway house roles: alcoholism professional, reformed drunk, abstainer, and drunk. The *alcoholism professional* is a former long-term resident of a halfway house who graduates to assume a post in the rehabilitation of alcoholics, sometimes in the very halfway house in which he regained his sobriety, but more often in a different one. Other posts include alcoholism counselor in a detox center, in an antipoverty program, in an inpatient or outpatient treatment center, or in industry. The *reformed drunk* regains sobriety but lacks membership in a group. Where the alcoholism professional acquires and maintains his status by working for the sobriety of "active" alcoholics, the reformed drunk seeks simply to proclaim and maintain his own status solely on the fact that he has achieved sobriety. In effect, he claims the status of a self-made man. The *abstainer* has rejoined conventional society. He acquires status by what he has achieved in the occupational world. The fact of his abstinence is strictly incidental.

When a person presents himself at the front door of a halfway house, he is a candidate for recovery from alcoholism. He brings with him a mixed set of motives and skills, and a characteristic way of doing things, in short, his own personality. These affect his candidacy. In addition, the nature of his candidacy is already well shaped in considerable degree by his alcoholic class and his past experiences as a member of that class—how others have reacted to him as an alcoholic, how much his drinking has impaired him, and what social resources he still has left. But whether or not he arrives at the stated goal of recovery depends on how he behaves in the process of social rehabilitation.

Social rehabilitation takes place in the halfway house. It constitutes a patterned set of experiences in a special kind of group. These experiences, necessarily, impose a good deal of strain on the candidate, the new resident. Experience in groups induced his drinking and drinking altered his experiences in groups. Thus, being a member of a human group came to mean for the average alcoholic more punishments and fewer rewards. Now he is placed in the difficult position of having to be a member of a group in order to achieve the highly valued goal of sobriety. He has to come to terms with the strain of these new group experiences in some fashion. His candidacy for recovery, thus, depends on how he goes about handling this difficult task.

The halfway house, as a helping organization, has a stated goal, recovery from alcoholism for its residents. This outcome becomes possible if and when social control leads to self-control. This can happen in several ways. Residents form an attachment to staff people and then later become committed to their sober norms. Or, residents, already committed to the sober norms of the staff, later become attached to them as people. Or, as can happen on occasion, residents may become alienated from staff as people, but may become committed to their sober norms through their attachments to those residents who do subscribe to the discipline of abstinence (Rubington, 1963). The upshot, in either instance, is that after halfway house residence, ex-residents continue to make these commitments their own. Thus, if they remained sober despite cravings for alcohol during their residence largely to retain the benefits of group membership, then after membership they are able to resist craving to drink by themselves.

This process of resocialization is quite complex and there are numerous obstacles in the way of successful outcomes of the kind just described. The principal ones have to do with any event or process that interferes with either attachment to staff people or commitment to their norms. The nature of the halfway house program of rehabilitation contains its own built-in obstacles. The principal obstacle, as already noted, is the authority system in the halfway house itself. The exercise of this authority, so necessary for the maintenance of the organization as well as for operation of its program, can very easily interfere with the process of attachment to sober staff people, commitment to sober norms, or both. The exercise of that authority in whatever manner is a problem that all residents have to come to terms with.

As already noted, the system of social types becomes the sources of responses to the authority problem. Selection of one of the four types constitutes four different ways of participating in the halfway house. It culminates in four different kinds of experiences in the halfway house. These types have implications, then, not only for the rehabilitation process but for the post-halfway house or recovery stage. Different kinds of careers are followed inside the halfway house, depending upon whether a resident becomes a mixer, company man, regular guy, or loner. Signs of these careers include length of stay, kind of discharge, conformity with house rules, and changes in attitude and in behavior.

Thus, a resident engaged in the process of social rehabilitation is a member of a group. How he experiences that group depends on the social type he has elected. Mixers experience the halfway house quite differently from company men. And, similarly, the experience of regular guys bears no similarity to that of the typical loner. The important point is that these variations in group membership styles prepare residents for alternate kinds of experience during the recovery stage or the post-halfway house period. Thus, there is some degree of consistency between the kinds of group membership exhibited during the halfway house career and the subsequent kinds of membership in which the ex-resident becomes involved. It is in this sense that there is a relationship between rehabilitation and recovery. The kinds of social types which characterize the resident's main adaptation during his career in the halfway house influences the kind of post-halfway house social role he is most likely to assume.

Halfway House Social Types and Ex-Resident Social Roles

The stated goal of the halfway house is to assist its residents to recover from alcoholism. Thus measures of its effectiveness, in part, must take into account the post-halfway house role of its ex-residents. Focus on social role, necessarily, requires information on the kinds of groups the ex-resident is now a member of and the kinds of experiences he is currently having in them. The halfway house equips its members, other things being equal, to develop motives and skills for maintaining membership in a group demanding abstinence as a condition of membership. Thus, the test of a halfway house's success is the number and kinds of sober roles ex-residents perform after membership and the kinds of groups in which they perform them.

The four principal post-halfway house social roles are alcoholism professional, reformed drunk, abstainer, and drunk. These roles are performed in different kinds of groups. The alcoholism professional, for example, has gone from a group in which he received help with his alcoholism to a helping group in which he has membership and in which he now helps others who may want help with their alcoholism. He validates his sober role in that kind of group membership. He has joined a helping group after his experience in a halfway house. By contrast, the reformed drunk has left the halfway house and yet continues in a sober role but without the benefits of any group membership (Rubington, 1964). Where the alcoholism professional justifies his past inebriety by his current sober service in a helping group, the reformed drunk justifies his past inebriety by his own current sobriety without reference to any group membership whatsoever. His sobriety may well be quite precarious because of the fact that he is not a member of a primary group. He may nonetheless experience fairly long-term sobriety in spite of its shaky foundation. The abstainer performs that sober role in a set of conventional groups that he has either joined or rejoined. That sober role is tangential to rather than contingent upon those group memberships. And, in the course of time, it will become hardly noticeable. Finally, the drunk has rejoined the actively drinking alcoholic category. For many, though not necessarily all, that connotes performing the inebriate's role in a deviant drinking group that the drunk has either joined or rejoined (Rubington, 1968).

Some Hypotheses on Social Types and Ex-Resident Social Roles

The theory presented here thus must answer two questions: (1) Which social types in the halfway house are most likely to produce what kinds of sober ex-resident roles? (2) What kinds of houses are more likely to produce which kinds of sober ex-resident roles?

To begin with, the frequency and kinds of opportunities open to ex-residents have a significant bearing on ex-resident roles. Fluctuations in the supply and demand for jobs, inside as well as outside the alcoholism movement, will have their influence on posttreatment outcomes. Similarly, shifts in the tolerance for relapse and the like will affect the climate of opinion surrounding ex-residents' adjustment. Along with variation in the array of opportunities available, there is the fact that relapse is always a possibility, regardless of social type. There is, then, no 100 percent correspondence between a given halfway house social type and a particular ex-resident social role, sober or drunk. Any of the social types described here can relapse at any time, during or after their residency, and all of them have. Any of the social types can move into one of the four post-halfway house social roles and, at some time, all of them have. But the theory outlined here predicts that certain of the social types are more likely to lead to one rather than the other kind of ex-resident social role. Hence, the theory predicts that: (1) a significantly greater percentage of mixers will end up as alcoholism professionals; (2) a significantly greater percentage of company men will end up as reformed drunks; (3) a significantly greater percentage of regular guys will end up as drunks; and (4) a significantly greater percentage of loners will end up as abstainers.

In adapting to the internal problems of authority in the halfway house, each of the four social types opens up certain kinds of experiences in the halfway house while closing off others. More to the point, these adaptations carry over into the postresident social world, closing off certain kinds of experiences in groups and at the same time opening up other kinds. There is then a rough goodness of fit such that social types of one kind fit the ex-resident best for one rather than another post-halfway house social role. Thus, the theoretical argument is that mixers are best fitted for post-halfway house social roles as alcoholism professionals, that company men are best fitted for the subsequent roles of reformed drunks, that regular guys are best fitted for continuing in the social roles of inebriates, and that, finally, loners are best fitted for the social roles of abstainers. The basis for these theoretical expectations follows.

Mixer to Alcoholism Professional

Mixers handle the problem of authority through direct contact with the social segment in whose company they are. When in mixed company, mixers appear as allies of both social segments in the halfway house. And when in the company of their fellow residents only, they present themselves as supporters of the residents' viewpoint. But when alone with staff members, they give the impression of accepting their authority. Thus mixers reduce whatever friction social interaction in a halfway house is likely to produce. Other things equal, mixers have the easiest time in their residence with both social segments. For mixers, as their title suggests, have the capacity to get along well with people. As a result, an important reciprocal arises out of their pattern of social relationships. As they can give their best as members of the halfway house group, so, similarly, are they able to take the best that the organization has to offer. Mixers reduce exposure to staff authority on the one hand, exposure to residents' appeals to solidarity on the other. The way they can take appropriate sides in face-to-face interaction implies an ability to understand how each segment views the world. Living in a halfway house, they cultivate this ability as a means of managing their own personal affairs. But the development of this ability as a means of coming to terms with the problems of halfway house existence is, at the same time, excellent on-the-job training for anyone who would pursue counseling work in an alcoholism service agency. By virtue of their adaptation, mixers are able to integrate staff as well as resident perspectives. Thus, by the time the average mixer completes his halfway house residency, he has already become an excellent candidate for work in an agency servicing alcoholics. Borrowing a term from medical education, a halfway house residency, for mixers, is the first step toward becoming an alcoholism professional.

Company Man to Reformed Drunk

Company men readily adopt the staff's viewpoint, in some instances right at the time of the initial intake interview. The ease with which they take on the staff's view is inversely correlated with their length of stay. In effect, there are short-term company men and long-term company men. Short-term company men are generally classed by staff and by residents alike as sycophants, manipulators, or "con artists." They are said to be using the halfway house as a "flop," or that they simply come in "to get out of the cold," or that they are only there to "get the wrinkles out of their bellies." Yet, during their very short stay, they manage to praise the staff while alienating their fellow residents. All halfway houses have a high attrition rate and company men contribute their fair share to the early dropouts. No small part of their early departure is the fact that all residents are onto their game and want little to do with them.

Long-term company men, by contrast, stand out by the fact that they have managed to form close ties with at least one staff member (Rubington, 1961) or have become committed to halfway house sobriety norms. These close ties become the basis for a relationship that enables the company man to withstand the implicit scorn of his fellow residents. The company man alienates fellow residents. The experience of alienation either thrusts him out of the house or into closer contact with staff or a stronger commitment to house norms. The company man's style induces abrasiveness with fellow residents which creates an internal problem which a staff member can resolve through counseling. Again, certain reciprocals in social relationships recur in the halfway house. The company man arouses antagonism among residents. Implicit abrasiveness can turn into explicit friction at any time. To cope with this internal adjustment problem, a company man can turn to a staff member. Low contact with potentially abrasive role-partners produces high contact with specialists in reducing any frictions in human relations that can become the pretext for a drinking bout. As a result, counselors help company men to remain residents in the halfway house without resorting to drink. The company man brings problems of halfway house adjustment to staff members. Staff members convert these human problems into alcohol problems and provide an in-house solution. In effect, then, the staff member becomes an inadvertent ally of the company man, making it possible for him to continue living in the halfway house without reducing his abrasive interactions with fellow residents. On the contrary, as the company man increases his length of sobriety relative to all the residents who are falling by the wayside, he actually becomes an important person to the halfway house and to the staff alike. Since he demonstrates the success potential of the program, staff can be indulgent with him.

The upshot is that should the company man enter the ranks of the alcoholism professionals which his long-term sobriety in part may entitle him to (scarcity of talent coupled with a high demand for recovering alcoholics may facilitate this kind of situation), he is most apt to continue the company man approach to rehabilitation. He will continue to antagonize residents in the exercise of his authority. Sobriety is the badge of his authority and he will wield that like a club over persons he was hired to counsel. But, most of the time, the long-term company man graduates into the ranks of the reformed drunks. These are special social types who exist on the borders of the alcoholic community, both its wet and its dry segments. The reformed drunk holds no active membership in any conventional primary groups during the period of his postresidency. What he is left with is exactly the same adjustment style that served him while a resident in the halfway house. During his residency, he remained aloof from his fellow residents and considered himself to be morally superior to them. In time, as he stayed on while many residents left, the staff confirmed this conception he held of himself, if only because of the hard-won sobriety which he had amassed in the face of so many failures. After his residency, he continues to lord it over people who are not yet sober or who are by no means as sober as long as he is. The paradox is that his behavior supports the norms of the halfway house though it is in no sense based upon any strong attachment to the staff as people. In the end, then, the company man experience fits the person for a parade of sobriety. The company man parades sobriety to his and to the staff's satisfaction during his residence. And, during the period of postresidency, he continues to parade it to the exclusion of all else. With the help of the halfway house and the staff, he has made himself into a reformed drunk.

Regular Guy to Drunk

As noted earlier, the regular guy is the company man turned inside out. Where the company man sides with the staff in most situations, the regular guy sides with fellow residents in most situations. The difficulty with this adaptation is that in the long run it does not reduce problems with authority. On the contrary, regular guys intensify their exposure to halfway house authority. As a result, over time, this style increases the chance of abrasiveness between staff and residents. As abrasiveness in interaction mounts, it only serves to confirm the regular guy's view of the halfway house as a social organization. He is particularly sensitive to issues that arise from time to time and the fact that some staff members may be playing favorites in one way or another. One way is to indulge one resident while depriving a second. Laxity in the enforcement of rules is one approach to this problem. Near confrontations with staff members coupled with backbiting among residents when staff are absent activates a cycle of alienation. Internal alienation of regular guys from staff members is stabilized and is made the fundamental basis of solidarity among regular guys. Necessarily, this only increases the antagonism, latent or manifest, between regular guys and staff. As they develop their adversary relationship to the staff, they simultaneously draw closer together with others in the segment of regular guys.

The fundamental symbol of acceptance of halfway house authority is "opening up." A resident opens up when he begins to talk about his drinking problem openly in all three of the halfway house situations. When staff see signs of residents opening up they begin to believe they are getting the program. Opening up is the most visible proof that residents are now looking at their alcohol problems through the eyes of the staff. Regular guys reverse this social process. To the extent that they become involved in a resistance movement against the staff, their norms, and their way of defining alcohol problem situations, they go the other way and reduce, if not avoid, talk about their drinking problem. As the saying goes, they "clam up." Their way of experiencing the halfway house as a group also influences the manner in which they leave it.

Attrition always is a fact of halfway house life because of the high rate of relapse. But relapse is much more noticeable when it involves regular guys. And this is so because of the fact that, at any given time, a preponderance of regular guys are in residence. Thus, more regular guys know more regular guys, and some times two or three establish a very close primary group relationship. Relapses, which of course are ruptures of halfway house group membership, take two forms when regular guys are involved. Relapses are either paired or sequential. In the paired relapse, two residents will leave the halfway house together and go on a drunk. In the sequential relapse, one person relapses, and then another feels compelled to follow his example shortly thereafter. And sometimes a sequence can become a train if enough regular guys follow each other's example. In the case of the paired relapse, two persons have become very close. Being close means that they have come to enjoy doing things together. In time, under special circumstances, that can come to mean getting drunk together. In the case of the sequential relapse, the persons involved do not have close ties with another resident. But they are friendly with, close to the whole segment of regular guys. This involves identification with them to some unknown though important extent. Thus, seeing one regular guy relapse makes another regular guy feel that his own relapse is imminent.

Staff members may seek to intervene and to point out the social signs of impending relapse. They may counsel a resident and indicate that what has happened to X and to Y is now very likely to happen to Z unless he takes some preventive action. However, the typical regular guy, no matter how much he longs for sobriety, by the nature of his adaptation to the halfway house, by the way he experiences it, is predisposed less and less to talk about his alcohol problems. Now though his urges toward sobriety have become stronger, and though he actually fears that what has happened to others may well happen to him, he lacks practice in doing the one thing that will most likely prevent if not postpone his own relapse, namely, talking to the staff about the way he feels. In addition to a lack of practice and of skill in talking about himself and his drinking problems (or other personal problems for that matter), there is a sense in which he regards such talk as morally wrong, inappropriate, out of place, and quite and most definitely out of character. Thus the wall that the regular guy has built between himself and members of the staff cannot come down as fast as the situation requires it, if he is to remain sober. As a result, resignation builds, he gives in to what he considers to be the inevitable, and goes out and gets drunk. Once again, an adaptation that is to well suited to being friendly and one of the guys while in the halfway house fits him quite well for rejoining the ranks of the actively drinking people. For that is precisely the kind of style that maximizes the fellowship of heavy drinking and minimizes the heavy toll it takes on those who participate in it.

Loner to Abstainer

The loner, because of his reticence in communication, is an enigma. He takes no sides if he can help it. He avoids alignment with both social segments of the halfway house. Hence, confrontations or displays of antagonism with either side, staff or residents, are reduced if not actually eliminated. The advantages of the loner's style of being in the halfway house is that it takes him out of the crossfire of social interaction. For him, this means interpersonal peace. But his adjustment has its costs. He does not become attached to persons in the halfway house, be they staff or residents. Similarly, it is unclear to all whether or not he is committed to halfway house norms, whether he is in process of becoming committed, or whether he has already become disenchanted with its rule and discipline of abstinence. Men size up other men by what they talk about and how they talk about it. Since the loner says so little and is so noncommittal, the loner cannot be sized or fitted to the prevailing system of social types in the halfway house. The staff doesn't know what to make of him but, then again, neither do his fellow residents.

An unknown number of loners—like mixers, company men, and regular guys—relapse and return to the ranks of the drunks. But, according to the theory presented here, a much higher percentage of loners successfully makes it into the ranks of the anonymous ex-alcoholics. Experiencing little group pressure during their period of halfway house residence because of their ability to sidestep the problem of authority, they are well suited after discharge to continue a style which enables them to navigate between groups rather than become a part of one. Loners make their own way by integrating their perspectives as they see fit. If they were committed to sobriety prior to their residence or developed a personal commitment to it in the course of their residence, then they can extend that same pattern of adaptation afterwards. They take what sobriety they can get on the margins of social groups, anonymous figures even in their abstinence. For loners are in but not of groups, conforming to while not being committed to their norms.

Types of Halfway Houses and Halfway House Social Types

Just as all halfway houses produce relapses at fairly high levels, so do they also engender the whole range of social types previously discussed. Thus, privately run, state-subsidized, and church-operated halfway houses all have their share of mixers, company men, regular guys, and loners. But, again, these houses vary among themselves in the proportions of social types they produce. Social types, as noted, always emerge as ways of coping with halfway house authority. Thus, any variation in the proportion of social types found across the spectrum of halfway houses will be in response to the variations in halfway house authority. These variations lead to the following predictions:

1. State-operated halfway houses will produce the highest proportion of mixers.
2. Privately run halfway houses will produce the highest proportion of company men.
3. Church-operated halfway houses will produce the highest proportion of regular guys.

Halfway houses have typical patterns of authority. The privately run halfway house is characterized by paternal authority, the state-subsidized halfway house by fraternal authority, (Janowitz, 1959), and the church-operated halfway house by impersonal authority. These patterns of authority engender somewhat different kinds of strain in the group experience and, in turn, evoke patterned responses on the part of the residents.

The privately run halfway house tends to be small and to be more often than not a one-man operation. Many directors of privately run halfway houses are truly charismatic leaders who set themselves up as exemplars of abstinence. The climate in these houses varies from benevolent to authoritarian paternalism. The response most conducive to handling the problems of paternal authority is that of the company man. The company man in the privately run halfway house is more apt to become a long-term resident, to become attached to both the director and his norms. In time, the company man will graduate and honor his director by seeking to emulate him by starting his own or working in someone else's halfway house. It is not the case that other types will not emerge in the privately run halfway house. It is just that mixers, regular guys, and loners are more apt to find living in the paternal halfway house a bigger strain.

The state-subsidized halfway houses are middle-sized. Authority is distributed among the counselors who are in more frequent and much closer contact with residents than with either director or manager. Distribution of authority lessens it somewhat and makes it more horizontal, whereas in the privately run halfway house it is vertical and much more concentrated. There are slightly more staff personnel, mainly counselors, who are in authority over residents. This authority is fixed by shift work. Residents in the state-subsidized halfway house have to come to terms with several authority figures, not just one. In addition, of course, there are their fellow residents. Given the array of staff and resident personalities that residents must somehow adapt to, the evidence suggests that mixers are in a better position to cope with the pattern of fraternal authority. Company men, in addition to antagonizing the bulk of the residents, are very likely to produce jealousy and antagonism among counselors. Regular guys are more apt to succeed in uniting staff members against them. Mixers, by the nature of their style, reduce the risk that counselors will have to act as authorities in their presence. Their interactions will be fluid in contrast to the abrasive interactions characteristic of company men and regular guys.

The church-operated halfway house tends to be the largest of the three kinds of halfway houses. A greater tendency towards specialization brings in its wake impersonal authority. Though all social types will appear in the church-operated halfway house, the regular guys have the best chance to stick it out in the face of the built-in kind of authority problems. In the face of this kind of authority, the lines are sharply drawn—are you one of us or one of them? One of the risks is that a mixer can easily be mistaken for a company man in this kind of a setup. The result is that, behind the scenes, regular guys make mixers, company men, and loners equally miserable. Their own abrasiveness with the other three draws their own segment closer together against impersonal authority. But it tends to shorten the length of time that mixers, company men, and loners can or want to remain in this kind of polarized atmosphere.

This theory of halfway house rehabilitation links the resident's antecedents with the halfway house and how he interacts in it to the kind of role he performs after he has left. The example of the alcoholism professional may clarify the argument. The alcoholism professional, though perhaps the least frequent in numbers, is the most successful of the ex-resident roles on any set of criteria sophisticated follow-up studies might use. According to the theory being advanced here, there are three ways of becoming an alcoholism professional. At a turning-point late in his drinking career, a white-collar alcoholic who has experienced mild societal reactions, has minimal impairments, but maximal social resources, enters a privately operated halfway house. Shortly thereafter he adopts the role of the company man. After a successful period of residence, he graduates from the halfway house and obtains a position in one of the many new organizations in the growing alcoholism treatment complex as an alcoholism counselor.

Similarly, a blue-collar alcoholic, at the same late point in his drinking career along with his own peculiar assortment of reaction, impairments, and resources, enters a state-subsidized halfway house. Shortly thereafter he becomes a mixer. And then, considerably later, he leaves the halfway house to take a job as a counselor in a detoxication center.

Finally, a skid row alcoholic, at the late stages of his drinking career, carrying his own experiences with severe societal reactions, maximal impairments, and meager social resources affiliates with a church-operated halfway house. Almost upon arrival he becomes a regular guy. And, given close attachment with other regular guys, all of whom are strongly committed to the abstinence norm, he manages to stay for the optimum period. Later, he graduates to become a paramedic in a detoxication center.

These three hypothetical illustrations indicate how alcoholics of certain kinds of social backgrounds are more apt to get into one rather than another kind of halfway house. And, once in and depending on the pattern of authority, they are more apt to become one rather than another kind of social type. The kind of social type they have become in the course of their residence fits them best for some rather than other kinds of ex-resident roles. Thus, the part ex-residents play in the social world after their halfway house residence depends on where they came from and how they managed the rehabilitation experience in the halfway house.

SUMMARY AND CONCLUSIONS

Experiences in groups produce alcoholics. They then go on to have certain types of experiences as alcoholics. These are shaped in part by the social class they begin in or end up in, and in part by the effects of societal reaction and of impairment on them. But, in any case, they come to exist on the margins of or outside of important groups. As they become concerned about their affliction, they seek to do something about it. The growing alcoholism treatment complex develops a variety of services in which to treat them. Whatever else these services are, they are also groups in their own right. Halfway houses have multiplied and have come to play an important part in the entire continuum of treatment for alcoholics. More and more alcoholics enter halfway houses, seeking rehabilitation and recovery. The characteristic ways in which halfway houses are organized and work present a set of problems, a number of strains to persons who have had less and less successful experience in groups over the years. Though all may seek recovery, how they handle these strains influences treatment outcome. Only a certain range of experiences in these groups depending again on the kind of halfway house are apt to result in recovery.

REFERENCES

"A Body of Knowledge. Halfway House Alcoholism Programs. Administration and Programming," 1975, Association of Halfway House Alcoholism Programs of North America, Inc., St. Paul, Minnesota.

Albrecht, G. L., 1969, The structure and dynamics of two halfway houses for Skid Road alcoholics, Proceedings, 4th Annual Conference of Association of Halfway House Alcoholism Programs of North America, Inc., Mayo Hotel, Tulsa, Oklahoma, October 12-15, pp. 62-77.

Baker, T. B., 1972, Halfway houses for alcoholics: Shelters or shackles, *International Journal of Social Psychiatry* 18:201.

Blacker, E., and Kantor, D., 1960, Halfway houses for problem drinkers, *Federal Probation,* 24:18.

Blumberg, L., Shipley, T. E., and Shandler, I. W., 1973, "Skid Row and Its Alternatives: Research and Recommendations from Philadelphia," p. 173, Temple University Press, Philadelphia.

Cahn, S., 1970, "The Treatment of Alcoholics," p. 155, Oxford University Press, New York.

Cannon, M. S., 1973, Alcoholism Halfway Houses—General Characteristics, U. S. Department of Health, Education and Welfare, Statistical Note 73.

Emrick, C. D., 1974, A review of psychologically oriented treatment of alcoholism. I. The use and interrelationships of outcome criteria and drinking behavior following treatment, *Q. J. Stud. Alcohol.* 35:523.

Emrick, C. C., 1975, A review of psychologically oriented treatment of alcoholism. II. The relative effectiveness of different treatment approaches and the effectiveness of treatment versus no treatment, *Q. J. Stud. Alcohol* 36:88.

Janowitz, M., 1959, Changing patterns of organizational authority: The military establishment, *Adm. Sci. Quart.* 3:473.
Jellinek, E. M., 1960, "The Disease Concept of Alcoholism," p. 112, Hillhouse Press, New Haven.
Katz, L., 1966, The Salvation Army's Men's Social Service Center II. Results, *Q. J. Stud. Alcohol* 27:641.
Myerson, D. J., and Mayer, J., 1966, Origins, treatment and destiny of Skid-Row alcoholic men, *N. Engl. J. Med.* 275:419.
Nash, D. T., 1962, Chronic alcoholism treated via halfway house and calcium carbimide, *N. Y. State J. Med.* 1962:3098.
Pattison, E. M., 1974, Rehabilitation of the chronic alcoholic, *in* "The Biology of Alcoholism" (B. Kissin and H. Begleiter, eds.), p. 633, Plenum, New York.
Pattison, E. M., Coe, R., and Rhodes, R. J., 1969, Evaluation of alcoholism treatment: A comparison of three facilities, *Arch. Gen. Psychiatry* 20:486.
Pittman, D. J., and Gordon, C. W., 1958, "The Revolving Door," Yale Center of Alcohol Studies, New Haven.
Rubington, E., 1957, The chronic drunkenness offender in Connecticut. III. The rehabilitation experiment evaluated, *Conn. Rev. Alc.,* September (no pagination).
Rubington, E., 1960, Relapse and the chronic drunkenness offender, *Conn. Rev. Alc.* 12:10.
Rubington, E., 1961, The alcohol offender and his treatment, *in* "Legal and Criminal Psychology" (H. Toch, ed.), p. 397, Holt, Rinehart & Winston, New York.
Rubington, E., 1963, Social cohesion in a halfway house, *in* "Toward Intensive Treatment for the Tuberculous Alcoholic Patient," p. 9, Firland Sanatorium and NIMH.
Rubington, E., 1964, Grady 'breaks out': A case study of an alcoholic's relapse, *Soc. Prob.* 11:372.
Rubington, E., 1965, Organizational strains and key roles, *Adm. Sci. Quart.* 9:350.
Rubington, E., 1968, The bottle gang, *Q. J. Stud. Alcohol* 29:943.
Rubington, E., 1970, Referral, past treatment contacts, and length of stay in a halfway house. Notes on consistency of societal reactions to chronic drunkenness offenders, *Q. J. Stud. Alcohol* 31:659.
Schmidt, W., Smart, R. G., and Moss, M. K., 1968, "Social Class and the Treatment of Alcoholism," University of Toronto Press, Toronto.
Trice, H. M., 1966, "Alcoholism in America," pp. 30–38, McGraw-Hill, New York.

CHAPTER 10

Evaluation of Treatment Methods in Chronic Alcoholism

Frederick Baekeland

Department of Psychiatry
Division of Alcoholism and Drug Dependence
State University of New York
Downstate Medical Center
Brooklyn, New York

INTRODUCTION

The past 20 years and the last 10 in particular have witnessed an impressive increase in interest both in treating and understanding chronic alcoholism. The early part of this period saw the introduction of disulfiram and citrated calcium carbimide. A little later, tranquilizers and antidepressants came to the fore. At the same time there was a rapid increase in the vogue for group as opposed to individual psychotherapy. (The latter, along with Alcoholics Anonymous, had been an established therapeutic mainstay since the Second World War.)

More recently, behavioral psychotherapeutic approaches have become popular. Each treatment method has its proponents, who tend to downgrade other techniques. Nonetheless, however understandable therapists' devotion to the methods they know and however human their suspicion of those they are less familiar with, well-designed research on methods for the treatment of chronic alcoholism, and, therefore, solid conclusions about their indications and relative efficacy are badly needed. Unfortunately, they are few and far between.

I hope the reader will understand that even if my critical review of the past 20 years' English language literature at times may sound nihilistic, its intention, embodied in the summary at the end of each section, is to offer whatever suggestions, however tentative, can be made on the basis of present knowledge. I have chosen to look at alcohol treatment both programmatically, in terms of inpatient and outpatient treatment programs, which include a variety of treatment regimens, and in terms of the currently popular single modality approaches of behavioral psychotherapy, A.A., and drug treatment.

The reader unacquainted with alcoholism may be surprised to discover how high a percentage of alcoholics receive substantial and lasting benefit from treatment and it may astonish him even more to be told that therapeutic outcome seems pretty much the same regardless of the kind of treatment, but depends to a much larger extent on the kind of patient being treated.

TREATMENT GOALS AND OUTCOME

First of all, what are the goals of alcoholism treatment and how should the patient's progress be assessed? It would hardly occur to the professional unfamiliar with alcoholism to question the goals of its treatment, even less to inquire how its outcome should be measured. Surely, he thinks, the goal of treatment must be abstinence, and drinking behavior the measure of its outcome. Yet, these are two of the most vexing questions in the field of alcoholism, questions that have generated a great deal of controversy, questions, it turns out, that are inextricably intertwined.

How, then, should the results of treatment be measured? Does it suffice to determine abstinence or some other measure of alcohol intake, or should a more comprehensive and broad-based assessment be made that includes the patient's role functioning at work and at home as well as his physical status and general psychiatric symptomatology? Published studies have varied widely with respect to the outcome measures they used. Despite its intuitive plausibility as *the* measure of treatment success or failure, there are several arguments against the use of abstinence as the sole criterion of outcome: (1) Because of patient denial, it is less reliable than other measures such as occupational status, stability, and performance, failure in which is not so fraught with shame, and which can be determined from employers, who may keep objective and accurate records on such matters. (2) A patient may improve on one measure but not on others. (3) Abstinence is no guarantee of a good life adjustment. Thus, Gerard *et al.* (1962) reported that of a group of alcoholic outpatients abstinent for at least a year, 54 percent were "overtly disturbed" (had excessive community or social activities), 24 percent were "conspicuously inadequate" (led meager lives), 12 percent were "A.A. successes" (had little or no social life apart from A.A.), while only 10 percent were "independent successes." Similarly, in an eight-clinic study, the same authors (Gerard and Saenger, 1966) found that 12–32 percent of patients whose drinking had improved functioned poorly in one or more of five other areas. Similarly, it has been reported that 30–40 percent of abstinent patients show only slight improvement in other respects (Moore and Ramseur, 1960; Pfeffer and Berger, 1957) and that some of them even deteriorate (Gerard and Saenger, 1966). (4) Some alcoholics can become normal drinkers. This was first reported by Davies *et al.* (1956) in 7.5 percent of his hospital patients and has been followed by similar reports by other investigators (Cahalan, 1970; Gerard and Saenger, 1966; Kendall, 1965; Kendall and Staton, 1966; Lemere, 1953; Moore and Ramseur, 1960; Penick *et al.*, 1969; Reinert and Bowen, 1968; Selzer and Holloway, 1957). In assessing the results of treatment, it hence seems best to use multifactorial outcome measures rather than alcohol intake alone.

Granted that the patient's performance in all areas should be followed, what should the goal of treatment be, abstinence or moderate drinking? Many, the author included, are unwilling to accept at face value published accounts of alcoholics who achieve normal drinking. Reluctance to accept such results has been in part based on adherence to the loss of control theory of alcoholism (Ludwig and Wikler, 1974; Ludwig *et al.*, 1974). Recently this has apparently been challenged by findings that the blind administration of disguised vodka does not increase alcoholics' desire to drink (Engle and Williams, 1972; Merry, 1966), that on an experimental alcoholization ward craving did not occur with the first drink but only after large quantities of whiskey were consumed over a period of many days (Mendelson, 1964), and that in a similar setting one-third of the patients never took a drink and that one-third began drinking but stopped (Gottheil *et al.*, 1972). However, all of these experiments were conducted in highly artificial settings with little resemblance to everyday life with its privation, stresses, and insecurity. They were, rather, protected supportive environments where the patient could seek out company and support if he wished. We should also be aware that the drinking behavior of patients in the above-mentioned "normal drinking" studies was sampled only once during their moderate drinking. In long-term longitudinal outcome studies it has been found that drinking status in a given individual can be quite variable over time even when group means do not change (Fitzgerald *et al.*, 1971; Wilby and Jones, 1962), so that the drinking status of these "normal drinkers" could quickly change in response to renewed life stress, be it occupational, marital, or other. Indeed, Rosenberg *et al.* (1973) reported that half the abstinent alcoholics in a halfway house returned to drinking within two weeks after it burned down. Furthermore, those studies with case histories (Bailey and Stewart, 1967; Davies, 1962; Kendall, 1965) provide no consistent information that would help us predict what kind of patient could reasonably achieve sustained moderate drinking. Thus, of 19 subjects, 7 changed to less vulnerable occupations, 4 embarked on less stressful sexual lives with new partners, and 3 lost their tolerance for alcohol. Five others were exposed to "treatment" influences (religious conversion, A.A., fear of medical consequences). Finally, only one of Davies's patients had a history of clear-cut withdrawal symptoms. [Apropos, Reinert and Bowen's (1968), three patients who reverted to normal drinking all had short histories of heavy drinking.] Hence, it seems safest to propose abstinence as a goal for all alcoholics. Reports that abstinent alcoholics may function poorly in a number of other areas (Gerard and Saenger, 1966; Gerard et al., 1962; Moore and Ramseur, 1960; Pfeffer and Berger, 1957; Rossi et al., 1963) should not be taken as evidence against this position. Rather, they should be taken to indicate that the alcoholic is often a person with many problems besides alcoholism, and it should not be forgotten that, on the average, abstinence has been found to be strongly related to work adjustment (Bateman and Petersen, 1971; Bowen and Androes, 1968; Clancy et al., 1967; Gillis and Keet, 1969; Rathod et al., 1966; Rohan, 1972) and to health status, interpersonal relationships, and social stability (Clancy et al., 1967; Gerard and Saenger, 1966; Gillis and Keet, 1969; Goldfried, 1969), as well. Therefore, ideally, the thrust of treatment should be to try to help the patient with all his major psychosocial problems, not just with his drinking.

TREATMENT LENGTH

How long should the alcoholic receive treatment? First of all, we should be aware that even if he becomes abstinent, he is notoriously prone to relapse. The following figures will give some idea of how rapidly he may lose the ground he seems to have gained in the hospital once he is again exposed to the slings and arrows of the outside world. Of those who relapse after discharge from hospital, 50 percent do so by one month (Pokorny et al., 1968), 66 percent do by three months (Rohan, 1972) [in the case of committed multiple drunkenness offenders, 60 percent within a week (Mindlin, 1960)], 67 percent and 88 percent by six months (Davies et al., 1956; Selzer and Holloway, 1957) and 95 percent by one year (Selzer and Holloway, 1957).

Davies et al. (1956) found that a patient's drinking status at 6 months correctly predicted his two-year outcome in better than 80 percent of cases. In outpatients, Charnoff et al. (1963) discovered a 100 percent failure rate in patients who attended clinic for 3 months with improvement but dropped out as opposed to a 23 percent failure rate in patients who had attended for 6 months with improvement. Other investigators have found that 6- and 12-month abstinence and improvement rates are similar (Ritson, 1969; Walton et al., 1966), while one and two-year rates are even closer to each other (Bill C., 1965; Davies et al., 1956; Gerard and Saenger, 1966; Gibbins and Armstrong, 1957; McCance and McCance, 1969; Willems et al., 1973a). These results imply that a patient could continue to be treated for at least 6 months after he has achieved sobriety. How long treatment should continue beyond this is an open question that should be tempered by consideration of the findings of Fitzgerald et al. (1971). In a four-year follow-up of alcoholics treated in hospital they reported only 30.4 percent of patients kept the same sobriety status from year to year. While 76.9 percent were abstinent at one year, only 22.5 percent were abstinent in each of the next three years. Hence, in practice, many clinics continue to see patients on an indefinite, open-ended, but less frequent basis long after they have become abstinent.

SPONTANEOUS IMPROVEMENT, OR WHAT HAPPENS TO THE UNTREATED ALCOHOLIC?

Any consideration of the effectiveness of various treatment regimens for alcoholism must consider what happens to the untreated alcoholic. Apparently, spontaneous improvement does take place. A variety of factors may play a role. Change of occupations seems particularly important as persons in certain occupations suffer unusually high risks of alcoholism (Hitz, 1973). As already noted, new and less stressful sexual relationships, decreased tolerance, religious conversion, A.A. attendance, and fear of the medical consequences of alcoholism all can help call a halt to drinking in some patients (Bailey and Stewart, 1967; Davies, 1962; Kendall, 1965). Thus, even if the alcoholic does not receive treatment, there is a small but definite chance that he will improve or recover due to nonmedical influences, something therapists assessing the effectiveness of the treatment regimens they espouse should remember in their more grandiose moments. In survey studies there is ample evidence that alcoholism may spontaneously abate without formal treatment. For example, at reinterview after three years, 26.7 percent of Bailey et al.'s (1966) subjects now denied alcoholism (Bailey and Stewart, 1967), and Cahalan (1970) discovered that drinking problems taper off sharply in men after age 50, especially in higher SES subjects. He found that changes in drinking behavior were highly susceptible to social influences. Similarly, Smart (1970) reported that from 1951 to 1961 2 percent of alcoholics in Frontenac County, Ontario, apparently recovered without formal treatment, and Goodwin *et al.* (1971) found that on eight-year follow-up, 18.4 percent of their alcoholic felons had been abstinent for at least two years while 45 percent were drinking moderately (at least once a month but no more than twice a week and without intoxication). Of the seven abstinent subjects, two had received formal treatment, thus reducing their figure to 13.2 percent. Improved subjects were more likely to be white (an indication of higher SES), older, and to have superior social adjustment and less general psychopathology, all predictors of better outcome in persons treated for alcoholism. Similarly, other studies of untreated alcoholics have reported spontaneous sustained abstinence figures of 11 percent over an unspecified interval (Lemere, 1953) and 15 percent at seven years (Kendall and Staton, 1966). However, in Kendall and Staton's (1966) study, which examined the fate of patients referred for hospitalization but refusing it, only one out of nine abstinent patients improved without medical or lay treatment. In all, 23 percent of their subjects showed marked improvement. They were probably sicker than those of Goodwin *et al.* (1971) since they had been forced to seek hospital treatment. Finally, Kissin *et al.* (1970) found a 4 percent one-year improvement rate in untreated lower class outpatients. It thus appears that depending on the patient's personal and social assets, there is a 2–15 percent spontaneous improvement rate in alcoholics who do not receive formal treatment.

INPATIENT TREATMENT

Many alcoholics continue to deny their alcoholism and to avoid treatment for it until it reaches a point where they have to be hospitalized. The first stage of hospitalization usually includes detoxification from the acute toxic effects of alcohol discussed elsewhere. This may or may not be followed by a longer stay, the goals of which are social and psychological rehabilitation. Such long-term inpatient facilities include state, Veterans Administration, and private hospitals. The former, inter alia, usually have special treatment units for alcoholics, while the latter often specialize in alcoholism treatment alone. After the alcoholic leaves the hospital, if he is too well to remain in it but still too marginally adjusted and too bereft of social resources to do well in an outpatient program, he may go to a halfway house, where he can live with other recovering alcoholics while he goes out to work. State hospitals have the lowest staff-to-patient ratios, but their staffing pattern is similar to that of VA and private hospitals, which employ doctors, nurses, psychologists, social workers, and recreational and vocational counselors. Halfway houses are often run by recovered alcoholics and depend almost exclusively on A.A. and other group methods. They usually have no medical facilities or staff. Group therapy is currently stressed in all hospitals, but more so in public institutions, which must operate on less money per patient. At present, the trend is toward specialized units or facilities with suggested patient stays ranging from 60 to 90 days.

The fundamental questions about hospital treatment are the following: (1) How effective is it? (2) How much of its reported effectiveness can be attributed to treatment and how much to the kind of patient treated? (3) Is any particular kind of pretreatment regimen better than any other? (4) Is hospital treatment better than outpatient treatment? (5) How necessary is posthospital outpatient follow-up treatment?

Effectiveness of Inpatient Treatment

When the reported improvement rates of 30 studies (Cowen, 1954; Davies et al., 1956; Dubourg, 1969; Dunne, 1973; Edwards, 1966; Edwards and Guthrie, 1966; Glatt, 1961; Hoff and Forbes, 1955; Jensen, 1962; Katz, 1966; Kish and Hermann, 1971; McCance and McCance, 1969; Mindlin, 1960; Moore and Ramseur, 1960; Myerson and Mayer, 1966; Nørvig and Nielsen, 1956; Pittman and Tate, 1969; Pokorny et al., 1968; Rae, 1972; Rathod et al., 1966; Rhodes and Hudson, 1969; Ritson, 1971; Rossi et al., 1963; Selzer and Holloway, 1957; Simpson and Webber, 1971; Tomsovic, 1968; Trice et al., 1969; Vallance, 1965; Walton et al., 1966; Willems et al., 1973a) with an average follow-up period of 2.2 years were corrected for sample attrition and spontaneous improvement, they amounted to 29.9 percent. Thus, about 30 percent of patients appear to improve with hospitalization, or, dispensing with the debatable correction for spontaneous improvement, 41.5 percent. Either figure seems a very substantial one for a chronic condition so beset with physical, psychological, and social complications.

Patient and Treatment

These figures seem good but should we applaud the treatment programs or rather the patients they treat? Comparisons of the populations of good and poor outcome programs suggest the latter, as do studies of prognostic factors in the treatment of alcoholism. Thus, it has repeatedly been found that social (residential/occupational/marital) stability is positively related to outcome (Bowen and Androes, 1968; Davies et al., 1956; Dubourg, 1969; Evenson et al., 1973; Gillis and Keet, 1969; Glatt, 1961; Kish and Hermann, 1971; Kurland, 1968; McCance and McCance, 1969; Moore and Ramseur, 1960; Pemberton, 1967; Pokorny et al., 1968, 1971, 1973; Rathod et al., 1966; Rosenblatt, et al., 1971; Selzer and Holloway, 1957). Similarly, in those studies where there was enough spread profitably to examine it, SES (itself related to social stability) has a strong positive relationship to outcome (Edwards et al., 1973; Gillis and Keet, 1969; Glatt, 1961; McCance and McCance, 1969; Mindlin, 1960; Pokorny et al., 1973; Trice et al., 1969). Conversely, a poor prognosis is associated with factors such as legal trouble (Corotto, 1963; Dubourg, 1969; McCance and McCance, 1969; Pokorny et al., 1968; Rubington, 1970; Trice et al., 1969; Willems et al., 1973a) and psychopathic features (Glatt, 1961; Muzekeri, 1965; Pokorny et al., 1968; Rathod et al., 1966), themselves interrelated.* With this in mind, let us look at the four studies with the best outcomes and the four with the worst. Papas (1971), Rae (1972), Willems et al. (1973a) and Davies et al. (1956), respectively, found (corrected) improvement rates of 32.4 percent, 46.4 percent, 55.8 percent, and 68 percent. Papas's (1971) patients were Air Force personnel with 10–18 years of service, and the other three study populations were heavily larded with Class 1, 2, and 3 patients. All four investigations, in addition, had rather extensive exclusion criteria and stressed group therapy. The four studies with the worst outcome were those of Mindlin (1960), Katz (1966), Rubington (1970), and Rhodes and Hudson (1969), which had (corrected) improvement rates of 18 percent, 7.9 percent, 2.2 percent (abstinence), and 0 percent. They had virtually no exclusion criteria. Mindlin's (1960) population was heavily seasoned with skid row type patients, while Katz's (1966) were all skid row alcoholics treated in a Salvation Army service center, and Rubington's (1970) were tuberculous alcoholics treated in a sanatorium. Group therapy was also used in all of these settings. The differences in success rates of about 50 percent between the good and poor outcome programs, much larger than that ever reported between different treatment regimens even in uncontrolled studies, strikingly affirms the dominant role played by patient factors. On the other hand, if we compare the results of inpatient programs with comparable populations (neither skid row nor upper class) done through 1963, when treatment

---

* Other prognostic indicators of good long-term outcome in inpatient treatment are the following: higher IQ (Heilbrun, 1971); better education (Heilbrun, 1971); older patient (Bateman and Petersen, 1971; Foulds and Hassall, 1969; Glatt, 1961; Rathod et al., 1966; Trice et al., 1969); later onset of heavy drinking (Foulds and Hassall, 1969; Rae, 1972; Selzer and Holloway, 1957; Trice et al., 1969); noninstitutional referral (Goodwin et al., 1969); no court convictions (Clancy et al., 1965); social stability (Edwards, 1966); at least six months' prior abstinence (Rossi et al., 1963); abstinence for at least one week before admission (Bateman and Petersen, 1971); fewer previous hospitalizations (Ellis and Krupinski, 1964; Kurland, 1968; McCance and McCance, 1969); lower current symptom levels (Goodwin et al., 1969; Trice et al., 1969); does not drink while in hospital (Gottheil et al., 1972); admits alcoholism on admission (Rossi et al., 1963); less self-esteem (Pokorny et al., 1968); high on extroversion and low neuroticism (Edwards, 1966); lower Sc and Ma scales scores on the MMPI (Heilbrun, 1971); moderately self-critical rather than grossly self-punitive (Walton et al., 1966); lower hostility scores (Ritson, 1971); capacity to maintain interpersonal relationships (Goodwin et al., 1969); and mother dead, or, if living, seen less than once a month (Bateman and Petersen, 1971).

was either custodial or less intense than later on, with those carried out from 1964 to the present, when more intensive treatment was the rule, the former (Cowen, 1954; Moore and Ramseur, 1960; Nørvig and Nielsen, 1956; Rossi et al., 1963; Selzer and Holloway, 1957; Trice et al., 1969; Vallance, 1965) have an average success rate of 26.6 percent (sd = 6.7 percent) and the latter (Kish and Hermann, 1971; Pokorny et al., 1968; Rohan, 1970; Tomsovic, 1968) one of 33.9 percent (sd = 8.6 percent), a relative increase of 25.6 percent. The small Ns involved demand that we hedge our bets, but yet it does seem that improved and more intensive inpatient treatment methods may have been responsible for a substantial gain in treatment effectiveness.

Treatment Length and Outcome

How long should a patient be hospitalized for his alcoholism? Is there a point of diminishing returns, and if so, when is it? Several investigators (Ellis and Krupinski, 1964; Moore and Ramseur, 1960; Rathod et al., 1966) reported that longer hospitalization gives a better prognosis, Pemberton (1967) that more intensive treatment favors a good outcome. Specifically, two studies found that patients who stay longer than four months do markedly better (Ferneau and Desroches, 1969; Katz, 1966). Other investigations have clouded the issue. Thus, Tomsovic (1970) found length of stay predictive of outcome, but only in nonresponders to his questionnaires (they tended to be dropouts), while Ritson (1969), who excluded dropouts from his outcome analyses, found that length of inpatient stay was not predictive of outcome. It is hard to know just what to make of outcome studies like the above since better-motivated patients tend to stay in treatment longer and motivation is probably related to outcome. Yet, since the last two studies were conducted in higher SES populations than the first two, there is the implication that higher SES patients require shorter hospitalizations (perhaps because they are subject to more favorable extramedical posthospital influences), while low SES patients need longer stays in order to better their ability to deal with more difficult life situations. Paradoxically, it seems that the dropout, the patient who leaves treatment earliest, is just the one who would most benefit from longer hospitalization.

The Effectiveness of Inpatient Psychotherapy

Since most inpatient programs subject the patient to a variety of therapeutic influences (individual and/or group psychotherapy, occupational and recreational therapy, A.A., vocational counseling), it is difficult and perhaps even artificial to ask which of these approaches is more effective in an inpatient setting. On the other hand, most programs tend to stress one of these modalities at the expense of the others and they vary widely in terms of their demands on personnel time and training.

Although most physicians involved in the treatment of chronic alcoholism feel that some kind of inpatient psychotherapy is helpful, there is no general concensus about what form it should take or how intensive it should be. To make things even worse, all controlled long-term follow-up studies done so far fail to show that any advantage accrues to inpatient psychotherapy as opposed to other approaches, whether it be self-confrontation by videotape (Schaefer *et al.*, 1972) or a combination of individual and group therapy (Wolff, 1968). Yet, four controlled studies of group therapy have demonstrated positive short-term changes in psychological test measures applied during or at the end of hospitalization (Armstrong and Hoyt, 1960; Ends and Page, 1957, 1959; Mindlin and Belden, 1965). Hence, it seems very possible that inpatient psychotherapy of alcoholics is ineffective unless it is supplemented by follow-up treatment after discharge.

Hospital versus Outpatient Treatment

Because hospitalization is so much more expensive and disruptive of the life of the patient and his family than outpatient treatment, it is important to know how important the former is beyond the absolutely necessary initial drying-out period. Only one study adequately examined this issue. Edwards and Guthrie (1966) randomly assigned well-matched patients to two months of either inpatient or outpatient treatment (patients received group therapy and A.A. in both cases) and found nonsignificant between group differences at six and ten months. Thus, so far, the evidence does not support the idea that prolonged inpatient treatment offers any special advantage for most patients.

How Necessary Is Aftercare?

Pokorny *et al.* (1973) approached this question by comparing a 60-day inpatient program supplemented by outpatient group therapy with 90-day hospitalization without aftercare (Pokorny *et al.*, 1968). Similar outcome figures were obtained in both programs, which used similar patients, the same facility, and the same personnel and treatment methods. Along somewhat similar lines, Stein *et al.* (1975) recently reported that patients who were simply detoxified and received aftercare did just as well as those who were hospitalized longer with intensive psychosocial treatment followed by aftercare. These studies are important for two reasons. First of all, they suggest that current inpatient programs may be too long. Secondly, they imply that aftercare confers additional benefit on the alcoholic patient. Indeed, it has repeatedly been reported that patients who got group therapy after discharge do better (Gillis and Keet,

1969; Hansen and Teilmann, 1954; Moore and Ramseur, 1960; Ritson, 1971). However, this might simply be because they were better motivated patients who tend to persist in treatment. On the other hand, Pittman and Tate (1969) executed a carefully designed study in which primarily lower-class patients were randomly assigned either to 7–10 days of hospitalization or to 3–6 weeks of inpatient care with subsequent outpatient treatment. (The main emphasis was on group therapy in both settings.) This meant in essence that the hospital-only patients were getting not a great deal more than detoxification. At one year there was no between group differences in outcome. Their results should tempered by the finding that of the 19 patients who remained abstinent for the whole follow-up period, 18 had extensive follow-up contact with the outpatient clinical staff or with some community group. On the other hand, Dubourg (1969), in a follow-up study of 79 hospital-treated middle- and upper-class alcoholics, found that among the 25 with a good outcome, only 3 had received regular outpatient treatment (as had 3 failures), while 6 got support from agencies and 13 from relatives only. Hence, it appears that outpatient follow-up treatment care can replace some or even most of the time patients currently spend in inpatient treatment after they are detoxified, and that it is especially indicated in the lower-class patient with poor social supports, while it may not be so important in the more socially competent upper-class patient.

OUTPATIENT TREATMENT

The number of outpatient alcohol clinics in the United States has doubtless grown since 1963, when Bahn *et al.* (1963) put their number at 39 out of 1429 outpatient psychiatric clinics. Their figures give a very cogent rationale for specialized alcohol clinics: 51 percent of alcoholics treated at alcohol clinics improved as opposed to only 29 percent of those seen in general psychiatric clinics, where they amounted to only 3 percent of patients seen. The negative bias of medical and paramedical personnel against alcoholics is well known (Cahn, 1970; Knox, 1971; Mulford, 1966; Pittman and Sterne, 1965).

Dropping Out of Treatment

Two of the biggest problems in the outpatient treatment of alcoholics are engaging the patient in treatment in the first place and, once in treatment, keeping him there, issues that are discussed in Chapter 4 of this volume. At this point, it is enough to be aware that both inpatient and outpatient programs are plagued with patients dropping out of treatment, but it is a much more severe problem in the latter and that the long term outlook of the dropout is not as good as that of the patient who persists in treatment.

Basic Issues of Outpatient Treatment

Some of the same basic questions apply to outpatient as to inpatient treatment: (1) How well does it work? (2) Which determines treatment outcome more, type and amount of treatment or patient characteristics? (3) Is one kind of treatment better than another?

Effectiveness of Outpatient Treatment

Taking sample attrition into account in those studies where the authors did not do so, the mean reported outcome for 18 clinics (Asma *et al.*, 1971; Bruun, 1963; Cellar and Grant, 1952; Clancy *et al.*, 1965; Gerard and Saenger, 1966; Lynn and Smith-Moorehouse, 1966; Mindlin, 1959; Pfeffer and Berger, 1957; Rankin *et al.*, 1967; Storm and Cutler, 1968) was one of 41.6 percent (sd = 15.3 percent) improvement, a figure that dropped to 36 percent if an assumed 5 percent per year spontaneous improvement rate is taken into account. Thus, here, as in the case of inpatient treatment, the results of treatment turn out to be quite respectable.

Treatment Length and Outcome

Treatment length repeatedly has been found to be positively related to outcome in outpatient treatment studies (Fox and Smith, 1959; Gerard and Saenger, 1966; Kissin *et al.*, 1968b; Ritson, 1969; Thomas *et al.*, 1959). Yet, since it and other prognostic factors were confounded, it is hard to know what to make of these findings. In this respect, it is interesting that Gibbins and Armstrong (1957), who excluded patients who attended less than three sessions, found no relationship between abstinence gain and extent of outpatient contact. If other authors had likewise eliminated rapid dropouts from their analyses, perhaps their results would have vanished. Gibbins and Armstrong's (1957) findings also imply that short-term intensive hospitalization (which their patients had) is enough for better-motivated, better-prognosis patients, but that it is just those who do not get outpatient aftercare who might benefit most from it. A very rough look at this question can be obtained by examining the relationship between treatment length and outcome in studies rather than individuals. Twenty-four inpatient studies (Bateman and Petersen, 1971; Devenyi and Sereny, 1970; Dubourg, 1969; Edwards and Guthrie, 1966; Farrar *et al.*, 1968; Fox and Smith, 1959; Gillis and Keet, 1969; Jensen, 1962; Knox, 1972; Lemere and Voegtlin, 1950; Levinson and Sereny, 1969; McCance and McCance, 1969; Mills *et al.*, 1971; Mindlin, 1960; Nørvig and Nielsen, 1956; Papas, 1971; Pittman and Tate, 1969; Quinn and Kerr, 1963; Rae and Drewery, 1972; Rohan, 1970; Rubington, 1970; Vallance, 1965; Walton *et al.*, 1966; Willems *et al.*, 1973a) and seven outpatient studies

(Bruun, 1963; Burton and Kaplan, 1968; Fox and Smith, 1959; Gerard and Saenger, 1966; Pfeffer and Berger, 1957; Rankin *et al.*, 1967; Stojiljkovic, 1969) gave information on either abstinence or improvement, or both, at long-term follow-up. Among 14 inpatient studies with good prognosis populations (Bateman and Petersen, 1971; Devenyi and Sereny, 1970; Dubourg, 1969; Edwards and Guthrie, 1966; Farrar *et al.*, 1968; Gillis and Keet, 1969; Jensen, 1962; Lemere and Voegtlin, 1950; McCance and McCance, 1969; Papas, 1971; Rae, 1972; Vallance, 1965; Walton *et al.*, 1966; Willems *et al.*, 1973a) and 9 with medium-prognosis populations (Fox and Smith, 1959; Jensen, 1962; Knox, 1972; Levinson and Sereny, 1969; McCance and McCance, 1969; Mills *et al.*, 1971; Nørvig and Nielsen, 1956; Pittman and Tate, 1969; Rohan, 1972), correlations between treatment length and abstinence and between length and improvement were respectively 0.46 and 0.06 ($N = 11$ and 10) and 0.40 and 0.27 ($N = 7$ and 7). Similarly, in the outpatient studies, which were too few for separate analyses for prognostic groups, the correlations were 0.46 and 0.27. These results, which are most tentative, suggest that treatment length is related more to abstinence than to lesser degrees of improvement which, perhaps, may depend more on environmental factors.

Patient versus Treatment as Outcome Predictor

A wide array of factors have been reported to favor good outcome in the outpatient treatment of alcoholism. They include: SES and social stability (Baekeland *et al.*, 1973; Gerard and Saenger, 1966; Goldfried, 1969; Kissin *et al.*, 1968b; Mayer and Myerson, 1971; Mindlin, 1959), abstinence on admission or length of abstinence in the year before treatment (Baekeland *et al.*, 1973; Goldfried, 1969; Platz *et al.*, 1970), motivation* (Baekeland *et al.*, 1973; Gerard and Saenger, 1966; Goldfried, 1969; Mayer and Myerson, 1971; Mindlin, 1959), spouse in treatment (Thomas *et al.*, 1959), 40-45 years old

---

* *Motivation* is a term commonly used by clinicians and its circular use has been rightly criticized by Pittman and Sterne (1965). It is not always clear what clinicians mean by the word motivation. Often they are not as explicit as Mindlin (1959), who stipulated that it include at least one of the following: (1) wanting to change beyond stopping drinking, (2) realizing that one must take an active part in treatment, and (3) being willing to make sacrifices for the sake of treatment. The reader will recognize these as the usual canons of psychoanalytically oriented psychotherapy. On the other hand, the term motivation is often used in connection with positive accepting attitudes toward treatment and those who render it. In addition, a distinction certainly should be made between extrinsic motivation (pressure from wife, boss, etc.), which is what initially drives most alcoholics to treatment, and intrinsic motivation, a desire to get better for one's own sake, into which extrinsic motivation is often converted in the course of treatment. Finally, motivation in the above sense and motivation as a personality trait in the sense of a general tendency to persevere in endeavors once they are undertaken may be only distantly related. The relative prognostic value of motivation in these various meanings of the word should be investigated.

(Kissin et al., 1968b), and periodic drinking (Kissin et al., 1968b).* On the other hand, sociopathic features confer a poor outcome (Baekeland et al., 1973; Kissin et al., 1968b; Mindlin, 1959; Ritson, 1971), as does having a punitive wife (Ritson, 1971).

Of the 19 clinics surveyed (Asma et al., 1971; Bruun, 1963; Cellar and Grant, 1952; Clancy et al., 1965; Gerard and Saenger, 1966; Lynn and Smith-Moorehouse, 1966; Mindlin, 1959; Pfeffer and Berger, 1957; Rankin et al., 1967; Storm and Cutler, 1968), the two with the highest improvement figures, 52 percent (Asma et al., 1971) and 72 percent† were, respectively, one where patients had first completed inpatient treatment at $100 per week and would accept disulfiram, hence highly motivated, higher SES patients, and another where patients were telephone company employees, 56 percent of whom had been employed by their firm for 10–29 years, thus a group of more than middle income with exceptional occupational stability. On the other hand, the two clinics with the poorest outcomes, 17.7 percent (Rankin et al., 1967) and 22.2 percent‡ were ones which catered largely to skid row type patients. Treatment methods, on the other hand, were different within each set of two clinics (disulfiram vs. psychotherapy and medication vs. psychotherapy). Again, it would seem that the nature of the patient is much more important than that of the treatment used on him.

The Relative Value of Different Treatments

Currently, outpatient treatment of alcoholism runs the gamut from classical individual psychotherapy through group and couple therapy to behavior therapy and drug treatment. Whatever kind of treatment the patient receives, it is usually supplemented by the practical social work assistance so many lower class alcoholics are badly in need of. Class 1 and 2 patients tend to get psychotherapy but not medication, while class 4 and 5 patients tend to receive medication but not psychotherapy. Middle-class patients often receive both. In alcoholism clinics the social worker, rather than the psychiatrist or clinical psychologist, has more and more come to be the mainstay of psychotherapy.

The outpatient, unlike the inpatient, is more likely to be subjected to only one kind of treatment rather than to a milieu which includes a number of them. Hence, the question of the relative value of and indication for different kinds of

---

* The last is probably an artifact of data analysis depending on the one-year follow-up interval used. Thus Willems et al. (1973b) found that periodic drinking was not a predictor of good outcome at six months but was at one year in their patients.
† Gerard and Saenger (1966), Clinic G.
‡ Gerard and Saenger (1966), Clinic C.

treatment is of much more practical importance in outpatient than in inpatient settings.

Unfortunately, there have been relatively few controlled studies which evaluated different treatment regimens in outpatient settings. So far, however, conventional individual psychotherapy has not fared well in comparison with other approaches. Thus, Bruun (1963) tried to assess the relative effectiveness of two programs which respectively featured either 10 sessions of individual therapy by a psychiatrist or a multidisciplinary approach (3 visits to an internist, 7 to a social worker, 2 to group therapy, and 20 to a nurse). When improvement due to extraneous therapeutic influences was taken into account, multidisciplinary treatment turned out to be more effective than individual psychotherapy. A more trenchant indication of the relative ineffectiveness of individual psychotherapy with the alcoholic is given in a report by Hayman (1956). In a survey of members of the Southern California Psychiatric Association, he found that among those who treated alcoholics (mostly by psychoanalytically oriented individual psychotherapy), over one-half had no success with any of their alcoholic patients, and of those who did report successes, these occurred in no more than 10 percent of cases. These results with higher SES patients, if at all representative, are a telling indictment of psychoanalytically oriented individual psychotherapy of alcoholics when it is not supplemented by other kinds of treatment.

An interesting and promising psychotherapeutic approach is that of couple therapy. It has long been felt by those who do psychotherapy with alcoholics that the patient's wife is often partly responsible for his alcoholism. However, Edwards *et al.* (1973), in a thorough and critical review of the literature, showed that the classical description of the alcoholic's wife advanced between 1937 and 1959 as an aggressive, domineering woman who married her husband to mother or control him is without foundation. Nevertheless, in some cases there seems to be a good rationale for couple treatment. Thus, Rae and Drewery (1972) found that wives of alcoholics who score high on the MMPI psychopathic deviate (Pd) scale, see themselves as more masculine than do wives of normals, as do their husbands, who see themselves as feminine. However, high Pd wives deny their husbands' feminine trends, an attitude that may work counter to therapists' efforts to get patients to recognize problems in this area. Pd scale scores also seem connected with alcoholics patients' wives' perception of their husbands' independence or dependence. Control husbands scored much more independent than dependent, control wives much more dependent than independent, as one might expect of those who conform to conventional sex-role stereotypes. However, alcoholic husbands saw themselves as equally independent and dependent. High Pd wives recognized this conflict, but low Pd wives denied it, something that may have negated these patients' efforts to achieve independence if they had low Pd wives. Hence, it is not surprising that in another study by Rae (1972), successfully treated inpatients had wives with lower Pd scores. However, having a low Pd wife was a prognostic indicator only in marriages marked by employment instability and sexual disturbance. A number of uncontrolled studies suggest how important it can be to include the alcoholic's wife in his treatment. Thus, it has been reported that male inpatients whose wives are willing to attend a wives' group do better (Smith, 1967) as do those who have their wives' assistance and support (Pemberton, 1967). Similarly, selected patients who engage in couple group therapy do well (Burton and Kaplan, 1968; Gallant et al., 1970). One controlled study (Corder et al., 1972) bears out these reports of the effectiveness of alcoholic couple group psychotherapy, though in a hospital setting. Controls received the usual four weeks of daily group therapy, didactic lectures, RT, and OT. Experimental subjects got the same program for three weeks, then an intensive four-day workshop with wives and their husbands which consisted of two-couple therapy sessions, session videotape analyses, group lectures and discussions about alcoholism, discussions of game-playing in alcoholism joint recreational counseling, A.A. and Al-Anon meetings, and meetings with representatives from follow-up treatment agencies. At 7-month follow-up, experimental subjects were significantly more abstinent (55 percent), and a higher proportion of them were attending follow-up treatment and had been employed one month. Obviously, in this experiment, couple treatment was effective, but it is difficult to know which aspect of it was most responsible for the observed treatment effects or to what extent subject selection, or simply the intensiveness of treatment, had a hand.

In summary, then, it seems that multidisciplinary treatment is more effective than individual psychotherapy and that in some cases active involvement of spouses in treatment is worthwhile.

The Issue of Compulsory Treatment

The typical alcoholic gets therapeutic attention relatively late in the course of his illness. One important reason for this sad state of affairs is that he tends to deny his alcoholism until he is forced to seek treatment. Sometimes it is DTs or a seizure that pushes him into a hospital, but much more often it is pressure from his wife or boss, who in effect, tell him to "shape up or ship out," or the more formal demands of a court or welfare agency that prompt him to show up at an outpatient facility. There, it is indeed the unusual alcoholic who is self-referred. The idea of exerting pressure on or coercing patients somehow goes against the grain of therapists, who in their training have been set to graze in Elysian fields populated by middle- and upper-middle-class neurotic patients and have been told that coercion, manipulation, and the like are untherapeutic. Yet, in the case of the alcoholic, social pressure not only helps get him into treatment but also helps keep him there until he can benefit from it. This is clearly shown in a recent report by Smart (1974) about alcoholics who were coerced by their employers to seek treatment. Under threat of job loss, they were forced into a therapeutic regimen that included three weeks of inpatient treatment followed by about a year of outpatient therapy. Twelve to 14 months after discharge, 56 percent showed great improvement in drinking behavior and 28 percent moderate improvement, figures that are not significantly different from those obtained in a group of voluntary patients. Similarly, Lemere et al. (1958) found a positive correlation between threatened loss of job or spouse and good outcome.

By extension, it certainly seems reasonable for any institution or agency that is in a position to reward the alcoholic with its services to require him to get treatment in return. Yet, it is only welfare agencies and municipal court systems that have done so, the former in an effort to keep their clients from drinking up the meager funds they receive for food and shelter, the latter in the hope of cutting down to size the costly and time-consuming problem of the multiple offense public drunkenness offender. One of the most difficult problems facing not only judges and policemen but also clinicians is the treatment of the skid row patient. His expense to society and his suffering are great. As already indicated, he does poorly with any kind of treatment. The problem is such a desperate one that clinicians and municipal authorities have experimented with compulsory treatment of this otherwise poorly motivated kind of patient who accounts for the bulk of arrests for drunkenness and vagrancy. Four such experiments have been reported. Three of them got good results, one a very poor one. Thus Maier and Fox (1958) found a 37.9 percent improvement rate with only a 44.8 percent parole violation rate in a three-month program involving 34 group meetings and 5 individual therapy sessions at an alcoholic clinic. In a similar program, Jackson (1958) reported a significant reduction in arrest rates, and Gallant et al. (1968b) got good long-term results with state penitentiary alcoholics paroled to compulsory individual treatment in an alcohol clinic. However, the same authors get only a one-year 6.1 percent success rate in a later and very well-designed larger-scale experiment with house of detention skid row alcoholics. At first glance, these widely different results are puzzling until we realize that the patients of Jackson et al. (1958) were screened so as to have a good prognosis, while Maier and Fox's (1958) 29 Georgia patients had to be white and only six of them were unskilled, and the prison inmates of Gallant et al. (1968a) had been imprisoned for at least four years and, hence, had been abstinent for at least that long and that the penalty for parole violation was a long sentence. On the other hand, their skid row patients were unselected and suffered only a short sentence if they relapsed. The conclusion seems to be that skid row alcoholics, if carefully selected, can benefit from compulsory treatment, especially if the sentence for parole violation (i.e., dropping out of treatment) is a stiff one.

ALCOHOLICS ANONYMOUS (A.A.)

General Considerations

Of all the treatment methods I have chosen to consider, A.A., a self-help organization founded in 1935, is one of the oldest and best established. However, from an objective, medical–sociopsychiatric point of view it is the worst understood. Hence, in order to try to answer a few basic questions about A.A. in the section that follows, I have tried to pull together hard-to-find material that is scattered throughout the research literature where it is likely to escape the attention of clinicians. I hope the reader will bear with the details for the sake of the conclusions, meager as they are. The importance of A.A. can best be appreciated if we consider a few statistics. It has been estimated that there are nine million alcoholics in the United States (National Council on Alcoholism Fact Sheet, 1972). In 1970, the last year about which complete information is available, 278,000 patients with a primary diagnosis of alcoholism were being treated in psychiatric facilities (Redick, 1973). (The number with a secondary diagnosis of alcoholism is not known.) Jones and Helrich (1972) estimated that about 70,000 alcoholics were being seen by private practitioners. Thus, at least 350,000 alcoholics received medical and/or psychiatric treatment for alcoholism per se. On the other hand, A.A.'s current membership in the United States is about 250,000 (Alcoholics Anonymous, 1972a), which reaches almost as many alcoholics as does the medical profession.

From a sociological point of view, Jones (1970) has pointed out that A.A. has much in common with religious sects. Indeed, it has a strongly religious orientation and its precepts (the so-called Twelve Steps) involve the alcoholic's admission that he cannot control his drinking by himself, his admitting his wrongs and making amends, his submission to and reliance an God, and his carrying the message to other alcoholics (the sponsor system).

Butts and Chotlos (1973) found that according to psychological tests, alcoholics are more apt to see their behavior as being externally controlled than normals. In other words, as every clinician dealing with them knows, they tend to be "alibi Ikes." Along the same lines, alcoholism specialists emphasize their patients' extraordinary propensity to overuse the psychological defense mechanism of denial which, along with dependency, is considered one of the hallmarks of the alcoholic. It is not surprising, then, that an MMPI study showed alcoholics with better insight to be more candid and less defensive than those with poor understanding of their problems (Button, 1956). More important, several investigators (Kurland, 1968; Moore and Ramseur, 1960) found that among hospitalized alcoholics extent of denial on admission did not predict long-term outcome but rather decrease in denial during hospitalization did so. More recently, Willems *et al.* (1973a) reached the identical conclusion.* They stressed that change in denial was the single most powerful prognostic indicator they could find in their patients.† Hence, it is not surprising that it is just the alcoholic's denial of his problems, or if he admits them, their attribution to others, that A.A. tries to change by insisting that he admit his alcoholism and stop blaming others for his troubles. Indeed, it is just A.A.'s emphasis on these points that may give it much of its therapeutic leverage. Besides stressing the patient's dropping the barriers of denial, personal responsibility is stressed. While the role of denial reduction in A.A.'s effectiveness is fairly clear, that of the sponsor system is not as well understood.

The use of A.A., either on a voluntary or compulsory basis (the former is much more common), is an explicit and integral part of most inpatient and many outpatient treatment programs. It has been called "probably the single most effective method of treatment that we have" (Fox, 1957) by a well-known expert in the field of alcoholism who was merely voicing an opinion widely shared by other workers in the field. Yet, surprisingly little is known about A.A. in a systematic way as it has consistently avoided scientific study. At a descriptive level, Hayman (1966) has pointed out that via A.A. the alcoholic becomes a member of an important subculture, belongs to an in-group which can function for him as an auxiliary, external superego. Important, he feels, is the destruction of fantasies of omnipotence, atonement of guilt, and the construction of powerful reaction formations in a setting that promises love, care, and help. A number of characteristic defense mechanisms seem to be turned into their opposites: denial into open acknowledgment and projection into self-blame. Heavy emphasis is put on the use of reaction-formation: Hostility and destructiveness are replaced by profession of love. The role of these and other psychological phenomena in A.A. deserves the systematic scientific study that has up to now been denied them.

---

* However, Rossi *et al.* (1963) did find that patients who admitted their alcoholism on admission had a better prognosis.

† The skeptic might object that reduction in denial is not a predictor but merely a concomitant of therapeutic change or of therapist–patient agreement about causes and goals. The important questions for the researcher should be: What kind of patient diminishes his denial under treatment, under what treatments, and with what kinds of therapists? Here AA's cooperation could be of enormous value.

A.A. Population Characteristics

Who seeks out A.A., who sticks with it, and who does not? Most studies of A.A. members have been carried out on hospital or clinic populations (Allen and Dootjes, 1968; Canter, 1966; Gynther and Brilliant, 1967; Jackson, 1958; Mindlin, 1964; Trice and Roman, 1970), hence, on A.A. members for whom it may be less helpful, who may be sicker, or who may be less strongly affiliated with it. A few have looked at nonpatients, i.e., A.A. volunteers, in themselves probably a biased sample which preferentially includes more successful A.A. members (Karp *et al.*, 1965; Machover *et al.*, 1959a,b; Mathias, 1955).

With these reservations in mind, we note that past or present A.A. members are less likely to be married (Gynther and Brilliant, 1967) and, if successful (at least two years sobriety), more likely to have been solitary drinkers (Jackson, 1958), a finding that emphasizes A.A.'s role in resocializing the alcoholic. Compared to non-A.A. members, A.A. attenders seem to have weaker aggressive needs, to be less dysphoric and more verbally productive and to have stronger hypomanic trends (Mathias, 1955), all features that would favor successful membership in socially oriented A.A. groups, as would the social dependence implied by the finding that people who join A.A. seem to be more field-dependent than those who do not (Karp *et al.*, 1965). Canter (1966) found that the hospital patient who preferred to participate in A.A. rather than other kinds of treatment had a higher IQ and was more authoritarian.

Trice (1957, 1959) devoted several studies to the process of affiliation with A.A. He made the important distinction between affiliates who stick with A.A. (attended at least twice a month over the past year) and nonaffiliates (attended less than once every six months in the past three years). He sees A.A. affiliation as consisting of three phases: (1) the pre-A.A. phase, (2) the phase of initial contact, and (3) the phase in which meetings have been attended for several weeks. He found the affiliate to be one who could share emotional reactions with others, had lost his drinking friends (and thus was increasingly socially isolated and had lost an important support for his drinking) and was exposed to favorable descriptions of A.A. He had no close friend or relative who had quit drinking on his own and hence had no competing willpower model of treatment. He had a better history of childhood churchgoing and was more likely to have a wife or girl friend who accompanied him to meetings and supported his affiliation with A.A. (We recall the wife's important role in the outcome of hospitalized patients and will find this an important factor in disulfiram treatment.) The affiliate was not found to be class-conscious (Trice, 1957). (However, it should be kept in mind that A.A. membership is mainly restricted to classes 2–4 and that 30 percent of its members have executive, professional, or technical jobs. Class 5 or skid row persons do not often seek out A.A. [Edwards *et al.*, 1967; Jones, 1970; Pittman and Gordon, 1958].)

The fact that A.A. appeals to socially isolated persons is highlighted by the finding that while affiliates scored higher on affiliation motives than did nonaffiliates, both groups scored relatively low on affiliation compared to controls (Trice, 1959). In another and more methodologically sophisticated study of A.A. affiliates and nonaffiliates, Trice and Roman (1970) found that psychological rather than social–demographic factors accounted for more of the experimental variance.* Implicated were guilt proneness, intensive labeling, and physical stability. Quite opposed to the accepted stereotypes about A.A. were the findings that 0/3 variables indicated social stability and only 1/5 middle-class background as predictors of affiliation. However, the authors were studying a state hospital population (thus one weighted away from middle-class status and possibly one consisting in part of persons who failed in A.A. because of high levels of somatic and/or psychological symptoms which drove them into a hospital). It may be that the socially, medically, and psychiatrically stable middle-class A.A. affiliate only infrequently ends up in state hospitals.

It is well known to clinicians but not well publicized by them that A.A. tends to discourage members' attendance at alcohol clinics, which offer alternative forms of treatment. A.A.'s attitudes toward alcohol clinics may be one factor which tends to select different kinds of patients for A.A. and clinic treatments. Trice and Roman's (1970) finding that affiliators were high on guilt proneness, intensive labeling, and physical stability goes along with this idea. Psychiatrists, psychologists, and social workers tend to adopt a much less directive posture with alcoholics than does A.A. and they attack the alcoholic's denial much more gently with an approach calculated not to make him feel guilty. (He is led to believe that he has "emotional problems" rather than that he is defective in "willpower.") Hence, patients low on denial may be overrepresented among A.A. members and underrepresented in alcohol clinic populations, a bias that in itself would give the A.A. member a better prognosis. Since A.A. does not offer medical treatment, it is possible that the patient with high levels of psychiatric and somatic symptoms may drop out of it for the sake of more specific treatment or else supplement his attendance with visits to medical clinics. The contrast between clinic and A.A. populations is further suggested by a study of outpatient alcoholics conducted by Allen and Dootjes (1968). They found that clinic attendance correlated negatively with autonomy and positively with deference (on the Gough Adjective Check List). Subjects who agreed with A.A. were higher on autonomy and lower on deference. The implication is that there was an inverse relationship between A.A. affiliation and clinic attendance. Hence, it is possible that A.A. members (past or present)

---

* Similarly, Overall *et al.* (1971) have shown that although both historical and psychological items help to account for the variance in outpatient psychopathology, it is more efficiently accounted for by the former.

who attend alcohol clinics are either A.A. nonaffiliates in Trice's sense of the word or else former affiliates who derived no benefit from it, a point that should be kept in mind in trying to sort out findings that relate A.A. membership to other variables in outpatient treatment studies. Much work needs to be done on the problem of the similarities and differences between A.A. members and alcoholics who go to medical clinics. In any case, when specialists who routinely deal with alcoholics ask patients about prior A.A. attendance, they should determine whether they were A.A. affiliates or nonaffiliates, as should the nonspecialist.

Predictors of Success in A.A.

Who is and who is not likely to benefit from A.A.? Machover et al. (1959b), who used an extensive battery of psychological tests on remitted A.A. members, unremitted alcoholics, normal controls, and homosexuals (Mattachine Society members), concluded that remitted A.A. members were less defensive. In other words, they either had been low on denial in the first place or had reduced it as the result of attending A.A. They also reported that remitted A.A. members were not socially inhibited (the best discriminant) and were more likely to be identified with their mothers, to be obsessive–compulsive, and to use overcontrol, rationalization, and reaction formation. These are valuable studies which deserve repetition with larger and less biased samples.

In summary, it seems that the A.A. affiliate is most likely to be a single, religiously oriented class 2–4 individual, who has lost his drinking friends. If he has a wife or girl friend, she supports the idea of his going to A.A. He is not highly symptomatic, and is a socially dependent guilt-prone person with obsessive–compulsive and authoritarian personality features, prone to use rationalization and reaction formation. Finally, he is a verbal person who can share his reactions with others and is not threatened by groups of people.

The Effectiveness of A.A.

Now that we have some idea about who goes to A.A. and who does best in it, how well does it work? We recall Fox's (1957) statement to the effect that A.A. is the most valuable treatment method available in the field of alcoholism. One often finds similar statements in medical and psychiatric papers. For example, Gellman (1964) estimated that as many as 75 percent of A.A. attenders are cured and A.A. itself claims that 60 percent of those attending A.A. achieve sobriety within one year (Alcoholics Anonymous, 1972b). What are the facts? Because A.A. has consistently declined to let itself be scientifically studied, they are, alas, pitifully few. The only available and useful source of hard data on the effectiveness of A.A. as a primary treatment method known to us is a report by an A.A. member, Bill C. (1965). We are indebted to him for providing the data he did but equally frustrated because he chose to report only on members who attended A.A. at least ten times, so that we can have no idea of the overall dropout rate in the A.A. group he reported on. Of the 393 members who attended at least ten meetings, 122 stayed sober for at least one year and were still sober, while another 14 had stayed sober for at least one year but had slipped and were currently attending. Thus, this gives an improvement rate of 34.6 percent, a cumulative 7.4 year attrition rate of 65.4 percent, and a yearly attrition rate of 8.8 percent among members who came at least ten times. On the face of it, then, allowing for probable sampling variation among A.A. groups, A.A. seems to do about as well as alcohol clinics. However, we should recall that class 5 patients, who have a poor prognosis, tend to avoid A.A. So do those who reject religious values and are especially apprehensive about social and group interaction and hence are probably unaffiliated, i.e., those who do not belong to any organization. Thus, at least three kinds of poorer prognosis persons tend to avoid A.A. or quickly drop out of it, something that makes A.A. populations much better prognosis ones than those in typical alcohol clinics. We also recall that A.A. members may be lower on denial than clinic patients, yet another favorable prognostic feature. Seen in this light, a 34.6 percent improvement rate becomes a little less impressive, especially if one keeps in mind the fact that the base from which the 41.6 percent outpatient figure was computed included 18 programs, 5/6 of which had typical poorer prognosis, predominantly lower-class clinic populations. The apparently better results of alcohol clinics is not so surprising if one considers that they offer a much wider range of services (medical, psychiatric, social work) and a variety of treatment methods (conventional individual and group psychotherapy, couple and family therapy, behavior therapy, and drug treatment). Clearly, the issue of the effectiveness of A.A. as a primary treatment method cannot easily be further clarified unless it opens its doors to researchers. Until that time, the researcher, whose professional stance must be the infuriating and unpopular one of the man from Missouri, will take with a grain of salt claims of very high success rates so often expressed by A.A. members.

The issue of the value of A.A. as a secondary treatment method is a separate one in which clear-cut results are lacking. However, as an adjunct to other kinds of treatment, it is likely that it is useful in a much wider range of patients.

In summary, then, as a primary treatment method, compared to alcohol clinic treatment, A.A. seems to be applicable to a narrower range of patients in whom it may not be as effective. However, as a supplementary treatment method, it may be applicable to a much wider range of alcoholics. In any event, the physician who diagnoses alcoholism in one of his patients should not forget to suggest that he try A.A. If the patient finds it incompatible, he will abandon it. On the other hand, if he finds it congenial, he will stick with it and may derive great benefit from it. However, the physician should not blindly recommend A.A. to patients who are unlikely to accept it (class 1 or 5, non- or antireligious, uncomfortable in groups) but rather to those likely to profit from it.

A.A. Attendance as a Predictor of Success in Other Settings

Many clinicians take past or present contact with A.A. to be one of several indicators of good motivation. In many studies patients were involuntary and usually they were asked whether they had ever attended A.A. rather than how frequently they had done so, a procedure that would not identify A.A. affiliates but rather patients' willingness voluntarily to label themselves as alcoholics. On the other hand, Bateman and Petersen (1971) and Rossi *et al.* (1963), who asked voluntary inpatients whether or not they had regularly attended A.A., found that regular attenders had a better long-term outcome. Posthospital A.A. attendance was also associated with better outcome in 3/4 clinical studies that looked at it (Bowen and Androes, 1968; Kish and Hermann, 1971; Tomsovic, 1970). It is interesting that they were conducted in VA hospitals, while the negative report (Selzer and Holloway, 1957) came from a state hospital, which tends to take sicker, lower-class, and more deteriorated alcoholics than VA programs, in other words, just the kind of patient who would tend not to attend A.A. anyway. In outpatients, the picture seems much the same. There we find that previous A.A. contact was not predictive of success in two studies with predominantly lower-class populations (Kissin *et al.*, 1968b; Mayer and Myerson, 1971) but was in three others that dealt with somewhat higher SES patients (Baekeland *et al.*, 1973; Gerard and Saenger, 1966; Robson *et al.*, 1965). Since frequently of prior A.A. attendance was not determined, these results suggest that, provided he has more environmental supports, or alternatively, that he belongs to the SES category that would go to A.A., a patient's willingness to label himself as an alcoholic in the past gives him a better prognosis in outpatient treatment. Here again, the role of prior A.A. affiliation and nonaffiliation remains to be determined. In any case, the physician who detects alcoholism in one of his patients should not forget to ask him whether he has had previous treatment for it, and he should also make it clear that he considers A.A. a form of treatment. Otherwise, the patient may fail to seek out a valuable form of treatment which may help him.

BEHAVIORISTICALLY ORIENTED PSYCHOTHERAPY

General Considerations

In my discussion of inpatient and outpatient treatment, without going into details, I have repeatedly mentioned the use of conventional individual and group psychotherapy and their outcome. Logically, in the same sections, I should have included studies that used psychotherapeutic approaches based on principles of learning theory and, hence, on experimental psychology rather than on clinical medicine. That I did not do so and instead have devoted a special section to them is not because, in my opinion, there is compelling evidence that they are more effective than other kinds of psychotherapy, but simply because the reader may be less familiar with their details.

Behavioral approaches to alcoholism are hardly new. The use of aversion therapy dates back over 40 years, of drugs even farther. The principal forms of behavioral therapy are: (1) aversion therapy; this includes the following variants: classical aversive conditioning, instrumental avoidance conditioning, instrumental escape conditioning, aversive imagery techniques, and relaxation-aversion; (2) systematic desensitization; and (3) operant conditioning techniques. Because these terms may be unfamiliar, I will briefly define them.

In *simple classical aversive conditioning,* not drinking alcohol in the presence of alcohol along with disgust at the sight, smell, taste, or thought of alcohol is the conditioned stimulus (CS) and an aversive stimulus (nausea and vomiting, paralysis, painful electric shock) is the unconditioned stimulus (UCS) (Hsu, 1965; Lemere *et al.,* 1942; Mills *et al.,* 1971); Sanderson *et al.,* 1963; Voegtlin, 1940). This and other aversive techniques depend on the establishing of classical (Pavlovian) conditioned responses. The CS precedes the UCS. In *instrumental avoidance conditioning* the patient can avoid shock if he turns off the stimulus (stimuli range from photographs of spirits to the sight of a bottle) within eight seconds of its presentation (MacCulloch *et al.,* 1966). In *instrumental escape* conditioning (Blake, 1965), the patient learns to spit out his favorite drink in order to terminate shocks. In *aversive imagery techniques* (Anant, 1966, 1967a,b, 1968; Cautela, 1966, 1967), the patient is first given relaxation training by deep breathing and inhaling, after which he is required to imagine various situations involving drinking, and finally is asked to visualize these situations and then various unpleasant consequences involving nausea, vomiting, and social shame and degradation. In *relaxation–aversion* (Blake, 1965), the procedure is the same as in instrumental escape conditioning except that the aversion training is preceded by training in progressive relaxation. In *systematic desensitization* (Wolpe, 1958), anxiety-evoking stimuli are graded according to the intensity of the anxiety they induce in the patient. He is then asked to imagine the least disturbing stimulus (event, feeling, fantasy, interpersonal situation) while at the same time undergoing a relaxation response in which he has been previously trained and which is thought to be incompatible with anxiety. The procedure is repeated until the patient reports that the stimulus no longer makes him feel anxious. In the next session it is repeated but with the next most anxiety provoking stimulus on the hierarchical list of graded stimuli. Relaxation procedures include progressive relaxation, hypnosis, or carbon dioxide inhalation. In *operant conditioning* techniques an alternative behavior to drinking is established by selectively rewarding it. Approaches based on classical conditioning have depended on the assumption that punishment leads to avoidance of the punished response and thus have stressed passive avoidance training, while those based on operant conditioning have emphasized instead active avoidance learning, in which learning rather than inhibiting a response is the aim. Theoretically, there is no reason both approaches cannot be combined, although this does not seem to have been done in large series of patients.

Aversive Conditioning

The oldest approach to aversive conditioning is through the use of drugs (apomorphine, emetine, and the like) to induce nausea and vomiting just after the subject smells, tastes, swills, and swallows an alcoholic beverage. The procedure is an arduous one for both therapist and patient and carries with it potential medical hazards. The reported results of investigations of this kind of aversive conditioning cannot be taken at face value. In 3 of 4 studies (Lemere and Voegtlin, 1950; Thimann, 1949; Voegtlin and Broz, 1949), the patients were upper-class persons who might have done well with any kind of treatment regimen, and in none were they randomly assigned to aversive conditioning, which was confounded with other kinds of treatment influences. When the trials with upper-class patients are corrected for retreatment of relapsed patients, they give long-term abstinence rates that range from 30 to 40 percent, certainly not even as good as those obtained with other treatment methods in similar patients. However, the results of Stojiljkovic (1969), who reported at 51.5 percent long-term abstinence rate, are more impressive since he dealt only with somewhat lower SES outpatients who had been unsuccessfully treated by other techniques. That they were so good may gave been due not only to the intensiveness of his six-month course of conditioning procedures but also to the fact that all his patients had to join an AA-like organization and get social work counseling. Obviously, patients willing to submit to this kind of treatment must be highly motivated, a supposition supported by the unusually low dropout rates, 10.9 percent in one of the studies with upper SES patients, and only 15.1 percent in Stojiljkovic's, who must have been especially carefully selected since this is an extraordinarily low dropout rate for any kind of outpatient program. In any case, this kind of conditioning has at least two drawbacks: (1) It is potentially hazardous. (2) It is unacceptable to most patients.

Since apomorphine and emetine, particularly the former, are CNS depressants which produce nausea of gradual onset, they are theoretically not well suited for aversive conditioning: central depressants inhibit the formation of conditioned responses and precise timing of the CS-UCS interval (it should be about 0.5 seconds) is important for optimal conditioning. Franks (1970) has outlined other requirements for effective conditioning. They include distributed rather than massed trials (to avoid the accumulation of reactive inhibition), prominence of the CS with respect to background stimuli, a realistic overall training situation and the use of a number of different forms and attributes of alcohol as of the CS. Finally, partial or intermittent reinforcement is most effective. The cogency of these recommendations is illustrated by the results of Quinn and Kerr (1963), who treated 22 poor prognosis alcoholics, of whom 60 percent were readmissions, with apomorphine at the rate of six days a week for three months. Of the 15 patients (68.2 percent) who completed 50 reinforcements, all but one resumed drinking within a short time, this despite the fact that they also received group and individual therapy. A later report (Quinn and Henbest, 1967) on patients who had received apomorphine aversion therapy and got a long-lasting aversion to whiskey showed that these patients had switched to other alcoholic beverages, clear evidence of a failure of generalization of aversion from one type of alcoholic beverage to another, something that may help to explain their poor results.

An even more drastic aversion conditioning procedure, but one in which the CS-UCS interval can be precisely timed and the UCS has a sudden onset, involves the use of the intravenously administered muscle relaxant succinyl choline as the UCS just after the presentation of the CS. Terrifying total paralysis with full consciousness immediately ensues for 60 to 90 seconds. Needless to say, the procedure is not without risk, requires the services of an anesthesiologist, and is acceptable only to most atypical patients. Good results have never been reported with it. Thus, two series without random assignment of patients reported improvement rates of 17.3 percent at four months (Holzinger et al., 1967) and 16.7 percent at one year (Farrar et al., 1968). Controlled studies have been even less encouraging. Thus Laverty (1966) found no significant long-term effects or abstinence, although there was a significant decrease in the amount of alcohol ingested when patients drank. On the other hand, in two other well-designed and controlled experiments (Clancy et al., 1967; Madill et al., 1966) no effect whatever of this kind of treatment on outcome was found.

By contrast, the results of electric aversion therapy have varied widely. Since some of these experiments not only have received a good deal of sensational and uncritical publicity but also had a goal of moderation rather than abstinence, I want to discuss them in detail. Thus Devenyi and Sereny (1970), who controlled for the conditioning procedure but not for volunteering and used reinforcement sessions, reported only a 16.6 percent abstinence rate with a dropout rate of 86.6 percent within the first three reinforcement sessions. On the other hand, it is not surprising that McCance and McCance (1969), who randomly assigned their much higher SES patients either to electric aversion or to group therapy, found them equally effective and almost three times as much so as had Devenyi and Sereny (1970). Blake (1965) has reported similar results in class 1 and 2 patients. An interesting refinement was supplied by Mills et al. (1971), who used a control group whose comparability to their experimental subjects was not shown. Building on their previous findings that when alcoholics and social drinkers are compared, the former are found to drink straight drinks in gulps and large quantities while the latter mixed drinks in small sips and small overall quantities (Sobell et al., 1971; Sobell and Sobell, 1972), they tried to train alcoholics to drink normally. This they did in a hospital setting in a simulated bar by shocking undesired responses (gulping rather than sipping, too frequent sipping or gulping). The intended course of treatment was one of 14 two-hour sessions. Four out of 14 subjects dropped out before the sixth session and 9 of the completers showed conditioning effects. Booster sessions after discharge from hospital were not used. At six weeks, 5 out of 13 experimental subjects were improved as opposed to 2 out of 13 controls, a nonsignificant difference. Three of the improved patients did not drink moderately, which they had supposedly been trained to do, but instead were abstinent, something that casts grave doubts on the highly publicized procedure which otherwise might have been faulted for omitting reinforcement sessions. In a refinement of this experiment (Sobell and Sobell, 1972) volunteer patients in a voluntary hospital, chosen either to be appropriate to a goal of abstinence (identification with A.A., requested abstinence, poor social support for controlled drinking) or of controlled drinking (previous history of controlled drinking, good social support for controlled drinking) were not only given aversive conditioning on a variable avoidance schedule but also were interviewed when drunk, looked at videotape replays of these episodes, were subjected to an artificial failure experience, and in the remaining sessions (the bulk of treatment), there was emphasis on defining prior setting events for heavy drinking and training subjects in alternative, socially acceptable responses to these situations. Patients were randomly assigned to these treatments or to the control condition, which included group therapy, A.A., and so forth, which experimental subjects also received. At six-month follow-up, 75 percent of the controlled drinker experimental patients were functioning well versus 31.6 percent of controls, and 80 percent of the abstinence experimental subjects were functioning well as opposed to only 20 percent of the controls, both significant differences. Equally striking results were reported by Lovibond and Caddy (1970), who trained alcoholics to discriminate their blood alcohol concentrations from 0 to 0.98 percent. They were then shocked on a variable intermittent schedule if they drank to levels above 0.065 percent. These largely self-referred patients were told that treatment alone could not ensure control of their drinking behavior but that its purpose was to provide them with a stop mechanism to help their own efforts at self-control. They were given 6–12 sessions and 30–70 shocks. Family members were actively involved in treatment. At 4–15 months, 21 of 31 (67.7 percent) patients were rarely exceeding 0.07 percent blood alcohol levels and only 3 of 31 subjects had failed to complete the full course of treatment. On the other hand, control subjects, who received shocks on a random basis for the first three sessions but contingent on blood alcohol levels thereafter, dropped out very rapidly (8 out of 13 before the third session) and quickly returned to their normal drinking pattern after initially reducing it. These seem like very impressive results. However, Sobell and Sobell's (1972) abstinence experimental subjects were better educated than their controls and one would like to know whether experimental and control subjects had equivalent periods of hospitalization and also whether the same therapists administered experimental and control treatments (details not mentioned in their long report). Furthermore, does the subject who had hoped for a novel kind of treatment but instead receives an old one feel that he is being given short shrift and hence become discouraged? Surely Lovidbond and Caddy's (1970) control subjects must quickly have realized that they were merely receiving painful shocks which were in no way therapeutic, so that one wonders in what sense their sessions filled the bill as control *treatments*. An even more questionable aspect of their experiment, however, is their reliance on potentially unreliable patient self-reports in their follow-ups. Indeed, a more recent experiment along the same lines found that patients were unable to maintain their blood alcohol levels within an acceptable range whenever external blood alcohol feedback was withdrawn (Silverstein *et al.*, 1974).

Although convincing proof of the specific value of electric aversion remains to be given, it is noteworthy that the experiments which reported the best results with it either had good prognosis patients or more comprehensive treatment programs. They may thus be indicative of a trend in behavior therapy away from simplistic procedures to ones more geared to the specifics of individual patients' clinical histories and personalities.

Blake (1965) pioneered in the introduction of instrumental escape conditioning with shock. He reported that at one year 27 percent of his patients were improved, not a striking result (Blake, 1967). A strong proponent of behavior therapy might object that Blake's subjects were not treated under realistic conditions. They should consult the experiment published by Vogler et al. (1970), who used a control group, gave instrumental escape conditioning with shock in a simulated public bar, and used a random reinforcement schedule with 20 sessions of 20 sips per session over a two-week period. At eight months, experimental subjects did no better than controls in terms of the number of relapses they had suffered, although they did have a significantly greater number of days to relapse. Poor results could be due to the fact that few patients returned for booster sessions, perhaps only another way of saying that subject characteristics (motivation) were much more important than specific treatment effects. That chemical and electric aversion therapies have not shown any special worth despite their apparently good theoretical foundation is cause for reflection. Their failure to live up to their initial promise may be due to the following: (1) They can generate high general levels of anxiety as opposed to the specific anxiety conditioned to the alcoholic beverage. If behavior (drinking, for example) is motivated by anxiety, *aversive conditioning can augment rather than reduce such behavior* (Eysenck and Rachman, 1965). (2) If a subject is habituated to receiving punishment together with a reward, punishment during extinction can actually increase resistance to extinction (Solomon, 1964). Many alcoholics learn to associate both reward and punishment with drinking. (3) Punishing a response originally established by punishment, such as drinking because of a painful experience, may strengthen rather than weaken the response. (4) Adaptation to punishment may take place. (5) In practice, aversive therapy has focused on only one aspect of excessive drinking, the actual act of drinking. The drinking response, as Franks (1970) points out, is but one item in a complex series of closely interrelated responses, which are established earlier and are essential to the act of drinking. (A behaviorist's way of saying, "Treat the whole man.") In other words, these techniques have tended to ignore the history of the symptom and the full social context in which it occurs, something that ordinary psychotherapy, which is on less firm theoretical grounds, has not neglected to do. Finally, many alcoholics (22.5 percent in one study—Keehn, 1970) don't like the taste of any alcoholic beverage very much. Hence, for such patients it would be inappropriate to make alcohol aversive.

An apparent drawback of all the aversive techniques discussed is the unpleasantness of the procedures, which tend to have low patient acceptability and to induce high dropout rates. On the face of it, then, verbal conditioning techniques would seem to hold more promise. Aversive imagery techniques include both so-called aversive imagery per se and the use of hypnosis. Anant (1967a) initially reported that 25 of 26 alcoholics who completed his program abstained for 8–15 months, while in a later study without booster sessions (Anant, 1968) only 20 percent were found to be abstinent at 6–23-month follow-up. Ashem and Donner (1968) used aversive imagery techniques in a controlled study of patients who had been previously unsuccessfully treated with other methods (A.A., clinic treatment, or private psychotherapy). Besides group therapy, they received nine formal sessions (as well as practicing by themselves). At six-month follow-up, 6 out of 15 (40 percent) were abstinent as opposed to none of the controls. Considering that patients were randomly assigned to treatments and that these were presumably poor prognosis patients, the procedure seems promising and warrants further and more extensive study, which it will no doubt receive in the future.

Hypnosis has been tried sporadically in the treatment of alcoholism for a long time. However, reasonably well designed experiments are rare. Smith-Moorehouse (1969) reported on his results with hypnosis in alcoholic outpatients. In his procedure he tells the patient that whenever he sees or smells alcohol he will remember that he is an alcoholic and that if he wants to keep fit, well, and happy, he should not take the first drink. He points out that suggesting to the patient under hypnosis not to drink again may produce severe conflicts in him, which in turn can lead to increased tension and anxiety and renewed drinking. He reported a six-month to two-year improvement rate in experimental subjects of 53.4 percent as compared with 37.5 percent in controls, a significant difference. The hypnotic subjects were self-selected, but on the other hand if they had been more highly motivated than the controls, one would expect that their dropout rate would have been lower than that of the latter. Dropout rates were in fact the same (37 percent). In a better designed study which used random assignment of patients to treatments, Edwards and Guthrie (1966), on the other hand, found no lasting benefit of hypnosis at one year, this despite the fact that six days of hypnotic suggestion with apparently adequate suggestions followed by weekly and then monthly reinforcement had been used. However, an overriding emphasis on absolute abstinence may have worked counter to the author's intentions.

In summary, it seems that chemical aversion does no better than other methods but that it has not yet fairly been tested because of the absence of controlled experiments, while succinyl choline aversion and instrumental escape conditioning with shock offer no special treatment advantages that outweigh their unpleasantness. However, aversive imagery techniques seem both more palatable and promising. In experiments that failed, one is struck by the small number of treatments. Both booster sessions and ancillary treatment methods are probably important for best results, and aversive techniques should not be used in patients who do not enjoy alcoholic beverages.

Systematic Desensitization

There has been little study of systematic desensitization in alcoholics. Storm and Cutler (1968), however, found it ineffective. They attributed their lack of success to their patients' relatively low anxiety–tension levels, which hindered identification of special foci of anxiety. Even so, one would hardly expect it to work in an unselected population of alcoholics but rather only in highly self-aware and self-observant patients who are not physically dependent on alcohol and do not drink to alleviate depression or to allow the release of inhibited sexual or aggressive drive expression under conditions of diminished self-criticism. However, the technique deserves reexamination in a suitable population as it is by now a well-established kind of psychotherapy.

Operant Conditioning

A major criticism of all the aversive techniques is that they rather shortsightedly focus on the extinction of a response (drinking) without giving the patient an alternative way of reducing anxiety of teaching him any behavior incompatible with drinking. This criticism is met by operant conditioning, which shapes desired alternative responses by rewarding them. Extensive studies of the effectiveness of this approach with alcoholics remain to be reported. However, a very promising small study was reported by Hunt and Azrin (1973), who based their treatment strategy in eight patients on the idea of time out from positive reinforcement, in other words, that individuals may be deterred from drinking because of its interference with other sources of satisfaction provided they are of maximum quality and frequency, are varied in nature, and occur regularly. Accordingly, they rearranged the vocational, family, and social reinforcers of these patients so that time out from these reinforcers would occur if they began to drink, with the result that their eight patients did significantly better than matched controls. Notable in this treatment program was its rather intensive involvement at all levels with every major area of the patient's life.

In summary, then, systematic desensitization may be applicable in self-aware patients whose drinking is triggered by special foci of anxiety and is not reinforced by the reduction of other dysphoric symptoms by alcohol. On the other hand, operant conditioning techniques seem much more generally applicable and are in many ways reminiscent of more conventional psychotherapy.

An obvious extension of behavioral techniques would be the combination of aversive and operant conditioning. This obviously calls for very careful study of the patient and the details of his history and doubtless will be done in the future.

DRUG TREATMENT

The use of drugs to modify alcoholism has a long history. Drug treatment of alcoholism has been one of the most carelessly and mechanically treated areas of research, careless because of the enormous proliferation of uncontrolled studies, mechanical because of the blind application of double blind techniques with utter disregard of many other equally important methodological requirements. Hence, the hitherto too easily accepted conclusions of three comprehensive reviews which decided that the drug treatment of alcoholism is ineffective must be questioned. To show that, on the contrary, it is effective in some patients will require a rather detailed discussion both of the assumptions underlying drug treatment and of the methodological problems that plague its evaluation.

Implicit Assumptions

The basic assumption of most drug treatments involving tranquilizers and antidepressants is the drive reduction theory of alcoholism, which runs something like this: (1) Alcoholics suffer from dysphoric symptoms such as anxiety, depression, and, if physically dependent, symptoms of partial withdrawal. Alcohol reduces such symptoms. Therefore, drugs that also diminish these symptoms will lower the patient's drive to consume alcohol. Alcoholics do indeed seem to be more anxious and depressed than normal controls (Kissin and Platz, 1968; Weingold et al., 1968), and two factor analytic studies suggest not only that their depression is like that of patients with a primary diagnosis of depression (Gibson and Becker, 1973) but also that alcohol reduces their depression, anxiety, and tension (MacAndrew and Garfinkel, 1962). Yet, a recent review (Cappell and Herman, 1972) concluded that the evidence in human studies for the tension reduction hypothesis of alcoholism is conflicting. Uninvestigated individual differences in tension reduction by alcohol may explain some of the contradictory reports reviewed by Cappell and Herman (1972). Thus Vannicelli (1972) studied the effects of divided doses of alcohol in hospitalized alcoholics and found that half of his subjects were anxious when intoxicated, while the reverse was true in the other half. Change in anxiety after the first drink was a good predictor of overall change in anxiety after several drinks. Subjects who changed the least in anxiety scored the highest on the California Psychological Inventory Cm Scale, indicating a greater tendency for them to be dependable, moderate, steady, etc., a result consistent with the finding of Cutter *et al.* (1973) that inhibition rather than power needs determined whether subjects did or did not take a second drink. Thus, it is possible that tranquilizers might only be effective in less inhibited and, perhaps, socially stable patients. Similarly, longitudinal studies of the relationship between stressful life events (Hore, 1971a,b) or depressive episodes (Mayfield and Coleman, 1968) and drinking bouts have found consistent relationships in only about a third of patients. Finally, alcohol has been reported to have an antidepressant effect only at lower doses (Mayfield, 1968a,b). All this suggests that antidepressants may be applicable only to a subgroup of depressed alcoholics who drink more or less continuously at moderate levels.

At first glance, the above suggestions might seem questionable because of reports that in alcoholics negative affects increase as they continue drinking (McGuire *et al.*, 1966; Mendelson *et al.*, 1964; Vanderpool, 1969). Since they may preferentially remember the short-term rather than the long-term effects of alcohol on them (Weingartner *et al.*, 1971) (a major problem in psychotherapy) or, put differently, since only the immediate effects of alcohol significantly reinforce alcohol consumption, such observations are probably irrelevant. Much more to the point are reports to the effect that alcohol causes an immediate, albeit transitory, reduction in dysphoric affects (Berg, 1971; Kissin and Platz, 1968; Mayfield, 1968a,b).

Negative Results in Drug Studies with Alcoholics

Earlier, I indicated that almost all controlled studies of the effects of psychotropic medication, either on psychological symptoms or on drinking patterns of chronic alcoholics, have proved negative. Yet, many specialists feel that such drugs have a valid, important, and useful place in the treatment of alcoholism. Who is right, the researcher or the practitioner? In my opinion as a researcher, the latter rather than the former, since most double blind drug studies conducted with alcoholics have probably been invalid on methodological grounds. Two things more than anything else have vitiated most such studies: high dropout rates and so-called nonspecific effects.

Reported dropout rates have tended to be lower (13.7–39.2 percent with a mean of 28 percent) in inpatient treatment settings (Bowen and Androes, 1968; Gross and Nerviano, 1973; Kamin and Caughlan, 1963; Knox, 1972; Moore and Ramseur, 1960; Pokorny *et al.,* 1968; Rathod *et al.,* 1966; Rhodes and Hudson, 1969, Rohan, 1970; Simpson and Webber, 1971; Tomsovic, 1968; Wilkinson *et al.,* 1971) than in outpatient treatment programs, where 52–75 percent of patients drop out before the fourth session (Baekeland *et al.,* 1973; Blane and Meyers, 1964; Ditman and Cohen, 1959; Gerard and Saenger, 1966; Storm and Cutler, 1968; Wilby and Jones, 1962). Dropout rates in outpatient drug studies (by far the most common kind) have not been much less. Here they present the following rather dismal picture: by the first visit, 36.5 percent (Bourne *et al.,* 1966); by one week, 46.8 percent (Ditman, 1961; Leivonen *et al.,* 1966; Mooney *et al.,* 1964), by two weeks, 57.6 percent (Charnoff *et al.,* 1963; Cohen *et al.,* 1961; Ditman and Mooney, 1959; Mooney and Ditman, 1965); by one month, 50.1 percent (Bourne *et al.,* 1966; Haden and Fowler, 1962; Ritter, 1956; Tyndel *et al.,* 1969), by three months, 61.3 percent (Charnoff *et al.,* 1963; Ditman, 1961; Ritter, 1956), and at six months, 89.2 percent (Charnoff *et al.,* 1963; Ditman, 1961; Ditman *et al.,* 1962; Ritter, 1956). Figures like these are characteristic of the typical outpatient drug study, which is usually conducted in clinics catering to predominantly lower-class patients. Such high dropout rates have had the effect of curtailing the time span of many of them to the point where they were in effect one-, two-, or three-week affairs or else of reducing sample size to a point where drug effects, even if they existed, would be difficult to pick up, that is, below the 50-patient group size usually considered necessary in drug research with heterogeneous populations (Overall and Hollister, 1967). In this regard, it is noteworthy that Swinson (1971), who had a low dropout rate in carefully selected class 1–3 patients, found a definite effect of metronidazole on drinking patterns at 12 months but not at 6 months. One wonders whether other drugs now felt (by researchers, not clinicians) to be ineffective in the treatment of alcoholism would have proved of benefit had they only been put to the test in longer-term studies in non-skid row and lower-class populations, in other words, with patients who would stay in treatment and thus allow a fair test of the drugs. We recently obtained some highly suggestive evidence in this regard (Baekeland and Lundwall, 1975a), in a double blind one-month study of the effects of the tranquilizer–antidepressant combination oxazepam–protriptyline on anxiety and depression in lower-class outpatient alcoholics. When we simply took difference scores between the last time the patient attended and his first visit, we found no drug effect. However, in a number of patients, this included intervals during which time they had missed one or more visits and hence had been without medication for varying lengths of time. However, if the analysis included only periods of continuous attendance (and hence of continuous medication) and excluded those patients with missed visits, a significant reduction in anxiety and depression was registered. Certainly, a corollary conclusion should be that it is a waste of time to give medication to patients who do not attend regularly.

General disregard of so-called nonspecific effects may have been even more crippling to the validity of studies of drug treatment of alcoholics by virtue of their vitiating real drug–placebo differences in a number of ways. First of all—and this is generally overlooked—many patients do not take medication as it is prescribed, either omitting it altogether, taking too little, or taking it at irregular intervals (Kearney and Bonime, 1966; Rickels, 1968; Rickels et al., 1968; Willcox et al., 1966) and patients who deviate from prescribed dosages are most likely to drop out of treatment (Rickels, 1968; Winkelman, 1964). Published drug studies for the most part do not seem to take this into account. Most apparently do not even try to correct for missed visits. Second, alcoholic patients on active medication may drop out of treatment more rapidly and more often than those on placebo depending on their SES characteristics (Charnoff et al., 1963; Kissin et al., 1968a). If not taken into account, such differential dropout rates can obscure true drug–placebo differences. Early dropping out of treatment can be especially misleading in studies evaluating the effectiveness of drugs with a slow onset of therapeutic action—such as antidepressants—a point especially relevant in a population which tends to suffer from depression. Third, physicians are usually resentful about having to give their patients a placebo and communicate this resentment of drug studies to their patients, another factor contributing to high dropout rates and the paradoxical relative ineffectiveness of active medication in such studies. Fourth, the effectiveness of active medications has been clearly shown in the case of antidepressants (Sheard, 1963). It is particularly important to take into account therapist differences since therapeutic style (detached vs. active, type A therapist vs. type B [Whitehorn and Betz, 1960]) determines both attrition rate (Howard et al., 1970; McNair et al., 1967) and placebo response (Rickels and Cattell, 1969) and differential assignment of patients to therapists, depending on patient and therapist characteristics, often takes place in supposedly well-designed studies (Rickels and Cattell, 1969). Fifth, placebo reactors may report so much improvement on placebo that drug–placebo differences are obscured (McNair et al., 1967, 1970). Sixth, patients with an established pattern of dropping out of treatment and who are most likely to drop out (Rickels, 1966) are included in drug studies. Finally, inadequate dosage regimens below levels affording adequate symptomatic relief for many patients have often been used, something that may contribute to high dropout rates within the first month of treatment (Baekeland et al., 1973). Hence, until drug studies with outpatient alcoholics that take these factors into account are forthcoming, the negative results with which the literature so far abounds must be taken with more than a grain of salt.

Outcome in Drug Studies

It is not surprising that few double blind studies have shown drugs to be effective in the treatment of chronic alcoholism. Rather it is amazing that any of them have panned out.

Disulfiram and Citrated Calcium Carbimide

Disulfiram (DS) and citrated calcium carbimide (CCC) are different from other drugs studied in the treatment of alcoholism in two respects: (1) Their effect on drinking is an indirect or potential one, and (2) neither has been subjected to a double blind study. If a patient takes either, he experiences a highly unpleasant physiological reaction, the knowledge or fear of which, it is hoped, will keep him from drinking. However, once having discontinued the medication (for 48 hours in the case of disulfiram, for 24 hours in that of CCC), he is free to drink again with impunity. Therefore, unless he takes either drug on a compulsory basis, he is not compelled not to drink and knows it. Thus, it can be argued that any effect of either drug in non-double blind studies simply reflects the patient's motivation not to drink or, at least, to control his drinking when he feels most driven to it, the strength of his bond with the therapist who administers it, the strength of the latter's belief in efficacy of DS, and the former's suggestibility. Clearly, a priori, it is hard to imagine any rationale for using either DS or CCC except in patients who have moderately good but not perfect control of their drinking and who are neither forgetful about taking medication nor fooling themselves that they want to stop drinking. Absolute medical contraindications to their use have been outlined elsewhere. In practice this usually only includes arteriosclerotic heart disease (Lundwall and Baekeland, 1971).

Studies with controls suggest that the patient who does better on disulfiram is older, better motivated, better able to form dependent relationships, and more likely to have had blackouts and to have compulsive personality traits (Baekeland et al., 1971; Winship, 1957). Depression confers a worse outcome in DS therapy (Baekeland et al., 1971; Winship, 1957). Uncontrolled studies suggest that more socially stable and better motivated patients do better, while those with sociopathic features do worse (Lundwall and Baekeland, 1971). None of these prognostic indicators is remarkable in itself; in fact, most of them also apply to other treatment modalities. Specific only, perhaps, are the contraindication of depression and the proviso of blackouts, which indicate rapid heavy intake (Goodwin et al., 1969). Hence, the use of disulfiram is probably contraindicated in patients with more than mild depression and also in those whose drinking is highly compulsive and seem to be unable, under any circumstances, to stop after a few drinks.

Only seven studies of disulfiram have used any kind of control group (Baekeland et al., 1971; Gerard and Saenger, 1966; Gerrein et al., 1973; Gibbins and Armstrong, 1957; Hoff, 1955; Hoff and McKeown, 1953; Winship, 1957). Patients were not randomly assigned to treatments, and in all but one, the control conditions were nondrug treatments (Gerrein et al., 1973). All these studies concluded that DS was effective, but in such designs any differences found between experimental and control groups cannot *only* be attributed to the drug per se. The important role of motivation is suggested by Hoff and McKeown's (1953) finding that, within their control group, patients who wanted DS but were denied it for various reasons did better than those who refused it.

A major problem in DS treatment is that the patient may take himself off it and drink. Indeed, Baekeland et al. (1971) found that the proportion of dry appointments their patients kept varied from 0.51 in the group with the poorest outcome to 0.89 in the one that did best. Hence, ideally, something else with strong reward value should be made contingent on the patient's taking his DS. In everyday terms, he should be gently forced to take his DS. Such a strategy was reported by Bourne et al. (1966) in a study of jailed skid row alcoholics put on probation and given DS daily for 30 or 60 days (after which they could either continue or stop treatment) either without legal obligation, by a family member, or else by the courthouse probation officer with return to jail as the penalty for missing appointments. Patients were also given vocational counseling, and A.A. and psychotherapy were urged upon them. Nine months after the start of treatment both groups had an abstinence rate of 50 percent, unusually high in this kind of patient. One-month dropout rates were similar (50 percent, 40.9 percent) and 8.1 percent of the patients drank while on DS and hence had DS-alcohol reactions. A related and promising approach to the currently serious problem of the development of alcoholism among heroin addicts (Baden, 1971; Goldstein, 1972) on methadone maintenance treatment was reported by Liebson et al. (1973), who found that patients among whom receiving their methadone was made contingent on their taking disulfiram had much better control of their drinking than those simply urged to take it on a noncontingent basis. One can imagine that such approaches could be extended to patients in alcohol clinics who are dependent on minor tranquilizers such as chlordiazepoxide.

A potentially useful and new technique is that of surgical implantation of DS to ensure long-term effective blood levels. Widely used in Europe, extensive trials in the English-speaking world have only recently been reported in two studies (Malcolm and Madden, 1973; Whyte and O'Brien, 1974). Both reported postoperative abstinence periods that were significantly longer than those following previous treatments. Contraindications were those for oral disulfiram (respiratory, cardiac or renal disease, cirrhosis, diabetes, epilepsy, mental subnormality, and psychosis) and the patient's living alone. In practice, the last had to be considered. (Many clinicians using oral DS have felt those patients not living alone do better on it.) The fact that infection was a problem in 14.9 percent of patients and that tablet extrusion occurred in another 14.9 percent in one of the two studies (Malcolm and Madden, 1973) emphasizes that this is a procedure best reserved for those who have failed by every other treatment method, thus truly a court of last resort.

CCC produces an alcohol reaction which is like that of DS but is milder (Armstrong, 1957; Brunner-Orne, 1962; Collins and Brown, 1960; Glatt, 1959; Lader, 1967; Levy et al., 1967; Marconi et al., 1960, 1961; Minto and Roberts, 1960; Mitchell, 1958; Nason, 1958; Peck, 1962; Smith et al., 1957, 1959), shorter-lived (Glatt, 1959; Marconi et al., 1960), and of more rapid onset (Collins and Brown, 1960; Ferguson, 1956). Its duration of action is only about half that of DS (Collins and Brown, 1960; Peck, 1962). In view of CCC's shorter duration of action, it is hard to imagine any rationale for its use except in some persons in whom DS is contraindicated. No properly controlled studies comparing its effectiveness with that of DS are known to me.

Metronidazole

The case of metronidazole is an excellent example of how poor experimental design can lead to the erroneous conclusion that a probably effective drug is useless. First introduced with the claim that it reduced craving for alcohol, it was found effective by a spate of uncontrolled studies. Then, when controlled studies were done, nine (Egan and Goetz, 1968; Gallant et al., 1968a; Lehmann and Ban, 1967; Linton and Hain, 1967; Penick et al., 1969; Platz et al., 1970; Rothstein and Clancy, 1970; Strassman et al., 1970; Tyndel et al., 1969) found it ineffective as opposed to one that did not (Swinson, 1971). The democratic procedure, if one were to blindly believe in the efficacy of double blind studies, would be to head-count and ignore the one paper that reported positive results. However, let us examine it and the others critically. Swinson (1971), first of all, elected to study a population not too likely to drop out or to be impossible to follow up: the first 60 male alcoholics discharged from an inpatient unit willing to be studied and likely to stay in the area for at least 12 months. His patients did not include class 4 or 5 patients and had an established relationship with the investigator, all factors militating against dropping out of treatment. Patients were told they might expect to feel less like drinking on the medication than previously. Drinking was assessed on a 5-point scale, weekly at first, and then at 3-month intervals by interview with patient and spouse/next of kin and patient. At one month there was no significant between-group difference but a trend in favor of metronidazole, which increased at 6 months and turned out to be significant at 1 year. Dropout rates were only 5 percent at 1 month and 35 percent at 3 months. (Class 1 and 2 patients were least likely to drop out.) Now let us consider the studies with negative results. All of them used treatment periods that were shorter and ranged from 2 to 6 months, and all but one of them (Penick et al., 1969) used skid row or largely lower-class populations. In one study (Tyndel et al., 1969) a 50 percent dropout rate in the first month reduced the effective sample size to only 23 patients. Two trials (Gallant et al., 1968a; Merry and Whitehead, 1968), which found metronidazole no better than chlordiazepoxide and neither better than routine treatment, suggest that the latter drug, too, may have been tested for too short a time in other published studies on it. Although Swinson's (1971) work should be replicated, it appears that metronidazole may be effective on a long-range basis. How does it work? Several investigators found that it had a tranquilizing effect (Gelder and Edwards, 1968; Lehmann and Ban, 1967) and that it interfered with the taste of or for alcohol (Lehmann and Ban, 1967; Platz et al., 1970; Strassman et al., 1970; Tyndel et al., 1969). Further work needs to be done in this area, especially in view of Lehmann and Ban's (1967) finding that clinical factors such as early onset of alcoholism were predictive of these effects in their patients. (Patients who had been drinking for less than 13 years had a significant decrease in anxiety and alcohol consumption.)

Lysergic Acid Diethylamide (LSD)

The history of LSD treatment of alcoholism starts with sensational anecdotal reports, followed by enthusiastic and almost always successful uncontrolled studies and, finally, by consistently unsuccessful controlled trials. A number of the latter (Bowen et al., 1970; Pahnke et al., 1970) showed transient short-term effects, but they, like the others (Cheek et al., 1966; Hollister et al., 1969; Johnson, 1969; Ludwig et al., 1969; O'Reilly and Funk, 1964; Smart et al., 1966; Tomsovic and Edwards, 1970), were not able to demonstrate any long-term efficacy of LSD. If nothing else, this research, which made a number of schizophrenic patients decompensate, has drawn attention to the hazards of using hallucinogens in borderline or compensated schizophrenics.

Antidepressants and Tranquilizers

Considering the pitfalls of routine double blind drug research in alcoholic populations, it is hardly surprising that, except for our oxazepam–protriptyline study cited above, the only solid results have been from short-term inpatient studies which have addressed themselves to symptoms of anxiety and depression in patients specially selected to exceed cutting scores on these symptoms. Thus, imipramine proved effective in a six-week trial on depressed female alcoholic alcoholic outpatients (Wilson *et al.*, 1970) and at three weeks in depressed male alcoholic inpatients (Butterworth, 1971), while doxepin was more effective than chlordiazepoxide in treating symptoms of anxiety and depression at three weeks in male alcoholic outpatients (Butterworth and Watts, 1971). Similarly, Overall *et al.* (1973) in a large-scale and methodologically sophisticated study of hospitalized alcoholics recently reported that chlordiazepoxide in low doses (10 mg t.i.d.) was ineffective in reducing symptoms of anxiety and depression while mesoridazine or amitryptaline did help. The issues of effective dosage and cross-tolerance were raised but not settled. Their findings are quite the reverse of ours in outpatients (Baekeland *et al.*, unpublished results), where we discovered that patients on phenothiazines did worse than equally symptomatic patients on chlordiazepoxide, which in turn retained in treatment many less symptomatic patients given vitamins who otherwise might have been expected to persist better in treatment (Baekeland and Lundwall, 1975b). The difference between our results and those of Overall *et al.* (1973) is not surprising since more of their patients were psychotic and they used lower doses of chlordiazepoxide. Several other studies have also reported the effectiveness of chlordiazepoxide in keeping alcoholic outpatients in treatment. Thus, Ditman (1961) found that unselected and largely skid row patients on chlordiazepoxide had a lower short-term dropout rather than those on placebo, imipramine, thioridazine, or diethylpropion. Similarly, Rosenberg (1974) discovered that patients on chlordiazepoxide stayed in treatment better than those on disulfiram, vitamins, or no medication, in that order. It is interesting that these differences disappeared after five months. Chlordiazepoxide is effective in controlling withdrawal symptoms (Kaim *et al.*, 1969), so that its working in outpatient alcoholics may be related to its action on such symptoms, which some think last as long as six months after withdrawal. Although phenothiazines do not appear indicated for many patients because they potentiate the action of alcohol, in certain populations they seem to help. Thus, Ditman (1961) found that thioridazine was most effective in keeping hostile, irritable, unsociable, and lonely patients in treatment. Similarly, Haden and Fowler (1962) found that outpatients with withdrawal symptoms were less likely to drop out before the fifth session if they were given promazine than if they were not. An intriguing recent double blind study by Kline *et al.* (1974) of the effects of lithium on alcoholics, who were first hospitalized and then discharged to outpatient treatment and who received the drug for one year, concluded that the patients on lithium did much better than those on placebo in terms both of drinking episodes and rehospitalizations. Because of the widespread uncritical publicity this paper has received, a careful critique of it is in order. In this preliminary study with a 58.9 percent one-year dropout rate, when dropouts are *included* in the analysis, the significance of the difference between the two groups' readmission rates *increases* to $p < 0.025$. Careful examination of the data shows that this happened because a much smaller proportion of the lithium dropouts than of the placebo dropouts were rehospitalized. Since both groups had the same dropout rate (59.0 percent vs. 58.8 percent), the conclusion is unavoidable that patients assigned to lithium treatment were a better prognosis group than those assigned to placebo, so that the between-group differences obtained could be (but may not be) due entirely to sampling bias rather than to drug effects. (The two groups were matched only on age and number of previous hospitalizations.) The finding that lithium program completers had significantly fewer drinking episodes in the follow-up interval than did patients on placebo must be tempered by this possibility. Furthermore, if dropouts are included, the between-group difference in number of drinking episodes vanishes. Finally, there was no between-group difference in change scores on two depression inventories, this notwithstanding the fact that the rationale of the study was the repeated report that the drug is effective in mania and depression.

In summary, a variety of drug treatments are available as aids in the treatment of chronic alcoholism. At this point, their respective indications have not been definitely mapped out, but on the basis of research to date the following tentative statements can be made: (1) Psychotropic medication is indicated only in those patients who attend regularly and can be trusted to take it in reasonably reliable fashion. (2) Chlordiazepoxide may be helpful in nonpsychotic patients who are anxious and/or suffer from withdrawal symptoms. There seems to be no justification for its use beyond three or four months. (3) On the other hand, phenothiazines may be more helpful in psychotic alcoholics. However, they are contraindicated in all others. (4) On a long-term basis metronidazole may be helpful in higher SES patients. (5) Disulfiram seems indicated in well-motivated, older, and nondepressed patients who have compulsive personality makeups, do not have heart disease, and have already demonstrated that they can control their drinking.

SUMMARY AND CONCLUSIONS

The English-language literature from 1953 to the present on methods for the treatment of chronic alcoholism was critically reviewed. It appears that multifactorial outcome measures are superior to abstinence alone as a criterion of success and that a six-month follow-up interval is the absolute minimum acceptable.

We were repeatedly impressed with the dominant role played by patient rather than treatment factors both in persistence in treatment and eventual outcome. Thus, in inpatient studies good prognosis patients (higher SES and social stability) had improvement rates varying from 32.4 to 68 percent while poor prognosis patients (largely skid row alcoholics) had rates ranging from 0 to 18 percent. Hospital treatment programs from 1960 to 1973 seemed more effective than those from 1953 to 1963. No differences were found in the effectiveness of different kinds of treatment regimens. There was some evidence that higher SES patients may need shorter hospitalizations and little aftercare.

Outpatient clinics had higher improvement rates (41.6 percent) than did inpatient programs despite their higher dropout rates (36.9 percent vs. 17.0 percent). Outpatient programs with the highest improvement rates had good prognosis patients (higher SES, better motivation, higher social stability), while those with the poorest outcomes catered largely to skid row patients. The evidence favors multidisciplinary approaches, involvement of the spouse in treatment, and forced treatment of selected skid row alcoholics.

On the basis of the available evidence, which is limited, it seems that the population served by A.A. is quite different from that which goes to hospitals and clinics; also, that the general applicability of A.A. as a treatment method is much more limited than has been supposed in the past. Available data do not support A.A.'s claims of much higher success rates than clinic treatment. Indeed, when population differences are taken into account, the reverse seems to be true. Of all the treatment methods reviewed, A.A. is most in need of further systematic study.

Behavioral approaches, if one takes into account that they are usually tried on highly selected volunteers, seem to give about the same results as other treatment methods. However, succinyl choline aversion is ineffective while aversive imagery techniques and those combining aversive and operant conditioning show promise.

Because of high dropout rates and the operation of uncontrolled nonspecific factors, double blind drug studies conducted in heterogeneous, unselected, and predominantly lower-class clinic populations have not fairly tested the utility of antidepressants or tranquilizers in the treatment of chronic alcoholism.

In the absence of controlled studies, it is still an open question whether the major determinant of success in the use of disulfiram is not simply the strength of the therapist–patient relationship. In any case, better results seem to be obtained if it is taken under supervision. The successful disulfiram patient, besides having characteristics which confer a good prognosis with any kind of treatment, is likely not to be depressed and to have compulsive personality features. Disulfiram may be particularly effective in heroin addict alcoholics maintained on methadone when receiving methadone is made contingent on taking disulfiram.

Metronidazole may be effective on a long-term basis.

Patients who do well on drugs, psychotherapy, or rehabilitation programs seem to have different characteristics, and success rates go up with the number of treatment options given the patient.

REFERENCES

Alcoholics Anonymous, 1972a, "The Fellowship of Alcoholics," New York.
Alcoholics Anonymous, 1972b, "Profile of an AA Meeting," Alcoholics Anonymous World Services, Inc., New York.
Allen, L. R., and Dootjes, I., 1968, Some personality considerations of an alcoholic population, *Percept. Mot. Skills* 27:707–712.
Anant, S. S., 1966, The treatment of alcoholics by a verbal aversion technique: A case report, *Manas* 13:79–86.
Anant, S. S., 1967a, A note on the treatment of alcoholics by a verbal aversion technique, *Can. J. Psychol.* 8:19–22.
Anant, S. S., 1967b, Treatment of alcoholics and drug addicts by verbal aversion technique, Paper read at the 7th International Congress of Psychotherapy, Wiesbaden.
Anant, S. S., 1968, Treatment of alcoholics and drug addicts by verbal aversion techniques, *Int. J. Addict.* 3:381–388.
Armstrong, J. D., 1957, The protective drugs in the treatment of alcoholism, *Can. Med. Assoc. J.* 77:228–232.
Armstrong, R. G., and Hoyt, D. B., 1960, Personality structure of male alcoholics as reflected in the IES test, *Q. J. Stud. Alcohol* 24:239–248.
Ashem, B., and Donner, L., 1968, Covert sensitization with alcoholics: A controlled replication, *Behav. Res. Ther.* 6:7–12.
Asma, F. E., Eggert, R. L., and Hilker, R. R., 1971, Long term experience with rehabilitation of alcoholic employees, *J. Occup. Med.* 13:581–585.
Baden, M. M., 1971, Methadone-related deaths in New York City, *in* "Methadone Maintenance" (S. Einstein, ed.), pp. 143–152, Marcel Dekker, New York.
Baekeland, F., and Lundwall, L. K., 1975a, Effects of discontinuity of medication on the results of a double-blind study in outpatient alcoholics, *Q. J. Stud. Alcohol* 36 (9):1268–1272.
Baekeland, F., and Lundwall, L., 1975b, Dropping out of treatment: A critical review, *Psychol. Bull.* 82 (5):738–783.
Baekeland, F., Lundwall, L., Kissin, B., and Shanahan, T., 1971, Correlates of outcome in disulfiram treatment of alcoholism, *J. Nerv. Ment. Dis.* 153:1–9.
Baekeland, F., Lundwall, L., and Shanahan, T., 1973, Correlates of patient attrition in the outpatient treatment of alcoholism, *J. Nerv. Ment. Dis.* 157:99–107.
Bahn, A. K., Anderson, C. L., and Norman, V. B., 1963, Outpatient psychiatric clinic services to alcoholics, *Q. J. Stud. Alcohol* 24:213–226.
Bailey, M. B., and Stewart, J., 1967, Normal drinking by persons reporting previous problem drinking, *Q. J. Stud. Alcohol* 28:305–315.
Bailey, M., Haberman, P., and Sheinberg, J., 1966, Identifying alcoholics in population surveys, *Q. J. Stud. Alcohol* 27:300–315.
Bateman, N. I., and Petersen, D. M., 1971, Variables related to outcome of treatment for hospitalized alcoholics, *Int. J. Addict.* 6:215–224.
Berg, N. L., 1971, Effects of alcohol intoxication on self-concept; studies of alcoholics and controls in laboratory conditions, *Q. J. Stud. Alcohol* 32:442–453.

Blake, B. G., 1965, The application of behavior therapy to the treatment of alcoholism, *Behav. Res. Ther.* 3:75–85.
Blake, B. G., 1967, A follow-up of alcoholics treated by behavior therapy, *Behav. Res. Ther.* 5:89–94.
Blane, H. T., and Meyers, W. R., 1964, Social class and the establishment of treatment relations by alcoholics, *J. Clin. Psychol.* 20:287–290.
Bourne, P. G., Alford, J. A., and Bowcock, J. Z., 1966, Treatment of skid-row alcoholics with disulfiram, *Q. J. Stud. Alcohol* 27:42–48.
Bowen, W. T., and Androes, L., 1968, A follow-up study of 79 alcoholic patients: 1963–1965, *Bull. Menninger Clin.* 32:26–34.
Bowen, W. T., Soskin, R. A., and Chotlos, J. W., 1970, Lysergic acid diethylamind as a variable in the hospital treatment of alcoholis; a follow-up study, *J. Nerv. Ment. Dis.* 150:111–118.
Brunner-Orne, M., 1962, Evaluation of calcium carbimide in the treatment of alcoholism, *J. Neuropsychiatry* 3:163–167.
Bruun, K., 1963, Outcome of different types of treatment of alcoholics, *Q. J. Stud. Alcohol* 24:280–288.
Burton, G., and Kaplan, H. M., 1968, Marriage counseling with alcoholics and their spouses—II. The correlation of excessive drinking behavior with family pathology and social deterioration, *Br. J. Addict.* 63:161–170.
Butterworth, A. T., 1971, Depression associated with alcohol withdrawal; imipramine therapy compared with placebo, *Q. J. Stud. Alcohol* 32:343–348.
Butterworth, A. T., and Watts, R. D., 1971, Treatment of hospitalized alcoholics with doxepin and diazepam; a controlled study, *Q. J. Stud.* 32:78–81.
Button, A. D., 1956, A study of alcoholics with the Minnesota Multiphasic Inventory, *Q. J. Stud. Alcohol* 17:263–281.
Butts, S. V., and Chotlos, J., 1973, A comparison of alcoholics and nonalcoholics on perceived locus of control, *Q. J. Stud. Alcohol* 34:1327–1332.
C., Bill, 1965, The growth and effectiveness of Alcoholics Anonymous in a Southwestern city, *Q. J. Stud. Alcohol* 26:279–284.
Cahalan, D., 1970, "Problem Drinkers: A National Survey," Jossey-Bass, San Francisco.
Cahn, S., 1970, "The Treatment of Alcoholics: An Evaluation Study," Oxford University Press, New York.
Canter, F., 1966, Personality factors related to participation in treatment of hospitalized male alcoholics, *J. Clin. Psychol.* 22:114–116.
Cappell, H., and Herman, P. C., 1972, Alcohol and tension reduction: A review, *Q. J. Stud. Alcohol* 33:33–64.
Cautela, J. R., 1966, Treatment of compulsive behavior by covert sensitization, *Psychol. Rec.* 16:33–41.
Cautela, J. R., 1967, Covert sensitization, *Psychol. Rep.* 20:459–468.
Cellar, F. A., and Grant, A. H., 1952, The treatment of alcoholism in an outpatient clinic, *J. Mich. Med. Soc.* 51:722–723, 729.
Charnoff, S. M., Kissin, B., and Reed, J. I., 1963, An evaluation of various psychotherapeutic agents in the long term treatment of chronic alcoholism. Results of a double blind study, *Am. J. Med. Sci.* 246:172–179.
Cheek, F. E., Osmond, H., Sarett, M., and Albahary, R. S., 1966, Observations regarding the use of LSD-25 in the treatment of alcoholism, *J. Psychopharmacol.* 1:56–74.
Clancy, J., Vornbrock, R., and Vanderhoof, E., 1965, Treatment of alcoholics; a follow-up study, *Dis. Nerv. Syst.* 26:551–561.
Clancy, J., Vanderhoof, E., and Campbell, P., 1967, Evaluation of an aversive technique as a

treatment for alcoholism; controlled trial with succinylcholine-induced apnea, *Q. J. Stud. Alcohol* 28:476–485.

Cohen, S., Ditman, K. S., Mooney, H. B., and Whittlesey, J. R. B., 1961, Prothiperidyle (Timovan) in the treatment of alcoholism: A preliminary report, *J. New Drugs* 1:235–237.

Collins, J. M., and Brown, L. M., 1960, Calcium carbimide a new protective drug in alcoholism, *Med. J. Aust.* 1:835–838.

Corder, B. F., Corder, R. F., and Laidlaw, N. D., 1972, An intensive treatment program for alcoholics and their wives, *Q. J. Stud. Alcohol* 33:1144–1146.

Corotto, L. V., 1963, An exploratory study of the personality characteristics of patients who volunteer for continued treatment, *Q. J. Stud. Alcohol* 24:432–442.

Cowen, J., 1954, A six-year follow-up of a series of committed alcoholics, *Q. J. Stud. Alcohol* 15:413–423.

Cutter, H. S. G., Kay, J. C., Rothstein, E., and Jones, W. C., 1973, Alcohol, power and inhibition, *Q. J. Stud. Alcohol* 34:381–389.

Davies, D. L., 1962, Normal drinking in recovered alcohol addicts, *Q. J. Stud. Alcohol* 23:94–104.

Davies, D. L., Shephard, M., and Myers, E., 1956, The two years prognosis of fifty alcohol addicts after treatment in hospital, *Q. J. Stud. Alcohol* 17:485–502.

Devenyi, P., and Sereny, G., 1970, Aversion treatment with electro-conditioning for alcoholism, *Br. J. Addict.* 65:289–292.

Ditman, K. S., 1961, Evaluation of drugs in the treatment of alcoholics. *Q. J. Stud. Alcohol Suppl.* 1:107–116.

Ditman, K. S., and Cohen, S., 1959, Evaluation of drugs in the treatment of alcoholism, *Q. J. Stud. Alcohol* 20:573–576.

Ditman, K. S., and Mooney, H. B., 1959, Effects of phenyltoxamine in alcoholics, *Q. J. Stud. Alcohol* 20:276–280.

Ditman, K. S., Mooney, H. B., and Cohen, S., 1962, New drugs in the treatment of alcoholism, in "Neuropsychopharmacology," Vol. 3, pp. 352–355, Elsevier, New York.

Dubourg, G. O., 1969, After-care for alcoholics—a follow-up study, *Br. J. Addict.* 64:155–163.

Dunne, J. A., 1973, Counseling alcoholic employees in a municipal police department, *Q. J. Stud. Alcohol* 34:423–434.

Edwards, G., 1966, Hypnosis in treatment of alcohol addiction; controlled trial, with analysis of factors affecting outcome, *Q. J. Stud. Alcohol* 27:221–241.

Edwards, G., and Guthrie, S., 1966, A comparison of inpatient and outpatient treatment of alcohol dependence, *Lancet* 1:467–468.

Edwards, G., Hensman, C., Hawker, A., and Williamson, V., 1967, Alcoholics Anonymous: The anatomy of a self-help group, *Soc. Psychiatry* 1:195–204.

Edwards, P., Harvey, C., and Whitehead, P. C., 1973, Wives of alcoholics: A critical review and analysis, *Q. J. Stud. Alcohol* 34:112–132.

Egan, W. P., and Goetz, R., 1968, Effect of metronidazole on drinking by alcoholics, *Q. J. Stud. Alcohol* 29:899–902.

Ellis, A. S., and Krupinski, J., 1964, The evaluation of a treatment program for alcoholics: A follow-up study, *Med. J. Aust.* 1:8–13.

Ends, E. J., and Page, C. W., 1957, A study of three types of group psychotherapy with hospitalized male inebriates, *Q. J. Stud. Alcohol* 18:263–277.

Ends, E. J., and Page, E. W., 1959, Group psychotherapy and concomitant psychological change, *Psychol. Monogr.* 73, No. 480.

Engle, K. B., and Williams, T. K., 1972, Effect of an ounce of vodka on alcoholics' desire for alcohol, *Q. J. Stud. Alcohol* 33:1099–1105.

Evenson, R. C., Altman, H., Sletten, I. W., and Knowles, R. R., 1973, Factors in the description and grouping of alcoholics, *Am. J. Psychiatry* 130:49–54.

Eysenck, H. J., and Rachman, S., 1965, "The Causes and Cures of Neurosis," Routledge & Kegan Paul, London.
Farrar, C. H., Powell, B. J., and Martin, L. K., 1968, Punishment of alcohol consumption by apneic paralysis, *Behav. Res. Ther.* 6:13–16.
Ferguson, J. K. W., 1956, A new drug for alcoholism treatment. I. A new drug for the treatment of alcoholism, *Can. Med. Assoc. J.* 74:793–795.
Ferneau, E. W., Jr., and Desroches, H. F., 1969, The relationship between abstinence and length of hospitalization, *Q. J. Stud. Alcohol* 30:447–448.
Fitzgerald, B. J., Pasewark, R. A., and Clark, R., 1971, Four-year follow-up of alcoholics treated at a rural state hospital, *Q. J. Stud. Alcohol* 32:636–642.
Foulds, G. A., and Hassall, C., 1969, The significance of age of onset of excessive drinking in male alcoholics, *Br. J. Psychiatry* 115:1027–1032.
Fox, R., 1957, Treatment for alcoholism, in "Alcoholism: Basic Aspects and Treatment" (H. E. Himwich, ed.), Publication #47 of the American Association for the Advancement of Science, Washington, D.C.
Fox, V., and Smith, M. A., 1959, Evaluation of a chemopsychotherapeutic program for the rehabilitation of alcoholism, *Q. J. Stud. Alcohol* 20:767–780.
Franks, C. M., 1970, Alcoholism, in "Symptoms of Psychopathology: A Handbook" (C. G. Costello, ed.), pp. 160–190, John Wiley, New York.
Gallant, D. M., Bishop, M. P., Camp, E., and Tisdale, C. A., 1968a, A six-month controlled evaluation of metronidazole (Flagyl) in chronic alcoholic patients, *Curr. Ther. Res.* 10:82–87.
Gallant, D. M., Bishop, M. P., Faulkner, M. A., Simpson, L., Cooper, A., Lathrop, D., Brisolara, A. M., and Bossetta, J. R., 1968b, A comparative evaluation of compulsory (group therapy and/or Antabuse) and voluntary treatment of the chronic alcoholic municipal court offender, *Psychosomatics* 9:306–310.
Gallant, D. M., Rich, A., Bey, E., and Terranova, I., 1970. Group psychotherapy with married couples: A successful technique in New Orleans alcoholism clinic patients, *J. La. State Med. Soc.* 122:41–44.
Gelder, M. G., and Edwards, G., 1968, Metronidazole in the treatment of alcohol addiction; a controlled trial, *Br. J. Psychiatry* 114:473–475.
Gellman, G. I. P., 1964, "The Sober Alcoholic: An Organizational Analysis of Alcoholics Anonymous," College and University Press, New Haven, Connecticut.
Gerard, D. L., and Saenger, G., 1966, "Outpatient Treatment of Alcoholism," University of Toronto Press, Toronto.
Gerard, D. L., Saenger, G., and Wile, R., 1962, The abstinent alcoholic, *Arch. Gen. Psychiatry* 6:83–95.
Gerrein, J. R., Rosenberg, C. M., and Manohar, V., 1973, Disulfiram maintenance in outpatient treatment of alcoholism, *Arch. Gen. Psychiatry* 28:798–802.
Gibbins, R. J., and Armstrong, J. D., 1957, Effects of clinical treatment on behavior of alcoholic patients, an exploratory methodological investigation, *Q. J. Stud. Alcohol* 18:429–450.
Gibson, S., and Becker, J., 1973, Alcoholism and depression; the factor structure of alcoholics' responses to depression inventories, *Q. J. Stud. Alcohol* 34:400–408.
Gillis, L. S., and Keet, M., 1969, Prognostic factors and treatment results in hospitalized alcoholics, *Q. J. Stud. Alcohol* 30:426–437.
Glatt, M. M., 1959, Disulfiram and citrated calcium carbimide in the treatment of alcoholism, *J. Ment. Sci.* 105:476–481.
Glatt, M. M., 1961, Treatment results in an English mental hospital alcoholic unit, *Acta Psychiatr. Scand.* 37:143–148.
Goldfried, M. R., 1969, Prediction of improvement in an alcoholism outpatient clinic. *Q. J. Stud. Alcohol* 30:129–149.

Goldstein, A., 1972, Blind controlled dosage comparisons with methadone in 200 patients, in "Proceedings of the Third National Conference on Methadone Treatment," Public Health Service Publication No. 2172, pp. 31–37, U.S. Government Printing Office, Washington, D.C.

Goodwin, D. W., Crane, J. B., and Guze, S. B., 1969, Alcoholic "blackouts": A review and clinical study of 100 alcoholics, *Am. J. Psychiatry* 126:191–198.

Goodwin, D. W., Crane, J. B., and Guze, S. B., 1971, Felons who drink: An 8-year follow-up, *Q. J. Stud. Alcohol* 32:136–147.

Gottheil, E., Murphy, B. F., Skoloda, T. E., and Corbett, L. O., 1972, Fixed interval drinking decisions. II. Drinking and discomfort in 25 alcoholics, *Q. J. Stud. Alcohol* 33:325–340.

Gross, W. F., and Nerviano, V. J., 1973, The prediction of dropouts from an inpatient alcoholism program by objective personality inventories, *Q. J. Stud. Alc.* 34:514–515.

Gynther, M. D., and Brilliant, P. I., 1967, Marital status, readmission to hospital, and intrapersonal and interpersonal perceptions of alcoholics, *Q. J. Stud. Alcohol* 28:52–58.

Haden, H. H., and Fowler, R. D., 1962, Prozine as an adjunct to psychotherapy with alcoholic outpatients in the withdrawal stage, *Q. J. Stud. Alcohol* 23:442–448.

Hansen, H. A., and Teilmann, K., 1954, A treatment of criminal alcoholics in Denmark, *Q. J. Stud. Alcohol* 15:245–287.

Hayman, M., 1956, Current attitudes to alcoholism of psychiatrists in Southern California, *Am. J. Psychiatry* 112:484–493.

Hayman, M., 1966, "Alcoholism: Mechanism and Management," Charles C Thomas, Springfield, Illinois.

Heilbrun, A. B., 1971, Prediction of rehabilitation outcome in chronic court-case alcoholics, *Q. J. Stud. Alcohol* 32:328–333.

Hitz, D., 1973, Drunken sailors and others; drinking problems in specific occupations, *Q. J. Stud. Alcohol* 34:496–505.

Hoff, E. C., 1955, The use of disulfiram (Antabuse) in the comprehensive therapy of a group of 1,020 alcoholics, *Conn. State Med. J.* 19:793–798.

Hoff, E. C., and Forbes, J. C., 1955, Some effects of alcohol on metabolic mechanisms with applications to therapy of alcoholics. III. The effect of poly-vitamin supplementation in the diet of alcoholics upon their clinical course, in "Origins of Resistance to Toxic Agents" (M. G. Sevag, R. D. Reid, and O. E. Reynolds, eds.), pp. 184–193, Academic Press, New York.

Hoff, E. C., and McKeown, C. E., 1953, An evaluation of the use of tetraethylthiuram disulfide in the treatment of 560 cases of alcohol addiction, *Am. J. Psychiatry* 109:670–673.

Hollister, L. E., Shelton, J., and Kriger, G., 1969, A controlled comparison of lysergic acid diethylamide (LSD) and dextroamphetamine in alcoholism, *Am. J. Psychiatry* 125:1352–1357.

Holzinger, R., Mortimer, R., and Van Dusen, W., 1967, Aversion conditioning treatment of alcoholism, *Am. J. Psychiatry* 124:246–247.

Hore, B. D., 1971a, Life events and alcoholic relapse, *Br. J. Addict.* 66:83–88.

Hore, B. D., 1971b, Factors in alcoholic relapse, *Br. J. Addict.* 66:89–96.

Howard, K., Rickels, K., Mock, J. E., Lipman, R. S., Covi, L., and Baum, N. C., 1970, Therapeutic style and attrition rate from psychiatric drug treatment, *J. Nerv. Ment. Dis.* 150:102–110.

Hsu, J. J., 1965, Electroconditioning therapy of alcoholics: A preliminary report, *Q. J. Stud. Alcohol* 26:449–459.

Jackson, J. K., 1958, Type of drinking patterns of male alcoholics, *Q. J. Stud. Alcohol* 19:269–302.

Jackson, J. K., Fagan, R. J., and Burr, R. C., 1958, The Seattle police department rehabilitation project for chronic alcoholics, *Federal Probation* 22:36–41.

Jensen, S. E., 1962, A treatment program for alcoholics in a mental hospital, *Q. J. Stud. Alcohol* 23:315–320.
Johnson, F. G., 1969, LSD in the treatment of alcoholism, *Am. J. Psychiatry* 126:481–487.
Jones, R. K., 1970, Sectarian characteristics of Alcoholics Anonymous, *Sociology* 4:181–195.
Jones, R. W., and Helrich, A. R., 1972, Treatment of alcoholism by physicians in private practice, *Q. J. Stud. Alcohol* 33:117–131.
Kaim, S. C., Klett, C. J., and Rothfield, B., 1969, Treatment of the acute alcohol withdrawal state: A comparison of four drugs, *Am. J. Psychiatry* 125:54–60.
Kamin, I., and Caughlan, J., 1963, Subjective experiences of outpatient psychotherapy, *Psychotherapy* 17:660–668.
Karp, S. A., Witkin, H. A., and Goodenough, D. R., 1965, Alcoholism and psychological differentiation: Effect of achievement of sobriety on field dependence, *Q. J. Stud. Alcohol* 26:580–585.
Katz, L., 1966, The Salvation Army men's social service center: II. Results, *Q. J. Stud. Alcohol* 27:636–647.
Kearney, T., and Bonime, J. C., 1966, Problems of drug evaluation in outpatients, *Dis. Nerv. Syst.* 27:604–606.
Keehn, J. D., 1970, Reinforcement of alcoholism; schedule control of solitary drinking, *Q. J. Stud. Alcohol* 31:28–39.
Kendall, R. E., 1965, Normal drinking by former alcohol addicts, *Q. J. Stud. Alcohol* 26:247–257.
Kendall, R. E., and Staton, M. C., 1966, The fate of untreated alcoholics, *Q. J. Stud. Alcohol* 27:30–41.
Kish, G. B., and Hermann, H. T., 1971, The Fort Meade alcoholism treatment program: A follow-up study, *Q. J. Stud. Alcohol* 32:628–635.
Kissin, B., and Platz, A., 1968, The use of drugs in the long term rehabilitation of chronic alcoholics, in "Psychopharmacology: A Review of Progress 1957–1967" (D. H. Efron, ed.), pp. 835–851, Public Health Service Publication 1836, Washington, D.C.
Kissin, B., Charnoff, S. M., and Rosenblatt, S. M., March 1968a, Drug and placebo responses in chronic alcoholics, *Am. Psychiatr. Assoc. Psychiatr. Res. Rep.* 24:44–60.
Kissin, B., Rosenblatt, S. M., and Machover, S., 1968b, Prognostic factors in alcoholism, Washington, D. C. *Am. Psychiatr. Assoc. Psychiatr. Res. Rep.* 24:22–43.
Kissin, B., Platz, A., and Su, W. H., 1970, Social and psychological factors in the treatment of chronic alcoholism, *J. Psychiatr. Res.* 8:13–27.
Kline, N. S., Wren, J. C., Cooper, T. B., Vargas, E., and Canal, O., 1974, Evaluation of lithium therapy in chronic alcoholism, *Clin. Med.* 81:33–36.
Knox, W. J., 1971, Attitudes of psychiatrists and psychologists toward alcoholism, *Am. J. Psychiatry* 127:1675–1679.
Knox, W. J., 1972, Four-year follow-up of veterans treated on a small alcoholism ward, *Q. J. Stud. Alcohol* 33:105–110.
Kurland, A. A., 1968, Maryland alcoholics: Follow-up study I., *Psychiatr. Res. Rep.* 24:71–82.
Lader, M. H., 1967, Alcohol reactions after single and multiple doses of calcium cyanamide, *Q. J. Stud. Alcohol* 28:468–475.
Laverty, S. G., 1966, Aversion therapies in the treatment of alcoholism, *Psychosom. Med.* 28:651–666.
Lehmann, H. E., and Ban, T. A., 1967, Chemical reduction of the compulsion to drink with metronidazole; a new treatment modality in the therapeutic program of the alcoholic, *Curr. Ther. Res.* 9:419–428.
Leivonen, P., Stenij, P., and Thesleff, C. J., 1966, Experience with three drugs in ambulatory treatment of alcoholism, *Acta Psychiatr. Scand.* 42 (Suppl. 192):177–181.

Lemere, F., 1953, What happens to alcoholics, *Am. J. Psychiatry* 109:674–676.
Lemere, F., and Voegtlin, W. L., 1950, An evaluation of the aversion treatment of alcoholism, *Q. J. Stud. Alcohol* 11:199–204.
Lemere, F., Voegtlin, W. L., Broz, W. R., O'Holloran, P., and Tupper, W. E., 1942, The conditioned reflex treatment of chronic alcoholism: VII. Technic, *Dis. Nerv. Syst.* 3:243–247.
Lemere, F., O'Holloran, P., and Maxwell, M. A., 1958, Motivation in the treatment of alcoholism, *Q. J. Stud. Alcohol* 19:428–431.
Levinson, T., and Sereny, G., 1969, An experimental evaluation of "insight therapy" for the chronic alcoholic, *Can. Psychiatr. Assoc. J.* 14:143–145.
Levy, M. S., Livingstone, B. L., and Collins, D. M., 1967, A clinical comparison of disulfiram and calcium carbimide, *Am. J. Psychiatry* 123:1018–1022.
Liebson, I., Bigelow, G., and Flamer, R., 1973, Alcoholism among methadone patients: A specific treatment method, *Am. J. Psychiatry* 130:483–485.
Linton, P. H., and Hain, J. D., 1967, Metronidazole in the treatment of alcoholism, *Q. J. Stud. Alcohol* 28:544–546.
Lovibond, S. H., and Caddy, G., 1970, Discriminated aversive control in the moderation of alcoholics' drinking behavior, *Behav. Ther.* 1:437–444.
Ludwig, A. M., and Wikler, A., 1974, "Craving" and relapse to drink, *Q. J. Stud. Alcohol* 35:108–130.
Ludwig, A., Levine, J., Stark, L., and Lazar, R., 1969, A clinical study of LSD treatment in alcoholism, *Am. J. Psychiatry* 126:59–69.
Ludwig, A. M., Wikler, A., and Stark, L. H., 1974, The first drink, *Arch. Gen. Psychiatry* 30:539–547.
Lundwall, L., and Baekeland, F., 1971, Disulfiram treatment of alcoholism; a review, *J. Nerv. Ment. Dis.* 153:381–394.
Lynn, L., and Smith-Moorehouse, P. M., 1966, A survey of treatment of alcoholism at an outpatient clinic, *Br. J. Addict.* 61:197–200.
MacAndrew, C., and Garfinkel, H., 1962, A consideration of changes attributed to intoxication as common-sense reasons for getting drunk, *Q. J. Stud. Alcohol* 23:252–266.
MacCulloch, M. J., Feldman, M. P., Orford, J. F., and MacCulloch, M. L., 1966, Anticipatory avoidance learning in the treatment of alcoholism: A record of therapeutic failure, *Behav. Res. Ther.* 4:187–196.
Machover, S., Puzzo, F. S., and Plumeau, F., 1959a, Clinical and objective studies of personality variables in alcoholism. I. Clinical investigation of the "alcoholic personality," *Q. J. Stud. Alcohol* 20:505–519.
Machover, S., Puzzo, F. S., and Plumeau, F., 1959b, Clinical and objective studies of personality variables in alcoholism. II. Clinical study of personality correlates of remission from active alcoholism, *Q. J. Stud. Alcohol* 20:520–527.
Madill, M. F., Campbell, D., Laverty, S. G., Sanderson, R. E., and Vanderwater, S. L., 1966, Aversion treatment of alcoholics by succinylcholine-induced apneic paralysis; an analysis of early changes in drinking behavior, *Q. J. Stud. Alcohol* 27:483–509.
Maier, R. A., and Fox, V., 1958, Forced therapy of probated alcoholics, *Med. Times/N.Y.,* 86:1051–1054.
Malcolm, M. T., and Madden, J. S., 1973, The use of disulfiram implantation in alcoholism, *Br. J. Psychiatry* 123:41–55.
Marconi, J., Solari, G., Gaete, S., and Piazza, L., 1960, Comparative clinical study of the effects of disulfiram and calcium carbimide. I. Side effects, *Q. J. Stud. Alcohol* 21:642–654.
Marconi, J., Solari, G., and Gaete, S., 1961, Comparative clinical carbimide. II. Reaction to alcohol, *Q. J. Stud. Alcohol* 22:46–51.

Mathias, R. E. S., 1955, An experimental investigation of the personality structure of chronic alcoholics, Alcoholics Anonymous, neurotic and normal groups, Doctoral dissertation, University of Buffalo.
Mayer, J., and Myerson, D. J., 1971, Outpatient treatment of alcoholics; effects of status, stability and nature of treatment, *Q. J. Stud. Alcohol* 32:620–627.
Mayfield, D. G., 1968a, Psychopharmacology of alcohol. I. Affective change with intoxication, drinking behavior and affective state, *J. Nerv. Ment. Dis.* 146:314–321.
Mayfield, D. G., 1968b, Psychopharmacology of alcohol. II. Affective tolerance in alcohol intoxication, *J. Nerv. Ment. Dis.* 146:322–327.
Mayfield, D. G., and Coleman, L. L., 1968, Alcohol use and affective disorder, *Dis. Nerv. Syst.* 29:467–474.
McCance, C., and McCance, P. F., 1969, Alcoholism in North-East Scotland; its treatment and outcome, *Br. J. Psychiatry* 115:189–198.
McGuire, M. T., Stein, S., and Mendelson, J. H., 1966, Comparative psychosocial studies of alcoholic and nonalcoholic subjects undergoing experimentally induced ethanol intoxication, *Psychosom. Med.* 28:13–26.
McNair, D. M., Kahn, R. J., Droppleman, L. F., and Fisher, S., 1967, Compatibility, acquiescence and drug effects, *in* "Neuropsychopharmacology" (H. Brill, J. O. Cole, P. Deniker, H. Hippins, and P. B. Bradley, eds.), pp. 536–542, Excerpta Medical Foundation, New York.
McNair, D. M., Fisher, S., Kahn, R. J., and Droppleman, L. F., 1970, Drug-personality interaction in intensive outpatient treatment, *Arch. Gen. Psychiatry* 22:128–135.
Mendelson, J. H., 1964, Experimentally induced chronic intoxication and withdrawal in alcoholics, *Q. J. Stud. Alcohol* Suppl. No. 2:1–126.
Mendelson, J. H., LaDou, J., and Solomon, P., 1964, Experimentally induced chronic intoxication and withdrawal in alcoholics. Part 3. Psychiatric findings, *Q. J. Stud. Alcohol* Suppl. No. 2:40–52.
Merry, J., 1966 (June 4). The "loss of control" myth, *Lancet* 1:1257–1258.
Merry, J., and Whitehead, A., 1968, Metronidazole and alcoholism, *Br. J. Psychiatry* 114:859–861.
Mills, K. C., Sobell, M. B., and Schaeffer, H. H., 1971, Training social drinking as an alternative to abstinence for alcoholics, *Behav. Ther.* 2:18–27.
Mindlin, D. F., 1959, The characteristics of alcoholics related to prediction of therapeutic outcome, *Q. J. Stud. Alcohol* 20:604–619.
Mindlin, D. F., 1960, Evaluation of therapy for alcoholics in a workhouse setting, *Q. J. Stud. Alcohol* 21:90–112.
Mindlin, D. F., 1964, Attitudes toward alcoholism and toward self. Differences between three alcoholic groups, *Q. J. Stud. Alcohol* 25:136–141.
Mindlin, D. F., and Belden, E., 1965, Attitude changes with alcoholics in group therapy, *California Mental Health Research Digest* 3:102–103.
Minto, A., and Roberts, F. J., 1960, "Temposil," a new drug in the treatment of alcoholism, *J. Ment. Sci.* 106:288–295.
Mitchell, E. H., 1958, Use of citrated calcium carbimide in alcoholism, *J. Am. Med. Assoc.* 168:2008–2009.
Mooney, H. B., Ditman, K. S., and Cohen, S., 1964, Cyphroheptadine in the treatment of alcoholics, *J. New Drugs* 4:46–51.
Mooney, H. B., and Ditman, K. S., 1965, Tybamate, a meprobamate analog, in the treatment of alcoholics, *J. New Drugs* 5:233–235.
Moore, R. A., and Ramseur, F., 1960, Effects of psychotherapy in an open-ward hospital in patients with alcoholism, *Q. J. Stud. Alcohol* 21:233–252.

Mulford, H. A., 1966, "Identifying Problem Drinkers," USPHS Publ. #1000, Washington, D.C.
Muzekeri, L. H., 1965, The MMPI in predicting treatment outcome in alcoholism, *J. Consult. Psychol.* 29:281.
Myerson, D. J., and Mayer, J., 1966, Origins, treatment, and density of skid-row alcoholic men, *N. Engl. J. Med.* 275:419–425.
Nason, Z. M., 1958. Temposil; new chemotherapy for treatment of alcoholism, *J. Kans. Med. Soc.* 59:391–392.
National Council on Alcoholism Fact Sheet, 1972 (August), "Facts on Alcoholism."
Nørvig, J., and Nielsen, B., 1956, A follow-up study of 221 alcohol addicts in Denmark, *Q. J. Stud. Alcohol* 17:633–642.
O'Reilly, P. O., and Funk, A., 1964, LSD in chronic alcoholism, *Can. Psychiatr. Assoc. J.* 9:258–261.
Overall, J. E., and Hollister, L. E., 1967, Psychiatric drug research: Sample size requirements for one vs. two raters, *Arch. Gen. Psychiatry* 16:152–161.
Overall, J. E., Henry, B. W., and Ford, H., 1971, Background variables and outpatient psychopathology, *Psychol. Rep.* 28:303–309.
Overall, J. E., Brown, D., Williams, J. D., and Neill, L. T., 1973, Drug treatment of anxiety and depression in detoxified alcoholic patients, *Arch. Gen Psychiatry* 29:218–221.
Pahnke, W. N., Kurland, A. A., Unger, S., Savage, C., and Grof, S., 1970, The experimental use of psychedelic (LSD) psychotherapy, *J. Am. Med. Assoc.*, 212:1856–1863.
Papas, A. N., 1971, An Air Force alcoholic rehabilitation program, *Mil. Med.* 136:277–281.
Peck, R. E., 1962, Use of CCC in treatment of alcoholic patients of Meadowbrook Hospital, *N.Y. State J. Med.* 62:1626–1629.
Pemberton, D. A., 1967, A comparison of the outcome of treatment in female and male alcoholics, *Br. J. Psychiatry* 113:367–373.
Penick, S. B., Carrier, R. N., and Sheldon, J. B., 1969, Metronidazole in the treatment of alcoholism, *Am. J. Psychiatry* 125:1063–1066.
Pfeffer, A. Z., and Berger, S., 1957, A follow-up study of treated alcoholics, *Q. J. Stud. Alcohol* 18:624–648.
Pittman, D. J., and Gordon, C. W., 1958, "Revolving Door: A Study of the Chronic Police Case Inebriate," Monographs of the Yale Center of Alcohol Studies #2, New Haven, Free Press, Glencoe, Illinois.
Pittman, D. J., and Sterne, M., 1965, Concept of motivation: Sources of institutional and professional blockage in the treatment of alcoholics, *Q. J. Stud. Alcohol* 26:41–57.
Pittman, D. J., and Tate, R. L., 1969, A comparison of two treatment programs for alcoholics, *Q. J. Stud. Alcohol* 30:888–899.
Platz, A., Panepinto, W. C., Kissin, B., and Charnoff, S. M., 1970, Metronidazole and alcoholism: An evaluation of specific and non-specific factors in drug treatment, *Dis. Nerv. Syst.* 31:631–636.
Pokorny, A. D., Miller, B. A., and Cleveland, S. E., 1968, Response to treatment of alcoholism: A follow-up study, *Q. J. Stud. Alcohol* 29:364–381.
Pokorny, A. D., Miller, B. A., Kanas, T. E., and Valles, J., 1971, Dimensions of alcoholism, *Q. J. Stud. Alcohol* 32:699–705.
Pokorny, A. D., Miller, B. A., Kanas, T., and Valles, J., 1973, Effectiveness of extended aftercare in the treatment of alcoholism, *Q. J. Stud. Alcohol* 34:435–443.
Quinn, J. T., and Henbest, R., 1967, Partial failure of generalization in alcoholics following aversion therapy, *Q. J. Stud. Alcohol* 28:70–75.
Quinn, J. T., and Kerr, W. S., 1963, The treatment of poor prognosis in alcoholics by prolonged apomorphine aversion therapy, *J. Irish Med. Assoc.* 53:50–54.

Rae, J. B., 1972, The influence of the wives on the treatment outcome of alcoholics: A follow-up study at two years, *Br. J. Psychiatry* 120:601-613.
Rae, J. B., and Drewery, 1972, Interpersonal patterns in alcoholic marriages, *Br. J. Psychiatry* 120:615-621.
Rankin, J. G., Santamoria, N. N., O'Day, D. M., and Doyle, M. C., 1967, Studies in alcoholism. I. A general hospital medical clinic for the treatment of alcoholism, *Med. J. Aust.* 1:157-162.
Rathod, N. H., Gregory, E., Blows, D., and Thomas, G. H., 1966, A two-year follow-up study of alcoholic patients, *Br. J. Psychiatry* 112:683-692.
Redick, R. W., 1973, Utilization of psychiatric facilities by persons diagnosed with alcohol disorders, *National Institute of Mental Health, Mental Health Statistics, Series B., No. 4*, U.S., Dept. of Health, Education and Welfare, Rockville, Maryland.
Reinert, R. E., and Bowen, W. T., 1968, Social drinking following treatment for alcoholism, *Bull. Menn. Clin.* 32:280-290.
Rhodes, R. J., and Hudson, R. M., 1969, A follow-up study of tuberculous Skid-Row alcoholics. I. Social adjustment and drinking behavior, *Q. J. Stud. Alcohol* 30:119-128.
Rickels, K., 1966, Drugs in the treatment of neurotic anxiety and tension: Controlled studies, *in* "Psychiatric Drugs" (P. Solomon, ed.), pp. 225-235, Grune & Stratton, New York.
Rickels, K., 1968, Non-specific factors in drug therapy of neurotic patients, *in* "Non-Specific Factors in Drug Therapy" (K. Rickels, ed.), pp. 3-26, Charles C Thomas, Springfield, Illinois.
Rickels, K., and Cattell, R. B., 1969, Drug and placebo response as a function of doctor and patient type, *in* "Psychotropic Drug Response" (P. R. A. May and J. R. Wittenborn, eds.), pp. 126-140, Charles C Thomas, Springfield, Illinois.
Rickels, K., Raab, E., Gordon, P. E., Laquer, K. G., DeSilverio, R. V., and Hesbacher, P., 1968, Differential effects of chlordiazepoxide and fluphenazine in two anxious patient populations, *Psychopharmacologia* (Berl.) 12:181-192.
Ritson, B., 1969, Involvement in treatment and its relation to outcome amongst alcoholics, *Br. J. Addict.* 64:23-29.
Ritson, B., 1971, Personality and prognosis in alcoholism, *Br. J. Psychiatry* 118:79-82.
Ritter, N. S., 1956, Experience with reserpine in the treatment of alcoholism, *Q. J. Stud. Alcohol* 17:195-197.
Robson, R. A. H., Paulus, I., and Clarke, G. G., 1965, An evaluation of the effect of a clinic treatment program on the rehabilitation of alcoholic patients, *Q. J. Stud. Alcohol* 26:264-278.
Rohan, W. P., 1970, A follow-up study of hospitalized problem drinkers, *Dis. Nerv. Syst.* 31:259-267.
Rohan, W. P., 1972, Follow-up study of problem drinkers, *Dis. Nerv. Syst.* 33:196-199.
Rosenberg, C. M., 1974, Drug maintenance in the outpatient treatment of chronic alcoholism, *Arch. Gen. Psychiatry* 30:373-377.
Rosenberg, C. M., Manohar, V., O'Brien, J., Cobb, F., and Weinberger, S., 1973, The Hello House fire: Response of alcoholics to crisis, *Q. J. Stud. Alcohol* 34:199-202.
Rosenblatt, S. M., Gross, M. M., Malenowski, B., Broman, M., and Lewis, E., 1971, Marital status and multiple psychiatric admissions for alcoholism, *Q. J. Stud. Alcohol* 32:1092-1096.
Rossi, J. J., Stach, A., and Bradley, N. J., 1963, Effects of treatment of male alcoholics in a mental hospital: A follow-up study, *Q. J. Stud. Alcohol* 24:91-108.
Rothstein, E., and Clancy, D. D., 1970, Combined use of disulfiram and metronidazole in treatment of alcoholism, *Q. J. Stud. Alcohol* 31:446-447.
Rubington, E., 1970, Referral, post treatment contacts and length of stay in a halfway house, *Q. J. Stud. Alcohol* 31:659-668.
Sanderson, R. E., Campbell, D., and Laverty, S. G., 1963, An investigation of a new aversive conditioning treatment for alcoholism, *Q. J. Stud. Alcohol* 24:261-275.

Schaefer, H. H., Sobell, M. B., and Sobell, L. C., 1972, Twelve month follow-up of hospitalized alcoholics given self-confrontation experiences by videotape, *Behav. Ther.* 3:283–285.

Selzer, M. L., and Holloway, W. H., 1957, A follow-up of alcoholics committed to a state hospital, *Q. J. Stud. Alcohol* 18:98–120.

Sheard, M. A., 1963, The influence of doctor's attitude on the patient's response to antidepressant medication, *J. Nerv. Ment. Dis.* 136:555–560.

Silverstein, S. J., Nathan, P. E., and Taylor, H. E., 1974, Blood alcohol level estimation and controlled drinking by chronic alcoholics, *Behav. Ther.* 5:1–15.

Simpson, W. S., and Webber, P. W., 1971, A field program in the treatment of alcoholism, *Hospital and Community Psychiatry* 22:170–173.

Smart, R. G., 1970, The evaluation of alcoholism treatment programs, *Addictions* (Toronto) 17:41–51.

Smart, R. G., 1974, Employed alcoholics treated voluntarily and under constructive coercion, *Q. J. Stud. Alcohol* 35:196–209.

Smart, R. G., Storm, T., Baker, E. F. W., and Solursh, L., 1966, A controlled study of lysergide in the treatment of alcoholism. I. The effects on drinking behavior, *Q. J. Stud. Alcohol* 27:469–482.

Smith, C. G., 1967, Marital influences on treatment outcome in alcoholism, *J. Irish Med. Assoc.* 60:433–434.

Smith, J. A., Wolford, J. A., Weber, M., and McLean, D., 1957, Use of citrated calcium carbimide (Temposil) in the treatment of chronic alcoholism, *J. Am. Med. Assoc.* 165:2181–2183.

Smith, J. A., Mansfield, E., and Herrick, H. D., 1959, Treatment of chronic alcoholics with citrated calcium carbimide (Temposil), *Am. J. Psychiatry* 115:822–824.

Smith-Moorhouse, P. M., 1969, Hypnosis in the treatment of alcoholism, *Br. J. Addict.* 64:47–55.

Sobell, L. C., Sobell, M. B., and Schaefer, H. H., 1971, Alcoholics name fewer mixed drinks than social drinkers, *Psychol. Rep.* 28:493–494.

Sobell, M. B., and Sobell, L. C., 1972, "Individualized Behavior Therapy for Alcoholics: Rationale, Procedures, Preliminary Results and Appendix," State of California Dept. of Mental Hygiene, Sacramento.

Solomon, R. L., 1964, Punishment, *Psychologist* 19:239–253.

Stein, L. I., Newton, J. R., and Bauman, R. S., 1975, Duration of hospitalization for alcoholism, *Arch. Gen. Psychiatry* 32:247–252.

Stojiljkovic, S., 1969, Conditioned aversion treatment of alcoholics, *Q. J. Stud. Alcohol* 30:900–904.

Storm, T., and Cutler, R. E., 1968, "Systematic Desensitization in the Treatment of Alcoholics: An Experimental Trial," Alcoholism Foundation of British Columbia, Vancouver.

Strassman, H. D., Adams, B., and Pearson, A. W., 1970, Metronidazole effect on social drinkers, *Q. J. Stud. Alcohol* 31:394–398.

Swinson, R. P., 1971, Long term trial of metronidazole in male alcoholics, *Br. J. Psychiatry* 119:85–89.

Thimann, J., 1949, Conditioned reflex treatment of alcoholism. I. Its rationale and technic, *N. Engl. J. Med.* 241:368–370.

Thomas, R. E., Gliedman, L. H., Imber, S. D., Stone, A. R., and Freund, J., 1959, Evaluation of the Maryland Alcoholic Rehabilitation Clinics, *Q. J. Stud. Alcohol* 20:65–76.

Tomsovic, M., 1968, Hospitalized alcoholic patients. I. A two-year study of medical, social and psychological characteristics, *Hospital and Community Psychiatry* 19:197–203.

Tomsovic, M., 1970, A follow-up study of discharged alcoholics, *Hospital and Community Psychiatry* 21:94–97.

Tomsovic, M., and Edwards, R. V., 1970, Lysergide treatment of schizophrenic and nonschizophrenic alcoholics: A controlled evaluation, *Q. J. Stud. Alcohol* 31:932–949.

Trice, H. M., 1957, A study of the process of affiliation with Alcoholics Anonymous, *Q. J. Stud. Alcohol* 18:39–54.

Trice, H. M., 1959, The affiliation motive and readiness to join Alcoholics Anonymous, *Q. J. Stud. Alcohol* 20:313–320.

Trice, H. M., and Roman, P. M., 1970, Sociopsychological predictors of affiliation with Alcoholics Anonymous: A longitudinal study of "treatment success," *Soc. Psychiatry* 5:51–59.

Trice, H. M., Roman, P. M., and Belasco, J. A., 1969, Selection for treatment: A predictive evalaution of an alcoholism treatment regimen, *Int. J. Addict.* 4:303–317.

Tyndel, M., Fraser, J. G., and Hartleib, C. J., 1969, Metronidazole as an adjuvant in the treatment of alcoholism, *Br. J. Addict.* 64:57–61.

Vallance, M., 1965, Alcoholism: A two-year follow-up study of patients admitted to the psychiatric department of a general hospital, *Br. J. Psychiatry* 111:348–356.

Vanderpool, J. A., 1969, Alcoholism and the self-concept, *Q. J. Stud. Alcohol* 30:59–77.

Vannicelli, M. L., 1972, Mood and self-perception of alcoholics when sober and intoxicated. I. Mood change. II. Accuracy of self-perception, *Q. J. Stud. Alcohol* 33:341–357.

Voegtlin, W. L., 1940, The treatment of alcoholism by establishing a conditioned reflex, *Am. J. Med. Sci.* 199:802–809.

Voegtlin, W. L., and Broz, W. R., 1949, The conditioned reflex treatment of chronic alcoholism. X. An analysis of 3125 admissions over a period of ten and a half years, *Ann. Intern. Med.* 30:580–597.

Vogler, R. E., Lunde, S. E., Johnson, G. R., and Martin, P. L., 1970, Electrical aversion conditioning with chronic alcoholics, *J. Consult. Clin. Psychol.* 34:302–307.

Walton, H. J., Ritson, E. B., and Kennedy, R. I., 1966, Response of alcoholics to clinic treatment, *Br. Med. J.* 2:1171–1174.

Weingartner, W., Faillace, L. A., and Markley, H. G., 1971, Verbal information retention in alcoholics, *Q. J. Stud. Alcohol* 32:293–303.

Weingold, H. P., Lechin, J. M., Bell, H. A., and Coxe, R. C., 1968, Depression as a symptom of alcoholism: search for a phenomenon, *J. Abnorm. Psychol.* 73:195–197.

Whitehorn, J. C., and Betz, B. J., 1960, Further studies of the doctor as a crucial variable in the outcome of treatment with schizophrenic patients, *Am. J. Psychiatry* 117:215–223.

Whyte, C. R., and O'Brien, P. M. J., 1974, Disulfiram inplant: A controlled trial, *Br. J. Psychiatry* 124:42–44.

Wilby, W. E., and Jones, R. W., 1962, Assessing patient response following treatment, *Q. J. Stud. Alcohol* 23:325–334.

Wilkinson, A. E., Prado, W. M., Williams, W. O., and Schnadt, F. W., 1971, Psychological test characteristics and length of stay in alcoholism treatment, *Q. J. Stud. Alcohol* 32:60–65.

Willcox, D. R. C., Gillan, R., and Hore, E. H., 1966, Do psychiatric outpatients take their drugs?, *Br. Med. J.* 2:790–792.

Willems, P. J. A., Letemendia, F. J. J., and Arroyave, F., 1973a, A two-year follow-up study comparing short with long stay in-patient treatment of alcoholics, *Br. J. Psychiatry* 122:637–648.

Willems, P. J. A., Letemendia, F. J. J., and Arroyave, F., 1973b, A categorization for the assessment of prognosis and outcome in the treatment of alcoholism, *Br. J. Psychiatry* 122:649–654.

Wilson, I. C., Lacoe, B. A., and Riley, L., 1970, Tofranil in the treatment of postalcoholic depressions, *Psychosomatics* 11:488–494.

Winkelman, N. W., 1964, A clinical and socio-cultural study of 200 psychiatric patients started on chlorpromazine 10½ years ago, *Am. J. Psychiatry* 120:861–869.

Winship, G. M., 1957, Antabuse treatment, *in* "Hospital Treatment of Alcoholism; A Comparative, Experimental Study" (R. S. Wallerstein, ed.), pp. 23–51, Menninger Clinic Monogr. Ser., No. 11, Basic Books, New York.

Wolff, K., 1968, Hospitalized alcoholic patients. III. Motivating alcoholics through group psychotherapy, *Hospital and Community Psychiatry* 19:206–209.

Wolpe, J., 1958. "Psychotherapy by Reciprocal Inhibition," Stanford University Press, Stanford, California.

CHAPTER 11

Factors in the Development of Alcoholics Anonymous (A.A.)

Barry Leach
Research Psychologist
formerly, Departments of Medicine and Psychiatry
The Roosevelt Hospital
New York, New York

and

John L. Norris
General Service Board of Alcoholics Anonymous
and
Eastman Kodak Company
Rochester, New York

INTRODUCTION: CRISIS FOR A HUNGOVER DOCTOR

The surgeon had the shakes. Bad.

Fortunately he was not scheduled to operate on anybody that Sunday afternoon in May 1935, in Akron, Ohio. In fact, as word of his excessive drinking had spread, his once-flourishing practice had been dwindling sharply for several years. He was even about to lose the home in which he and Anne had reared the children.

At 55, he was discouraged almost to despair. For the thousandth time, his foggy brain searched the past for reasons. His childhood had been spent in the almost completely "dry" town of St. Johnsbury, Vermont, where drinking had been looked upon with great disfavor by the "good" townspeople. His stern but loving Presbyterian parents, Judge and Mrs. W. P. Smith, had seen to it that their only son attended church as many as five times a week.

Upon leaving home to attend Dartmouth College, he had rebelled. Not only had he withdrawn almost totally from religious activity, but he had also become a heavy drinker while still an undergraduate. At first, among his tippling classmates, he had seemed to have a unique ability to swill copiously without even so much as a headache the next day.

However, his drinking had got so bad at the University of Michigan, where he had gone on for premed work in 1905, he had been asked to leave. He had been able to get his medical degree only by transferring to Rush Medical College, Chicago, and keeping a promise to stay "bone dry" for two years.

But once he had set up his own Akron practice, in 1912, he had become a daily drinker. In retrospect, it was hard to understand how he had functioned successfully at all, since for nearly 20 years the routine had been a downward, nightmarish cycle of earning money to get liquor, smuggling it home, getting drunk every night (he had a phobia about being unable to sleep), the morning jitters which necessitated the taking of large doses of sedatives every day, to make it possible to earn more money to get more liquor—and on and on down the spiral, punctuated by more than a dozen hospitalizations for detoxification in a local sanitarium. Hundreds of desperately sincere promises to swear off had lasted less than a day.

Although he had not really wanted to, he had even "tried religion." Late in 1932, Dr. Bob and his wife, Anne, had been attracted to the "seeming poise, health, and happiness" of members of the Akron chapter of the Oxford Groups,* the religious movement led by Frank N. Buchman, a Lutheran minister from Pennsylvania, which was extremely popular among middle- and upper-class Protestants in the United States, England, and South Africa in the 1920s and 1930s.

Because it had awakened memories of his strict religious upbringing, the physician had been somewhat repelled by the spiritual nature of the Oxford Groups. Nevertheless, he had studied their teachings diligently, searching for relief from his dipsomania. But he still got drunk every night.

An Oxford Grouper, Mrs. Henrietta Seiberling, had called Anne the day

---

* The history and assessments of Buchman's Oxford Groups movement, more recently known as Moral Rearmament, can be found in several works, such as Begbie (1923), Shoemaker (1936), and Ferguson (1936). A sharp analysis of the movement by Cantril (1941) points out several features of Buchmanism which are also features of Alcoholics Anonymous.

before the 1935 Sunday that was to change Dr. Bob's life. She asked Anne to bring Bob to her house to meet a fellow from New York who might be able to help the doctor with his drinking problem.

Dr. Bob had come home that afternoon bearing a plant as a guilt-easing Mother's Day gift. Potted himself, he had then passed out. But Anne promised Henrietta she would try to bring her husband over the following day.

Riding to the Seiberling estate that Sunday afternoon, Dr. Bob felt no real hope. He had read everything published about alcoholism, he was sure, and had consulted all the experts. Nothing, nothing at all, had helped. He expected no help now. Almost all he could think of was how to get a drink to calm his nearly unbearable jitters and dull the horrors in his mind. So he extracted a promise from Anne that they would visit with Henrietta and the New York man no more than 15 minutes.

At 5:00 p.m. Mrs. Seiberling introduced the hungover doctor to the New Yorker, whose name was Bill, and the two men went into a room by themselves. It was 11:15 p.m. Sunday, May 11, 1935, that Dr. Bob and his new friend emerged from their first private session. Except for one bender about three weeks later, Dr. Bob never again had another drink during his remaining 15 years, nor did the other former drunkard, Bill W.,* for his last 37 years.

What happened between these men in those six hours?

They were to become known a few years later as the cofounders of Alcoholics Anonymous, and that first session between them is usually given as the start of the movement (e.g., Trice, 1958a).

Before presenting their respective versions of that individual encounter, and what had led up to that very first "A.A. meeting," this chapter shows the present size and influence of its outgrowth, the history of the movement is outlined, evaluations of its effectiveness are offered, the literature on A.A. is reviewed, and questions are raised about some features of the movement.

THE GROWTH AND SIZE OF A.A.

The names of, and other information about, all known local chapters (in A.A. parlance, groups) are listed by the A.A. General Service Office (G.S.O.)

---

* Upon their deaths—Dr. Bob of cancer, November 16, 1950, Bill of pneumonia, January 28, 1971—their full identities were made public. Dr. Bob was Robert Holbrook Smith, M.D., proctologist, and Bill W. was William Griffith Wilson, investment analyst, Bedford Hills, New York. In this chapter, however, we follow the still prevailing A.A. custom of calling them Dr. Bob and Bill, or Bill W.

in New York in publishing annually the [A.A.] *World Directory,** a practice which began with a mimeographed listing of 22 cities in which groups were said to be "well established and holding weekly meetings" in 1940 (A.A. Bulletin #1, 1940).

In English-Speaking Lands

Figure 1 shows the growth from 1940 through 1970 in the number of *Directory*-listed autonomous A.A. groups per one million population in four English-speaking lands. Only Australia, the British Isles, Canada, and the United States are included, not only because all the necessary data, of almost equivalent reliability, are available, but also because those countries alone share, along with other similarities of culture, the tongue in which the A.A. ideology was originally conceptualized and in which the original A.A.-published literature has consistently been available in the native language of their residents.

Autonomous A.A. groups are those formed spontaneously by the alcoholics themselves (The A.A. Group, 1965). Institutional groups, formed for patients, clients, and inmates of health and correctional facilities, are usually somewhat less spontaneous in origin and have less autonomy, a difference discussed in detail by Leach *et al.* (1969). Institutional groups are not included in Figure 1, but are discussed later (see pages 450–451).

The exact number of A.A. groups is relatively verifiable, by correspondence with G.S.O. and by personal visits to group meetings which are made by the delegates to the annual General Service Conference for the United States and Canada and the biennial A.A. World Service Conference, who, in cooperation with local area General Service Committee members and their overseas counterparts, can verify the existence of groups in their respective areas.

Why the Canadian rise in Figure 1 is steeper than that shown for the United States, why Australian growth apparently slowed down and leveled off after 1960, and why the British Isles growth began in 1965 to accelerate slightly, we do not know. These variations could be artifacts of A.A. record and reporting systems, but the present authors prefer to speculate no further than that.

---

* As of 1973, 4 volumes have been issued. *World Directory Part I: United States and Canada,* and *World Directory Part II: Africa, Asia, Australia, Europe, Central and South America and all other areas not covered in World Directory Part I* are annuals. Biannually G.S.O. also publishes a *Directory of A.A. Groups in Correctional Facilities* and a *Directory of A.A. Groups in Hospitals and Rehabilitation Centers.* Because all the directories contain full names and addresses of A.A. members, which is privileged information, they are labeled "Confidential. For A.A. Members Only." A complete collection of these volumes is the only published listing of the numbers and locations of A.A. groups throughout the world, and is the source of all such information in this chapter, unless otherwise noted.

The Development of Alcoholics Anonymous

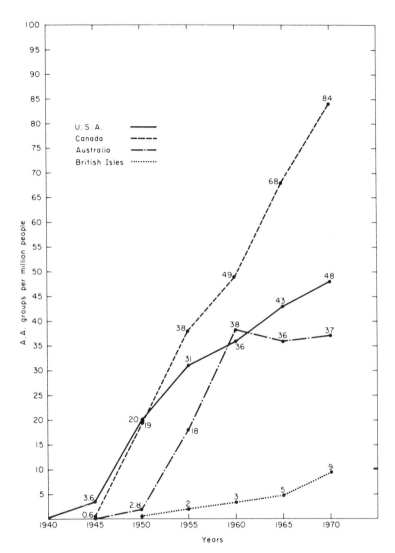

FIGURE 1. 1940–1970 growth of A.A. in number of known autonomous groups per million people in four English-speaking lands.

At any rate, the available evidence shows that from 1945 through 1970, the number of autonomous A.A. groups in Australia for each million of its general population increased an average of 1.4 groups (57.1 percent) annually. The yearly increase in the British Isles, 1950 through 1970, was an average of 0.4 groups (20.5 percent). For Canada, the average annual increment from 1945 through 1970 was 3.2 groups (125.3 percent), and an average yearly increase of 1.5 groups (72.8 percent) per one million population occurred in the United States, 1940 through 1970. Clearly, the number of autonomous A.A. groups has been growing in these countries faster than their respective general populations.

No study has been published on the life span of A.A. groups, but some evidence can be seen in the *Directories* of the longevity of groups. Of 22 in existence in 1940, 86 percent were still extant in 1970. Of 626 groups in 1945, 81 percent were alive in 1970. Of 4052 groups in 1950, 58 percent survived in 1970, and 58 percent of the 8615 groups of 1960 were active in 1970. As will be seen, the total number of groups has increased consistently since 1940.

This would not necessarily mean steady growth in the number of A.A. *members,* of course, if growth in the number of groups resulted simply from splitting of large groups into tinier fragments. Therefore the reported active membership estimates for autonomous groups in A.A. *Directories* were totaled and averaged, with the results shown in Table 1.

In A.A.'s earlier years, some groups may have been inclined to inflate their counts a bit, anxious to demonstrate their success at attracting new members. Afterward, however, pressure developed which could have influenced groups to underreport, as will be shown.

At any rate, the only records available show that the mean number of

TABLE 1. Average Estimated Membership of Autonomous A.A. Groups

Year	Average estimated number of members per group	Year	Average estimated number of members per group
1941	40.8	1957	19.1
1942	39.2	1958	19.1
1943	43.3	1959	17.7
1944	38.1	1960	17.3
1945	35.1	1961	17.8
1946	35.4	1962	17.3
1947	31.5	1963	17.4
1948	30.2	1964	17.0
1949	28.6	1965	16.7
1950	28.2	1966	16.7
1951	30.1	1967	16.7
1952	25.2	1968	16.8
1953	22.6	1969	16.9
1954	21.9	1970	17.2
1955	20.6	1971	17.3
1956	19.9	1972	17.3

active members reported per group has had an overall tendency to decline, until it began slight rises in 1968. This may indeed indicate that large A.A. groups tend to spin off smaller units, one of the chief ways new A.A. groups get started (Leach *et al.,* 1969). This has been described (The A.A. Group, 1965) as a measure to make possible smaller, more closely knit groups in which each alcoholic receives more personal attention than would be possible in larger groups.

Figure 1 and Table 1 leave unanswered the question: Has Alcoholics Anonymous actually grown in its absolute number of members, in proportion to the general population, in the first three decades for which A.A. records are available?

The exact number of A.A. members is unknowable, presumably forever, because membership in A.A. is a subjective state capable of verification only by self-report. The most nearly formal, or official, A.A. statement on the matter is the Fellowship's Third Tradition, "The only requirement for A.A. membership is a desire to stop drinking" (Wilson, 1952). It is commonly said in A.A. that "anyone is a member who says he is." There is no provision whatsoever for expelling from the Fellowship any self-proclaimed member, even if he gets drunk. If he disrupts a meeting, he may be escorted outside, but this is rare and it does not deny in any way his claim on A.A. membership. He is welcome to return to any A.A. meeting, drinking or not, as long as he does not disrupt it.

Aware of the impossibility of establishing precise, verifiable membership statistics, G.S.O. nevertheless every year asks each listed group to report its number of active members. Almost no groups keep membership lists, so it is believed each group generally reports to G.S.O. an estimate of the number of members commonly seen at its meetings.

Just as it is known that A.A. groups exist which have never made their existence known to G.S.O., it is known that a few groups fail to report membership figures, despite efforts by General Service Conference delegates and their local Committees to achieve accurate reporting by all groups. They may be motivated, in part, to underreport, because of intergroup, interstate, and interprovincial competitiveness to see which group's or area's financial contribution for support of G.S.O. is higher per capita. The lower the number of members reported, the higher average contribution per member is recorded for each group. In addition, there is evidence in Table 15 (see also Bailey and Leach, 1965) that after about five years' sobriety many members tend to attend fewer A.A. meetings, so are not likely to be included in group membership estimates. For these reasons, the estimated A.A. membership figures used here seem conservative, possibly too low by as many as 100,000 members (Leach, 1973). However, all membership estimates in this chapter are those recorded in the A.A. *Directories* as reported by the groups themselves.

Figure 2 shows the totals of active A.A. membership at five-year intervals,

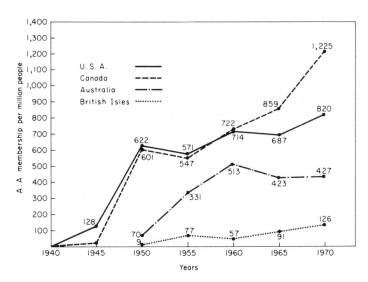

FIGURE 2. 1940-1970 growth of A.A. in reported estimated active group membership per million people in four English-speaking lands.

normalized for general population growth, in the same four English-speaking global areas previously discussed.

Figure 2 shows growth in estimated active A.A. membership similar to the growth in the number of A.A. groups shown in Figure 1, and to that extent they tend to confirm each other.

Perhaps coinciding with improved accuracy of reporting to Australian A.A. service offices, Australian estimated membership, like its number of groups, apparently declined after 1960, but has again begun to climb. Nevertheless, overall, from 1945 through 1970, there was an increase of 17.0 active A.A. members (19.6 percent) per one million population in Australia. In the British Isles, the same average annual increment, 1950 through 1970, has been 5.6 members (40.2 percent). Canada, in membership as in groups, has shown the most growth, from 1945 through 1970, with an average yearly addition of an estimated 46.3 members (114.6 percent).

The 1950-1955 decline shown for Canada and the United States coincides with establishment of the General Service Conference for those countries, and it seems likely this improved the accuracy of group reporting. Even so, from 1945 through 1970, in the United States, there was apparently an average yearly increase of an estimated 26.6 active A.A. members (16.6 percent) per one million of the general population. It seems a safe conclusion that actual A.A. membership, like the number of A.A. groups, has been growing in these countries faster than their respective general populations.

This is an important finding since it demonstrates the danger of any generalizations about A.A., or any other movement of such size and rate of growth. It also underscores the fact that the results of experiments, surveys, or any other studies based on only small, localized samples of the total A.A. membership may not be representative of the entire Fellowship.

A.A. Growth Worldwide

In other parts of the world, also, A.A. has increased in number of groups since a published report showing the 1965 size of A.A. by nation and continent (Leach et al., 1969).

Table 2 shows the increase in the absolute number of known autonomous groups worldwide, *not* normalized for population growth, from 1965 through 1970.

The biggest percents of increase in those five years are those of Latin America—170.5 percent in Central America, 108.8 percent in South America. There was a growth of almost four-fifths in the number of European A.A. groups, and a more than 50.0 percent growth in Asia. In sum, the worldwide increase in the absolute number of known autonomous A.A. groups was 40.7 percent.

TABLE 2. Comparison of Number of Known Autonomous A.A. Groups Worldwide in 1965 and 1970

Location	1965[a]	1970[b]	Increase Number	Increase Percent
United States	7,143	9,541	2,128	28.7
Canada	1,245	1,667	422	33.9
(Subtotals	8,658	11,208	2,550	29.5)
Africa	166	199	33	19.9
Asia	39	61	22[c]	56.4
Australasia and Oceania	418	482	64	15.3
Europe	726	1,298	572	78.8
North and Central America[d]	536	1,450	914	170.5
South America	204	426	222	108.8
(Subtotals	2,089	3,916	1,827	87.5)
Totals	10,747	15,124	4,377	40.7

[a] From Leach et al. (1969).
[b] From [A.A.] *World Directory Part I Spring 1971* and [A.A.] *World Directory Part II Summer 1971.*
[c] This may reflect in part the large number of United States armed forces personnel then in Southeast Asia.
[d] Exclusive of the United States and Canada.

Table 3 shows the worldwide increase in the number of listed institutional A.A. groups.

Recent dramatic increase in the number of hospital and rehabilitation facilities for alcoholics, especially in the United States, probably accounts for the fact that the percent of increase in the number of A.A. groups in such centers is higher than that in correctional institutions. That impression is supported by the numbers of groups in such facilities listed in the 1972 *Directories*. In health facilities, the 1972 listing is 914, an increase of 333 groups (57.3 percent) since 1965, or 147 groups (19.2 percent) since 1970.

For correctional institutions, the 1972 count is 986 A.A. groups, an increase of 278 groups (39.3 percent) since 1965, but an increase of only 61 groups (6.2 percent) since 1970.

The 1972 totals show dramatic growth since 1947, when the [A.A.] *World Directory* issued in August of that year—the first to carry such separate numbers—showed only 22 A.A. groups in hospitals, 29 in prisons.

These findings are important in perhaps two ways. First, they may illustrate in part some shift in emphasis away from the penal approach to alcoholism toward the health approach. Second, the total of 1900 A.A. groups which meet at least once weekly inside institutions means that more than 98,800 times each year there are demonstrated (a) the regard in which A.A. is apparently held by institutional officials, and (b) one kind of cooperation possible between A.A. members and the professional community.

The reader is here spared other evidence of the growth in size and influence of A.A. such as figures on increases in distribution of A.A.-published literature (e.g., see Leach *et al.*, 1969), since they could be expected as a logical reflection of the increase in the number of A.A. groups and members.

One other fact, however, gives a somewhat unusual measure of the

TABLE 3. Comparison of Number of Known Institutional A.A. Groups Worldwide in 1965 and 1970

Kind of institution	Number of groups		Increase	
	1965[a]	1970	Number	Percent
Hospitals and rehabilitation centers	581	767[b]	186	32.0
Correctional facilities	708	925[c]	217	30.6
Totals	1,289	1,692	403	31.0

[a] From Leach *et al.* (1969).
[b] From *Directory of A.A. Groups in Hospitals and Rehabilitation Centers 1970*.
[c] From *Directory of A.A. Groups in Correctional Facilities 1970*.

increasing influence of Alcoholics Anonymous. It is the number of often frankly imitative, at least similar, movements, especially in fields other than alcoholism, which have begun as A.A. has become bigger, better known, and better understood.

As of July 1972, A.A. World Services, Inc., knew of 27 such organizations actually using the word *anonymous* as the last part of their names, 2 using the suffix *-anon,* and 12 others using largely A.A. principles, and sometimes A.A. literature, in their own programs. None of them is in any way endorsed by, or affiliated with, A.A. They include programs for people (and their relatives) afflicted with narcotics addiction, alcoholism, asthma, obesity, cancer, various kinds and degrees of criminality, divorce, compulsive gambling, domestic problems, teenage delinquency, headaches, neuroses, excessive smoking, psychoses, schizophrenia, and acne. A few are openly commercial, money-making enterprises, but most, like A.A., are not.

In the field of alcoholism, the oldest and best known of these is the Al-Anon Family Groups, the youth division of which is called Alateen, which is for the relatives and friends of alcoholics (Living with an Alcoholic with the Help of Al-Anon, 1966; Al-Anon Faces Alcoholism, 1965). Al-Anon, although modeled closely after A.A., has no official ties with it, and the two maintain completely separate services.

Perhaps it is true that "the major contribution of Alcoholics Anonymous has been not only in the rehabilitation of alcoholics, but also in the dramatization that alcoholics can be helped" (Chafetz and Demone, 1962). However, the imitators of A.A. are evidence it may be of value in solving some other problems, too.

To the degree that emulation represents influence, then the widespread use of A.A. concepts in many differing areas and ways indicates a scope of influence far beyond the concerns of Dr. Bob S. and Bill W. in their first conversation in Akron, Ohio. They were only trying to refrain from their own pathological drinking. What had led up to their 1935 encounter, and highlights of subsequent A.A. history, are sketched next, after one final look at the latest available numbers of A.A. groups and members worldwide.

As of this writing, the 1972 A.A. *Directories* show a grand total of 20,829 groups (including both autonomous and institutional ones), an increase of 4008 A.A. groups (23.8 percent) in two years, and an estimated grand total of 395,244 reported active A.A. members* in 92 countries, an increase of 78,767 members (24.8 percent) in two years. Such was the outgrowth of the two "members" composing the only such "group" in the world, 37 years before.

---

* Including 502 "Lone Members" and 400 seagoing "Internationalists," categories which have not been included among previous membership estimates in this chapter, since they are counted separately from group members.

THE DEVELOPMENT OF A.A. AND ITS STRUCTURE

A.A. was born in the land among whose discoverers was the Englishman Henry Hudson, who gave some Holland gin to the Algonquian Indians living on the island of Manhattan (the Indian word *Manahachta-nienk* means "place where we all got drunk"), and among whose founders were the Mayflower's passengers who quarreled with their sailors over the beer aboard (Chidsey, 1969).

The founding fathers of the nation where A.A. started also included Calvinists who inveighed against any use of "ardent spirits," French, Spanish, and Italian vintners, German brewmeisters, and Quakers who used rum not only for breakfast but also for buying in Africa black men and women they sold into slavery. Its culture fostered Benjamin Rush, the Whiskey Rebellion, the Maine Law, "blind pigs," the American Society for the Promotion of Temperance and its far-from-temperate propaganda, and the Baptist clergyman who "invented" Bourbon whiskey; also Carry Nation, Frances Willard's Women's Christian Temperance Union, the Anti-Saloon League, and the Sunday School movement (Furnas, 1965; McCarthy and Douglass, 1949; Roueche, 1960). Plus the Washingtonian, Father Mathew's, and the Catch-My-Pal movements, Blue Ribbon and Red Ribbon Clubs (Maxwell, 1949; Sagarin, 1969), Wets, Drys, and National Prohibition, speakeasies, bootleggers, rumrunners, bathtub gin, moonshine, home brew, Al Capone, Dutch Schultz (Allsop, 1961; Sann, 1971), Izzy and Moe, the Wickersham Commission, and the St. Valentine's Day Massacre (Chidsey, 1969).

Earliest A.A. Origins

On December 5, 1933, Repeal became the law of the United States—about 17 months before Bill W. met Dr. Bob. That encounter was the first time Bill W.'s efforts to help another alcoholic brought enduring relief to the subject of his attentions. He had tried with others before, but without apparent effect. So the history of A.A. is commonly reported to date from 1935.

However, Bill W. reported often* that the roots of A.A. stretched back into the Zurich consulting room of psychoanalyst C. G. Jung. In a letter to

---

* Bill told this story at international A.A. conventions in St. Louis in 1955, in Long Beach, California, in 1960, and in Toronto, in 1965, and at several A.A. banquets in New York. Recordings of those addresses are in the archives of the A.A. General Service Office in New York. This account is based on those, on personal interviews with Bill and with Mrs. Henrietta Seiberling; on W. G. Wilson (1939, 1957, 1968a,b,c), Wilson and Jung (1963), the Dr. Bob (January 1951) and Bill W. (March 1971) memorial issues of *The A.A. Grapevine,* on "The Co-Founders of Alcoholics Anonymous" (1972), and Smith (1973). An authorized biography of Bill W. (Thomsen, 1957) was published by Harper & Row in 1975.

Jung written in 1961, Bill (Wilson and Jung, 1963) recounted the pre-Dr. Bob A.A. story. It is given here in Bill's words, along with Jung's reply;* both are slightly abridged but neither was ever before published outside an A.A. publication, because many features of this story are still being recapitulated daily in the case history of every sober A.A. member. Bill wrote:

> My dear Dr. Jung:
>
> This letter of great appreciation has been very long overdue. . . . Though you have surely heard of us, I doubt if you are aware that a certain conversation you once had with one of your patients, a Mr. Roland H., back in the early 1930's, did play a critical role in the founding of our fellowship. . . .
>
> Having exhausted other means of recovery from his alcoholism, it was about 1931 that he became your patient. I believe he remained under your care for perhaps a year. His admiration for you was boundless, and he left you with a feeling of much confidence.
>
> To his great consternation, he soon relapsed into intoxication. Certain that you were his "court of last resort," he again returned to your care. Then followed the conversation between you that was to become the first link in the chain of events that led to the founding of Alcoholics Anonymous. . . .
>
> First of all, you frankly told him of his hopelessness, so far as any further medical or psychiatric treatment might be concerned. This candid and humble statement of yours was beyond doubt the first foundation stone upon which our society has since been built.
>
> Coming from you, one he so trusted and admired, the impact upon him was immense.
>
> When he then asked you if there was any other hope, you told him there might be, provided he could become the subject of a spiritual or religious experience—in short, a genuine conversion. You pointed out how such an experience, if brought about, might re-motivate him when nothing else could. But you did caution, though, that while such experiences had sometimes brought recovery to alcoholics, they were, nevertheless, comparatively rare. You recommended that he place himself in a religious atmosphere and hope for the best. . . .
>
> Shortly thereafter, Mr. H. joined the Oxford Groups, an evangelical movement then at the height of its success. . . . You will remember their large emphasis upon the principles of self-survey, confession, restitution and the giving of oneself in service to others. They strongly stressed meditation and prayer. In these surroundings, Roland H. did find a conversion experience that released him for the time being from his compulsion to drink.
>
> Returning to New York, he became very active with the "O.G." here, then led by an Episcopal clergyman, Dr. Samuel Shoemaker. Dr. Shoemaker had been one of the founders of that movement, and his was a powerful personality that carried immense sincerity and conviction.
>
> At this time (1932–1934) the Oxford Groups had already sobered a number of alcoholics, and Roland, feeling that he could especially identify

---

* Reprinted by permission of the copyright owners, The A.A. Grapevine, Inc., New York, N.Y. 10017

with these sufferers, addressed himself to the help of still others. One of these chanced to be an old schoolmate of mine, named Edwin T. He had been threatened with commitment to an institution, but Mr. H. and another ex-alcoholic O.G. member, procured his parole and helped to bring about his sobriety.

Meanwhile, I had run the course of alcoholism and was threatened with commitment myself. Fortunately I had fallen under the care of a physician—a Dr. William D. Silkworth—who was wonderfully capable of understanding alcoholics. But just as you had given up on Roland, so had he given me up. It was his theory that alcoholism had two components—an obsession that compelled the sufferer to drink against his will and interest, and some sort of metabolism difficulty which he then called an allergy.[*] The alcoholic's compulsion guaranteed that the alcoholic's drinking would go on, and the "allergy" made sure that the sufferer would finally deteriorate, go insane, or die. Though I had been one of the few he had thought it possible to help, he was finally obliged to tell me of my hopelessness; I, too, would have to be locked up. To me, this was a shattering blow. Just as Roland had been made ready for his conversion experience by you, so had my wonderful friend, Dr. Silkworth, prepared me.

Hearing of my plight, my friend Edwin T. came to see me at my home where I was drinking. By then, it was November 1934. I had long marked my friend Edwin for a hopeless case. Yet here he was in a very evident state of "release" which could by no means be accounted for by his mere association for a very short time with the Oxford Groups. Yet this obvious state of release, as distinguished from the usual depression, was tremendously convincing. Because he was a kindred sufferer, he could unquestionably communicate with me at great depth. I knew at once I must find an experience like his, or die.

Again I returned to Dr. Silkworth's care. . . .

Clear once more of alcohol, I found myself terribly depressed. This seemed to be caused by my inability to gain the slightest faith. Edwin T. again visited me and repeated the simple Oxford Groups' formulas. Soon after he left me I became even more depressed. In utter despair I cried out, "If there be a God, will He show Himself." There immediately came to me an illumination of enormous impact and dimension, something which I have since tried to describe in the book, "Alcoholics Anonymous," and also in "A.A. Comes of Age. . . ."

My release from the alcohol obsession was immediate. At once I knew I was a free man.

Shortly following my experience, my friend Edwin came to the hospital, bringing me a copy of William James' "Varieties of Religious Experience." This book gave me the realization that most conversion experiences, whatever their variety, do have a common denominator of ego collapse at depth. . . .

[* The notion of alcoholism as an allergy had been set forth before, as early as 1896 (see Jellinek, 1960), and was later published by Silkworth (1937a). Although it has been discredited scientifically, it continues to be used by A.A. members and by psychiatrists as an elucidating analog, and probably "is as good as or better than anything else for their purposes" (Jellinek, 1960).—*B.L., J.L.N.*]

> In the wake of my spiritual experience there came a vision of a society of alcoholics, each identifying with, and transmitting his experience to the next—chain style. . . . This concept proved to be the foundation of such success as Alcoholics Anonymous has since achieved. . . .

In reply, Jung wrote:

> Dear Mr. W.:
>
> Your letter has been very welcome indeed.
>
> I had no news from Roland H. anymore and often wondered what has been his fate. Our conversation which he had adequately reported to you had an aspect of which he did not know. The reason that I could not tell him everything was that those days I had to be exceedingly careful of what I said. I had found out that I was misunderstood in every possible way. Thus I was very careful when I talked to Roland H. But what I really thought about, was the result of many experiences with men of this kind.
>
> His craving for alcohol was the equivalent, on a low level, of the spiritual thirst of our being for wholeness, expressed in medieval language: the union with God.
>
> How could one formulate such an insight in a language that is not misunderstood in our days?
>
> The only right and legitimate way to such an experience is, that it happens to you in reality and it can only happen to you when you walk on a path which leads you to higher understanding. You might be led to that goal by an act of grace or through a personal and honest contact with friends, or through a higher education of the mind beyond the confines of mere rationalism. I see from your letter that Roland H. has chosen the second way, which was, under the circumstances, obviously the best one.
>
> I am strongly convinced that the evil principle prevailing in this world leads the unrecognized spiritual need into perdition, if it is not counteracted either by real religious insight or by the protective wall of human community. . . . But the use of such words arouses so many mistakes that one can only keep aloof from them as much as possible.
>
> These are the reasons why I could not give a full and sufficient explanation to Roland H. but I am risking it with you because I conclude from your very decent and honest letter that you have acquired a point of view above the misleading platitudes one usually hears about alcoholism.
>
> You see, Alcohol in Latin is "spiritus" and you use the same word for the highest religious experience as well as for the most depraving poison. The helpful formula therefore is: *spiritus contra spiritum.*
>
> Thanking you again for your kind letter
>
> > I remain
> > yours sincerely,
> > C. G. Jung

But Bill W.'s vision "of a society of alcoholics" took some time to become the reality known today as A.A. During the five months immediately after his own "release," he frequently visited other alcoholics under the care of Dr. Silkworth at New York's Towns Hospital, pressing upon them the need for a conversion like his own. No one seemed interested.

He finally undertook the small business venture which took him to Akron in May 1935. It failed, and Bill headed for the cocktail lounge of his hotel. "Then I panicked," he wrote later. "I remembered that in trying to help other people, I had stayed sober myself. For the first time I *deeply* realized it. I thought, 'You need another alcoholic to talk to. You need another alcoholic just as much as he needs you!'" (Wilson, 1957).

Choosing at random from a church directory, he telephoned a clergyman to whom he told his predicament, and was in due course put in touch with another Oxford Group member, Mrs. Seiberling. She said she knew someone he might help—thinking of Dr. Bob—and invited him to her house, where the meeting with Dr. Bob took place the following day.

Dr. Bob later explained the meeting in these words: "[Bill] was a man who had experienced many years of frightful drinking, who had had most all the drunkard's experience known to man. . . . *[H]e was the first living human with whom I had ever talked who knew what he was talking about in regard to alcoholism from actual experience. In other words, he talked my language*" (Wilson, 1939). The emphasis is Dr. Bob's.

Bill later recalled that, instead of harping on the conversion experience in talking to Bob, he had followed some advice offered by Dr. Silkworth, and first talked at length about his own experience until Dr. Bob began to say, "Yes, that's me, I'm like that." He had also borne down hard on the "verdict of inevitable annihilation" for the alcoholic, quoting both Dr. Jung and Dr. Silkworth to Dr. Bob. "Our talk was a completely *mutual* thing. I had quit preaching. I knew that I needed this alcoholic as much as he needed me. *This was it,*" Bill realized (Wilson, 1957).

The day after his last drink on June 10, 1935, Dr. Bob asked Bill if they would not be safer if they began working on other alcoholics. Bill agreed, so at Akron City Hospital they found the case who became A.A. No. 3, and more followed in the several weeks Bill stayed in Akron.

From "The Big Book" to the G.S. Conference (1938–1955)

As the movement's chief architect and historian, Bill, in *Alcoholics Anonymous Comes of Age* (Wilson, 1957) traced the subsequent developments. They are not gone into here, except as they shaped the present-day structure of the movement.

After lengthy discussions among the first members in Akron and New York, in 1938 Bill W. wrote the first section of the basic textbook, *Alcoholics Anonymous* (Wilson, 1939), which also contained 28 other members' histories in their own words. For the book he adapted the Oxford Groups' formulas (see seventh paragraph of Bill's letter to Jung), which had until then been the "word-of-mouth program" of "the alcoholic squad" or "society of souses," as they called themselves, into the Twelve Steps of A.A.

The Twelve Steps

Finally agreed upon as a historical record of the first 100 members' experience, and "suggested as a program of recovery," they are:

1. We admitted we were powerless over alcohol—that our lives had become unmanageable.
2. Came to believe that a Power greater than ourselves could restore us to sanity.
3. Made a decision to turn our will and our lives over to the care of God *as we understood Him*.
4. Made a searching and fearless moral inventory of ourselves.
5. Admitted to God, to ourselves, and to another human being the exact nature of our wrongs.
6. Were entirely ready to have God remove all these defects of character.
7. Humbly asked Him to remove our shortcomings.
8. Made a list of all persons we had harmed, and became willing to make amends to them all.
9. Made direct amends to such people wherever possible, except when to do so would injure them or others.
10. Continued to take personal inventory, and when we were wrong, promptly admitted it.
11. Sought through prayer and meditation to improve our conscious contact with God *as we understood Him* praying only for knowledge of His will for us and the power to carry that out.
12. Having had a spiritual awakening as the result of these steps we tried to carry this message to alcoholics, and to practice these principles in all our affairs.*

Bill W. and his fellow members, after intense and lengthy discussion, agreed with great care on the wording of the Steps (Wilson, 1957) as the most exact, precise terms for recording the essence of their own subjective experience. No one was better qualified for that task, of course, and Bill's writings (1939, 1952) are the most nearly "official" (A.A. members resist that word) interpretation of them.

---

* Reprinted by permission of the copyright owners, A.A. World Services, Inc., New York, N.Y. 10017.

"Power greater than ourselves" and "as we understood Him" echo something Edwin T. had said to Bill when first he visited Bill to acquaint him with the Oxford Groups program. Edwin said, "Why don't you choose your own conception of God?" (Wilson, 1939). These three statements are charters that sanction a tremendous range of almost all possible conceptualizations of experience similar to those of the first A.A. members, ranging from the orthodox religious view (e.g., Taylor, 1953) through the psychiatric formulations of Tiebout (1944, 1946, 1961) to the somewhat more humanistic interpretation suggested by Maxwell (1962). Each member is encouraged to translate the Steps, if desired, into terms most acceptable to himself or herself. The *A.A. Grapevine,* for example, repeatedly publishes various members' personal explications (e.g., J. E., 1961), as well as those of non-A.A. members (e.g., Shoemaker, 1964). The same privilege is available to the present authors and to others who will be cited in a modest attempt to synthesize some of the insights of scientists with those of A.A. The following comments are offered principally to suggest illustrations of the broad spectrum of interpretations and applications of the Twelve Steps possible.

Step 1. As we have seen, Jung and Silkworth had told Roland H. and Bill W. they were hopeless cases if they continued to drink. This facilitated their own admissions of inability to drink, and enabled them to infer the concomitant need for abstinence.

It is notable that abstinence is not specifically called for; in fact, the first mention in the book (Wilson, 1955a) of members' urging it upon each other comes in a case history where Bill and Dr. Bob suggested it to A.A. No. 3 as only a temporary measure: "You can quit twenty-four hours, can't you?" (p. 188). Whether this 24-hour idea is a benign use of rationalization (Clancy, 1961), or represents an "optimal goal discrepancy" (Blumberg *et al.,* 1966), the idea works (Emrick, 1970; Becker and Israel, 1961).

Jung's and Silkworth's statements also produced profound despair in their patients, but right on the heels of those feelings both men were offered hope, a way out, as the next Step suggests.

Some such admission of powerlessness* over alcohol, or its equivalent, is crucial in motivating the alcoholic to take steps toward recovery, in our opinion. The questionnaire (Big Light Can Be Shed on Alcoholism, 1945) on which Jellinek's (1946, 1952) seminal phaseology of alcohol addiction was based, suggested three possible aspects of such an admission: admission to self, admission to someone else, and one's "lowest point." Matkom (1965) referred to the alcoholic's "pride in past excessive drinking" as a powerful motivating agent, the structure on which the success of A.A. largely rests. Jellinek (1952) described it as "bankruptcy of alibis" and "admission of defeat." Tiebout (1949, 1951a,

---

* It is striking that the term *power* appears in Steps 1, 2, and 11, anticipating by more than three decades the conclusions of McClelland *et al.* (1972) about feelings of power in regard to drinking.

1953, 1954, 1961, 1963a, 1965) wrote extensively about the phenomenon of "hitting bottom" and "surrender" as necessary to recovery.

Many studies have demonstrated the importance of this recognition, understanding, or at least absence of self-deception about the drinking as a turning point in recovery (e.g., Bailey and Leach, 1965; Bell, 1970; Brook, 1962; Connor, 1962; Emrick, 1970; Gerard et al., 1962; Gynther and Brilliant, 1967; Latimer, 1953; Maxwell, 1954; Palola et al., 1961; Roth et al., 1971; Trice, 1957a,b).

It has not escaped scientific notice that the very First Step in the A.A. attack against alcoholism is aimed directly at its distinguishing symptom, drinking (e.g., Chafetz and Demone, 1962; Tiebout, 1951b, 1958a,b). It has also been suggested that A.A. may appeal less to some alcoholics than it does to those who have suffered more egregious, severe consequences or greater complications of their drinking (e.g., Bailey and Leach, 1965; Chandler et al., 1971; Gibbins et al., 1959; McMahan, 1942; Miller et al., 1970a; Trice and Wahl, 1958), and the thrust of the First Step may in part account for this. Alcoholics with most to lose, such as jobs, family, and social ties, may, because of such dire consequences or the prospect of them, be more highly motivated toward recovery than those alcoholics who have apparently less to lose, such as the indigent, isolated, unemployed, skid row, or institutionalized alcoholics (e.g., Wood, 1969; Trice, 1959a,b), so it would not be surprising to find A.A. less attractive to some skid row alcoholics (e.g., Katz, 1966; Miller et al., 1970b; Pittman and Gordon, 1958) than it is to some middle-class victims of alcoholism.

No valid claim to a "success rate" can be established for A.A., in the present authors' opinion, because no records are kept either of how many alcoholics are exposed to A.A. only to go away and die unremitted, or of how many alcoholics who, after only minimal exposure to A.A., go away and recover. Nevertheless, the success that A.A. has had may be in part accounted for by the fact that the First Step may in effect turn away some unmotivated alcoholics— at least temporarily. We suggest, too, that the wording of the Step may be highly significant in what it does *not* include. The absence of a direct call for immediate abstinence leaves open the choice to drink or not to drink, and the A.A. group is relatively unpunitive, uncritical, and accepting if the choice is to drink (Alexander, 1963; Chafetz, 1959; Hartcollis, 1968), as illustrated by the number of members who keep returning to A.A. meetings after slips (see Table 10). The First Step does not use the term *alcoholic*, or ask the prospective member to decide initially whether or not he or she is an alcoholic. Such a decision can of course be extremely difficult in a culture which often stigmatizes anyone who bears the label *drunkard* or *alcoholic* as well as one who bears the label *teetotaler*. Smith (1961) declared that A.A. had made the public realize that "alcoholics" stay sober, and help others, and that alcoholics are no longer considered alcoholics, but "A.A.'s"; and Turfboer (1959) found A.A. membership so prestigious in one locale that nonalcoholics sought to join. Nevertheless, it seems likely that the stigmatization described by Straus (1956), which drives alcoholics to join A.A. in communities other than their own to avoid becoming known in their own home towns as alcoholics, is much more common in places where A.A. exists. Roman and Trice (1968) and Trice and Roman (1970a,b) have described the delabeling and relabeling processes the First Step makes possible, so that within A.A. the discrepancy between the alcoholic and his environment can be reduced, making him or her, in the terms of Koumans (1969), reachable and motivated.

Before making a decision about being an alcoholic, the problem drinker who begins to attend A.A. meetings has an opportunity to learn the A.A. definition (Pfeiffer, 1968) and "model" of alcoholism (Siegler et al., 1968), and to learn the A.A. value system (Murphy, 1953; Pope, 1956) and the A.A. norms (Trice, 1957a; Edwards et al., 1966), as he or she previously shared those of the drinking peers (Bacon, 1958, 1973; Trice, 1957a). While this is going on, he or she is subjected to the impact of the identification resulting from the other members' empathy (Esser, 1961b; Ripley and Jackson, 1959; Sommer et al., 1963; Stewart, 1955, 1960). There is a chance to substitute thinking for drinking (Simmel, 1948).

Step 2. As previously noted, this Step offers hope quickly after the First Step can produce despair. It happened to Roland H. when Jung told him that some recoveries from alcoholism occurred, and it happened to Bill W. when he saw his friend Edwin, who he had thought was hopeless, sober. Undoubtedly it starts to occur in the experience of many A.A. newcomers when they first see and hear recovered alcoholics at an A.A. meeting.

The Step avoids locating the source of the helping power, simply describing its strength—"greater than ourselves." This, like an admission of powerlessness, can be viewed as undermining the omnipotent, defensive, immature, egocentric, dependent, unstoppable aspects of the alcoholic personality described so often in the logorrhea of alcoholism literature (e.g., Abraham, 1927; Adler, 1941; Alexander, 1956; Armstrong, 1958; Armstrong, 1959; Armstrong and Wertheimer, 1950; Blackburn, 1955; Blane, 1968; Chafetz and Demone, 1962; Chafetz et al., 1970; Clinebell, 1968; Cohen, 1962; Delehanty, 1966; Dunlap, 1961; Fenichel, 1945; Ferenczi, 1916, 1926; Fox and Lyon, 1955; Freud, 1912, 1917, 1930; Gerard, 1959; Jellinek, 1942; Jellinek and Haggard, 1942; Knight, 1937a,b; Landis, 1945; Lisansky, 1960; Lolli, 1960; Manson, 1949; Menninger, 1938; Mowrer, 1959; Palola et al., 1961; Pfeffer, 1958; Rado, 1933; Schilder, 1941; Seiden, 1960; Sherfey, 1955; Sutherland et al., 1950; Syme, 1957; Tähkä, 1966; Tiebout, 1954, 1955b, 1961, 1965; Wexberg, 1959; Witkin et al., 1959; Van Suetendael, 1959; Zwerling, 1959).

A new A.A. member who expresses unease about the Power concept is often advised to use an A.A. "sponsor"—a member of some standing who especially befriends a newcomer—or to recognize the A.A. group or the Fellowship as a whole as an entity possessing power greater than his or her own.

Step 3. This Step, like the first two, suggests that the first A.A. members felt it important that they stop trying to run their own lives. It is not unlike the decision by a diabetic or coronary patient to abdicate direction of some aspects of his own life in favor of following the regimen prescribed by his physician.

Some observers, as well as some A.A. members, clearly view this Step as similar to or identical with a religious conversion (Benson, 1960; Clinebell, 1963; Jones, 1970; Keller, 1966; Lawrence, 1962; Paster, 1948; Tiebout, 1951a), and some of the original enthusiasm for the use of LSD in treating alcoholics (e.g., Fox, 1967b; Hoffer and Osmond, 1968; Jensen and Ramsay, 1963; Smith, 1959) was based on an apparent similarity between its effect and that of a religious conversion. Haertzen *et al.* (1968) and Machover *et al.* (1962), however, cast doubt on its *religious* nature. Chafetz (1959) described the effect of Step 3 as that of a *spiritual* conversion with an implied maternal reunion. Clagett (1962) also distinguished between a religious conversion and a spiritual one, as did Oates (1966).

In their sample of New York alcoholics in A.A., Bailey and Leach (1965) found 15.1 percent of the men, 14.1 percent of the women reporting no religious affiliation, and only 55.6 percent of the no-slip subgroup and 44.0 percent of the frequent relapsers reported atending religious services regularly. Among 50 Merseyside, England, A.A. members, Jones (1970) found 90.0 percent reporting they felt "regenerated" in A.A., an experience described by 88.0 percent as "conversion." Sixty-four percent of the sample claimed to be Christian, but 26.0 percent said they found the Steps too religious in context.

It is suggested here that the exact nature of God to A.A. members is, like the pink elephant and the "typical alcoholic personality," visible only to the eyes of the beholder, which is precisely what the authors of Step 3 intended.

Steps 4 and 5. Although insight alone rarely has produced abstinence (Gerard *et al.*, 1962), the value of insight, intraception, catharsis, and ventilation in recovery from alcoholism seems fairly well established (e.g., Connor, 1962; Davis, 1947; Gynther and Brilliant, 1967; Mindlin, 1964).

In addition to lightening the alcoholic's load of guilt (Maxwell, 1951), however, Step 5 has another value which has apparently been overlooked in the scientific literature. Whether it is taken with another A.A. member, with a clergyman, psychiatrist, or counselor—and Bill W. (1952) suggested all three possibilities—the action represents a further commitment to recovery. If Bill's suggestions are followed, the person taking the Step clearly identifies himself to another person as *an alcoholic* trying to recover. There is reason to believe such self-identification affects his or her future behavior (Lewin and Grabbe, 1948).

Steps 6 and 7. Evidently many alcoholics who approach A.A. have severe inner conflicts because they fall so short of their own demands for perfection (Alexander, 1963; Carroll and Fuller, 1969; Connor, 1962; Gynther and Brilliant, 1967; Mindlin, 1964). These Steps recognize the discouragement this produces. The authors of the Steps apparently felt that a complete willingness to try for improvement, and acknowledgment of the need for it, were important in their own recovery. Becker and Israel (1961) pointed out that A.A. provides an organized set of symbols and rituals, and Chafetz (1959) directed attention to the fact that A.A. is an "action" or "doing" group. Maxwell (1951) reported that A.A. members he interviewed felt they had given up fears, ego-inflation, hostility, and intolerance to some degree, and had gained through membership in A.A. some ability to interact more satisfyingly with other persons.

It is suggested here that the A.A. member is helped to give up qualities or behavior which trouble him or her by doing the actions—symbolic, ritual, or otherwise—suggested by these Steps. It should also be noted that Step 7, in the phrase "humbly asked," calls for an additional relinquishment of the omnipotent, unstoppable ego.

Steps 8, 9, and 10. The phrase "became willing" in Step 8 echoes the insistence by religious leaders that actions which belie feelings are insufficient for "salvation," that "a willing heart" must precede the action.

It has not been mentioned often in the psychological alcoholism literature that "making a list," as suggested in Step 8, helps to objectify in a finite, concrete way some of the problems of guilt and fear which often plague the alcoholic who approaches A.A.

The Ninth Step, if carried out as Bill W. (1939, 1952) suggested, calls once again for open, face-to-face acknowledgment by the A.A. member to other people of being an alcoholic trying to recover, as Step 5 did.

Step 10, as Tiebout (1954) noted, can be seen as continuance of the ego reduction and maintenance of the humility necessary for recovery.

Step 11. This one, too, reflects concern of the Steps' authors that the old "self-will run riot" patterns (Wilson, 1939, 1952) of the drinking behavior not be permitted to flourish again.

In Edwin T.'s initial description of his own remission to Bill W., he said he had learned "that I should try to pray to whatever God I thought there was . . . and if I did not believe there was any God, then I had better try the experiment of praying to whatever God there *might* be" (Wilson, 1957, p. 59).

Bill W. and his fellow members could not have known when they wrote Step 11, of course, of the physiological changes said to be produced by certain meditation methods (Swinyard *et al.,* 1973), but they evidently felt meditation valuable in their own recovery.

Step 12. The A.A. newcomer seeking recovery beyond abstinence can be shown that this Step indicates a "spiritual awakening" is the result of the previous Steps. Whether or not this is precisely what James (1902) would describe as a religious experience, it does seem to be a kind of personality change. The 12th Step then suggests that the alcoholic try to help other alcoholics, and that members in order to sustain recovery continue new behavior and thinking patterns in all areas of their life.

Talking to newcomers, or replying to an inquiry about A.A., is called 12th Step work, or sponsorship, by A.A. members. Although the activity is the source of much of the fame of A.A., it is not original with A.A. Members of the Washingtonian, Catch-My-Pal (Maxwell, 1949, 1950), Oxford Groups (Begbie, 1923; Cantril, 1941; Ferguson, 1936; Shoemaker, 1936), and Father Mathew's (Maguire, 1864) movements felt under the same obligation. Requet (1951) described alcoholics as proselytizers "by definition and by nature," and pointed out the need for "conspirators" in helping alcoholics. The empathy a recovered alcoholic shows for other alcoholics (Alexander, 1963; Bird, 1949; Davis, 1947; Esser, 1961b; Gordon and Hooker, 1969; Gellman, 1962; Ripley and Jackson, 1959; Sommer et al., 1963; Stewart, 1955, 1960) has also been realized by antialcoholism workers in several nations where A.A. has very few if any known chapters (e.g., Borghesi and Medaglini, 1965; Durzynski, 1971; Haasz, 1967; Morice, 1963; Nyárádi, 1970; Raiteri, 1969).

Whether it is compulsive (Chafetz, 1959) or just a form of countertransference (Moore, 1965), "practical experience shows that nothing will so much insure immunity from drinking as intensive work with other alcoholics. It works when other activities fail," Wilson (1939) wrote. Such activity may be described as "gratification leading to acceptance and status in the group" (Moore, 1965), "gratifying dependent needs without inner conflict"* (Alexander, 1963), "escape from inner reality" (Bird, 1949), or "increased concern for others" (Eckhardt, 1967); it also is of some value to the alcoholic who is the subject of such concern. As Mindlin (1964) pointed out, and Chafetz et al. (1970) subsequently demonstrated, the attitude of helping people can be a strong motivation to the alcoholic to recover. In A.A., the newcomer being helped finds hope developing (Maxwell, 1954), pessimism reducing (Mogan et al., 1969), and benign changes in the self-concept commencing (Connor, 1962; Trice, 1958a,b).

It should be observed that the alcoholic being helped in A.A. often voices gratitude to the member doing the 12th Stepping, only to be told, surprisingly, that no debt is incurred by accepting such help, since the reward to the helper is immunity from drinking whether or not the person being helped recovers (Wilson, 1952). Presumably, this knowledge that he or she can ask for and

---

* Undoubtedly the personal gratification many A.A. members find in 12th Step work is what has led many of them to seek non-A.A. careers in alcoholism, as laymen, paraprofessionals, or professionals (Beacham, 1968; Moore and Buchanan, 1966; Staub and Kent, 1973; Schultz, 1961; Wilson, 1952).

accept support without incurring an unbearable, monumental moral indebtedness makes it easier for the new A.A. member to utilize constructively the concern of other members than it is to accept the same help from the family, old friends, the employer, or the fee-charging professional.

The newcomer also perceives a new model for his own behavior—that of the nurturant person. Two studies of drinking which did not purport to study A.A. or even alcoholism per se reached similar conclusions in this regard. In their cross-cultural study of drinking, Bacon *et al.* (1965) found that indulgence of dependency needs in adulthood is correlated with a low frequency of drunkenness, and they suggested that much of what happens in A.A. is just such indulgence. "A new member becomes a part of a large group of former alcoholics who wish to help him. He is encouraged to tell his troubles and to ask for help at any time of the day or night that he feels the need. He is protected by anonymity and forgiven for backsliding. Eventually he is helped in his turn to become nurturant toward others like him" (p. 46). In their opinion, this accounts for "the remarkable success of Alcoholics Anonymous in arresting alcoholism."

McClelland *et al.* (1972) concluded that socializing the alcoholic's power drive by getting him or her to do things for other people "cures" excessive drinking, and said, "Alcoholics Anonymous does precisely that by insisting that everyone who joins must spend his time trying to cure other alcoholics" (p. 336).

Numerous studies (Alexander and Gudeman, 1965; Allen and Dootjes, 1968; Bailey *et al.*, 1961; Bateman and Petersen, 1971, 1972; Belasco, 1971; Belasco and Trice, 1969; Bell, 1970; Brook, 1962; Carothers, 1971; Canter, 1966; Carroll and Fuller, 1969; Chandler *et al.*, 1971; Clancy, 1962; Connor, 1962; Cornutt, 1953; Dunlap, 1961; Emrick, 1970; Fuller, 1965; Gerard *et al.*, 1962; Gibbins *et al.*, 1959; Gynther and Brilliant, 1967; Haberman, 1966; Jackson and Connor, 1953; Jackson and Kogan, 1963; James, 1950; Karp *et al.*, 1965; Kish and Hermann, 1971; Kurland, 1968; Lahey, 1950; Machover and Puzzo, 1959a,b; Machover *et al.*, 1962; Machover *et al.*, 1959; Mathias, 1955; McGinnis, 1963; McMahan, 1942; Mindlin, 1964; Mogan *et al.*, 1969; Palola *et al.*, 1962; Pinardi, 1963; Plumeau *et al.*, 1960; Robson *et al.*, 1965; Roth *et al.*, 1971; Seiden, 1960; Sommer *et al.*, 1963; Trice, 1957a,b, 1959a,b; Trice and Roman, 1970a,b; Trice and Wahl, 1958; Viaille, 1963; Voth, 1963, 1965; Wood, 1969) have reported psychological and sociological differences between alcoholics who have successfully affiliated with A.A., presumably exhibiting abstinence and/or other indices of recovery, and alcoholics who have not successfully affiliated with A.A.—yet. But almost no one has explored the differences between the two groups in certain kinds of behavior after their initial exposure to A.A.

Since the Steps constitute the program for recovery suggested to all A.A.

members, we find it curious that so little scientific curiosity has been displayed, even by behaviorists, in experiments designed to measure the differences between A.A. "successes" and "not-yet-successes" in behavior with regard to Steps 3, 4, 5, 6, 7, 8, 9, 10, 11, and 12. For instance, while the importance of *receiving* sponsorship has been noted by Gordon and Hooker (1969), Matkom (1965), Trice (1959a,b), and Wood (1969), almost no one has examined whether or not *giving* sponsorship is significant in the recovery of A.A. members. Davis (1947) and others already cited have observed its apparent value, and Bailey and Leach (1965) offered evidence that various forms of 12th Step work, including attending meetings, speaking at meetings, visiting institutions, holding group office, etc., are correlated with recovery. But no well-designed experiment has tested the efficacy of the Steps in recovery from alcoholism.

One more facet of the 12th Step deserves notice. It, too, requires the alcoholic to reveal frankly, usually to strangers, the facts of his or her own alcoholism and membership in A.A. Therefore the A.A. member who takes Steps 5, 9, and 12 as suggested (Wilson, 1939, 1952) openly reveals his alcoholism as well as his A.A. membership to many people. If he or she speaks at A.A. meetings, the same revelation is made in a quasi-public way to hundreds of people. It is important to understand that this does not violate the anonymity Tradition of the Fellowship, because the Tradition proscribes such revelations only in public print and broadcast media.

Anonymous, Not Nameless

When the book was titled *Alcoholics Anonymous* (Wilson, 1957), the members who had called themselves "the nameless bunch of drunks" happily accepted Alcoholics Anonymous as the movement's name, and the textbook itself became known to them as something else. "We fixed the retail price of the book at $3.50," Bill (1957) wrote. "This figure was the result of long and heated arguments. . . . As a consolation to the contestants, we directed Mr. [Edward] Blackwell [of Cornwall Press] to do the job on the thickest paper in his shop. The original volume[*] proved to be so bulky that it became known as the 'Big Book.' Of course, the idea was to convince the alcoholic purchaser that he was indeed getting his money's worth!"

The rhetoric of the book, Bill's charismatic leadership (Bales, 1944), and the movement's first national publicity (Markey, 1939; Alexander, 1941) left

---

[* A revised edition was published in 1955 (Wilson, 1955a) with new introductory matter, additional appendices, and nine additional case histories, but the first 11 basic chapters were left in their original form. They are also intact in the Third Edition (Wilson, 1976), which contains 44 case histories, 13 of them new.]

the impression that the movement was more religious than most members experience it, and steps were taken to remedy this. For the second (1941) printing of the original edition, Bill wrote an appendix pointing out that "spiritual awakenings" of the mystical, sudden, spectacular sort he had had were in the decided minority; that most members in the Fellowship experience that William James (1902) called the "educational variety," which develop slowly over a period of time in the company of one's peers in A.A. He also suggested that the "Power greater than ourselves" might be an "unsuspected inner resource." The similarity between these ideas and Jung's "higher education of the mind" induced "through a personal and honest contact with friends" is striking.

But publishing even a book about steps "on a path which leads you to higher understanding" is a business venture, and the question of money had to be faced. In 1938, a noncommercial corporation, originally called the Alcoholic Foundation, was established, consisting of a majority of reputable, nonalcoholic friends of the movement and a minority of A.A. members. It and its successors have continuously owned the book, and maintained whatever central office at the national level (G.S.O. in New York) was required for answering inquiries.

The Twelve Traditions

By 1946, it was apparent that more principles than those of the Twelve Steps were involved in the operations of A.A., and the Twelve Traditions were developed (Wilson, 1952, 1955b, 1957; The Twelve Traditions—Illustrated: A Distillation of A.A. Experience, 1971). At the movement's first international convention in Cleveland, Ohio, in 1950, for all members and their families who could get there, the Traditions were "officially" adopted. They are:

1. Our common welfare should come first; personal recovery depends upon A.A. unity.
2. For our group purpose there is but one ultimate authority—a loving God as He may express Himself in our group conscience. Our leaders are but trusted servants; they do not govern.
3. The only requirement for A.A. membership is a desire to stop drinking.
4. Each group should be autonomous except in matters affecting other groups or A.A. as a whole.
5. Each group has but one primary purpose—to carry its message to the alcoholic who still suffers.
6. An A.A. group ought never endorse, finance, or lend the A.A. name to any related facility or outside enterprise, lest problems of money, property, and prestige divert us from our primary purpose.

7. Every A.A. group ought to be fully self-supporting, declining outside contributions.
8. Alcoholics Anonymous should remain forever nonprofessional, but our service centers may employ special workers.
9. A.A., as such, ought never be organized; but we may create service boards or committees directly responsible to those they serve.
10. Alcoholics Anonymous has no opinion on outside issues; hence the A.A. name ought never be drawn into public controversy.
11. Our public relations policy is based on attraction rather than promotion; we need always maintain personal anonymity at the level of press, radio, and films.
12. Anonymity is the spiritual foundation of our traditions, ever reminding us to place principles before personalities.*

The present authors here resist the almost irresistible temptation to comment at length on the enormous but overlooked importance of each of the A.A. Traditions, especially to try to clarify some misunderstandings which persist among some A.A. members as well as in the professional community. We simply direct attention to the fact that anonymity, in its spiritual sense, warns that public glory and acclaim to the A.A. member could result not only in dangerous reinflation of the ego (Tiebout, 1954, 1961, 1965), as several bitter experiences have proved, but also in divisive quarrels within the Fellowship. The Traditions do not prevent private or public identification as a "recovered alcoholic," but do help to prevent use of the Fellowship for public prestige and other personal gain. Traditions 1, 2, 6, 7, 9, 11, and 12 emphasize this, and all twelve help to protect the amateur, lay character of the movement. It should be noted that there is no provision or authority for disciplining or expelling any A.A. member who does not abide by the Traditions; instead, each A.A. member has the responsibility for his own behavior. For further understanding of the value and applications of the Traditions, see the references already cited.

The A.A. General Service Conference (United States and Canada)

In 1955, in St. Louis, Missouri, the members at the second international convention of A.A. "officially adopted" the idea of the General Service Conference as a representative assembly linking the Fellowship to the Foundation, which was given its present name, the General Service Board (Wilson, 1957). In effect, this turned over to the Fellowship itself the responsibilities previously shouldered by Bill W., Dr. Bob, and the other initial founders. This

* Reprinted by permission of the copyright owners, A.A. World Services, Inc., New York, N.Y. 10017.

responsibility became known as "Service—the Third Legacy" of A.A.'s pioneers to their succeeding members. Recovery (the Twelve Steps) was the first such legacy, and unity (the Twelve Traditions), the second legacy. At subsequent international conventions every five years, members have taken no "official" actions of any sort, because the Conference now meets annually to perform such tasks.

The local or neighborhood group is still, as it was in the earliest Akron days of A.A., the primary operating unit of the Fellowship (The A.A. Group, 1965; Leach et al., 1969). Their links with each other through the General Service Office are detailed in the A.A. Service Manual (1969).

Developments Since 1965

In 1966, the composition of the Board was changed to include 14 A.A. members and only 7 nonalcoholics, all unpaid. Eight of the A.A. members, whose terms are limited to four years, are chosen by groups within regions of the U.S. and Canada, 6 for specific skills and geographical proximity to the General Service Office. The Board operates through a business corporation (A.A. World Services, Inc.) which handles publishing and operations of the General Service Office, and a separate corporation for the *A.A. Grapevine*.

Started in 1944 by six New York members as a local monthly A.A. newspaper, the *Grapevine* in 1945 was adopted (by mail poll of the known groups) as the national monthly journal of the movement (Wilson, 1957). Its monthly circulation in 1976 was over 84,000, and all its contents are contributed free by members, except for an occasional guest contribution (e.g., Aharan, 1971; Bloom, 1958; Churan, 1965; D'Alonzo, 1960; Dancey, 1966, 1968; Deering, 1970; de Kruif, 1959; Dole, 1972; Dowling, 1966; Edwards, 1965; Flesch, 1961; Fox, 1966; Fox, 1973a,b; Gitlow, 1968; Haggard and Jellinek, 1944; Heard, 1958; Hopkins, 1944; Keith-Lucas, 1970; Krouse, 1965; Lincoln, 1958; Logan, 1961; Magnum, 1965; Maxwell, 1971, 1975; McQueen, 1960; Menninger, 1965; Norris, 1965, 1968a,b; 1972; Oursler, 1944; Perelman, 1944; Powdermaker, 1945; Sessions, 1966; Shoemaker, 1964; Silkworth, 1947; Strunk, 1966; Thurber, 1971; Tiebout, 1955a, 1956, 1963a,b, 1965; Waldron, 1965; Wylie, 1944), many of which are original contributions to understanding of alcoholism and recovery.

The *Grapevine* accepts no advertising, being supported entirely by sales of copies and of a few collateral items (e.g., "A.A. Today," 1960; "Best Cartoons from the Grapevine," 1970). Four A.A. members are its paid staff, and there are seven nonalcoholic employees. The *Grapevine* does not speak with authority as "the voice of A.A." (that role is filled by the General Service Conference), and each issue cautions on Page 1 that "opinions expressed herein are not to be attributed to Alcoholics Anonymous as a whole, nor does publication of any article imply any endorsement, either by Alcoholics Anonymous or the A.A. *Grapevine*." It therefore publishes pieces reflecting a tremendously wide range of viewpoints on such subjects as atheism, medication, psychiatry, and almost all aspects of A.A. activities.

"The A.A. Preamble," a statement read at many A.A. meetings throughout the world, originated in the *Grapevine*. It is:

> Alcoholics Anonymous is a fellowship of men and women who share their experience, strength, and hope with each other that they may solve their common problem and help others to recover from alcoholism.
>
> The only requirement for membership is a desire to stop drinking. There are no dues or fees for A.A. membership; we are self-supporting through our own contributions. A.A. is not allied with any sect, denomination, politics, organization or institution; does not wish to engage in any controversy, neither endorses nor opposes any causes. Our primary purpose is to stay sober and help other alcoholics to achieve sobriety.*

Partially summarizing the Twelve Traditions, the Preamble somewhat paraphrases the original introduction to the Big Book (Wilson, 1939), and is the nearest thing to an "official" definition of itself ever published by the movement.

Neither the Board nor any of its subsidiaries nor the General Service Office nor the General Service Conference has any formal authority over any A.A. member or group. They have no enforcement powers, only advisory ones. Both the Board and the Conference utilize committees on these specialized interests: trustees (of the Board), movement policy, literature, institutions, public information, finance, relations with the professional community, and Conference admissions and agenda. The General Service Office is staffed by 12 A.A. members and in 1976 had about 70 nonalcoholic employees.

Financially, the movement declines all contributions from any source except its own members, and there is a $300 per year per member limit on them. All bequests from nonmembers are declined. Each group passes the hat to pay for its rent, meeting refreshments, literature, and whatever contributions it may make for the support of a local central office (sometimes in nonurban areas these are simply telephone-answering services) and for the support of G.S.O. For 1973, it budgeted for expenses of service to A.A. groups and free literature distribution a sum slightly in excess of $1 million, of which approximately three-fifths are expected to be met by groups' and members' contributions, with profits from sales of A.A. publications used to make up the balance.

Largely underwritten by the General Service Board of the United States and Canada, in 1969 representatives of A.A. in 12 overseas countries, Central

---

* Reprinted by permission of the copyright owners, The A.A. Grapevine, Inc., New York, N.Y. 10017.

and North America held the First A.A. World Service Meeting. The Second, like the first, met in New York, in 1972, and the Third was held London in 1974, with A.A. in eighteen nations represented.

Such international gatherings, and the 1973 size of the G.S.O. budget, stand in stark contrast to the financial ruin and indebtedness facing Bill W. and Dr. Bob when, alone, they first faced each other in Akron in 1935. However, only the quantity of such mutual communication among alcoholics as theirs has changed—by magnification—in the subsequent years. The exact quality of their service to each other is still maintained in more than one million A.A. meetings annually on six of the world's continents (Leach *et al.*, 1969).

As A.A. has grown and inevitably had to develop offices, more publications, budgets, and other somewhat more sophisticated structural elements, one question has steadfastly remained the touchstone for measuring the value of all suggested new A.A. developments. It is: Will this help, for all who approach A.A. in the future, to protect the basic characteristics of that initial moment when Bill W. first honestly shared his own experience with Dr. Bob? Anything that would alter the nature of that moment—such as professionalism, vainglory, politics, publicity, any deceit, sectarian religion, unlicensed professional advice, unsubstantiated biochemical, psychological, or other scientific theories outside the personal experience of the A.A. member—is strongly and widely disapproved within the Fellowship. A.A. remains, essentially, "one drunk talking to another" (Wilson, 1939), in confidence, candor, and trust.

EVALUATIONS OF A.A. EFFECTIVENESS

In 1968 the General Service Board made its first try at evaluating the effectiveness of A.A. on a large scale by a "sobriety census" of a sample of members attending A.A. meetings in June and July in the United States and Canada. In 1971 and 1974* it similarly surveyed a smaller sample. Earlier attempts to measure the effectiveness of A.A. have been discussed in detail elsewhere (Leach, 1973).

The method and major findings of the 1968 Board survey have already been reported in some detail (Norris, 1970; Leach, 1973), and confirmation by independent investigators of principal Board findings in 1968 has also been demonstrated (Leach, 1973).

Summaries of the 1968 data are compared here with those of 1971. In addition, hitherto unreported findings of the 1968 study are shown, as are answers to some questions used in 1971 but not in 1968.

* Findings for 1974 were not available for preparation of this chapter. However, cursory examination has shown them to be roughly similar to those of 1971.

Survey Method

Delegates to the A.A. General Service Conference for the United States and Canada, a representative assembly of A.A. in the two nations, were mailed a supply of brief questionnaires—a copy is shown in the A.A. Survey (1970). Each was asked to contact a certain number of the A.A. groups by which he or she had been chosen. In 1968, this corresponded to approximately 5.0 percent of the groups; in 1971, 3.0 percent, but a minimum of three for each delegate both years. Specifically excluded were A.A. groups that meet in hospitals, rehabilitation centers, and correctional facilities, eliminating a total of 1475 groups in 1968 (Leach, 1973), nearly 1900 in 1971.*

During the months of June and July, each delegate took questionnaires to one meeting of each of the groups selected at random. Questionnaires were completed anonymously on the spot, and returned by the delegate to the General Service Office.

Representativeness of the Sample

In 1968, 11,355 usable questionnaires received as of September 1 were utilized in compiling analyses of the results, but in 1971 only the first 6000 usable questionnaires received (as of July 26) were analyzed. Table 4 compares the distribution of responses in 1968 with that of 1971.

In 1968, the number of assigned groups, 518, was 5.2 percent of the known (that is, listed)† 9895 United States and Canadian A.A. groups. The 438 groups actually covered in the survey represented 84.6 percent of the hoped-for coverage, and 4.4 percent of the listed groups. In 1971, the survey design called for coverage of 377 groups, 3.4 percent of the listed‡ 11,208; but only 262 groups were actually covered, or 69 percent of the assigned coverage, 2.3 percent of the listed U.S. and Canadian groups.

By comparison, Bailey and Leach (1965) reported coverage of 73.2 percent of the New York groups in 1962, Edwards et al. (1966, 1967) had responses from 88.9 percent of the London groups contacted in 1964, and replies reported on by Kiviranta (1969) came from 70.6 percent of the Finnish groups in 1960–1961. These were three non-A.A. studies of A.A., and they and the Board's investigations were all entirely independent of one another.

Most answers furnished by 6000 A.A. meeting-goers to the Board in 1971 were not very different from those of 11,355 members in 1968, as will be

---
* According to *Directory of A.A. Groups in Correctional Facilities 1972* and *Directory of A.A. Groups in Hospitals and Rehabilitation Centers 1972.*
† According to the [*A.A.*] *World Directory Part I, 1968.*
‡ According to the [*A.A.*] *World Directory Part I, 1971.*

TABLE 4. Distribution of Responses in 1968 and 1971 A.A. Surveys by General Service Board

	Number of A.A. groups			
	1968		1971	
Region	Assigned	Covered	Assigned	Covered
United States				
New England	35	35	28	25
Maine	3	3	3	3
New Hampshire	3	3	3	3
Vermont	3	3	3	0
Massachusetts	15	17	9	9
Rhode Island	3	1	3	3
Connecticut	8	8	7	7
Middle Atlantic	57	53	37	28
New York	29	26	19	19
New Jersey	13	12	8	3[a]
Pennsylvania	15	15	10	6
East North Central	97	41	62	51
Ohio	24	14	15	27
Indiana	9	9	6	?[a]
Illinois	34	7	21	14
Michigan	19	8	12	3[a]
Wisconsin	11	3	8	7
West North Central	43	40	31	26
Minnesota	14	5	8	7
Iowa	5	5	4	5
Missouri	9	13	6	4
North Dakota	3	2	3	3
South Dakota	3	3	3	0
Nebraska	4	4	3	3
Kansas	5	8	4	4
South Atlantic	54	52	42	33
Delaware	3	2	3	3
Maryland	5	5	5	5
District of Columbia	4	5	3	2
Virginia	9	8	5	0
West Virginia	3	3	3	3
North Carolina	8	8	5	5
Georgia	7	7	5	4
South Carolina	4	4	3	3
Florida	11	10	10	8
East South Central	18	21	13	8
Kentucky	5	5	4	4
Tennessee	4	7	3	1
Alabama	6	6	3	3
Mississippi	3	3	3	0
West South Central	33	23	21	20
Arkansas	3	6	3	3
Louisiana	5	6	3	3
Oklahoma	6	3	3	3
Texas	19	8	12	14

TABLE 4. (Continued)

Region	Number of A.A. groups			
	1968		1971	
	Assigned	Covered	Assigned	Covered
Mountain	28	29	26	13
Montana	3	4	3	0
Idaho	3	3	3	3
Wyoming	3	3	3	3
Colorado	4	4	4	3
New Mexico	3	3	3	0
Arizona	6	6	4	1[a]
Utah	3	3	3	3
Nevada	3	3	3	0
Pacific	78	82	59	50
Washington	3	9	7	0
Oregon	6	6	4	4
California	63	61	42	42
Alaska	3	3	3	1
Hawaii	3	3	3	3
Territories	3	1	3	0
Canal Zone	0	0	0	0
Puerto Rico	3	1	3	0
U.S. Totals	443	377	322	254[a]
Canada				
Alberta	7	6	5	5
British Columbia, Yukon Territory	10	6	7	0
Manitoba	4	4	3	0
New Brunswick, Prince Edward Island	3	3	3	0
Newfoundland, Nova Scotia, Labrador	4	3	3	0
Northwest Territory	0	0	0	0
Ontario	24	9	13	3[a]
Saskatchewan	7	7	5	0
Canada Totals	75	61	55	8
Grand Totals	518	438[b]	377	262[a]

[a] It could not be determined from which or how many A.A. groups 322 responses used in 1971 were returned: an additional 10 from Arizona; all 105 from Indiana; an additional 181 from Michigan; an additional 36 from New Jersey; and an additional 43 from Ontario. No 1971 responses were received in time to be used from groups in Louisiana, Mississippi, Montana, Nevada, New Mexico, South Dakota, Vermont, Virginia, Washington, and Puerto Rico. From Canadian groups completed 1971 questionnaires were returned in time for use only from Alberta and Ontario.

[b] Probably at least 28 additional A.A. groups were represented, bringing this total to 466. Only the numbers of groups clearly identifiable by group name is used in this table, however.

shown, so a look at the 1968 distribution is in order. Table 5 shows the regional distribution of respondents in 1968.

It can be seen that distribution of respondents in 1968 was only roughly comparable to that of the known A.A. groups. For example, Canada, with 1411 groups (14.3 percent of the total) was represented by 11.9 percent of the respondents. Regions apparently overrepresented by 1968 questionnaire returns were New England, the South Atlantic, the East South Central, the Mountain, and the Pacific states, with the remaining regions and Canada underrepresented.

The largest overrepresentation, that of New England, may have been due in part to the density per square mile of the number of A.A. groups there, where a total of 566 groups were the responsibility of seven delegate-interviewers (about 81 groups each) who actually contacted 6.2 percent of the region's groups (an average of five groups each). Delegates in the over-represented regions tended to have been elected by smaller numbers of groups (an average of 98.5 each) than those of regions underrepresented in the survey (an average of 115.6 each).

Large urban centers, the most likely sites of the possible 20.0 percent membership overlap suggested by Chafetz and Demone (1962), were probably not overrepresented. No groups were contacted in Atlanta, Boston, Chicago, Cincinnati, Houston, Indianapolis, Louisville, Milwaukee, Minneapolis, Omaha, Richmond, San Antonio, or Toronto; only one each in Buffalo, Dallas, Denver, Des Moines, Memphis, Miami, Newark, Philadelphia, Seattle, St. Louis, and Wilmington; only two each in Cleveland, Fort Worth, New

TABLE 5. Regional Distribution of Respondents in 1968 A.A. Survey

Region	Percent of region's groups covered	Number of respondents	Percent of total respondents ($n = 11,355$)
United States	4.4	10,006	88.1
New England	6.2	1,585	14.0
Middle Atlantic	4.7	1,216	10.7
East North Central	2.2	991	8.7
West North Central	4.8	880	7.8
South Atlantic	5.0	1,434	12.6
East South Central	7.0	337	3.0
West South Central	3.4	673	5.9
Mountain	6.5	600	5.3
Pacific	5.1	2,260	19.9
Territories	2.2	30	0.0
Canada	4.3	1,349	11.9

Orleans, Rochester, and Syracuse; only three each in Los Angeles and Pittsburgh; only four in Akron; and only seven in New York City.

The largest underrepresentation, that of the East North Central U.S. region, was no doubt due in part to the fact that Ohio, Illinois, and Michigan—three of the five states in that region—were incompletely covered, and no groups in the Chicago area were covered at all, as noted.

At any rate, East North Central delegates did contact 41 groups, and the responding delegates returned an average of 24.2 usable questionnaires per group contacted.

In the Pacific region of the United States, by far the largest geographically and the second largest in number of A.A. groups, nine delegates responsible for 1597 groups (an average of 177.4 groups each) covered 82 (5.1 percent) of them for the survey—or an average of 9.1 groups each—returning an average of 27.6 usable questionnaires for each group contacted. California alone supplied 74.0 percent of these questionnaires.

The most obvious imbalance in the 1968 Canadian figures is, of course, the difference between Ontario and Quebec. Ontario was incompletely covered, and Toronto not at all, so only nine (2.1 percent) of the province's 420 groups were contacted. However, 237 questionnaires were returned—an average of 26.3 respondents per group. Quebec, on the other hand, with only 323 groups, was represented in the 1968 survey by 23 (7.1 percent) of them, with a total of 466 usable questionnaires—an average of 20.3 respondents per group.

By states, territories, and provinces, the percentage of groups covered in 1968 ranged from 1.1 percent of Illinois's 622 groups to the 12.0 percent of the 25 groups in Wyoming. By U.S. regions, the arithmetical mean percentage of group coverage was 4.7. By states (including Puerto Rico and the District of Columbia, excluding the Canal Zone), the mean percentage of group coverage was 5.8; by Canadian provinces (excluding the Northwest Territory), 4.6.

Each of the 77 U.S. delegates was responsible at the 1968 Conference for a mean of 110.4 groups, and covered an average of 4.9 groups for the survey. The 12 Canadian delegates, responsible for a mean of 118.0 groups each at the Conference, covered for the survey an average of 5.1 groups each. The mean number of all groups covered per responding interviewer-delegate was 5.3, each returning an average of 140.2 questionnaries.

Distribution of the 1971 responses tends to redress, in part, some of the 1968 imbalances. For example, regions overrepresented in 1968—New England, South Atlantic, East South Central, Mountain, and Pacific—were all slightly underrepresented in relation to the study design in the 6000 responses used for 1971. Those underrepresented in 1968—Middle Atlantic, East North Central, West North Central, West South Central, and the Territories—were still underrepresented in 1971. However, whereas only 42.3 percent of the survey design was achieved in the East North Central in 1968, the percent of the design coverage actually fulfilled in 1971 rose to 82.3. For the West South Central, percent of design coverage achieved rose from 69.7 percent in 1968 to 95.2 percent in 1971.

Since, as noted, in 1968 Quebec was better covered than Ontario, the improved showing of Ontario and absence of Quebec respondents in the 1971 results tend somewhat to rectify that imbalance.

Anyhow, in neither 1968 nor 1971 were any statistically significant interstate, interprovincial, interregional, or international differences discernible in respondents' answers to any questions, so combined United States and Canadian totals are used hereinafter, unless otherwise specifically noted, for reporting the findings.

Moreover, it should be noted that results reported here are representative only of the A.A. members at meetings of the surveyed groups who chose to complete questionnaires. Undoubtedly some of those present chose not to do so. There is also evidence that the longer-sober A.A. members are probably underrepresented in such a sample because they apparently attend fewer A.A. meetings than newer members (Bailey and Leach, 1965; Norris, 1970). Thus no attempt is made to assess what proportion of either the total A.A. membership, or that of each group, is represented by the survey results.

Findings of the A.A. Board Surveys

Gender and Age of A.A. Members

Table 6 compares the gender of A.A. members responding in the 1968 survey with that of 1971 respondents.

That the percent of women among A.A. members rose almost 3.0 percent 1968 and 1971 is not particularly surprising. Analyses of the 1968 returns indicate that the ratio of women to men among newcomers to A.A. has tended to rise as A.A. in North America has grown older. Of those who first attended A.A. before 1958, 81.4 percent of the 1968 respondents were male, only 17.4 percent female. But of those who first affiliated with A.A. after 1967, 75.1 percent of the 1968 respondents were male, 24.2 percent female. Further, it was found that 29.0 percent of the 1968 respondents under the age of 30 were women, compared with 22.0 percent of those over 30.

Bailey and Leach (1965) reported findings tending in the same direction. Among their New York respondents in A.A. less than one year, 64.4 percent were male, 32.5 percent female, as contrasted with those in A.A. 10 or more years: 72.4 percent male, 24.0 percent female. Overall, in their 1058-member sample in 1962, the percents were 66.7 percent male, 29.9 percent female, with 3.4 percent not specified. In 1962, Cooper and Maule (1962) reported among 152 A.A. members in the United Kingdom 82.0 percent men, 18.0 percent

TABLE 6. Gender of A.A. Members by Percent

Gender	1968	1971	Change
Male	75.3	74.3	−1.0
Female	22.4	25.3	+2.9
No answer	2.3	0.4	−1.9

women. In 1964, Edwards *et al.* (1966, 1967) reported among London members 81.0 percent male, 19.0 percent female. However, it should be noted that A.A. in England was only 17 years old in 1964, whereas the New York findings involved an A.A. history of 27 years, and the Board's 1971 report reflects a 36-year history.

We know of no reason to believe the ratio of female to male members in A.A. will not continue to rise.

Why the percent of A.A. members under the age of 30 in the summer of 1971 was smaller than the 1968 percent, we do not know. If this reflects a bias in sampling, we are unable to pinpoint it. It is the universal impression of the G.S. Board, the Conference delegates, the A.A. General Service Office staff, and local A.A. intergroup offices that the number of younger people attending A.A. meetings is growing steadily. Also, we have learned from hitherto unpublished analyses of the 1968 data that among new members (under one year of affiliation with A.A.) in that year, 11.0 percent were under 30. A planned 1974 survey could reveal whether the reduction shown in Table 7 has reversed or continues. Since each generation of Americans has a larger proportion of drinkers and most remain drinkers (Cahalan *et al.,* 1969), the percent of both younger and older people in A.A. presumably could increase.

In the 1960–1961 study of Finnish A.A. members (Kiviranta, 1969), 13.7 percent were reported 30 years old or younger. In the 1962 study of New York members (Bailey and Leach, 1965), 4.1 percent of the men and 5.4 percent of the women were reported under 30, and Edwards *et al.* (1966, 1967) found 4.0 percent of their London respondents under 30 in 1964.

TABLE 7. Comparison between 1968 and 1971 of Ages of A.A. Members (%)

Age	1968	1971	Change
30 and under	7.1	4.5	−2.6
31–50	57.1	52.1	−5.0
Over 50	34.0	42.2	+8.2
No answer	1.8	1.2	−0.6

Although the 5.0 percent drop of those in the 31–50 age bracket is, like that of the under-30s, puzzling, it is certainly not unexpected to find more than half the 1971 responding A.A. members within that age range. The 1960–1961 Kiviranta (1969) study reported 74.4 percent of its Finnish respondents in that bracket; the 1962 New York study (Bailey and Leach, 1965) reported 45.9 percent of its male and 57.6 percent of its female respondents between the ages of 30 and 49, and the 1964 London (Edwards et al., 1966, 1967) investigation reported 64.0 percent in the 30–49 age bracket. Cooper and Maule (1962) had found 66.0 percent of their United Kingdom sample between 35 and 54 years old.

The percent of members over 50 among Board respondents increased 8.2 percent between 1968 and 1971, a bigger gain than loss of those 50 and under. This could reflect the aging of that portion of the A.A. population which attends many A.A. meetings regularly. (Frequency of meeting attendance is discussed elsewhere.) Kiviranta (1969) found 11.9 percent of the Finnish members 51 or over; Bailey and Leach (1965) reported 35.4 percent of their New York sample, and Edwards et al. (1966, 1967) 32.0 percent of their London members, over 50.

Occupational Status of A.A. Members

Only in the 1971 G.S. Board survey were respondents asked to indicate their occupations. Results are shown in Table 8.

Almost 1900 A.A. groups which meet in prisons, state hospitals, and other institutions were deliberately eliminated from the 1971 survey, as noted previously, and delegate-interviewers may not have visited for survey purposes a representative sample of A.A. groups on or near skid rows or other communities of the lower socioeconomic levels. Also, A.A. members of that standing might be less likely to complete survey questionnaires than those in higher socioeconomic situations. As a result, the 1971 Board sample is very likely biased toward members of middle or higher socioeconomic status, and it is known that Canadian groups are severely underrepresented.

Even so, it is of interest that the percent of A.A. respondents in the highest classes of employment (first three lines of Table 8) is 41.2 for A.A. members, compared with 31.0 percent of the United States employed population in those occupational categories.*

For male A.A.s, 42.8 percent were found in those classes compared with 34.0 percent for employed United States males. However, the percent among A.A. women is 20.7 percent, as against 26.0 percent for the United States norm.

---

* All United States norms used here are from U.S. Bureau of the Census (1970).

TABLE 8. Occupational Categories (%) of A.A. Members in 1971 (All Ages)

Occupation	Total (n = 6000)		Male (n = 4459)		Female (n = 1518)		Sex not specified (n = 23)	
1. Professional, technical	12.6		13.6		10.1		—	
2. Proprietary, managerial, official	12.3		14.6		5.9		4.3	
3. Sales	12.0		14.6		4.7		—	
(Subtotals)		36.9		42.8		20.7		4.3
4. Craftsman	14.6		19.3		0.7		4.3	
5. Clerical	7.5		3.8		18.2		17.4	
6. Semiskilled workers, operators	7.4		9.3		1.9		—	
7. Service workers	5.8		4.8		8.5		—	
8. Laborers	2.2		2.8		0.2		4.3	
9. Alcoholism counselors (paraprofessional)	0.8		0.8		0.8		—	
(Subtotals)		38.3		40.8		30.3		26.0
10. Housewives, widows	10.3		—		40.0		4.3	
11. Unemployed, retired	8.2		9.9		4.0		13.0	
12. Students	0.6		0.6		0.7		—	
(Subtotals)		19.1		10.5		44.7		17.3
13. No answer	5.7		5.9		4.3		52.4	

In the next highest list of occupations (lines 4–9 in Table 8), the percent of A.A. members is consistently lower than the norms for the United States, as follows: 64.3 percent for the A.A. total, with 69.0 percent of the national norm; 40.8 percent of the A.A. men, 66.0 percent of the United States norm; and 30.3 percent for A.A. women as against 74.0 percent of employed United States women. Nearly one-fifth of the A.A. sample makes up the housewife/widow, unemployed/retired, and student category.

Since occupational functioning is sometimes considered an index to the level of recovery from alcoholism, it is interesting to see these 1971 Board findings are so nearly consistent with what has been reported by independent investigators about A.A. members in New York (Bailey and Leach, 1965), London (Edwards *et al.*, 1966, 1967), and Finland (Kiviranta, 1969), as follows. Apparently holding jobs at the time of the studies were 80.9 percent of the 1971 Board total (excluding housewives and students), 91.7 percent of the men and 97.8 percent of the women (housewives excluded) in New York, 96.0 percent of London A.A.s, and 80.6 percent of those in Finland.

Why Alcoholics Go to A.A.

Motivating problem drinkers to seek help, from A.A. or anywhere else, is a complex, subtle, and even misunderstood problem, as Sterne and Pittman (1965) and Chafetz *et al.* (1970), among others, have so clearly shown, but the Board surveys contribute some possibly worthwhile knowledge. Board questionnaires contained a list of possible sources of referral to, and information about, A.A. Respondents were asked, "Which *two* of the following do you feel were most responsible for your coming to your first A.A. meeting? (*Please check two*)." Table 9 shows the distribution of replies.

Among the 11,355 respondents in 1968, 6225 (54.8 percent) designated an A.A. member as an important motivating factor in getting them to an A.A. meeting, 3865 (34.0 percent) the family, and 1840 (16.2 percent) the physician. No other factor received even half as many "votes," though almost all respondents checked at least two factors, some more.

These 1968 findings stress the strong effect a recovered alcoholic, making a "Twelfth Step call," can have on a still-suffering alcoholic. They also support the finding of Bailey and Leach (1965) of the weight of "social diagnosis" (by the family, for instance) in persuading an alcoholic to seek help. And they indicate the strong influence of the physician's referral of an alcoholic to A.A.

It was also found among the hitherto unpublished data that males were proportionately more responsive than females to other A.A. members, their own families, and employers, but that females tended to be more responsive to doctors, A.A. literature, newspapers, magazines, and television, and social workers.

TABLE 9. 1968–1971 Comparison of Factors (%) Most Responsible for First Attendance at A.A. Meeting[a]

Factor responsible	1968	1971	Change
A.A. Member	54.8	51.6	−3.2
Family	34.0	35.2	+1.2
Doctor	16.2	16.2	—
A.A. literature	6.9	7.2	+0.3
Clergyman	6.5	5.3	−1.2
Employer	5.5	5.8	+0.3
Newspaper	4.9	4.5	−0.4
Magazine	4.2	3.1	−1.1
Friend	4.1	4.2	+0.1
Counseling agency	3.9	5.3	+1.4
Hospital	1.9	3.4	+1.5
National Council on Alcoholism	1.6	—	−1.6
Television	1.5	2.2	+0.7
Clubs, social organizations	1.2	0.8	−0.4
Radio	1.1	0.8	−0.3
Attorney/jail	1.0	2.4	+1.4
Psychologist	1.0	0.1	−0.9
Social worker	0.4	0.1	−0.3

[a] The total exceeds the base since many respondents named more than one factor.

The 1968 data also showed a difference by length of sobriety in factors most responsible for first A.A. meeting attendance. Among A.A.'s earliest members—those with more than 20 years of sobriety—27.0 percent cited newspapers and magazine stories as responsible for getting them to their first meeting, as opposed to only 2.6 percent for other A.A. members. Those early members, of course, had begun their recovery in A.A. in 1948 or before, when A.A. was still new enough to fascinate journalists, when television was still in its early childhood, and before there were very many already-sober A.A. members either to demonstrate sustained recovery to the professional community or to "Twelfth Step" other alcoholics.

The newest (not youngest) members in 1968 also showed the sharply increasing influence of the medical profession in getting alcoholics into A.A. Of those sober less than one year, 18.5 percent reported a doctor's influencing them toward A.A., whereas of those with 15 or more years of sobriety, only 12.0 percent noted a physician's influence—which means more than a 50.0 percent increase.

Some cross tabulations of the 1971 responses reveal the following:

1. While the A.A. member declined 3.2 percent as an overall influence, the member's importance increased from 50.0 percent to 52.0 percent

among the 18–30 age group, and declined from 55.0 percent to 51.0 percent in the 31–50 age bracket.
2. Among the 18–30 age group, the family influence rose from 37.0 percent to 39.0 percent.
3. Although the doctor declined from 13.0 percent to 11.0 percent among the 18–30-year-olds, the physician's importance increased in the 51–65-year bracket.
4. A friend was more important in getting the 18–30-year-old members to an A.A. meeting (designated by 5.0 percent) than in overall importance (designated by 4.0 percent).
5. The counseling agency increased sharply, from 4.0 percent to 7.0 percent as a factor responsible for getting the 18–30 group to attend A.A.
6. The influence of a hospital rose from 2.0 percent to 4.0 percent among the 18–50 group.
7. There was also a notable increase in the influence of the attorney/jail factor among younger members, up from 5.0 percent to 7.0 percent for the 18–30 group, from 1.0 percent to 3.0 percent for those 31–50.

In their 1962 examination of professional treatment histories of New York A.A. members, Bailey and Leach (1965) reported only 18.0 percent of those who sought help from clergymen, 18.7 percent of those treated by psychiatrists, and 30.0 percent of those treated by nonpsychiatric physicians were referred to A.A. Cooper and Maule (1962), having surveyed by questionnaire 152 A.A. members in the United Kingdom, reported 23.0 percent had been advised to join A.A. by medical specialists, 20.0 percent by other A.A. members, and 13.0 percent by general practitioners. Asked to rate the pre-A.A. treatment for alcoholism they had received as satisfactory or harmful, 64.6 percent of the Bailey and Leach sample of A.A. members rated referral to A.A. as satisfactory, but 14.9 percent rated it as harmful! Professional contacts were most often rated harmful if the treater told respondent he was *not* an alcoholic (46.8 percent), if he told the respondent he *was* an alcoholic (42.6 percent), and if tranquilizers were prescribed (36.2 percent). Professional treatment which continued one year or more was more likely to be rated satisfactory (40.7 percent) than harmful (36.2 percent).

Many more, and other, interpretations of these data are, naturally, possible in any such matter involving so many variables. The G.S. Board of A.A. is hopeful that interpretations within A.A. of the Table 9 data are translated into even closer cooperation between A.A. members and others in the community concerned about the welfare of alcoholics. But the present authors ask the professional reader to make his or her own inferences.

Relapses after Joining A.A.

Instant and enduring abstinence does not, of course, always set in upon an alcoholic's first contact with A.A. But for 7434 (42.8 percent) of the Board's 17,355 respondents, it did—a fact worth underscoring. Table 10 discloses that the "instant abstinence" proportion rose 5.3 percent between 1968 and 1971, and also shows how long the other responding members kept having slips before their present terms of sobriety in A.A. began.

Becoming an abstinent A.A. member can perhaps best be viewed as a process involving several stages, ranging from initial knowledge of the existence of the Fellowship through firm, acted-out commitment to its ideals, such as speaking of one's alcoholism and recovery at a public A.A. meeting.

The approximate direction in which the stages occur might be:

1. Learning of the existence of A.A.
2. Perceiving A.A. as relevant to one's need.
3. Being referred to A.A. by a helping agency.
4. Making first personal contact with A.A., perhaps by telephone or by letter, but more likely face to face, involving (a) visiting an A.A. office, (b) visiting a clubhouse for A.A. members, (c) visiting, or being visited by, an individual A.A. member or two, or (d) attending an "open" (anyone welcome) A.A. meeting.

TABLE 10. Comparison Between 1968 and 1971 Findings on Time Lapsed from First Meeting to Last Drink for A.A. Members (%)

Time lapsed	1968		1971		Change	
Sober before first visit	—		0.7		+0.7	
Sober from first visit	41.0		46.3		+5.3	
1 year or less	22.9		16.3		−6.6	
Cumulative subtotals		63.9		63.3		−0.6
1 to 5 years	18.5		17.6		−0.9	
Cumulative subtotals		82.4		80.9		−1.5
5 to 10 years	8.2		8.1		−0.1	
Cumulative subtotals		90.6		89.0		−1.6
10 to 15 years	4.0		4.7		+0.7	
Cumulative subtotals		94.6		93.7		−0.9
15 to 20 years	2.6		2.1		−0.5	
Cumulative subtotals		97.2		95.8		−1.4
Over 20 years	0.9		1.7		+0.8	
Don't know/no answer	1.9		2.5		+0.6	
	100.0		100.0			

5. Attending a "closed" (for members or prospective members only) A.A. meeting.
6. Participating in various other A.A. activities, such as those related to the Twelve Steps and others which enable the alcoholic to internalize the norms of the movement, especially that of abstinence and those codified in the Twelve Traditions.
7. Taking the last drink.
8. Making a "Twelfth Step" visit (to help another alcoholic).
9. Speaking at an A.A. meeting, possibly resulting in disclosure to nonalcoholic acquaintances of one's A.A. membership.

Not every abstinent member has proceeded through every single phase, of course (for instance, referral to A.A. by a helping agency does not happen in every case). And perhaps the crisis generally described as "hitting bottom," or other significant events, should be inserted in the list at some point for some cases.

Nor is the above order followed exactly by every member.

Moreover, some stages may occur almost simultaneously for some members, while in other cases months or years may elapse between stages. "Joining A.A." may thus be seen as a learning process during which *relapses can and do occur for some members* at almost any point before the last drink (Stage 7 above), and, for a minority, even after the Stage 9 above—the "final" commitment. In those cases, the process is, roughly, recapitulated at least partly in somewhat similar fashion.

It is notable that 11,054 (over three-fifths) of the Board's respondents in 1968 and 1971 reported they had stopped drinking either immediately after their first A.A. meeting or within their first A.A. year.

But some other members, in both surveys, although still considering themselves A.A. members, continued to have relapses. Perhaps these are alcoholics of the Epsilon or Pseudoepsilon species described by Jellinek (1960). A few reported continuing relapses for as long as 20 years before starting their terms of sobriety current at the time of the surveys (see Length of Abstinence of A.A. Members, p. 497).

To study the possibility that gender or age influence the time lapse between first A.A. contact and last drink, some of the 1968 data have been reanalyzed. It was found that fewer women reported no slips (40.6 percent) than men (41.1 percent), but 67.5 percent of the women had stopped drinking in that first A.A. year, as compared with 62.5 percent of the men. Among 1968 respondents who had their last drink *more* than one year after their first A.A. meeting, 35.5 percent were men, 30.6 percent women.

Among those 30 years of age and under in 1968, 42.4 percent reported no slips, 31.3 percent had their last drink within their first year, and for 25.0

percent it had taken more than one year since their first A.A. meeting for them to have their last drink. Among those over 30, 40.8 percent reported no slips, 22.1 percent none since their first year—a total of 62.9 percent whose drinking stopped within the first year. But 35.1 percent had not had their last drink until more than one year after first attending an A.A. meeting.

In addition, reported dates of last drink among 1968 respondents were tabulated according to gender, age, and factors reported most responsible for joining A.A. (see Table 11).

A larger proportion of men (80.7 percent) than women (18.5 percent) reported their last drink occurred before 1958. Among those whose last drink had occurred within one year before the 1968 survey, 76.6 percent were men, 22.8 percent women.

Among 1968 respondents reporting their last drink before 1958, only 0.3 percent were 30 years old or younger, with 98.5 percent over that age. But among those whose last drink occurred between 1963 and 1967 (inclusively), the percent of those 30 or under rose to 5.2 percent, while among those reporting the last drink since 1967—less than one year before the Board's first survey—11.2 percent were in the younger age bracket, with 87.8 percent over 30. These findings seem consistent with other 1968 data on gender and age.

Caution must be used in interpreting findings on dates of last drink according to factors reported most responsible for first attending an A.A. meeting, because most respondents, as requested, reported more than one factor.

One other variable was analyzed in relation to slips. Among factors the 1968 respondents reported responsible for their first attending A.A. (see Table 9), four were selected as possible "pressure" factors. That is, some alcoholics may have perceived mentions of, or referrals to, A.A. by certain people as external coercion, and felt they first attended A.A. somewhat under duress. The four factors possibly perceived as pressure are the attorney/jail, the employer, the physician, and the family. By observation, it is known that many alcoholics sent to the Fellowship by such resources are sometimes quite reluctant to approach A.A. and often react initially with resistance to the A.A. program.

By contrast, the alcoholic who is self-referred, or who reported he or she first attended A.A. because of a television show or a newspaper story, for example, would presumably feel under no such duress. In A.A. terms, he or she would more likely "be ready" to accept A.A. help. Table 12 shows a possible difference in the "slips" record between those 1968 respondents who perhaps felt pressured into A.A. and those presumably more self-motivated.

As predicted, among those reporting the "pressure factors" (first four lines of Table 12) as most responsible for their first going to A.A.—those who may have felt somewhat coerced into it—the percent of No Slips was smaller than it was among those attributing their first A.A. meeting to 11 other factors (lines 5–15 of Table 12). Psychologists, the National Council on Alcoholism (perhaps

TABLE 11. Dates Reported in 1968 for Last Drink According to Gender, Age, and Factors Responsible for Joining A.A.

	Subjects reporting last drink									
	Since 1967		1963–1967		1958–1962		Before 1958		Don't know/no answer	
	Number	Percent	Number	Percent	Number	Percent	Number	Percent	Number	Percent
By Sex										
Male	3298	76.6	2965	75.0	1172	77.6	1102	80.7	136	74.7
Female	983	22.8	964	24.4	328	21.7	252	18.5	35	19.2
No answer	26	0.6	24	0.6	10	0.7	11	0.8	11	6.0
Total	4307	100.0	3953	100.0	1510	100.0	1365	100.0	182	100.0
By Age										
15–30 years	481	11.2	204	5.2	13	0.9	4	0.3	7	3.8
Over 30 years	3782	87.8	3694	93.4	1474	97.6	1344	98.5	159	87.4
No answer	44	1.0	55	1.4	13	1.5	17	1.2	16	8.8
By Factors Most Responsible for Joining A.A.										
A.A. member	2368	55.0	2236	56.6	866	57.4	787	57.7	91	50.0
Family	1581	36.7	1395	35.3	521	34.5	419	30.7	57	31.3
Doctor	791	18.4	626	15.8	257	17.0	177	13.0	32	17.6

A.A. literature	303	7.0	279	7.1	98	6.5	98	7.2	14	7.7
Clergyman	264	6.1	280	7.1	107	7.1	81	5.9	12	6.6
Employer	232	5.4	219	5.5	80	5.3	81	5.9	10	5.5
Newspaper	184	4.3	186	4.7	81	5.4	103	7.5	4	2.2
Friend	170	3.9	171	4.3	63	4.2	73	5.3	5	2.7
Magazine	137	3.2	154	3.9	80	5.3	95	7.0	4	2.2
Counseling agency	211	4.9	160	4.0	37	2.5	28	2.1	5	2.7
National Council on Alcoholism	73	1.7	62	1.6	21	1.4	18	1.3	4	2.2
Television	70	1.6	66	1.7	22	1.5	12	0.9	2	1.1
Hospital	60	1.4	66	1.7	22	1.5	13	1.0	2	1.1
Clubs, Social organizations	46	1.1	63	1.6	23	1.5	9	0.7	1	0.5
Radio	47	1.1	35	0.9	14	0.9	18	1.3	2	0.5
Attorney/jail	42	1.0	47	1.2	11	0.7	14	1.0	2	1.1
Psychologist	36	0.8	45	1.1	17	1.1	9	0.7	2	1.1
All others	12	0.3	9	0.2	4	0.3	3	0.2	—	—
None	55	1.3	70	1.8	18	1.2	33	2.4	4	2.2
Don't know/no answer	19	0.4	23	0.6	13	0.9	11	0.8	3	1.6
Total	4307	100.0[a]	3953	100.0[a]	1510	100.0[a]	1365	100.0[a]	182	100.0[a]

[a] Adds up to more than 100% because of multiple mentions.

TABLE 12. Relationship Between Slips and Factors Reported in 1968 Most Responsible for Joining A.A.

Factor reported most responsible for attending first A.A. meeting	No slips		Last drink within 1 year after first meeting		Last drink more than 1 year after first meeting		No answer	
	Number	Percent	Number	Percent	Number	Percent	Number	Percent
1. Family (n = 3973)	1534	38.6	930	23.4	1438	36.2	71	1.8
2. Doctor (n = 1883)	727	38.6	501	26.6	617	32.8	38	2.0
3. Employer (n = 622)	245	39.4	147	23.6	219	35.2	11	1.8
4. Attorney/jail (n = 116)	35	30.2	25	21.6	54	46.5	2	1.7
5. Television (n = 172)	87	50.6	36	20.9	45	26.2	4	2.3
6. Radio (n = 116)	51	44.0	28	24.1	34	29.3	3	2.6
7. Self-referred (n = 180)	78	43.4	40	22.2	56	31.1	6	3.3
8. Friend (n = 482)	208	43.2	109	22.6	153	31.7	12	2.5
9. Club or social organization (n = 142)	61	42.9	39	27.5	39	27.5	3	2.1
10. A.A. member (n = 6348)	2667	42.0	1341	21.1	2222	35.0	118	1.9
11. Counseling agency (n = 441)	183	41.5	127	28.8	124	28.1	7	1.6
12. Magazine (n = 470)	195	41.4	83	17.7	188	40.0	4	0.9
13. Newspaper (n = 558)	230	41.2	114	20.4	207	37.1	7	1.3
14. A.A. literature (n = 792)	319	40.3	203	25.6	252	31.8	18	2.3
15. Clergyman (n = 744)	299	40.2	142	19.1	287	38.5	16	2.2
16. Psychologist (n = 109)	42	38.6	41	37.6	25	22.9	1	0.9
17. Nat'l Council on Alcoholism (N.C.A.) (n = 178)	65	36.5	43	24.2	66	37.0	4	2.3
18. Hospital (n = 163)	55	33.7	61	37.5	46	28.2	1	0.6
19. All others (n = 28)	8	28.6	6	21.4	14	50.0	—	—
20. Don't know or no answer (n = 69)	33	47.9	13	18.8	19	27.5	4	5.8

this means a local N.C.A. affiliate, since many of these agencies are known to be highly motivated to make strong referrals to A.A.), hospitals, and all other factors also rank low in the No Slips percentages, along with the preselected, supposed "pressure" factors.

The largest percentages reporting no relapses occurred in the categories which presumably reflect a higher degree of self-motivation.

Generalizations about these A.A. members and other alcoholics, relapses, and referring agencies would be unwise based on this sampling and such unsophisticated statistical work as comparison of simple mathematical averages. Moreover, in such a complex problem it is likely that some influential variables have been entirely overlooked. For example, there were suggestions in the 1968 data that many of those A.A. members reporting no slips after the first year appeared also to be under some continuing professional therapy, such as services of a physician, a hospital, a psychologist, or some counseling agency. Certainly it is not concluded here that pressure to join A.A., from any source, necessarily results in more relapses, or relapses over a longer period, than self-referral does. Instead, attention is directed to the fact that no matter whether the referring agency was perceived as coercion or not, large percents of *all* respondents could, apparently, begin sooner or later to maintain the unbroken abstinence reported at survey time (see Length of Abstinence, p. 497).

It can be seen from Table 12 that regardless of the factor reported most responsible for joining A.A., an average of 39.7 percent of the respondents reported no slips, an average of 24.5 percent no slips after the first year, with an average of 34.0 percent continuing to relapse more than one year. These findings correspond closely with those in Table 10.

They are also in close agreement with findings and observations from non-A.A. sources, such as industrial programs on alcoholism and correctional officials, that many alcoholics manifest initial hostility to any therapy for alcoholism, but after being even unwillingly exposed to some such help, the alcoholic often can and does begin to recover, despite some relapsing. As Clancy (1964) has noted, A.A. makes therapeutic use of slips.

A number of differences between those reporting no slips and "frequent relapsers" were found by Bailey and Leach (1965) among a subsample of 450 of the New Yorkers who had been in A.A. five or more (average, about ten) years. Among the frequent relapsers, the percent of females was smaller than that of males, educational level was lower, a smaller percent of marriages were intact, and age level tended to be lower. Participation in all A.A. activities reported on except meeting attendance was lower among the relapsers, and larger proportions of relapsers than nonrelapsers had taken tranquilizers and barbiturates since joining A.A. Also, the frequent relapsers proportions in ongoing professional psychotherapy, taking disulfiram (Antabuse) and attending three or more weekly A.A. meetings, were larger than those of the nonslippers. It seems possible the last three activities represented stepped-up, intensified efforts to achieve recovery.

"It would be valuable," Bailey and Leach (1965) wrote, "to know what holds . . . [frequent relapsers] . . . in the Fellowship despite continued lack of success in achieving A.A.'s principal goal of sobriety. Perhaps it is hope that they may yet become sober, perhaps a sense of belonging to a group and of participation in meaningful human relationships . . ." (pp. 40–41).

Edwards *et al.* (1966) noted that emphasis on therapeutic successes in A.A. "has inevitably led to the neglect of a less dramatic but perhaps more important achievement—A.A. has created a supportive organization which accepts and continues to tolerate the relapsing alcoholic who has little ability to maintain long-term sobriety. . . ." They called for greater emphasis on A.A.'s "possibly larger and more unique role as a supporting fellowship for the relapsing alcoholic . . ." (pp. 383–384).

These statements were occasioned by the respective authors' findings on relapses among A.A. members in New York and London, which closely parallel the 1968 Board findings on relapses, as Leach (1973) has shown.

For the therapist and the alcoholic, however, the present authors believe the findings on slips are useful chiefly to the extent they prevent discouragement should a relapse occur. Clearly, the odds are in favor of abstinence being maintained by the alcoholic who perseveres (and is so encouraged) at being an A.A. member.

A unique study by Bill C. (1965) shows the mathematical probabilities of gaining sobriety in A.A. The author kept records on meeting attendance and sobriety of members of a large A.A. group and did a seven-year follow-up on 393 members. (This was an unusual A.A. situation. As far as we know, in only one other case—Bohince and Orensteen (1953)—have such group records been kept. By far most A.A. groups prefer to maintain no such records.)

Bill C. reported 31.0 percent of his sample had stayed sober for at least one year and were sober at the time of the follow-up, and 12.0 percent had stayed sober at least one year, but slipped. He found that "about 70 percent of those who stay sober for 1 year continue on to 2 years, and about 90 percent of those who stay sober for 2 years continue on to 3 years." From his data, Bill C. plotted the curve in Figure 3 showing the possibility of continuation of sobriety at least one year for those who attend at least ten A.A. meetings is approximately 0.43. Thereafter, the curve rises fairly steeply until two years is attained, then begins to flatten.

What A.A. Members Do

Rokeach (1960) and colleagues found that it is easier for people with "closed" thinking systems to exchange total sets of old ideas all at once for total

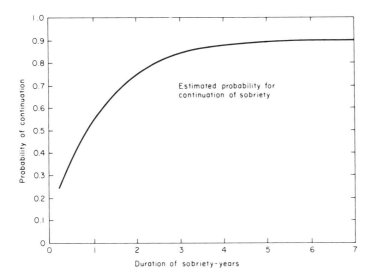

FIGURE 3. Probability of continuation of sobriety. Reproduced by permission from *Quarterly Journal of Studies on Alcohol*, Vol. 26, p. 283, 1965; also by permission of the author. Copyright by Journal of Studies on Alcohol, Inc., New Brunswick, N.J. 08903.

sets of new ideas than it is to change old beliefs into new systems of belief piecemeal, or gradually. It seems fair to describe the rationalizing and denying (Jellinek, 1952) of the drinking alcoholic as a closed thinking system. A.A. members do remind each other with great frequency to "keep an open mind" (e.g., Wilson, 1952), and likewise report (e.g., Wilson, 1939) that piecemeal, gradual efforts to "cut down" on drinking have failed for them as a treatment of alcoholism, whereas successful recoveries in A.A. are characterized by total abstinence, among other things.

The Rokeach (1960) conclusion that "formation of new systems is facilitated when the new beliefs are presented all at once, in which case the new beliefs do not have to be reconciled with old ones" is often demonstrated by what happens to those alcoholics who, upon joining A.A., plunge fully into the highly numerous, all-encompassing activities which can characterize the role of A.A. member.

The basic A.A. textbook (Wilson, 1939) says, "Some of us have tried to hold on to our old ideas and the result was nil until we let go absolutely.... Half measures availed us nothing" (pp. 70–71). Alcoholics Anonymous is frequently described (e.g., Maxwell, 1962) as an entirely "new way of life."

A.A. members' belief and action systems in their "new way of life" can be divided into at least two kinds. First are those which are intrapersonal, affecting the interior life of the alcoholic. These can be heard discussed at any A.A. meeting and can be read about on almost any page of the *A.A. Grapevine* and in the writings of Wilson (1939, 1952, 1967). The second kind of activity participated in by A.A. members as such are the overt behaviors described in detail by Bird (1949), Trice (1958a), Maxwell (1962), Gellman (1964), Bailey and Leach (1965), in The A.A. Group (1965), and by Leach *et al.* (1969), among other places.

They include both informal social activities and those more formal, specific actions which have become virtually institutionalized in A.A. and are immediately recognizable to both the A.A. member and to students of A.A. as "A.A. activities." We suggest the network of interpersonal relationships formed by these activities can be described in the phrases used by Jung (Wilson and Jung, 1963), "personal and honest contact with friends," and "the protective wall of human community," as possible paths to the experience of recovery from alcoholism.

These A.A. behaviors all center around one chief action, which is attending weekly (or more often) meetings of an A.A. group (Leach *et al.,* 1969). Perhaps because it is the A.A. activity most easily quantifiable, respondents in the Board surveys were asked about the frequency with which they attend such meetings. Table 13 compares the 1968 and 1971 findings.

Members are free to attend as many meetings as they wish. Although most

TABLE 13. Comparison of 1968 and 1971 G.S. Board Findings on Frequency of Attending A.A. Meetings

Frequency	1968		1971		Change	
3 or more times a week	49.7%		55.2%		+5.5	
Once a week or more (but less than 3 times)	44.7%		40.4%		−4.3	
		94.4%		95.6%		+1.2
Once a month or more (but less than once a week)	2.1%		1.9%		−0.2	
		96.5%		97.5%		+1.0
3 times a year or more (but less than once a month)	1.2%		0.2%		−1.0	
		97.7%		97.7%		+0
Once a year	0.2%		—		−0.2	
		97.9%		97.7%		−0.2
Less than once a year	0.5%		—		−0.5	
		98.4%		97.7%		−0.7
No answer	1.6%		2.3%		−0.7	
	100.0%		100.0%			

members have a "home" group and tend to attend its meetings with most regularity, they can also go to almost any other group meetings. Groups hold meetings at least weekly, but many meet as often as daily, as Leach et al. (1969) showed.

In some larger urban areas, there are groups listed with the local central A.A. office as, or generally understood to be, only or primarily for women, for men, for doctors, for priests, for policemen, or for homosexuals. Members unsuited for meetings of these groups would not be likely to return again after an accidental appearance, or would, in most cases, be asked to go, instead, to another nearby meeting.

Virtually no change shows between 1968 and 1971 in the percent of respondents (regardless of length of sobriety) who attend an A.A. meeting once a week or more—about 95.0 percent. However, there was a 5.5 percent increase in the number of members reporting they attend meetings three or more times a week.

Among 1968 respondents, it has also now been found that younger members (under 31) tend to go to more meetings than older ones, as Table 14 shows.

Of those 30 years of age and younger, 78.6 percent reported going to three to seven meetings weekly in 1968, as compared to 49.7 percent for the total sample, 51.6 percent for the 31–50 age group, 45.2 percent for those in the 51–65 age bracket, and 43.7 percent for those over 65.

By 1971, the percent of respondents in the 18–31 age bracket reporting attendance at meetings thrice or more weekly had risen from 60.8 percent to 68.0 percent, and the percent of this age group attending daily meetings was up from 6.9 percent to 9.0 percent.

The 31–50 age group in 1971 followed closely the total membership. The percent of those reporting three or more meetings a week was up from 51.6 percent in 1968 to 58.0 percent, while the percent of those attending daily meetings had doubled, from 3.5 percent to 7.0 percent of the 1971 respondents of those ages.

In 1968, newly sober members, as could be expected, reported attending more meetings than old-timers, but the difference was small, as Table 15 shows.

For instance, 94.0 percent of those with less than one year of sobriety in 1968 reported at least one meeting per week, but 92.8 percent of those with 15–20 years of sobriety also reported at least one per week. The main difference between newer members and those with longer sobriety was in the *number* of meetings attended each week. Of those with less than a year of sobriety, 52.6 percent reported from three to seven meetings weekly. Of those sober 1 to 5 years, 53.4 percent reported a similar frequency, but this number

TABLE 14. Frequency of Attending A.A. Meetings in 1968 by Age

	Total		Under 18		18–30		31–50		51–65		Over 65		Age not shown	
	Number	Percent	Number	Percent	Number	Percent	Number	Percent	Number	Percent	Number	Percent	Number	Percent
Total respondents	11,355	100.0	90	100.0	718	100.0	6481	100.0	3407	100.0	449	100.0	210	100.0
Every day	444	3.9	1	1.1	50	6.9	225	3.5	106	3.1	25	5.6	37	17.6
3–6 times a week	5,202	45.8	16	16.7	387	53.9	3108	48.0	1435	42.1	171	38.1	85	40.4
1–2 times a week	5,080	44.7	16	16.7	228	31.8	2858	44.1	1678	49.3	233	51.7	68	32.4
2–3 times a month	183	1.6	—	—	8	1.1	94	1.5	68	2.0	10	2.2	3	1.4
Once a month	59	0.5	1	1.1	3	0.4	26	0.4	25	0.7	3	0.7	3	1.4
4–5 times a year	46	0.4	17	18.9	2	0.3	15	0.2	9	0.3	2	0.4	1	0.5
2–3 times a year	85	0.8	39	43.3	7	1.0	17	0.3	13	0.4	3	0.7	6	2.9
Once a year	21	0.2	1	1.1	—	—	13	0.2	5	0.1	—	—	2	1.0
Less frequently	56	0.5	1	1.1	6	0.8	12	0.1	32	0.9	1	0.2	4	1.9
No answer	179	1.6	—	—	27	3.8	113	1.7	36	1.1	2	0.4	1	0.5

TABLE 15. Frequency of Attending A.A. Meetings in 1968 by Length of Sobriety

	Total		Less than 1 year		1–5 years		5–10 years		10–15 years		15–20 years		Over 20 years		No answer or don't know	
	Number	Percent	Number	Percent	Number	Percent	Number	Percent	Number	Percent	Number	Percent	Number	Percent	Number	Percent
Total respondents	11,355	100.0	4320	100.0	3962	100.0	1513	100.0	740	100.0	417	100.0	229	100.0	174	100.0
Every day	444	3.9	256	5.9	125	3.2	27	1.8	15	2.0	10	2.4	4	1.7	7	4.0
3–6 times a week	5,202	45.8	2012	46.7	1990	50.2	657	43.4	274	37.0	133	31.9	74	32.3	62	35.6
1–2 times a week	5,080	44.7	1787	41.4	1684	42.5	760	50.2	401	54.2	244	58.5	136	59.5	68	39.1
2–3 times a month	183	1.6	49	1.1	55	1.4	32	2.1	20	2.7	17	4.1	4	1.7	6	3.5
Once a month	59	0.5	13	0.3	25	0.6	9	0.6	5	0.7	2	0.5	2	0.9	3	1.7
4–5 times a year	46	0.4	14	0.3	14	0.3	10	0.7	3	0.4	2	0.5	1	0.4	2	1.2
2–3 times a year	85	0.8	43	1.0	23	0.6	9	0.6	5	0.7	2	0.5	—	—	3	1.7
Once a year	21	0.2	9	0.2	4	0.1	3	0.2	—	—	3	0.7	1	0.4	1	0.6
Less frequently	56	0.5	14	0.3	31	0.8	2	0.1	4	0.5	—	—	2	0.9	3	1.7
No answer	179	1.6	123	2.8	11	0.3	4	0.3	13	1.8	4	0.9	3	2.2	19	10.9

of weekly meetings was reported by only 45.2 percent of those sober 5 to 10 years, 39.0 percent of those sober 11 to 15, and about 34.0 percent of those sober 15 or more years at the time of the 1968 survey.

As Table 16 shows, almost no difference was found in 1968 between males and females in frequency of meeting attendance in the United States and Canada.

On the other hand, the independent London study by Edwards et al. (1966), did show a difference between men and women in meeting attendance in 1964. Twenty-one percent of the men, as compared with only 15.0 percent of the women, reported attending meetings "most nights" but reporting meeting attendance frequency as simply "most weeks" were 80.0 percent of the women and only 71.0 percent of the men. "Most months" was the report of 4.0 percent of the men, 2.0 percent of the women; and "occasionally," said 4.0 percent of the men and 3.0 percent of the women. Of the total sample, 20.0 percent reported "most nights"; 73.0 percent, "most weeks"; 3.0 percent "most months"; and 4.0 percent, "occasionally."

Another independent study of A.A. showing frequency of meeting attendance, the Bailey and Leach (1965) report on 1058 New York members surveyed in 1962, also gives evidence of such differences, and others, as Table 17 indicates.

In their total New York sample, Bailey and Leach (1965) found nearly one-third (31.8 percent) reported attending *more* than three meetings a week in 1962, and almost half (48.7 percent) said *at least* three times a week, with only 9.2 percent reporting attendance less than once a week.

TABLE 16. Frequency of Attending A.A. Meetings in 1968 by Gender

	Total		Male		Female		Sex not mentioned	
	Number	Percent	Number	Percent	Number	Percent	Number	Percent
Total	11,355	100.0	8549	100.0	2542	100.0	271	100.0
Every day	444	3.9	325	3.8	75	2.9	44	16.2
3–6 times a week	5,202	45.8	3925	45.9	1194	47.0	89	32.8
1–2 times a week	5,080	44.7	3889	45.5	1132	44.5	61	22.5
2–3 times a month	183	1.6	138	1.6	41	1.6	4	1.4
Once a month	59	0.5	44	0.5	14	0.5	1	0.5
4–5 times a year	46	0.4	10	0.2	8	0.3	20	7.4
2–3 times a year	85	0.8	28	0.3	21	0.8	36	13.4
Once a year	21	0.2	8	0.1	—	—	13	4.8
Less frequently	36	0.3	23	0.3	32	1.9	1	0.5
No answer	179	1.6	159	1.8	25	1.0	1	0.5

TABLE 17. Differences Among New York A.A. Members Reporting Meeting Attendance Three or More Times Weekly in 1962[a]

Level of education		
High school graduation or less	Men ($n = 358$)	Women ($n = 154$)
	57.3%	46.1%
Some college or more	Men ($n = 342$)	Women ($n = 159$)
	49.7%	32.1%
Length of continuous sobriety	Total[b] ($n = 1005$)	
Less than six months ($n = 276$)	56.2%	
Six months, less than 2 years ($n = 248$)	57.3%	
2 years, less than five ($n = 235$)	48.1%	
5 years and over ($n = 246$)	39.0%	
Incidence of slips among those in A.A. 5 or more years[c]		
No relapses ($n = 144$)	42.4%	
Frequent relapses[d] ($n = 84$)	52.4%	

[a] Drawn from Bailey and Leach (1965).
[b] 42 did not report meeting attendance frequency.
[c] Data on relapses were not reported by 34 of the 484 members in A.A. for 5 or more years.
[d] Defined, in respondents' terms, as "many," "dozens," or "more than 20." Meeting frequency was not shown for "occasional relapses."

Lengths of Abstinence of A.A. Members

"A.A. is of course in part a treatment organisation achieving 'cure' or 'arrest,' and it is these therapeutic successes which have given it prestige," Edwards et al. (1966) noted. However, that these "therapeutic successes," presumably abstinence, had occurred in large numbers was simply asserted for many years (e.g., Belden, 1962; Block, 1962a,b; Clinebell, 1968; Fox and Lyon, 1955; Hoffer and Osmond, 1968; McCarthy, 1946; Pfeffer, 1958; Pittman and Gordon, 1958; Stone, 1962; Strachan, 1968; Streeseman, 1962; Trice, 1958a; G. C. Wilson, 1968), often with such claims based only on A.A.'s own pronouncements (e.g., Chafetz and Demone, 1962; Silkworth, 1939a,b; Tiebout, 1944).

The abstinence of many A.A. samples of varying sizes has been cited as a criterion of "A.A. success" in numerous studies on aspects of alcoholism other than abstinence per se (such as those of Bateman and Petersen, 1971; Beaubrun, 1967; Bräutigam, 1967; Brown, 1963; Byrne, 1968; Connor, 1962; Corder et al., 1972; Davis, 1970; Emrick, 1970; Haberman, 1966; Jackson and Kogan, 1963; Kamner and Dupong, 1969; Karp et al., 1965; Robson et al., 1965; Rohan, 1970; Soden, 1960; Stein, 1966; Swegan, 1957; Thorpe and Perret, 1959; Timmens, 1965; Trice, 1957a; Trice et al., 1969; Trice and Roman, 1970a,b; White, 1965).

Smith (1941), Bohince and Orensteen (1953), Bailey and Leach (1965), and Edwards *et al.* (1966, 1967) offered independent findings on length of abstinence on relatively small A.A. samples, as will be shown. But the results of the 1968 and 1971 surveys by the G.S. Board are the most extensive evidence so far that A.A. members by the thousands do remain abstinent for years. A comparison of the 1968 and 1971 Board findings on "time since last drink" is shown in Table 18.

In 1968, the Board found 6861 A.A. members (60.3 percent) sober one year or more; in 1971, 3636 (60.6 percent) sober one or more years.

Reporting on the first experiment by a state mental hospital of "trying" Alcoholics Anonymous as a treatment of alcoholic patients, Smith (1941) told of 111 alcoholics exposed to A.A. in Rockland State Hospital, Orangeburg, New York, beginning in August 1939. As of March 1941, of 56 discharged patients, 6 had been abstinent 12 to 18 months, 16 for 6 to 12 months, 6 for 3 to 6 months, and 14 for 1 to 3 months. Forty-two (75.0 percent) of the discharged had had no relapses, 11 had had one relapse, and 3 had had 2 relapses.

Bohince and Orensteen (1953), reporting on 231 Minneapolis A.A. members studied in 1950, rated 149 (63.0 percent) as "successfully abstinent."

In 1962, Bailey and Leach (1965) found among 1058 New York A.A.s 57.6 percent reporting one year or more of sobriety, and 45.7 percent reporting at least two or more years since their last drink. In 1964, Edwards *et al.* (1966, 1967) found 46.2 percent of 306 London A.A. members sober two or more years.

The Board found sober less than one year 38.1 percent of its 1968 respondents; three years later, 37.8 percent. Bailey and Leach (1965) found 37.1 percent of their New York sample with abstinence terms under one year, 49.0 percent sober one year or less, and Edwards *et al.* (1966, 1967) found 52.5 percent of their London respondents sober one year or less.

TABLE 18. Length of Abstinence of A.A. Respondents in 1968 and 1971 Board Surveys

	1968		1971		% Change
	Number	Percent	Number	Percent	
Less than one year	4320	38.1	2268	37.8	−0.3
1 to 5 years	3962	34.9	2016	33.6	−1.3
6 to 10 years	1513	13.3	834	13.9	+0.6
11 to 15 years	740	6.5	408	6.8	+0.3
16 to 20 years	417	3.6	198	3.3	−0.3
Over 20 years	229	2.0	180	3.0	+1.0
Don't know/no answer	174	1.6	96	1.6	±0
Totals	11,355	100.0	6000	100.0	

As could be expected from the respective ages of A.A. in England and the United States, the London sample (Edwards *et al.*, 1966, 1967) showed the smallest percent of longer-term sobrieties—only 20.8 percent sober six or more years. The Board found 25.4 percent in 1968, 27.0 percent three years later, reporting six or more years' abstinence. In New York, Bailey and Leach (1965) found 23.4 percent sober *five* or more years. It should be recalled, moreover, that the Board sampling, as well as those in London and New York, was probably somewhat biased against those sober longest who tend to go to fewer A.A. group meetings (see Table 15).

None of the foregoing reported investigations meets all the requirements for rigorous, scientific follow-up or evaluative studies of alcoholism treatment propounded by Bowman and Jellinek (1941), Bruun (1963), and Chafetz *et al.* (1970). Very few studies of any alcoholism treatment agency or modality do meet many such criteria, as Belasco (1971), Boggs (1967), Gillespie (1967), Hill and Blane (1967), and Voegtlin and Lemere (1942) have pointed out.

With such qualifications borne in mind, it is still notable that the G.S. Board found such small percents of change in lengths of abstinence between 1968 and 1971, and that the Board findings are so similar to those reported by two sets of independent investigators who had studied different A.A. populations in New York and London. Perhaps the major Board finding is that covering only 4.4 percent of the United States and Canadian groups in 1968, 2.3 percent in 1971, the Board found 10,497 respondents reporting abstinence for one or more years, 4519 for five or more years. Until better-quality studies are published, the assertion that A.A. members by the thousands do remain abstinent for years seems reasonable and modest.

Other Life Changes Associated with A.A. Membership

As previously indicated, not all A.A. members sustain extended abstinence, although many do and others try. So, for many A.A. members, as for many professionals treating alcoholics, it is sometimes difficult to understand why anyone should discourage or belittle abstinence or its value for people whose addiction to ethanol, or record of repeated difficulties connected with drinking, has brought them to seek or need treatment for a drinking problem.

Yet abstinence as a goal, or criterion of success, in alcoholism treatment has been questioned. For example, Chafetz and Demone (1962) objected to the A.A. emphasis on abstinence, noting that "employment, marital status and interaction with others" can also be indices to successful recovery from alcoholism. (They are also possible for many active, drinking alcoholics.) McElroy (McElroy and O'Brien, 1963) predicted doom for A.A. because its emphasis on abstinence "forsakes the virtue of temperance." Pattison (1966) sharply questioned the concept of abstinence, quoting several studies (Davies,

1962; Fox and Smith, 1959; Moore and Ramseur, 1960a; Nørvig and Nielsen, 1956; Pfeffer and Berger, 1957; Selzer and Holloway, 1957; Lemere, 1953; Shea, 1954) that report "social drinking" among purported alcoholics. In addition, Rossi *et al.* (1963), Wilby and Jones (1962), Pfeffer and Berger (1957), Flaherty *et al.* (1955), Wellman (1955), Wallerstein (1956), Gerard *et al.* (1962), Shea (1954), and Foster *et al.* (1972) have questioned the "cost" of abstinence in such terms as deteriorated health and personality disintegration, involving constricted social and psychological functioning which Pittman and Snyder (1962) suggest are "dysfunctionalities [of A.A.] from a therapeutic point of view."

Considering the size of A.A., the autonomy of each A.A. group, and the vigor with which each individual A.A. member is encouraged to set his own therapeutic goals (Wilson, 1939, 1952), it is no doubt possible to find A.A. members, and A.A. dropouts, who would confirm almost all the observations made in the studies cited just above.

However, abstinence is not "the sole measure of success" (Chafetz and Demone, 1962) in A.A. Nor is A.A. concerned exclusively with "the symptom rather than the underlying problems which resulted in alcoholism" (Chafetz and Demone, 1962), although A.A. has been commended for focusing attention on the "symptom," i.e., drinking, by Chafetz and Demone (1962) as well as by Tiebout (1951a, 1956, 1958a,b, 1963b).

Among the Twelve Traditions of A.A., abstinence is referred to in only number 3, which says that the *desire* to stop drinking (not its actual cessation) is the only requirement for A.A. membership. In this context, abstinence as an ideal, or goal, represents a unifying principle on which those whose experience has repeatedly demonstrated the desirability of their own not-drinking can agree.

The A.A. Twelve Steps "suggested as a program of recovery," as explained by their chief articulator (Wilson, 1939, 1952, 1967) and frequently elucidated by other members in almost every issue of the *A.A. Grapevine*, make it clear that abstinence is only a starting point for the full recovery which is the A.A. ideal. Leach (1968) has offered evidence that drinking, intoxication, and abstinence are less popular topics at A.A. discussion meetings than other life problems. "Not drinking" is nowhere specifically mentioned in the Twelve Steps, where emphasis is placed on other measures recommended for recovery. For example, Wilson (1952), writing about Step 3, noted that an A.A. newcomer's "life is still unmanageable even though he is sober, that after all, only a bare start on A.A.'s program has been made. More sobriety brought about by the admission of alcoholism and by attendance at a few meetings is very good indeed, but it is bound to be a far cry from permanent sobriety and a contented, useful life" (p. 41). About Step 4, he wrote, "[U]nless he [the newcomer] is now willing to work hard at the elimination of the worst of these [character] defects, both sobriety and peace of mind will still elude him . . . all the faulty foundation of his life will have to be torn out and built anew on bedrock" (p. 51). Discussing Step 8, Wilson (1952) wrote, "Defective relations with other human beings have nearly always been the immediate cause of our woes, including our alcoholism . . ." (p. 82). In the foreword to the original book, *Alcoholics Anonymous* (Wilson, 1939), it was pointed out that A.A. membership is only an avocation, that it is necessary for members to carry on their occupations in addition to participating in A.A. activities. The member whose A.A. behavior is compulsive (Chafetz, 1959), obsessive, and fetishistic (Shea, 1954) is nowhere described in the A.A. literature as the ideal A.A. example of recovery. Even so, it may of course be the only level of recovery possible for some alcoholics who go to A.A.

A.A. membership has been associated by some investigators with little or no change in some alcoholics. For example, Machover *et al.* (1962) reported that 23 A.A. members sober two or more years had no greater investment in social and religious* values than 23 relapsing alcoholics. On the other hand, Eckhardt's (1967) value analyses of 12 alcoholics' life stories found that membership in A.A. had produced less egocentric, more ethical values, including more concern for others and for long-run consequences. Haertzen *et al.* (1968) reported that membership in A.A. was "nonsignificant" in a sample of 608 alcoholics still hospitalized. Zwerling (1959) found no significant differences between 23 A.A. males sober two years or more and 23 drinking alcoholics—both groups showing schizoid adaptation (isolation), dependence, depression, hostility, and sexual immaturity. Mayer and Myerson (1971) concluded that attendance at A.A. meetings was not related to stability, status, or reduction of drinking in 222 alcoholism clinic patients. Belasco (1971), too, reporting on 378 alcoholics discharged from a hospital, found that high ratings on his behavioral/social adjustment criteria of success were not associated with frequent A.A. attendance. Williams (1970) investigated the "loss of control" concept in alcoholism by giving alcohol to 20 recovered alcoholics in A.A. and 20 active alcoholics, and telling only half of each group they had ingested ethanol. All of those so informed said they wanted more, but the uninformed expressed no such desire. He found no difference between A.A.-oriented alcoholics and the others regarding loss of control. Robson *et al.* (1965), comparing rehabilitation rates among alcoholics with long A.A. association and those who received professional clinic treatment in addition, found 71.0 percent of the former, 70.0 percent of the latter, showing evidence of rehabilitation. Oakley and Holden (1971) reported no significant difference in A.A. attendance between 58 followed-up patients sober six months or more and the total group

---

* A.A. literature (e.g., Wilson, 1939, 1952, 1957, 1967) insists that A.A. is "spiritual" in nature, not "religious."

of 115. Karp *et al.* (1965) reported that remission did not significantly affect field dependence scores among 80 alcoholic patients, although "A.A. patients," both those sober at least 15 months and relapsers, showed significantly less field dependence than clinic patients.

Other investigators have associated A.A. membership with supposedly undesirable traits or changes in some alcoholics. For instance, Strayer (1961) seemed to suggest that 29 patients in prolonged group therapy (up to eight years) did better after somewhat disassociating themselves from A.A. McGinnis (1963) reported that 21 patients exposed for six weeks both to professionally directed group therapy and to A.A. grew more in ego strength than 21 alcoholics exposed only to A.A. On the other hand, Seiden (1960), using Bender-Gestalt performances as an index, found 25 Denver A.A. members possessing greater ego strength and self-reliance than 25 still-hospitalized alcoholics. Of course, it is possible that the "ego reduction" mechanisms operative in A.A., according to Tiebout (1944, 1946, 1949, 1954, 1955b, 1961, 1963a, 1965) could reduce scores on the ego strength measurements used by McGinnis.

Mathias (1955), after administering MMPIs, Rorschachs, and two new projective tests to 100 alcoholics, including 30 members of A.A., reported that the principal change which the chronic alcoholic experiences as a member of A.A. occurs in the altered direction of his aggressive impulses. The A.A. members tended to direct aggression outward in the form of hypomanic and paranoid trends, while the non-A.A. alcoholics tended to direct such impulses inward, making attempts to counteract them by drinking and self-debasement. Mindlin (1964) reported on a sample of 155 nonpsychotic alcoholic patients in California hospitals, of whom 55 had had no help, 38 had attended 10 or more A.A. meetings but had had no professionally guided psychotherapy, and 40 had had five or more psychotherapy sessions, with "some" A.A. He found a high degree of information about alcoholism in 58.0 percent of the psychotherapy group, only 31.0 percent of the A.A. group. The A.A.s also had lower self-esteem than the other patients. Palola *et al.* (1961) reported that although alcoholics tend to stop denying alcoholism after joining A.A., they tend to begin denying unhappiness. And Ludwig (1972), studying "reasons for drinking" among 176 male alcoholic relapsers, reported that of those abstinent, 7.0 percent gave A.A. as the reason for their most recent resumption of drinking.

Whether or not some A.A. members do achieve supposedly desirable changes in life other than abstinence was explored in the G. S. Board 1971 survey. Respondents were asked, "If having joined A.A. has affected your life in any way, will you use the space below to describe any changes that have taken place?" Space on the questionnaire was provided for reporting changes in family life, business or occupation, leisure activities, and community interests. Results are shown in Table 19.

TABLE 19. Life Changes Since Joining A.A. Reported in 1971 G.S. Board Survey[a]

	Total (%)	Men (%)	Women (%)
Changes in family life			
Family-oriented responses, such as "happy, more harmonious home life," "family reunited," "better communications," "regained acceptance," including respect and understanding, etc.	77.1	76.1	80.6
References exclusively to self	17.3	15.5	22.2
Negative responses	2.9	2.7	3.2
No family	2.3	2.4	2.0
Changes in business or occupational life			
Improved	54.6	58.2	45.9
No change	2.1	2.2	1.7
Negative responses	0.8	0.8	0.8
No business or occupation	3.8	3.8	3.4
Changes in leisure activities			
More fulfilling, active social life	32.4	32.4	31.5
Most leisure spent in A.A.	11.9	11.5	12.9
Very little or no leisure activity, too busy	3.4	3.4	2.9
No change	1.5	1.6	1.2
Changes in community interest			
More community involvement, including broader interests and more concern about helping other people	22.6	23.0	21.6
Less time now for community interests	1.5	1.3	2.1
No change	4.5	5.0	3.1
No community interests	8.3	8.6	7.2

[a] Responses in all columns add to more or less than 100% because of multiple responses in some cases, no responses in others. (Number of respondents = 6000.)

The sample reported on in Table 19—6000 A.A. members—is larger, and drawn from a much wider geographic distribution (see Table 4) than the A.A. samples reported on by any of the other authors cited in this section, and none of the A.A.s participating in the G.S. Board surveys were in hospital, but several of the other samples reported on above were currently hospitalized. It should also be noted again that, in sampling only A.A. members in attendance at meetings of autonomous A.A. groups, the G.S. Board findings are probably less representative of the longer-sober members than of those with shorter terms of abstinence, as Table 15 shows, and there is evidence that members who are frequent relapsers may be somewhat more likely to be included in this sample than others, as Bailey and Leach (1965), Edwards et al. (1966, 1967), and Miller et al. (1970) suggested.

It is also inaccurate to attribute all reported positive improvements in the lives of A.A. members solely to their A.A. membership, since it is known that many A.A. members, in addition to A.A. help, also receive continuing help from professional people and agencies and other non-A.A. resources.

Nevertheless, the findings reported in Table 19 are interesting in that they correspond with findings by so many other investigators. Similarities of the findings on job and marital status by the G.S. Board, Bailey and Leach (1965), Edwards et al. (1966, 1967), and Kiviranta (1969) have already been shown. The Table 19 data on changes in family life adjustment tend to support the impression that A.A. membership is associated with improved marital status. Ancillary evidence that A.A. membership may indeed improve family life came from three studies of alcoholics' wives. Bailey et al. (1962) found 23 wives (among 69) with abstinent husbands, 17 of whom had been helped by A.A.; but among the separated families, only 4 husbands had become abstinent in A.A. Jackson and Kogan's (1963) study of wives of alcoholics included wives of 29 A.A.s sober more than 12 consecutive months (median sobriety: 4.5 years), and these families differed from others in having used help more effectively and having resolved problems more rapidly than the non-A.A. families. An MMPI study of 50 wives of A.A. members and 30 wives of nonalcoholic controls by Corder et al. (1964) found the former not markedly neurotic or disturbed.

As to occupational adjustment, Belasco (1971), after a two-year follow-up study, reported that among 26 alcoholics who increased frequency of posthospital A.A. attendance, 10 had more, 16 less, occupational stability. He further reported that 33 of 35 persons with good occupational adjustment also had good ecological (type of residence) stability, and 24 had high sobriety maintenance; 33 of 53 with good ecological stability also had good occupational adjustment and 24 had high sobriety maintenance. Examining job performance by 340 employees "treated" primarily by A.A. and disulfiram (Antabuse), Davis (1970) found 220 "recovered," and 75 had had relapses but remained employable. Kish and Hermann (1971) followed up 173 alcoholic patients discharged from a Veterans Administration hospital and found that patients who were married, employed, and active A.A. members showed more improvement than those who were not. They found group psychotherapy, age, education, personality, vocational interest and aptitudes, and commitment status (vs. voluntary patient status) not related to improvement. However, Pokorny et al. (1968), after following up 13 months after discharge on 88 alcoholic patients exposed to A.A. at another Veterans Administration hospital, found 51.0 percent distinctly improved, and those with highest degrees of abstinence had also improved in vocational, mental, emotional, social, and legal adjustment.

Belasco (1971), in addition to the findings noted above, also reported that among "frequent or moderate" attenders at A.A. meetings, 6 of 19 had good occupational adjustment, 8 of 20 good ecological stability, and 10 of 20 good sobriety maintenance. Pittman and Tate (1969), after following up on patients of a mental health center, reported that those whose treatment and exposure to A.A. were longer, and who got aftercare, improved in many areas in addition to abstinence. Palola et al. (1962) found A.A. members sober at least two years much more willing to admit to pre-A.A. suicidal thoughts and behavior than unremitted alcoholics.

Other reported changes in A.A. members seem to link improvements with length of remission. For example, Carothers (1971), reporting on 60 alcoholics tested with two psychometric instruments, found that the 20 active A.A. members sober two or more years had a "locus of control" more internal than that of "intemperate" (sober less than one year) alcoholics, and the former were also found more conscientious, imaginative, forthright, placid, conservative, and self-sufficient than the intemperates. Carroll and Fuller (1969) concluded that length of sobriety and voluntary participation in A.A. interact to reduce significantly the self–ideal self discrepancy. Connor (1962) reported the self-acceptance of alcoholics as influenced by length of sobriety in A.A. nearly doubles in such areas as use of secondary relationship qualities and neurotic word choice declines. The self-acceptance index rose from zero for those sober less than three months up to .421 for those sober 7 or more years, which was a score indistinguishable from that of nonalcoholic controls. White (1965), having compared 25 A.A. members sober 3 months or less with 25 members with 3 or more years of verifiable abstinence, pointed to the meaningfulness of length of sobriety in evaluating psychological characteristics of such subjects. Machover and Puzzo (1959a) compared 23 male A.A.s sober at least two years with 23 equally "exposed" to A.A. (average A.A. contact, 5 years; range, 1–14 years), and reported the clinical impression of "higher level of ego tonus" correlated with remission.

The Board findings shown in Table 19 can all be described as relating to some kind of interpersonal relations, and as such they tend in the same direction as reports by other investigators. However, where the Board found only 11.9 percent of its sample volunteering they spent most of their leisure time "in A.A.," a factor found significant by Lahey (1950) in maintaining A.A. abstinence, the New York sample reported on by Bailey and Leach (1965), asked specifically to estimate the percent of "social life" spent in A.A., found that it increased from 32.2 percent ($n = 276$) of those sober less than six months to 40.4 percent ($n = 235$) for those sober two years but less than five, then decreased to 39.4 percent ($n = 246$) for those sober five years and over. Mindlin (1964), in addition to findings mentioned previously, found that 68.0 percent of his psychotherapy alcoholic sample felt isolated and ill at ease socially, whereas only 39.0 percent of the A.A. group reported such feelings. Maxwell (1951) found 150 alcoholics reporting feelings of less hostility and intolerance, and more ability to interact satisfyingly with others, all attributed to A.A. membership. Latimer (1953), comparing six alcoholics sent to A.A. as sole treatment with six alcoholics not referred to A.A., found the A.A.s displayed more self-understanding, stability, and ability to handle social situations.

Two studies of A.A. members in prisons have reported on findings other than abstinence. Dellinger (1953) said that the 200 A.A. inmates were "model prisoners," the first to be reestablished because of good behavior after a prison riot. Roth *et al.* (1971) compared 50 inmate members of A.A. in a federal prison with 50 alcoholic inmates not in A.A. and 50 nonalcoholic inmates. They reported the combined samples of alcoholics were more intelligent and older than the nonalcoholics, and fewer had records of homosexual acts (6.0 percent vs. 22.0 percent). But the inmates who belonged to the prison A.A. group showed better prison adjustment overall, better disciplinary indices, better work reports, and better records of participation in the prison's rehabilitation programs than the non-A.A. alcoholics and the nonalcoholic inmates.

Evidence that A.A. can also produce beneficial changes in some alcoholics' perceptions of time has been presented by Emrick (1970), whose experiment demonstrated the value in recovery from alcoholism of the A.A. 24-hour concept, which is similar to the emphasis on the "here and now" in reality therapy (Glasser, 1965), and to the "endurance of temporary discomfort" practiced in Recovery, Inc. (Low, 1950). As far as is known, Recovery, Inc. developed entirely apart from and independently of A.A., encouraged only informally by some early A.A. supporters. There are some similarities in the two and in their separate evolutions there may have occurred some cross-fertilization of ideas,* since some members of each have attended meetings of the other.

As the Rev. Edward Dowling, S. J., pointed out to Bill W. (Wilson, 1957), the Twelve Steps of A.A. resemble the spiritual exercises of St. Ignatius of Loyola, founder of the Jesuit order, so presumably practice of the Twelve Steps could produce results which Jesuits would consider spiritual growth. Cohen (1962, 1964), whose experiment demonstrated mental health improvement in A.A. members, recommended the use of A.A. principles in treatment of a broad range of disorders besides alcoholism, as did Beeley (1959), Hobbs (1960), Frank (1961), Mowrer (1961, 1964, 1967), and Zusman (1969).

That benefits other than abstinence can result from participation in such A.A.-type activities as the group meetings, practice of the Twelve Steps, and shared social activities can also be inferred from the continued flourishing of the

---

* Unpublished private correspondence between January 14, 1954, and July 29, 1958, between Bill W. and C.D., a Louisville, Kentucky, member of Recovery, Inc.

various organizations previously discussed which apply A.A. principles and practices to problems other than alcoholism. Some of them have simply "borrowed" the Twelve Steps (which are protected by copyright), adapting them by changing only one or two words. Some of these adaptations are discussed in publications by Becker and Israel (1961), Casriel (1963), Casson (1969), Cohen (1965), Endore (1968), Gamblers Anonymous (1962), Gartner and Riessman (1968), Glaser (1971), Hagie (1971), Hurvitz (1970), Patrick (1963), Peer Teacher-Leaders Aid in Self-Help Groups (1970), Professional Committee of the Schizophrenia Foundation of New Jersey (1970), Quinn (1967), Stefan (1965), and Yablonsky (1965).

In conclusion, it can be said that many people approach A.A. wanting only to stop drinking, while others arrive wanting almost everything but that. One, or the other—or both—can come about.

THE LITERATURE ON A.A.

At least six kinds of literature on A.A. can be found: (1) books, pamphlets, and the magazine and bulletins published by the movement itself; (2) journal articles, chapters of books, and books of professional, scientific, or technical character; (3) authoritative books and articles on alcoholism for the lay reader; (4) news or feature items in the popular mass media, as well as semiprofessional periodicals; (5) pamphlets and magazine articles published by special interest groups such as religious organizations; and (6) belles lettres such as novels, plays, and biographies.

With few exceptions, primarily those of historic or other importance, only selections from the first three categories are considered here. But this is not meant to disparage the quality or value of all the others. Astor (1959), Berryman (1973), Boyington (1958), Jackson (1944), Lowry (1947), Mainc (1947), Newlove (1974), Only (1974), Randall (1957), Scott (1947), Seabrook (1935), Taintor (1945), Thomsen (1975), Westheimer and Miller (1963), and Willis (1963), for example, all give unique or universal insights into facets of the alcoholisms which are not provided by statistical tables, just as do such O'Neill plays as *The Iceman Cometh* (1946) and *Long Day's Journey into Night* (1956), and Shakespeare's Cassio in *Othello* and the porter in *Macbeth*.

Material Published by A.A.

In addition to the three editions of the basic text of the movement (Wilson, 1939, 1955a, 1976), five other volumes have been published by A.A.: *Twelve Steps and Twelve Traditions* (Wilson, 1952); *A.A. Comes of Age* (Wilson,

1957), a brief history of the Fellowship through 1955; *The A.A. Way of Life, as Bill Sees It* (Wilson, 1967), brief readings from Bill W.'s writings for the Grapevine, from his voluminous private correspondence, and from his other works; *AA/30* (1965), published for the international convention in Toronto; *Came to Believe* (1973), a collection of pieces by 75 A.A. members telling their personal versions of what the phrases "Power greater than ourselves" and "spiritual awakening" mean; and *Living Sober,* discussing 31 techniques members use for maintaining abstinence. In English, material is available in Braille and on tape cassettes.

The Big Book, or major portions of it plus parts of other A.A. publications are also available in Afrikaans, Finnish, French, German, Norwegian, Portuguese, Spanish, and Swedish. Some A.A. literature is also published in Chinese, Dutch, Flemish, Hungarian, Isixhosa, Italian, Japanese, Sotho, Swahili, Vietnamese, Zulu, and several Western Indian tongues. G.S.O. also publishes 17 pamphlets on recovery, 5 *Directories,* 11 pamphlets on "unity and service," 6 pamphlets for professional and business people, 15 items for use in public information activity about A.A. (plus taped radio and television announcements), 4 bimonthly bulletins, and 12 miscellaneous items such as wallet cards and plaques for display. By arrangement with A.A. World Services, Inc., in New York, A.A. material is distributed overseas by 20 literature distribution centers. All material is continually under review by Conference and Board committees which recommend additional publications as needs arise, or oversee revisions and updating of older publications. All of the above are formally approved by the G.S. Conference, and bear a legend noting that fact. G.S.O. also publishes limited quantities of service material for A.A. members, such as "Guidelines" on specific A.A functions in cooperation with various professional alcoholism agencies.

In addition, local A.A. offices, groups, and individual members publish many bulletins, newsletters, pamphlets and other materials which are often extremely popular within certain areas or among certain segments of the A.A. population, but they are not Conference-approved material.

The *A.A. Grapevine* and its publications have been discussed earlier.

In addition to his writings published by A.A., Bill W. authored some other published material. A paper he read at the 1944 meeting of the Section on Neurology and Psychiatry of the Medical Society of the State of New York was carried in its journal (Wilson, 1944). A short article (Wilson, 1945a) about future prospects for A.A. was published in the *Quarterly Journal of Studies on Alcohol,* and his 1944 lecture at the (then) Yale (now Rutgers) Summer School of Alcohol Studies was carried in the compendium of its proceedings (Wilson, 1945b). "Is A.A. for Alcoholics Only?" (Wilson, 1947) appeared in *Guideposts.* A paper (Wilson, 1949) given at the 105th annual meeting of the American Psychiatric Association appeared in its journal, and another (Wilson, 1950) appeared the following year in a state medical journal. In 1963, he published "A.A.'s Debt to Its Friends" (Wilson, 1963), and in 1968, two books on alcoholism carried pieces by Bill (Wilson, 1968b,c).

Bibliographies

No truly exhaustive bibliography on A.A. has been published, and the list of references following this chapter is not such, but we believe it is representative of the main kinds of scholarly and other material published by and about the Fellowship and related phenomena.

In the *International Bibliography of Studies on Alcohol* (I.B.S.A.), Keller et al. (1968) listed 69 entries relating to Alcoholics Anonymous which had been published as of 1950, beginning with Wilson (1939). As of December 1972, the Rutgers Center of Alcohol Studies, which maintains the invaluable Classified Abstract Archive of the Alcohol Literature, could furnish a bibliography of 98 "selected references" on A.A. It includes 12 articles from local, state, or provincial journals; 4 articles in languages other than English; studies presented for obtaining graduate degrees, 5 items published in the United Kindgom, 4 in other European nations, 3 in Australia, 2 in Canada, and 1 in Chile; and 2 A.A.-published books, but not the basic text of the Fellowship (Wilson, 1939).

The National Clearing House for Alcohol Information (N.C.A.L.I.), opened in 1972 by the National Institute on Alcohol Abuse and Alcoholism (N.I.A.A.A.), early in 1973 furnished the present authors a printout of abstracts produced by a computer search on A.A. which contains 227 items. Thirty-four relate to alcoholism with no mention of A.A., and 17 contain the initials *AA* but are not about either alcoholism or Alcoholics Anonymous. Fifty-seven refer to A.A. primarily as a therapeutic resource for physicians, clinics, or hospitals; 29 are abstracts of items published outside the United States and Canada; 18 are about organizations other than A.A. but in some way analogous to it; 10 refer to the historical importance of A.A.; 10 mention A.A. as a resource for police, prison, or court programs on alcoholism; 8 refer to attendance at A.A. meetings as a gauge of recovery from alcoholism; 7 are about A.A. as a therapeutic resource for industry; 4 are critical of A.A.; 3 mention A.A. as a resource for clergymen or religious organizations; 3 are A.A.-published items, but the basic A.A. textbook (Wilson, 1939) is not included; and 20 make other references to A.A. Seven describe the use of A.A. members as subjects in studies.

In searching the literature on A.A., the present authors found a total of 9941 members of A.A. have been used as subjects of reported scientific or professional studies, experiments, or surveys on alcoholism. (This does not include the 17,355 who participated in the G.S. Board surveys previously discussed.) Although A.A. itself does not engage in research on alcoholism (see Traditions 5, 6, and 8), individual A.A. members are of course free to cooperate in such projects if they choose, and thousands do.

Leach *et al.* (1969) offered 53 references, of which 16 are published by A.A. itself, and six are general reference works without specific mentions of A.A., and Leach (1973) listed 114 bibliographic items, of which 12 are A.A. publications and 7 are general reference works, leaving only 95 articles on or in reference to A.A. not published by the Fellowship itself.

Phillips (1973) has produced (and privately published*) *Alcoholics Anonymous: An Annotated Bibliography, 1935-1972.* Compiled in partial fulfillment of requirements for a Master of Library Science degree, including only material published in English, it is selective but includes 26 books on history and descriptions of A.A., including 8 published by A.A. itself, and 52 magazine and journal articles on those subjects; 7 periodical and journal articles on A.A. programs in penal institutions; 16 on A.A. programs in hospitals; 16 unpublished theses and dissertations which are psychological and sociological studies of A.A. and its members; and 36 periodical articles of similar nature. Twenty-one books and articles are on members of the family of the alcoholic, including materials on the Al-Anon Family Groups and Alateen, which, though similar to A.A., prefer to be considered officially unrelated to A.A. An appendix simply lists 65 items published by A.A. and 52 by Al-Anon and Alateen. Thus the strictly A.A.-related annotated part of this bibliography consists of 153 entries. Phillips clearly described in her preface the purpose and scope of this work, and it is a unique contribution to study of Alcoholics Anonymous.

Significant Early Publications in Mass Media

In the autumn of 1939, when there were about a score of sober A.A.s in Cleveland, Ohio, the *Cleveland Plain Dealer* ran a series of newspaper feature articles and editorials endorsing A.A. The result was such a deluge of requests for help that members sober only a week were "sponsoring" newcomers. A year later, Cleveland had nearly 30 A.A. groups and several hundred members. It was the first indication that massive A.A. growth was possible (Wilson, 1957).

Liberty magazine that year also carried a story about A.A. (Markey, 1939), which brought about 800 inquiries (Wilson, 1957) for help to the New York office of the Alcoholic Foundation.

---

* It may be ordered from Julianne Phillips, 406 North Main Street, London, Ohio 43140. 63 pp., $2.50.

John D. Rockefeller, Jr., gave a dinner February 8, 1940, at the Union Club in New York "in the interest of Alcoholics Anonymous." Of the 400 tycoons and philanthropists invited, 75 appeared. Because his father had become ill, Nelson A. Rockefeller presided. Among the speakers were Bill W., who told his own story; Foster Kennedy, M.D., of Bellevue Hospital and Cornell Medical University; and the Rev. Harry Emerson Fosdick, D.D., of New York's Riverside Church, both of whom endorsed the movement.* Eventually news of the dinner hit the wire news service (Digest of Proceedings, 1940; Alcoholics Anonymous, 1940a,b) and the size of the Fellowship doubled within the year (Wilson, 1957).

However, it was not until the *Saturday Evening Post* ran an article on A.A. (Alexander, 1941) that the movement really became nationally known. At the end of 1940, membership had been estimated at 2000; by the end of 1941, it was 8000 (Wilson, 1957).

One other story (Alcoholics Anonymous Spurns a Trap, 1950) in the nonprofessional press contributed significantly to A.A.'s public image. It told that the Fellowship had declined a bequest of $10,000 from a nonalcoholic woman, and A.A. announced a policy of never accepting any gifts of money whatsoever from anyone except A.A. members (Wilson, 1952, 1957). It has adhered to this position. It will accept a single bequest of not more than $300 from an A.A. member, and that is also the maximum size contribution G.S.O. will accept from a member in one year.

Professional, Scientific, and Technical Publications

Most publications of this sort already cited herein or described elsewhere (e.g., Leach, 1973) will not be discussed again here, but some such literature on A.A. deserves attention.

In 1937, before A.A. had its name, Silkworth reported Bill W.'s recovery and said, "[H]e . . . help[s] others in his former condition and has gathered about him a group of over fifty men, all free from their former alcoholism through the application of this method of treatment and 'moral psychology'" (Silkworth, 1937b). This description of what came to be known as the A.A. experience was the first mention of it in a medical publication. In an article two years later in the *Journal-Lancet* of Minneapolis, the name Alcoholics Anony-

---

* Despite the high hopes of Bill W. and other early members, John D. Rockefeller, Jr., declined to subsidize A.A. As his son Nelson said, "[I]t is our belief that Alcoholics Anonymous should be self-supporting so far as money is concerned" (Wilson, 1957), and so it has been. The elder Rockefeller did, however, give to the treasury of Riverside Church a total of $1,000 which was doled out in $30 weekly amounts to the destitute Bill and Lois and Dr. Bob and his wife in 1940–1941 when neither family had any other income. The Rockefeller "loan" was eventually paid back in full from income from the Big Book, and Bill's only A.A. income throughout his life was the royalties for his writings (Wilson, 1957).

mous made its debut in a medical publication, and Silkworth (1939a) reported there were 150 members. Two years following that, Silkworth (1941) reported 3500 members and quoted endorsements of the A.A. approach from six other reputable physicians who had seen alcoholic patients recover. In effect, all these were simply physicians' clinical reports on their own patients, not reports of controlled experiments. No one reported follow-ups on these cases.

A brief anonymous description of A.A. appeared in the *Illinois Medical Journal* in 1940 (Correspondence. Alcoholics Anonymous, 1940). It invited inquiries. The next year the *New York State Journal of Medicine* termed the A.A. book "empirical but useful" (Book Review. Alcoholics Anonymous, 1941).

But first mentions in national, more prestigious medical publications were different. In 1939 the *Journal of the American Medical Association* greeted the book with this anonymous review:

> The seriousness of the psychiatric and social problem represented by addiction to alcohol is generally underestimated by those not intimately familiar with the tragedies in the families of victims or the resistance addicts offer to any effective treatment. Many psychiatrists regard addiction to alcohol as having a more pessimistic prognosis than schizophrenia. For many years the public was beguiled into believing that short courses of enforced abstinence and catharsis in "institutes" and "rest homes" would do the trick, and now that the failure of such temporizing has become common knowledge, a considerable number of other forms of quack treatment have sprung up. The book under review is a curious combination of organizing propaganda and religious exhortation. It is in no sense a scientific book, although it is introduced by a letter from a physician who claims to know some of the anonymous contributors who have been "cured" of addiction to alcohol and have joined together in an organization which would save other addicts by a kind of religious conversion. The book contains instructions as to how to intrigue the alcoholic addict into the acceptance of divine guidance in place of alcohol in terms strongly reminiscent of Dale Carnegie and the adherents of the Buchman ("Oxford") movement. The one valid thing in the book is the recognition of the seriousness of addiction to alcohol. Other than this, the book has no scientific merit or interest (Book Notices. Alcoholics Anonymous, 1939).

The following year, an anonymous reviewer for the *Journal of Nervous and Mental Disease* wrote:

> As a youth we attended many "experience" meetings more as an onlooker than as a participant. We never could work ourselves up into a lather and burst forth in soapy bubbly phrases about our intimate states of feeling. That was our own business rather than something to brag about to the neighbors. Neither then nor now do we lean to the autobiographical, save occasionally by allusion to point a moral or adorn a tale, as the ancient adage puts it.
>
> This big book, i.e. big in words, is a rambling sort of camp-meeting confession of experiences, told in the form of biographies of various alcoholics

who had been to a certain institution and have provisionally recovered, chiefly under the influence of the "big brothers get together spirit." Of the inner meaning of alcoholism there is hardly a word. It is all on the surface material.

Inasmuch as the alcoholic, speaking generally, lives a wish-fulfilling infantile regression to the omnipotency delusional state, perhaps he is best handled for the time being at least by regressive mass psychological methods, in which, as is realized, religious fervors belong, hence the religious trend of the book. Billy Sunday and similar orators had their successes but we think the methods of Forel and of Bleuler infinitely superior (Book Reviews. Alcoholics Anonymous, 1940).

By 1951, however, the methods of A.A. had earned it the Lasker Award of the American Public Health Association (Wilson, 1957).

In addition to materials already cited, there exists a wealth of miscellaneous publications about, or partially about, A.A., and no survey of the important literature on A.A. could pretend to be representative without mention of a few examples of some of the major types of such publication. Among several notable studies, interpretations, or discussions of A.A. itself, or specific A.A. phenomena, are those of Bacon (1957, 1959, 1970), Bales (1942, 1945), Bayne (1963), Bean (1975), Bircher (1946), Edwards (1967), Fletcher (1944), Furman (1949), Glatt (1967), Groves (1972), Hanfmann (1951), Krystal and Moore (1963), Lofland and Lejeune (1960), Madsen (1974), Mann (1947), McAfee (1952), McCarthy (1964), Metcalfe (1944), Meyers (1941), Minogue (1948), Nelson (1941), Ritchie (1948), Rosenbarger (1971), Rotman (1945), Rusterholz (1946), Sapir (1957), Sariola (1962, 1963), Silcott (1971), Thompson (1952), Tiebout (1958b), Trice (1956, 1957a,b,c), Trice and Roman (1970b), and Williams (1958).

Some books not already mentioned on alcoholism and other alcohol problems also give information about A.A., including Bissell (1973), Blum and Blum (1967), Collins (1966), Coppolino and Coppolino (1965), Cross (1968), D'Alonzo (1959), Ford (1951, 1955), Free (1955), Fullam (1950), Hayman (1966), Jackson (1960), Kant (1954), Kennedy (1966), Kessel and Walton (1965), Kessel (1962), King (1953), Lovell (1953), Mann (1958, 1970), Milt (1967), Mullan and Sangiuliano (1966), Pittman (1967), Rea (1956), Regan (1962), Smithers (1968), Steiner (1971), Sullivan (1969), Trice (1966), Ullman (1960), Voldeng (1962), Whitney (1965, 1970), and Williams (1960), as do books not solely about alcohol problems, such as Shibutani (1961), and Toch (1965).

Many surveys on alcohol problems have also produced findings on A.A., such as Alcoholism (1961), Borg (1964), Dorsch *et al.* (1969), Dunn and Clay (1971), Grover (1967), Hart *et al.* (1964), Hayman (1956), Johnson (1965), Knox (1969, 1971), Knudsen (1959), Martin *et al.* (1962), Moore (1971), Moore and Wood (1963), Morales-de-Flores (1970), New York City Alcoholism Study (1962), Park (1962), Perceval (1969), Plaut (1967), Rhodes *et al.* (1969), Ritson and Thompson (1970), Seaborn (1970), Slater and Smith (1961), and Williams (1950).

Alcoholics Anonymous as a treatment and research resource (social or medical, in many kinds of health setting) has been discussed in Alcoholics Anonymous (1948), Alcoholics Anonymous in London (1948), Alford (1942), All Nippon Sobriety Association (1968), Alstrup (1950), Abstinencia jedno z meradiel narodnej kultury? (1969), Andelković *et al.* (1969), Behrendt (1961), Bell (1960, 1964, 1971), Bell *et al.* (1969), Bissell (1969), Boczán (1970), Bowen and Androes (1968), Bradfer (1964), Britton (1967), Cabrera (1961), Canter (1969), Chambers (1953), Chambers *et al.* (1962), Chambers and Schultz (1965), Chegwidden (1968), Cheney and Kish (1970), Clavreul (1971), Cohen *et al.* (1961), Conroy *et al.* (1971), Cuccia (1970), Curlee (1971a,b), Davis (1948), Delehanty (1956), Desroches *et al.* (1969), Edwards (1966), Einstein *et al.* (1970), Ellis and Krupinski (1964), Ends and Page (1959), Esser (1961a), Fitzgerald *et al.* (1971), Fort and Porterfield (1961), Fox (1967a,c, 1973b), Fox and Leach (1970), Glatt (1959), Haggard and Jellinek (1942), Heersema (1942), Hewitt (1943), Hoover (1960), Hunter (1947), Ignatia (1951), Jackson *et al.* (1960), Jellinek (1962), Jensen (1962), Johnson *et al.* (1966), Jones *et al.* (1969), Kelly (1968), Kissen (1961), Krzysztof (1970), Lawrence (1961), Lister (1971), Mann (1946), Manson (1948), Marquis (1971), Mascarenhas *et al.* (1970), Matkom (1969), McInerney (1970), Meer and Amon (1963), Miller *et al.* (1970), Minogue (1946), Mitchell (1958, 1961), Mitchell (1941), Moore and Ramseur (1960b), Nunan (1944), Pacy (1968), Parry (1961), Prothro (1961), Quaranta (1947), Quehl (1971), Ramsay *et al.* (1963), Reinert (1965), Reed (1970), Ritson (1968), Rohan *et al.* (1969), Ross (1965), Rossi (1970), Rossi and Bradley (1960), Rothstein *et al.* (1966), Selzer (1958, 1968), Silver Award (1971), Silverman (1968), Solms and de Meuron (1969), Steinhilber *et al.* (1967), Tähkä (1966), Tessmann (1971), Thompson (1942), Thompson and Kolb (1953), Tolentino (1967), Tuckwiller (1947), U.K. National Health Service Memorandum (1963), Valles and Sikes (1964), Victor (1970), West and Swegan (1956), White and Gaier (1965), Williams (1964), Wilson (1970), X (1954), Zanko and Kozlina (1970), and Zimberg *et al.* (1971).

The social work literature, including that on family counselling, has also included some references to A.A., such as Alateen (1973), Bailey (1965, 1968, 1970), Bailey and Fuchs (1960), Burton (1960), Dilemma of the Alcoholic Marriage (1971), Esser (1968), Jackson (1956), Lee (1960), Scott (1970), Shipp (1963), Valles (1965), Whitney (1968), and Wilding (1949).

Corrections and criminal justice publications with reference to A.A. include Alcoholics and Alcoholics Anonymous in Prisons (1969), Alcoholics Anonymous in Prisons (1954), Anderson (1944), Bacon (1945), Brown (1955),

Cozart (1952), Curlee (1968), Ditman and Crawford (1966), Ditman et al. (1967), Dolan (1953), Easy Does It (1948), Evrard (1971), Florida Corrections (1971), Green (1968), I Think You Guys Mean It! (1947), MacCormick (1963), McKinney (1946), Pinardi (1966), St. Clair (1958), Sell (1946), Tatham (1969), Truax (1972), Upton (1944), and Wolff (1949).

Religious publications with reference to A.A., besides those already listed, include Echols (1947), Hunter (1945), Middlebrooks (1945), Nace (1949), Roth (1965), Tiemeyer (1948) and Walton (1941).

Mentions of A.A. are also extensive in the literature on skid row alcoholism, such as Bourne et al. (1966), Block (1962b), and Bahr (1969), in addition to citations already made. Several authors have also spoken of A.A. in relation to halfway houses for alcoholics (e.g., Donohue, 1971; Richards, 1968; and Simonds, 1969). Ruprecht (1961) errs, however, in stating that A.A. actually operates such facilities. Neither the Fellowship itself nor any of its component units (offices, groups, committees, etc.) owns or operates day centers, detox centers, clubs, rehabilitation centers, hospitals or hospital wards, or any other such agency, in accordance with the Traditions of the Fellowship. Some Fellowship members, acting in their capacities as private citizens, not in their A.A. roles, are often instrumental in establishing and operating such facilities (just as they serve on various alcoholism committees and commissions or in professional or paraprofessional jobs—as private citizens, never as representatives of A.A.), but technically and in actuality agencies of this sort are not started, financed, or run by A.A. or by A.A. members per se (Wilson, 1952, 1957).

Mentions of A.A. also abound in material on industrial alcoholism, such as Dealing with the Drinking Problem (1970), F. (1947), Fellowship of Alcoholics Anonymous (1948), Habbe (1969, 1970), It's Fun to be Sober (1945), Leggo (1967), Maurer (1968), Maxwell (1960), Norris (1948, 1968a,b), Slotkin et al. (1971), and Trice (1958b, 1959b) in addition to materials already mentioned.

Features common to both A.A. and various modalities of professional therapy have led to descriptions of A.A. as "forms of" or "nothing but" various treatments, including cybernetics (Bateson, 1971), psychoanalysis (Blum, 1966), behavior therapy (Geary, 1967), logotherapy (Holmes, 1970), role theory (Meyer, 1967), transactional analysis (Steiner, 1971), systems theory (Ward and Faillace, 1970), and others. Such views can of course be quite enlightening as to the dynamics of both A.A. and the professional therapy under discussion. But oversimplification can also obscure vital differences between A.A. as experienced by its members and professional therapy modalities.

Various other approaches to alcohol problems have also been described as analogous to A.A., such as attitudes toward alcohol use in Polynesia (Lemert,

1962). Some such claims, however, can be questioned, such as descriptions of the Swedish Link organization as "like A.A." (Källström, 1959; Lambert, 1958; Simm, 1965; Tillgren *et al.*, 1958; and Wegerman, 1968). Link crusades for "Temperance" and is government-supported. For reasons explained by Wilson (1952, 1957), neither is true of A.A.

There are also crucial differences between A.A. and the Japanese ANSA (All Nippon Sobriety Association, 1968), Ring i Ring (Alstrup, 1950), the Bratislava temperance club (Abstinencia, 1969), the "prayomatic groups" of Lyon (Clavreul, 1971), A.A. groups led or "interpreted" by psychiatrists, as in Holland (Esser, 1961a), the government-funded clubs for alcoholics in Poland (Krzysztof, 1970) and other Socialist countries (Nyárádi, 1970; Zanko and Kozlina, 1970). To fail to understand the genuine, not superficial, distinctions between these approaches and the traditional A.A. approach does disservice to both A.A. and various good programs which, although somewhat similar to, are also different from A.A. in critical ways.

Some observers (e.g., Button, 1956; Canter, 1966; Chafetz and Demone, 1962; Curlee, 1968; Dozier, 1966; Goodman, 1970; Jackson and Connor, 1953; Lemert, 1954; Madsen, 1964; Parker *et al.*, 1960; Pittman and Gordon, 1958; Reinert, 1968; and Stone, 1962) have hypothesized that A.A. could not be successful with certain alcoholics on grounds of gender, ethnicity, personality types, age, socioeconomic or other status. Jellinek (1960) felt A.A. would not appeal to victims of his *delta* species of alcoholism, and, as pointed out previously (Leach, 1973), the relatively slow growth of A.A. in France tends to confirm Jellinek's prediction. Otherwise, however, no substantial body of evidence supporting any theories of immunity to A.A. has been published. Trice's (1957c, 1959a) findings of factors associated with nonaffiliation are valuable, but no follow-up revealed whether or not nonaffiliates may have become abstinent A.A. members at a later stage of their alcoholism. The General Service Board and others, as shown before, found many alcoholics who continued to drink after their first contact with A.A. but later began sustained sobriety. Perhaps only longitudinal studies of the size and type undertaken by Cahalan *et al.* (1969) and Cahalan (1970) will clearly delineate a body of alcoholics with true, lasting immunity to A.A. Studies of smaller samples, investigated psychometrically at only one point in their drinking history, have produced a logorrhea of findings, many of them conflicting, as reflected in "Keller's Law": "The investigation of any trait in alcoholics will show that they have either more or less of it" (Keller, 1972). His conclusion that "alcoholics are different in so many ways that it makes no difference" is a terse, accurate collective abstract of many studies of A.A. members.

Obsession with the phenomenon of nonaffiliation ignores and does little to explain the fact that so many alcoholics have successfully affiliated with A.A.

Criticisms of A.A.

A.A.'s Twelve Steps, its suggested program of recovery, and the ideals and standards of conduct described in the Traditions, have received very little serious analysis in the professional literature, so they have received almost no thoughtful criticism. Instead, most criticism of A.A. has been in the form of attacks on sick alcoholic patients who turn to A.A. for help, none of whom "has been able to maintain anything like perfect adherence to these [A.A.] principles" (Wilson, 1939).

Aharan (1970, 1971) deplored some members' denigration of disulfiram (Antabuse) and other means of treatment they have found ineffective, and proposed that A.A. establish "centers for training in human relations and group dynamics." Bennett (1970) noted the tension that can develop between some physicians and A.A. members (or other lay groups). Bonacker (1958) described A.A. as repressive toward hostility, with the "virtues of passivity and placidity too highly exalted." In the Twelve Steps he found exhibitionism, dependence, and rebellion against authority. Cahn (1970) noted the difference between the traditional mental health approach to alcoholism and that of A.A. Cain (1963, 1964a,b, 1967) criticized A.A. members for being anti-intellectual, cultish, ritualistic, antiprofessional, and not always truthful. Chafetz and Demone (1969), too, noted cultlike aspects and the A.A. focus on drinking. The Committee on Prisons, Probations and Parole, District of Columbia (1957) was critical of A.A. members who gave up holding A.A. meetings in the local workhouse. Dinitz (1949) called attention to euphemisms in the A.A. "argot," and to illicit sexual behavior by some A.A. members. Elkins (1966) criticized A.A. members for ignorance, fear, omnipotence, anger, dishonesty, idolatry, and indolence. Ellison (1964), describing himself as a consultant to A.A., was critical of the growth of the G.S.O. budget, the neuroticism of some members, the ethnic distribution of the G.S.O. staff, and the political ideology of a nonalcoholic former trustee of the General Service Board. Haverstick (1955) insisted that the movement feuded with the Roman Catholic church and deliberately tried to recruit a member from among European nobility for snob appeal. Krystal (Krystal and Moore, 1963) criticized the nonprofessionalism of A.A. and belittled the special empathy alcoholics manifest for each other in A.A. Lindt (1959) said an A.A. member remains "improved" only so long as he or she remains on good terms with an A.A. "sponsor." (He did not mention the fact that many alcoholics sober in A.A. never had an A.A. sponsor.) Lovald and Neuwirth (1968) said A.A. "perpetuates the sick role" and thus prevents the alcoholic's freeing himself from stigma. Ottenberg (1969) pointed to friction between A.A. members and professional and paraprofessional therapists. Pfau (Pfau and Hirshberg, 1958) said the structural setup of

A.A. "tends far too much to organization and this, in spiritual entities, could prove disastrous." Pfau was a Roman Catholic priest. Pittman and Snyder (1962) were critical that some A.A. groups do not "reach out enough" to welcome newcomers, that some members have low tolerance for members who have slips, that some alcoholics are not reached by A.A., and that some members do not achieve "relative autonomy from special external therapeutic supports." Boyes (Ripley and Jackson, 1959) criticized some A.A. members for drinking, some for lack of patience with newcomers, some for lack of tolerance, and some for "craving only sobriety." Sagarin's (1969) criticism is based largely on the works of Cain, Dinitz, and Ellison already noted. He is also highly critical of the loose organizational structure of the movement, which makes absolutely accurate statistics on recovery virtually impossible, and of its lay folklore on alcoholism because it is not, at all points, congruent with some scientific theories on alcoholism.

The movement itself declines to answer public criticism, to avoid public controversy. But all of the above criticisms and many even more severe have also been voiced by A.A. members—both drinking and sober ones—at A.A. meetings and in the A.A. literature, as any investigator can find. Perhaps this is one of the movement's most overlooked strengths and a reason for its continued growth.

Recently, A.A. has been attacked implicitly, if not formally, on two other grounds. The first charge leveled is that its continued use of the concept of anonymity helps to perpetuate the stigma on alcoholism.

It seems only fair to point out that anonymity was originally used by A.A. as a response to the stigma placed on alcoholics by our society. A.A. did not originate the stigma, but has helped considerably to diminish it.

The term *anonymous* in the name of the Fellowship still has value in promising confidentiality to the sick alcoholic, just as any client has a right to expect a physician, psychologist, nurse, clergyman, lawyer, or social worker to respect the privacy of privileged information.

The A.A. Tradition of anonymity in no way inhibits a member from publicly identifying himself or herself as "an alcoholic" or "a recovered alcoholic," as the A.A. literature clearly shows. But it does discourage the public announcement of A.A. membership for purposes of personal gain or aggrandizement. This custom helps prevent the appearance of affiliation with any governmental, commercial, health, religious, educational, or other institution toward which a prospective A.A. member might have an aversion.

However, the chief value of the anonymity concept for alcoholics in A.A. may well be its effect on the egocentricity of alcoholism, as explained by Tiebout (1944, 1946, 1949, 1951a, 1953, 1954, 1955b, 1961, 1963a,b, 1965).

It takes little imagination to foresee what would happen to Alcoholics Anonymous if its members relinquished the search for humility occasioned by the concept of anonymity; to see what would happen to medical science if all health practitioners decided no longer to keep privileged information privileged; or to see what would happen to the biological and other sciences if their practitioners forsook the search for truth as their chief motivation in favor of personal grandiosity.

The second serious criticism of A.A. can be heard in those voices denying that alcoholism is a disease and faulting A.A. for use of this idea. That discussion is beyond the scope of this paper, but in the opinion of the present writers it seems to have little if any impact on A.A.

A young woman A.A. member, told that some experts are saying alcoholism is not a disease, looked blank for a moment, shrugged slightly, and said, "Well, it sure isn't a picture of health." She then turned back to the A.A. newcomer she was trying to help, who had the shakes almost as bad as those of the surgeon, Dr. Bob, nearly 40 years before.

REFERENCES

A.A. Bulletin #1, 1940, Alcoholic Foundation, New York.
A.A. Group, The, 1965, A.A. World Services, New York.
A.A. Service Manual, The, 1969, A.A. World Services, New York.
A.A. Survey, The, 1970, A.A. World Services, New York.
"AA/30," 1965, A.A. World Services, New York.
"A.A. Today," 1960, A.A. Grapevine, New York.
Abraham, K. L., 1927, "Selected Papers of Psychoanalysis," Hogarth, London.
Abstinencia jedno z meradiel narodnej kultury? 1969, *Protialkoholicky Obzor* (Bratislava) 4(1):19.
Adler, A., 1941, The individual psychology of the alcoholic patient, *Journal of Criminal Psychopathology* 3:74.
Aharan, C. [H.], 1970, AA and other treatment programs: Problems in cooperation, *Addictions* (Toronto) 17(4):25.
Aharan, C. H. 1971, A.A. and other treatment programs, *The A.A. Grapevine* 27(11):17.
"Al-Anon Faces Alcoholism," 1965, Al-Anon Family Group Headquarters, New York.
"Alateen: Hope for Children of Alcoholics," 1973, Al-Anon Family Group Headquarters, New York.
Alcoholics and Alcoholics Anonymous in prisons, 1969, *Prison Service Journal* 8(31):35.
Alcoholics Anonymous, 1940a, *Newsweek*, February 19:47.
Alcoholics Anonymous, 1940b, *Time*, February 19:56.
Alcoholics Anonymous, 1940c, *Illinois Medical Journal*, February: 1.
Alcoholics Anonymous, 1948, *British Medical Journal* 1:664.
Alcoholics Anonymous in London, 1948, *Lancet* 254:31.
Alcoholics Anonymous in prisons, 1954, *Federal Probation* 18:17.
Alcoholics Anonymous spurns a trap, 1950, *Christian Century*, February 22:229.
Alcoholism, 1961, *J. Am. Med. Assoc.* 177:275.
Alexander, F., 1956, Views on the etiology of alcoholism II. The psychodynamic view, *in* "Alcoholism as a Medical Problem" (H. D. Kruse, ed.), pp. 40–46, Hoeber-Harper, New York.

Alexander, F., 1963, Alcohol and behavioral disorder—alcoholism, *in* "Alcohol and Civilization" (S. P. Lucia, ed.), pp. 130–141, McGraw-Hill, New York.

Alexander, J., 1941, Alcoholics Anonymous; freed slaves of drink, now they free others, *Saturday Evening Post* 213(35):9.

Alexander, J. B., and Gudeman, H. E., 1965, Perceptual and interpersonal measures of field dependence, *Percept. Mot. Skills* 20:79.

Alford, L. B., 1942, Symposium on medical emergencies; management of psychoses and alcoholism in general practice, *Med. Clin. North Am.* 26:401.

All Nippon Sobriety Association (ANSA), 1968, Zen Nippon Danshu Renmei (Zen Dan Ren), ANSA, Tokyo.

Allen, L. R., and Dootjes, I., 1968, Some personality considerations of an alcoholic population, *Percept. Mot. Skills* 27:707.

Allsop, K., 1961, "The Bootleggers: The Story of Chicago's Prohibition Era," Arlington House, New Rochelle.

Alstrup, K., 1950, Group treatment of alcoholics in the organization "Ring i Ring," the Danish A.A., *Ugeskr. Laeg.* 112:807.

Andelković, A., Milosavčević, V., and Mijušković, N., 1969, Neke karakteristike alkoholicara koji traže anonimnost u lečenju, *Alkoholizam*, Beograd 9(2):43.

Anderson, D., 1944, The process of recovery from alcoholism, *Federal Probation* 8(4):14.

Armstrong, J. D., 1958, The search for the alcoholic personality, *Ann. Amer. Acad. Pol. Soc. Sci.* 315:40.

Armstrong, R. G., 1959, A review of the theories explaining the psychodynamics and etiology of alcoholism in men, *Psychological Newsletter* 10:159.

Armstrong, R. G., and Wertheimer, M., 1950, Personality structure in alcoholism, *Psychological Newsletter* 10:341.

Astor, M., 1959, "My Story, An Autobiography," Doubleday, New York.

Bacon, M. K., Barry, H., III, and Child, I. L., 1965, A cross-cultural study of drinking. *II*. Relations to other features of culture, *Q. J. Stud. Alcohol* Suppl. 3:29.

Bacon, S. D., 1945, Alcoholism and social isolation, in "Cooperation in Crime Control" (M. Bell, ed.), pp. 209–234, National Probation and Parole Association, New York.

Bacon, S. D., 1957, A sociologist looks at Alcoholics Anonymous, *Minnesota Welfare* 10:35.

Bacon, S. D., 1958, Alcoholics do not drink, *Ann. Amer. Acad. Pol. Soc. Sci.* 315:55.

Bacon, S. D., 1959, Prevention can be more than a word, *in* "Realizing the Potential in State Alcoholism Programs. Proceedings of the Northeast States Conference on Alcoholism," pp. 5–18, Connecticut Commission on Alcoholism, Hartford.

Bacon, S. D., 1970, Meeting the problems of alcoholism in the United States, *in* "World Dialogue on Alcohol and Drug Dependence" (E. D. Whitney, ed.), Beacon Press, Boston.

Bacon, S. D., 1973, The process of addiction to alcohol: Social aspects, *Q. J. Stud. Alcohol* 34:1.

Bahr, H. M., 1969, Lifetime affiliation patterns of early- and late-onset heavy drinkers on Skid Row, *Q. J. Stud. Alcohol* 30:645.

Bailey, M. B., 1965, Al-Anon Family Groups as a resource for wives for alcoholics, *in* "Al-Anon Faces Alcoholism," pp. 33–37, Al-Anon Family Group Headquarters, New York.

Bailey, M. B., 1968, "Alcoholism and Family Casework," Community Council of Greater New York, New York.

Bailey, M. B., 1970, Attitudes toward alcoholism before and after a training program for social caseworkers, *Q. J. Stud. Alcohol* 31:669.

Bailey, M. B., and Fuchs, E., 1960, Alcoholism and the social worker, *Social Work* 5(4):14.

Bailey, M. B., and Leach, B., 1965, "Alcoholics Anonymous, Pathway to Recovery. A Study of 1,058 Members of the AA Fellowship in New York City," National Council on Alcoholism, New York.

Bailey, W., Hustmyer, F., and Kristofferson, A., 1961, Alcoholism, brain damage, and perceptual dependence, *Q. J. Stud. Alcohol* 22:387.
Bailey, M. B., Haberman, P., and Alksne, H., 1962, Outcomes of alcoholic marriages: endurance, termination, or recovery, *Q. J. Stud. Alcohol* 23:610.
Bales, R. F., 1942, Types of social structure as factors in "cures" for alcohol addiction, *Applied Anthropology* 1(3):1.
Bales, R. F., 1944, The therapeutic role of Alcoholics Anonymous as seen by a sociologist, *Q. J. Stud. Alcohol* 5:267.
Bales, R. F., 1945, Social therapy for a social disorder—compulsive drinking, *Journal of Social Issues* 1(3):14.
Bateman, N. I., and Petersen, D. M., 1971, Variables related to outcome of treatment for hospitalized alcoholics, *Int. J. Addict.* 6:215.
Bateman, N. I., and Petersen, D. M., 1972, Factors related to outcome of treatment for hospitalized white male and female alcoholics, *J. Drug Issues*, Tallahassee 2:66.
Bateson, G., 1971, The cybernetics of "self": A theory of alcoholism, *Psychiatry* 34(1):1.
Bayne, J. R., 1963, Chronic disease and custodial care. III. A program of treatment for alcoholics. *Med. Serv. J.,* Canada 19:256.
Beacham, E. G., 1968, First-year review of alcoholism program, Tuberculosis Division, Baltimore City Hospitals, *Md. State Med. J.* 17(8):108.
Bean, N., 1975, "Alcoholics Anonymous: A.A.," Insight Communications, New York.
Beaubrun, M. H., 1967, Treatment of alcoholism in Trinidad and Tobago, 1956–1965, *Br. J. Psychiatry* 113:643.
Becker, G. S., and Israel, P., 1961, Integrated drug- and psychotherapy in the treatment of alcoholism, *Q. J. Stud. Alcohol* 22:610.
Beeley, A. L., 1959, Alcoholism, social work and mental hygiene, *Mental Hygiene* 43:577.
Begbie, H., 1923, "More Twice-Born Men," Putnam's, New York.
Behrendt, V. M., 1961, Treatment of alcoholism, *Journal Lancet* 81:321.
Belasco, J. A., 1971, The criterion question revisited, *Br. J. Addict.* 66:39.
Belasco, J. A., and Trice, H. M., 1969, "The Assessment of Change in Training and Therapy," McGraw-Hill, New York.
Belden, S., 1962, A program for the treatment of alcoholics in a mental hospital, *Q. J. Stud. Alcohol* 23:650.
Bell, A. H., 1970, The Bell Alcoholism Scale of Adjustment; a validity study, *Q. J. Stud. Alcohol* 31:965.
Bell, A. H., Weingold, H. P., and Lachin, J. M., 1969, Measuring adjustment in patients disabled with alcoholism, *Q. J. Stud. Alcohol* 30:634.
Bell, R. G., 1960, A method of clinical orientation to alcohol addiction, *Can. Med. Assoc. J.* 83:1346.
Bell, R. G., 1964, Who is qualified to treat the alcoholic? Comment on the Krystal-Moore discussion, *Q. J. Stud. Alcohol* 25:562.
Bell, R. R., 1971, Alcohol, *in* "Social Deviance: A Substantive Analysis," pp. 171–190, Dorsey Press, Homewood, Illinois.
Bennett, A. E., 1970, Problems in the rehabilitation of the alcoholic, *in* "Summaries, Vol. I: Third International Congress of Social Psychiatry, Zagreb," pp. 200–202.
Benson, P. H., 1960, "Religion in Contemporary Culture," Harper & Bro., New York.
Berryman, J., 1973, "Recovery," Farrar, Straus & Giroux, New York.
"Best Cartoons from the Grapevine," 1970, A.A. Grapevine, New York.
Big light can be shed on alcoholism, 1945, *A.A. Grapevine* 1(11):1.
Bircher, R., 1946, "A.A." Alcoholics Anonymous oder das Bündnis der "Nomenloser Trinker," *Wendepunkt,* Zür. 23:214.

Bird, B., 1949, One aspect of causation in alcoholism, *Q. J. Stud. Alcohol* 9:532.
Bissell, L., 1969, Suicide in the alcoholic population, *in* "Identifying Suicide Potential (Conference Proceedings, Teachers College, Columbia University, New York, 1969)," pp. 69–74, Behavioral Publications, New York.
Bissell, L., 1973, "Understanding Alcoholism," Claretian Press, Chicago.
Blackburn, J. M., 1955, Report of the section on psychological research, *Q. J. Stud. Alcohol* 16:546.
Blane, H. T., 1968, "The Personality of the Alcoholic, Guises of Dependency," Harper & Row, New York.
Block, M. A., 1962a, "Alcoholism: Its Facets and Phases," John Day, New York.
Block, M. A., 1962b, A program for the homeless alcoholic, *Q. J. Stud. Alcohol* 23:644.
Bloom, H., 1958, Non-alcoholics need to listen, *A.A. Grapevine* 15(2):26.
Blum, E. M., 1966, Psychoanalytic views of alcoholism: A review, *Q. J. Stud. Alcohol* 27:259.
Blum, E. M., and Blum, R. H., 1967, "Alcoholism: Modern Psychological Approaches to Treatment," Jossey-Bass, San Francisco.
Blumberg, L., Shipley, T. E. Jr., Shandler, I. W., and Niebuhr, H., 1966, The development, major goals and strategies of a Skid Row program: Philadelphia, *Q. J. Stud. Alcohol* 27:242.
Boczán, J., 1970, Registráló módszereink az idült alkoholizmus szociopszichoterápiájában, *Alkohologia,* Bp. 1:27.
Boggs, S. L., 1967, Measures of treatment outcome for alcoholics: A model analysis, *in* "Alcoholism" (D. J. Pittman, ed.), pp. 174–197, Harper & Row, New York.
Bohince, E., and Orensteen, A. C., 1953, An evaluation of the services and program of the Minneapolis chapter of Alcoholics Anonymous, Doctoral dissertation, University of Minnesota.
Bonacker, R. D., 1958, Alcoholism and Alcoholics Anonymous viewed symptomatologically, *Mental Hygiene* 42:562.
Book Notices. Alcoholics Anonymous, 1939, *J. Am. Med. Assoc.* 113(16):1513.
Book Review. Alcoholics Anonymous, 1941, *N.Y. State J. Med.* 41(24):2392.
Book Reviews. Alcoholics Anonymous, 1940, *J. Nerv. Ment. Dis.* 92(3):399.
Borg, V., 1964, Klinikkbehandling av alkoholskadede, *Tidsskr. norske Laegeforen* 84:1117.
Borghesi, R., and Medaglini, E., 1965, A proposito di una forma di psicoterapia di gruppo nell'alcoholismo cronico; considerazioni critiche su l'alcoholics anonymous, *Rass. Studi Psychiatr.* 54:79.
Bourne, P. G., Alford, J. A., and Bowcock, J. A., 1966, Treatment of skid-row alcoholics with disulfiram, *Q. J. Stud. Alcohol* 27:42.
Bowen, W. T., and Androes, L. R., 1968, A follow-up study of 79 alcoholic patients: 1963–1965, *Bull. Menninger Clin.* 32:26.
Bowman, K. M., and Jellinek, E. M., 1941, Alcohol addiction and its treatment, *Q. J. Stud. Alcohol* 2:98.
Boyington, G., 1958, "Baa, Baa, Black Sheep," Putnam's, New York.
Bradfer, J., 1964, Observation de 102 cas d'alcooliques suivis en cure ambulatoire dans un centre de consultation urbain, *Acta Neurol. Belg.* 64:657.
Bräutigam, W., 1967, Neuere Erfahungren bei der Behandlung des Alkoholikers, *Münch. Med. Wschr.* 109:2698.
Britton, W. L. S., 1967, The National Society on Alcoholism of New Zealand, Inc., *Alcoholism* (Zagreb) 3(1):56.
Brook, R. R., 1962, Personality correlates associated with differential success of affiliation with Alcoholics Anonymous, Doctoral dissertation, University of Colorado.
Brown, P. R., 1955, The problem drinker and the jail, *Q. J. Stud. Alcohol* 16:474.
Brown, R. F., 1963, An aftercare program for alcoholics, *Crime and Delinquency* 9(1):77.
Bruun, K., 1963, Outcome of different types of treatment of alcoholics, *Q. J. Stud. Alcohol* 24:280.

Burton, G., 1960, The alcoholic and his community search for help, *American Journal of Public Health* 50 (July): 960.
Button, A. D., 1956, Psychodynamics of alcoholism, *Q. J. Stud. Alcohol* 17:456.
Byrne, M., 1968, Resocialization of the chronic alcoholic, *Am. J. Nurs.* 68:99.
C., B., 1965, The growth and effectiveness of Alcoholics Anonymous in a southwestern city, *Q. J. Stud. Alcohol* 26:279.
Cabrera, F. J., 1961, Group psychotherapy and psychodrama for alcoholic patients in a state hospital rehabilitation program, *Group Psychotherapy* 14(3-4):154.
Cahalan, D., 1970, "Problem Drinkers," Jossey-Bass, San Francisco.
Cahalan, D., Cisin, I. H., and Crossley, H. M., 1969, "American Drinking Practices," Rutgers, New Brunswick.
Cahn, S., 1970, "The Treatment of Alcoholics; an Evaluative Study," Oxford, New York.
Cain, A., 1963, Alcoholics Anonymous, cult or cure? *Harper's Magazine* 226(1353):48.
Cain, A., 1964a, Speaking out: Alcoholics can be cured despite AA, *Saturday Evening Post*, September 19:6.
Cain, A., 1964b, "The Cured Alcoholic," John Day, New York.
Cain, A., 1967, "Paul King's Rebellion," John Day, New York.
"Came to Believe . . . the Spiritual Adventure of A.A. as Experienced by Individual Members," 1973, A.A. World Services, New York.
Canter, F., 1966, Personality factors related to participation in treatment of hospitalized male alcoholics, *J. Clin Psychol.* 22:114.
Canter, F. M., 1969, A self-help project with hospitalized alcoholics, *Int. J. Group Psychother.* 19:16.
Cantril, H., 1941, "The Psychology of Social Movements," Wiley, 1941.
Carothers, C., 1971, A discriminatory analysis of personality characteristics of the intemperate and the rehabilitated alcoholic. Doctoral dissertation, Texas Technical University.
Carroll, J. L., and Fuller, G. B., 1969, The self and ideal-self concept of the alcoholic as influenced by length of sobriety and/or participation in Alcoholics Anonymous, *J. Clin. Psychol.* 25:363.
Casriel, D., 1963, "So Fair a House: The Story of Synanon," Prentice-Hall, Englewood Cliffs, New Jersey.
Casson, F. R. C., 1969, The gamblers. *Mental Health* (London) Winter:15.
Chafetz, M. E., 1959, Practical and theoretical considerations in the psychotherapy of alcoholism, *Q. J. Stud. Alcohol* 20:281.
Chafetz, M. E., and Demone, H. W. Jr., 1962, "Alcoholism and Society," Oxford, New York.
Chafetz, M. E., and Demone, H. W., 1969, Alcoholics Anonymous, *in* "Deviance" (S. Dinitz, ed.), pp. 264-272, Oxford, New York.
Chafetz, M. E., Blane, H. T., and Hill, M. J., 1970, "Frontiers of Alcoholism," Science House, New York.
Chambers, F. T. Jr., 1953, Analysis and comparison of three treatment measures for alcoholism: Antabuse, the Alcoholics Anonymous approach, and psychotherapy, *Br. J. Addict.* 50:29.
Chambers, J. F., and Schultz, J. D., 1965, Double-blind study of three drugs in the treatment of acute alcoholic states, *Q. J. Stud. Alcohol* 26:10.
Chambers, J. F., D'Agostino, A. Sheriff, W. H. Jr., and Schultz, J. D., 1962, Comprehensive care of the acute alcoholic in a municipal hospital, *Can. Med. Assoc. J.* 86:1112.
Chandler, J., Hensman, C., and Edwards, G., 1971, Determinants of what happens to alcoholics, *Q. J. Stud. Alcohol* 32:349.
Chegwidden, M. J., 1968, The management of alcoholism in general practice, *Med. J. Aust.* 55-2(10):445.
Cheney, T. M., and Kish, G. B., 1970, Job development in a Veterans Administration hospital, *Vocational Guidance Quarterly* 19(1):61.

Chidsey, D. B., 1969, "On and Off the Wagon," Cowles, New York.
Churan, C. A., 1965, Cure—the questionable word in alcoholism, *A.A. Grapevine* 21(10):38.
Clagett, A. F., 1962, Primary essentials for organizing and administering a successful institutional A.A. program, *Ment. Hyg.* 46:438.
Clancy, J., 1961, Procrastination: A defense against sobriety, *Q. J. Stud. Alcohol* 22:269.
Clancy, J., 1962, The use of intellectual processes in group psychotherapy with alcoholics, *Q. J. Stud. Alcohol* 23:432.
Clancy, J., 1964, Motivation conflicts of the alcohol addict, *Q. J. Stud. Alcohol* 25:511.
Clavreul, J., 1971, L'alcoolisme est une maladie, *Information Psychiatrique* (Lyon) 47(1):17.
Clinebell, H. J. Jr., 1963, Philosophical-religious factors in the etiology and treatment of alcoholism, *Q. J. Stud. Alcohol* 24:473.
Clinebell, H. J. Jr., 1968, "Understanding and Counseling the Alcoholic, Revised and Enlarged," Abingdon, Nashville.
Co-Founders of Alcoholics Anonymous, The: Brief Biographies of A.A.'s First Two Members, 1972, A.A. World Services, New York.
Cohen, F., 1962, Personality changes among members of Alcoholics Anonymous, *Ment. Hyg.* 46:427.
Cohen, F., 1964, Alcoholics Anonymous principles and the treatment of emotional illness, *Ment. Hyg.* 48:621.
Cohen, F., 1965, Prescriptive group therapy, *Journal of Religion and Health* 4(2):188.
Cohen, S., Meighan, S. S., Hoover, M. P., and Bell, R. G., 1961, Alcoholism, *Can. Med. Assoc. J.* 84:341.
Collins, M. C., 1966, "Defeating Alcoholism the Fairview Way," Whitmore, Philadelphia.
Committee on Prisons, Probation and Parole, District of Columbia, 1957, "Prisons, Probation and Parole in the District of Columbia," Board of Commissioners, District of Columbia.
Connor, R. G., 1962, The self-concepts of alcoholics, *in* "Society, Culture, and Drinking Patterns" (D. J. Pittman and C. R. Snyder, eds.), pp. 455–467, Wiley, New York.
Conroy, R. W., Friedberg, B., and Krizaj, P., 1971, A community plan for military alcoholics, *Am. J. Psychiatry* 128:774.
Cooper, J., and Maule, H. C., 1962, Problems of drinking: An enquiry among members of Alcoholics Anonymous, *Br. J. Addict.* 58:45.
Coppolino, C. A., and Coppolino, C. M., 1965, "The Billion Dollar Hangover," A. S. Barnes, New York.
Corder, B. F., Hendricks, A., and Corder, R. F., 1964, An MMPI study of a group of wives of alcoholics, *Q. J. Stud. Alcohol* 25:551.
Corder, B. F., Corder, R. F., and Laidlaw, N. D., 1972, An intensive treatment program for alcoholics and their wives, *Q. J. Stud. Alcohol* 33:1144.
Cornutt, R. H., 1953, The use of the Bender-Gestalt with an alcoholic and nonalcoholic population, *J. Clin. Psychol.* 9:287.
Correspondence. Alcoholics Anonymous, 1940, *Ill. Med. J.* 77(2):1.
Cozart, R., 1952, Release preparation of the prisoner, *Federal Probation* 16:13.
Cross, J. N., 1968, "Guide to the Community Control of Alcoholism," American Public Health Association, New York.
Cuccia, E., 1970, When you're drunk it don't count, *Mental Health* (London) Spring:11.
Curlee, J., 1968, Women alcoholics, *Federal Probation* 32(1):16.
Curlee, J., 1971a, Combined use of Alcoholics Anonymous and outpatient psychotherapy, *Bull. Menninger Clin.* 35:368.
Curlee, J., 1971b, Sex differences in patient attitudes toward alcoholism treatment, *Q. J. Stud. Alcohol* 32:643.

D'Alonzo, C. A., 1959, "The Drinking Problem—And Its Control," Gulf, Houston.
D'Alonzo, C. A., 1960, The problem of alcoholism in industry, *A. A. Grapevine* 17(6):16.
Dancey, T. E., 1966, Do A.A.'s need psychiatry? *A.A. Grapevine* 23(6):6.
Dancey, T. E., 1968, Are psychiatry and A.A. incompatible? *A.A. Grapevine* 24(5):28.
Davies, D. L., 1962, Normal drinking in recovered alcohol addicts, *Q. J. Stud. Alcohol* 23:94.
Davis, C. N., 1947, Alcoholics Anonymous, *Arch. Neurol. Psychiatry* 57:516.
Davis, C. N., 1948, An approach to the problem of alcoholism, *Del. State Med. J.* 20:85.
Davis, W. W., 1970, Practical experience with an alcoholism program in industry, *Ohio State Med. J.* 66:84.
Dealing with the drinking problem, 1970, *Manpower* 2(12):2.
Deering, G. E., 1970, Doctor, why can't I stop drinking? *A.A. Grapevine* 27(3):7.
de Kruif, P., 1959, God is not yourself, *A.A. Grapevine* 16(1):29.
Delehanty, E. J., 1956, Motivation in the alcoholic, *Dis. Nerv. Syst.* 17:224.
Delehanty, E. J., 1966, Treatment of alcoholism with help from Alcoholics Anonymous, *Modern Treatment* 3:537.
Dellinger, J. B., 1953, Alcoholics Anonymous operating in a prison setting. Mimeographed, 6 pp. N.P.
Desroches, H., Ferneau, E., and Cobble, J., 1969, Factors associated with successful responsiveness to an alcoholism treatment unit, *Newsletter for Research in Psychology* 11(1):1.
Digest of Proceedings at Dinner Given by Mr. John D. Rockefeller, Jr., in the Interest of Alcoholics Anonymous, at Union Club, New York City, February 8, 1940. Privately printed, 14 pp. N.P.
"Dilemma of the Alcoholic Marriage," 1971, Al-Anon Family Group Headquarters, New York.
Dinitz, S., 1949, The role of Alcoholics Anonymous as a therapeutic agent, Master's thesis, University of Wisconsin.
Ditman, K. S. and Crawford, G. G., 1966, The use of court probation in the management of the alcohol addict, *Am. J. Psychiatry* 122:757.
Ditman, K. S., Crawford, G. G., Forgy, E. W., Moskowitz, H., and MacAndrew, C., 1967, A controlled experiment on the use of court probation for drunk arrests, *Am. J. Psychiatry* 124:160.
Dolan, G., 1953, Convicts take the cure, *American Mercury*, October:130.
Dole, V. P., 1972, A.A., drug addiction, and pills, *A.A. Grapevine* 29(5):2.
Donahue, J., 1971, A halfway house program for alcoholics, *Q. J. Stud. Alcohol* 32:468.
Dorsch, G., Talley, R., and Bynder, H., 1969, Response to alcoholics by the helping professions and community agencies in Denver, *Q. J. Stud. Alcohol* 30:905.
Dowling, E., 1966, Three dimensions of A.A., *A.A. Grapevine* 22(9):28.
Dozier, E. P., 1966, Problem drinking among American Indians. The role of sociocultural deprivation, *Q. J. Stud. Alcohol* 27:72.
Dunlap, N. G., 1961, Alcoholism in women: Some antecedents and correlates of remission in middle-class members of Alcoholics Anonymous, Doctoral dissertation, University of Texas.
Dunn, J. H., and Clay, M. L., 1971, Physicians look at a general hospital alcoholism service, *Q. J. Stud. Alcohol* 32:162.
Durzynski, B., 1971, Wplw Leczniczy Spolecznosci Klubu AA, *Problemy Alkoholizmu* (Warszawa) 6(6):10.
E., J., 1961, The twelve steps revisited, *A.A. Grapevine* 47(4):2.
Easy does it, 1948, *Federal Probation* 12(11):25.
Echols, W., 1947, Here's hope for alcoholics, *Christian Century*, November 26:1451.
Eckhardt, W., 1967, Alcoholic values and Alcoholics Anonymous, *Q. J. Stud. Alcohol* 28:277.
Edwards, G., 1965, The puzzle of A.A., *A.A. Grapevine* 21(11):26.

Edwards, G., 1966, Hypnosis in treatment of alcohol addiction. Controlled trial, with analysis of factors affecting outcome. *Q. J. Stud. Alcohol* 27:221.
Edwards, G., 1967, The alcoholic. *Journal of the Royal Institute of Public Health and Hygiene* (London) 30:28.
Edwards, G., Hensman, C., Hawker, A., and Williamson, V., 1966, Who goes to Alcoholics Anonymous? *Lancet* 11 (Aug. 12):382.
Edwards, G., Hensman, C., Hawker, A., and Williamson, V., 1967, Alcoholics Anonymous: The anatomy of a self-help group. *Social Psychiatry* 1:195.
Einstein, S., Wolfson, E., and Geght, D., 1970, What matters in treatment: Relevant variables in alcoholism, *Int. J. Addict.* 5:43.
Elkins, H. K., 1966, Our mutual sins, *A.A. Grapevine* 22(11):28.
Ellis, A. S., and Krupinski, J., 1964, The evaluation of a treatment programme for alcoholics; a follow-up study, *Med. J. Aust.* 51:8.
Ellison, J., 1964, Alcoholics Anonymous: Dangers of success, *Nation* 198(2):212.
Emrick, C. D., 1970, Abstinence and time perception of alcoholics, *Q. J. Stud. Alcohol* 31:384.
Endore, G., 1968, "Synanon," Doubleday, Garden City.
Ends, E. J., and Page, C. W., 1959, Group psychotherapy and concomitant psychological change, *Psychological Monographs* 73 (480).
Esser, P. H., 1961a, Group psychotherapy with alcoholics, *Br. J. Addict.* 57:105.
Esser, P. H., 1961b, Group psychotherapy with alcoholics. Its value for significs, *Q. J. Stud. Alcohol* 22:646.
Esser, P. H., 1968, Conjoint family therapy for alcoholics, *Br. J. Addict.* 63(3/4):177.
Evrard, F. H., 1971, "Successful Parole," Charles C Thomas, Springfield, Illinois.
F., J. I., 1947, Alcoholism—an occupational disease of seamen. Approaches to a solution of the problem in the Port of New York, *Q. J. Stud. Alcohol* 8:498.
Fellowship of Alcoholics Anonymous, The, 1948, Proc. Industr. Conf. Alcsm., N.P.
Fenichel, O., 1945, "The Psychoanalytic Theory of Neurosis," Norton, New York.
Ferenczi, S., 1916, "First Contributions to Psychoanalysis," Hogarth, London.
Ferenczi, S., 1926, "Further Contributions to the Theory and Technique of Psychoanalysis," Basic Books, New York.
Ferguson, C. W., 1936, "Confusion of Tongues," Doubleday Doran, Grand Rapids.
Fitzgerald, B. J., Pasewark, R. A., and Clark, R., 1971, Four-year follow-up of alcoholics treated at a rural state hospital, *Q. J. Stud. Alcohol* 32:636.
Flaherty, J. A., McGuire, H. T., and Gatski, R. I., 1955, The psychodynamics of the "dry drunk," *Am. J. Psychiatry* 112:460.
Flesch, R., 1961, The open mind, *A.A. Grapevine* 17(11):36.
Fletcher, A. N., 1944, Alcoholics Anonymous, *Trained Nurse* 112:194.
Florida Corrections respond to opportunities, challenges, crises, 1971, *American Journal of Correction* 33(4):18.
Ford, J. C., 1951, "Depth Psychology, Morality, and Alcoholism," Weston College, Weston, Massachusetts.
Ford, J. C., 1955, "Man Takes a Drink," Kennedy and Sons, New York.
Fort, T., and Porterfield, A. L., 1961, Some backgrounds and types of alcoholism among women, *J. Health Hum. Behav.* 2:283.
Foster, F. M., Horn, J. L., and Wanberg, K. W., 1972, Dimensions of treatment outcome: A factor-analytic study of alcoholics' responses to a follow-up questionnaire, *Q. J. Stud. Alcohol* 33:1079.
Fox, R., 1967a, Disulfiram (Antabuse) as an adjunct in the treatment of alcoholism, *in* "Alcoholism: Behavioral Research, Therapeutic Approaches" (R. Fox, ed.), pp. 242–255, Springer Publishing, New York.

Fox, R., 1967b, Is LSD of value in treating alcoholics? *in* "The use of LSD in Psychotherapy and Alcoholism" (H. A. Abramson, ed.), pp. 477–495, Bobbs-Merrill, Indianapolis.

Fox, R., 1967c, A multidisciplinary approach to the treatment of alcoholism, *Am. J. Psychiatry* 123:7.

Fox, R., 1973a, What a doctor learned from A.A. *A.A. Grapevine* 29(8):26.

Fox, R., 1973b, Treatment of the problem drinker by the private practitioner, *in* "Alcoholism: Progress in Research and Treatment," (P. G. Bourne and R. Fox, eds.), pp. 227–243, Academic Press, New York.

Fox, R., and Leach, B., 1970, Office treatment of the alcoholic by the private physician, *in* "World Dialogue on Alcohol and Drug Dependence" (E. D. Whitney, ed.), pp. 173–198, Beacon Press, Boston.

Fox, R., and Lyon, P., 1955, "Alcoholism: Its Scope, Cause, and Treatment," Random House, New York.

Fox, V., 1966, Escape, *A.A. Grapevine* 23(7):14.

Fox, V., and Smith, M. A., 1959, Evaluation of a chemopsychotherapeutic program for the rehabilitation of alcoholics; observations over a two-year period, *Q. J. Stud. Alcohol* 20:767.

Frank, J. D., 1961, "Persuasion and Healing: A Comparative Study of Psychotherapy," Johns Hopkins, Baltimore.

Free, J. L., 1955, "Just One More," Coward-McCann, New York.

Freud, S., 1912, Contributions to the psychology of love. The most prevalent form of degradation in erotic life (1912), *in* "Collected Papers" Vol. IV, pp. 203–216, Hogarth, London, 1925.

Freud, S., 1917, Mourning and melancholia, *in* "Collected Papers" Vol. IV, pp. 152–170, Hogarth, London, 1925.

Freud, S., 1930, Three contributions to the theory of sex (4th ed.), Nervous Mental Diseases Publishing House, Washington D.C.

Fullam, T., 1950, "Here's to Sobriety; A Plain Approach to Understanding the Compulsive Drinker and His Problems," Abelard Press, New York.

Fuller, G. B., 1965, Validity of the Alcadd Test in identifying alcoholism and some of its attributes, *Current Conclusions* 3:29.

Furman, J. C., 1949, The Alcoholics Anonymous organization in the United States of America, *in* International Congress on Alcoholism 23:209.

Furnas, J. C., 1965, "The Life and Times of the Late Demon Rum," Putnam's Sons, New York.

Gamblers Anonymous, 1962, G. A. Publishing Co., Los Angeles.

Gartner, A., and Riessman, F., 1968, "Law and Order, A New Approach," New York University, New York.

Geary, J. R., 1967, Alcoholics Anonymous: An effective form of behavior therapy, Read at the First Annual Meeting of the Association for Advancement of the Behavioral Therapies, Washington, D.C., September 2, 1967. 20 pp., mimeographed.

Gellman, I. P., 1962, Alcoholics Anonymous: An autonomous organization of social deviants. Doctoral dissertation, New York University.

Gellman, I. P., 1964, "The Sober Alcoholic: An Organizational Analysis of Alcholics Anonymous," College and University Press, New Haven.

Gerard, D. L., 1959, Personality and social factors in intoxication and addiction, *in* "Drinking and Intoxication" (R. G. McCarthy, ed.), pp. 298–305, Free Press, Glencoe.

Gerard, D. L., Saenger, G., and Wile, R., 1962, The abstinent alcoholic, *A.M.A. Arch. Gen. Psychiatry* 6:83.

Gibbins, R. J., Smart, R. G., and Seeley, J. R., 1959, A critique of the Manson Evaluation test, *Q. J. Stud. Alcohol* 20:257.

Gillespie, D. G., 1967, The fate of alcoholics: An evaluation of follow-up studies, *in* "Alcoholism" (D. J. Pittman, ed.), pp. 159–173, Harper & Row, New York.

Gitlow, S. E., 1968, A pharmacological approach to alcoholism, *A.A. Grapevine* 24(5):12.
Glaser, F. B., 1971, Gaudenzia, Inc.: Historical and theoretical background of a self-help addiction treatment program, *Int. J. Addict.* 6(4):615.
Glasser, W., 1965, "Reality Therapy: A New Approach to Psychiatry," Harper & Row, New York.
Glatt, M. M., 1959, An alcoholic unit in a mental hospital, *Lancet* 2:397.
Glatt, M. M., 1967, Treatment of alcohol dependency, *Lancet* (London) No. 7493:791.
Goodman, H. T. Jr., 1970, D.D.T.—the drinker's dilemma and treatment, a psychiatrist's view, *Ohio State Med. J.* 66:684.
Gordon, H. L., and Hooker, C. A., 1969, Opinions of alcoholics concerning effectiveness of various treatment methods, *Newsletter for Research in Psychology* 11(1):24.
Green, K. M., 1968, Don't blame the policeman, *Canada's Mental Health* (Ottawa) 16(5):16.
Grover, M. L., 1967, Drinking patterns in California; a special study in one community, *Alcoholism*, Zagreb 3:22.
Groves, D. H., 1972, charismatic leadership in Alcoholics Anonymous; a case study, *Q. J. Stud. Alcohol* 33:684.
Gynther, M. D., and Brilliant, P. J., 1967, Marital status, readmission to hospital, and intrapersonal and interpersonal perceptions of alcoholics, *Q. J. Stud. Alcohol* 28:52.
Haasz, I., 1967, Alkoholicari su alkoholicarima upalili avjetlo, *Al-Klub* (Zagreb) 4(1–2):6.
Habbe, S., 1969, The drinking employee—management's problem? *Conference Board Record* 6(2):27.
Habbe, S., 1970, Company controls for drinking problems (Personnel Policy Study #218) National Industrial Conference Board, New York.
Haberman, P. W., 1966, Factors related to increased sobriety in group psychotherapy with alcoholics, *J. Clin. Psychol.* 22:229.
Haertzen, C. A., Hooks, N. T., and Monroe, J. J., 1968, Nonsignificance of membership in Alcoholics Anonymous in hospitalized alcoholics, *J. Clin. Psychol.* 24:99.
Haggard, H. W., and Jellinek, E. M., 1942, "Alcohol Explored," Doubleday, Doran, Garden City.
Haggard, H. W., and Jellinek, E. M., 1944, Two Yale savants stress alcoholism as true disease, *A.A. Grapevine* 1(1):1.
Hagie, F. E., 1971, Young people initiate Richmond drug council, *J. Indiana State Med. Assoc.* 64(5):391.
Hanfmann, E., 1951, The life history of an ex-alcoholic, *Q. J. Stud. Alcohol* 14:468.
Hart, W. T., Gardner, E. A., and Zax, M., 1964, Community facilities for alcoholics in Rochester, N.Y., *Q. J. Stud. Alcohol* 25:747.
Hartcollis, P., 1968, Denial of illness in alcoholism, *Bull. Menninger Clin.* 32:47.
Haverstick, J., 1955, Big book: Bible for alcoholics, *Saturday Review*, August 27:17.
Hayman, M., 1956, Current attitudes to alcoholism of psychiatrists in Southern California, *Am. J. Psychiatry* 112:485.
Hayman, M., 1966, "Alcoholism: Mechanism and Management," Charles C Thomas, Springfield, Illinois.
Heard, G., 1958, The search for ecstasy, *A.A. Grapevine* 14(12):15.
Heersema, P. H., 1942, Present role of Alcoholics Anonymous in the treatment of chronic alcoholism, *Minn. Med.* 25:204.
Hewitt, C. C., 1943, A personality study of alcohol addiction, *Q. J. Stud. Alcohol* 4:368.
Hill, M. J., and Blane, H. T., 1967, Evaluation of psychotherapy with alcoholics: A critical review, *Q. J. Stud. Alcohol* 28:76.
Hobbs, A. H., 1960, The consumption of alcohol and the hypothesis of reciprocal complementarity, *Am. J. Psychiatry* 117:228.

Hoffer, A., and Osmond, H., 1968, "New Hope for Alcoholics," University Books, New Hyde Park, New York.
Holmes, R. M., 1970, Alcoholics Anonymous as group logotherapy, *Pastoral Psychology* 21(202):30.
Hoover, M. P., 1960, Management of acute alcoholic intoxication, *Can. Med. Assoc. J.* 83:1352.
Hopkins, A., 1944, Alcoholics give famous producer moving experience, *A.A. Grapevine* 1(3):1.
Hunter, A. A., 1945, Sees A. A. rise, Los Angeles, *Christian Century,* January 3:22.
Hunter, O. B., 1947, Treatment of alcoholism in Washington and the Yale Plan, *Med. Ann. D. C.* 16:633.
Hurvitz, N., 1970, Peer self-help psychotherapy groups and their implications for psychotherapy, *Psychotherapy* 7(1):41.
I think you guys mean it! 1947, *Federal Probation* 11(1):8.
Ignatia, M., 1951, The care of alcoholics, *Hospital Progress* 32:293.
It's fun to be sober, 1945, *Newsweek,* January 1:74.
Jackson, C., 1944, "The Lost Weekend," Farrar Straus, New York.
Jackson, D., 1960, "Stumbling Block," Abingdon Press, New York.
Jackson, J. K., 1956, The adjustment of the family to alcoholism, *Marriage and Family Living* 18:361.
Jackson, J., and Connor, R., 1953, The Skid Road alcoholic, *Q. J. Stud. Alcohol* 14:468.
Jackson, J. K., Mykut, M., Burr, R. C., and Fogan, R. J., 1960, The alcoholism training program at the University of Washington School of Medicine, *Q. J. Stud. Alcohol* 21:298.
Jackson, J. K., and Kogan, K. L., 1963, The search for solutions: Help-seeking patterns of active and inactive alcoholics, *Q. J. Stud. Alcohol* 24:449.
James, B. E., 1950, Study of the personal and social adjustment characteristics of members of Alcoholics Anonymous, Master's thesis, Kent State University.
James, W. H., 1902, "The Varieties of Religious Experience: A Study in Human Nature," Longmans Green, New York.
Jellinek, E. M., 1942, "Alcohol Addiction and Chronic Alcoholism," New Haven Press.
Jellinek, E. M., 1946, Phases in the drinking history of alcoholics; analysis of a survey conducted by the official organ of Alcoholics Anonymous, *Q. J. Stud. Alcohol* 7:1.
Jellinek, E. M., 1952, Phases of alcohol addiction, *Q. J. Stud. Alcohol* 13:673.
Jellinek, E. M., 1960, "The Disease Concept of Alcoholism," College and University Press, New Haven.
Jellinek, E. M., 1962, Cultural differences in the meaning of alcoholism, *in* "Society, Culture, and Drinking Patterns" (D. J. Pittman and C. R. Snyder, eds.), pp. 382–388, Wiley, New York.
Jellinek, E. M., and Haggard, H. W., 1942, "Alcohol Explored," Doubleday Doran, Garden City.
Jensen, S. E., 1962, A treatment program for alcoholics in a mental hospital, *Q. J. Stud. Alcohol* 23:315.
Jensen, S. E., and Ramsay, R., 1963, Treatment of chronic alcoholism with lysergic acid diethylamide, *Can. Psychiatr. Assoc. J.* 8:182.
Johnson, M. W., 1965, Physicians' views on alcoholism: With special reference to alcoholism in women, *Nebr. State Med. J.* 50:378.
Johnson, M. W., DeVries, J. C., and Houghton, M. I., 1966, The female alcoholic. *Nursing Research* 15:343.
Jones, K. L., Shainberg, L. W., and Byer, C. O., 1969, Alcoholism, *in* "Drugs and Alcohol" (K. Jones, ed.), pp. 99–109, Harper & Row, New York.
Jones, R. K., 1970, Sectarian characteristics of Alcoholics Anonymous, *Sociology* (Oxford) 4:181.
Källström, B., 1959, Om verksamheten vid Statens vardanstalt för alkoholmissbrukare a Venngarn, *Sven. Läkartidn.* 56:19.

Kamner, M. E., and Dupong, W. G., 1969, Alcohol problems: study by industrial medical department, *N.Y. State J. Med.* 69(24):3105.

Kant, F., 1954, "The Treatment of the Alcoholic," Charles C Thomas, Springfield, Illinois.

Karp, S. A., Witkin, H. A., and Goodenough, D. R., 1965, Alcoholism and psychological differentiation: Effect of achievement of sobriety on field dependence, *Q. J. Stud. Alcohol* 26:580.

Katz, L., 1966, The Salvation Army Men's Social Service Center II. Results, *Q. J. Stud. Alcohol* 27:636.

Keith-Lucas, A., 1970, Helping people is a ticklish business, *A.A. Grapevine* 27(7):8.

Keller, J. E., 1966, "Ministering to Alcoholics," Augsburg Publishing, Minneapolis.

Keller, M., 1972, The oddities of alcoholics, *Q. J. Stud. Alcohol* 33:1147.

Keller, M., Jordy, S. S., and Efron, V., 1968, "International Bibliography of Studies on Alcohol Vol. II: Indexes, 1901-1950," Rutgers, New Brunswick, New Jersey.

Kelly, C. M., 1968, Fighting poverty with health care, *Am. J. Nurs.* 68(2):282.

Kennedy, R. J. H., 1966, "Steps to Sobriety," Joseph F. Wagner, New York.

Kessel, N., and Walton, H., 1965, "Alcoholism," Penguin, Baltimore.

Kessell, J., 1962, "The Road Back," Knopf, New York.

King, A. R., 1953, "Basic Information on Alcohol, Revised Edition," Cornell College Press, Mt. Vernon, Iowa.

Kish, G. B., and Hermann, H. T., 1971, The Fort Meade alcoholism treatment program; a follow-up study, *Q. J. Stud. Alcohol* 32:628.

Kissen, M. D., 1961, The treatment of alcoholics, *Q. J. Stud. Alcohol* Suppl. 1:101.

Kiviranta, P., 1969, "Alcoholism Syndrome in Finland," Finnish Foundation for Alcohol Studies, Helsinki.

Knight, R. P., 1937a, The dynamics and treatment of chronic alcohol addiction, *Bull. Menninger Clin.* 1:233.

Knight, R. P., 1937b, The psychodynamics of chronic alcoholism, *J. Nerv. Ment. Dis.* 86:538.

Knox, W. J., 1969, Attitudes of psychologists toward alcoholism, *J. Clin. Psychol.* 25:446.

Knox, W. J., 1971, Attitudes of psychiatrists and psychologists toward alcoholism, *Am. J. Psychiatry* 127:1675.

Knudsen, R., 1959, Public and private measures in Norway for the prevention of alcoholism (15 pp.), International Bureau against Alcoholism, Lausanne.

Koumans, A. J. R., 1969, Reaching the unmotivated patient, *Ment. Hyg.* 53:298.

Krouse, W., 1965, Doctor's report, *A.A. Grapevine* 22(3):25.

Krystal, H., and Moore, R. A., 1963, Who is qualified to treat the alcoholic? *Q. J. Stud. Alcohol* 24:705.

Krzysztof, B., 1970, Z pola walki z alkoholizmem w stolicy: Wywiad z przewodniczacym prezydium rady narodowej m. St. Warszawy, Inz. Jerzym Majewskim, *Problemy Alkoholizmu* (Warszawa) 5(1):1.

Kurland, A. A., 1968, Maryland alcoholics: Follow-up study 1, *Psychiatric Research Reports of the American Psychiatric Association* 24:71.

Lahey, W. W., 1950, A comparison of personal and social factors identified with selected members of Alcoholics Anonymous, Doctoral thesis, University of Southern California.

Lambert, B., 1958, Länkrörelsen och dess resultat, *Social-Med. Tidskr* 35:275.

Landis, C., 1945, Theories of the alcoholic personality, *in* "Alcohol, Science, and Society" pp. 129-142, Quarterly Journal of Studies on Alcohol, New Haven.

Latimer, R., 1953, The social worker and the A.A. program in a state hospital, *Journal of Psychiatric Social Work* 22:175.

Lawrence, F. E., 1961, The outpatient management of the alcoholic, *Q. J. Stud. Alcohol* Suppl. 1:117.

Lawrence, F. G., 1962, Alcoholics Anonymous: A clergyman's viewpoint, *in* "Problems in Addiction: Alcoholism and Narcotics" (C. W. Bier, ed.), pp. 127–132, Fordham University Press, New York.

Leach, B., 1968, Alcoholics Anonymous: Its effectiveness, nature, and availability, *in* "Procedures of the 28th International Congress on Alcohol and Alcoholism, Vol. I: Abstracts" (M. Keller and M. Majchrowicz, eds.), pp. 56–57, Rutgers, New Brunswick, New Jersey.

Leach, B., 1973, Does Alcoholics Anonymous really work? *in* "Alcoholism: Progress in Research and Treatment" (P. Bourne and R. Fox, eds.), pp. 245–284, Academic Press, New York.

Leach, B., Norris, J. L., Dancey, T., and Bissell, L., 1969, Dimensions of Alcoholics Anonymous: 1935–1965, *Int. J. Addict.* 4:507.

Lee, J. P., 1960, Alcoholics Anonymous as a community resource, *Social Work,* N.Y. 5(4):20.

Leggo, C., 1967, The supervisor and the alcoholic employee, *J. Occup. Med.* 9:96.

Lemere, F., 1953, What happens to alcoholics? *Am. J. Psychiatry* 109:674.

Lemert, E. M., 1954, "Alcohol and Northwest Coast Indians," University of California Press, Berkeley.

Lemert, E. M., 1962, Alcohol use in Polynesia, *Trop. Geogr. Med.* 14:183.

Lewin, K., and Grabbe, P., 1948, Conduct, knowledge, and acceptance of new values, *in* "Resolving Social Conflicts" (G. Lewin, ed.), pp. 56–68, Harper and Brothers, New York.

Lincoln, V., 1958, I was made to be me, *A.A. Grapevine* 15(4):12.

Lindt, H., 1959, The "rescue fantasy" in group treatment of alcoholics, *Int. J. Group Psychother.* 9:43.

Lisansky, E., 1960, The etiology of alcoholism: The role of psychological predisposition, *Q. J. Stud. Alcohol* 21:314.

Lister, L. M., 1971, Inpatient alcoholic rehabilitation care, *Md. State Med. J.* 20(5):33.

"Living Sober," 1975, A.A, World Services, New York.

"Living with an Alcoholic with the Help of Al-Anon," 1966, Al-Anon Family Group Headquarters, New York.

Lofland, J. R., and Lejeune, R. A., 1960, Initial interaction of newcomers in Alcoholics Anonymous: A field experiment in class symbols and socialization, *Social Problems* 8:102.

Logan, A. B., 1961, Can we permit justice to remain blind? *A.A. Grapevine* 17(4):26.

Lolli, G., 1960, "Social Drinking," World, New York.

Lovald, K., and Neuwirth, G., 1968, Exposed and shielded drinking; drinking as role behavior and some consequences for social control and self-concept. *Arch. Gen. Psychiatry* 19:95.

Lovell, H. W., 1953, "Hope and Help for the Alcoholic," Doubleday, Garden City.

Low, A. A., 1950, "Mental Health Through Will Training," Christopher Publishing House, Boston.

Lowry, M., 1947, "Under the Volcano," Reynal and Hitchcock, New York.

Ludwig, A. M., 1972, On and off the wagon; reasons for drinking and abstaining by alcoholics, *Q. J. Stud. Alcohol* 33:91.

MacCormick, A. H., 1963, Correctional views on alcohol, alcoholism, and crime, *Crime and Delinquency* 9(1):15.

Machover, S., and Puzzo, F. S., 1959a, Clinical and objective studies of personality variables in alcoholism. I. Clinical investigation of the "alcoholic personality," *Q. J. Stud. Alcohol* 20:505.

Machover, S., and Puzzo, F. S., 1959b, Clinical and objective studies of personality variables in alcoholism II. Clinical study of personality correlates of remission from active alcoholism, *Q. J. Stud. Alcohol* 20:520.

Machover, S., Puzzo, F. S., Machover, K., and Plumeau, F., 1959c, Clinical and objective studies of personality variables in alcoholism III. An objective study of homosexuality in alcoholism, *Q. J. Stud. Alcohol* 20:528.

Machover, S., Puzzo, F. S., and Plumeau, F. S., 1962, Values in alcoholics, *Q. J. Stud. Alcohol* 23:267.

Madsen, W., 1964, The alcoholic agringado, *Am. Anthropol.* 66:355.

Madsen, W., 1974, "The American Alcoholic," Charles C Thomas, Springfield, Ill.

Magnum, E. K., 1965, Court class on alcoholism, *A.A. Grapevine* 21(12):33.

Maguire, J. F., 1864, "Father Mathew," Sadlier & Co., New York.

Maine, H. [pseud.; Winslow, Walker], 1947, "If a Man Be Mad," Doubleday, New York.

Mann, M., 1946, Alcoholics Anonymous; new partner for hospitals, *Mod. Hosp.* 66(1):77.

Mann, M., 1947, What shall we do about alcoholism? *Vital Speeches* 13:253.

Mann, M., 1958, "New Primer on Alcoholism," Holt, Rinehart & Winston, New York.

Mann, M., 1970, "Marty Mann Answers Your Questions about Drinking and Alcoholism," Holt, Rinehart & Winston, New York.

Manson, M. P., 1948, A psychometric differentiation of alcoholics from nonalcoholics, *Q. J. Stud. Alcohol* 9:175.

Manson, M. P., 1949, A psychometric analysis of psychoneurotic and psychosomatic characteristics of alcoholics, *J. Clin. Psychol.* 5:77.

Markey, M., 1939, Alcoholics and God, *Liberty Magazine*, September 30:6.

Marquis, P. A., 1971, Le traitement des alcooliques, *Laval Médical* (Quebec) 42(6):604.

Martin, M., Best, E. W. R., Josie, G., and Leblanc, R., 1962, Survey of organizatons in the field of alcoholism in Canada—1960, *Laval Médical* (Quebec) 33:586.

Mascarenhas, E., Hanrahan, F. R., and Plucinsky, J. J., 1970, Rosary Hall; an AA-oriented hospital alcoholic care unit, *Ohio State Med. J.* 66:812.

Mathias, R., 1955, An experimental investigation of the personality structure of chronic alcoholic, Alcoholics Anonymous, neurotic and normal groups, Doctoral dissertation, University of Buffalo.

Matkom, A. J., 1965, The alcoholic in the state mental hospital, *Q. J. Stud. Alcohol* 26:499.

Matkom, A. J., 1969, An alcoholic treatment center in a general hospital, *Q. J. Stud. Alcohol* 30:453.

Maurer, H., 1968, The beginning of wisdom about alcoholism, *Fortune* 77(5):177.

Maxwell, M. A., 1949, Social factors in the Alcoholics Anonymous program, Doctoral dissertation, University of Texas.

Maxwell, M. A., 1950, The Washingtonian movement, *Q. J. Stud. Alcohol* 11:440.

Maxwell, M. A., 1951, Interpersonal factors in the genesis and treatment of alcohol addiction, *Social Forces* 29(4):443.

Maxwell, M. A., 1954, Factors affecting an alcoholic's willingness to seek help, *Northwest Sci.* 28:116.

Maxwell, M. A., 1960, Early identification of problem drinkers in industry, *Q. J. Stud. Alcohol* 21:655.

Maxwell, M. A., 1962, Alcoholics Anonymous: An interpretation, *in* "Society, Culture, and Drinking Patterns" (D. J. Pittman and C. R. Snyder, eds.), pp. 577–585, John Wiley and Sons, New York.

[Maxwell, M. A.], 1971, Anonymity's spiritual dividends, *A.A. Grapevine* 28(7):10.

Maxwell, M. A., 1975, Changing patterns of sponsorship, *A.A. Grapevine* 32(4):12.

Mayer, J., and Myerson, D. J., 1971, Outpatient treatment of alcoholics; effects of status, stability, and nature of treatment, *Q. J. Stud. Alcohol* 32:620.

McAfee, W. T., 1952, Alcoholics Anonymous: An evaluative study, Doctoral dissertation, University of Chicago.

McCarthy, R. G., 1946, Group therapy in an outpatient clinic for the treatment of alcoholism, *Q. J. Stud. Alcohol* 7:98.

McCarthy, R. G., 1964, The fellowship of Alcoholics Anonymous, in "Alcohol Education for Classroom and Community" (R. G. McCarthy, ed.), pp. 226–234, McGraw-Hill, New York.
McCarthy, R. G., and Douglass, E. M., 1949, "Alcohol and Social Responsibility," Thomas Y. Crowell Co., New York.
McClelland, D. C., Davis, W. N., Kalin, R., and Wanner, E., 1972, "The Drinking Man," Free Press, New York.
McElroy, C., and O'Brien, J., 1963, Alcoholism and total abstinence, *America,* January 5:18; January 12:52.
McGinnis, C. A., 1963, The effect of group therapy on the ego-strength scale scores of alcoholic patients, *J. Clin. Psychol.* 19:346.
McInerney, J., 1970, The use of Alcoholics Anonymous in a general hospital alcoholism treatment program, *Medical Ecology and Clinical Research,* Park Ridge, Ill. 3(1):21.
McKinney, G. F., 1946, Alcoholics Anonymous at New York's Wallkill Prison, *Prison World* 8(2):20.
McMahan, H. G., 1942, The psychotherapeutic approach to chronic alcoholism in conjunction with the Alcoholic Anonymous program, *Illinois Psychiatry Journal* 2:15.
McQueen, S. D., 1960, A.A. is the chaplain's ally, *A.A. Grapevine* 17(1):14.
Meer, B., and Amon, A. H., 1963, Age-sex preference patterns of alcoholics and normals, *Q. J. Stud. Alcohol* 24:417.
Menninger, K. A., 1938, "Man Against Himself," Harcourt, Brace, New York.
Menninger, K. A., 1965, The vital balance, *A.A. Grapevine* 22(2):13.
Metcalfe, G. E., 1944, Alcoholics Anonymous, *J. Indiana State Med. Assoc.* 37:684.
Meyer, W. T., 1967, The application of role theory in an investigation of affiliation with Alcoholics Anonymous, Master's thesis, Western Michigan University.
Meyers, T. J., 1941, What solution for the problem of chronic alcoholism? *J. Am. Osteopath. Assoc.* 41:153.
Middlebrooks, A. E., 1945, Urges treatment for alcoholics, *Christian Century,* November 14:1264.
Miller, B. A., Pokorny, A. D., Valles, J., and Cleveland, S. E., 1970a, Biased sampling in alcoholism treatment research, *Q. J. Stud. Alcohol* 31:97.
Miller, B. A., Pokorny, A. D., and Kanas, T. E., 1970b, Problems in treating homeless, jobless alcoholics, *Hospital and Community Psychiatry* 21:98.
Milt, H., 1967, "Basic Handbook on Alcoholism," Scientific Aids Publications, Fair Haven, New Jersey.
Mindlin, D. F., 1964, Attitudes toward alcoholism and toward self: Differences between three alcoholic groups, *Q. J. Stud. Alcohol* 25:136.
Minogue, S. J., 1946, Alcoholism, *Med. J. Aust.* 33:271.
Minogue, S. J., 1948, Alcoholics Anonymous, *Med. J. Aust.* 35:586.
Mitchell, E. H., 1958, Use of citrated calcium carbimide in alcoholism, *J. Am. Med. Assoc.* 168:2008.
Mitchell, E. H., 1961, Rehabilitation of the alcoholic, *Q. J. Stud. Alcohol* Suppl. 1:93.
Mitchell, T. B., 1941, Effective therapy in chronic alcoholism, *South. Med. Surg.* 103:656.
Mogan, R. E., Helm, S. T., Snedeker, M. R., Snedeker, M. H., and Wilson, W. M., 1969, Staff attitudes toward the alcoholic patient, *Arch. Gen. Psychiatry* 21:449.
Moore, R. A., 1965, Some countertransference reactions in the treatment of alcoholism, *Psychiatry Digest* 26(11):35.
Moore, R. A., 1971, Alcoholism treatment in private psychiatric hospitals; a national survey, *Q. J. Stud. Alcohol* 32:1083.
Moore, R. A., and Buchanan, T. K., 1966, State hospitals and alcoholism. A nation-wide survey of treatment techniques and results, *Q. J. Stud. Alcohol* 27:459.

Moore, R. A., and Ramseur, F., 1960a, Effects of psychotherapy in an open-ward hospital on patients with alcoholism, *Q. J. Stud. Alcohol* 21:233.
Moore, R. A., and Ramseur, F., 1960b, A study of the background of 100 hospitalized veterans with alcoholism, *Q. J. Stud. Alcohol* 21:51.
Moore, R. A., and Wood, J. T., 1963, Alcoholism and its treatment in Yugoslavia, *Q. J. Stud. Alcohol* 24:128.
Morales-de-Flores, I., 1970, Organization of local services for alcoholism control, in "Alcohol and Alcoholism" (R. E. Popham, ed.), pp. 368–371, University of Toronto Press.
Morice, A., 1963, Les rapports de l'alcoolisme et des accidents du travail, *Bull. Acad. Nat. Méd.* 147:86.
Mowrer, H. R., 1959, A psychocultural analysis of the alcoholic, in "Drinking and Intoxication" (R. G. McCarthy, ed.), pp. 287–297, Free Press, Glencoe.
Mowrer, O. H., 1961, "The Crisis in Psychiatry and Religion," Van Nostrand, New York.
Mowrer, O. H., 1964, "The New Group Therapy," Van Nostrand, New York.
Mowrer, O. H., (ed.), 1967, "Morality and Mental Health," Rand McNally, Chicago.
Mullan, H., and Sangiuliano, I., 1966, "Alcoholism: Group Psychotherapy and Rehabilitation," Charles C Thomas, Springfield, Illinois.
Murphy, M. M., 1953, Values stressed by two social class levels at meetings of Alcoholics Anonymous, *Q. J. Stud. Alcohol* 14:576.
Nace, R. K., 1949, Alcoholics Anonymous speaks to the church, *J. Clin. Pastoral Work* 2:124.
Nelson, A., 1941, Alcoholics Anonymous, *Med. J. Aust.* 28(2):43.
New York City Alcoholism Study: A Report, 1962, National Council on Alcoholism, New York.
Newlove, D., 1974, "The Drunks," Saturday Review Press/E. P. Dutton, New York.
Norris, J. L., 1948, Cost and remedy of the alcoholic hangover in industry, *Ind. Med.* 17:129.
Norris, J. L., 1965, A look at A.A., *A.A. Grapevine* 22(6):17.
Norris, J. L., 1968a, Doctors and A.A., *A.A. Grapevine* 24(5):4.
Norris, J. L., 1968b, Alcoholism in industry, *Arch. Environ. Health* 17(3):436.
Norris, J. L., 1970, Alcoholics Anonymous, in "World Dialogue on Alcohol and Drug Dependence" (E. D. Whitney, ed.), pp. 155–172, Beacon Press, Boston.
Norris, J. L., 1972, Acting within the Traditions, *A.A. Grapevine* 28(9):15.
Nørvig, J. and Nielsen, B., 1956, A follow-up study of 221 alcohol addicts in Denmark, *Q. J. Stud. Alcohol* 17:633.
Nunan, T. R., 1944, A note on the handling of alcoholics in a Navy camp, *Q. J. Stud. Alcohol* 5:426.
Nyárádi, É., 1970, A klubtherápia jelentösége az alkoholista beteg utógondozásában, *Alkohologia*, Bp. 1:24.
Oakley, S., and Holden, P. H., 1971, Alcoholic rehabilitation center; follow-up survey 1969, *Inventory*, N.C. 20(3):2.
Oates, W. E., 1966, "Alcohol In and Out of the Church," Broadman Press, Nashville.
O'Neill, E., 1946, "The Iceman Cometh," Random House, New York.
O'Neill, E., 1956, "Long Day's Journey into Night," Yale University Press, 1956.
Only, M. [pseud.], 1974, "High: A Farewell to the Pain of Alcoholism," Prentice-Hall, Englewood Cliffs, N.J.
Ottenberg, D. J., 1969, The Eagleville interdisciplinary rehabilitation program for alcoholics: Lessons after two years, *Q. J. Stud. Alcohol* 30:449.
Oursler, F., 1944, Charming is the word for alcoholics, *A.A. Grapevine* 1(2):1.
Pacy, H., 1968, The management of alcoholism in general practice, *Med. J. Aust.* 2(7):335.
Palola, E. G., Jackson, J. K., and Kelleher, D., 1961, Defensiveness in alcoholics; measures based on the Minnesota Multiphasic Personality Inventory, *Journal of Health and Human Behavior* 2:185.

Palola, E. G., Dorpat, T. L., and Larson, W. R., 1962, Alcoholism and suicidal behavior, *in* "Society, Culture, and Drinking Patterns" (D. J. Pittman and C. R. Snyder, eds.), pp. 511–534, Wiley, New York.

Park, P., 1962, Drinking experience of 806 Finnish alcoholics in comparison with similar experiences of 192 English alcoholics, *Acta Psychiatr. Scand.* 38:227.

Parker, J. B. Jr., Meiller, R. M., and Andrews, G. W., 1960, Major psychiatric disorders masquerading as alcoholism, *South. Med. J.* 53:560.

Parry, A.A., 1961, Doctors and alcoholism, *Med. Times* 89:659.

Paster, S., 1948, The treatment of chronic alcoholism, *Memphis Med. J.* 23:88.

Patrick, S. W., 1963, Our way of life: A short history of Neurotics Anonymous, *in* "Drug Addiction in Youth" (E. Harms, ed.) Vol. 3, pp. 148–157, Pergamon Press, Oxford.

Pattison, E. M., 1966, A critique of alcoholism treatment concepts, with special reference to abstinence, *Q. J. Stud. Alcohol* 27:49.

Peer teacher-leaders aid in self-help groups, 1970, *Public Health Reports* 85(4):356.

Perceval, R., 1969, Alcoholism in Ireland; the role of the Irish National Council on Alcoholism, *Journal of Alcoholism* London 4:241.

Perelman, S. J., 1944, The meditation of old Mr. Perelman on "dynamic drunks," *A.A. Grapevine* 1(6):1.

Pfau, R., and Hirshberg, A., 1958, "Prodigal Shepherd," Lippincott, Philadelphia.

Pfeffer, A. Z., 1958, "Alcoholism," Grune & Stratton, New York.

Pfeffer, A. Z., and Berger, S., 1957, A follow-up study of treated alcoholics, *Q. J. Stud. Alcohol* 18:624.

Pfeiffer, E., 1968, "Disordered Behavior: Basic Concepts in Clinical Psychiatry," Oxford, New York.

Phillips, J., 1973, "Alcoholics Anonymous: An Annotated Bibliography, 1935–1972." 63 pp., privately published, London, Ohio.

Pinardi, N., 1966, The chronic drunkenness offender; what one city is doing about the revolving door, *Crime and Delinquency* 12:339.

Pinardi, N. J., 1963, Helping alcoholic criminals; a pilot study, *Crime and Delinquency* 9(1):71.

Pittman, D. J. (ed.), 1967, "Alcoholism," Harper & Row, New York.

Pittman, D. J., and Gordon, C. W., 1958, "Revolving Door: A Study of the Chronic Police Case Inebriate," Free Press, Glencoe.

Pittman, D. J., and Snyder, C. R., 1962, Responsive movements and systems of control, *in* "Society, Culture, and Drinking Patterns" (D. J. Pittman and C. R. Snyder, eds.), pp. 547–552, Wiley, New York.

Pittman, D. J., and Tate, R. L., 1969, A comparison of two treatment programs for alcoholics, *Q. J. Stud. Alcohol* 30:888.

Plaut, T. F. A., 1967, "Alcohol Problems: A Report to the Nation," Oxford, New York.

Plumeau, F., Machover, S., and Puzzo, F., 1960, Wechsler-Bellevue performances of remitted and unremitted alcoholics and their normal controls, *J. Consult. Psychol.* 27:240.

Pokorny, A. D., Miller, B. A., and Cleveland, S. E., 1968, Response to treatment of alcoholism: a follow-up study, *Q. J. Stud. Alcohol* 29:364.

Pope, B., 1956, Attitudes toward group therapy in a psychiatry clinic for alcoholics, *Q. J. Stud. Alcohol* 17:233.

Powdermaker, F., 1945, Dr. Powdermaker discusses the role of a psychiatrist in Alcoholics Anonymous, *A.A. Grapevine* 1(9):1.

Professional Committee of the Schizophrenia Foundation of New Jersey, The, 1970, "The Schizophrenias: Yours and Mine," Pyramid, New York.

Prothro, W. B., 1961, Alcoholics can be rehabilitated, *Am. J. Public Health* 51:450.

Quaranta, J. V., 1947, Alcoholism: A study of emotional maturity and homosexuality as related factors in compulsive drinking, Master's thesis, Fordham University.
Quehl, T. M., 1971, The M.D. and the alcoholic, *J. Fla. Med. Assoc.* 58(4):43.
Quinn, W. F., 1967, Narcotic and dangerous drug problems: Current status of legislation, control, and rehabilitation, *Calif. Med.* 106(2):108.
Rado, S., 1933, Psychoanalysis of pharmacothymia, *Psychoanalytical Quarterly* 2:1.
Raiteri, M., 1969, Il problema dell' alcoolismo considerato dal punto di vista associativo, *Alcoholism* (Zagreb) 5(2):122.
Ramsay, R., Jensen, S., and Sommer, R., 1963, Values in alcoholics after LSD-25, *Q. J. Stud. Alcohol* 24:443.
Randall, T., 1957, "The Twelfth Step," Scribner's, New York.
Rea, F. B., 1956, "Alcoholism: Its Psychology and Cure," Philosophical Library, New York.
Reed, L. K., 1970, Let's take another look at alcoholism, *Ohio State Med. J.* 66(8):816.
Regan, J. R. Jr., 1962, "What About Alcohol?" National Council of the Churches of Christ, New York.
Reinert, R. E., 1965, The alcoholism treatment program at Topeka Veterans Administration Hospital, *Q. J. Stud. Alcohol* 26:674.
Reinert, R. E., 1968, The concept of alcoholism as a disease, *Bull. Menninger Clin.* 32:21.
Requet, A., 1951, Aspect actuel de la cure de l'alcolisme psychiatrique masculin dans la région lyonnaise, *Progres Medical,* Paris 59:1516.
Rhodes, R. J., Hames, G. H., and Campbell, M. D., 1969, The problem of alcoholism among hospitalized tuberculous patients; report of a national questionnaire survey, *Am. Rev. Respir. Dis.* 99:440.
Richards, T. B., 1968, The A.A. halfway house and service center, *in* "Alcoholism: The Total Treatment Approach" (R. Catanzaro, ed.), pp. 373-382, Charles C Thomas, Springfield, Illinois.
Ripley, H. S., and Jackson, J. K., 1959, Therapeutic factors in Alcoholics Anonymous, *Am. J. Psychiatry* 116:44.
Ritchie, O. W., 1948, A sociohistorical survey of Alcoholics Anonymous, *Q. J. Stud. Alcohol* 9:119.
Ritson, B., 1968, The prognosis of alcohol addicts treated by a specialized unit, *Br. J. Psychiatry* 114:1019.
Ritson, E. B., and Thompson, C. P., 1970, Planning a rural alcoholism program, *Br. J. Addict.* 65:199.
Robson, R. A. H., Paulus, I., and Clarke, C. G., 1965, An evaluation of the effect of a clinic treatment program on the rehabilitation of alcoholic patients, *Q. J. Stud. Alcohol* 26:264.
Rohan, W. P., 1970, A follow-up study of hospitalized problem drinkers, *Dis. Nerv. Syst.* 31:259.
Rohan, W. P., Tatro, R. L., and Rotman, S. R., 1969, MMPI changes in alcoholics during hospitalization, *Q. J. Stud. Alcohol* 30:389.
Rokeach, M., 1960, "The Open and Closed Mind," Basic Books, New York.
Roman, P. M., and Trice, H. M., 1968, The sick role, labelling theory, and the deviant drinker, *International Journal of Social Psychiatry* 14:245.
Rosenbarger, M., 1971, An investigation of selected community alcoholism resources and facilities and their relationship to posthospitalization adjustment of male alcoholics, Doctoral dissertation, University of Michigan.
Ross, J. L., 1965, Alcoholics Anonymous: A neglected adjunct to hospital treatment, *Kans. Med. Soc. J.* 56:23.
Rossi, J., 1970, A holistic treatment program for alcoholism rehabilitation, *Medical Ecology and Clinical Research,* Park Ridge, Ill. 3 (1):6.

Rossi, J. J., and Bradley, N. J., 1960, Dynamic hospital treatment of alcoholism, *Q. J. Stud. Alcohol* 21:432.
Rossi, J. J., Stach, A., and Bradley, N. J., 1963, Effects of treatment of male alcoholics in a mental hospital. A follow-up study, *Q. J. Stud. Alcohol* 24:91.
Roth, L. H., Rosenberg, N., and Levinson, R. B., 1971, Prison adjustment of alcoholic felons, *Q. J. Stud. Alcohol* 32:382.
Roth, R. J., 1965, William James and alcoholics, *American*, July 10:38.
Rothstein, E., Norton, B. A., Lahage, E. H., and Mueller, S. R., 1966, An experimental alcoholism unit in a psychiatric hospital, *Q. J. Stud. Alcohol* 27:513.
Rotman, D. B., 1945, Alcoholism, a social disease, *J. Am. Med. Assoc.* 127:564.
Roueche, B., 1960, "The Neutral Spirit," Little Brown, Boston.
Ruprecht, A. L., 1961, Day-care facilities in the treatment of alcoholics, *Q. J. Stud. Alcohol* 22:461.
Rusterholz, A., 1946, A.A.—Namenlose Alkoholiker, *Gesnd. Woholfahrt* 26:626.
Sagarin, E., 1969, "Odd Man In: Societies of Deviants in America," Quadrangle, Chicago.
Sann, P., 1971, "Kill the Dutchman! The Story of Dutch Schultz," Arlington House, New Rochelle.
Sapir, J. V., 1957, The A.A. story in Connecticut, *Connecticut Review of Alcoholism* 8:25.
Sariola, S., 1962, Anomia, vieraantuminen ja sosiaaliset orgelmat, *Alkoholipolitiikka*, Hels. 27:145.
Sariola, S., 1963, Anomi, alienation och de sociala problemen,*Alkoholpolitik*, Hels. 26:3.
Schilder, P., 1941, The psychogenesis of alcoholism, *Q. J. Stud. Alcohol* 2:277.
Schultz, J. D., 1961, Treatment of alcoholism in a general hospital, *Q. J. Stud. Alcohol* Suppl. 1:85.
Scott, E. M., 1970, "Struggles in an Alcoholic Family," Charles C Thomas, Springfield, Illinois.
Scott, N. A., 1947, "The Story of Mrs. Murphy," Dutton, New York.
Seaborn, R., 1970, Alcoholism in Australia, in "World Dialogue on Alcohol and Drug Dependence" (E. D. Whitney, ed.), pp. 68–69, Beacon Press, Boston.
Seabrook, W., 1935, "Asylum," Harcourt Brace, New York.
Seiden, R. H., 1960, The use of Alcoholics Anonymous members in research on alcoholism, *Q. J. Stud. Alcohol* 21:506.
Sell, J. T., 1946, An experience with Alcoholics Anonymous, *Federal Probation* 10 (4):42.
Selzer, M., 1958, On involuntary hospitalization for alcoholics; suggestions for an inpatient treatment program, *Q. J. Stud. Alcohol* 19:660.
Selzer, M. L., 1968, Michigan Alcoholism Screening Test (*MAST*): Preliminary report, *Univ. Mich. Med. Cent. J.* 34:143.
Selzer, M. L., and Holloway, W. H., 1957, A follow-up study of alcoholics committed to a state hospital, *Q. J. Stud. Alcohol* 18:98.
Sessions, P. M., 1966, Self-indulgence vs. self-destruction, *A.A. Grapevine* 22 (10):30.
Shea, J. E., 1954, Psychoanalytic therapy and alcoholism, *Q. J. Stud. Alcohol* 15:595.
Sherfey, M. D., 1955, Psychopathology and character structure in chronic alcoholism, in "Etiology of Chronic Alcoholism" (O. Diethelm, ed.), pp. 16–42, Charles C Thomas, Springfield, Illinois.
Shibutani, T., 1961, "Society and Personality," Prentice-Hall, Englewood Cliffs, New Jersey.
Shipp, T. J., 1963, "Helping the Alcoholic and His Family," Prentice-Hall, Englewood Cliffs, New Jersey.
Shoemaker, S. M., 1936, "National Awakening," Harper, New York.
Shoemaker, S. M., 1964, Those twelve steps, *A.A. Grapevine* 20 (8):26.
Siegler, M., Osmond, H., and Newell, S., 1968, Models of alcoholism, *Q. J. Stud. Alcohol* 29:571.

Silcott, E. J., 1971, The correspondence between Alcoholics Anonymous and the adaptive capacities of its members, Doctoral dissertation, Smith College.
Silkworth, W. D., 1937a, Alcoholism as manifestation of allergy, *Med. Rec.* 145:249.
Silkworth, W. D., 1937b, Reclamation of the alcoholic, *Med. Rec.* Apr. 21:1.
Silkworth, W. D., 1939a, A new approach to psychotherapy in chronic alcoholism, *Journal-Lancet* 59 (7):312.
Silkworth, W. D., 1939b, Psychological rehabilitation of alcoholics, *Med. Rec.* 150:65.
Silkworth, W. D., 1941, Highly successful approach to alcoholic problem confirmed by medical and sociological results, *Med. Rec.* 154 (3):105.
Silkworth, W. D., 1947, Doctor calls "slips" more normal than alcoholic, *A.A. Grapevine* 3 (8):1.
Silver Award: Emergency care for alcoholics—the Alcoholic Detoxification Center, Washington, D.C., 1971, *Hospital and Community Psychiatry* 22 (10):302.
Silverman, M., 1968, Community medicine in relation to a comprehensive general hospital centered psychiatric service, *in* "The Treatment of Mental Disorders in the Community" (G. Daniel, ed.), pp. 58–67, Williams and Wilkins, Baltimore.
Simm, U., 1965, Länkrörelsen i Storbrittanien, *Social-Med. Tidskr.* 42:206.
Simmel, E., 1948, Alcoholism and addition, *Psychoanal. Q.* 17:6.
Simonds, J., 1969, The halfway house for alcoholics, *Alcoholism Treatment Digest* 14 (1):1.
Slater, W. J. B., and Smith, J. A., 1961, Alcoholism and treatment facilities in the Cape Province, *Rehabilitation South Africa* 5 (2):95.
Slotkin, E. J., Levy, L., Wetmore, E., and Rank, F. N., 1971, "Mental Health Related Activities of Companies and Unions; a Survey Based on the Metropolitan Chicago Area," Behavioral Publications, New York.
Smith, C. M., 1959, Some reflections on the possible therapeutic effects of the hallucinogens, with special reference to alcoholism, *Q. J. Stud. Alcohol* 20:292.
Smith, J. A., 1961, Problems in the treatment of the alcoholic, *Q. J. Stud. Alcohol* Suppl. 1:129.
Smith, P. L., 1941, Alcoholics Anonymous, *Psychiatr. Q.* 15:554.
[Smith, R. H.], 1973, Dr. Bob's last major talk, *A.A. Grapevine* 30 (1):2.
Smithers, The Christopher D., Foundation, 1968, "Understanding Alcoholism," Scribner's, New York.
Soden, E. W., 1960, How a municipal court helps alcoholics, *Federal Probation* 24 (3):45.
Solms, H., and de Meuron, M., 1969, Thérapie de groupe pour alcooliques en milieu psychiatrique hospitalier; résultats préliminaires d'une expérience pratique faite dans des conditions dificiles, *Toxicomanies*, Québec 2:201.
Sommer, R., Ramsay, R. [W.], and Jensen, S. [E.], 1963, The social relations of hospitalized alcoholics. *Can. Psychiatr. Assoc. J.* 8:51.
Staub, G. W., and Kent, L. M., 1973, "The Para-Professional in the Treatment of Alcoholism," Charles C Thomas, Springfield, Illinois.
St. Clair, O., 1958, Alcoholic delinquency, *Med. Arts Sci.* 12:34.
Stefan, G. [pseud.], 1965, "In Search of Sanity," University Books, New Hyde Park, New York.
Stein, D. R., 1966, Linguistic analysis of the speech of a selected group of former alcoholics: A research note, *Q. J. Stud. Alcohol* 27:106.
Steiner, C., 1971, "Games Alcoholics Play: The Analysis of Life Scripts," Grove Press, New York.
Steinhilber, R. M., Kuluvar, V. D., Anderson, D. J., Heilman, R. O., and Hansen, P. L., 1967, Symposium on the problem of the chronic alcoholic, *Mayo Clin. Proc.* 42 (11):705.
Sterne, M. W., and Pittman, D. J., 1965, The concept of motivation: A source of institutional and professional blockage in the treatment of alcoholics, *Q. J. Stud. Alcohol* 26:41.
Stewart, D. A., 1955, The dynamics of fellowship as illustrated in Alcoholics Anonymous, *Q. J. Stud. Alcohol* 16:251.
Stewart, D. A., 1960, "Thirst for Freedom," Hazelden Foundation, Center City, Minnesota.

Stone, G. P., 1962, Drinking styles and status arrangements, in "Society, Culture, and Drinking Patterns" (D. J. Pittman and C. R. Snyder, eds.), pp. 121–140, Wiley, 1962.
Strachan, J. G., 1968, "Alcoholism: Treatable Illness," Mitchell Press, Vancouver.
Straus, R., 1956, Survey of the problems of alcoholism in Auburn and Cayuga County, May 1956, p. 10; mimeographed, N.P.
Strayer, R., 1961, Social integration of alcoholics through prolonged group therapy, Q. J. Stud. Alcohol 22:471.
Streeseman, A. E., 1962, Alcoholics Anonymous: A psychiatrist's viewpoint, in "Problems in Addiction: Alcoholism and Narcotics" (W. C. Bier, ed.), pp. 133–137, Fordham University Press, New York.
Strunk, J. H., 1966, Potentiation: New top killer? A.A. Grapevine 23 (1):9.
Sullivan, V. F., 1969, "How to Stop Problem Drinking," Fell, New York.
Sutherland, E. H., Shroeder, H. G., and Tordella, C. L., 1950, Personality traits and the alcoholic: A critique of existing studies, Q. J. Stud. Alcohol 11:547.
Swegan, W. E., 1957, Emotional aspects of alcoholism and economy of rehabilitation, Med. Tech. Bull. 8:177.
Swinyard, C. A., Chaube, S., and Sutton, D. B., 1973, Neurological and behavioral aspects of transcendental meditation relevant to alcoholism: A review, Paper read at the 4th annual meeting, American Medical Society on Alcoholism, Valley Forge, Pennsylvania, September 30, 1973.
Syme, L., 1957, Personality characteristics of the alcoholic: A critique of recent studies, Q. J. Stud. Alcohol 18:288.
Tähkä, V., 1966, "The Alcoholic Personality: A Clinical Study," Finnish Foundation for Alcohol Studies, Helsinki.
Taintor, E. [pseud.; Mason, G., and Mason, R. F.], 1945, "September Remember," Prentice-Hall, New York.
Tatham, R. J., 1969, Detoxification center; a public health alternative for the "drunk tank," Federal Probation 33 (4):46.
Taylor, G. A., 1953, "A Sober Faith: Religion and Alcoholics Anonymous," Macmillan, New York.
Tessmann, R., 1971, Erkennung von Alcohol- und Tablettenabusus in Klinik und Praxis, Med. Welt (Stuttgart) 22 (1):20.
Thompson, H. S. [M. A. Maxwell, ed.], 1952, An experience of a nonalcoholic in Alcoholics Anonymous leadership, Q. J. Stud. Alcohol 13:271.
Thompson, C. E., and Kolb, W. P., 1953, Group psychotherapy in association with Alcoholics Anonymous, Am. J. Psychiatry 110:29.
Thompson, W. A., 1942, Treatment of chronic alcoholism, Am. J. Psychiatry 98:846.
Thomsen, R., 1975, "Bill W.," Harper & Row, New York.
Thorpe, J. J., and Perret, J. T., 1959, Problem drinking: A follow-up study. A.M.A. Arch. Ind. Health. 19:24.
Thurber, J., 1971, The bear who let it alone, A.A. Grapevine 28 (1):24.
Tiebout, H. M., 1944, Therapeutic mechanisms of Alcoholics Anonymous, Am. J. Psychiatry 100:468.
Tiebout, H. M., 1946, Psychological factors operating in Alcoholics Anonymous, in "Current Therapies of Personalitity Disorders" (B. Glueck, ed.), pp. 145–165, Grune & Stratton, New York.
Tiebout, H. M., 1949, The act of surrender in the therapeutic process, with special reference to alcoholism, Q. J. Stud. Alcohol 10:48.
Tiebout, H. M., 1951a, Conversion as a psychological phenomenon in the treatment of the alcoholic, Pastoral Psychology 2 (13):28.

Tiebout, H. M., 1951b, The role of psychiatry in the field of alcoholism, with comment on the concept of alcoholism as symptom and as disease, *Q. J. Stud. Alcohol* 12:52.

Tiebout, H. M., 1953, Surrender versus compliance in therapy with special reference to alcoholism, *Q. J. Stud. Alcohol* 14:58.

Tiebout, H. M., 1954, The ego factors in surrender in alcoholism, *Q. J. Stud. Alcohol* 15:610.

Tiebout, H. M., 1955a, The pink cloud and after, *A. A. Grapevine* 12 (4):2.

Tiebout, H. M., 1955b, A factor in the psychology of the alcoholic. Mimeographed, 8 pp. Read at 10th anniversary meeting of the National Council on Alcoholism, New York, March 18, 1955.

Tiebout, H. M., 1956, Why psychiatrists fail with alcoholics, *A.A. Grapevine* 13 (4):5.

Tiebout, H. M., 1958a, Direct treatment of a symptom, in "Problems of Addiction and Habituation," (P. H. Hoch and J. Zubin, eds.), Grune & Stratton, New York.

Tiebout, H. M., 1958b, Alcoholism: Its nature and treatment. Read at Southern Regional Conference on Alcoholism, Birmingham, Alabama, November 1958, National Council on Alcoholism, New York.

Tiebout, H. M., 1961, Alcoholics Anonymous—an experiment of nature, *Q. J. Stud. Alcohol* 22:52.

Tiebout, H. M., 1963a, What does "surrender" mean? *A.A. Grapevine* 19 (11):30.

Tiebout, H. M., 1963b, Treating the *causes* of alcoholism, *A.A. Grapevine* 20 (6):9.

Tiebout, H. M., 1965, When the big "I" becomes nobody, *A.A. Grapevine* 22(4):2.

Tiemeyer, T. N., 1948, What Yale teaches about alcohol, *Christian Century,* September 29:1004.

Tillgren, J., Odencrants, G., and Ljungstrom, C., 1958, Alkoholdispensaren vid Nykterhetsfolkets rådgivningbyrå, *Sven. Läkartidn.* 55:2692.

Timmens, J. M., 1965, A program for the rehabilitation of alcoholics in a combat division overseas, *Med. Bull. U.S. Army, Europe* 22:382.

Toch, H., 1965, "The Social Psychology of Social Movements," Bobbs-Merrill, New York.

Tolentino, I., 1967, Rapporti patodinamici tra alcoolismo e stati depressivi, *Anali Bolnice "Dr. M. Stojanovic" (Zagreb)* 6 (2–3):393.

Trice, H. M., 1956, The "outsider's" role in field study, *Sociology and Social Research* 41:27.

Trice, H. M., 1957a, Alcoholism: Group factors in etiology and cure, *Human Organization* 15 (Summer):33.

Trice, H. M., 1957b, Social factors in association with A.A., *Journal of Criminal Law, Criminology, and Police Science* 48 (Nov.–Dec.):378.

Trice, H. M., 1957c, A study of the process of affiliation with Alcoholics Anonymous, *Q. J. Stud. Alcohol* 18:39.

Trice, H. M., 1958a, Alcoholics Anonymous, *Annals of the American Academy of Political and Social Sciences* 315:108.

Trice, H. M., 1958b, Absenteeism among high-status and low-status problem drinkers, *Industrial and Labor Relations Research* 4 (1):10.

Trice, H. M., 1959a, The affiliation motive and readiness to join Alcoholics Anonymous, *Q. J. Stud. Alcohol* 20:313.

Trice, H. M., 1959b, The problem drinker on the job, New York State School of Industrial and Labor Relations, Bulletin No. 40, Cornell University, Ithaca, New York.

Trice, H. M., 1966, "Alcoholism in America," McGraw-Hill, New York.

Trice, H. M., and Roman, P. M., 1970a, Delabeling, relabeling, and Alcoholics Anonymous, *Social Problems* 17:538.

Trice, H. M., and Roman, P. M., 1970b, Sociopsychological predictors of affiliation with Alcoholics Anonymous. A longitudinal study of "treatment success," *Social Psychiatry* 5 (1):51.

Trice, H. M., Roman, P. M., and Belasco, J. H., 1969, Selection for treatment; a predictive evaluation of an alcoholic treatment regimen, *Int. J. Addict.* 4:303.

Trice, H. M., and Wahl, J. R., 1958, A rank order analysis of the symptoms of alcoholism, *Q. J. Stud. Alcohol* 19:636.
Truax, L. H., 1972, Judge's guide for alcohol offenders, National Council on Alcoholism, New York.
Tuckwiller, P. A., 1947, Medical perspective, *W. Va. Med. J.* 43:261.
Turfboer, R., 1959, The effects of in-plant rehabilitation of alcoholics. An analysis of 160 alcoholic petroleum workers on a Caribbean island, *Med. Bull. Standard Oil, N.J.* 19:108.
Twelve Traditions, The (Illustrated): A Distillation of A.A. Experience, 1971, A.A. World Services, New York.
Ullman, A. D., 1960, "To Know the Difference," St. Martin's Press, New York.
U.K. National Health Service Memorandum on Hospital Treatment of Alcoholism, 1963, *Q. J. Stud. Alcohol* 24:191.
U.S. Bureau of the Census: Statistical Abstract of the United States: 1970 (91st edition), Washington, D.C.
Upton, C. H., 1944, Alcoholics Anonymous, *Federal Probation* 8 (3):29.
Valles, J., 1965, "How to Live with an Alcoholic," Simon and Schuster, New York.
Valles, J. and Sikes, M. P., 1964, A program for the treatment and study of alcoholism in a Veterans Administration hospital, *Q. J. Stud. Alcohol* 25:100.
Van Suetendael, P. T., 1959, The alcoholic patient—his needs as met by Alcoholics Anonymous and the caseworker, *Mil. Med.* 124:851.
Viaille, H. D., 1963, Prediction of treatment outcome of chronic alcoholics in a state hospital, Doctoral dissertation, Texas Technological College.
Victor, M., 1970, Some remarks on the physician's role in the treatment of alcoholism, *Ohio State Med. J.* 66 (8):808.
Voegtlin, W. L., and Lemere, F., 1942, The treatment of alcohol addiction: A review of the literature, *Q. J. Stud. Alcohol* 2:717.
Voldeng, K. E., 1962, "Recovery from Alcoholism," Regnery, Chicago.
Voth, A. C., 1963, Group therapy with hospitalized alcoholics. A twelve-year study, *Q. J. Stud. Alcohol* 24:289.
Voth, A. C., 1965, Autokinesis and alcoholism, *Q. J. Stud. Alcohol* 26:412.
Waldron, K. R., 1965, A.A. and the drug addict, *A.A. Grapevine* 22 (5):38.
Wallerstein, R. S., 1956, Comparative study of treatment methods for chronic alcoholism, *Am. J. Psychiatry* 113:228.
Walton, O. M., 1941, Alcoholics hold Cleveland dinner, *Christian Century*, October 22:1313.
Ward, R. F., and Faillace, L. A., 1970, The alcoholic and his helpers: a systems view, *Q. J. Stud. Alcohol* 31:684.
Wegerman, A., 1968, Riksdagskronika: Ur motionsfloden M.M. *Alkoholfragan* (Stockholm) 62 (3):95.
Wellman, M., 1955, Fatigue during the second six months of abstinence, *Can. Med. Assoc. J.* 72:338.
West, L. J., and Swegan, W. H., 1956, An approach to alcoholism in the military service, *Am. J. Psychiatry* 112 (12):431.
Westheimer, D., and Miller, J P, 1963, "Days of Wine and Roses," Bantam Books, New York.
Wexberg, L. E., 1959, Psychodynamics of patients with chronic alcoholism, *J. Clin. Psychopathol.* 10:147.
White, W. F., 1965, Personality and cognitive learning among alcoholics with different intervals of sobriety, *Psychol. Rep.* 16:1125.
White, W. F., and Gaier, E. L., 1965, Assessment of body image and self-concept among alcoholics with different intervals of sobriety, *J. Clin. Psychol.* 21:374.

Whitney, E. D., 1965, "The Lonely Sickness," Beacon, Boston.
Whitney, E. D., 1968, "Living with an Alcoholic," Beacon, Boston.
Whitney, E. D. (ed.), 1970, "World Dialogue on Alcohol and Drug Dependence," Beacon, Boston.
Wilby, W. E., and Jones, R. W., 1962, Assessing patient response following treatment, *Q. J. Stud. Alcohol* 23:325.
Wilding, D. J., 1949, Alcoholics Anonymous in Melbourne, Australia. Observations of a welfare worker, *Q. J. Stud. Alcohol* 9:609.
Williams, L., 1950, Some observations on the recent advances in the treatment of alcoholism, *Br. J. Addct.* 47:62.
Williams, L., 1958, An experiment in group therapy, *Br. J. Addct.* 54:109.
Williams, L., 1960, "Tomorrow Will Be Sober," Harper, New York.
Williams, L., 1964, Treatment aspects of alcoholism, *Practitioner* 192:114.
Williams, T. K., 1970, The ethanol-induced loss of control concept in alcoholism, Doctoral dissertation, Western Michigan University.
Willis, G. W., 1963, "The Bottle Fighters," Random House, New York.
Wilson, G. C., 1968, The management of the alcoholic, *Med. J. Aust.* 55-2 (20):875.
Wilson, G. C. 1970, The medical management of the alcoholic, *Alcoholism* (Zagreb) 6 (2):153.
[Wilson, W. G.], 1939, "Alcoholics Anonymous: The Story of How More Than One Hundred Men and Women Have Recovered from Alcoholism," Works Publishing Co., New York.
W[ilson, W.] Bill [G.], 1944, Basic concepts of Alcoholics Anonymous, *N.Y. State J. Med.* 44 (16):1805.
W[ilson, W.] B. [G.], 1945a, Alcoholics Anonymous in a postwar emergency, *Q. J. Stud. Alcohol* 6:239.
W[ilson, W.] Bill [G.], 1945b, The fellowship of Alcoholics Anonymous, in "Alcohol, Science, and Society," Lecture 28, pp. 461–463, Quarterly Journal of Studies on Alcohol, New Haven.
W[ilson, W.] Bill [G.], 1947, Is A.A. for alcoholics only? *Guideposts* 2 (7), Section 2, unpaged, September.
W[ilson, W.] Bill [G.], 1949, The society of Alcoholics Anonymous, *Am. J. Psychiatry* 106 (5):370.
W[ilson, W.] Bill [G.], 1950, Alcoholics Anonymous, *N.Y. State J. Med.* 50:1697.
[Wilson, W. G.], 1952, "Twelve Steps and Twelve Traditions," A.A. Publishing, Inc., New York.
[Wilson, W. G.], 1955a, "Alcoholics Anonymous: The Story of How Many Thousands of Men and Women Have Recovered from Alcoholism. New and Revised Edition." A.A. World Services, New York.
[Wilson, W. G.], 1955b, A.A. Tradition: how it developed, A.A. World Services, New York.
[Wilson, W. G.], 1957, "Alcoholics Anonymous Comes of Age: A Brief History of A.A.," A.A. Publishing, Inc., New York.
[W]ilson, [W.] Bill [G.], 1963, A.A.'s debt to its friends, *Faith at Work* July–August:24.
[Wilson, W. G.], 1967, "The A.A. Way of Life," A.A. World Services, New York.
[Wilson, W. G.], 1968a, Dr. Jung, Dr. Silkworth, and A.A., *A.A. Grapevine* 24 (8):10.
W[ilson], W. G., 1968b, Foreword, in "Understanding Alcoholism" (The Christopher D. Smithers Foundation, ed.), pp. vii–viii, Scribner's, New York.
W[ilson, W.] Bill [G.], 1968c, The fellowship of Alcoholics Anonymous, in "Alcoholism: The Total Treatment Approach" (R. J. Catanzaro, ed.), pp. 116–124, Charles C Thomas, Springfield, Illinois.
[Wilson, W. G.], 1976, "Alcoholics Anonymous: The Story of How Many Thousands of Men and Women Have Recovered from Alcoholism," Third Edition, A.A. World Services, New York.
W[ilson], Bill, and Jung, C., 1963, The Bill W.—Carl Jung Letters, *A.A. Grapevine* 19 (8):2.

Witkin, H. A., Karp, S. A., and Goodenough, D. R., 1959, Dependence in alcoholics, *Q. J. Stud. Alcohol* 20:493.

Wolff, P. O., 1949, Conceptos modernos sobre la lucha contra el alcoholismo, *Rev. Psiquiatr. Criminol.* 14:327.

Wood, P., 1969, The selection of members in Alcoholics Anonymous, Master's thesis, Kent State University.

Wylie, P., 1944, Philip Wylie jabs a little needle into complacency, *A.A. Grapevine* 1 (4):1.

X, [pseud.], 1954, Alcoholics Anonymous, *Kans. Med. Soc. J.,* 55:451.

Yablonsky, L., 1965, "The Tunnel Back: Synanon," Macmillan, New York.

Zanko, D., and Kozlina, G., 1970, Die Rolle des Klubs geheilter Alkoholiker in der Kommune, *Alcoholism,* Zagreb 6:70.

Zimberg, S., Lipscomb, H., and Davis, E. B., 1971, Sociopsychiatric treatment of alcoholism in an urban ghetto, *Am. J. Psychiatry* 127:1670.

Zusman, J., 1969, "No-therapy": A method of helping persons with problems, *Community Ment. Health J.* 5:482.

Zwerling, I., 1959, Psychiatric findings in an interdisciplinary study of 46 alcoholic patients, *Q. J. Stud. Alcohol* 20:543.

CHAPTER 12

Role of the Recovered Alcoholic in the Treatment of Alcoholism

Sheila B. Blume

Central Islip Psychiatric Center
Central Islip, New York

School of Medicine
State University of New York at Stony Brook

and

Caribbean Institute on Alcoholism

INTRODUCTION

In December of 1963 the Quarterly Journal of Studies on Alcohol published a discussion entitled "Who is Qualified to Treat the Alcoholic?" One of the participants (Krystal, 1963) held the extreme position that only a person able to "diagnose all emotional and physical problems" and trained in psychotherapy, ideally one who has undergone a personal psychoanalysis, is truly qualified to treat alcoholism. He refers to treatment, however, as individual psychotherapy. On the other extreme has been the position that "only an alcoholic can help an alcoholic," an assumption made in the early prehistory of Alcoholics Anonymous (A.A., 1955). The implication of the second position is that a given number of years addicted to the bottle will provide better preparation for treating alcoholism than the same number of years on an analyst's couch.*

Today, the conflict between these two extremes seems anachronistic. It is reminiscent of the old "wets" versus "drys" debate in its counterproductive impact on the severe alcoholism problems now facing society. The issue of who will treat the alcoholic tends to rephrase itself today as "What are the roles of the various professionals and nonprofessionals (including recovered alcoholics) in the treatment of alcoholism?" Reference to the 1963 debate, however, is useful in pointing out the necessity to sharpen definitions. What exactly do we mean by "recovered," "alcoholism," "treatment"?

DEFINITIONS

The term *alcoholism* refers here to a disease or disorder, characterized by pathological dependency on alcohol, leading to functional impairment.

The physiological and behavioral features of this disorder and the criteria for its diagnosis have been spelled out by a committee of the National Council on Alcoholism (Kaim *et al.*, 1972).

For those who prefer non-medical models, alcoholism can be defined as a habitual pattern of alcohol ingestion which results, over an extended period of time, in interference with functional behavior in one or more important areas of an individual's life.

The word "recovery" refers to a return to the healthy and functional state of a person afflicted with alcoholism. This includes, at the present state of our knowledge, abstinence from alcohol. It also includes restoration of the ability to function in all areas of life, in a subjective state free from undue distress, and without the substitution of another addictive drug. Criteria for recovery are also discussed by Kaim *et al.* (1972). The term *recovered alcoholic* is used here in preference to others suggested. "Ex-alcoholic" seems to imply an absolute cure, which is not now available. "Recovering alcoholic" (Falkey, 1971), although it conveys the dynamic nature of the recovery process, blurs the distinction between those alcoholics in the initial stages of their treatment and recovery, and those who have achieved a stable readjustment.

SCOPE OF CHAPTER

The best known and most successful setting in which recovered alcoholics have helped others recover is Alcoholics Anonymous. Since this fellowship is

---

* Alcoholics Anonymous does not now hold to this position (A.A. 1972a).

the subject of another chapter, A.A. will be discussed here only in relation to professionally headed treatment programs in which A.A. members may take part.

The role of the recovered alcoholic who is also a professional (e.g., social worker, physician, nurse, health educator, clergyman, psychologist) will be discussed in general terms only, since this role in the treatment of alcoholism differs in very few ways from that of the nonalcoholic professional. The recovered alcoholic who works in a nonprofessional or a paraprofessional role in the treatment of alcoholism will be covered in greater detail. It is in the motivation, selection, training, job responsibility, and effectiveness of therapists in this category (referred to here as alcoholism counselor or lay therapist) that the major issues arise. Although the alcoholism counselor may be male or female, the term *he* is used for convenience to substitute for *he* or *she*.

ROLES OF THE RECOVERED ALCOHOLIC, PAST AND PRESENT

Independent Lay Therapists and Group Programs

In *The Other Side of the Bottle,* Dwight Anderson (1950) discusses the history of recovered alcoholics working as lay therapists in the United States. The first of these to be widely known was Courtnay Baylor, who began to work with Dr. Elwood Worcester, rector of the Emmanuel Church in Boston, in 1912. This same Dr. Worcester was an important contributor to the early history of group psychotherapy in this country (Mullan and Rosenbaum, 1962). Baylor developed a concept of the interaction of physical and mental tensions in causing the craving for alcohol, and made use of a form of deep relaxation in his therapy. As a participant in the "Emmanuel Church health classes" or "Emmanuel Movement," as it was later called, Baylor was particularly effective in treating alcoholics and training other lay therapists.

Richard R. Peabody, one of Baylor's patients and students, became an independently working lay therapist. His treatment included a very detailed schedule of daily activities, recorded in a diary, and a series of 80 to 100 individual counseling sessions (Anderson, 1944). Francis Chambers, a lay therapist who studied with Peabody, represents a third generation in Baylor's line. Chambers also worked primarily in individual counseling sessions and relaxation exercises, in conjunction with a psychiatrist, Dr. Edward Strecker. Treatment extended over the course of a year or more. Their method of treatment is well described (Strecker and Chambers, 1939).

A colorful lay therapist, independent of those mentioned above, was Sam Leake of San Francisco. Although he overcame his own alcoholism with the help of a Christian Science practitioner, he worked independently as a counselor in treating others (Anderson, 1944).

Edward J. McGoldrick, Jr., a New York attorney, devoted his life to the treatment of alcoholism following his own recovery. He founded Bridge House in 1945, as an official agency of the then New York City Department of Welfare. Believing that few alcoholics need medical help (McGoldrick, 1964), he staffed this small residential facility entirely with recovered alcoholics trained in the use of his technique. The McGoldrick method, based on the premise that "what the alcoholic needs is information rather than reformation," is basically an educational approach. He states (1966) that over 40 percent of the alcoholics who entered his program reached graduation (including one year of sobriety). Since McGoldrick's death in 1967, the Bridge House program has been carried on by members of his staff.

A group self-help movement which was instrumental in the recovery of many alcoholics flourished in the United States during the 1840s and 1850s. The Washington Temperance Society was founded in Baltimore, Maryland, in 1840 by a group of six friends who, after a temperance lecture, decided to give up drinking. To that end, they composed and signed a pledge of total abstinence. They began to meet regularly for the purpose of persuading others to sign and embrace sobriety. Millions of Americans signed this pledge as the movement spread. Although the majority of these people were not alcoholics, the Washingtonians made the "reformation of drunkards" a major goal. This was accomplished through kindness, concern, and respect for the drinker—a departure from earlier approaches. It was customary for recovered alcoholics to tell the stories of their drinking and recovery at weekly meetings and to become active in recruiting new members from among their still-drinking companions. Although valid statistics are lacking, it seems likely that many thousands of alcoholics were helped by these groups. The history of the rise and decline of the Washingtonian Movement is documented by Maxwell (1950).

Religious Programs

The recovered alcoholic who becomes active in religious approaches to the treatment of alcoholism has usually gained his own sobriety following a conversion or some other religious experience.

One such story is that of William Raws, an Australian-English artist who emigrated to the United States as an attempt at "geographic cure" of his alcoholism in 1889. Here he recovered following a religious experience and six years later, in 1897, he founded the Keswick Colony of Mercy in Whiting, New Jersey. The Colony is now directed by the founder's grandson. It provides a program of "spiritual therapy" through group discussions, Bible study, worship, sharing meetings, and personal counsel. No medical or psychotherapy program is included. Recovery occurs through "spiritual regeneration." One successful graduate of Keswick has founded another similar program in another state (Raws, 1972).

Recovered alcoholics have also played an active part in the skid row religious mission. Several of these establishments, set up to provide food, clothes, shelter, and spiritual comfort to the down-and-out in the areas where they live, have developed specific treatment programs for alcoholics.

One such organization, the Bowery Mission, has been active on New York City's Bowery since 1879. The mission has been traditionally staffed almost entirely by recovered alcoholics. The alcoholism program consists of three components, referred to as physical treatment, mental treatment, and spiritual treatment. Alcoholics Anonymous meetings, work therapy, and recreation are included. Recovered alcoholics function as providers of "mental treatment" in the form of information-giving groups, therapy groups, and individual counseling. In addition, they are active in "spiritual treatment" as leaders of Bible study groups, prayer meetings, and seminars.

Men are hired in various capacities by the mission directly after completing a three-month program, and may work their way up the ladder to a counseling position. Training for the latter may include study at a Bible Institute and attendance at an alcoholism summer school. Although the present director of the Bowery Mission is not an alcoholic, all of his recent predecessors and all of his counseling staff are recovered alcoholics (Lockwood, 1971.)

Independent Facilities

A large number of small, independently operated facilities have been established and/or staffed by recovered alcoholics throughout the United States during the past few decades. Some, calling themselves farms, lodges, or rest homes, provide essentially a friendly, alcohol-free atmosphere in which an alcoholic may rest, repair nutritional deficiencies, and receive practical advice in a protected setting. Alcoholics Anonymous meetings may be held in the home or nearby. Other facilities have developed formal programs of evaluation, psychotherapy, and rehabilitation. An example of the latter is Chit Chat Farms in Wernersville, Pennsylvania. This 60-bed facility has a mixed staff of professionals and alcoholism counselors working as a team. Approximately half of the therapists are recovered alcoholics. The program stresses therapeutic community concepts in treating alcohol and sedative addictions (Shulman, personal communication).

Medically and Psychiatrically Sponsored Programs

Hospitals and clinics specializing in the treatment of alcoholism provide a long and interesting history of the successful collaboration between recovered alcoholic counselors and professionals in the field. Perhaps the first was the Washingtonian Hospital in Boston, Massachusetts, founded in 1857 by members of the Washingtonian movement. This hospital continues to operate today as a center for treatment of the addictions. Shadell Sanitarium in Seattle was founded, according to Anderson (1950), by a recovered alcoholic businessman and a physician, in 1935. Their treatment for alcoholism consisted of a combination of medically supervised aversion therapy in the hospital, and follow-up care by recovered alcoholic counselors on an outpatient basis.

The collaboration of Dr. Stecker and Mr. Chambers in the 1930s has already been mentioned. The Yale Plan Clinics, in Hartford and New Haven, were established as model treatment facilities in 1944 by the Yale Laboratory of Applied Physiology (later the Yale Center of Alcohol Studies). These outpatient clinics employed recovered alcoholics as counselors under the supervision of professional staff members (Mann, 1976).

Many other inpatient and outpatient programs have followed this lead or developed models of their own. These have included general hospitals, specialized clinics, and psychiatric hospitals. The Alcoholism Rehabilitation Center at Lutheran General Hospital, Park Ridge, Illinois, for example, employs and trains recovered alcoholics to function as treatment team members. These counselors are given "duties and responsibilities identical to those of the teams' social workers and clergymen." (McI., 1970). At Roosevelt hospital in New York City, another general hospital, alcoholism counselors have been employed since 1968. Their duties include individual and group work with alcoholics admitted to the general medical and surgical wards under a variety of primary diagnoses. The counselors attempt to involve these patients in Alcoholics Anonymous and outpatient follow-up. At Eagleville Hospital, in Eagleville, Pennsylvania, which treats both alcoholism and drug addiction, about one-third of the therapists are nonprofessional recovered alcoholics. Some of them are recovered ex-patients trained by the hospital (Ottenberg, 1969). At Central Islip Psychiatric Center in New York, the Charles K. Post Rehabilitation Unit has employed recovered alcoholics as counselors since its inception in 1962. Counselors are hired under the New York State Civil Service classification "Alcoholism Rehabilitation Assistant." The job has no specified educational requirement nor does it specify that its holder must be a recovered alcoholic. It does require five years of experience in a field requiring "communication skills" and at least a year of "paid or volunteer experience in an organized program on alcoholism."

One of the more interesting concepts in the employment of alcoholics as therapists has been developed at Mendocino State Hospital in California. This program has trained "senior" alcoholic patients as individual counselors and group leaders to work with other patients (Slaughter and Torno, 1968). Whether this method has been found successful and whether graduates of this training have been hired as staff members after discharge, rather than working simply as volunteers, is not yet clear.

Although the above-mentioned examples indicate that recovered alcoholic counselors have gained considerable acceptance in medically run facilities, this phenomenon is far from universal. A survey of alcoholism treatment facilities published in 1967 makes no mention of recovered alcoholics in any of the hospital inpatient or clinic outpatient programs surveyed. In only one halfway house and one court program was mention made of recovered alcoholics providing care (Glasscote *et al.*, 1967).

Industrial Programs

A number of pioneering industrial alcoholism programs were begun by recovered alcoholic employees (Warren, personal communication; Vaughn, personal communication). Initially, knowing of their recovery from alcoholism, the firm's personnel department had consulted with these individuals informally for advice concerning other, actively alcoholic employees. Later on, this relationship developed into a formal program, involving the medical department of the firm, but continuing to utilize the talents of recovered alcoholics. These counselors very effectively encourage, advise, and motivate alcoholic employees to enter and continue treatment. The success of such industrial programs, with their early case-finding, job-centered motivation and careful follow-up, has been outstanding (D'Alonzo, 1969).

Antipoverty Programs

Under the Federal Economic Opportunity Act of 1964 and later amendments, the administrative structure and funding mechanism for alcoholism programs for the poor was established.

The services suggested by the O.E.O. as a model program form an integral part of the local Community Action Agency serving each identified poverty group (O.E.O., 1971a). The major staff resource for this service is the alcoholism counselor. His duties include direct services to families with alcohol problems; liaison to A.A., Al-Anon, and Alateen; liaison to medical and other resources in the community; and education of both the C.A.A. staff and the neighborhood residents in alcohol problems. Training programs have been set up specifically for these counselors, counselor-coordinators, and their aides (O.E.O., 1971b). In a truly historic legislative action, Section 222 (a) (8) of the amended law encourages "the use of the services of recovered alcoholics as counselors." This recommendation is reflected in the qualifications for candidates for training programs and the requirement that each local project "develop specific written policies and procedures for encouraging the employment of recovered alcoholics as staff."

Courts and Correctional Facilities

Recovered alcoholics have made many significant contributions to the programs aimed at rehabilitating the group known as "chronic drunkenness offenders" or "revolving door alcoholics." Two rather dissimilar examples of this type of effort, staffed by recovered alcoholics, are the Miami court project and Connecticut's Compass Club (Glasscote *et al.,* 1967).

In the state of Indiana, alcoholism counselors are part of the regular staffing pattern of state correctional institutions. These counselors are recovered alcoholics. They are trained by the state's Division on Alcoholism after passing a civil service examination (Falkey, 1971).

Special programs for persons convicted of driving while intoxicated have been designed to identify, motivate, and guide into treatment the many alcoholics who fall into this group. The Phoenix, Arizona, program, begun in 1966, employs recovered alcoholics as course assistants for this purpose (Stewart and Malfetti, 1970). Other subsequently established drunk-driver programs have employed such counselors as group leaders in preference to professional driving instructors (Adams, 1976).

Public Education and Information Agencies

Issues concerning education and prevention are dealt with in another chapter in this volume. It is of interest here, however, that the nation's major voluntary health agency concerned with alcoholism, the National Council on Alcoholism, was founded by a recovered alcoholic, Mrs. Marty Mann.

Many of the local affiliates of the National Council, which carry out on the local level the functions of public education, community development, referral, and counseling services, are also staffed by recovered alcoholics. Both the National Council on Alcoholism and the local councils and their functions are well described in Mrs Mann's book (1958).

THE ALCOHOLISM COUNSELOR AS MEMBER OF A TREATMENT TEAM

Unique Advantages of the Recovered Alcoholic as Counselor

One of the most frequent arguments in favor of the employment of nonprofessional counselors in treating alcoholism is the scarcity of trained professionals (Rosenberg et al., 1971), specifically psychiatrists (Edwards, 1970). Since there will never be enough professionals to treat all the alcoholics, the reasoning goes, we must learn to utilize others. Consequently, there emerges the image of the alcoholism counselor as an ersatz psychiatrist or junior social worker, filling in to meet a real but unfortunate shortage of the genuine article. I am personally in complete disagreement with this position. My view is based on our program's* ten years of experience with a joint professional and nonprofessional staff. Although some functions of team members are interchangeable, each member also brings a special expertise (the counselor no less than the psychiatrist) that makes a necessary contribution to the total program. I would not be willing to trade our four recovered alcoholic counselors for an equal number of psychiatrists. They are representatives of an important discipline and should be recognized as such.

Among the special contributions of the counselor are the following. The counselor serves as a living example of hope for each new entrant into the program. Most alcoholics approach treatment with extreme ambivalence. Although they are unhappy with the present situation and therefore want help, the prospect of making a commitment, accepting the diagnosis, and aiming for sobriety is a terrifying one. Fear of failure is expressed as a reluctance to try, a denial of illness, and a mistrust of the program and staff. The common background of alcohol addiction in patient and counselor encourages a healthy identification on the part of the patient, bringing hope that he too has a chance to recover. With the emergence of hope, defensiveness is relaxed, and the fears of failure can be brought into open discussion.

Day-to-day contact with the counselor provides the patient with an opportunity to observe the counselor's handling of major and minor anxieties and frustrations, both within and outside the formal therapy situation, without a drink. Although *all* team members are closely watched by their patients to see whether they "practice what they preach" outside of their therapy groups, the counselor represents a special role model for the alcoholic patient. He is also in a unique position to reinforce this kind of learning in the patient by expressing

---

* The Charles K. Post Rehabilitation Unit at Central Islip Psychiatric Center.

his feelings honestly and pointing out the contrast between his present and former way of reacting to stress.

The counselor can often communicate better, especially in the early stages of treatment, with those alcoholics who have special problems with authority figures. These patients view professional staff members as distant parent figures, with a mixture of awe, fear, and hate. This distance may be magnified by differences in education or social class. By "speaking the same language," backed up with common experience, the counselor is often able to effect the first breakthrough in such patients. He is then able to lead the patient into realistic relationships with the other members of the team.

The common history of alcoholism provides a further advantage to the counselor in making him able to identify very deeply with the patient. Although underlying personality characteristics may vary widely, most alcoholics have shared a great many common experiences, predicaments, and emotional reactions, as a result of the behavior of drinking itself. The counselor's ability to identify in this way generates in him a patient and tolerant approach—an attitude much less prone to moralistic judgment than might otherwise be so. This identification on the basis of common experience also makes the counselor harder to fool or "con," a source of great pride to both counselor and patient.

Another special advantage of the counselor is his experience in handling the very important day-to-day social mechanics of living abstinent in the midst of a drinking society. He can provide practical advice on handling various unusually difficult situations which the nonalcoholic professional may never have encountered.

Lastly, the counselor, if he is a member of Alcoholics Anonymous, can help the patient understand the A.A. program and how it can help him. Many patients entering our own program have had earlier disappointing experiences with A.A. They are initially unwilling to try again. An opportunity for these patients to talk about their experiences with a staff member who is knowledgeably involved in A.A. but not defensive about the A.A. program often leads to a new and more successful A.A. participation. The counselor can help interpret the A.A. program to the patient, and interpret the methods and goals of the treatment program to local and institutional A.A. groups.

All of the above advantages also apply to the professional mental health worker who is a recovered alcoholic as well. These people bring to their job *two* special areas of expertise.

Motivation of Counselors

In my opinion, most recovered alcoholics seeking to enter the counseling field are primarily motivated by a wish to spare others some of the pain and suffering they personally endured during their illness, especially during their search for help. Often they had been rejected or given bad advice by people they had approached in this search, and are painfully aware of the shortage of adequate treatment resources. They have often had some success at "12th step" work in A.A. or as volunteers in an agency or program of some kind. They find their own usual occupation unrewarding by contrast, and wish to help others full time. The prospective counselor also hopes that such a career will afford him some safety in his own recovery. Seldom have I found a desire for profit or control over others the primary motivation. The wish for social status as an "expert on alcoholism," and the wish to learn by working with professionals is often present. Less often is there a wish to "show up" the failure of medicine to help alcoholics by beating the professionals at their own game.

Selection of Candidates to Be Counselors

The motivation of the prospective counselor, as discussed above, is an important criterion in selection. If the wish to help is accompanied by an unrealistic set of expectations of success, the applicant should be discouraged. Underlying competitive or hostile feelings toward professionals would also mitigate against successful performance as a counselor.

Most programs select candidates for training and employment on the basis of personal qualities. Such terms as *warm, strong, stable, good at communicating, adaptable, flexible, interested,* and *people-oriented* appear frequently in the description of the ideal (Augustus, 1971; Mann, 1976; Wilson, 1968; Shulman, personal communication). Perhaps the best criteria are those studied and documented by Truax and Carkhuff (1967) as the most important characteristics of any effective therapist: genuineness, nonpossessive warmth, and accurate empathy.

In selecting recovered alcoholics in particular, length of sobriety is a factor often given considerable weight. Most training programs require a minimum of two years of continuous sobriety, although the O.E.O. programs require only one year (O.E.O., 1971a), and the Mendocino program mentioned above (Slaughter and Torno, 1968) begins training while the alcoholic is still hospitalized. In our own program at Central Islip we will not hire an ex-patient to work on the unit in any capacity, even as kitchen worker, until he has been employed elsewhere and sober for at least six months. We do not consider training for a counseling role until three to five years later. This measure is partly designed to assure some stability in sober adjustment, but primarily it aims at preserving the therapeutic value of the program and staff to the newly discharged patient. Early in his posthospital adjustment, the staff can be of great help to him in handling his problems, including those at work. This ability is much curtailed if his employer and therapist are one. Experience working outside of the unit gives both ex-patient and staff confidence that the "umbilical cord" has been cut and that the new employee's sobriety does *not* depend only on his working in the unit.

In my opinion, the existence of serious underlying psychopathology would make a candidate unsuitable for work as a counselor in most cases. Therefore, the already-offered history of alcoholism in an applicant should not preclude a sensitive examination for indications of other pathological trends. Psychological testing has been employed by at least two counselor training programs but results have not yet been correlated with later job performance (Willmar brochure, n.d.; Baltimore brochure, n.d.).

The question of whether "low bottom drunks" (alcoholics who reached skid row during their drinking) versus "high bottom drunks" (those who did not) are more effective as counselors depends on the type of patient population to be served. Unless the program is aimed specifically at the homeless, skid row, or chronic drunkenness offender alcoholic (in which case the "low bottom" is preferable), either type of life history will be alcoholic enough for purposes of identification between counselor and patient.

The final criterion used in selection of counselors is affiliation with A.A. or a similar organization. Since A.A. keeps no list and membership is self-defined, no job requirement can specify a "card-carrying A.A. member." However, the unique role of the counselor as liaison between the therapy program and A.A. will be operative only if the counselor is an active participating member. A husband or wife active in Al-Anon and/or Alateen is occasionally mentioned as a fringe benefit to the employer of a counselor, with the husband and wife functioning as a team in some programs (Vaughn, 1972).

Training

A survey conducted by the National Council on Alcoholism in 1970 listed six formal programs in the United States for the training of lay counselors (Seixas and Sutton, 1971). These programs vary greatly in educational prerequisites, from none to a bachelor's degree. All tend to cover a considerable amount of academic material and most rely heavily on supervised field work. In addition to these, a number of O.E.O. training programs have been established. Many treatment centers continue to train their own counselors, usually with a combination of on-the-job training and attendance at the Rutgers Center of Alcohol Studies or a similar alcoholism institute.

Other issues in alcoholism training are covered elsewhere in this volume.

Job Responsibility

The special experience and expertise a recovered alcoholic counselor brings to the job, discussed above, defines in advance some of his special usefulness. Certain of his functions will be interchangeable with those of other treatment team members. Such functioning may include interviewing, evaluating candidates for treatment, working with individual alcoholics and family members, and providing information to members of the public. They may include participating in the development of treatment plans for individuals and the development of the total program. They may include functioning as a leader of education-informative groups, discussion groups, psychotherapy groups, and role-playing or psychodrama groups (Blume et al., 1968).

The counselor also has functions different from those of other team members. As liaison to Alcoholic Anonymous he should attend all intramural meetings and have a voice in the selection of A.A. member volunteers working with the program. He helps patients make contact with A.A. groups in their local communities and helps them find potential A.A. sponsors where indicated.

The counselor may organize and assist volunteers who drive patients to neighboring A.A. meetings, make 12th step calls, or help patients find lodgings, jobs, and recreation. He serves as an important resource for staff members and professionals training at the treatment facility, in helping them understand A.A. and how it works.

It is important for programs wishing to employ recovered alcoholics as counselors to have a clear understanding of the differences between alcoholism counseling and A.A. 12th step work, and between counselor-led group psychotherapy and A.A. meetings. The term *12th step* refers to the 12th of the suggested steps to recovery of A.A. This step involves "carrying the message" personally to other alcoholics, sometimes in the form of a home visit. The technique is described in "Alcoholics Anonymous" (A.A., 1955). The essence of a 12th step call is an A.A. member relating the story of his own alcoholism to the alcoholic in need of help, in whatever form he is most likely to listen to it, until the listener becomes curious about how the caller recovered. At that point, the A.A. member describes his own personal recovery through the A.A. program. All conversation is kept at this personal level of the member's own history. The "12th stepper" may also leave A.A. literature, will invite or take the alcoholic to an A.A. meeting, and may even given him considerable personal help over an extended period of time.

Open A.A. meetings run on similar principles. Members of the group listen to and identify with the personal stories of drinking and recovery of a number of speakers. These talks are followed by discussions among group members on how to follow the A.A. program. The program itself approaches alcoholism as a "three-fold illness; physical, mental and spiritual," and lays heavy emphasis on faith in a power greater than oneself. Tiebout (1961) and Leach *et al.* (1969) discuss the history and working of A.A.

Techniques of counseling, on the other hand, involve listening to the patient far more than talking. The counselor's personal experiences enter the process only so far as they help him understand the patient, establish a basic identification by the patient, and ensure accurate empathy. He rarely directs or suggests but rather helps the patient explore and understand the patient's own feelings and behavior. Since the processes involved in taking the first drink that leads to a relapse is basically a psychological one, motivated by a feeling of distress and search for relief (Jellinek, 1960), the counseling process seeks to help each alcoholic identify those emotional states from which he has habitually sought relief through drinking. Having correctly identified these feelings, both past and present, and the kinds of life events which tend to arouse them, the patient is helped to find new solutions to the old problems. Usually this involves accepting previously unacceptable and denied feelings, and developing new and more flexible attitudes toward himself, others, and life. The exploration of deeply unconscious material is not involved. The counseling process is conducted in a warm, nonjudgmental atmosphere, encouraging honesty and frankness. It is combined with constant reinforcement of motivation toward sobriety. The same principles apply in both individual and group work. By following these principles the counselor may be quite effective in helping the patient find his own alcohol-free solutions to a great variety of problems, including sexual ones. Thus alcoholism counseling and A.A. "carrying the message" are complementary to one another, but entirely separate methods and techniques. For this reason the terms *A.A. counselor* or *A.A. counseling* are inaccurate ones. A counselor does not take the role of A.A. sponsor or 12th stepper to the patients in the treatment program in which he works. On the other hand, in his personal life he may certainly, as an A.A. member, serve in this capacity for other alcoholics.

Other job responsibilities for which the counselor is particularly suited are the various outreach activities of the rehabilitation program in which he works. Precare, which involves education and efforts to develop motivation in individuals and families, is part of this job. Another part involves contacting and following up patients who miss appointments or fail to return from passes. A third involves maintaining contact with ex-patients of the program through whatever means are available. The program at Central Islip Psychiatric Center, for example, publishes a quarterly newsletter which is sent free of charge to any ex-patient who wishes it. In return, the ex-patient must write, call, or visit the unit at least once a year. A counselor edits and sends out this newsletter and handles the correspondence, including the sending of sobriety anniversary cards. If he has not heard from the ex-patient for a year, he includes a reminder to remain in touch. The same counselor at Central Islip organizes two annual reunions for ex-patients at the hospital. One of these is an A.A. anniversary meeting, the other a large dinner-dance. Additional aftercare outreach activities might include recruiting and supervising volunteers outside of the hospital, and making home visits.

Although the job responsibilities listed above are both extensive and intensive, nonprofessional alcoholism counselors have been successfully performing these duties for many years. The success of the counselor in these difficult functions will depend partially upon his ability and training, but also to a large degree on the extent of his backing and supervision by other, professional members of the treatment team. A counselor will feel quite secure in helping an individual alcoholic work through emotional problems if he has adequate supervisory time in which to discuss his work, and if he is certain he can refer patients with problems he feels are beyond his competence to a psychiatrist or other professional. He can lead a psychotherapeutic group in which an atmosphere of free communication prevails, as long as he feels secure in the respect and backing of his professional colleagues. He must be able to come to them freely for advice (as they will also come to him). No alcoholism counselor should ever be asked to carry on individual or group work in isolation, nor should his work be looked upon as something apart from that done by other staff members. Respect, backup, and supervision are the most important factors in assuring optimal effectiveness.

Problems Unique to the Recovered Alcoholic as Counselor

The most commonly cited problems of the recovered alcoholic as counselor involve competition and conflict with professional staff members. Differences in ideas on the nature of alcoholism, treatment goals, and administrative philosophies have tended at times to polarize staff members. In my experience, professionals, especially those new to the field of alcoholism, tend to set very high life-goals for their patients with very little real expectation that these will be attained. Therefore, they are content with small gains and tend to blame failures on the disease, or on conditions, other people, or similar externals. The counselor, on the other hand, tends to set more restricted goals, chiefly abstinence, and has every expectation that the patient will succeed. He tends to place the blame for failures on the patient, on himself, or on both. This difference may be illustrated by the counselor's satisfaction with an alcoholic lawyer who maintains his sobriety driving a truck. The professional might take equal satisfaction from the same lawyer maintaining his law practice, even though periodic drinking bouts continue. Prevention of this type of conflict is partly a function of the training period, when the issues of goals, expectations, and treatment methods should be thoroughly explored. Communication between treatment team members, good leadership, and the development of a common treatment philosophy will also help resolve such disagreements. Sensitivity training (Lowry, 1971) and marathon sessions (Ottenberg, 1969) have been suggested as methods of facilitating such staff communication.

A second problem occasionally encountered is the attempt on the part of a counselor to overcompensate for his lack of professional training, or for a personal feeling of inadequacy by overworking. He may put in extra hours, donate large amounts of money or equipment to individual patients or to the unit, and make other personal sacrifices. This behavior may be welcomed, or even encouraged, by fellow staff members who are unaware of its meaning. The counselor in such a situation should be helped by his supervisor or team leader to understand and alter this pattern of behavior. If this is not done, a sense of dissatisfaction and resentment usually develops, accompanied by a very real fatigue, which may lead the counselor to resign with ill feeling.

A third problem occasionally noted grows out of the identification of the counselor with his patient. Although this identification is in general a positive and desirable factor, it occasionally leads the counselor to make hasty and incorrect assumptions about the reactions of his patient to events and situations. The assumption, "Because we are both alcoholics, and I reacted this way, he must also react the same way," may be a dangerous one, especially when the counselor is working with alcoholics who have underlying schizophrenia or personality disorders. It is here that the special training and supervision of a counselor is needed, not primarily to diagnose his patients, but to listen carefully to them and be sensitive to their feelings and reactions. A counselor so trained will never feel, "If it was good for me, it has to be good for him," but will take a flexible approach to treatment.

I have never personally confronted the problem of a drinking episode in a counselor. Inquiries of directors of other programs seem to indicate that this is in fact a very rare situation. It is probably best handled, first, by helping the counselor with his problem, as the program would help any alcoholic in trouble. Second, he should be assigned to limited or noncounseling duties until the program director feels sure his sobriety is again stable enough to resume counseling. Effort should be made to help the counselor identify the emotional factors leading up to the drinking episode. Occasionally such investigation will lead to a decision to leave the counseling field. Decisions of this type will vary greatly with the individual involved.

The final set of problems apply to the counselor who is both an A.A.

member and the liaison agent between his treatment program and A.A. This counselor may find himself set apart as an "expert," in a way which threatens to interfere with the personal help he derives from attending A.A. meetings. He may be repeatedly asked for medical, marital, and personal advice. He may be asked for special consideration by the treatment program for group members or friends. He may be envied by other group members, or criticized for "making money by doing A.A. work." It is important in this regard for the counselor to keep his counseling and A.A. roles separate in his own thinking. He knows he is not being paid his salary to stay sober, belong to A.A., or follow its suggested steps. It is up to the counselor to communicate, by his attitudes and in discussion, that he considers himself a member of his A.A. group like any other. Members of his group will be helped to understand that his work in counseling is not 12th stepping or sponsoring, and that he comes to A.A. like other group members, to "share his experience, strength and hope." Seldom have problems of this sort interfered seriously with a counselor's effectiveness.

It is not considered a breach of the tradition of anonymity for the counselor to identify himself as an A.A. member to staff and patients, as this tradition applies only to the public media (A.A., 1955).

The General Service Office of Alcoholics Anonymous has prepared guidelines for A.A. members working as counselors and for professional programs wishing the cooperation of A.A. (A.A., 1972a,b).

Remuneration and Status

Many alcoholism treatment programs utilize volunteers, usually A.A. members and often also ex-patients of the program, who perform many useful services. These may include leading intramural A.A. meetings, raising funds for the program, sponsoring newly discharged patients in A.A., making 12th step calls, and doing other volunteer work. Their activities should be carefully distinguished from those of the counselor, who must be adequately paid and accorded status in proportion to the considerable responsibilities he assumes in the treatment process. His expertise, skills, and relationships with professional staff members should not be expected from untrained volunteers, nor should the counselor be regarded as a category of "cheap labor." Because many treatment programs are publicly funded, and salaries tend to rise with educational attainments and numbers of degrees, the problem of obtaining realistic salary levels for counselors has been a difficult one. Efforts are being made by some programs to arrange for credit toward associate or bachelor's degrees to be given for on-the-job counseling experience (Ottenberg, 1969). Other efforts must be made to define and develop the role of recovered alcoholic as counselor into well-recognized and well-established category of mental health worker.

THOUGHTS ON THE FUTURE OF RECOVERED ALCOHOLICS AS COUNSELORS

Alcoholism counselors have both an interesting and colorful past and an active and vital present. As interest in the treatment of alcoholism has grown in this country, their value has achieved increasing recognition. The future of the recovered alcoholic as counselor, however, is not at all as clearly defined as his past. As more programs for training paraprofessionals (physicians' assistants, mental health technicians, alcoholism counselors, and others) are developed, the unique contribution of the recovered alcoholic runs the risk of being submerged in the large paraprofessional group. Educational standards may tend to replace personal qualities and past life experience as qualifications for positions in alcoholism counseling.

Questions of policy and philosophy will continue to arise. Should job qualifications include such statements as "bachelor of arts degree and/or history of recovery from alcoholism," or "recovery from alcoholism or a high school diploma, plus training and experience"? Can any amount of education or counseling training be equated with a personal history of recovery from the disease? Since educational requirements are easier to specify, the recovered alcoholic may find himself on equal ground with the untrained nonalcoholic in competition for counselor training positions and/or counseling jobs which include on-the-job training. If he is hired or trained, what weight should be given the history of recovery in setting salary, grade, and title? Should he be considered equal to nonalcoholic paraprofessionals?

The answers to these questions are not simple, and may perhaps vary with the type of work setting. I believe, however, that recovered alcoholics in the counseling field ought to make an intense effort to preserve the separate identity of their work role. This might be done by organizing on a state or national level to develop standards for training, job functions, remuneration, and other matters. Such an organization would be separate from other alcoholism or mental health workers' groups. It might work to resist the exclusion of capable people through irrelevant educational standards, and to preserve recognition of the value of the life experience of recovery. No large-scale organization of this type exists today. Consequently, vital issues concerning the future of alcoholism counseling are being decided without organized input from recovered alcoholic counselors.

Addendum

Since this chapter was submitted for publication, the field of alcoholism counseling has experienced a significant move toward professionalism. In a project funded by the National Institute on Alcohol Abuse and Alcoholism, a Proposed National Standard for Alcoholism Counselors was developed by Roy Littlejohn Associates. The final Littlejohn report, dated August 1974, reflected the combined effort of hundreds of people in the alcoholism field. It was designed for use in a future nation-wide voluntary certification program, to be developed in cooperation with examining boards at the state level.

The report lists six areas of competency required for certification: communication, knowledge of alcohol use and alcoholism, evaluation and assessment, planning, information and referral, counseling and treatment. Other certification requirements and a code of ethics are included in the report. Twenty-two professional tasks which counselors can perform are also listed. Although no specific mention of the recovered alcoholic as counselor is made, the first requirement for certification is, "No history of alcohol or drug misuse for a minimum of two years immediately prior to certification." This, of course, would apply to alcoholics and nonalcoholics alike.

In some state, efforts have already begun to develop certification for alcoholism counselors. These include New York, North Carolina, Kansas, Minnesota, North Dakota, and others. In some cases the certification will be granted by a voluntary organization, in others by a state governmental body, and in others by an independent board set up primarily for this purpose. In some states, efforts to develop certification for addiction counselors competent in both alcoholism and drug abuse care are being made. Until these diverse developmental efforts reach fruition, the field will remain somewhat fluid and not well defined. It is to be hoped that the professionalization movement does not overlook the unique contribution of the recovered alcoholic as counselor.

REFERENCES

Adams, J. R., 1973, The driver rehabilitation group leader: Who qualifies?, unpublished.
"Alcoholics Anonymous," 1955, Alcoholics Anonymous World Services, Inc., New York.
Alcoholics Anonymous, 1972a, "A.A. Wants to Work With You," brochure, Professional Relations Committee, General Service Office, Alcoholics Anonymous, Box 459, New York 10017.
Alcoholics Anonymous, 1972b, "A.A. Guidelines for Members Employed in the Alcoholism Field" (those who wear "two hats"), brochure, Professional Relations Committee, General Service Office, Alcoholics Anonymous, Box 459, New York 10017.
Anderson, D., 1944, The place of the lay therapist in the treatment of alcoholics, *Q. J. Stud. Alcohol* 5:257.
Anderson, D., 1950, "The Other Side of the Bottle," A. A. Wyn, New York.
Augustus, G., 1971, The Maryland program for training alcoholism counselors, in "Professional Training on Alcoholism" (F. A. Seixas and J. Y. Sutton, eds.) Vol. 178, Annals of the New York Academy of Sciences, pp. 58–60, New York Academy of Sciences, New York.
Baltimore City Health Department Alcoholism Center, n.d., "Training Program for Alcoholism Counselors," brochure, Alcoholism Center, 2221 St. Paul Street, Baltimore, Maryland.

Blume, S. B., Robins, J., and Branston, A., 1968, Psychodrama techniques in the treatment of alcoholism, *Group Psychotherapy* 21:241.

Burlington Northern, 1971, "A Rehabilitation Program for Problem Drinkers," brochure, Burlington Northern, Inc., Personnel Department, St. Paul, Minnesota.

D'Alonzo, C. A., 1969, Testimony before the Senate sub-committee on alcoholism and narcotics, New York (transcript from Medical Division, E. I. DuPont DeNemours and Company).

Edwards, G., 1970, Place of treatment professions in society's response to chemical abuse, *Br. Med. J.* 11:195.

Falkey, D. B., 1971, "Standards, Recruitment, Training and Use of Indigenous Personnel in Alcohol and Drug Misuse Programs," presented at the 22nd annual meeting of the North American Association of Alcoholism Programs, Hartford, Connecticut.

Glasscote, R. M., Plaut, T. F. A., Hammersley, D. W., O'Neill, F. J., Chafetz, M. E., and Cumming, E., 1967, "The Treatment of Alcoholism; A Study of Programs and Problems," Joint Information Service of the American Psychiatric Association and the National Association for Mental Health, Washington, D.C.

Jellinek, E. M., 1960, "The Disease Concept of Alcoholism," College and University Press, New Haven.

Kaim, S. C., Brill, H., Cloud, L. A., Knott, D. H., Lieber, C. S., McIsaac, W. M., Mendelson, J. H., Rankin, J., Reading, A., Shore, R. S., Willard, H. N., and Wolin, S. J., 1972, Criteria for the diagnosis of alcoholism, *Am. J. Psychiatry* 129:127

Krystal, H., 1963, Who is qualified to treat the alcoholic? II Advantages of the professional psychotherapist, *Q. J. Stud. Alcohol* 24:706.

Leach, B., Norris, J. L., Dancey, T., and Bissel, L., 1969, Dimensions of Alcoholics Anonymous: 1935-1965, *Int. J. Addict.* 4:507.

Lockwood, J. W., 1971, "Syllabus, Report and Summary Appraisal: The Bowery Mission Program to Date," brochure, Bowery Mission and Young Men's Home, 227 Bowery, New York.

Lowry, L., 1971, A.A. members as alcoholism counselors, *Michigan Alcoholism Review* 16:10 (Published by Department of Public Health Alcoholism Program 3500 North Logan, Lansing, Michigan).

Mann, M., 1958, "New Primer on Alcoholism," Holt, Rinehart & Winston. New York.

Mann, M., 1973, Attitude: Key to successful treatment, *in* "The Paraprofessional in the Treatment of Alcoholism: A New Profession" (G. E. Staub and L. M. Kent, eds.), Charles C Thomas, Springfield, Illinois.

Maxwell, M., 1950, The Washingtonian movement, *Q. J. Stud. Alcohol* (Suppl) 11:410.

McGoldrick, E. J., 1964, Who is qualified to treat the alcoholic? Comment on the Krystal-Moore discussion, *Q. J. Stud. Alcohol* 25:351.

McGoldrick, E. J., 1966, "The Conquest of Alcohol," Delacorte Press, New York.

McI., J., 1970, "How Can Alcoholics Anonymous Members Work Effectively as Alcoholism Counselors?," Presented at the 35th Anniversary International Convention of Alcoholics Anonymous, Miami Beach, Florida.

Mullan, H., and Rosenbaum, M., 1962 "Group Psychotherapy Theory and Practice," Free Press, Glencoe, Illinois.

O.E.O., 1971a, "O.E.O. Guidance, Alcoholic Counseling and Recovery Program," No. 6136-1a, Office of Economic Opportunity, Washington, D.C.

O.E.O., 1971b, "Training Alcoholism Counselors for Poverty Communities; First Annual Report," Rutgers Center for Alcohol Studies, New Brunswick, New Jersey.

Ottenberg, D. J., 1969, The Eagleville interdisciplinary rehabilitation program for alcoholics: Lessons after two years, *Q. J. Stud. Alcohol* 30:449.

Raws, W., n.d., "Saved by Grace; Testimony of an Artist," pamphlet, Keswick Colony of Mercy, Keswick Grove, Whiting P.O., New Jersey.

Raws, W. A., 1972, Keswick Colony of Mercy, personal communication.
Rosenberg, S. S., Keller, M., Bellicha, T. C., Katz, J. W., Light, L., and Spiegler, D. L. (eds.), 1971, "Alcohol and Health," Department of Health, Education and Welfare, Washington D.C.
Seixas, F. A., and Sutton, J. Y. (eds.), 1971, Professional training on alcoholism, *Ann. N.Y. Acad. Sci.* 187:132–138.
Slaughter, L. D., and Torno, K., 1968, Hospitalized alcoholic patients IV the role of patient-counselors, *Hospital and Community Psychiatry* 19:209.
Stewart, E. I., and Malfetti, J. L., 1970, "Rehabilitation of the Drunken Driver," Teachers College Press, Columbia University, New York.
Strecker, E. A., and Chambers, F. T., 1939, "Alcohol—One Man's Meat," Macmillan, New York.
Tiebout, H. M., 1961, Alcoholics Anonymous—an experiment of nature, *Q. J. Stud. Alcohol* 22:52.
Truax, C. B., and Carkhuff, R. R., 1967, "Toward Effective Counseling and Psychotherapy: Training and Practice," Aldine, Chicago.
Willmar State Hospital, n.d., "Counselor on Alcoholism Training Program," brochure, Willmar State Hospital, Willmar, Minnesota.
Wilson, W. M., 1968, Hospitalized alcoholic patients VI Training personnel to work with alcoholic patients, *Hospital and Community Psychiatry* 19:211.

CHAPTER 13

Training for Professionals and Nonprofessionals in Alcoholism

Edward Blacker

Division of Alcoholism
Massachusetts Department of Public Health
Boston, Massachusetts

INTRODUCTION

All professional workers in the field of mental health are thoroughly taught about the classical types of psychosis and neurosis. Other general health workers in the course of their schooling are also familiarized with these problems. Few people complete medical, social work, psychology, nursing, and similar degrees without learning about schizophrenia, depression, phobias, and the like. Clinical experience is provided on these problems and doctoral dissertations on them have been produced in the thousands. In contrast, training and education on the problem of alcoholism has been sketchy at best and in most cases nonexistent, despite the fact that its prevalence in the United States and throughout the world is extensive.

Relatively speaking, formal training programs in alcoholism until a few years ago have been few in number. But in the late 1960s a new national surge of interest in the alcoholism problem produced a corresponding expansion in training programs. For example, in 1971 the National Institute on Alcohol Abuse and Alcoholism (NIAAA) funded approximately 25 training programs in the United States. In 1972, NIAAA funded 48 training programs, practically doubling their effort in one single year! Clearly, the long extant training gaps in schools and clinical settings on the problem of alcoholism are beginning to be corrected.

The thinly spread existence of alcoholism training programs has resulted in a lack of information and communication on the subject of training in alcoholism. This shortage is compounded by the fact that directors of training programs, NIAAA supported and others, simply have not published descriptions of their programs or evaluations of their methods. A search of the *Abstract Archive on Alcohol Literature,* the main bibliographic reference in the field of alcoholism, yielded less than a handful of references. Therefore, the information on training in this chapter is based on the author's personal experience with several training programs, on unpublished materials privately supplied to the author by several directors of training programs, and on very brief abstracts of 46 NIAAA-funded programs made available by the training branch of NIAAA. As of 1972, half of the latter programs had barely begun.

Despite the relative dearth of published material on training in alcoholism, there is sufficient information available to provide an assessment of some of the essential elements of training. These will be described below and will serve to provide those interested in developing training programs with some guidelines on factors that need to be considered, such as objectives, target groups, settings, teaching methods and models, and subject matter.

SIGNIFICANT COMPONENTS OF TRAINING

Objectives

Those familiar with the field of alcoholism know that the role-players often seem to fall into two categories: (1) those who are strong advocates of the alcoholic, fighting to reduce discrimination, hostility, and rejection against him while promoting his access to the health delivery system, and (2) those who are indifferent, unaware, or downright hostile to the alcoholic, resisting his access to the health delivery system. The latter group far outnumbers the former group, but over the last 30 years the advocates of helping the alcoholic have been increasingly successful in overcoming negative attitudes and practices of many groups of caretakers in our society. This prevailing set of circumstances has necessitated making attitude change one of the main objectives of training. The specific content of the attitude change sought varies with the target group being trained.

Typically, many training programs knowingly seek out target groups of one kind or another who hold negative attitudes toward the alcoholic ranging from neutrality to therapeutic nihilism. People in positions of authority in hospitals (including trustees, physicians, hospital administrators, etc.) are one example. They may hold beliefs and myths that the alcoholic is a difficult patient—refusing to accept a doctor's orders, causing trouble, unacceptable to other patients, and having little chance of treatment success. As a result, the facilities they govern have more often than not been closed to the alcoholic. Some hospitals, for example, have actual closed-door policies written into their bylaws, and many refuse treatment by throwing up various forms of informal, unspoken barriers. Therefore, for these target groups, one important goal of attitude change is to persuade those in key institutional positions to make services available to alcoholics without discrimination.

For clinician target groups, the objective of attitude change has a different purpose. Patients are sensitive to how therapists feel about them, and this obviously affects client behavior and the results of treatment. If the therapist feels the client won't or can't improve his condition, in all likelihood this will be transmitted to the client and result in a self-fulfilling prophecy. For clinicians, then, training seeks to convert their feelings and attitudes from therapeutic nihilism to feelings of positive hope for a favorable treatment outcome.

A third attitude widely held by members of our society is that the alcoholic is an immoral failure and ought to be punished. Witness the fact that public drunkenness has always been treated as a crime and the penalty for this behavior has been a jail sentence. Less than half of the states have repealed public drunkenness as a crime, replacing punishment facilities with treatment facilities. Because of this widely held attitude, training programs make every effort to eliminate this belief in anyone who will listen, from the average citizen to the highest officeholders.

In short, there are three pervasive expressions of negative attitudes toward the alcoholic: (1) "Keep out of my house." (2) "You can't or won't get better even if I try to help you." (3) "You are a moral failure and ought to be punished." A primary objective of training is to change these attitudes because little else can be accomplished when they exist. This is why training programs and alcoholism agencies hammer away on the theme that alcoholism is a treatable illness.

Assuming the presence or accomplishment of attitude change, a second major objective is the transmission of information on alcoholism and associated problems. The content of the information may include basic knowledge of the problem, its epidemiology, and matters of application. For example, basic knowledge might cover symptoms and theories; epidemiology would deal with the size and distribution of the problem; applied information might cover types of treatment resources. These are but a few examples of subject matter which will be discussed in greater detail in a section below.

A third major training objective is teaching technical skills. The particular skills taught depend on the role of the target group. If one is training high school teachers, attempts would be made to teach them about prevention and how to lead group discussions. For clinicians, skill training might include counseling techniques, medical management, and referral procedures. Research specialists might learn about evaluation techniques, the design of experiments, and statistical procedures. Obviously the particular content will vary with the type of trainee and the level of the trainee's prior experience.

Attitude change, knowledge acquisition, and skill development are, then, three fundamental objectives of training. These objectives have been sought sometimes individually and often in different combinations. In developing any training program, clarity of objectives is essential, for this is the basis for choosing teaching strategies. Moreover, only when objectives are clearly stated can evaluation be possible.

Target Groups

There are one national and four regional training centers which specialize in alcoholism. However, the main responsibility for training is carried by many directly funded NIAAA programs, short-term summer schools, a few academic programs, some voluntary agencies, and official state alcoholism programs. These are the main agents of training and they choose to train and educate different target groups.

Target groups range from the general public to the specialist in alcoholism. Teaching objectives and strategies vary, depending on the target group. Efforts are made by the training agents to reach all kinds of target groups which are enumerated and described as follows.

General Public

There are two main subgroups in this category: (a) the general lay citizen of a community and (b) community leaders holding various positions of importance.

The main training agents for the general public are the voluntary committees on alcoholism (affiliated with the National Council on Alcoholism) and the official state alcoholism programs. The efforts of these agents are to promote alcoholism as an illness to the general public. Obviously, the power to create and fund service programs for alcoholics and their families lies in the hands of the general public and its official leaders. If they are not convinced of the merits of helping the alcoholic there will be little or no effort to create the necessary treatment facilities. Therefore, the general public is a primary target of education.

Another reason for education and training of the general public is to assist them in getting help when they encounter the problem of alcoholism in their families. There is a need for the public, as in the case of any illness, to recognize the symptoms of alcoholism so that early identification of the problem can be followed by appropriate action. Through such means as the mass media the public is provided with information on the problem and resources that they can use for assistance.

Sometimes community citizens, both laymen and officials of government, form committees to examine the problem in their communities and to take action steps to meet these problems. Training agents will make special efforts to organize such groups and provide them with more intensive information than the general public. Members of such groups may include recovered alcoholics, their wives, public officials, clergymen, business leaders, and facility directors and planners. The purpose of these groups generally is to assist in organizing the community to develop resources for assisting the alcoholic. The members of the committee are not necessarily themselves agents of treatment, but rather instruments for the development of services. For example, a member of the school committee might be instrumental in advocating an alcohol education program in his school system, but would not be responsible for its implementation. Or, a hospital trustee might arrange for a ten-bed alcoholism unit to be opened in his hospital. Or, as a third example, a city mayor might provide an unoccupied city building to be used as a detoxification facility or a halfway house.

Obviously the general public constitutes the largest target group of alcoholism education. But even though this audience is vast and sometimes difficult to reach, its support is vital to the field of alcoholism.

Students

A major target group for purposes of prevention is composed of elementary and high school students. Although drug education has predominated in schools in recent years, efforts are continually made to introduce alcohol education into school curricula. There is even one summer school of alcohol studies, the Nebraska school, which invites high school-age youngsters to attend.

College- and graduate-level students form another target group. As part of their general course curricula, the subject of alcoholism is introduced, generally in the social and behavioral science courses. However, the subject matter is more often than not sparsely treated, and most students complete their degree training without learning very much about alcoholism. Efforts are now being made to correct this deficiency and these efforts will be discussed in more detail in the section below on settings.

Finally, professional school students—as, for example, in medicine, law, and social work—are another target group. Their education on alcoholism, or rather lack of it, is parallel to that of graduate students.

General Caretakers

The alcoholic in the course of his treatment is in contact with many different agencies and institutions and the people who work in them. Many alcoholics, especially the skid row type, frequently encounter the workers and officials in the whole spectrum of law enforcement. This would include the police, lawyers, judges, probation officers, and prison personnel.

Other alcoholics never intersect with the law enforcement agencies, except perhaps in the divorce courts. Instead, most commonly they might seek help for their problems from clergymen or from physicians. They might also try to obtain assistance from mental health resources or social and welfare agencies.

Of the major institutions in our society dealing with social or health problems—hospitals, courts, the police, social and welfare agencies—very few fail to have some type of contact with the problem drinker. Yet, many of them fail to deal directly with the problem, consciously or unconsciously overlooking the core problem and dealing only with its effects. For example, a physician might treat a patient in a hospital for all kinds of medical complications of alcoholism, but will not confront the patient with suggestions for dealing with the root cause of the complications, namely the excessive, uncontrolled drinking. This behavior, coupled with the alcoholic's own inclination to deny his problem, in effect creates a conspiracy of avoidance.

Another group that falls in the general caretaker category is business and industry. Although not involved in direct clinical care, they are affected by virtue of the fact that their employees' work performance is impaired if they have alcoholism problems.

All of these caretaker groups have generic work responsibilities or functions. They are not specialists in the treatment of alcoholism. As a matter of fact most of these workers have little or no knowledge of the nature of alcoholism. Since it would be impossible to provide treatment for the vast number of alcoholics in specialized settings, it is essential for general caretakers to learn how to deal with the alcoholism problem. This is not to say that everyone has to be an expert, but at least they should know enough about the problem to identify it and to make appropriate referrals to knowledgeable resources. They should also be sensitive to the fact that hiding or avoiding the problem is detrimental to the alcoholic. Other general caretakers, however, can acquire sufficient knowledge to deal with the alcoholic directly as part of their general practice or responsibilities without having to refer the client to other resources.

Training agents have long recognized the need for training general caretakers. This training activity is very actively fostered by official state alcoholism agencies and also by the National Institute on Alcohol Abuse and Alcoholism. As a matter of fact approximately 25 percent of the formal training programs funded by NIAAA in 1972 were targeted for trainees from a wide variety of backgrounds (Mitnick, 1972). These included: school officials such as teachers, counselors, and administrators; medical personnel such as physicians, nurses and public health officials; law enforcement workers such as police officers and probation officers; social agency personnel such as social workers, welfare workers, and vocational counselors; and community development workers such as program administrators or planners, clergymen, mental health workers, and industry representatives.

This wide spectrum of general caretaker target groups mirrors the wide impact of the alcoholism problem in our society. It is most important for these groups to be knowledgeable about alcoholism since most alcoholics receive their primary treatment from them rather than from specialists.

Specialists

Most mental health and other health facilities in the United States are designed to deal with many different types of problems. A typical mental health facility or agency will accept patients with a wide variety of diagnoses. People with a problem of depression go to the same treatment center as a person with an anxiety neurosis or a phobia. In medical facilities to a large extent the same is true. A hospital is equally available to the person in need of a hernia repair and to the person requiring appendix removal.

Unhappily, problem drinkers have not been viewed as having an illness meriting access to many of the multipurpose facilities or hospitals. Somehow the alcoholic has been excluded for one reason or another. As a result there have been many specialized services that have been designed and organized especially for the alcoholic. For example, in Massachusetts there are 25 outpatient clinics and 40 halfway houses exclusively for the alcoholic. People of different professional backgrounds work in these types of facilities and so do nonprofessionals. Many of these workers devote their efforts to the alcoholism field on a full-time basis; others do so part time. But full time or part time, they have an in-depth interest in the problem.

Training for professional specialists requires providing them with a full range of information on the alcoholism problem and teaching them the skills necessary to deal with it. Professional groups range from the physician to the social worker, and skills range from appropriate medical treatment to counseling. As experience grows, professional groups also require opportunities to share these experiences, and this is often accomplished in special conferences or meetings.

Nonprofessionals or paraprofessionals also work in the field of alcoholism, and lately in expanding numbers. Unfortunately, the term *nonprofessional* is not clearly defined or uniformly applied. Sometimes it seems that the label is assigned to anyone who is not a member of one of the health professions—for example, to clergymen or to probation officers. At other times the label is assigned to people without college or advanced degrees. At still other times the label might apply to someone who has professional degree training in a field unrelated to dealing with personal problems of human behavior—for example, a chemist or an economist who decides to become an alcoholism counselor. Perhaps the future will bring clearer definitions of the term, ideally a label defined in its own right rather than as the negative entity of the classification of professionals.

In any case, in the context of this discussion the term is used to apply to less-than-college-educated individuals and to college-trained individuals whose education has not been directly relevant to providing counseling or other clinical services to emotionally disturbed individuals. Although in other health problem areas nonprofessionals come from a wide variety of backgrounds, in the field of alcoholism the nonprofessional worker is most often a recovered alcoholic or the spouse of an alcoholic. Their desire to work in the field of alcoholism stems from a combination of personal gratitude in having overcome the problem and their unique experience with self-help as members of Alcoholics Anonymous (A.A.) or Al-Anon. Through A.A. many alcoholics have learned to control their own problem drinking and in turn have assisted others in this accomplishment.

In the last few years many recovered alcoholics have begun to wear two hats. They continue as members of A.A. for the sake of their own sobriety, but they have also sought and found employment as paid workers in the field of alcoholism. They may work as alcoholism counselors in various settings such as hospitals, outpatient programs, and halfway houses. In many states they also operate and manage private, nonmedical treatment facilities. More recently they have become community workers, serving communities as advocates for the alcoholic and ombudsmen assisting the alcoholic to treatment resources.

Most of these nonprofessional workers do not have academic knowledge of the alcoholism problem and its ramifications. Rather, they know the problem from the perspective of personal experience. Sometimes this is a great advantage, but at other times not. The problem of alcoholism manifests itself in many different ways and not all alcoholics are cut from the same cloth.

Therefore, the alcoholic who becomes a paid worker has to expand his perspective.

As paid workers, nonprofessionals are also called upon to carry out duties and functions that are independent of their experience as alcoholics. For example, how to appropriately evaluate a person's problems and decide what to advise the person; or, how to organize and manage the services of a halfway house; or, how to organize a community to plan for resources for the alcoholic. All of these functions require special skills which professional groups often acquire in their academic training. If the nonprofessional is lacking these skills, then training programs have to be provided to teach them. Such programs are springing up throughout the country. There are programs in Baltimore, Boston, and Los Angeles, to mention but a few.

To summarize our discussion of target groups several generalizations can be made: (1) The problem of alcoholism in one form or another confronts a wide range of the public, professionals, and nonprofessionals, generalists and specialists; (2) knowledge and experience with the alcoholism problem by these groups is not widespread; (3) interest in acquiring knowledge of the problem and motivation to help the alcoholic varies with professional disciplines and individuals in these disciplines; (4) accordingly, training objectives and teaching strategies vary for different target groups depending on their attitude, discipline, and prior education and experience.

Settings

Training programs in alcoholism are generally carried out in three main types of settings: (1) short-term (one to three weeks) summer schools operated on college or university campuses; (2) academic year programs conducted at colleges and universities; and (3) various length programs conducted at clinical treatment centers and facilities.

Summer schools on alcohol studies are available in all regions of the United States and in Canada. There are at least a dozen or more of these schools and they have provided training for tens of thousands of people. They are always conducted in university settings, often in conference centers which are part of the university. The university involvement is usually under the auspices of a Division of Continuing Education in cooperation with the official state alcoholism program.

The Rutgers School of Alcohol Studies, which runs for three weeks, is the oldest specialized school on alcohol studies. It was established originally at Yale University in 1943 and moved to Rutgers University in 1962. The other summer schools, which run for one week, are located throughout the country. In many cases the summer schools operate from the same university campus every year. In other cases of regionally sponsored schools—for example, the New England School of Alcohol Studies—the campus settings are rotated each year, or every other year.

In the case of the summer schools, the role of the university (except for the Rutgers School) is more or less that of hotel host. The educational input is minimal. Recently, however, in some cases the university has become a setting for training programs which are fully integrated with its education and degree-granting purposes. All of these programs have been launched with the financial support of NIAAA and about 15 to 20 such programs are now under way at different universities throughout the country. The programs are, with few exceptions, geared to graduate-level training.

The purposes and the target groups of the university-based training programs vary with the program. Some programs offer specialized training on alcoholism for a selected and limited number of students from the general graduate student body. In these programs, the specialized alcoholism training is patched onto the generic degree training, and degrees are earned in a general discipline rather than in alcoholism specifically. Other students at the universities who are not participants in the special training program are not generally exposed to any course work on alcoholism. An example of this type of university-affiliated training program is one on research training conducted by the author (Blacker, 1968). Another example is a program in the Department of Special Education and Rehabilitation at the University of Pittsburgh to train counselors and leaders in the field of alcoholism (Mitnick, 1972).

A second type of university training program is designed so as to integrate training in alcoholism very closely with general degree training. Students are expected to take courses in the core curriculum required for a basic generic degree plus extra courses and practicum training in alcoholism. The degree earned may be a general degree (Master of Health Science, Master of Mental Health, Master or Doctor of Science), as in the case of the comprehensive alcoholism services program at The Johns Hopkins University, School of Hygiene and Public Health (Mandell, 1972), or the degree earned may be specifically for alcoholism, as in the case of the University of Arizona, which offers a Master's degree in rehabilitation counseling in alcoholism (Mitnick, 1972). The general student body at the schools in which this type of training program is offered is not necessarily exposed to the alcoholism curriculum. But in a few cases a basic course provided for the special trainees is available also to other graduate students.

A third type of university program is geared to providing some training on alcoholism to all students at a particular school or subdivision of a school, as contrasted to training for a select few. One such program was developed at the Graduate School for Social Work at Boston University in Massachusetts. In this program the emphasis is on training the social work student who may *not* work in specialized alcoholism programs. As we have previously pointed out, more alcoholics seek help from generic treatment facilities than from specialized resources. The Boston University program, recognizing this fact, is therefore attempting to train all of their students to be better equipped to meet problems of alcoholism in their professional activities, no matter what setting they work in.

Another example of this third type of program was started at the Bowman Gray School of Medicine of Wake Forest University in Winston-Salem, North Carolina. This educational program is designed to introduce and include material (information, concepts, and techniques) on alcoholism and alcohol abuse into the existing curricula of this teaching and training complex. The program is unique in that its trainee target group is not confined to students alone. Rather, the program considers all of their 314 medical students, 137 house staff of all departments, 188 attending staff, the local physicians in surrounding communities, and all paramedical personnel in surrounding communities to legitimately qualify as trainees of the program. Quite an ambitious program!

A fourth type of university program combines the intensive training of a selected group of graduate students and at the same time offers some general training on alcoholism to all students enrolled in the particular department. An example of this type of training was the addition to the Clinical Psychology Training Program in the Department of Psychology at the University of Nebraska in Lincoln. The primary purpose of this program is to provide training for clinical psychology specialists in alcoholism. But in addition, the program provides training opportunities in alcoholism for all students in the clinical program. The program is based on the joint efforts of both the university and existing community projects.

A third major type of setting for training programs is located at clinical treatment centers and facilities. These facilities run the gamut of clinical centers where the alcoholic obtains services—outpatient clinics, inpatient hospitals, mental health centers, and various community agencies.

These programs are designed for the practitioner and tend to stress direct clinical experience, although didactic lectures are also included. However, theory is not emphasized as much as with trainees in university settings.

An interesting example of a training program with the emphasis on practicum settings was launched by the East Los Angeles Health Task Force in California. This organization is providing paraprofessional training to 20 community residents to develop skills in alcoholism rehabilitation. The trainees are specially selected *barrio* residents, the majority recovered alcoholics, who are bilingual and bicultural. They train and work in different community agencies and in the community itself with an emphasis on assisting Spanish-speaking problem drinkers and their families.

A rather novel training program uses the community itself as the training site. The Western Regional Office of the American Public Health Association has developed a program which travels through seven different states. The purpose of the program is to help agency and community leaders to decide, after examining the alcoholism problem in their own area, what to do about controlling it at community and state levels, and to improve their capability to develop and manage alcoholism control programs.

Although the above-described types of settings have been discussed separately, it would be misleading to interpret this as reflecting isolated operations. Many times the universities and practicum centers work together cooperatively and trainees move back and forth between the university and the clinical setting. However, this kind of combined training is more characteristic of those training programs where the grantee is a university. When the grantee is a clinical setting, backup university support may or may not be present.

The setting of a training program of course tends to correlate with particular target groups. Medical schools are oriented to medical students and physicians, social work schools to social workers, and so on. But the clinical settings generally provide training for members of a variety of professions and also for nonprofessionals. Of the close to 50 NIAAA-funded training programs previously mentioned (some examples of which have been given in this section) social workers, paraprofessionals, and varied caretakers constitute about two-thirds of the target groups. The remaining target groups consist of physicians, clergy, rehabilitation counselors, public health workers, behavioral scientists, and psychiatrists.

On the surface, it would appear that the setting of a training program is a straightforward matter. However, this is not the case, for the setting can carry with it the support, endorsement, power, and charisma of the sponsor of the training program. When such sponsors have widespread status and influence, the impact on trainees, trainers, agencies, and the community is enhanced.

Teaching Methods and Models

A teaching method can be defined as a technique or strategy for transmitting information and knowledge. A model is the particular or unique organization of methods. Methods of alcoholism training programs are similar to those used in familiar teaching situations. However, the selection and emphasis of particular methods vary from program to program. Models also vary, but their teaching components can be classified basically into three categories: (1) academic subject matter, (2) practicum experience, and (3) time involvement.

All training programs try to teach about the subject matter on the nature of alcoholism—its characteristics, causes, epidemiology, and treatment. Typical methods of conveying subject matter include lectures, seminars, films, videotapes, closed-circuit TV, group discussions, and panel presentations.

Most training programs also provide practicum experience. Typical methods used for this include live patient interviews, counseling techniques, field trips, professors' rounds, experience groups, encounter groups, one-way mirror observation, case conferences, direct treatment, community liaison work, and research work.

Trainees' time involvement can range from a very brief exposure to an intensive experience. Some target groups may simply learn about alcoholism from listening to one or a few lectures. Others might participate in a week-long program, as in the case of the typical summer school on alcohol studies. Still others might receive training for a full three months or even for as much as a year. No programs are structured for more than a year except perhaps for programs that are part of formal generic degree training. In the latter programs the trainee may spread his alcoholism speciality over several years on a part-time basis while simultaneously meeting the requirements for his general degree.

Content material in lectures and seminars is, of course, provided by knowledgeable instructors. The same is true of fieldwork where the trainee is generally under the supervision of experienced fieldwork instructors. Sometimes trainees will rotate their fieldwork through different types of services and clinical activities.

Training program models are shaped from the three major components mentioned and methods are chosen from the wide array available to suit the training objectives for a particular target group. For example, the Institute of the Pennsylvania Hospital in Philadelphia in 1972 began a training program for nonpsychiatric physicians in the Greater Delaware Valley (Mitnick, 1972). The program is designed to improve the care of alcoholic patients by educating individual physicians and physicians holding teaching or special organizational roles. The training model includes formal as well as informal approaches. The formal program involves: (1) weekly seminars, (2) experience-based learning, (3) three-session courses on specific aspects of alcohol abuse, and (4) single sessions on alcoholism included as part of continuing education programs devoted to other topics. The informal approaches are rather innovative and include: (1) circuit riders, (2) programed correspondence education using patient management problems, and (3) teaching special groups such as medical school faculties and industrial physicians. The particular methods used in this training program include case presentations, sensitivity training groups, live patient interviews, didactic presentations, group discussions, audiovisual presentations, visits to treatment facilities, and consultations in physicians' offices.

Another example of an interesting training program model was carried out at the Massachusetts General Hospital in Boston. This program was designed to provide leadership training to social workers, nurses, psychologists, and psychiatrists. All trainees had already received their advanced degrees and had past work experience. The training offered was for one year on a full-time basis. Components of training included an academic seminar, an experience group, a didactic group on group therapy, field visits and consultation to other types of facilities, and supervised direct clinical practice. Included in the clinical practice was an especially innovative technique that involved having an interdisciplinary team composed of a social worker, a nurse, a psychologist, and a psychiatrist operate as a unit to service the alcoholics coming to the emergency ward and the outpatient clinic. The team worked together for a full year and each three months the leadership of the team was rotated so that each member would have an opportunity to lead the work responsibilities of the team. This procedure was a courageous break with tradition where the physician generally holds the superordinate status and responsibility.

Of the many methods available and used in training one might well ask which are most effective and with whom. Unhappily there are very few, if any, published accounts of evaluations of the effectiveness of particular methods or strategies of teaching. If at all, this type of information tends to be transmitted informally by word of mouth. Teaching techniques are tried, tested, and revised based on personal experience. For example, in the first year of the research training program directed by the author, the seminar component involved parading expert after expert before the trainees to lecture on their particular special area of alcoholism. Although this faculty of lecturers was very impressive and did an outstanding job of communicating information to the students, it became clear that some serious liabilities were connected with this type of teaching approach. The small seminar group of students were obtaining their knowledge through passive learning and therefore it did not "stick" as well as it should. In addition group discussion and group solidarity did not develop over the course of time. So in the subsequent years of the training program the experts were scrapped, except for an occasional lecturer. Instead the students themselves were assigned the responsibility of teaching the core seminar subject matter. On a rotation basis each student chose a topic area and presented it to the seminar. Experience has shown that these presentations are not as precise, comprehensive, or erudite as those of the experts, but the sacrifice of some expert information is a small loss compared to the gain of active student learning and greater liveliness and participation in the seminar.

Probably experiences with training methods similar to this example can be commonly found in other training programs. One film may be superior to another; it may be better to cover a particular topic before another; emphasis on certain clinical experiences might be better for one type of target group as compared to another, etc. However, as indicated previously, this type of information is not readily available. If it is available, it is based on personal experience only and not on systematic efforts to compare and evaluate methods. There should be more of the latter, but this is very rare. However, one such attempt is being made in a training program at the Malcolm Bliss Mental Health Center in St. Louis, Missouri (Mitnick, 1972).

The target groups of the Malcolm Bliss training program are practicing care-givers representing various disciplines and settings at the local level. In providing the training on treatment of the alcoholic to these groups, the program seeks to evaluate the effectiveness of two different teaching methods in terms of acquisition of new knowledge, change of attitude, and retention of new information and attitudes. The two methods compared are lecture-discussion-demonstration versus more condensed informational sessions plus small "gut-level" discussion groups. This program illustrates the principle of designing a training program which provides needed training and simultaneously permits the systematic evaluation of different methods.

Some of the examples given above on training methods and models are intended to be illustrative only. There is nothing sacrosanct about any one method or model. The particular combination of elements and methods should be carefully chosen by training program designers in accordance with the objectives of training for any particular target group so as to ensure a positive learning experience.

Subject Matter

All training programs cover more or less the subject matter related to alcohol, drinking, and alcoholism. Some typical topics covered are listed as follows under basic information, special areas, and service and treatment.

Basic Information

    Alcohol as a substance and its biochemistry.
    Epidemiology of alcoholism.
    Drinking in primitive societies and anthropological theories of alcoholism.
    Drinking attitudes and practices in civilized societies and sociological theories of alcoholism.
    Psychological and psychiatric theories of alcoholism.
    Psychological effects of alcohol on performance and personality.
    Classification and typology systems of alcoholics.
    Stages in the development of alcoholism.
    Survey of treatment approaches to alcoholism and its complications.
    Education and prevention of alcoholism.

Special Areas

    Alcoholism and the family.
    Alcoholism and skid row subculture.
    Alcohol, problem drinking, and traffic safety.
    The Churches and alcohol.
    Teen-age drinking.
    Alcoholism and industry.
    Alcoholism in women.
    Legal issues and government controls of alcohol.

Service and Treatment

    Patient motivation and therapist attitudes toward the alcoholic.
    Residential treatment approaches.
    Pharmacological treatment of alcoholism and its complications.
    Emergency ward and outpatient treatment.
    Special techniques such as behavior therapies and LSD therapy.
    Counseling goals and strategies in treating the alcoholic.
    Alcoholics Anonymous, Al-Anon, and Al-Ateen.
    How to evaluate and refer patients.
    Resources for treatment of the alcoholic.
    Community consultation approaches.
    Planning and organizing a treatment delivery system for alcoholics.

It is obvious from the above outline of basic information, special areas, and service and treatment topics that there is a wide range of subject matter. Training programs generally choose from among these topics (and from others not listed) to cover the discursive subject matter whether in lectures, seminars, or small group discussions. Depending on the intensity of the training program and the interests of the trainees, the number and emphasis of the topics will vary. Also, the level on which any of these topics are presented has to be adjusted in accordance with the level of sophistication and experience of the trainees.

For each of the topics mentioned there is appropriate reading material available in the literature. Formal training programs generally have prepared reading lists which are made available to students. Samples of these reading lists can often be obtained by writing directly to the directors of training programs. There is an extensive literature in the field of alcoholism (as in most other fields ranging from very good to very bad) offering different and sometimes questionable perspectives and information. It is the responsibility of trainers and educators to expose trainees to the best that is available and to provide a balanced and unbiased account of any topic.

EXAMPLES OF PROGRAMS

The components of training that have been discussed as separate elements are of course melded together in any training program. A few examples of training programs with different objectives will now be described treating them as complete entities.

Research Program

The author (Blacker, 1968) has directed a program to train behavioral and social science graduate students in alcoholism research which is now in its 15th year of operation. The objectives of the program are: (1) to train the student in an interdisciplinary approach; (2) to train and recruit students for operating more effectively in research, clinical, and academic settings either wholly or partly concerned with alcoholism; and (3) to contribute knowledge in the field of alcoholism during and after the trainee's experience in the program.

The training program (supported by NIAAA) is a cooperative effort of the Massachusetts Division of Alcoholism, universities (Boston University, Harvard, and Tufts) in the Boston area, and various field settings. Field settings include inpatient and outpatient alcoholism services, and any other suitable agencies. The selected field settings are staffed by highly qualified researchers and clinicians competent to provide expert supervision.

Toward the end of the third quarter of each academic year, graduate students are recruited for the succeeding academic year from the social and behavioral science departments of the universities mentioned. Seven doctoral candidates, majoring in sociology or any branch of psychology, participate in the program each year. To be eligible a student must be enrolled in the regular doctoral program in his school. Students at different levels of graduate training have been accepted, but more advanced students are preferred. Because we wanted to train as many students as possible, with few exceptions, the students are in the program for only one academic year. Each student receives a stipend plus payment for tuition. The students are expected to spend from $2\frac{1}{2}$ to 3 days a week on the program.

The training model consists of three components—a seminar, a field placement, and an original research project. Throughout the academic year 30 two-hour seminars are held at one of the participating universities. The seminars of the first semester are devoted to coverage of basic content information, ranging from the chemistry and physiology of alcohol to theories of alcoholism. The basic information is covered by the students' own presentations, with each student responsible for two major topics. For some topics, experts follow the student presentations, going into the finer points in each area and discussing related research studies. In the second semester the seminars include presentations by the students of their research proposals and research findings.

The teaching staff present at the seminars includes the author as seminar leader and a faculty member from each of the participating universities. The faculty includes sociologists and psychologists, assuring a good interdisciplinary exchange. The participation of university faculty is an exceedingly important feature, for it keeps the student allied with university supervision and results in an academic atmosphere. All of the schools have provided some form of academic credit for the seminar.

The second training component is a field placement. Each student is usually assigned to a single field placement but in some cases they have split placements. In the event that a student's research requires a setting different from this initial placement, he can readily be switched. In the field stations the student becomes acquainted with the details of the clinical setting and its operations—staffing patterns, clinical meetings, treatment approaches, and research activities. Here the student obtains a concrete exposure to the subjects he will investigate. The student spends about two days at the field station.

In the field setting the student is supervised by a social or behavioral scientist with a Ph.D. or by a physician with research sophistication. The supervisor is responsible for involving the trainee in clinical activities, for helping him to develop a research project, and for implementing the collection of data.

The third component of the training model is a research project. The student, with the seminar and the field experience as background, is required to develop and carry out an original research project during his year in the program. The student develops his questions or hypotheses from theory and experience, designs the methodology, collects the data, and writes a final report covering all of these matters.

Taken together, the seminar, the field placement, and the research project activities provide a comprehensive and constructive training experience which gives the student in-depth knowledge of the problem of alcoholism.

The program has been evaluated twice by means of a mailed questionnaire to graduates of the program. The first evaluation reported on the trainees in the first four years of the program and the very positive results have been published elsewhere (Blacker, 1968). More recently a second evaluation was made of the students who had completed their training sometime during the program's first nine years. This second follow-up survey is as yet unpublished but preliminary findings are consistent with the first study. The findings show that: (1) approximately 62 percent of the students have worked in the field of alcoholism, part or full time, (2) twelve students have published doctoral dissertations in alcohol-related research, (3) thirteen of the students have written over 75 publications in the field of alcoholism of which 3 are books, (4) 80 percent of the students rate all aspects of the program and the program as a whole as good or excellent.

Clearly, the results of the two follow-up studies show that the training program has quite successfully accomplished its objectives.

University Program

At The Johns Hopkins University in Maryland there is a university based training program directed by Mandell (1972). The general purpose of this NIAAA supported program is to prepare program directors, planners, and evaluators of alcoholism programs to work in medical care, mental health, public health, and social welfare settings on local, state, and national levels.

The program operates out of the School of Hygiene and Public Health and is a joint program of the departments of behavioral sciences, mental hygiene, and psychiatry. Students require a bachelor's degree to be eligible for the program. They are expected to spend one year in the program and on successful completion can earn a master's degree in health science, science, or mental health. If desired, students can continue their training beyond one year in order to earn a Doctor of Science degree.

Settings for the program include the university and various field settings. The academic courses and the supervised field placement constitute the two major training components. A preceptor is assigned to each student to assist him throughout the course of study.

The academic portion of the program provides the background information and theory necessary for the development of skills in planning, administration, and evaluation of alcoholism services. The student attends courses and seminars, some required and some elective. Many general courses and a few specifically on alcoholism are available and can be chosen according to the student's goals. The courses cover epidemiological and etiological aspects of alcoholism, material on organizations and administration, and research and program evaluation. For example, a student interested in research can take courses in biostatistics, research methods in community health, and advanced research methods in medical care and behavioral sciences.

The field placement possibilities cover the range of services and include a community-based outpatient alcoholism clinic, an inpatient alcoholism facility at a mental hospital, a city hospital-based comprehensive alcoholism service, an area council on alcoholism, the state agency on alcoholism, and a university-affiliated hospital with a comprehensive alcoholism program. Each student has a short-term placement in each of four agencies and a long-term placement in one agency. Each short-term placement is one month in duration on a half-time

(100 hours) basis. The long-term placement is for five months on a half-time basis.

The field placements are closely supervised by highly skilled leaders in the field of alcoholism. In the field placements the students are expected to accomplish dual tasks: (1) to acquire or refurbish skills in interviewing, referral, and group work techniques with alcoholics, and (2) to come to understand the constraints and advantages of health delivery services as affected by historical, organizational, and administrative factors.

Evaluation is planned for this program and will include assessments of the trainees' knowledge, skills, and attitudes for organizing methods and approaches for solving the problems of alcoholism. Trainees will be followed up to assess their success in adopting models of administrative behavior and alcoholism treatment to practical settings and circumstances.

Community Leaders Program

Under the auspices of the Social Science Institute at Washington University in St. Louis, Missouri, a program has been operating to train upper-echelon policy makers and professionals. This program, under the direction of Pittman (1972), was begun in 1968.

The goal of the program is to train people in community organization in order to enable them to establish and improve comprehensive alcoholism programs at the state and local levels. Trainees include high-level government officials and agency leaders from many fields and disciplines such as judges, legislators, law enforcement personnel, and health and welfare workers. The program is nationally oriented so that trainees are eligible from anywhere in the nation. The program aims to equip the trainees so that they can serve as consultants and catalytic agents in their home communities.

The training program model involves six days of concentrated sessions in which the trainees (1) hear lectures on various aspects of alcoholism and the practical applications of this information in local Missouri agencies, (2) make field visits to various local agencies, and (3) draft simulated programs of their own communities which are discussed and evaluated by fellow trainees and staff. In each six-day session there are approximately 10 trainees. The six-day sessions are repeated with new trainees eight times a year so that about 80 people are trained each year by the program. The training program staff also provides follow-up consultation to trainees after they return to their local communities.

The curriculum content of the program emphasizes a community organization approach to the alcoholism problem. The community organization strategy involves four phases: study, analysis, plan formulation, and implementation. Phase one delineates the scope of the alcoholism problem, community attitudes, power structures, and resources. Phase two evaluates the results of the study and defines unmet needs and obstacles. In phase three a plan is formulated to meet the needs in a particular community. And in phase four techniques and strategies are outlined for implementing a community program.

An evaluation of this program was conducted on the 71 trainees who had attended the program during 1970–1971. The trainees came from 22 different states and included people of diverse backgrounds, with many high-level agency and community leaders. A questionnaire was mailed to the trainees who had been back in their communities for six months or more. The trainees wrote short essays in answer to questions related to their current work. In some cases the questionnnaire responses were supplemented by telephone interviews.

The results showed that 70 percent of the respondents had made significant progress in developing alcoholism resources in their communities, some of which occurred as a result of their training and some of which would have occurred in the routine course of work. Achievements of trainees involved a full range of enterprises from the expansion of detoxification and comprehensive treatment services to submission of legislation to repeal public intoxication as a criminal offense.

What is especially unusual and rewarding about this program is its tremendous leverage effect. The program reaches a relatively small number of community leaders by means of intensive short-term training who return to communities throughout the nation, and by their activities create programs and resources which benefit thousands of alcoholics and their families.

Paraprofessionals Program

Informally for a number of years a training program primarily for paraprofessionals was carried out by the Alcoholism Division of the Department of Psychiatry at Boston City Hospital (Hyde, 1972). Considerable experience was gained from this effort which has led to the development of a more formal program supported by NIAAA (Rosenberg, 1972).

The new program under the direction of Dr. Rosenberg uses the Boston City Hospital as the primary training site. Other secondary training facilities are also used, including a free-standing detoxification facility, a long-term illness hospital, and halfway houses. The program aims at providing paraprofessionals with: (1) a broad experience in alcohol-related problems, (2) an orientation to participate as members of interdisciplinary therapeutic teams, and (3) the skills to undertake treatment tasks which do not require advanced training.

The program is not directed solely to employees of the Boston City Hospital. Rather the program is conceived so as to provide a central training location primarily for the employees of dozens of alcoholism agencies in the Greater Boston area. The trainees are mainly alcoholism counselors who continue to work at their agencies while they are on the training program. Emphasis is placed on selecting trainees who are members of minority groups or who are serving minorities. Training is provided to 16 people each year by agreement with their parent agencies. The trainees are given a stipend of $1250 a year to compensate for their overtime hours in the program.

The training program model involves a combination of alcoholism education and practicum experience. The 16 trainees are divided into two teams, each of 8 members, and all trainees spend a total of 12 hours per week over a 10-month training period. They spend two half-days and one evening in training, which is interspersed with their ordinary working hours on their primary jobs. In the training program one half-day is spent at the alcoholism clinic, one half-day in the inpatient or emergency wards, and one half-day in group dynamics or educational seminars.

At the alcoholism clinic they interview patients and discuss them with senior supervising therapists. In addition, they meet together as a team to discuss what they have observed with a team supervisor (a psychiatrist or psychiatric social worker). The clinical activity of the trainees is designed so as to gradually increase their responsibilities in assisting patients. They progress from doing intake interviews only, to cotherapist activities in groups, to counseling a small number of patients themselves.

A second type of practicum experience is gained by the trainees working as assistants to psychiatric residents, staff psychiatrists, and social workers in the inpatient and emergency wards of Boston City Hospital. Approximately one-third of the patients in the emergency ward and one-quarter of the patients in the medical wards have alcohol-related problems. The trainees assist the staff members in aiding these patients to find posthospital care and rehabilitation for their alcoholism.

In addition to the practicum experience the trainees acquire knowledge on alcoholism and how to deal with it by means of lectures and seminars, field trips to other facilities, and group dynamics sessions. Subject matter is taught at a level suited to the trainees and always with an emphasis on practical application. The group dynamic sessions are an important component of the training program for they serve as an experience group for ventilating and examining personal feelings and reactions to the program and for discussing how individual and group attitudes can assist or impede therapeutic effectiveness.

Evaluation of the program, when it is carried out, will include assessing the knowledge gained by the trainees and their attitudes toward alcoholics. Also, a year after the completion of the program, the trainees will be followed up to determine how many continue to work in the field of alcoholism, changes in their income and occupational status, and the impact they have had on their agencies.

EVALUATION

Each of the four programs described in the previous section included evaluation of performance. However, more often than not training activities take place with little or no attention to evaluation. Training, after all, is an action-oriented goal, and trainers are too busy satisfying training demands to do the necessary research required to assess their effectiveness. They assume, as is the case with other types of practitioners, that their actions result in positive outcomes. No doubt in many cases they do, but there are other instances where training is partially or totally ineffective. Unless some effort is made to evaluate training programs, one can never know if objectives are being achieved or if certain teaching strategies are more effective than others. Therefore, it is a good idea to assess training efforts, if not continuously, at least intermittently. The basic purpose of evaluation is to determine what is working, with whom, and how well. The resultant data of an evaluation will allow the trainer to make appropriate changes in the training program.

Designing an evaluation study requires some sophistication in research skills, but instruction in such skills is beyond the scope of this chapter. However, a general sketch of the factors that need to be considered can be briefly described.

To begin with it is essential to clearly define the objectives of training. Is it simply knowledge acquisition? Is it attitude change? Is it skill training? Is it an attempt to recruit people to work in the field of alcoholism? These and other objectives, singly and in combinations, are often sought. Strategies of evaluation will vary depending on the objectives. For example, knowledge acquisition may be measured simply by an objective test before and after training, but recruitment to work in the field of alcoholism would require a long-term follow-up procedure.

Secondly, it is important to build evaluation into a program right from the start. The status of the trainees (knowledge, attitudes, etc.) has to be pretested before they are exposed to the training in order to establish a baseline from which to measure change. If there is no baseline one is never sure whether the change resulted because of the training program or for some other reason.

Third, the trainees should be evaluated after they have completed the training program. This can be done at different time intervals which are chosen on the basis of the stated objectives. Immediate follow-up evaluation may determine whether the trainee "got the message" of the training. But if one wants to determine if the immediate change in knowledge, attitudes, and skills actually modified behavior, a longer-term follow-up is generally required. Unfortunately, trainers are too often satisfied with evaluating changes in knowledge or attitudes only, and fail to measure actual behavior. The two are not the same.

When possible, it is also desirable to compare trainee target groups with control or comparison groups. This strategy helps to delineate how many of the changes found (if any) are due to the training program or to other factors.

Finally, the evaluation should allow for pinpointing the factors in the training process that are affecting or not affecting the outcome. It is of little value to the trainer to learn only that his teaching strategies are producing no results or only partial results. Rather, he needs to know specifically what is wrong and what is right with his teaching methods so that revisions can be made accordingly.

These are but a few of the issues that need to be considered in setting up an evaluation strategy. Details can become complicated, so training agents unfamiliar with evaluation procedures should obtain the assistance of a research consultant, preferably prior to the start of training.

GUIDELINES FOR DESIGNING A TRAINING PROGRAM

In examining the experiences of a wide variety of training programs, there are certain factors and principles which stand out above others. For the reader who is interested in designing a training program the following factors and suggestions should be considered:

1. Survey the community to ascertain the need for training and set target group priorities.
2. Define clearly the objectives of training.
3. Select from the many teaching methods available the components most suitable for a particular target group or groups.
4. Make the content of the training relevant to trainees' interests and needs.
5. Try to schedule the training to interfere as little as possible with the career activity of the trainee or with his work.
6. Provide incentives to the trainee to encourage his full participation.
7. Provide the best teaching staff available to ensure identification with positive role models.

8. Provide for small group discussion, reaction, and feedback opportunities for the trainees and the staff.
9. Have a properly balanced mixture of academic information and practical experience.
10. Select training sites and facilities which are highly respected.
11. Try to keep the costs of training within reasonable limits.
12. If training resources are limited, allocate them to target groups who can leverage their training to others.
13. Build evaluation plans into programs from the start.
14. Don't hesitate to experiment with different methods and models for different groups.
15. Provide an interdisciplinary experience if possible.

CONCLUSION

Training in the field of alcoholism has for the most part been provided as a speciality. That is, the subject is not ordinarily covered in the course of basic professional education, but rather in postgraduate studies. Rarely is the subject matter of alcoholism and alcohol abuse problems covered in undergraduate education. And the same is true of graduate and professional school education. The reasons for this omission are somewhat elusive, but the reality of it is inexcusable inasmuch as the problems of alcoholism and alcohol abuse are widespread.

As we have pointed out, major efforts have been under way in recent years to correct deficiencies in professional education, especially through the efforts of the National Institute on Alcohol Abuse and Alcoholism. As a first step it seems that it has been necessary to "play catch-up," so to speak, by training caretakers who had already completed their formal education. More recently, however, as we have shown by a number of examples in this chapter, training programs in alcoholism have begun to be introduced into the curricula of colleges and universities. Clearly, this is the desired direction, for only when all members of professional disciplines are knowledgeable about alcoholism will there be sufficient manpower to help the alcoholic.

REFERENCES

Blacker, E., 1968, Training predoctoral social and behavioral science students in alcoholism research, *Am. J. Psychiatry* 124(12):108–113.
Hyde, A. P., 1972, A program for training paraprofessional counselors, Paper read at the 23rd annual meeting of the Alcohol and Drug Problems Association of North America, Atlanta, Georgia (September 10–15, 1972).

Mandell, W. M., 1972, Personal communication: unpublished document, training grant application number MH 12809, The Johns Hopkins University, Baltimore, Maryland.
Mitnick, L., 1972, Personal communication: abstracts of training grant applications supported by the National Institute on Alcohol Abuse and Alcoholism, National Institute of Mental Health, Rockville, Maryland.
Pittman, D. J., 1972, Personal communication: unpublished document, training grant application number MH 11639, Washington University, St. Louis, Missouri.
Rosenberg, C. M., 1972, Personal communication: unpublished document, training grant application number MH 13105, Boston City Hospital, Boston, Massachusetts.

CHAPTER 14

Public Health Treatment Programs in Alcoholism

Morris E. Chafetz and Robert Yoerg

National Institute on Alcohol Abuse and Alcoholism
National Institute of Mental Health
Rockville, Maryland

ALCOHOL PROBLEMS IN THE UNITED STATES

Alcohol has long been a public health problem in the United States. It permeates every area of life—industry, transportation, the courts, the home—and cuts across social, cultural, and economic lines. Historically, public attitudes toward drinking have reflected the intellectual climate of the times, as have the public health approaches to the problem.

In Colonial days (Furness, 1965), frequent and relatively heavy imbibing was socially accepted, even the Puritans referring to rum as "a Good Creature of God," and preachers were often paid for their services with bottles of brandy. Doctors relied upon distilled spirits to ward off fevers and chills and to deaden pain. In a predominantly rural society, taverns provided convenient meeting places. Public drunkenness, however, was considered a criminal offense, and was punished by jailing, fines, or a session in the stocks.

The public health aspect of the abuse of alcohol was recognized by Dr. Benjamin Rush (Furness, 1965), a public-spirited physician, a signer of the Declaration of Independence, and a Surgeon-General of the Continental Army. Although, like his professional contemporaries, he used alcohol in treatment, he published in 1785 *An Inquiry into the Effects of Ardent Spirits on the Human Body and Mind,* in which he defined alcoholism as "a disease" that killed some 4000 persons annually out of a population of 6 million. He also referred to it as "an addiction."

In 1798, the American Congress authorized a Marine Hospital Service for merchant seamen which later became the nucleus of the U.S. Public Health Service. In light of today's knowledge of the part heavy drinking can play in a wide range of physical problems, and since sailors on shore are notoriously given to overindulgence, problems of alcoholism were therefore partially responsible for the start of a public health service in the United States (Hanlon, 1963).

The cosmopolitan Benjamin Franklin (U.S. Department of Health, Education and Welfare, 1971) deplored, in print, the evil effects of heavy drinking, as did the 18th-century Quaker philanthropist, Anthony Benezet (U.S. Department of Health, Education, and Welfare 1971).

Fashions in drinking (Furness, 1965) in America are largely the history of the social customs of the various ethnic groups that populated the New World. The British, French, and Dutch drank distilled spirits freely, but were critical of overindulgence, particularly among the poor. The Irish immigrants that swarmed into the United States as the potato blight ravaged their homeland in the mid-19th Century used whiskey as an emotional outlet and a necessary ingredient for social occasions. They accepted drunkenness as a natural consequence. A major change in American drinking patterns occurred in 1848 with the influx into the United States of traditionally beer-drinking European peoples. Until the mid-19th Century, about 85 percent of the alcohol consumed in America was distilled spirits. During the latter half, more than 50 percent of the consumption was beer. The proportion of wine drinking remained about the same.

With the Industrial Revolution, the eastern American states began to change from a rural to an urban economy and pioneers started moving west to open up the continent. They took distilled spirits with them since beer was both bulky and perishable. "Hard liquor" and hard drinking became a characteristic of the frontier, usually in a saloon, as a panacea for sickness, loneliness, bad food, and an unremittingly hard life. It was also an important commodity to barter with the Indians, whose problems were compounded by it.

In the cities, where industries used cheap immigrant labor, alcoholism was a problem, for a drunken or hungover worker was less productive. The evangelical approach of the 19th-century employer was to elevate "temperance" to a virtue and consider excessive drinking a moral offense (Furness, 1965). And, as in the 1970s the movement toward equal rights for women is equated with that of upgrading minority groups, so, in the 19th century, "temperance" and later "total abstinence" provided a training ground in community action for the women's suffrage movement. Women know the dire effect of a heavy-drinking breadwinner upon his dependents. It was a short step to seeking legislative power to curb the sale of alcohol. "When women walk to the polls, goodby, Mr. Booze," said a state president of the national Womens Christian Temperance Union in 1915 (Furness, 1965).

The whole question of drinking became as obfuscated by emotionalism as is the Equal Rights Amendment today. The churches lent their prestige to abstinence. The distinction between moderate social drinking and alcoholism was lost. The popular image of anyone who took a drink was that of a skid row derelict, down and out through sin. Law was invoked to resolve the problem— to control the purchase of alcohol. Massachusetts, for instance, tried limiting sales to lots of 15 gallons as early as 1838, but repealed this legislation as unworkable. In 1844 the new State of Oregon included an antialcohol clause in its Constitution (U.S. Department of Health, Education, and Welfare, 1971). Maine brought in statewide prohibition in 1851 and within the next four years 13 out of the (then) 31 States of the Union followed suit. The U.S. Navy abolished spirits rations for enlishted men in 1862 and for officers in 1914. "Anti-Saloon Leagues" and the ubiquitous Womens Christian Temperance Union caused whole areas to adopt "dry" legislation. In 1890, Athens, Georgia, was the first town to experiment with making the sale of liquor a municipal monopoly (Furness, 1965).

Although none of this legislation proved very effective, and a great many local restrictions were abandoned as unworkable after a few years, the Prohibition Movement intensified up to and throughout World War I, and in 1919 imposed the concept of the nation through the passage of the Eighteenth Amendment (U.S. Department of Health, Education, and Welfare, 1971). This remained in force from 1920 until 1933, when it was superseded and canceled by the Twenty-first Amendment. The Eighteenth is the only constitutional amendment ever to be repealed.

Although the rate of admissions to mental hospitals for alcohol-related psychoses diminished as a result of this legislation, most observers felt that Prohibition was not the answer to the problem of alcoholism in this country. The chief reason for the failure of Federal Prohibition was that the nationwide law could not be made effective any more than had been the case with local laws. Plenty of illicitly distilled or imported liquor remained, sold by bootleggers to individuals and to the innumerable speakeasies of the exuberant postwar twenties. Distribution of alcohol passed into the hands of criminals. There were no manufacturing standards, and many highly toxic beverages were drunk. Federal authorities, moreover, could not but realize that valuable sources of revenue were escaping untapped.

As the general public came to appreciate that the laws were not working (Furness, 1965), many civic-minded persons and churchmen found themselves in the embarrassing position of being on the same platform as criminals in trying to maintain Prohibition legislation. Responsible citizens grew tired of the lack of social freedom to have an occasional relaxing drink, and it became increasingly obvious that most hard-core drinkers and alcoholics were no more being weaned from their addiction by the Eighteenth Amendment than today's youth are restrained from smoking "pot" because it is illegal.

Significantly, however, although the Eighteenth Amendment was ratified in the traditional manner, through voting in State legislatures, the Twenty-first Amendment was put up to popular referendum in each individual state, indicating that the desire for repeal was not unanimous (U.S. Department of Health, Education, and Welfare, 1971). In areas where Prohibitionist feeling was strong, repeal did not become law. Three states kept their antialcohol legislation for many years, Mississippi until 1966. Wide variations in local regulation of the purchase and consumption of alcoholic beverages remain across the continent to this day.

THE RISE OF ORGANIZATIONAL INTEREST IN THE PROBLEM OF ALCOHOLISM

During the 1920s and throughout the Depression years, the alcoholic person was characterized and stigmatized as incorrigible. Like Alfred Doolittle in George Bernard Shaw's play *Pygmalion* (Shaw, 1962), which was first produced in 1912, the heavy drinker as well as the alcoholic was considered "undeserving," or as Doolittle expresses it, "up against middle class morality. ... If there's anything going and I put in for a bit of it, it's always the same story. 'You're undeserving, so you can't have it.' But ... I don't need less than a deserving man, I need more. I don't eat less hearty than him; and I drink a lot more."

As a residue of Prohibition, all but one state required some instruction in schools on the dire effects of drinking, but the educational materials supplied seldom made any distinction between moderate and problem consumption. A typical attitude was expressed in a Nebraska syllabus: "When a dipsomaniac or alcohol addict persists in his mania for alcohol, so that his brain is repeatedly 'soaked' . . . the surface layers of brain cells (the very last to come in race history) finally break down completely. . . . Such people are known as 'alcoholics' and are found in considerable numbers in our asylums" (Roe, 1943). The implication was that the alcoholic was incapable of rehabilitation and probably to blame for his situation anyway.

The first quarter of the 20th century, however, coincided with an upsurge of public interest in psychology. Since alcoholism and mental illness frequently overlap, and since attempts were being made actually to treat mental illness, medical investigators returned to the study of alcoholism as a disease state. Starting with Freud, psychoanalytical research has been concerned with why certain persons drink to excess (U.S. Department of Health, Education, and Welfare, 1971).

Many difficulties have been encountered in the pursuit of research into alcoholism. For example, there has been a good deal of personal resistance to treating alcoholic persons on the part of the medical profession and little recognition of the early signs of problem drinking, so that by the time a patient reaches a professional in the field of alcoholism, the disease is fully developed. Another problem has been that alcohol addiction is unique to humans and it has been difficult to use animals as models for research in this field. Yet another problem in actual treatment has been (and still is) the fact that the health insurance plans (Hallan, 1972) covering hospitalization and medical care that were developed in the late 1930s (as medical costs began to escalate) showed wide variation in benefits for treatment of alcohol-induced illness and alcoholism itself.

Public attitudes also began to shift, returning to the concept that alcoholism is a legitimate illness rather than a failure of morals or willpower. Much of the credit for the beginning of this changed attitude is due to Alcoholics Anonymous. Started in 1935, when relatively little help was available for alcoholics, A.A. was formed loosely on the Oxford Group religious approach. It was initiated by a stockbroker and a surgeon who realized their own lives were being ruined by uncontrollable drinking and who had found that the treatments then available were ineffective. In desperation, they evolved their own form of group therapy. Those who participated had, first, to admit and accept that alcoholism was a disease only controllable by permanent abstinence from drinking. Moreover, members aimed, realistically, to remain sober from one day to the next. There was no "signing the Pledge" as in the 19th century, nor was normal censure or blame attached to remissions. Members were encouraged to seek help from a nondenominational Higher Power, and fellow members were also expected to offer help in emergency situations.

The movements escalated as success stories began appearing in magazines and newspapers showing that recovery from alcoholism was possible. Members were identified only by first names to minimize self-aggrandizement and maintain anonymity.

Recognition that families of alcoholic people also need constructive help led to the development of Al-Anon, where spouses, relatives, or concerned friends could pool their experiences and learn from one another. Alateen, a similar group for children of alcoholic parents, was formed in 1957 (Bacon and Jones, 1968). In recognizing that for each alcoholic person an average of four other lives could be deeply affected, a public health attitude toward the problem started to grow (Al-Anon Family Group, 1964). The structure of A.A. was flexible enough to permit members to work within established social agencies and hospitals, and the group has always emphasized the need for more public education on the whole problem of alcoholism.

Meanwhile, in the third decade of the century, a Research Council on Problems of Alcohol got under way, and a Laboratory of Applied Physiology at Yale University, established in 1923, started, for the first time, to investigate the physiological aspects of the disease, using modern scientific methods of research (Plaut, 1967). Archives were set up, and in 1940 the *Quarterly Journal of Studies on Alcohol* came into being. Yale opened a Center for Alcohol Studies in 1943 and the following year started two outpatient clinics which became models for many others across the country.

The director of the Yale group, D. E. M. Jellinek, organized, in 1943, a Summer School of Alcohol Studies for public health and social workers, educators, churchmen, and concerned members of the medical profession. The first lecture that was open to the public was "Alcohol as a Public Health Problem," delivered by Dr. Lawrence Kolb, then assistant surgeon-general of the U.S. Public Health Service.

The following year, a National Council on Alcoholism (at first known as the National Committee for Education on Alcoholism) grew out of the Yale seminars. This national voluntary health organization has consistently taken the approach that control of alcoholism is a public health problem.

Another volunteer group—the North American Association of Alcoholism Programs—started in 1949 with emphasis on communication at state and local levels. Since 1972 this group has been known as the Alcohol and Drug Problems Association of North America.

The Yale Plan clinics later became part of the Connecticut State Commission on Alcoholism within the State's Department of Mental Health. In 1962, the Center of Alcohol Studies moved to Rutgers State University of New Jersey, where its work continues, aided by federal grants and financial assistance from private groups, such as the Christopher D. Smithers Foundation.

During the 1940s, too, a number of large industrial corporations, such as the General Electric Company and Eastman Kodak (Ayerst Laboratories, 1967), started experimenting with treatment programs for workers who had drinking problems, often with the help of members of Alcoholics Anonymous.

SOCIAL INVOLVEMENT WITH PROBLEMS OF ALCOHOLISM

The American Medical Association maintained for some years a Committee on Alcoholism (which today includes drug abuse) as part of its Council on Mental Health. In 1956, this Committee reminded physicians that "alcoholic symptomalogy and complications which occur in many personality disorders come within the scope of medical practice. Acute alcoholic intoxication can and often is a medical emergency "in the category of a sick individual." Most recently, in November 1972, the AMA urged its members to break the "conspiracy of silence" surrounding doctors and hospital personnel who had alcohol-related problems, abused drugs, or had serious psychological difficulties. Colleagues and medical societies were advised to refer these cases, if need be, to the appropriate state body which could invoke disciplinary action, such as the suspension of a doctor's license to practice.

The reluctance of many psychiatrists to treat alcoholic people was deplored by the American Psychiatric Association in 1956 when members were told to abandon "therapeutic pessimism" and contribute to "an adequate national attack on alcohol problems which necessarily requires the application of the knowledge of many professionals reinforced by broad citizen support"—a public health program in essence.

As early as 1944, the American Hospital Association (Plaut, 1967) had told members that "the primary point of attack on alcoholism should be through the general hospital," but, in 1957, they admitted ruefully that many hospitals still refused to admit alcoholic patients although "such a policy denies to the alcoholic patient benefits which would be available to him were his acute poisoning from another source, such as food."

Another important social factor leading to public concern with excessive drinking was the increasing popularity of the automobile as a method of transportation. Drinking and drink-related problems have always been most intense in urban areas (Bacon and Jones, 1968). Now these problems began to spread outwards, along the highways, as middle-income families started to move from the cities to the suburbs in great numbers. Interestingly, the first federal involvement with alcoholism as a medical and public health problem was the highway Safety Act of 1966.

CHANGING LEGAL PATTERNS IN THE PUBLIC APPROACH TO ALCOHOLISM

Public intoxication had been considered a criminal offense since Colonial days, and currently accounts for one-third of all arrests reported annually.

(This does not include thousands of cases involving excess drinking which are described as disorderly conduct, vagrancy, and driving while under the influence of alcohol. If these categories are included, the proportion of arrests rises to between 40 and 49 percent.)

This concept of public drunkenness as a criminal offense was first challenged in the 1960s. In a number of well-publicized cases, attorneys sought to establish the essential principle that many of these all-too-frequently arrested persons were alcoholics, and as much suffering from a disease that was more properly handled by public health, welfare, or social rehabilitative agencies than by the law.

In 1964, test cases involving two public inebriates, Walter Bowles and DeWitt Easter, resulted in charges of public drunkenness being dismissed, on the grounds that they were sick persons rather than lawbreakers. Shortly after, another chronic alcoholic, Joe Driver—who had been arrested more than 200 times and spent two-thirds of his life in jail because of his public drunkenness—appealed a prison sentence on the grounds that his drinking was involuntary and thus he was not responsible, and as a homeless person, he could only drink in public places.

Following a number of similar cases, in different courts, in 1969, the matter reached the U.S. Supreme Court when LeRoy Powell appealed a conviction for public drunkenness. His sentence was confirmed, but solely on the grounds that he was not a homeless person. The contention that he was a sick man rather than a criminal was upheld by all nine justices. The nationwide publicity given to the case stimulated two Crime Commissions, appointed by President Johnson to investigate crime in the District of Columbia and the nationwide administration of Justice. Both Commissions upheld the fact that alcoholics were sick persons rather than criminals. In civil law, also, test cases led to the acceptance of the concept that chronic drunkenness is a disease state rather than a criminal offense.

In 1971, the National Conference of Commissioners on Uniform State Laws adopted a Uniform Alcoholism and Intoxication Treatment Act (Hirsh, 1967) that has been called "A Bill of Rights" for alcoholic people (Chafetz, 1972). Legislation promulgated by the National Conference of Commissioners on Uniform State Laws is suggested for introduction and enaction by state legislatures, and in this particular instance, every state governor also received a letter signed by the Secretary of the Department of Health, Education and Welfare explaining the importance of the Act and urging its early adoption.

The Uniform Act is a landmark, for it removes the crime of public intoxication and the illness of alcoholism from the criminal code and places these in the public health area, where they rightly belong. Under the Uniform Act, alcoholics are required to be given necessary emergency medical treatment and follow-up care, such as medical, psychological, and social services rather than jail sentences or arbitrary "drying out." The Act guarantees, in the few instances where civil commitment might be necessary, that the individual has the right to treatment "which is likely to be beneficial." It is aimed at substituting understanding and humane treatment for moralism and punishment and restores responsibility for handling alcoholic persons to medical and public health authorities.

The Uniform Act recommends that each state set up a Division of Alcoholism with facilities for public education, research, and alcoholism treatment programs of all kinds. The confidentiality of records and voluntary participation is stressed, although protective custody of law enforcement agencies is permitted. All previous legislation on public drunkenness is to be repealed by the states and no new laws could make alcoholism a criminal offense. Regulations covering the treatment of drunken drivers, however, are not affected.

As the Eighteenth Amendment reflected the nationalization of a good deal of legislation already in force in the states, so the Uniform Act can be seen as the climax to varying degrees of success in alcoholism programs already under way in different parts of the country. The pattern has varied with the degree of local urbanization, predominant fashions in drinking, and, in particular, the lobbying strength of special interest groups either in the beverage industry or philanthropy or among those concerned with community action.

During the 1950s, a number of states had instituted plans to help alcoholic people and by 1971 (U.S. Department of Health, Education, and Welfare, 1971), 44 states had programs in action, often within the structure of mental health or welfare agencies. Apart from a plethora of licensing laws to regulate the sale of liquor, the states' main efforts were through state mental hospitals. Only about 10 percent of these had special facilities for treating alcoholism— despite the fact that between 20 and 40 percent of the males in these hospitals had alcohol-related problems. All too frequently, the only help given to them for their alcohol addiction came from volunteer A.A. members, and they had to share the same quarters and therapy as the mentally disturbed. There was also little follow-up upon discharge, and no recognition of the fact that, as in any chronic disease, there were liable to be remissions.

INVOLVEMENT OF THE FEDERAL GOVERNMENT IN ALCOHOLISM PROBLEMS

Alcoholism was first singled out for special treatment as a health problem during the Johnson Administration when, in 1966, a study and report to Congress on state and local programs for the treatment of alcoholism grew out of investigations made on behalf of the Highway Safety Act.

This investigation was followed, in 1967, by the inclusion of a special program for the prevention and rehabilitation of alcoholism in the Economic Opportunity Amendments. An Alcoholic Counseling and Recovery Program aimed at early case-finding was added in 1969. It emphasized that the programs should be community based, to underscore the restoration of the alcoholic person to normal life in society rather than institutionalization.

President Johnson specifically mentioned alcoholism as a national problem in several Reports to Congress—in his 1966 Report on Health and in his 1968 Message on Crime and Law Enforcement which recommended the Alcoholic Rehabilitation Act. The Act became law in 1968, and followed the White House recommendation that alcoholism be considered a major social problem, preferably handled by community services, and using both public and private agencies. The objective was to prevent and control and spread of the illness through public health measures and remove alcoholic people as far as possible from law enforcement agencies. The Community Mental Health Centers Act and its 1968 and 1970 Amendments also contained provisions for special treatment services for alcoholic persons.

Another landmark was the Comprehensive Alcohol Abuse and Alcohol Prevention, Treatment and Rehabilitation Act of 1970, which authorized the creation of a National Institute on Alcohol Abuse and Alcoholism (NIAAA) within the structure of the Department of Health, Education, and Welfare's National Institute of Mental Health. The NIAAA coordinates all federal alcohol programs, including those of DHEW and other Federal agencies. As its goals, the Institute has put its priorities on mobilizing, strengthening, and expanding all existing resources for the treatment and rehabilitation of alcoholic persons at the community level, and the developing of effective methods of prevention for alcoholic persons and problem drinkers. So important is this last area that the NIAAA is unique among federal health agencies in having a special Division of Prevention within its administrative structure.

Special priority areas of the NIAAA are the provision of help for employed alcoholics, programs for the rehabilitation of public inebriates, and comprehensive services for individuals identified as problem drinkers through traffic safety programs. Another priority is the American Indian, due to the very high rate of alcoholism in this minority group. The NIAAA also supports a wide range of alcoholism research and training programs.

As part of the Office of Public Affairs of the NIAAA, a National Clearinghouse for Alcohol Information was organized in 1972. The Clearinghouse provides, for the first time, a national focal point for the collection and dissemination of all information about alcohol and alcohol-related problems, and this information is available both to specialists and to the general public. Using computer systems, the Clearinghouse is gathering together the available knowledge on the problem and building a comprehensive library. Regular newsletters and similar publications publicize the latest findings on rehabilitation programs and research in the field. The Clearinghouse is attempting to fill the wide information gap on the whole problem of alcoholism and problem drinking, particularly among the general public.

PATTERNS OF ALCOHOL USE AND ABUSE WITHIN COMMUNITIES

A series of studies of American Drinking Practices (ADP), made in the 1960s by a social research group at the George Washington University and analyzed by the School of Public Health at the University of California in Berkeley, found that 68 percent of the adult population drank at least once a year (U.S. Department of Health, Education, and Welfare, 1971). A higher percentage of men drank, but women drinkers are catching up, statistically, reflecting perhaps increasing tensions as women move up in the job scale, and also more public acceptance of drinking as part of the feminine life-style.

In these surveys, 77 percent of the males queried and 60 percent of the women said they took a drink but were not habitual drinkers. In both sexes, there tended to be more drinking among the younger generation than among the older groups.

A wide range in drinking fashions can be found across the American continent, reflecting the different backgrounds and cultural attitudes of the ethnic groups that make up the citizenry. The ADP surveys found that the rate of alcohol-related problems and of alcoholism tended to be low in those groups where drinking customs were clearly defined and adhered to, and also where drinking was primarily associated with eating or with religious ceremony rather than as a leisure-time occupation (U.S. Department of Health, Education, and Welfare, 1971).

Many abstainers were found to be older people, often below average in earning capacity. More abstainers lived in the South than in the North, and they were usually native-born Americans, conservative in attitude, and belonging to fundamentalist forms of Protestant religions.

Young people, a great many investigators have found, reproduce the attitudes and drinking styles of their families (Bacon and Jones, 1968). In cultures where alcohol is used as a natural component in religious or festive ceremonies, and children are brought up to sip wine from an early age in connection with these events, few alcohol-related problems result. The Chinese, the Jews, and the Italians, who traditionally include wine or beer with their meals as well as with religious or social occasions, have the lowest rates of alcoholism among the many ethnic groups in the United States. The children of these groups are taught to drink moderately, and are brought up to consider overindulgence and drunkenness socially unacceptable. The family structure and the closeness of the members is also well defined by tradition in these cultures.

On the other hand, in families where heavy drinking is usual, or where one or both parents is alcoholic, the children, in later life, are more likely to suffer from alcohol-related difficulties. This is also true of families where there are few clearly defined standards regarding alcohol use or where the taking of a drink is associated with guilt or hostility. Individuals who grow up in such surroundings may have ambivalent attitudes about drinking, whether from religious, philosophic, or socioeconomic reasons, and they too have a higher rate of problems involving the abuse of alcohol.

Members of minority groups, particularly those living in poverty areas frequently have a high rate of alcohol-related problems, possibly a reflection of the general frustration of their living conditions—poor housing, lack of educational and employment opportunities, and the life (U.S. Department of Health, Education, and Welfare, 1971).

Certain classifications of persons also seem more prone to abuse alcohol than others (U.S. Department of Health, Education, and Welfare, 1971). The highest rates of alcohol-related difficulties are found in young men under 25, usually from urban areas and in the lower socioeconomic levels. People who move from rural areas or small towns to the cities, single or divorced men, and those who do not go to church are also considered "high risk" groups. Among church attenders, Catholics and liberal Protestants drink more than other sects. The children of alcoholic parents or those from broken homes are also considered high risks where alcohol is concerned.

A public health approach to the problem of alcoholism, therefore, has to take into consideration the ethnic and cultural background of the community involved, and the life-style of the individuals who make up that community—in particular, their motivation for drinking and the extent to which their drinking patterns are established and condoned.

THE SCOPE OF PUBLIC HEALTH PROBLEMS INVOLVED IN ALCOHOLISM

Alcohol-related problems crop up in a great many areas of life. On the physical side, excessive drinking adds complications to many illnesses and may even trigger certain disease states. Alcoholic persons are prone to malnutrition, and to the neurological diseases that often occur in association with vitamin deficiencies. They also seem to have a low resistance to infection and many may suffer from tuberculosis. Gastrointestinal disorders are also common in alcoholic individuals. The range of illnesses to which they are prone runs the gamut from gastritis, esophagitis, malabsorption, chronic diarrhea, pancreatitis, alcoholic hepatitis, and cirrhosis of the liver through cancers of the liver and the esophagus. Other characteristic physical problems are alcoholic cardiomyopathy, skeletal muscle myopathy, and metabolic disorders. Since many heavy drinkers are also heavy smokers they suffer the health hazards that this habit can intensify (U.S. Department of Health, Education, and Welfare, 1972d). Alcoholic people are estimated to have their life expectancy reduced by 10 to 12 years (Kissin and Begleiter, 1972).

The physical problems of alcoholics and problem drinkers do not all come from illness. Over half of the casualties in fatal traffic accidents have been found to have significant amounts of alcohol in their blood (U.S. Department of Health, Education, and Welfare, 1971). The injured number over half a million annually and, as is well known, road accidents are most frequent on weekends and holidays when people are in drinking situations. The hazard is not confined to drivers; drunken pedestrians add their toll. One California study found that 40 percent of ambulatory persons hurt in traffic accidents had been drinking (American Public Health Association), as opposed to 62 percent of the drivers. Psychological studies of accident victims are finding that these individuals often have emotional problems which perhaps motivate or contribute to their excessive drinking.

The NIAAA has collaborated with the National Highway Traffic Safety Administration of the Department of Transportation (DOT) in developing programs to reduce alcohol-related road accidents. The NIAAA provides consultation and assistance to the Department of Transportation's community-oriented Alcohol Safety Action Project (ASAP) which treats problem-drinking drivers identified in ASAP road safety programs. The information and educational programs of the NIAAA and DOT are coordinated, and the two agencies also share their research findings in the field of rehabilitation of the drinking driver. One of the most useful results of this cooperative effort has been the early identification of problem drinkers and their incorporation into treatment programs.

The Department of Transportation is also working with the NIAAA in trying to change the attitudes of the police and the legal profession toward drinking drivers in the hope that judges and probation officers will take increasing responsibility in directing those individuals who have alcohol-related problems on the road into treatment programs within their community (National Institute on Alcohol Abuse and Alcoholism, 1972).

Alcoholics and problem drinkers are also prone to violent behavior. Some investigators have found that the percentage of patients injured in fights or assaults who had significant levels of alcohol in their blood was as high as 64 percent (U.S. Department of Health, Education, and Welfare, 1971). Nor were these patients necessarily the aggressors. Often they had been attacked by muggers while in an intoxicated condition. Another study of criminal homicides found that more than half of those arrested had been drinking at the time of the offense.

The handling of public inebriates and skid row alcoholic people, as a result of the legislation passed in the 1960s, is now moving from the law courts to health and social welfare authorities. Many times, to avoid a "revolving door" of arrest, discharge, and rearrest, some form of cheap housing must be found for the individual, and, later, some kind of work that, realistically, he can handle (U.S. Department of Health, Education, and Welfare, 1972b). The imaginative use of such community-founded facilities as halfway houses or even hostels may be part of the answer to the rehabilitation of chronic public inebriates. However, goals are necessarily limited since these individuals have limited social and personal resources and are very difficult to treat (U.S. Department of Health, Education, and Welfare, 1972c).

In the area of marital and family troubles, divorced persons are among those groups identified as having a high percentage of alcoholism. Children from broken homes, particularly if a parent is an alcoholic, are also considered prone to problem drinking and alcoholism in adult life. Here, community services are required to sort out the variety of family problems and provide intervention and treatment as needed.

COMMUNITY ALCOHOLISM TREATMENT SERVICES

Before any community can organize an effective program to treat alcoholic people in its midst, it is necessary to evaluate every kind of agency operating within that community—the available medical facilities such as hospitals, detoxification and emergency care centers, nursing homes, inpatient and outpatient services, halfway houses, follow-up treatment clinics, and the available family services, welfare units, civic, legal, industrial, philanthropic, and religious facilities, as well as the volunteer personnel that can be found in groups like A.A.

State and local planning committees which have undertaken surveys of communities with guidance from the NIAAA have often found useful agencies in operation that needed coordination. To utilize an agency or facility that is already established and working well can often save a community money. Given some public education through the media and through group meetings, taxpayers may even be agreeably surprised to learn that (a) even skid row alcoholics may be restored to gainful employment and cease to be a financial burden and (b) costs of treatment for alcoholism, if it is caught early enough, may be less than expected. For example, although there are wide variations across the continent, the per diem cost of hospitalization for alcoholism is usually lower than for other diseases, sometimes only half the price (U.S. Department of Health, Education, and Welfare, 1972c). Also, since the majority of alcoholics are employed, health insurance may underwrite a good deal of the cost of therapy.

However, to be effective, programs must be coordinated on a long-term basis. It is not enough simply to provide emergency care modalities and hope that they are maximally effective in terms of cost and management (Cross, 1962). Managers of alcoholism programs have, increasingly, to demonstrate to the community, state, and now the federal agencies that their programs are providing quality service. This calls for regular evaluation and flexibility in priorities and funding. It is also necessary to keep "selling" the idea to the local taxpayers that it can be more economical to treat a problem drinker before he becomes a "regular" at the public health clinics, welfare offices, or police stations. Solid facts and figures on the relative cost of rehabilitation as opposed to nontreatment or incomplete treatment can produce telling arguments for continued public funding of suitable agencies.

Over the past two years, the NIAAA has developed a multimodal evaluation and monitoring system for community treatment programs that has two major elements—data collection and a site visit system.

For proper data collection—a valuable source of research and evaluation material—a system of patient tracking has been developed. Each patient entering a treatment center is identified as to demographic characteristics, clinical status, referral source, job and marital status, and arrest record (Research Triangle Institute, 1973a,b). Patients are followed up at regular intervals to evaluate their progress, the yardstick being the individual's own report on himself, his subsequent drinking pattern, interpersonal, job, and marital adjustment. A side benefit of this tracking system is that it automatically gathers information on the costs of treatment, the staffing pattern of the agency, and the effectiveness of its management.

The second basic element in the NIAAA approach is the site visit program (Research Triangle Institute, 1973a,b). Trained teams from the Institute, composed of representatives from state alcoholism authorities, the National Institute of Mental Health regional offices, grants managers, and outside experts go for three days to the alcoholism treatment centers where they observe and then submit independent subjective evaluations, as well as team reports, to the NIAAA. An essential part of each visit is a public hearing, to listen to constructive criticism of the program by members of the community.

Further, the Institute has conducted cost benefit studies on a cross-section of private and public treatment programs to determine the actual costs of the various elements of service and how these can be integrated for best advantage to the community and the patient (Research Triangle Institute, 1973a,b).

USING AVAILABLE AGENCIES: THE EXPERIENCE OF INDUSTRY

The soundness of taking a community health approach to alcoholism rather than a specialized one has been dramatically illustrated by the success of rehabilitation programs that have been organized in large corporations, usually in a combined effort of management and labor to uncover and treat alcoholic workers.

Here, the economic stake in curbing problem drinking is enormous. Alcohol abuse has been estimated as costing American business over $15 billion annually (Hanlon, 1963; U.S. Department of Health, Education, and Welfare, 1971), through absenteeism, heightened accident rates, lost work time, lowered departmental morale due to the disruption caused by drunken or hungover employees, and related illnesses which raise medical expenses.

Between 5 and 10 percent of the nation's workforce are estimated to have drinking problems, and about 75 percent of this group is male.

How a community-oriented approach can create a successful and relatively inexpensive program for handling problems of alcohol in a company has been illustrated by the Kennecott Copper Corporation, the largest single employer in Utah. While investigating the possibility of setting up a rehabilitative program, representatives of both management and labor at Kennecott studied the experience of other corporations and based their program on the premise that "*all* the problems of employees and their dependents are cause for concern and reason for help." They hired a full-time psychiatric social worker, reachable to anyone by dialing the letters INSIGHT on any telephone. His department (composed of the professional, his secretary, and a couple of assistants) was open around the clock seven days a week, and complete confidentiality of records was promised. The social worker would meet or talk with the seeker of help at any time or in any place, and although he himself did not offer therapy, he did supply constructive advice on where this could be obtained and would make arrangements for treatment through any of some 220 existing community agencies in the state.

Kennecott has found that although every species of problem has cropped up—family, financial, occupational, physical, and emotional—more than half of these problems related somehow to the abuse of alcohol either in the seeker of help or within his immediate family. Through early identification of the drinking difficulty and its treatment, the recovery rate from alcoholism at Kennecott has been high. In a special investigation of a group of hard core drinkers within the corporation, absenteeism has been found to have dropped by 50 percent, medical costs have gone down by 49 percent, and weekly indemnity payments by 64 percent.

The cost of operating this Project Insight has worked out to less than a dollar a year for each employee. It has already been copied by a number of other corporations and also by some federal agencies where alcoholism rates were high.

The experience of the Kennecott Copper Corporation, in Project Insight, bears out the results of studies made at the NIAAA which indicate that the essential reason for the high rate of successful treatment of alcoholism in industry lies in the facts that (a) there is one easily identified agency or individual to whom a person can turn for help with personal problems, sometimes before it is realized that these come from problem drinking, (b) assistance is easily available and given promptly, and (c) records are kept confidential.

A great deal has been written about the psychology of those who drink and efforts have been made to identify a characteristic "alcoholic personality" without much success. From the community health standpoint, this research has found that alcoholic people, even more than other troubled people, are easily discouraged in their efforts towards self-help and they cannot cope well with frustration. They will, for example, not wait for long periods of time in clinics, and if they are referred to other agencies or departments they will often not go there unless they are taken by someone, preferably someone they know and trust. They need to identify with one person or one team, and who this is—doctor, nurse, volunteer, or paramedical helper—is less important than the attitude this person takes toward the alcoholic. If the person or team can meet the alcoholic individual's dependency needs, the relationship can serve as the catalyst to motivate the individual towards accepting further treatment (Hartford, personal communication).

When the initial contact with the patient is skillfully handled, even cases hitherto considered "untreatable" may be persuaded to stay the course of therapy, according to studies made in large city hospitals (Hartford, personal communication). In this connection, A.A. members or required alcoholic people are often good personnel on admissions teams, for they have gained firsthand understanding of alcoholism problems.

A MODEL ALCOHOLISM TREATMENT PROGRAM

The two most important considerations, therefore, when a community has allocated funds to treat alcoholism and studied the people who are to be served by the program are (a) a system that will involve the alcoholic promptly, upon the first visit or call to the treatment center, and (b) coordinated social and medical services for "whole person" treatment, possibly over a long period of time.

Ideally a "model" center needs someone analogous to the psychiatric social worker of Project Insight to handle cases in the earlier stages of the illness, and also an emergency care center for those in active phases of the disease, including those picked up by the police or involved in traffic accidents. Some experts (Kissin and Begleiter, 1972) have suggested a triage system in hospitals to separate those individuals who might be simply on a bender from chronic alcoholic persons in need of a broader program of therapy. This is the treatment area where A.A. volunteers or other recovered alcoholic people make useful workers, for, if trained in treatment, they can recognize the symptoms from their personal experience and approach the patient through this experience and their additional training.

An emergency care center is only the first step in a treatment program and needs to be well coordinated with other services. The ideal location for such a center is in or near a medical center or general hospital. These facilities can provide departments for the treatment of most medical problems associated with alcoholism. They can also offer social service units and maintain a good working relationship and liaison with agencies such as family services, housing authorities, public health clinics, and psychiatric units, both public and private.

A seminar sponsored by the NIAAA on emergency care in March 1972 also found that any such unit had to be part of a comprehensive program which, in turn, needed to be integrated into the total health delivery system of the community. Full rehabilitation of an alcoholic can take several years and there is a need for adequate follow-up, both to maintain the patient's physical and emotional well-being and to provide data for periodic reevaluation of the treatment facilities by local, state, and federal agencies. This followup material can also provide important data for research into the various aspects of alcoholism (U.S. Department of Health, Education, and Welfare, 1972c).

As with other medical specialties, there is a shortage of trained personnel to treat alcoholic people, and hospitals provide convenient training grounds in this field. They can be especially useful in orienting doctors, nurses, paramedical workers, and volunteers in the best therapeutic approaches to take with alcoholic or suspected alcoholic patients. It is important, too, that anyone likely to be in contact with alcoholic people be trained to recognize the early symptoms of the illness. Public health nurses and social case workers, clergymen, volunteers, and teachers who are in contact with teen-agers and students all should know how to guide them to the proper treatment center before the illness reaches an advanced stage.

PREVENTION OF ALCOHOLISM

Prevention is the ultimate objective of any public health program in a community, whether it be vaccination against a specific disease-causing agent or curbing the abuse of alcohol. History has demonstrated the ineffectiveness of this country's first large-scale attempt at preventing alcohol problems—legal prohibition of alcohol. There is little evidence that the various legal control systems employed in the United States to curb alcohol usage bear any relationship to the extent or nature of alcohol use or alcohol problems. One survey found no consistent relation between excessive drinking—as measured by drunk-driving arrests, public intoxication arrests, admissions to mental hospitals, or reported alcoholism rates—and sales of alcoholic beverages through state monopoly or private liquor stores (U.S. Department of Health, Education, and Welfare, 1972a). Minimum age laws have proved equally ineffective, since many young people start experimenting with alcohol when considerably under the legal age—usually around 14. Moral exhortations and scare tactics have also for the most part met with failure.

The Division of Prevention of the National Institute on Alcohol Abuse and Alcoholism has taken a positive approach: prevention of alcohol abuse through education for responsible decision-making. Personal attitudes toward responsible behavior, or the lack thereof, are first developed within the family setting. Parents are the most influential teachers of responsible or irresponsible drinking patterns, since children tend to follow the drinking customs of their families and cultures. Therefore, parents have a responsibility to be aware of and to reexamine their own values and attitude before imposing them—often unknowingly—on their children. If their own self-perceptions are clarified and accepted, parents will be in a more authentic position to influence and educate their children about alcohol and responsible decision-making, and the goal of preventing alcohol abuse can ultimately be achieved.

Enlightened education efforts—those which encourage individuals to assume responsibility for their own behavior—incorporate these general principles:

1. It is not essential to drink. An individual, youth, or adult who decides to abstain from alcohol for moral, medical, economic, or any other reason should not be placed under pressure to drink by other members of his society.
2. Uncontrolled drinking, or alcoholism, is an illness. Young people, including the children of alcoholic parents, should be aware that alcoholism is not perversity, not a character defect, and not even the direct result of drinking. They should know that an alcoholic person,

like a victim of diabetes or tuberculosis, is a sick person who can and should be helped.
3. Safe drinking depends on both specific physiological and psychological factors. These factors include: (a) early development of healthy attitudes toward drinking, within a strong family environment; (b) prevention of high intoxication levels by restricting beverage consumption to small amounts, in appropriate dilution, and preferably in combination with food; (c) recognition that drinking is dangerous when used in an effort to solve emotional problems; (d) universal agreement that drunkenness will not be sanctioned by the group. Most effective is to engender a public attitude that drinking to the point of intoxication is socially unacceptable.
4. "Alcohol education" should not be restricted to alcoholism or alcohol. Instead, education on alcoholism and excessive drinking should be considered as only one specific topic of education about living, coping with life, and developing self-respect (U.S. Department of Health, Education, and Welfare, 1972a).

Ultimately, the prevention of alcohol problems rests with the private decisions and behavior of each individual. Effective prevention of alcohol abuse and alcoholism involves not only the avoidance of actions or behavior which are harmful to the individual; it also means the promotion of positive changes which are more satisfying and personally rewarding for the individual. The goal of prevention through education is thus to provide individuals with information and experience which will lead them to make decisions consistent with effective and rewarding patterns of living.

SUMMARY

Although early identified in America as a "disease," alcoholism later became associated in the public mind with sin and crime. Today it is again recognized as an illness, and the country's leading drug problem. A "model" community service for alcoholics calls for a "whole person" approach, including carefully coordinated networks of emergency care units; diagnostic and referral services; inpatient, outpatient, and intermediate care facilities; rehabilitation and family counseling services; social services such as emergency housing, financial assistance, job counseling, and legal help; and training facilities for personnel. Alcoholic people need one clearly identified place or person to go to for help, and this help needs to be available around the clock, seven days a week, and delivered promptly, without loss of time or human dignity.

Guidelines for treatment coordination programs and their evaluation are available through the NIAAA, the spearhead of the federal attack on the problem. More public education on the need for early diagnosis and treatment is required, as well as changed public attitudes, and much information and help is being supplied through interrelated action on the part of federal, state, and local agencies, working in concerted action.

REFERENCES

American Public Health Association (archives), Washington, D.C.
Al-Anon Family Group, 1964, "Living With an Alcoholic," Al-Anon Family Group Headquarters, Inc., New York.
Ayerst Laboratories, 1967, Kodak gives major role to supervisors: A special report, *Recovery*. 1(2).
Bacon, M., and Jones, M. B., 1968, "Teenage Drinking," Crowell, New York.
Chafetz, M. E., 1972, "A Bill of Rights for Alcoholic People," *J. Am. Med. Assoc.* 219(11):1471.
Cross, J. N., 1962, "Guide to the Community Control of Alcoholism," American Public Health Association, Washington, D.C.
Furness, J. C., 1965, "The Life and Times of the Late Demon Rum," G. P. Putnam and Sons, New York.
Hallan, J. B., 1972, "Health Insurance Coverage for Alcoholism," Institute of Human Ecology, Raleigh, North Carolina.
Hanlon, J. J. (ed.), 1963, "Principles of Public Health Administration," C. V. Mosby, Co., St. Louis, Missouri.
Hirsh, J., 1967, "Opportunities and Limitations in the Treatment of Alcoholics," Charles C Thomas, Springfield, Illinois.
Kissin, B., and Begleiter, H., 1972, "The Biology of Alcoholism," Plenum Press, New York.
National Institute on Alcohol Abuse and Alcoholism, 1972, "Fiscal Year 1972 Annual Report," Prepared by the National Institute on Alcohol Abuse and Alcoholism.
Plaut, T. F. A., 1967, "Alcohol Problems: A Report to the Nation by the Cooperative Commission on the Study of Alcoholism," Oxford University Press, New York.
Research Triangle Institute, 1973a, "Analysis of 12 Site Visit Reports," Prepared for the National Institute on Alcohol Abuse and Alcoholism by the Research Triangle Institute, Research Triangle Park, North Carolina.
Research Triangle Institute, 1973b, "Establishment of a Task Force for Monitoring Community Assistance Grants and Development, Implementation, and Evaluation of a Site Visit Model," Prepared for the National Institute on Alcohol Abuse and Alcoholism by the Research Triangle Institute, Research Triangle Park, North Carolina.
Roe, A., 1943, A study of alcohol education in the USA, *Q. J. Stud. Alcohol* 3(4):574–662.
Shaw, G. B., 1962, "Six Plays of George Bernard Shaw," Constable & Company, London.
U.S. Department of Health, Education, and Welfare, National Institute on Alcohol Abuse and Alcoholism, 1971, "First Special Report to the U.S. Congree on Alcohol and Health," DHEW Publication No. (HSM) 72-9099, Superintendent of Documents, U.S. Government Printing Office, Washington, D.C.
U.S. Department of Health, Education, and Welfare, National Institute on Alcohol Abuse and Alcoholism, 1972a, "Alcohol and Alcoholism: Problems Programs and Progress," DHEW Publication No. (HSM) 72-9127, Superintendent of Documents, U.S. Government Printing Office, Washington, D.C.
U.S. Department of Health, Education, and Welfare, National Institute on Alcohol Abuse and

Alcoholism, 1972b, "Proceedings of the Joint Conference on Alcohol Abuse and Alcoholism," DHEW Publication No. (HSM) 73-9051, Superintendent of Documents, U.S. Government Printing Office, Washington, D.C.

U.S. Department of Health, Education and Welfare, National Institute on Alcohol Abuse and Alcoholism, 1973c, "Proceedings: Seminar on Alcoholism Emergency Care Services," DHEW Publication No. (HSM) 73-9024, Superintendent of Documents, U.S. Government Printing Office, Washington, D.C.

U.S. Department of Health, Education, and Welfare, 1972d. "The Health Consequences of Smoking," A Report of the Surgeon General, Superintendent of Documents, U.S. Government Printing Office, Washington, D.C.

Index

A.A., *see* Alcoholics Anonymous
A.A. Comes of Age, 507
A.A. counselors, 556-558
   see also Counseling; Counselors
A.A. General Service Board, 468-469
A.A. General Service Conference (1955), 467-468
A.A. General Service Office, 443-447, 561
A.A. Grapevine, 468-469, 492, 500, 508
A.A. Way of Life, The, 508
A.A. World Directory, 444-446, 450, 508
A.A. World Services, Inc., 451
Abstainer, in halfway house, 378-379
Abstinence
   craving during, 7
   in psychotherapy program, 122
   total, 248, 308
Abstract Archive on Alcohol Literature, 568
Acceptance, of alcoholic problem in psychotherapy, 126
Acetaldehyde, Antabuse and, 79
Acting out, 250-251
Activity groups, treatment through, 223-224
Addiction
   defined, 5-6
   heroin, *see* Heroin addiction
Addictive cycle, 5-8
   breaking of, 24-26
   loss of control in, 6-8

Aftercare, importance of, 394-395
Age grading, in problem drinking, 213-214
Agitation, phenothiazines and, 90
Al-Anon Family Groups, 74, 215, 451, 574, 598
   defined, 289
   as dropout remedy, 186
   family problems and, 30
   group psychotherapy in, 255
   group structure of, 290
   outpatient treatment and, 400
   psychosocial work of, 41
   Twelve Steps of, 291
Alateen, 74, 215, 451
   family problems and, 30
   psychosocial work of, 41
Alcohol
   addictive cycle in, 5-8
   anxiety level and, 302
   blood level, *see* Blood alcohol level
   craving for, 5-8, 10
   decreased reinforcing properties of, 328
   depression and, 12, 89-90, 164, 302
   first experience with, 5
   in learning theory, 226
   physical dependence on, 5-8, 10, 22-23, 27-28, 80-87
   powerlessness over, 457-460

Alcohol (cont.)
  as reward, 226
  tolerance to, 5
Alcohol abuse pattern, 78–80
Alcohol addiction, see Addiction
Alcoholic(s)
  see also Alcoholic patient
  aggressive, 109
  as "alibi Ikes," 402
  ambivalence toward physician, 53–54
  antisocial personality as, 12–13
  black, 16, 212
  blood alcohol level and, 321–328, 334, 339–341, 413
  blue collar, 18–20
  brain function impairment in, 166
  characteristic disorder type, 11–14
  compulsory treatment for, 253, 400–402
  confrontation techniques with, 124–130
  continuity of treatment for, 45–46, 161–187
  contraindicated drugs and, 85, 90–91
  "conversion" of to nonalcoholic, 24
  coping by, 31–32, 94
  danger of sedatives and tranquilizers for, 84–86, 90–91
  defined, 200, 597
  delta, 516
  denial in, 165–166
  dependency behavior in, 108–109
  depression and anxiety in, 12, 79, 89–90, 164, 302
  desire to stop drinking in, 99–100, 216, 469, 500
  detection of, 62, 162–174
  as drug addict, 91
  drugs and, 84–86, 90–91, 417–426
  dyssocial, 15
  education in therapy for, 32
  employment status of, 91–92
  epsilon or pseudoepsilon, 484
  essential, 4, 9, 108
  family disruption among, 21–22
  first clinic interview of, 176
  ghetto, 4, 18, 20–21, 30
  group therapy for, 240–254
  hospital admission for, 167
  hostile, 132–133, 255
  identification of, 62, 162–174
  impulse disorders in, 13

Alcoholic(s) (cont.)
  inner-city, 4, 15, 18, 20–21, 30
  insight in, 95, 403
  interpersonal relationship problem of, 36
  in jails and prisons, 164
  Jewish vs. Irish, 2, 16
  loneliness of, 165
  lower-middle-class, 18–20
  maladaptive coping in, 94–95
  marriage pattern of, 185, 260, 269–270, 278
  memory impairment in, 166
  neurotic, 15
  "normal," 15
  number of, 162
  occupations of, 165, 208–209
  passive-aggressive personality in, 13
  pathogenic classification of, 9–22
  physical problems of, 605
  private treatment of, by physicians, 33
  vs. problem drinkers, 8, 57, 79
  projection in, 136
  psychiatric diagnoses among, 11–17
  psychoanalytic view of, 244
  psychopathy in, 12
  psychotic, 14–15
  rationalization in, 136
  reactive, 108
  recovered, see Recovered alcoholic
  referral for, 260
  refusal of treatment by, 246
  schizoid personality in, 13
  self-destructive impulsivity in, 13, 88, 109
  self-reports by, 333
  as sex-role stereotype, 185
  situational, 4
  skid-row, 18, 20–21, 30, 166, 360–361, 381, 606
  socially stable, 29
  socially unstable, 29
  spontaneous improvement in, 389–390
  spouse of, 244, 270–275
  suicide in, 164
  susceptible individual as, 5
  suspension of thought in, 136–137
  as term used in A.A., 459, 460
  untreated, 389–390
  upper- and middle-class, 18–19
  in urban hospitals, 163

Index

Alcoholic(s) (*cont.*)
   violent behavior of, 605–606
   welfare workers' avoidance of, 169
   white- and blue-collar, 18–20, 29
   World Health Organization definition of, 200
Alcoholic Counseling and Recovery Program, 602
Alcoholic employee, third-party information on, 173
Alcoholic family, 279–284
   see also Family therapy
Alcoholic marriage, 185, 260, 269–270
   as homeostatic mechanism, 278
Alcoholic patient
   acceptability of, 167–170
   acting-out behavior in, 152
   continuity of treatment for, 177–184
   as dropout from treatment, 161, 177–184
   engaging of in treatment program, 161–187
   goals possible for, 127
   "good" vs. "bad," 149–150
   homework for, 128, 141–142
   hostile, 132–133
   indulgent attitude toward, 151
   insight in, 140
   medical management of, 53–100
   reality distortions in, 138–139
   at referral stage, 174–176
   relationship building with, 36, 113–115, 148–154
   therapist and, 148–150
   treatment for, *see* Treatment
   ventilation by, 135
Alcoholic personality, concept of, 12
Alcoholic professional, in halfway house, 370, 374
   see also Recovered alcoholic
Alcoholic rehabilitation, *see* Rehabilitation
Alcoholic Rehabilitation Act, 602
Alcoholics Anonymous
   acceptance in, 463
   affiliation with, 404–405
   age of members in, 476–478
   Al-Anon cooperation with, 290–293
   and alcoholism clinic attendance, 180
   anonymity in, 465, 467, 518
   antipsychotherapy attitude of, 237
   articles in mass media by, 510–511

Alcoholics Anonymous (*cont.*)
   bibliographies of, 509–510
   "Big Book" of, 456–457, 465, 501, 508, 557
   "bottom" in, 85, 458–459
   changes associated with, 491–492, 499–507
   characteristic structure of, 74
   co-founders of, 443, 451, 456
   Compass Club and, 354
   confrontation in, 41, 237–238
   counselors and, 557, 561
   criticism of, 41, 517–519
   defined, 469
   development of, 441–519
   developments in since 1965, 468–470
   dropout rate for, 407
   effectiveness of, 40, 403, 406–408, 470–507
   ego-enhancing techniques in, 238
   evangelistic emphasis in, 168
   founding of, 441–443, 452–456, 597
   gender of members in, 476–477
   General Service Office of, 443–447, 561
   goals of, 216, 466, 469
   Grapevine of, 468–469, 492, 500, 508
   gratitude in, 463–464, 574
   as group, 238–239
   as group therapy, 37, 40–41, 216, 237–240, 385
   growth of, 443–451
   guilt in, 238
   halfway houses and, 357–358
   Higher Power in, 458–460, 597
   hospital visits from, 176
   identifying in, 217
   imitators of, 451, 507
   importance of, 402
   individual psychotherapy and, 156
   in institutions, 450, 506
   interpersonal relationships and, 501, 505
   length of sobriety in, 483–489, 497–499
   life membership in, 240
   literature on, 507–519
   love in, 403
   as "loving family," 239
   major goal of, 216, 466, 469
   material published by, 507–509
   meetings of, 217–218, 408, 492–497
   membership of, 162, 215, 404–406, 443–446
   most suitable patients for, 74–75
   motivation for joining, 480–482

Alcoholics Anonymous (*cont.*)
   in multidisciplinary approach, 45
   multiple family group therapy and, 285–289
   as "new way of life," 491–492
   number of groups in, 449–450, 482–473
   occupational adjustment and, 504
   occupational status of members in, 478–480
   outpatient treatment and, 400
   patients as members of, 58
   physician and, 57
   population characteristics of, 404–406
   Preamble of, 469
   predictions of success in, 406
   primary purpose of, 469
   problem drinking and, 215–219
   professional and scientific publications on, 511–516
   professional counselors and, 556–557
   psychology of, 237–238
   in psychotherapeutic approach, 134
   reaction-formation in, 403
   recovery data for, 402–403
   recovery program in, 490–497, 500–501
   referrals to, 63, 174–176
   rehabilitation and, 363
   requirements for membership in, 216, 469, 500
   in rest homes and lodges, 549
   skid row missions and, 549
   "slips" or relapses in, 483–490
   socioeconomic levels in, 30
   spiritual growth in, 506
   sponsorship in, 218, 464
   success rate in, 459
   surrender in, 216, 365, 458–459
   as therapeutic network, 239–240
   total abstinence as goal of, 308
   tranquilizers in relation to, 54
   as treatment model, 40–41, 47–48
   12th Step work in, 217–463–465, 480–481, 484, 557, 561
   Twelve Steps of, 217, 357–358, 457–465, 500–501, 506–507, 517
   Twelve Traditions of, 217, 466–467
   "willingness" in, 462
   worldwide growth of, 449–451
Alcoholics Anonymous, 456–465, 501, 508, 557

Alcoholism
   see also Alcohol; Alcoholic; Drinking; Drinking behavior; Medical management; Problem drinking; Treatment
   acceptance of, 126, 463
   adaptive consequences of, 281
   agitation and, 90–91
   aversive conditioning in, *see* Aversive conditioning
   behavioral criteria in, 170–171
   behavioral theories of, 301–302
   behavior modification treatment in, 34–35
   biological mechanisms in, 9–11
   bio-psycho-social dimensions of, 23
   broad-spectrum behavioral approach to, 327–335
   causes of, 109–110
   changing legal patterns in public approach to, 599–601
   as character neurosis, 243
   compared to diabetes, 80
   conditioned stimulus and conditioned response in, 6–7
   core problem in, 24, 31–32, 76, 93–100
   core therapies in, 24–25
   "cure" for, 35
   defined, 170, 200, 546
   depression and, 12, 78, 89–90, 164, 302
   depsychologizing of treatment in, 38
   detection through third party, 172–173
   development of, 1–9
   diagnostic criteria for, 62–67
   as disease, 1–2, 8–9, 57, 80, 168, 237, 267, 291
   drugs in treatment of, 84–86, 90–91, 417–426
   epidemics of, 4
   factors screening identification in, 165–170
   family history of, 4
   family problems in, 30–31, 98–99, 268
   family therapy in, 259–295
   father relationship in, 110
   genetic and prenatal biological predisposition in, 9–10
   group psychotherapy in, *see* Group psychotherapy
   hereditary factors in, 9
   heroin addiction and, 2–4, 161–164
   history of in United States, 593–596

Index

Alcoholism (cont.)
    hospitalization vs. outpatient treatment for, 394
    as incurable condition, 168
    industrial research on, 598
    inpatient treatment for, 390–395
    internal medical complications of, 162
    loss of control in, 6–7, 108, 171, 308
    loss of self-respect in, 17
    manic-depressive illness in, 90
    medical complications of, 10–11, 28
    medical model in treatment of, 32–34, 54
    as medical problem, 267
    multidisciplinary approach to, 53
    organizational interest in, 596–598
    physical dependency and, 5–8, 10, 22–23, 80–87
    physicians' shunning of, 46–47, 168
    physiological dependency in, 64
    predisposing factors in, 1–4
    in pregnancy, 9–10
    prevention of, 611–612
    vs. problem drinking, 8, 57, 79
    professional and nonprofessional workers in, 567–591
    professionals' negative attitude toward, 167–170
    progression in, 8
    proximal and distal causes of, 109–110
    psychiatric disorders and, 163–164
    psychological consequences of, 17–18
    psychological mechanisms in, 11–22, 56
    psychopathology in, 16–17, 28–29, 94
    psychosocial equation in, 16
    public health problems involved in, 604–606
    public health systems and, 46–48
    public health treatment programs in, 593–613
    questionnaire for, 171
    research in, 597
    schizophrenia and, 88–89, 163
    shunning of by physicians, 46–47
    social disruption caused by, 18
    social forces in, 4, 29–31, 37, 599
    social model of, 54
    socioeconomic classification in, 18–21
    socioeconomic factors in, 92–93
    specialized treatment programs for, 47–48
    stigma of, 569

Alcoholism (cont.)
    "susceptibility" to, 1
    as "symptom," 8–9
    training programs in, see Training programs
    tranquilizers and, see Tranquilizers
    treatment of, see Treatment
    tuberculosis and, 162
    vicious cycles in, 23
    vitamin deficiency and, 10–11
Alcoholism clinic attendance, A.A. and, 180
Alcoholic counselor, see Counselor; see also Recovered alcoholic
Alcohol education, target groups in, 570–571
    see also Education; Training programs
Alcoholism problems, Federal government involvement in, 601–603
Alcoholism programs
    company-sponsored, 173
    physician's role in, 60–62
Alcoholism Rehabilitation Assistant, 550
Alcoholism Rehabilitation Center, 550
Alcohol-related problems, scope of, 604–606
Alcohol Safety Action Project, 605
Alcohol studies, summer schools in, 575
Alcohol taste rating task, 304–305, 309
Alcohol use and abuse, in communities, 603–604
Alienation, problem drinking and, 199
Alpha alcoholic, 23
Alzheimer's disease, 65
Amblyopia, toxic, 65–66
American Drinking Practices studies, 603
American Hospital Association, 599
American Indian, drinking problems of, 4, 213
American Medical Association, 1, 106, 599
American Psychiatric Association, 599
American Public Health Association, 578–605
Anemia, alcoholism and, 65, 67
Antabuse
    in behavioral treatment, 337–338
    chlordiazepoxide and, 86
    as clinical adjunct, 79–80
    denigration of by A.A. members, 517
    dropout rates with, 184
    effectiveness of, 421–423
    group therapy and, 243–244
Anticipatory socialization, 181

Antidepressants, in alcoholic depression, 90, 424–426
Antipoverty programs, recovered alcoholic in, 551–552
Anti-Saloon Leagues, 595
Anxiety level, alcohol and, 201–202, 302
Apomorphine, 316
Arizona, University of 576
Attitudinal signs, in diagnosis, 65–66
Authority, response to in halfway house, 366–367
Aversive conditioning, 310–320, 409–416
  booster conditioning in, 313
  chemical, 316–318
  conditioned stimulus in, 409
  covert aversion in, 318–319
  electrical, 310–316
  instrumental escape conditioning and, 414
  failure of, 315–316, 340–341
  unpleasantness of, 414–415
Aversive imagery techniques, 34, 409, 415

BAL, see Blood alcohol level
Baptist Church, abstinence efforts of, 205
Behavioral assessment, 302–308
  broad-spectrum and multifaceted therapies in, 327–335
  in laboratory, 303–307
  in natural environment, 307–308
Behavioral "contracts," 339
Behavioral problems, systematic desensitization and, 327
Behavioral signs, in diagnosis, 65–66
Behavioral theories, 301–302
Behavioral treatment, 308–343, 409–417
  Antabuse in, 337–338
  blood alcohol level discrimination training in, 323–328
  broad-spectrum, 327–335
  community-reinforcement counseling in, 335–340
  contingency management in, 335–340
  "contracts" in, 338–339
  controlled drinking and, 309–310, 341–342
  counseling in, 335–340
  evaluation of, 340–343
  future research in, 342
  goals of, 308–310, 338
  natural environment modification in, 335–340

Behavioral treatment (cont.)
  significant others in, 337
  systematic desensitization in, 327
Behavior modification, 34–35
Belmont Hospital, London, 219
Benzodiazepine tranquilizers, 26, 33
Bignami's disease, 65
Blackouts, 64
Blacks, drinking problems of, 16, 212
Blood alcohol level
  in behavioral contracts, 339–340
  in electric aversion therapy, 413
  in operant conditioning, 321
Blood alcohol level discrimination training, 323–328, 334
Boston City Hospital, 587–588
Boston University, 576–577, 583
"Bottom," in Alcoholics Anonymous, 85, 458–459
Bowery Mission, 549
Brain function, impairment of, 166
Bridge House, 548
Business and industry, training programs for, 572–573

California, University of, 603
Cardiac arrhythmia, in diagnosis, 67
Casework, in psychotherapy, 105
Catecholine metabolism, Antabuse and, 79–80
Catholics, drinking behavior of, 205
Cattell Anxiety Scale, 12
CCC, see Citrated calcium carbimide
Central Islip State Hospital, 550, 553 n., 535, 558
Central nervous system depressants, in aversive conditioning, 411
Cessation of drinking
  changes following, 111–112
  substitutive treatment following, 112
Character disorder alcoholic, 11–14
Character neurosis, alcohol as, 243
Charles K. Post Rehabilitation Unit, 550, 553 n.
Chemical aversion, 316–318, 337–338, 410–412
Chit Chat Farms, 549
Chlordiazepoxide (Librium), 26, 33, 58, 77, 424–426
  Antabuse and, 86

Index

Chlordiazepoxide (Librium) (cont.)
  danger of, 85-86
  dropout rates with, 184
  as drug of choice, 85
  effectiveness of, 84
  prolonged treatment with, 91
  withdrawn from, 89
Cirrhosis, 55, 65
Citrated calcium carbimide, treatment with, 421-423
Clergy, in medical management, 71
Community Action Agency, 551
Community alcoholism treatment services, 606-608
Community Mental Health Centers Act (1968), 602
Community Programs, 609-610
Community-reinforcement counseling, in behavioral treatment, 335-340
Company alcoholism programs, 173
Company man, in halfway house, 368-369, 375-376
Compass Club, 353-356
Complementary role functioning, 279
Comprehensive Alcohol Abuse and Alcohol Prevention, Treatment, and Rehabilitation Act (1970), 602
Compulsory treatment, 253, 400-402
Conditioned stimulus, in aversive conditioning, 409
  see also Operant conditioning
Confrontation
  in Alcoholics Anonymous, 237-238
  in psychotherapy, 124-130
Conjoint family therapy, 282-284
  see also Family therapy
Connecticut Commission on Alcoholism, 353
Contingency management, in behavioral treatment, 335-340
Contracts, therapeutic, 130-132
Controlled drinking, 307-308
  alcohol as reinforcing consequence in, 321
  in behavioral treatment, 338, 341-342
  blood alcohol level discrimination training and, 324, 328
  electrical aversion and, 329-332
  as inappropriate goal, 309
  as treatment goal, 309
Coping mechanisms, 77
  maladaptive, 94-95

Coping mechanisms (cont.)
  in medical management, 94-98
Core problem, 24, 31-32
  defined, 93
  in treatment program, 76-77, 93-100
Correctional facilities, recovered alcoholic and, 552
Counseling
  see also Counselor; Psychotherapeutic approach; Psychotherapy
  in behavioral treatment, 335-340
  family, 98-99
  in medical management, 96-97
  in psychotherapy, 96, 105-106
Counselors
  certification for, 563
  future of, 562
  job responsibility of, 557-559
  as liaison to Alcoholics Anonymous, 556-558
  motivation of, 554-555
  paraprofessional, 70-71
  patient identification with, 560
  recovered alcoholic as, 61, 553-554
  remuneration and status of, 561
  selection of, 555-556
  training for, 556
  in treatment programs, 553-561
  12th Step work and, 557, 561
  unique problems of, 559-561
Counterconditioning, 327
Counterdependence, 180
Countertransference, 148, 150-154
Covert aversion or sensitization, 318-319
Craving
  addiction and, 6-8
  physical dependency and, 10
Crisis intervention approach, 110-111
Cybernetics, 264-265

Delirium tremens, 7, 39, 167
  in diagnosis, 64
  impending, 81-82
  treatment in, 82
  as withdrawal phenomenon, 55, 81
Denial
  in detection of alcoholics, 165-166
  hospitalization and, 403
  in psychotherapy, 136-137, 155

Department of Health, Education, and
    Welfare, U.S., 602
Dependency
  in diagnosis, 64
  forms of, 108-109
  physical, *see* Physical dependency
  primary psychological, 5
  reaction-formation against, 180
Depression
  alcoholism and, 12, 78, 89-90, 164, 302
  in periodic drinker, 78
Desensitization, systematic, *see* Systematic
    desensitization
Detection of alcoholics, 162-174
  factors hindering, 165-170
Detoxification
  addictive cycle in, 24
  necessity for, 82-83
  nonmedical, 61
  repeated, 116
Detoxification center, first U.S., 61
Diabetes, alcoholism compared to, 80
Diagnosis, 162-174
  criteria in, 62-67
  factors hindering detection in, 165-170
  high-risk groups in, 162-165
  physician's failure in, 167
  methods of identification in, 170-174
Diazepam, 26, 33, 58
Disease concept of alcoholism, 1-2, 8-9, 57
    80, 168, 237, 268, 291
Disulfiram, *see* Antabuse
Doctor, *see* Physician
Dreams, in group psychotherapy, 251
Drinking
  cessation of, *see* Cessation of drinking
  controlled, *see* Controlled drinking
  desire to stop, 99-100, 216, 500
  during psychotherapy, 121-124
  loss of control in, 6-7, 108, 171, 308
  modification of, 127
  problem, *see* Problem drinking
  substitute activities for, 112
  U.S. Constitution and, 595
Drinking behavior
  see also Alcoholism; Behavioral
    assessment; Problem drinking
  age factors in, 603
  alpha and other typologies in, 202-203
  anxiety reduction in, 201-202

Drinking behavior (*cont.*)
  change in, 270-273
  educational achievement and, 208
  ethnic factors in, 594, 603-604
  family influences in, 209-211, 604
  in family therapy, 266
  Federal research in, 603
  gulping vs. sipping in, 305-306, 334
  in learning theory, 226
  national, 199-200
  occupational activity and, 208
  religious aspects of, 204-206
  social and ethnic customs in, 594
  social class influences in, 207-209
  social etiological factors in, 203-214
  social influences and, 389
  social systems theory in, 200-203
  specific-culture orientation in, 203
  stress in, 304
  substructural orientation in, 203
  supracultural orientation in, 201-203
Drinking customs, alcoholism and, 202
Drinking disposition, 307
Drinking episodes, in psychotherapeutic
    approach, 143-145
Drinking habits, change in, 328
  see also Drinking behavior
Drinking Man, The (McClelland), 254
Drinking response
  see also Drinking behavior
  associated behavioral problems and, 327-335
  aversive conditioning in, 310-320
  blood alcohol level discrimination and,
    323-327
  modifying of, 320-327
Dropping out
  by alcoholic patient, 177-184
  from Alcoholics Anonymous, 180, 407,
    483-490
  from aversion therapy, 414-415
  causes of, 178-179
  counterdependence in, 180-181
  from drug treatment, 418-419, 427
  from group psychotherapy, 252
  MMPI scores of, 178
  from outpatient treatment, 395
  remedies for, 184-187
  socioeconomic status and, 173, 178,
    181

Index

Drugs
  see also Antabuse; Chlordiazepoxide; Sedatives; Tranquilizers
  in aversive conditioning, 410
  in medical management, 77
Drug treatment, 417–426
  assumptions in, 417–418
  negative effects in, 418–426
  dropouts in, 418–419, 427
Drunkenness
  as criminal offense, 164, 569, 599–600, 606
  stigma of, 569
DS (disulfiram), see Antabuse
DTs, see Delirium tremens
Dynamic techniques, in medical management, 94–96
Dyssocial alcoholic, 15

Eagleville Hospital, 550
East Los Angeles Health Task Force, 577
Eastman Kodak Company, 598
Economic Opportunity Act (1964), 551
Education, in problem drinking treatment, 227–228, 612
  see also Training programs
Educational achievement, drinking behavior and, 208
Eighteenth Amendment, 595, 601
Electric aversion therapy, 310–316, 412–414
  operant methods in, 321–323
Emetine, 316
Emmanuel Movement, 547
Employment status, in medical management, 91–92
  see also Occupational (adj.)
Episcopalians, drinking behavior of, 205–206
Equal Rights Amendment, 595
Essential alcoholics, 4, 9, 108
Ex-alcoholic, vs. "recovered alcoholic," 546
  see also Recovered alcoholic
Ex-alcoholic counselors, see Counselors
Existential techniques, in medical management, 94–95
Ex-resident social roles, in halfway houses, 372–379

Family
  as "alcoholic system," 263–264, 280–282
  drinking behavior and, 604
Family counseling, 98–99
Family disruption, 21–22

Family dynamics, in psychotherapy, 154
Family groups, in group psychotherapy, 254–256
Family influences, on drinking behavior, 209–211
Family system, homeostasis and, 264–265
Family therapy, 259–295
  adaptation of to alcoholism therapy, 275–282
  alcoholism and, 267–289
  behavioral context in, 266
  boundaries in, 266–267
  communication patterns in, 265–266
  family as alcoholic system in. 280
  future of, 292
  group psychotherapy and, 272
  implications of, 293–295
  interaction model in, 289
  for lower-class families, 274
  major studies in, 276–277
  in problem drinking, 224–225
  spouses in, 270–275
  as treatment modality, 261–267
Father relationship, in alcoholism, 110, 210
Fears, drinking and, 7
Federal government, alcoholism programs of, 601–603
Feedback loop, in homeostasis, 264
Female Alcoholic, The (Kinsey), 201
Fetal alcohol syndrome, 10
FIDD (Fixed-Interval Drinking Decision) program, 306
Frontier, drinking on, 594–595

Games Alcoholics Play (Steiner), 253
General Electric Company, 598
General hospitals, physician's role in, 59–60
  see also Hospitals
General practitioners, avoidance of alcoholics by, 167–170
  see also Physician
General Service Office, A. A., 443–444
Generic counselor, in alcoholic rehabilitation, 46
Geographic cure, 548
George Washington University, 603
Ghetto alcoholic, 4, 15, 18, 20–21, 30
Goals
  of Alcoholics Anonymous, 469
  of halfway houses, 371–372

Goals (cont.)
  homework and, 141–142
  proximal or distal causative, 129
  of psychotherapy, 128–130
  of treatment, 386–388
Gough Adjective Check List, 405
Great Depression, 596
Group programs, recovered alcoholic in, 547–548
Group psychotherapy, 221–223, 235–256
  see also Group therapy; Psychotherapy
  authentic cohesion in, 250
  Alcoholics Anonymous as, 237–240
  conjoint and family groups in, 254–256
  dropouts from, 252
  family therapy and, 272
  history of, 241–242
  patient selection in, 249
  subjective response in, 249
Groups, number of in Alcoholics Anonymous, 237–240, 449–450, 472–473
Group therapy, 37, 221–223
  see also Group psychotherapy
  Alcoholics Anonymous as, 237–240
  Antabuse and, 243
  dreams in, 251
  husband and wife in, 246
  multiple-couples and multiple-family, 284–289
  survey of, 240–254
  transference in, 245
Guilt, in Alcoholics Anonymous, 238
Gulping, vs. sipping, 305, 334

Halfway houses, 220–221, 351–382
  authority and, 366–367, 379–388
  Bridge House (N.Y.C.) as example of, 548
  church-operated, 380
  Compass Club as example of, 353–356
  culture of, 356–358
  defined, 352–353
  detoxification and, 360–361
  ecology of, 356, 359
  ex-resident social roles in, 370–379
  goal of, 371–372
  major characteristics of, 356–357
  origins of, 351–352
  physical plant in, 356–358
  rehabilitation and, 361–381
  relapses in, 377–378

Halfway houses (cont.)
  resident careers in, 370–371
  resident types in, 359–361
  resocialization in, 371–372
  social organization of, 356, 358–359
  social types in, 367–381
  types of, 357–359, 379–381
Health, Education, and Welfare Department, U.S., 602
Heavy drinker, physical dependence in, 22–23
Hematological disorders, 65, 67
Heroin addiction, 2–4
  compared with alcohol addiction, 161, 163
  methadone in, 26, 89
  physical dependency and, 6
  race factor in, 2
Higher Power, in Alcoholics Anonymous, 458–461, 597
Highway Safety Act (1966), 599, 601
Homeostasis, in family systems, 264–265, 278
Honeymoon period, in residential programs, 120
Hospitals
  refusal to admit alcoholics, 167
  training programs in, 579–580
"Hot seat" technique, 246
Hyperreflexia, 66
Hypnotism, as treatment, 310, 327, 415

Iceman Cometh, The (O'Neill), 13
Identification
  in alcoholic diagnosis, 170–174
  through third party, 172–173
Identified patient, concept of, 265
Identity problem, in psychotherapy, 140
Impulse disorders, alcoholism and, 13
Inadequacy feelings, 110
Indians, American, see American Indian
Industrial programs, 551, 598–599, 608–609
Industrial Revolution, 594
Infant, withdrawal symptoms in, 10
Inner city alcoholics, see Ghetto alcoholic
Inpatient psychotherapy, 393–394
Inpatient treatment, 390–395
  length and outcome of, 393
Insight, coping mechanisms and, 95
Institutions, Alcoholics Anonymous in, 450, 506
  see also Correctional facilities

Index

Instrumental avoidance conditioning, 409
Instrumental escape conditioning, 34, 414
Interpersonal relationships, problem of in psychotherapy, 36, 113–115, 148–154
Intoxication, acute, 78
 see also Drunkenness
Irish alcoholics, vs. Jewish, 16
Isoniazid, 80

Jews, drinking behavior of, 2, 16, 204–206
Johns Hopkins University, 576, 585

Kennecott Copper Corp., 608
Keswick Colony of Mercy, 548–549
Kings County Hospital Center, 59–60
Korsakoff syndrome, *see* Wernicke-Korsakoff syndrome

Lay therapist, recovered alcoholic as, 547–548
Learning theory, problem drinking in, 225–226
Librium, *see* Chlordiazepoxide
Liver abnormality, in diagnosis, 34, 67
Loner, in halfway house, 369–370, 378–379
Loss of control, in drinking, 6–7, 108, 171, 308
Love, in Alcoholics Anonymous, 403
LSD (lysergic acid diethylamide), treatment of alcoholism with, 424
"Lush," alcoholic as, 109
Lutheran General Hospital, 550
Lutherans, drinking behavior of, 205

McGoldrick method, 548
"Maintenance" drinking, 303–304
Malcolm Bliss Mental Health Center, 581
Manic-depressive illness, in alcoholism, 90
MAO (monoamine oxidase) inhibitors, 80
Marathon sessions, 560
Marchigava, 65
Marriage, group therapy and, 246
 see also Alcoholic marriage
Massachusetts General Hospital, 580
MAST, *see* Michigan Alcoholism Screening Test
Medically sponsored programs, 550–551
Medical management, 53–100
 see also Treatment
 Alcoholics Anonymous and, 71

Medical Management (*cont.*)
 characteristic structure of, 72
 clergy and, 71
 complications in, 87–88
 core problem in, 93–100
 counseling in, 96–97
 depression in, 89–90
 detoxification in, 82–83
 diagnosis criteria in, 62–63
 dynamic techniques in, 94–96
 employment status in, 91–92
 existential techniques in, 94–95
 family counseling in, 98–99
 inpatient vs. outpatient treatment in, 97–98
 mental health professionals in, 69–70
 motivation in, 99–100
 paraprofessional counselors in, 70–71
 physical dependency in, 80–87
 physician vs. psychiatrist in, 68–69
 "scare factors" in, 87–88
 sedatives in, 83–87, 90–91
 socioeconomic factors in, 92–93
 therapist of choice in, 68
 tranquilizers in, 83–87, 90–91
 treatment model of choice in, 71–75
 treatment plan design in, 75–100
Mendocino State Hospital, 551
Mental health professionals, in medical management, 69–70
Mental hospital admissions, alcoholism in, 162
Methadone detoxification and maintenance, 26, 89
Methodist Church, total abstinence position of, 205
Metronidazole, 423–424
Michigan Alcoholism Screening Test, 171
Minnesota Multiphasic Personality Inventory, 12, 172, 178, 185–186, 307, 392 n., 399, 403, 502
Minority groups, drinking problem and, 604
Mixers, in halfway house, 367–368, 374–375
MMPI, *see* Minnesota Multiphasic Personality Inventory
Mormons, drinking behavior of, 206
Multidisciplinary approach, 45
Multimodal approach, 44
Multiple-couples group therapy, 284–289
Multiple-family group therapy, 284
Myelinolysis, central pontine, 65

Narcissism, pathological, 13
National Clearinghouse for Alcoholic Information, 602
National Council on Alcoholism, 59, 62, 173, 552, 556, 570, 598
National Highway Traffic Safety Administration, 605
National Institute on Alcohol Abuse and Alcoholism, 46–48, 259, 286–288, 563, 568, 570, 573, 578, 583, 585, 587, 591, 602, 605–607, 610–611
Natural environment, modification of, 335–340
Nausea-producing chemicals, in aversive conditioning, 316–317
  see also Antabuse
Nebraska, University of, 577
Negative feedback loops, homeostasis and, 264
Negroes, see Blacks
Neurotic alcoholic, 15
Neuropathy, peripheral, 65
NIAAA, see National Institute on Alcohol Abuse and Alcoholism
Nonprofessional workers, training for, 567–591
"Normal" alcoholic, 15–16

Occupational activity, drinking behavior and, 165, 208–209
Occupational adjustment, in Alcoholics anonymous, 504
Occupational status, in Alcoholics Anonymous, 478–480
Office of Economic Opportunity, 551, 555–556
Operant conditioning
  aversive techniques in, 416–417
  in drinking response, 320–323
Operant drinking behavior, in behavioral treatment, 320, 338
Other Side of the Bottle, The (Anderson), 547
Outpatient treatment, 395–402
  Alcoholics Anonymous and, 400
  basic issues of, 396–402
  dropping out of, 395
  effectiveness of, 396
  length and outcome of, 396–398
  relative values in, 398–400

Overpermissiveness, by therapist, 151
Oxazepam-protriptyline combination, 180
Oxford Group, 442, 597

Pancreatitis, 55, 65
Paraprofessional counselors, 70–71
  training program for, 587–589
Parental absence, problem drinking and, 210
Parental drinking, children and, 210–211
Past treatment, success or failure in, 116–118
Patient, see Alcoholic patient
Patient-therapist roles, reversal of, 143
Patient tracking, in community treatment services, 607
Pavlovian treatment, aversive conditioning as, 310, 409
Periodic drinker, 78
Permanent sobriety, 248, 308
Pennsylvania Hospital, 579
Phenothiazines, 77, 84–85
  in agitation, 90
  in schizophrenic alcoholics, 89
Physical dependency
  craving and, 5–8, 10
  in heavy drinker, 22–23
  medical management of, 80–87
  persistent, 27–28
Physician
  Alcoholics Anonymous and, 57
  attitude of toward alcoholism, 46–47, 53–54
  aversion to working with alcoholics, 167–170
  in general hospitals, 59–60
  ignorance of alcoholism in, 167
  medical specialties of, 55–56
  as primary therapist, 61
  vs. psychiatrist in medical management, 68–69
  role of in alcoholism treatment, 54–62
  shunning of alcoholic patient by, 46–47, 53–54, 167–170
  training programs for, 579
  tranquilizers and, 58
Physiological dependency, 64
Pittsburgh, University of, 576
Polyneuropathy, 55
Pregnancy, heavy drinking in, 9–10
Prevention programs, 611–612

Index

Primary psychological dependence,
    development of, 5
    see also Physical dependency
Problem drinker
    activity groups and, 223–224
    rejection of by father, 210
    social drinking and, 78–79
Problem drinking
    see also Alcoholism; Drinking behavior
    age, sex, and urbanization in, 213–214
    alcoholics as, 8
    Alcoholics Anonymous and, 215–219
    vs. alcoholism, 57, 79
    alienation and, 199
    biological model of, 198
    didactic approach to, 228
    educational approaches in, 227–228
    ethnic aspects of, 212–213
    family influences in, 209–211
    family therapy in, 224–225
    group psychotherapy in, 221–223
    male role and, 210
    among Negroes, 212
    psychological model of, 198–199
    social factors in, 197–229
    social learning approach in, 225–227
    therapeutic communities in treatment of, 219–220
    treatment approaches in, 214–229
Professional workers
    target groups and, 569–575
    training for, 567–591
Progression, in alcoholism, 8
Prohibition Amendment, 595
Prohibition movements, 452, 595
Project Insight, 608–609
Projection, in psychotherapeutic process, 136–138
Protestants, drinking behavior of, 205
Pseudomutuality, 262
Psychiatrically sponsored programs, 550
Psychiatric diagnoses, 11–17
Psychiatrist, role of in treatment, 55, 68–69
Psychoanalysis-in-group approach, 248
Psychoanalytic psychopathy, for select group, 139
Psychoanalytic research, 597
Psychodrama, 248
Psychological deterioration, 17
Psychological mechanisms, 11–22

Psychological model
    in alcoholic rehabilitation, 35–37
    characteristic structure of, 73
Psychological signs, in diagnosis, 65–66
Psychopathology
    medical treatment of, 28
    two positions of, 94
Psychopathy, in alcoholics, 12
Psychotherapeutic approach
    crisis intervention in, 110–111
    drinking episodes and, 143–144
    goals and confrontation in, 124–130
    homework in, 141–143
    patient-therapist role reversal in, 143
    pretherapy factors in, 115–124
    repeated detoxification and, 116
    successful and unsuccessful past treatment and, 116–118
    therapeutic contracts in, 130–132
    transfer in, 118–119
Psychotherapy
    see also Psychotherapeutic approach
    abstinence during, 135
    Alcoholics Anonymous and, 156
    behavioristically oriented, 409–417
    confrontation in, 114–115, 124–130
    counseling in, 96, 105–106
    countertransference in, 150–154
    couple treatment in, 185
    defined, 106
    denial in, 136–137, 155
    detoxification and, 155
    drinking episodes in, 143–145
    drinking or not drinking in, 121–122
    dropping out of, 177–184
    early termination in, 147
    family members and, 154
    general considerations in, 106–113
    goals in, 128–130
    group, *see* Group psychotherapy
    homework in, 142
    hostile patient in, 132–133
    individual, 107–108
    initial phase of, 113–133
    inpatient, 393–394
    late phase and termination of, 145–147
    length of treatment in, 145–146
    lying in, 123
    male–female patient–therapist combinations in, 153

Psychotherapy (*cont.*)
  middle phase in, 133–145
  permissiveness in, 151
  relationships in, 36, 113–115, 148–154
  role playing in, 142
  techniques and practices in, 107
  treatment methods in, 145–156
  uncongenial, 183
Psychotic alcoholic, 14–15
Public drunkenness, as crime, 164, 569, 599–560, 606
Public health treatment programs, 46–48, 593–613
Public information agencies, recovered alcoholic and, 552
Pygmalion (Shaw), 596

Quarterly Journal of Studies on Alcoholism, 598
Questionnaires, on alcoholism, 171

Rationalization, in psychotherapeutic treatment, 136
Recovered alcoholic
  as counselor, 551–554
  in courts and correctional facilities, 552
  defined, 546
  and Economic Opportunity Act, 551–552
  future of as counselors, 562–563
  in independent facilities, 549
  in industrial programs, 551
  in medically and psychiatrically sponsored programs, 550–551
  past and present roles of, 547–552
  in public information programs, 552
  in religious programs, 548–549
  role of, 545–563
  unique problems of, 559–561
Recovering alcoholic, defined, 546
Referral
  case history in, 260
  criteria for, 63–75
  failures and remedies at, 174–176
  following residential program, 119–121
"Reformed drunk," in halfway house, 371, 376–376
"Regular guy," in halfway house, 368–369, 376–378
Rehabilitation
  see also Treatment

Rehabilitation (*cont.*)
  Alcoholics Anonymous and, 363
  defined, 361
  "generic counselor" in, 46
  halfway houses and, 361–381
  indices of, 362
  medical model in, 32–34
  multivariant model in, 42–43
  psychological model in, 35–37
  research findings in, 363–364
  as social process, 362–365
  special problems in, 26–31
  surrender in, 365
  theory of, 364–381
Rehabilitation centers, 549
  see also Halfway houses
Relationships, reestablishing of in psychotherapy, 36, 113–115, 148–154
Relaxation aversion, 34, 409
Religious affiliation, drinking and, 204–206
Research Council on Problems of Alcohol, 598
Residential program, discharge from, 119–121
Rest homes, 549
  see also Halfway houses
Role-playing, in group therapy, 142
Rorschach test, for A.A. members, 502
Roy Littlejohn Associates, 563
Rutgers Center of Alcohol Studies, 556, 575, 598
Rutgers University, Alcohol Behavior Research Laboratory, 306

Scapegoat, concept of, 265
Schizoid personality, 13
Schizophrenia, alcoholism and, 88–89, 163
Sedatives
  see also Tranquilizers
  dangers of, 85–86, 91
  in medical management, 83–87
Self-confidence, loss of, 17
Self-destructive tendencies
  reinforcing of, 88
  stages of, 13, 109
Self-respect, loss of, 17
Sensitivity training, 560
SES, *see* Socioeconomic status
Sex factors, in problem drinking, 213–214, 476–477

Index

Sex factors, in problem drinking (*cont.*)
  see also Women
Shadell Sanatorium, 316, 550
"Shakes," addiction and, 7
Significant other, in behavioral treatment, 337
Sipping, vs. gulping, 305–306, 334
Situational alcoholic, 4
Skid row alcoholic, 18, 20–21, 23, 30, 166, 360–361, 381, 606
Skid row religious missions, 549
Smoking, alcoholism and, 605
Social agencies, in alcoholism treatment, 47
Social drinking, 78–79
Social institutions, educational programs for, 572
Social isolation, 179
Social learning theory
  behavioral theory and, 302
  in problem drinking therapy, 225–227
Social model
  in alcoholic rehabilitation, 37–40, 54, 197–229
  origins and perspectives of, 199–200
  social systems theory in, 200–203
Social pressures, recognition of, 37
Social Science Institute, 586
Social systems theory, 200–203
Socioeconomic status
  Alcoholics Anonymous and, 30
  dropping out and, 173, 178, 181
Southern California Psychiatric Association, 399
Spouses, therapy for, 270–275
  see also Family therapy; Wives
Spree drinking, 303
Stress, in drinking behavior, 304
Substitutive treatment, following cessation of drinking, 112
Suicide, in alcoholics, 164
Supracultural orientation, drinking in, 201–203
Surrender, 216, 365, 458–459
Suspension of thought, in psychotherapy patient, 136
Systematic desensitization, 35, 409, 416

Taste tests, 304–305, 309
Temperance movements, 452, 463, 548
Therapeutic community, goals of, 219–220

Therapeutic contracts, in psychotherapy, 130–132
Therapeutic network, Alcoholics Anonymous as, 239–240
Therapist
  countertransference and, 150–151
  cross-sex combinations and, 153
  hostility toward alcoholic patient, 151–152
  identifying role of, 152
  male–female relations and, 153
  as parental substitute, 141
  relationship building by, 113–114
Therapy
  see also Family therapy; Group therapy; Psychotherapy; Treatment
  for alcoholic family, 282–285
  core problem and, 93
  counseling as, 96–97
  defined, 93
  drinking during, 121–124
  family theory in, see Family therapy
  psychotherapeutic approach in, see Psychotherapeutic approach
  self-knowledge in, 140
  for spouses, 270–278
Third party, diagnosis or identification through, 172–173
Training programs, 567–591
  see also Education
  basic information in, 51–52
  for community leaders, 586–587
  evaluation of, 589–590
  examples of, 583–589
  for general caretakers, 572–573
  for general public, 570–571
  guidelines for, 590–591
  for paraprofessionals, 587–589
  research type, 583–585
  service and treatment in, 582
  settings for, 575–581
  for specialists, 573–575
  for students, 571–572
  subject matter of, 581–582
  target groups in, 570–575
  teaching methods and models in, 578–581
  in universities, 576–577
Tranquilizers
  see also Chlordiazepoxide; Drugs; Sedatives
  Alcoholics Anonymous antagonism to, 54

Tranquilizers (*cont.*)
  dangers of, 85–86, 90–91
  in medical management, 84
  physician and, 58
  in treatment, 26, 33, 424–426
  withdrawal from, 89
Transfer, in psychotherapeutic approach, 118–119
Transference, 140
  defined, 148
  in group therapy, 245
Transportation Department, U.S., 605
Treatment
  see also Alcoholics Anonymous; Group therapy; Medical management; Psychotherapy; Therapy
  acting out as basic resistance in, 250
  aftercare in, 394–395
  Alcoholics Anonymous model in, 40–41, 74–75
  alcoholism counselor in, 553–561
  aversive conditioning in, 310–320
  behavioral, *see* Behavioral treatment
  behavior modification in, 34–35
  community, 606–610
  compulsory, 253, 400–402
  continuity of care in, 45–46, 161–187
  criteria for, 63–75
  depsychologizing of, 38
  design of plan in, 75–100
  drinking during, 121–124
  dropping out from, 161, 177–184
  drugs in, 33, 77, 417–427
  educational group in, 227–228
  education for alcoholic in, 32
  empathy and warmth in, 36
  engaging alcoholic in process of, 161–187
  evaluation of, 385–428
  family therapy in, 261–267
  four systems of, 54
  generic counselor in, 46
  goals in, 386–388
  group therapy in, 221–223
  halfway houses in, 220–221
  hospitalization vs. outpatient forms of, 394
  hostile alcoholic in, 255
  hypnotism in, 310, 327, 415
  implications for, 22–32
  inpatient, *see* Inpatient treatment
  length of, 145–146, 388–389

Treatment (*cont.*)
  medical model of, 32–34, 54, 72–73
  model of choice for, 71–72
  motivation in, 99–100
  multimodal multidisciplinary approach to, 43–48
  for multiple couples or multiple families, 284–289
  multivariant model in, 42–43
  outcome of, 386–388
  outpatient, *see* Outpatient treatment
  physician as therapist in, 33, 54–62
  pretherapy factors affecting, 115–124
  in problem drinking, 214–229
  psychotherapeutic approach in, 105–157
  psychodrama in, 248
  psychological model in, 35–37
  public health services in, 46–48, 593–613
  recovered alcoholic in, 545–463
  relative values in, 398–400
  refusal of, 246
  reward and punishment in, 38
  social learning approach in, 225–227
  social model in, 37–40, 54
  specialized programs in, 47–48
  special problems in, 76–93
  for spouses of alcoholics, 270–275
  systematic desensitization in, 416
  theory and practice in, 1–48
  therapeutic community in, 219–220
  therapist of choice in, 68
  in training programs, 582
Tree Grows in Brooklyn, A (Smith), 13
Tuberculosis, alcoholism and, 162
12th Step work, in Alcoholics Anonymous, 217, 463–465, 480–481, 484, 557, 561
Twelve Steps, of Alcoholics Anonymous, 217, 357–358, 456–465, 500–501, 506–507, 517
Twelve Steps and Twelve Traditions, 507
Twelve Traditions, of Alcoholics Anonymous, 217, 466–467
Twenty-first Amendment, 595

Under the Volcano (Lowry), 13
Uniform Alcoholic and Intoxication Treatment Act (1971), 600–601
United States, alcohol problems in, 593–596

Universities, alcoholism training programs in, 576–577, 583–586
U.S. Public Health Service, 594

Veterans Administration, 390
Videotapes, in multifaceted therapy approach, 329–330
Vitamin deficiency, 10–11

Wake Forest University, 577
Washingtonian Hospital, 550
Washington Temperance Society, 463, 548
Washington University, 586
Wernicke–Korsakoff syndrome, 65
Wino, alcoholic as, 109
Withdrawal symptoms, 5–6
    acute, 81–82
    craving and, 7
    delirium tremens and, 55
    in infant, 10
    panic in, 120
    subclinical conditioned, 24, 26

Wives
    see also Spouses
    as Al-Anon members, 292
    participation of in group meetings, 273, 275
Women
    drinking problems of, 214
    number of in A.A., 476–478
    as prohibitionists, 595
Women alcoholics
    behavior of, 303–304
    as maintenance drinkers, 304
    membership in A.A., 476–478
Womens Christian Temperance Union, 595
World Health Organization, 200

Yale Center of Alcohol Studies, 575, 598
Yale Hope Mission, 353, 355

Zimberg Interview Scale, 60
Zung Depression Scale, 12